terra australis 32

Terra Australis reports the results of archaeological and related research within the south and east of Asia, though mainly Australia, New Guinea and island Melanesia — lands that remained *terra australis incognita* to generations of prehistorians. Its subject is the settlement of the diverse environments in this isolated quarter of the globe by peoples who have maintained their discrete and traditional ways of life into the recent recorded or remembered past and at times into the observable present.

Since the beginning of the series, the basic colour on the spine and cover has distinguished the regional distribution of topics as follows: ochre for Australia, green for New Guinea, red for South-East Asia and blue for the Pacific Islands. From 2001, issues with a gold spine will include conference proceedings, edited papers and monographs which in topic or desired format do not fit easily within the original arrangements. All volumes are numbered within the same series.

List of volumes in Terra Australis

Volume 1: *Burrill Lake and Currarong: Coastal Sites in Southern New South Wales.* R.J. Lampert (1971)
Volume 2: *Ol Tumbuna: Archaeological Excavations in the Eastern Central Highlands, Papua New Guinea.* J.P. White (1972)
Volume 3: *New Guinea Stone Age Trade: The Geography and Ecology of Traffic in the Interior.* I. Hughes (1977)
Volume 4: *Recent Prehistory in Southeast Papua.* B. Egloff (1979)
Volume 5: *The Great Kartan Mystery.* R. Lampert (1981)
Volume 6: *Early Man in North Queensland: Art and Archaeology in the Laura Area.* A. Rosenfeld, D. Horton and J. Winter (1981)
Volume 7: *The Alligator Rivers: Prehistory and Ecology in Western Arnhem Land.* C. Schrire (1982)
Volume 8: *Hunter Hill, Hunter Island: Archaeological Investigations of a Prehistoric Tasmanian Site.* S. Bowdler (1984)
Volume 9: *Coastal South-West Tasmania: The Prehistory of Louisa Bay and Maatsuyker Island.* R. Vanderwal and D. Horton (1984)
Volume 10: *The Emergence of Mailu.* G. Irwin (1985)
Volume 11: *Archaeology in Eastern Timor, 1966–67.* I. Glover (1986)
Volume 12: *Early Tongan Prehistory: The Lapita Period on Tongatapu and its Relationships.* J. Poulsen (1987)
Volume 13: *Coobool Creek.* P. Brown (1989)
Volume 14: *30,000 Years of Aboriginal Occupation: Kimberley, North-West Australia.* S. O'Connor (1999)
Volume 15: *Lapita Interaction.* G. Summerhayes (2000)
Volume 16: *The Prehistory of Buka: A Stepping Stone Island in the Northern Solomons.* S. Wickler (2001)
Volume 17: *The Archaeology of Lapita Dispersal in Oceania.* G.R. Clark, A.J. Anderson and T. Vunidilo (2001)
Volume 18: *An Archaeology of West Polynesian Prehistory.* A. Smith (2002)
Volume 19: *Phytolith and Starch Research in the Australian-Pacific-Asian Regions: The State of the Art.* D. Hart and L. Wallis (2003)
Volume 20: *The Sea People: Late-Holocene Maritime Specialisation in the Whitsunday Islands, Central Queensland.* B. Barker (2004)
Volume 21: *What's Changing: Population Size or Land-Use Patterns? The Archaeology of Upper Mangrove Creek, Sydney Basin.* V. Attenbrow (2004)
Volume 22: *The Archaeology of the Aru Islands, Eastern Indonesia.* S. O'Connor, M. Spriggs and P. Veth (2005)
Volume 23: *Pieces of the Vanuatu Puzzle: Archaeology of the North, South and Centre.* S. Bedford (2006)
Volume 24: *Coastal Themes: An Archaeology of the Southern Curtis Coast, Queensland.* S. Ulm (2006)
Volume 25: *Lithics in the Land of the Lightning Brothers: The Archaeology of Wardaman Country, Northern Territory.* C. Clarkson (2007)
Volume 26: *Oceanic Explorations: Lapita and Western Pacific Settlement.* S. Bedford, C. Sand and S. P. Connaughton (2007)
Volume 27: *Dreamtime Superhighway: Sydney Basin Rock Art and Prehistoric Information Exchange.* J. McDonald (2008)
Volume 28: *New Directions in Archaeological Science.* A. Fairbairn and S. O'Connor (2008)
Volume 29: *Islands of Inquiry: Colonisation, Seafaring and the Archaeology of Maritime Landscapes.* G. Clark, F. Leach and S. O'Connor (2008)
Volume 30: *Archaeological Science Under a Microscope: Studies in Residue and Ancient DNA Analysis in Honour of Thomas H. Loy.* M. Haslam, G. Robertson, A. Crowther, S. Nugent and L. Kirkwood (2009)
Volume 31: *The Early Prehistory of Fiji.* G. Clark and A. Anderson (2009)
Volume 32: *Altered Ecologies: Fire, Climate and Human Influence on Terrestrial Landscapes.* S. Haberle, J. Stevenson and M. Prebble (2010)

terra australis 32

Altered Ecologies

Fire, climate and human influence on terrestrial landscapes

THE AUSTRALIAN NATIONAL UNIVERSITY
E PRESS

ANU E PRESS

© 2010 ANU E Press

Published by ANU E Press
The Australian National University
Canberra ACT 0200 Australia
Email: anuepress@anu.edu.au
Web: http://epress.anu.edu.au

National Library of Australia Cataloguing-in-Publication entry

Author: Haberle, Simon G.

Title: Altered ecologies : fire, climate and human influence on terrestrial landscapes / Simon G. Haberle, Janelle Stevenson, Matthew Prebble.

ISBN: 9781921666803 (pbk.) 9781921666810 (ebook)

Series: Terra Australis ; 32

Notes: Includes bibliographical references.

Subjects: Human ecology--Australasia.
Environmental archaeology--Australasia.
Landscape assessment--Australasia.
Landscape changes--Australasia.
Nature--Effect of human beings on--Australasia.

Other Authors/Contributors:
Stevenson, Janelle.
Prebble, Matthew.

Dewey Number: 333.72099

Series Editor: Sue O'Connor

Typesetting and design: Adam Black

Cover image: Lake Gwam and treefern grasslands at 3500m altitude in the Sarawaget Ranges, northern Papua New Guinea (photo by Simon Haberle)

Back cover map: Hollandia Nova. Thevenot 1663 by courtesy of the National Library of Australia.
Reprinted with permission of the National Library of Australia.

Papers in honour of Geoff Hope

Mungo Lunette

by Mark O'Connor
(in honour of Geoff's leadership on departmental excursions)

Among saltbush greyer than seaweed
we boardwalk the yeasting waves of the old golden-perch pond.

Rabbit dung dots a lake bottom dry since their coming

– only once, last century, after seven inches of rain
a farmer had to rescue his sheep by boat.

The roo bounces down-slope,
a brown branch crackling.

The professor's eye, noting the local runoff
and what the westerlies do, observes
seas that rush in and out of a continent's middle,
salt lakes spreading and emptying,
hills pimpling up and down with cartoon-like speed,
or gullying themselves into extinction,
rainforests and deserts zipping back and forth like Tom and Jerry.
A human mouse plague swiftly swarms, perhaps soon gone.

Impermanent planet.
Impermanent surface.
All under air and water.
The lunette is a wave, a dumper of sand,
a collapsing tower in the mad rabbit years when its top blew off
and old bones winked through,
with a vault's worth of fine stone cores.
Once so simply mislaid in sand,
now so obligingly sluiced out
to the un-subincised world
that smiles at stone knives.

You might as well offer your table knife
to a carpenter.

Cane grass and the Dreamtime's stone
fight the eroding rivulets together.
In today's rare downpour each lost core, or blunted scraper
crowns some tiny eroding hillock,
plops its anchor in the dissolving silts.

The bettong whose skin took off the stone tool's edge
has vanished – sacrifice to future fertility.

Through the softening soil Homo sapiens's
skeleton bumps up like a carp's back.

Waits for the wind to shave it off in pieces.

Preface

The impetus for this volume was the retirement of Professor Geoffrey Hope from the Department of Archaeology and Natural History at the Australian National University in June 2009 after a distinguished career in teaching and research in the fields of physical geography, and more specifically, palaeoecology. His pioneering efforts and long term commitment to research in Australia and in neighboring countries such as Papua New Guinea, Indonesia and Vanuatu exemplifies his tenacity and unwavering fascination for the natural world as well as the human dimension of our environment. This is illustrated in the variety and depth of the subjects he has tackled in the numerous scientific publications outlined in Chapter 1. Throughout his career Geoff has been a mentor and friend to many students and colleagues and has instilled a deep appreciation of the small and large dimensions of global change science. His unwavering support and encouragement of all who show an interest in the study of our environment has been an inspiration and guiding light to all of us who work with Geoff. In this volume contributors were asked to write on a theme that would reflect the impact Geoff has made on the many disciplines he has influenced. The title Altered Ecologies: Fire, climate and human influence on terrestrial landscapes reflects this wide-ranging influence and engagement with the past as well as the present. All papers were refereed and we are grateful to the authors for their participation and the referees for their time and dedication to producing this collection of papers. We also thank Kirsten Lawson, Adam Black and Duncan Beard for their generous help with volume production. Publication was possible with financial assistance from the Department of Archaeology and Natural History (ANU) and the Australasian Quaternary Association.

Simon Haberle, Janelle Stevenson and Matthew Prebble (Editors)

Publication supported by the Australasian Quaternary Association

Australasian Quaternary Association Inc.
AQUA

Contents

Introduction

1. A D-section and a tin whistle: A tribute to Prof. Geoff Hope
Matthew Prebble, Janelle Stevenson and Simon Haberle 1

Ecosystem responses to long and short term climate change

2. The contrasting biology of tropical versus temperate *Nothofagus* species and its relevance to interpretations of Cenozoic rainforest history in southeast Australia
Jennifer Read, Robert S. Hill, Geoffrey S. Hope and Raymond J. Carpenter 15

3. Beneath the peat: A refined pollen record from an interstadial at Caledonia Fen, highland eastern Victoria, Australia
A. Peter Kershaw, G. Merna McKenzie, Jonathan Brown, Richard G. Roberts and Sander van der Kaars 33

4. The vegetation cover of New Zealand during the Last Glacial Maximum: Do pollen records under-represent woody vegetation?
Matt S. McGlone, Rewi M. Newnham and Neville T. Moar 49

5. Holocene vegetation history of a high-elevation (1200 m) site in the Lake Heron Basin, inland Canterbury, New Zealand
J. M. Pugh and J. Shulmeister 69

6. Last Glacial Maximum habitat change and its effects on the grey-headed flying fox (*Pteropus poliocephalus* Temminck 1825)
J. G. Luly, David Blair, Jennifer G. Parsons, Samantha Fox and Jeremy VanDerWal 83

7. Observations on feeding frequencies among native and exotic birds and fruit bats at *Erythrina variegata* and *Dysoxylum* trees on American Samoa
Julie A. Sherman and Patricia L. Fall 101

Human colonisation and ecological impacts

8. Megafaunal extinctions and their consequences in the tropical Indo-Pacific
Richard T. Corlett 117

9. Marsupials as introduced species: Long-term anthropogenic expansion of the marsupial frontier and its implications for zoogeographic interpretation
Thomas E. Heinsohn 133

10. The empty coast: Conditions for human occupation in southeast Australia during the late Pleistocene
Sandra Bowdler 177

11. Early Holocene human occupation and environment of the Southeast Australian Alps: New evidence from the Yarrangobilly Plateau, New South Wales
Ken Aplin, Fred Ford and Peter Hiscock 187

12. Holocene lowland vegetation change and human ecology in Manus Province, Papua New Guinea
Matthew Prebble, Jean Kennedy and Wendy Southern 213

13. Geomorphic and archaeological consequences of human arrival and agricultural expansion on Pacific islands: A reconsideration after 30 years of debate
Matthew Spriggs 239

14. Pollen evidence for plant introductions in a Polynesian tropical island ecosystem, Kingdom of Tonga
Patricia L. Fall 253

15. Integrating social and environmental change in prehistory: A discussion of the role of landscape as a heuristic in defining prehistoric possibilities in northeast Thailand
William E. Boyd and Nigel Chang 273

Fire and its role in transforming our environment

16. A 40,000 year wood charcoal record from Carpenter's Gap 1: New insights into palaeovegetation change and indigenous foraging strategies in the Kimberley, Western Australia
Susan Frawley and Sue O'Connor 299

17. The burning question: Claims and counter claims on the origin and extent of buttongrass moorland (blanket moor) in southwest Tasmania during the present glacial-interglacial
Mike Macphail 323

18. Ecological drift or stable fire cycles in Tasmania: A resolution?
Ian Thomas, Phil Cullen and Michael-Shawn Fletcher 341

19. Restoration of mires of the Australian Alps following the 2003 wildfires
Roger Good, Genevieve Wright, Jennie Whinam and Geoff Hope 353

20. Post-fire experimental trials of vegetation restoration techniques in the peatlands of Namadgi (ACT) and Kosciuszko National Parks (NSW), Australia
Jennie Whinam, Geoffrey Hope, Roger Good and Genevieve Wright 363

Methodological advances and applications in environmental change research

21. The archaic and puzzling record of Lake Xere Wapo, New Caledonia
Janelle Stevenson, Richard Gillespie, Geoff Hope, Geraldine Jacobsen, Stewart Fallon and Vladimir Levchenko 381

22. Comparative AMS ^{14}C dating of plant macrofossils, beetles and pollen preparations from two Late Pleistocene sites in southeastern Australia
Nick Porch, A. Peter Kershaw 395

23. Can Myrtaceae pollen of the Holocene from Bega Swamp (New South Wales, Australia) be compared with extant taxa?
Andrew H. Thornhill 405

24. The evolution of a coastal peatland at Byron Bay, Australia: Multi-proxy evidence from the microfossil record
Kathryn H. Taffs, Brendan Logan, Jeff F. Parr and Geraldine E. Jacobsen 429

25. Development of mountain peatlands in stable equilibrium with open-channel hydraulics: A new concept in peatland formation and maintenance
Rachel Nanson 443

26. Glacier crippling and the rise of the snowline in western New Guinea (Papua Province, Indonesia) from 1972 to 2000
Michael L. Prentice and S. Glidden 457

27. Altitudinal limits of 230 economic crop species in Papua New Guinea
R. Michael Bourke 473

1

A D-section and a tin whistle: A tribute to Prof. Geoff Hope

Matthew Prebble

Department of Archaeology and Natural History, Research School of Pacific and Asian Studies, Australian National University, Canberra, Australia
matthew.prebble@anu.edu.au

Janelle Stevenson

Department of Archaeology and Natural History, Research School of Pacific and Asian Studies, Australian National University, Canberra, Australia

Simon Haberle

Department of Archaeology and Natural History, Research School of Pacific and Asian Studies, Australian National University, Canberra, Australia

After more than 40 years of academic research and teaching in Quaternary Science, Prof. Geoff Hope retired in June 2009. As an undergraduate student at the University of Melbourne in the 1960s, Geoff studied the natural sciences under such greats as Ray Specht. He later became one of a pioneering band of researchers to pursue palynology in Australia, starting with a Master's project on the peatlands of Wilsons Promontory, investigating the history of *Nothofagus* forest and aboriginal plant use. For his PhD research, he made the big leap into New Guinea, exploring glaciation as a driver of vegetation and climate change. Obtaining a Queen Elizabeth Fellowship, Geoff completed his PhD at the Department of Biogeography and Geomorphology at the ANU, under Donald Walker and Joe Jennings. To undertake his field work in 1972, Geoff lived for the whole year in a small ANU field hut (that still stands to this day) at the base of Mount Wilhelm, next to Lower Pindaunde Lake. From there, he made regular trips down to Kundiawa, the capital of Simbu Province, to obtain supplies and to submit samples of 'black sludge' for radiocarbon dating, as Richard Gillespie recalls:

> 'The year 1972 is probably best remembered by some Australians for the first Sunbury music festival, for others the Aboriginal Tent Embassy established on the lawns of Parliament House, the election of Gough Whitlam, or that catalogue sales for a Tutankhamun exhibition in London realised more than gate takings from the five-test Ashes series in England. Very few would have expected that intrepid Australians were

interrogating sediments in the high mountains of western New Guinea, for reasons no doubt best known to themselves. My knowledge of this event came from several bags of non-descript dirt, mostly black, some with woody bits, which landed on my desk at the Sydney University Radiocarbon Laboratory.

Not long after construction of that lab began in 1970, by MSc student Ian Fraser and PhD student Gillespie, our supervisor Assoc. Prof. Richard Temple attached a small typed notice to the door, respectfully dedicating the venture to former student Glynn Parfitt, a fundamentalist Plymouth Brethren devotee who had unsuccessfully tried to build a gas counting radiocarbon system to prove that the creation date calculated by Bishop Ussher was correct. We eventually got everything working and the first SUA-dates were published in late 1972, containing mainly modern standard measurements and cross-check samples provided as blind tests by Henry Polach from the ANU Radiocarbon Research Laboratory.

Among the many other samples arriving at the lab in 1972, but not making the first date list, were the bags of dirt from Jim Peterson and Geoffrey Hope. I was never sure why those samples came to Sydney, I had never heard of Hope or Peterson, and later suspected that Henry Polach (busy on the biggest game in town, the Mungo dates in Bowler et al. 1972) had farmed out the black sludge from New Guinea to me for a shot at making an impression on clients willing to pay good money for dates. I think we were charging $60 per sample.

Results were sent back (by Telex!) less than two weeks after they arrived at the lab, while the Peterson and Hope team was still in the field. Not that such service was on the same scale available to Americans: samples and radiocarbon dates were shuttled by car, on a weekly turn-around, between Rainer Berger's UCLA lab and massive excavations at Tule Springs, Nevada, in 1969. And scintillation counting pioneers Jerry Stipp and Murry Tamers built a mobile radiocarbon lab-on-a-truck, providing an overnight onsite service; later they built a radiocarbon date factory called Beta Analytic.

Some of the samples submitted by Hope and Peterson were among the first couple of dozen unknowns processed in the Sydney lab, although publication of these and later samples from the same region was delayed for four years (Gillespie and Temple 1976; Hope et al. 1976). There were more exciting samples in the same datelist, from archaeological sites at Malakunanja, Deaf Adder Gorge and Devil's Lair, but the tropical glacier and pollen-related samples from New Guinea were given the best available treatment in 1972. Thirty-seven years on, and I'm still dealing with sediments collected by Geoffrey Hope and others in New Caledonia (Stevenson et al. this volume), so I have to ask: when will this black sludge obsession cease, and what might the uncertainty on that date be?'

Geoff's primary research outcomes from his PhD resulted in some of his most widely acclaimed research. According to Paul Hesse (2006), his 1976 *Journal of Ecology* paper sits in the Top 10 best-ever Quaternary papers from Australasia. The paper traces tropical tree-line movement through dated records extending to the late Pleistocene. According to Paul, the substantial depression of temperatures (~8C°) 'has remained as an inspiration and frustration to palaeoclimatologists' attempts to understand climate change in the tropics ever since'.

After his PhD, Geoff was appointed a Lecturing Fellow in the Department of Geography, ANU, in 1978. Since that time, Geoff has been instrumental in engaging researchers from different fields in the wonders of New Guinea's natural and cultural history. He first introduced Tim Flannery to these marvels in 1982 on an expedition he organised to Kosipe Mission to explore the Pleistocene palaeoecology of the Ivane basin. In Tim's fascinating account of his adventures in New Guinea, *Throwim Way Leg* (1998), he describes Geoff as the most inspiring teacher of natural history he has known. As Flannery's book illustrated, being introduced to New Guinea by Geoff not only involved delving knee deep into swamps in the rugged mountains or mosquito-plagued lowlands, but also led to real adventure and irresistible opportunities for students to experience one of the most remote and rugged regions on earth. Also joining Geoff in New Guinea in the 1980s was plant ecologist Jenny Read, researching *Nothofagus* ecology (e.g. Read and Hope 1989). Jenny recalls one incident as a not uncommon occurrence in the highlands:

Figure 1. Geoff coring on Hunter Island, northwest Tasmania in 1976. Photograph Sandra Bowdler

'Geoff was collecting sediment cores from Nurenk Swamp, and some of us who had been conducting vegetation surveys in the area headed back to our vehicle early. When we arrived we were greeted by a crowd who informed us that they would kill us all

Figure 2. Geoff trudging through his beloved bogs on the Kemabu Plateau, a high-altitude area covered with thick glacial deposits to the north of Mount Jaya, West Papua, 1993

Figure 3. Geoff on Mount Wilhelm in 2007 reminiscing and resting with members of the clan who worked with him during his PhD fieldwork on the mountain in the early 1970s. Photograph Simon Haberle

> unless we gave them $100. I went back into the swamp to get Geoff (for his pidgin skills) and when he arrived he spent a couple of hours negotiating them down to $16! I didn't catch all of his arguments, but foreign aid, Australian development assistance, etc all played a part. I think we were all just sitting quietly (keeping our heads down). There are a few morals in this, depending on how you look at it. Somehow, though, I imagine that nearly all of us [who have worked with Geoff] have a similar story. Mostly when I think of Geoff I sadly think of a take-away stand [*kai* bar] in Mount Hagen that had a tank full of boiling fat and brisket and similar delights, and Geoff's excitement at such.'

Everyone will be pleased to know that Geoff has been a successful vegetarian for many years, mainly at the insistence of partner Bren, who is clearly aware of the perils of *kai* bars and Ox & Palm Bully Beef!

Geoff has often described himself as a 'plant historian', contrary to how some of his colleagues have seen him. The botanist Jim Croft, when working with Geoff on the botany and vegetation of highland New Guinea in the 1970s and '80s, found that:

> 'He was more interested in getting knee deep in high-altitude bogs and he had patiently explained that he wasn't really interested in and did not see plants in the field - he was after vegetation, a thing that botanists never saw.'

Geoff has continued to work in Papua New Guinea ever since, perhaps in part for his interest in the natural history and the impressive diversity of people and cultures in the region, but also because of his family heritage. His mother came from a sugar-cane growing family who lived in New Guinea during the early colonial days. To Jack Golson, Geoff represents part of a strong historical lineage of great multidisciplinary scientists who pioneered research in Papua New Guinea. His most recent work has been refocusing on the Owen Stanley Range, to his relief, publishing data which he obtained back in 1974 (Hope 2009). In 2007, Geoff led a group of international scientists on a two-week trip through the highlands of Papua New Guinea, visiting many of the old haunts from his PhD days. This included walking, perhaps a little slower this time, to the alpine grasslands of Mount Wilhelm in the Chumbu Province. As word went around the valley that he was there, it was with little surprise that the group soon met up with the people with whom Geoff had worked so closely in the 1970s (see Figure 3), many remembering the work that he did in opening up the region for researchers to return through the past 40 years. He returns once more to Mount Wilhelm this year, with Tim Barrows and Brad Pillans, in their attempt to further substantiate the timing of peak glacial conditions during the Last Glacial Maximum in Papua New Guinea.

He joined the Department of Biogeography and Geomorphology in the Research School of Pacific Studies, ANU, led by John Chappell, as Senior Fellow, in 1990 and was head of the Department of Archaeology and Natural History from 1998 to 2003. Five years as head of an archaeology department still hasn't earned him the title 'archaeologist'. Geoff often recalls working on Kangaroo Island in the 1970s with Ron Lampert while conducting geomorphological surveys and swamp coring. On one occasion, Geoff picked up what he thought was clearly a waisted blade. Ron explained that when in Geoff's hands 'this is not an artefact it's merely a stone, in the hands of an archaeologist it's an artefact!'.

Geoff's research interests are broad and they have taken him across Australia, to the rainforests of Southeast Asia and to the remote parts of the Pacific Islands. Patrick Nunn writes of Geoff's contribution to the palaeoecology of Fiji and endurance in the field:

Figure 4. Geoff teaching undergraduate students from the ANU in an area affected by the 2003 fires in the Namadgi National Park, ACT. Photograph Simon Haberle

Figure 5. Geoff with partner Bren and a harvest of garlic at their Bunga property on the south coast of NSW

> 'In the Vanua Balavu islands in 2000, the site on which the punishment squad [Patrick's students] focused their attention was a sinkhole named Cavaura. Geoff recognized at once the extraordinary potential of the sediment sequence likely to exist in the sediments in the centre of Cavaura and was determined to core them. With our eyes almost popping out in horror, we contemplated mutiny as we followed Geoff out to the very centre. With water up to our chests (for some of us), we dutifully pushed the hated corer further and further down into the scarcely-yielding sediments beneath. Down, down, down and finally, turn, then pull out. A process accompanied by a large subterranean belch, the black mud oozed out of the D-section corer. With astonishing equanimity for someone in his position, Geoff would have us hold the D-section horizontally while he cut the mud carefully into sections and bagged them.
>
> One day in Cavaura was unusually hot. Three of us were with Geoff in the centre of the lake coring. The water and the mud were burning us, even to the point of losing feeling in our legs and midriffs. Mentioned this to Geoff and he laughed (as though he were not even uncomfortable!), saying that in Papua New Guinea you have all this plus a couple of eight-foot crocodiles!
>
> Lunch at Cavaura, precariously seated in the hornet-infested rock shelters that lined the side of the sinkhole, was invariably enlivened by Geoff's piping. Just when we thought nothing else could possibly impact our sensibilities, out would come a penny whistle and Geoff would play a jig. Enclosed by limestone cliffs, framing the blue sky above us, an eerie quiet all around, Geoff's music would instantly enthuse us. Reminding us that there was another world beyond Cavaura into which, after a few more hours in that watery hell, we would joyfully re-enter, the music gave us new hope and the willpower that we needed to finish the job.'

Geoff continues to pull out his tin whistle at odd occasions. On one of these occasions when Geoff was heading out of Brisbane airport he left his whistle in a carry-on bag and was abruptly asked by the security officer 'what's this for?' Geoff replied 'This!' and proceeded to play a jig to the joy of all.

Geoff's practical attitude to life and work is captured by this anecdote from Sue O'Connor and fellow archaeologist Matthew Spriggs, on an expedition to East Timor with Geoff in 2000. United Nations personnel had warned them they would be unable to get to Tutuala village, which is on the eastern tip of the island, because Lake Ira Lalaro had flooded, cutting off the road and isolating the region for months. UN vehicles were reluctant to venture on to the flooded roads. There was no stopping Geoff, 'We're Australian,' he said. 'No wozzas, we can get through this easy.' He then opened out his tent, took out the flysheet and proceeded to wrap it around the front grille of the vehicle. They triumphantly made the crossing and the three were saluted by locals as they drove down Tutuala's only street.

One of the most admirable characteristics of Geoff is his inclusiveness, both of his family, students and colleagues. Geoff's family have joined him on many expeditions to remote and often precarious locations, as well as welcoming many of us into their home for one of those not to be missed Greenhood Place gatherings. His 'can do' approach to life has meant that research gets done and everyone gets involved. In an interview by The Australian (17.2.2010), Geoff suggested that he was not retiring but entering a new phase as 'a gentleman scientist'. Geoff was recently awarded emeritus status at the ANU and all of us wish him well, and expect him to continue to probe the depths of Australasia's landscape history. To answer Richard's query, the quest for more 'black sludge' will never end!

Other papers mentioned in the text

Bowler, J.M., A.G. Thorne and H.A. Polach 1972. Pleistocene man in Australia: Age and Significance of the Mungo Skeleton. *Nature* 240:48-50.

Flannery, T.F. 1998. Throwim Way Leg: Tree-Kangaroos, Possums, and Penis Gourds – On the track of unknown mammals in wildest New Guinea. New York: Atlantic Monthly Press.

Gillespie, R. and R.B. Temple 1976. Sydney University Natural Radiocarbon Measurements III. *Radiocarbon* 18:96-109.

Hesse, P. 2006. The Australian Quaternary top 10. *Quaternary Australasia* 23:12-15

Stevenson, J., Gillespie, R., G.S. Hope, Jacobsen, G., Fallon, S. and V. Levchenko 2010. The archaic and puzzling record of Lake Xere Wapo, New Caledonia. In Simon G. Haberle, Janelle Stevenson and Matthew Prebble (editors), *Altered Ecologies: Fire, Climate and Human Influence on Terrestrial Landscapes*, Terra Australis 32. ANU E-Press, Canberra, The Australian National University, this volume.

Publications of Geoffrey S. Hope

Howard, T.M. and G.S. Hope 1970. The present and past occurrence of beech (*Nothofagus cunninghamii* Oerst.) at Wilsons Promontory, Victoria, Australia. *Proceedings of the Royal Society of Victoria* 83:199-210.

Hope, G.S. and P.J.F. Coutts 1971. Past and present aboriginal food resources at Wilson's Promontory. *Mankind* 8:104-114.

Hope, G.S. and G.K. Thomson 1971. The vegetation of Cliffy Island, Victoria, Australia. *Proceedings of the Royal Society of Victoria* 84:121-128.

Peterson, J.A. and G.S. Hope 1972. A lower limit and maximum age for the last major advance of Carstensz glaciers, West Irian. *Nature* 240:36-37.

Galloway, R.W., G.S. Hope, E. Loffler and J.A. Peterson 1973. Late Quaternary glaciation and periglaciation phenomena in Australia and New Guinea. In E.M. van Zinderen Bakker (ed.) *Palaeoecology of Africa, the surrounding Islands and Antarctica*, pp125-138. Capetown, South Africa: Balkema.

Hope, G.S., J.A. Peterson and R. Mitton 1973. Recession of the minor ice fields of Irian Jaya. *Zeitschrift fur Glestcherkunde and Glazialgeologie* 9:73-87.

Scarlett, N.H., G.S. Hope and D.M. Calder 1974. Natural history of the Hogan Group. 3. Floristics and plant communities. *Proceedings of the Royal Society of Tasmania* 107:83-98.

Hope, G.S. 1974. The vegetation history from 6000 B.P. to present of Wilsons Promontory, Victoria, Australia. *New Phytologist* 73:1035-1053.

Hope, G.S. and J.A. Peterson 1975. Glaciation and vegetation in the high New Guinea mountains. *Bulletin of the Royal Society of New Zealand* 13:153-162.

Schodde, R., J. van Tets, C.R. Champion and G.S. Hope 1975. Birds of Carstensz glaciers. *Emu* 75:65-72.

Hope, G.S. 1976. Palaeoenvironments for man in New Guinea. In R.L. Kirk and R.F. Thorne (eds) *The Origins of the Australians*, pp29-53. Canberra: Institute of Aboriginal Studies.

Hope, G.S. 1976. Vegetation. In G.S. Hope, J.A. Peterson, I. Allison and U. Radok (eds) *The Equatorial Glaciers of New Guinea (Results of the 1971-1973 Australian Universities' Expeditions to Irian Jaya: survey, glaciology, meteorology, biology and palaeoenvironments*, pp113-172. Rotterdam: A.A. Balkema.

Hope, G.S. 1976. Mt Jaya: the area and its exploration. In G.S. Hope, J.A. Peterson, I. Allison and U. Radok (eds) *The Equatorial Glaciers of New Guinea (Results of the 1971-1973 Australian Universities' Expeditions to Irian Jaya: survey, glaciology, meteorology, biology and palaeoenvironments*, pp1-14. Rotterdam: A.A. Balkema.

Hope, G.S. and J.H. Hope 1976. Man on Mt Jaya. In G.S. Hope, J.A. Peterson, I. Allison and U. Radok (eds) *The Equatorial Glaciers of New Guinea (Results of the 1971-1973 Australian Universities' Expeditions to Irian Jaya: survey, glaciology, meteorology, biology and palaeoenvironments*, pp225-238. Rotterdam: A.A. Balkema.

Hope, G.S. and J.A. Peterson 1976. Palaeoenvironments. In G.S. Hope, J.A. Peterson, I. Allison and U. Radok (eds) *The Equatorial Glaciers of New Guinea (Results of the 1971-1973 Australian Universities' Expeditions to Irian Jaya: survey, glaciology, meteorology, biology and palaeoenvironments*, pp173-206. Rotterdam: A.A. Balkema.

Hope, G.S. 1976. The vegetational history of Mt Wilhelm, Papua New Guinea. *Journal of Ecology* 64:627-663.

Bowler, J.M., G.S. Hope, J.N. Jennings, G. Singh and D. Walker 1976. Late Quaternary climates of Australia and New Guinea. *Quaternary Research* 6:359-394.

Bowler, J.M., G.S. Hope, J.N. Jennings, G. Singh and D. Walker 1978. Reply to comment by G. E. Williams. *Quaternary Research* 9:114-115.

Hope, G.S. 1978. The Late Pleistocene and Holocene vegetational history of Hunter Island, North-Western Tasmania. *Australian Journal Botany* 26:493-514.

Hughes, P.J., G.S. Hope, M. Latham and M. Brookfield 1979. Prehistoric man-induced degradation of the Lakeba landscape: evidence from two inland swamps. In H.C. Brookfield (ed.) *Lakeba: environmental change, population and resource use*, pp93-111. Island Reports, UNESCO/UNFPA.

Hope, G.S. 1980. Historical influences on the New Guinea flora. In P. van Royen (ed.) *Alpine flora of New Guinea*, pp223-248. Vaduz: Cramer Verlag.

Hope, G.S. 1980. Tropical mountains in retrospect and prospect. In J.N. Jennings and G. Linge (eds) *Of time and place*, pp153-169. Canberra: ANU Press.

Hope, G.S. 1980. New Guinea mountain vegetation communities. In P. van Royen (ed.) *Alpine flora of New Guinea*, pp153-222. Vaduz: Cramer Verlag.

Hope, G.S. 1982. Pollen from archaeological sites: a comparison of swamp and open archaeological site pollen spectra at Kosipe Mission, Papua New Guinea. In W. Ambrose and P. Duerden (eds) Archaeometry: an Australian perspective, pp211-219. Canberra: Department of Prehistory, Research School of Pacific Studies, Australian National University.

Hope, G. and M. Spriggs 1982. A preliminary pollen sequence from Aneityum Island, southern Vanuatu. *Bulletin of the Indo-Pacific Prehistory Association* 2:88-94.

Walker, D. and G.S. Hope 1982. Late Quaternary vegetation history. In J.L. Gressitt (ed) *Biogeography and ecology of New Guinea*, pp263-287. The Hague: Junk.

Hope, G.S. 1983. Coastal southern Australia. In J. Chappell and A. Grindrod (eds) *CLIMANZ: Quaternary climatic history of Australia*, pp12, 34, 56, 74, 98. Canberra: Department of Biogeography and Geomorphology, Australian National University.

Hope, G.S. 1983. New Guinea. In J. Chappell and A. Grindrod (eds) CLIMANZ: Quaternary climatic history of Australia, pp 12, 35, 55, 75, 97. Canberra: Department of Biogeography and Geomorphology, Australian National University.

Hope, G.S. 1983. The vegetation changes of the last 20,000 years at Telefomin, Papua New Guinea. *Journal of Tropical Geography* 4:25-33.

Hope, G.S., J. Golson and F.J. Allen 1983. Palaeoecology and prehistory in New Guinea. *Journal of Human Evolution* 12:37-60.

Hope, G.S., J. Golson and J. Allen 1983. Palaeoecology and prehistory in New Guinea. *Journal of Human Evolution* 12:37-60.

Hope, G.S. 1984. Australian environmental change: timing, directions, magnitudes and rates. In P.S. Martin and R.G. Klein (eds) *Pleistocene extinctions: a prehistoric revolution*, pp681-690. Tucson, Arizona: University of Arizona.

Gillieson, D., P. Gorecki and G. Hope 1985. Prehistoric agricultural systems in a lowland swamp, Papua New Guinea. *Archaeology in Oceania* 20:32-37.

Hope, G.S., P.J. Hughes and J. Russell-Smith 1985. Geomorphological fieldwork and the evolution of the landscape of Kakadu National Park. In R. Jones (ed.) *Archaeological research in Kakadu National Park*, pp229-240. Canberra: Australian National University & National Parks and Wildlife.

MacPhail, M.K. and G.S. Hope 1985. Late Holocene mire development in montane southeastern Australia: a sensitive climatic indicator. *Search* 15:344-348.

Gibson, N. and G.S. Hope 1986. On the origin and evolution of Australasian alpine cushion plants. In B.A. Barlow (ed) Flora and fauna of alpine Australasia, pp 62-91. Melbourne: CSIRO.

Hope, G.S. 1986. Development of present day biotic distributions in the New Guinea mountains In B.A. Barlow (ed). Flora and fauna of alpine Australasia, pp129-145. Melbourne: CSIRO.

Hope, G., D. Gillieson and J. Head 1988. A comparison of sedimentation and environmental change in New Guinea shallow lakes. *Journal of Biogeography* 15:603-618.

Gillieson, D. and G.S. Hope 1989. Development of the Mountain Ok environment. In B. Craig and D. Hyndman (eds) Children of Afek: tradition and change among Mountain Ok of Central New Guinea, pp27-34. Sydney: Oceania Monographs.

Gillieson, D., G. Hope and J. Luly 1989. Environmental change in the Jimi Valley. In P. Gorecki and D.S. Gillieson (eds) *A crack in the spine, prehistory and ecology of the Jimi-Yuat Valley, Papua New Guinea*, pp105-122. Townsville: James Cook University.

Hope, G.S. 1989. Climatic implications of timberline changes in Australasia from 30 000 BP to present. In T. Donnelly and R. Wasson (eds) *CLIMANZ 3*, pp91-99. Canberra: CSIRO Division of Water Research.

Hope, G.S. and J.B. Kirkpatrick 1989. The ecological history of Australian forests. In K. Frawley and N. Semple (eds) *Australia's ever changing forests*, pp3-22. Canberra: Department of Geography and Oceanography, Australian Defence Force Academy.

Read, J. and G.S. Hope 1989. Foliar Frost-Resistance of Some Evergreen Tropical and Extratropical Australasian Nothofagus Species. *Australian Journal Botany* 37:361-373.

Haberle, S.G., G.S. Hope and Y. DeFretes 1991. Environmental change in the Baliem Valley, montane Irian Jaya, Republic of Indonesia. *Journal of Biogeography* 18:25-40.

Swadling, P. and G.S. Hope 1992. Environmental change in New Guinea since human settlement. In J. Dodson (ed.) The naive lands - prehistory and environmental change in the South West Pacific, pp13-42. Cheshire: Longman.

Hope, G., T. Flannery and Boeardi 1993. A preliminary report of changing Quaternary mammal faunas in subalpine New Guinea. *Quaternary Research* 40:117-126.

Hope, G.S. 1994. Quaternary vegetation. In R.S. Hill (ed.) *History of the Australian vegetation. Cretaceous to recent*, pp368-389. Cambridge: Cambridge University Press.

Thomas, I. and G.S. Hope 1994. An example of Holocene stability from Camerons Lagoon, a new treeline site on the Central Plateau, Tasmania. *Australian Journal of Ecology* 19:150-158.

Hope, G. and J. Tulip 1994. A long vegetation history from lowland Irian Jaya, Indonesia. *Palaeogeography, Palaeoclimatology, Palaeoecology* 109:385-398.

Hope, G.S. 1995. Australian Quaternary palaeoecology and its relevance to current issues of biodiversity. *Australasian Biotechnology* 5:220-224.

Hope, G.S. and J. Golson 1995. Late Quaternary change in the mountains of New Guinea. *Antiquity* 69:818-830.

Markgraf, V., M. McGlone and G. Hope 1995. Neogene paleoenvironmental and paleoclimatic change in southern temperate ecosystems -- a southern perspective. *Trends in Ecology and Evolution* 10:143-147.

Hope, G.S. 1996. History of Nothofagus in New Guinea and New Caledonia. In T.T. Veblen, R.S. Hill and J. Read (eds) *The ecology and biogeography of Nothofagus forests*, pp257-270. New Haven: Yale University Press.

Hope, G.S. 1996. Quaternary change and historical biogeography of the Pacific Islands. In A. Keast and S.E. Miller (eds) *The origin and evolution of Pacific Island biotas, New Guinea to Eastern Polynesia: patterns and processes*, pp165-190. Amsterdam: SPB Academic Publishing.

McGlone, M.S., G.S. Hope, J.M.A. Chappell and P. Barrett 1996. Past climatic change in Oceania and Antarctica. In J. Bouma, G.I. Pearman and M.R. Manning (eds) *Greenhouse: coping with climate change*, pp81-99. Melbourne: CSIRO.

Read, J. and G.S. Hope 1996. Ecology of Nothofagus in New Guinea and New Caledonia. In T.T. Veblen, R.S. Hill and J. Read (eds) *The ecology and biogeography of Nothofagus forests*, pp200-256. New Haven: Yale University Press.

Kershaw, A.P., M.B. Bush, G.S. Hope, K.F. Weiss, J.G. Goldammer and R. Sanford 1997. The contribution of humans to past biomass burning in the tropics. In J. Clark, H. Cachier, J.G. Goldammer and B. Stocks (eds) Sediment records of biomass burning and global change, pp413-442. Berlin: Springer-Verlag.

Hope, G. 1998. Early fire and forest change in the Baliem Valley, Irian Jaya, Indonesia. *Journal of Biogeography* 25:453-461.

Dickinson, W.R., D.V. Burley, P.D. Nunn, A. Anderson, G.S. Hope, A.D. Biran, C. Burke and S. Matararaba 1998. Geomorphic and archaeological landscapes of the Sigatoka Dune Site, Viti Levu, Fiji: Interdisciplinary investigations. *Asian Perspectives* 37:1.

Hope, G. and J. Pask 1998. Tropical vegetational change in the late Pleistocene of New Caledonia. *Palaeogeography, Palaeoclimatology, Palaeoecology* 142:1-21.

Hope, G. 1999. Vegetation and fire response to late Holocene human occupation in island and mainland north west Tasmania. *Quaternary International* 59:47-60.

Read, J., T. Jaffre, E. Godrie, G.S. Hope and J.M. Veillon 2000. Structural and floristic characteristics of some monodominant and adjacent mixed rainforests in New Caledonia. *Journal of Biogeography* 27:233-250.

Haberle, S.G., G.S. Hope and S. van der Kaars 2001. Biomass burning in Indonesia and Papua New Guinea: natural and human induced fire events in the fossil record. *Palaeogeography, Palaeoclimatology, Palaeoecology* 171:259-268.

Hope, G. 2001. Environmental change in the Late Pleistocene and later Holocene at Wanda site, Soroako, South Sulawesi, Indonesia. *Palaeogeography, Palaeoclimatology, Palaeoecology* 171:129-145.

Hope, G.S. 2002. The Late Quaternary of Kiritimati (Christmas) Island, Kiribati. In A.J. Anderson, H. Martinsson Wallin and P. Wallin (eds) *The Prehistory of Kiritimati (Christmas) Island, Republic of Kiribati Excavations and analyses*, pp111-118. Oslo: Kon-Tiki Museum Occasional papers.

Nunn, P.D., C. Ollier, G. Hope, P. Rodda, A. Omura and W.R. Peltier 2002. Late Quaternary sea-level and tectonic changes in northeast Fiji. *Marine Geology* 187:299-311.

Hope, G.S. 2003. The mountain mires of southern New South Wales and the Australian Capital Territory: their history and future. In B. Mackay (ed.) *Proceedings of an International Year of the Mountains Conference, Jindabyne Nov 25-28, 2002*, pp67-79. Canberra: Australian Alps Liaison Committee.

Hope, G.S., R. Hnatiuk and J.B. Smith 2003. Asia y Oceania: Los Ecosistemas Alpinos De Nueva Guinea Y El Sudeste De Asia (The alpine ecosystems of New Guinea and Southeast Asia). In R. Hofstede, P. Segarra and P.M. Vasconez (eds) *Los Paramos del Mundo*, pp245-253. Quito: IUCN/EcoCiencia.

Whinam, J., G.S. Hope, B.R. Clarkson, R.P. Buxton, P.A. Alspach and P. Adam 2003. Sphagnum in peatlands of Australasia: Their distribution, utilisation and management. *Wetlands Ecology and Management* 11:37-49.

Bird, M.I., G. Hope and D. Taylor 2004. Populating PEP II: the dispersal of humans and agriculture through Austral-Asia and Oceania. *Quaternary International* 118-119:145-163.

Hope, G.S. 2004. Glaciation of Malaysia and Indonesia, excluding New Guinea. In J. Ehlers and P.L. Gibbard (eds) *Quaternary glaciations - extent and chronology, Part III, Developments in Quaternary Science*, pp211-214. Amsterdam: Elsevier.

Hope, G., A.P. Kershaw, S.v.d. Kaars, S. Xiangjun, P.-M. Liew, L.E. Heusser, H. Takahara, M. McGlone, N. Miyoshi and P.T. Moss 2004. History of vegetation and habitat change in the Austral-Asian region. *Quaternary International* 118-119:103-126.

Thomas, F.R., P.D. Nunn, T. Osborne, R. Kumar, F. Areki, S. Matararaba, D. Steadman, D. O'Day and G. Hope 2004. Recent archaeological findings at Qaranilaca Cave, Vanuabalavu Island, Fiji. *Archaeology in Oceania* 39:42-49.

Hope, G.S. 2005. The Quaternary in Southeast Asia. In A. Gupta (ed.) *The physical geography of Southeast Asia*, pp24-37. Oxford: Oxford University Press.

Hope, G.S. and K. Aplin 2005. Environmental change in the Aru Islands. In S. O'Connor, M. Spriggs and P. Veth (eds) *The Archaeology of the Aru Islands, Eastern Indonesia, Terra Australis*, pp25-40. Canberra: Pandanus Books.

Hope, G.S. and S.G. Haberle 2005. The history of the human landscapes of New Guinea. In A. Pawley, R. Attenborough, J. Golson and R. Hide (eds) *Papuan pasts: cultural, linguistic and biological histories of Papuan-speaking peoples*, pp541-554. Canberra: Pacific Linguistics, The Australian National University.

Hope, G., J. Whinam and R. Good 2005. Methods and preliminary results of post-fire experimental trials of restoration techniques in the peatlands of Namadgi (ACT) and Kosciuszko National Parks (NSW). *Ecological Management and Restoration* 6:214-217.

O'Connor, S., K. Aplin, J. Pasveer and G.S. Hope 2005. Liang Nabulei Lisa: A Late Pleistocene and Holocene sequence from the Aru Islands. In S. O'Connor, M. Spriggs and P. Veth (eds) *The Archaeology of the Aru Islands, Eastern Indonesia, Terra Australis*, pp125-162. Canberra: Pandanus Books.

Prentice, M.L., G.S. Hope, K. Maryunani and J.A. Peterson 2005. An evaluation of snowline data across New Guinea during the last major glaciation, and area-based glacier snowlines in the Mt. Jaya region of Papua, Indonesia, during the Last Glacial Maximum. *Quaternary International* 138-139:93-117.

Read, J., G.S. Hope and R.S. Hill 2005. Phytogeography and climate analysis of Nothofagus subgenus Brassospora in New Guinea and New Caledonia. *Australian Journal Botany* 53:297-312.

Hope, G., U. Chokkalingam and S. Anwar 2005. The stratigraphy and fire history of the Kutai Peatlands, Kalimantan, Indonesia. *Quaternary Research* 64:407-417.

Stevenson, J. and G. Hope 2005. A comparison of late Quaternary forest changes in New Caledonia and northeastern Australia. *Quaternary Research* 64:372-383.

Anderson, A., R. Roberts, W. Dickinson, G. Clark, D. Burley, A.d. Biran, G. Hope and P. Nunn 2006. Times of sand: Sedimentary history and archaeology at the Sigatoka Dunes, Fiji. *Geoarchaeology* 21:131-154.

Heinsohn, T.E. and G.S. Hope 2006. The Torresian connections: Zoogeography of New Guinea. In J.R. Merrick, M. Archer, G.M. Hickey and M.S.Y. Lee (eds) *Evolution and Biogeography of Australasian Vertebrates*, pp71-93. Sydney: Auscipub.

Hope, G.S. 2006. Histories of wetlands in the Australian Capital Territory and the bog recovery program. In K. McCue, S. Lenz and S. Freidrich (eds) *Caring for Namadgi Science and People. Proceedings of the NPA ACT Symposium Canberra May 2006*, pp129-144. Canberra: National Parks Association ACT.

Hope, G.S., J. Coddington and D. O'Dea 2006. Estuarine development and human occupation at Bobundara Swamp, Tilba Tilba, New South Wales, Australia. In M. Lillie and S. Ellis (eds) *Wetland Archaeology and environments: regional issues, global perspectives*, pp258-274. Oxford: Oxbow.

Whinam, J. and G.S. Hope 2006. The peatlands of the Australasian Region. In G.M. Steiner (ed.) *Mires from Siberia to Tierra Del Fuego*, pp397-434. Stapfia 85, Zugleich Kataloge der Oberösterreichischen Landsmuseen, N.S.

Read, J., Jaffré, T., J.M. Ferris, S. McCoy and G.S. Hope 2006. Does soil determine the boundaries of monodominant rain forest with adjacent mixed rain forest and maquis on ultramafic soils in New Caledonia? *Journal of Biogeography* 33:1055-1065.

Fairbairn, A., G. Hope and G. Summerhayes 2006. Pleistocene occupation of New Guinea's highland and subalpine environments. *World Archaeology* 38:371-386.

Hope, G.S. 2007. Palaeoecology and palaeoenvironments of Papua. In A.J. Marshall and B.M. Beehler (eds) *The ecology of Papua*, pp255-266. Singapore: Periplus editions.

Hope, G.S. 2007. The history of human impact on New Guinea. In A.J. Marshall and B.M. Beehler (eds) *The ecology of Papua*, pp1087-1097. Singapore: Periplus editions.

Hope, G.S. and K.P. Aplin 2007. Paleontology. In A.J. Marshall and B.M. Beehler (eds) *The ecology of Papua*, pp246-254. Singapore: Periplus editions.

Hope, G.S. and A.E. Hartemink 2007. Soils of Papua. In A.J. Marshall and B.M. Beehler (eds) *The ecology of Papua*, pp165-176. Singapore: Periplus editions.

Prentice, M. and G.S. Hope 2007. Climate of Papua. In A.J. Marshall and B.M. Beehler (eds) *The ecology of Papua*, pp177-195. Singapore: Periplus editions.

Donders, T.H., S.G. Haberle, G. Hope, F. Wagner and H. Visscher 2007. Pollen evidence for the transition of the Eastern Australian climate system from the post-glacial to the present-day ENSO mode. *Quaternary Science Reviews* 26:1621-1637.

Hope, G.S. and R.L. Clark 2008. A tale of two swamps: subalpine peatlands in the Kelly-Scabby area of Namadgi National Park. In K. McCue and S. Lenz (eds) Corridors for survival in a changing world, pp61-76. Canberra: National Parks Association, ACT.

Power, M., J. Marlon, N. Ortiz, P. Bartlein, S. Harrison, F. Mayle, A. Ballouche, R. Bradshaw, C. Carcaillet, C. Cordova, S. Mooney, P. Moreno, I. Prentice, K. Thonicke, W. Tinner, C. Whitlock, Y. Zhang, Y. Zhao, A. Ali, R. Anderson, R. Beer, H. Behling, C. Briles, K. Brown, A. Brunelle, M. Bush, P. Camill, G. Chu, J. Clark, D. Colombaroli, S. Connor, A.L. Daniau, M. Daniels, J. Dodson, E. Doughty, M. Edwards, W. Finsinger, D. Foster, J. Frechette, M.J. Gaillard, D. Gavin, E. Gobet, S. Haberle, D. Hallett, P. Higuera, G. Hope, S. Horn, J. Inoue, P. Kaltenrieder, L. Kennedy, Z. Kong, C. Larsen, C. Long, J. Lynch, E. Lynch, M. McGlone, S. Meeks, S. Mensing, G. Meyer, T. Minckley, J. Mohr, D. Nelson, J. New, R. Newnham, R. Noti, W. Oswald, J. Pierce, P. Richard, C. Rowe, M. Sanchez Goñi, B. Shuman, H. Takahara, J. Toney, C. Turney, D. Urrego-Sanchez, C. Umbanhowar, M. Vandergoes, B. Vanniere, E. Vescovi, M. Walsh, X. Wang, N. Williams, J. Wilmshurst and J. Zhang 2008. Changes in fire regimes since the Last Glacial Maximum: an assessment based on a global synthesis and analysis of charcoal data. *Climate Dynamics* 30:887-907.

Hope, G.S., J. Stevenson and W. Southern 2009. Vegetation histories from the Fijian Islands: Alternative records of human impact. In G. Clark and A. Anderson (eds) *The early prehistory of Fiji, Terra Australis*, pp64-86. Canberra: ANU Press.

Hope, G. 2009. Environmental change and fire in the Owen Stanley Ranges, Papua New Guinea. *Quaternary Science Reviews* 28:2261-2276.

2

The contrasting biology of tropical versus temperate *Nothofagus* species and its relevance to interpretations of Cenozoic rainforest history in southeast Australia

Jennifer Read

School of Biological Sciences, Monash University, Victoria, Australia
jenny.read@sci.monash.edu.au

Robert S. Hill

Faculty of Sciences, University of Adelaide, South Australia
South Australian Museum, Adelaide, South Australia

Geoffrey S. Hope

Department of Archaeology and Natural History, Research School of Pacific and Asian Studies, Australian National University, Canberra, Australia

Raymond J. Carpenter

Faculty of Sciences, University of Adelaide, South Australia

Introduction

Given the uncertain impacts of global climate change, it is becoming increasingly important to understand the factors controlling vegetation change, species distributions and vagility. The fossil record provides the only real source of long-term data for changes in distributions and characteristics of species and vegetation types. However, while making an enormous contribution to our knowledge of vegetation history, interpretation is limited by the incomplete fossil record, understanding of the biology of fossil species, and insufficient precision regarding the timing of key environmental events.

One advantage of the macrofossil record, particularly that relating to leaves, is that better indications can be obtained about the biology of fossil species than from microfossils alone, allowing clearer insights into possible controls of vegetation change in the distant past. For example, trends in leaf shape, size, margins, texture and anatomical features such as stomatal density have been described or modelled in relation to a range of environmental factors (Givnish 1979; Beerling and Kelly 1996; McElwain and Chaloner 1996; Wilf et al. 1998; Beerling et al. 2002; Royer and Wilf 2006). Some leaf traits also allow the systematic affinities of leaf fossils to be determined with greater rigour than for most microfossils. Nearest Living Relative (NLR) analysis is the most compelling approach to the reconstruction of past environments from botanical evidence, especially for the time since angiosperms assumed dominance (Hill and Scriven 1997). However, this approach relies on a detailed understanding of the biology of living relatives, for which data are often lacking. Here, we use studies of extant *Nothofagus* (Nothofagaceae) species to explore effects of changing climate on the distribution of evergreen *Nothofagus* during the Cenozoic in southeast Australia, focusing on the Palaeocene to Pliocene as a case study to explore effects of climate change on rainforest. We focus in particular on the changing composition of the rainforest, rather than on its extent.

Nothofagus – key genus in time and space[1]

Nothofagus is a key genus in the fossil record of the Southern Hemisphere, represented through an 80 Myr record across much of Gondwana (southern South America, Antarctica, New Zealand and Australia) in the form of pollen, leaves and reproductive organs (Hill and Dettmann 1996; Hill 2001). This evidence, together with the broad geographic range of living species, provides the opportunity of interpreting the distribution and ecology of different *Nothofagus* taxa in response to environmental changes over time. *Nothofagus* is now most often found in rainforests, but some deciduous species extend into mediterranean climates in central Chile (Donoso 1996) and into cool, dry climates of the eastern Andes (Veblen et al. 1996). Within rainforests, habitats range from equatorial montane forests in New Guinea (here taken to comprise the West Papuan region of Indonesia and Papua New Guinea), to the cold southernmost forests of Tierra del Fuego, with deciduous species extending to the treeline in Tasmania and southern South America. *Nothofagus* even occurs across a wide range of soils, including relatively fertile soils, as well as old, leached, nutrient-poor (e.g. Read and Brown 1996) and ultramafic soils (Read and Hope 1996). Moreover, the pollen and often the macrofossils of *Nothofagus* allow distinction between tropical and temperate taxa, and among some temperate taxa (Hill 1991; Hill and Read 1991), although some caution is necessary in interpreting fossils as belonging to extant clades (Jordan and Hill 1999; Sauquet et al. 2009). Species typically show relatively localised seed dispersal, with most seed-fall occurring close to parent trees (e.g. Ogden et al. 1996), but long-distance dispersal does seem to occur (e.g. Ogden et al. 1996) and may have contributed significantly to migration in the past (Hill and Dettmann 1996). A better knowledge of the biology of these species and the factors limiting their distribution should allow better interpretation of the fossil record, and hence of factors limiting tree distributions and vagility.

The fossil record of *Nothofagus* is unquestionably one of the best of any plant group worldwide. However, one problem with using *Nothofagus* as a case study of species distribution and diversity through time is that the genus is apparently well past its time of maximum diversity (Hill 2001). This means that extant species might not represent the full range of ecological responses that *Nothofagus* has exhibited throughout its history. Hence, it may be difficult to fully reconstruct the environmental changes to which members of this genus have responded in the past. Furthermore, this approach rests on the assumption that key features of the biology of fossil species that have influenced their distribution have been retained, at least in part, in living relatives (discussed in Read et al. 1990a). Leaves of some fossil

[1]van Steenis (1971)

Nothofagus species are morphologically very similar to those of living species (Hill 2001), suggesting that aspects of their biology have been conserved to some degree. In addition, some *Nothofagus* taxa that once coexisted diverged greatly in their distribution after the early Miocene, suggesting there was already ecophysiological differentiation in place. Nevertheless, caution must be applied in use of this approach.

Historical changes in climate and vegetation responses, focusing on *Nothofagus*

The first pollen grains attributed to *Nothofagus* appear at very high latitudes at about 80 Ma (Hill and Dettmann 1996). We know little about the palaeoecology of these early species in part because the records are predominantly from microfossils and confidence in the NLR approach decreases with the age of the fossils. *Nothofagus* species were present in the region through the cooling of the late Cretaceous, and during the mild, wet Palaeocene (with periods of winter darkness at high latitudes) (Truswell 1993; McGowran et al. 2000). During the non-seasonal/megathermal early Eocene, *Nothofagus* may have been common only in the Tasmanian highlands, becoming more widespread across Australia during the middle to late Eocene when conditions were cooler (mesothermal) with little or no seasonal variation (McGowran et al. 2000). Some assemblages recorded at that time, such as suggested deciduous *Nothofagus* species in diverse angiosperm forests in the middle Eocene (Scriven et al. 1995), have no modern analogues.

The development of what appear to be modern climatic responses in the genus begins with its rise to dominance at high southern latitudes from about the late Eocene (~40 Ma) onwards (Hill 2001). This was more or less coincident with the end of the early Palaeogene 'Greenhouse' interval of global warmth, high atmospheric CO_2 levels (Pagani et al. 2005) and high water availability, and the initiation of the major Cenozoic cooling trend, as Australia, and later South America, separated from Antarctica, resulting in the development of the circum-Antarctic ocean current. *Nothofagus* then assumed dominance over much of Australia, Antarctica, southern South America and New Zealand, probably reaching its greatest diversity and widest distribution during the Oligocene (Hill 2001). Some extraordinary coexistence of *Nothofagus* taxa occurred at this time, most notably at Little Rapid River in northwest Tasmania, where several species from three of the four extant subgenera (*Brassospora*, *Nothofagus* and *Lophozonia*), and possibly even the remaining subgenus (*Fuscospora*), coexisted within a small catchment area (Hill 1991, 1994, 2001). Today, subgenus *Brassospora* occurs only in New Guinea and New Caledonia, subgenus *Nothofagus* occurs only in southern South America, and the other subgenera occur in each of Australia, New Zealand and South America (Hill and Dettmann 1996). Given the considerable latitudinal and longitudinal separation of some subgenera today, their coexistence at any time in the past is remarkable. This poses two related questions that we will focus on here: What conditions allowed these subgenera to coexist in the past? And, in particular, which environmental changes forced the wide separation of some subgenera today? The most obvious environmental changes to explore are as follows: photoperiod regime and sunlight angle; atmospheric CO_2 levels; water availability, seasonality of rainfall and drought frequency; mean annual temperature, seasonality of temperature and temperature extremes (both high and low); and disturbance regimes, including fire. This paper does not focus on disturbance, but, for example, increased charcoal levels from the middle Miocene in southeast Australia are indicative of increased burning, which must have affected rainforest composition and distribution (Kershaw et al. 2002).

The differing biology of tropical vs temperate *Nothofagus* species

There are relatively few data that allow clear interpretations of evolved physiological traits associated with latitudinal position. To avoid confounding influences of contrasting growth conditions eliciting plastic responses, the plants must be grown in common conditions, usually requiring study of seedlings in pot trials. Even the simplest comparisons can sometimes indicate susceptibilities of genotypes to particular conditions, which may at least allow hypotheses to be refined, and at best signal key traits (Grime 1965). Unfortunately, mature plants, including reproductive traits, cannot easily be examined in this way in slow-growing trees such as *Nothofagus*. Field-based studies provide valuable adjuncts in this regard. Here, we review data that may be informative for interpreting past changes in distribution and abundance of evergreen *Nothofagus* taxa, focusing on the comparative biology of species from Australia and New Guinea, for which we have most information.

Current distributions and climate range

Nothofagus has a wide latitudinal distribution through the Australasian region, from northwest New Guinea to southeast New Caledonia, and from southeast Queensland along the eastern highlands to southernmost Tasmania and New Zealand (i.e. c. 0-46° S). The tropical group (subgenus *Brassospora*) of 19 species is confined to New Guinea (14 species, c. 600-3100 m asl) and New Caledonia (five species, 160-1350 m asl). Just three *Nothofagus* species occur in Australia, the evergreen species *N. moorei* in the eastern highlands of mid-NSW to southeast Queensland (c. 500-1550 m asl) and *N. cunninghamii* in Victoria and Tasmania (0-1450 m asl) (subgenus *Lophozonia*), and Australia's only winter (cold) deciduous species, *N. gunnii* (subgenus *Fuscospora*), restricted to the Tasmanian highlands (550-1500 m asl). The main climate differences between the tropical and temperate species relate to winter minimum temperatures, seasonality of temperature, summer precipitation and precipitation of the wettest period (Table 1, Figure 1). The Australian species separate along a gradient influenced most strongly by temperature of the warmest month, temperature of the wettest period, and winter rainfall (Table 1, Figure 1). In particular, *N. moorei* experiences high rainfall through the warmest time of the year. The New Guinea species experience little seasonal variation in temperature, and although rainfall is seasonal, monthly rainfall is usually high even in the 'dry' season (Table 1). Hence, these species effectively experience an ever-wet climate (New Caledonian species experience a more seasonal climate that includes a short, warm dry season: Read et al. 2005). In contrast, for *N. cunninghamii*, the warmest time of the year is on average the driest (Table 1).

Comparative responses of tropical vs temperate Nothofagus species to light regimes

Variation in rainfall regime and cloudiness, together with latitudinal position, influence the light regimes experienced by *Nothofagus* species. Two components of light regimes are relevant in this context: irradiance and photoperiod. The limited data available suggest that the Australian evergreen species have higher light requirements than some of the New Guinea species, as judged by slightly higher light compensation points, dark respiration rates and light saturation points (Figure 2). However, the tropical species are quite variable: *N. resinosa* shows similar photosynthetic light-response curves to those of the temperate species, but some tropical species appear to be a little more shade-tolerant (Figure 2). Other factors such as biomass allocation patterns will also influence whole-plant shade tolerance. Shade tolerance can influence population dynamics and the degree of dependence on exogenous disturbance (e.g. Read et al. 1990b). It is likely that Palaeogene forests in southeast Australia had a high leaf area index (total leaf area per unit ground area), higher than in the current *Nothofagus* forests of southeast Australia, given the suggested high year-round rainfall and warm to mild

Table 1. Climate variables derived by BIOCLIM (Nix 1986) for evergreen *Nothofagus* species from New Guinea (all species combined) and Australia. Temperature variables (°C) and precipitation (mm) are given as mean ± standard deviation (5-95 percentile in brackets) for *n* location records. The data for New Guinea species are updated from Read et al. (2005). The data for Australian species are updated from Read et al. (2010). Seasonal temperature range is calculated as the hottest-month mean temperature minus the coldest-month mean temperature. Precipitation seasonality is calculated as the coefficient of variation across monthly values

Climate variable	New Guinea species (n = 301)	*N. moorei* (n = 139)	*N. cunninghamii* (n = 318)
Annual mean temperature	16.1 ± 2.9 (12.0-21.5)	12.2 ± 1.4 (10.1-14.6)	9.5 ± 2.0 (5.8-12.4)
Hottest-month mean temp.	16.6 ± 3.0 (12.5-22.1)	17.5 ± 1.6 (15.3-19.9)	15.1 ± 2.3 (11.0-18.4)
Coldest-month mean temp.	15.3 ± 2.8 (11.2-20.2)	5.2 ± 2.1 (2.6-9.1)	4.8 ± 2.1 (1.2-7.8)
Hottest-month max. temp.	21.4 ± 3.1 (16.5-27.2)	23.2 ± 1.4 (21.0-25.7)	20.8 ± 2.8 (15.9-25.3)
Coldest-month min. temp.	10.7 ± 3.0 (6.9-16.4)	0.9 ± 1.6 (-0.9-5.1)	1.6 ± 1.8 (-1.5-4.2)
Diurnal temperature range	10.4 ± 1.6 (7.1-12.1)	11.1 ± 1.1 (8.1-12.5)	11.9 ± 1.5 (10.0-14.6)
Seasonal temperature range	1.3 ± 0.8 (0.8-1.9)	12.4 ± 0.8 (10.8-13.4)	10.3 ± 1.6 (7.9-12.8)
Annual precipitation	2891 ± 743 (2073-4439)	1633 ± 325 (1229-2282)	1816 ± 523 (1127-2950)
Precipitation of hottest quarter	888 ± 249 (666-1159)	564 ± 142 (390-844)	304 ± 92 (191-516)
Precipitation of coldest quarter	526 ± 254 (301-1069)	315 ± 49 (232-388)	551 ± 158 (351-887)
Precipitation of wettest quarter	955 ± 348 (689-1491)	611 ± 179 (396-959)	570 ± 160 (364-899)
Precipitation of driest quarter	474 ± 166 (301-869)	249 ± 35 (194-302)	304 ± 95 (191-516)
Precipitation seasonality	28 ± 11 (14-44)	34 ± 11 (18-46)	25 ± 5 (18-35)

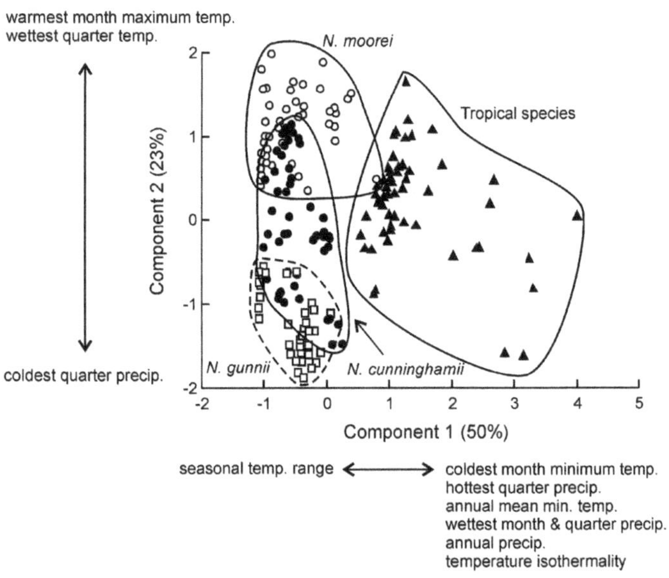

Figure 1. Ordination (PCA) of site climate variables for *Nothofagus* species from New Guinea and Australia. A subset of c. 50 locations was selected randomly for each Australian species and for the New Guinea species as a whole, to avoid bias towards well-surveyed species. The percentage contribution of the first two components to the total variation is indicated, and the variables contributing most strongly to each axis (component loadings ≥ 0.80) are listed. Climate data are from Read et al. (2005) and Read and Brown (1996), using 20 of the 24 BIOCLIM-derived variables (excluding some temperature variables)

temperatures. If so, opportunities for regeneration of some shade-intolerant species may have been relatively limited, possibly associated with exogenous disturbance, as found in many *Nothofagus* species in both tropical (Read and Hope 1996) and temperate (Ogden et al. 1996; Veblen et al. 1996) regions today. Under these conditions, some *Nothofagus* species may have been confined to disturbance-prone sites, such as ridge tops, or to suboptimal sites.

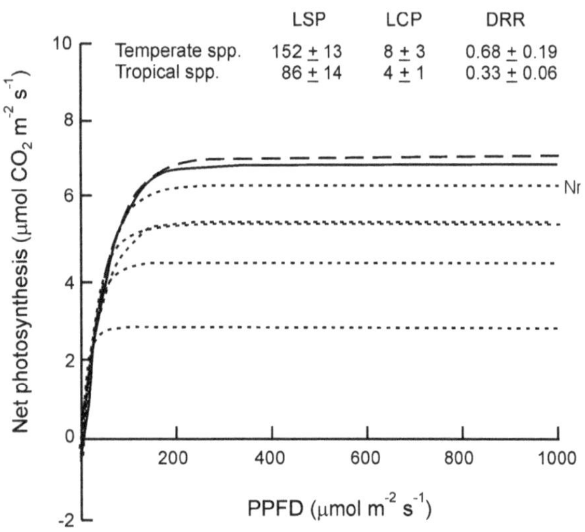

Figure 2. Instantaneous light-dependence curves of net photosynthesis of shade-grown evergreen *Nothofagus* species from New Guinea and Australia. *Nothofagus cunninghamii* is shown by a solid line, *N. moorei* by long dashes, and tropical species (*N. pullei, N. resinosa* (Nr), *N. crenata* var. *sapei, N. grandis* and *N. brassii*) by short dashes. Plants were grown at a photosynthetic photon flux density (PPFD) of 50-60 µmol m^{-2} s^{-1} in a growth chamber at 20°C with 12 h days. The average dark respiration rate (DRR, µmol CO_2 m^{-2} s^{-1}), light compensation point (LCP, µmol m^{-2} s^{-1}) and light saturation point (LSP µmol m^{-2} s^{-1}) are given for tropical (n = 5 species) and temperate (n = 2) species

Nothofagus at high latitudes – the constraints of polar winters

Nothofagus has a long history at high latitudes. It currently extends south to c. 56° on the islands of Cape Horn, but during the late Cretaceous and early Palaeogene, occurred even further south in the Antarctic region (Hill 2001). Exchange of many taxa between Australia and South America, including *Nothofagus*, is likely to have occurred across latitudes of at least 80° S during this time. Today, high-latitude forests experience severe conditions – extremely low minimum temperatures, short and cool growing seasons and low availability of soil nutrients and water – with growth and survival potentially limited by factors including frost damage, physiological drought and carbon balance (Woodward 1987; Givnish 2002). However, a range of evidence, including plant fossils, suggests that milder temperatures prevailed at southern high latitudes during the Cretaceous and Palaeogene. Cool winters, rather than severe freezing conditions, occurred along the coastal margins of Antarctica (Otto-Bliesner and Upchurch 1997; Hayes et al. 2006; Poole and Cantrill 2006), and there was even an abundance of now mostly tropical taxa (e.g. cycads, *Nypa*, Lauraceae) at a coastal early Eocene site in Tasmania estimated to have been at 65° S (Pole and Macphail 1996; Carpenter et al. 2007).

While milder winters may have allowed species without highly developed cold-adaptation to occupy polar forests, the dark and mild winters may have constrained tree growth and survival (by primary means or via competition) through negative effects on carbon balance. In addition, the low angle of light may have reduced rates of carbon assimilation during summer, particularly for species of the understorey and lower canopy (Specht et al. 1992). No modern analogue exists for these high-latitude environments, especially given the likely high atmospheric CO_2 concentrations, but interacting effects of temperature and light regime on modern vegetation limits and species mixtures have been reported at high latitudes in Scotland and Ireland (Bannister 1976). The typically thin leaves of deciduous plants often do not fossilise well, but there is abundant evidence that deciduous taxa were common in some parts of southern Australia during the Palaeogene (Hill 2001). The deciduous habit may have been advantageous for canopy species at high latitudes (e.g. Axelrod 1984; Spicer 1987), for example in terms of carbon balance, although the role of the latter has been questioned (Royer et al. 2003). It is anyway likely that these unusual high-latitude conditions acted as a sieve to species movements between and within land masses during the Cretaceous and Palaeogene (Read and Francis 1992). They may also have contributed to the sometimes greater than expected abundance of gymnosperms and *Nothofagus* in mild-to-warm climates and to the abundance of deciduous species at this time in southeast Australia, where only one winter-deciduous species occurs today.

Responses to freezing temperatures

As temperatures decreased and seasonality increased during the middle Miocene-Pliocene in southeast Australia, plants would have been exposed to lower minimum temperatures, with temperatures continuing to decline into the Pleistocene. In addition, the drying climate with increasingly seasonal rainfall may have led to a higher incidence of frosts as the cooler season became drier. There is certainly some correspondence of changes in species distribution with the levels of foliar frost resistance recorded in extant relatives: tropical *Nothofagus* species have a lower frost resistance than southern evergreen species (Figure 3), and Tasmanian rainforest species of several other genera are typically more frost resistant than related species native to NSW and Queensland (Read and Hill 1989). However, the frost resistance of New Guinea *Nothofagus* species native to cool climates is close to that of *N. moorei*, despite the estimated higher (by 8-11°C) mean minimum temperature of the coldest month (MTCM) experienced by the tropical species (Read and Hope 1989, 1996) (Figure 3). This comparison suggests there is no fundamental difference in capacity to develop frost resistance between the New Guinea species and *N. moorei*, i.e. the genetic difference is potentially small.

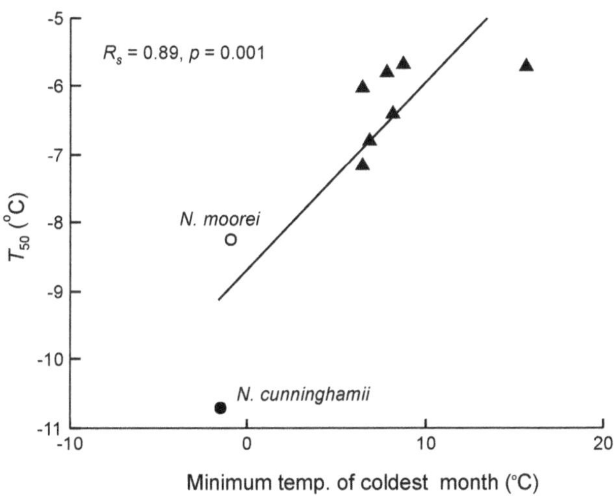

Figure 3. The relationship between frost resistance (T_{50}, the temperature producing 50% damage) and the minimum temperature of the coldest month (5-percentile values of species profiles) across evergreen *Nothofagus* species from New Guinea and Australia (updated from Read and Hope 1989). Climate data were estimated by BIOCLIM (Nix 1986), updated from Read et al. (2005). Tropical species are indicated by triangles, temperate species by circles. The result of Spearman correlation is given, using average T_{50} values for species sampled from multiple populations (*N. pullei*, *N. moorei* and *N. cunninghamii*)

Like some of the New Guinea species, *N. cunninghamii* has a higher frost resistance than expected from the relationship with estimated MTCM values, relative to *N. moorei* (Figure 3). These suggested anomalies in frost resistance may reflect the importance of key climate variables not derived by the climate estimation model, including annual minima, number of frost days, or rare (e.g. decadal) severe frost events. However, they may also reflect the retention of traits resulting from greater exposure to colder conditions during the Pleistocene glacial periods in *N. cunninghamii* (Read and Hope 1989) and possibly some of the New Guinea species. *Nothofagus* was more common near the treeline in the late Pleistocene in New Guinea, in cooler (although possibly cloudier and so less exposed to frost) conditions (Walker and Hope 1982; Hope 2009). The particularly high frost resistance recorded in *N. cunninghamii* may reflect substantial increases in frost resistance that evolved during glacial periods, particularly if refugia were colder and drier than is currently experienced (McKenzie 1995, 2002; Worth et al. 2009), and might involve more complex genetic change. However, frost resistance is the outcome of a range of physiological and morphological traits (Larcher 2003, 2005), and these traits have multiple functions. For example, osmotic adjustment can contribute to resistance to both freezing and drought; tissues providing densely lignified or cutinized barriers (as in scleromorphic leaves of species such as *N. cunninghamii*) can potentially contribute to frost resistance (Larcher 2005) as well as drought resistance, and to protection from herbivores (Turner 1994) and possibly from damaging levels of light (Jordan et al. 2005). Hence, some leaf traits of *N. cunninghamii* may reflect efficient simultaneous resistance to the multiple stresses that prevail in seasonal cool temperate climates.

Photosynthesis and growth responses to temperature

Little information is available about growth responses to temperature for tropical and temperate *Nothofagus* species. Leaf flushes in the New Guinea species can occur throughout the year (Ash 1982), consistent with the low seasonal temperature range experienced by the species, together with high year-round rainfall. In contrast, both photosynthesis and growth are constrained by low winter temperatures in the southern species, with flushes of shoot growth largely confined to spring, sometimes with a second flush in summer (Howard 1973; Lowman 1992). However, the temperate evergreen species have a high acclimatory capacity, at least for photosynthesis. Under experimental conditions, leaves developed at 20°C adjusted to changing growth temperatures so that they maintained 80% of their maximum rate of net photosynthesis over a span of 14-17°C; the tropical species showed an acclimatory capacity of only 8-11°C (Read 1990). Notably, no difference was recorded between tropical and temperate evergreen species in the acclimation temperature at which the maximum rate of photosynthesis was reached (Figure 4). These trends are consistent with the occurrence of tropical *Nothofagus* species in a climate that experiences relatively little daily and seasonal variation in maximum temperatures, and the restriction of temperate species to climates with higher daily and seasonal variability in temperature, but with similar hottest-month maximum temperatures (on average) (Table 1). The trends are also consistent with the changes in distribution recorded from the mid-Miocene, with subgenus *Brassospora* declining in abundance in southern Australia as temperature seasonality increased.

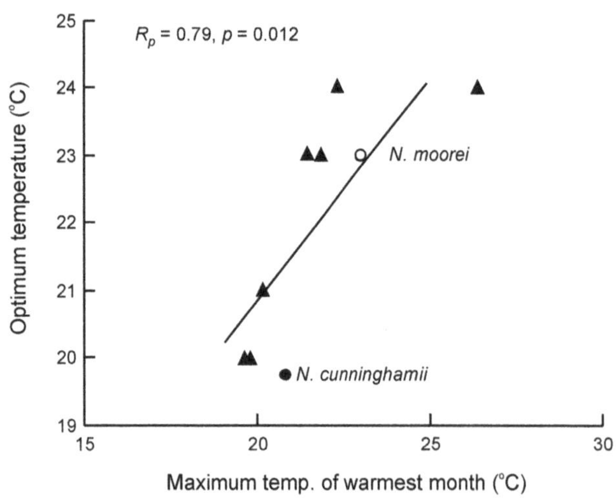

Figure 4. The relationship between the optimum acclimation temperature for photosynthesis and the maximum temperature of the warmest month (mean values of species profiles) across evergreen *Nothofagus* species from New Guinea and Australia (updated from Read and Hope 1996). Photosynthesis data were taken from Read (1990) and climate data were estimated by BIOCLIM (Nix 1986), updated from Read et al. (2005). Tropical species are indicated by triangles, temperate species by circles. The result of Pearson correlation is given, using average values of the optimum temperature for species sampled from multiple populations (*N. pullei, N. moorei* and *N. cunninghamii*)

Similar trends have been recorded among other Australian rainforest species, with temperate species better able to maintain maximum rates of photosynthesis across a greater range of acclimation and growth temperatures than tropical species (Cunningham and Read 2002, 2003a), but with a lower optimum temperature for photosynthesis (Cunningham and Read 2002). Hence, the 'strategy' of the temperate species appears to be to produce a growth flush in spring, the leaves then having a high capacity to adjust to the changing daily and seasonal temperatures, while the tropical species rely less on acclimation since they experience less variation in daily maxima across the year and potentially produce new leaves across a longer period (Read 1990; Cunningham and Read 2002, 2003a).

Whole-plant responses (e.g. carbon balance and growth rate) are likely to influence competitive success and contribute most to distributional changes. Little is known about comparative growth responses of tropical and temperate *Nothofagus* species to temperature. A study of some Australian tropical lowland rainforest trees showed higher temperature optima for growth than in temperate species, as expected from their occurrence in climates with higher mean annual maximum temperatures; however, the capacity to explain finer-scale latitudinal patterns of rainforest zonation based on growth responses was limited (Cunningham and Read 2003b). Much more research is needed in this area.

Responses to water deficits

In New Guinea, *Nothofagus* is typically associated with high rainfall and cloudiness in montane rainforests, where even the dry season seems to have relatively high and predictable rainfall (Ash 1982; Read et al. 2005). The responses of the New Guinea species to water deficits reflect this humid environment in which any water deficits are likely to be generally relatively short lived. Rates of leaf water loss tend to be low in the tropical *Nothofagus* species under experimental conditions of water deficit (Figure 5a), and higher long-term water-use efficiency has been recorded than in temperate species grown under common conditions (Read and Farquhar 1991), possibly due in part to higher stomatal sensitivity to water deficits. While this response seems paradoxical given the wet environment in which New Guinea species occur (and appears to contrast with some data from southern conifers: Brodribb and Hill 1998), stomatal sensitivity may be a relatively low-cost response to infrequent or short-lived water deficits. The main opportunity cost of foliar water conservation by reduced stomatal conductance is the potential reduction in photosynthetic rates. In ever-wet environments, this conservative response to water deficits may impose a relatively small opportunity cost. *Nothofagus moorei,* native to a summer-wet climate at mid-latitudes in Australia, shows a somewhat similar response (Figure 5a), but has better-developed mechanisms for enhancing water uptake than the New Guinea species (Read et al. 2010).

The very different response of *N. cunninghamii,* growing in southernmost Australia where the warmest time of the year is on average the driest (Table 1), indicates a more complex 'strategy'. Assimilation rates are limited in the cool-cold winters, so traits may be selected that maximise assimilation when temperatures are higher, in spring through to autumn. This may include mechanisms that enhance uptake of soil water, allowing high rates of stomatal conductance and thereby CO_2 uptake and assimilation through drier periods. *Nothofagus cunninghamii* maintains higher rates of stomatal conductance, but develops low water potentials for a given decrease in tissue water content (Figure 5b) and a high root mass:leaf mass ratio (Read et al. 2010), traits that potentially enhance water supply to photosynthetic tissues. Hence, it seems to have a suite of traits that together should enhance assimilation and growth in a winter-dominated rainfall regime in which some degree of water deficit may be experienced in summer. However, such a strategy would limit growth in a summer-wet or ever-wet climate. That is, the water-spending habit when paired with costly diversion of resources to roots at the expense of leaves would probably only be competitively advantageous to plants occurring in climates in which photosynthesis is limited to a period that may experience dry conditions.

This difference in behaviour compared with *N. moorei* and the tropical species is consistent with the late appearance of *N. cunninghamii* (early Pleistocene, perhaps c. 1 Ma) (Jordan 1997) in the fossil record of southeast Australia. However, small-leafed forms of *Nothofagus* appear earlier in the fossil record, including during periods in which rainfall is likely to be high year-round (Hill 1983), and may reflect evolved responses to other environmental conditions. The traits shown by *N. cunninghamii* would be particularly advantageous when conditions deteriorated in the Neogene, with temperatures becoming more seasonal and rainfall decreasing (potentially with higher evaporative loads in summer), with the later development of a winter-high rainfall regime.

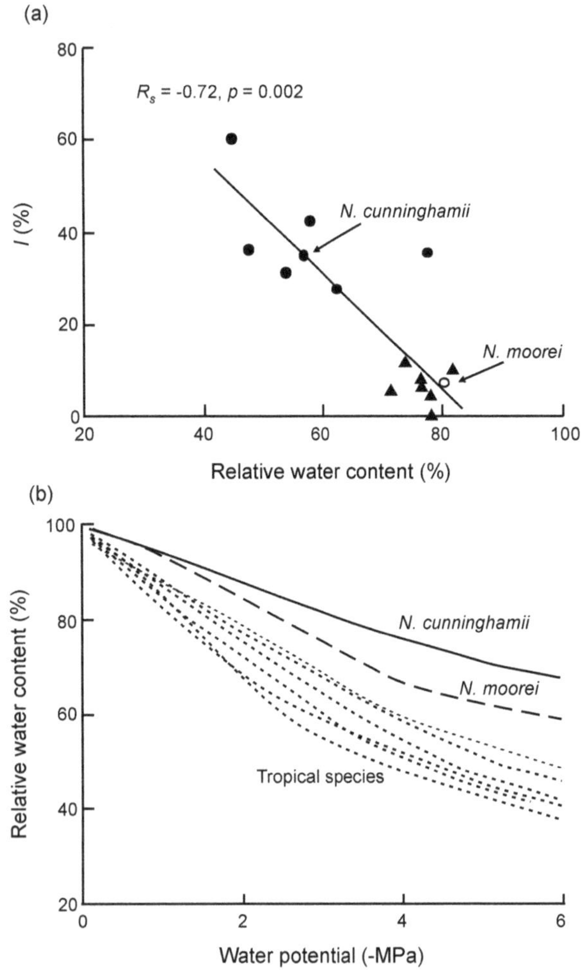

Figure 5. Comparisons of water relations traits between tropical and temperate evergreen *Nothofagus* species. (a) The relationship between leaf tissue damage (*I*, index of injury, calculated from leaching of solutes in damaged tissues) and relative water content in *Nothofagus* leaves 2 h after detachment (after Read et al. 2010). Tropical species and *N. moorei* maintain high water content and suffer little injury. Tropical species, triangles; temperate evergreen species from Australia, New Zealand and South America, circles. (b) The relationship between leaf relative water content and shoot water potential in detached leaves of evergreen species from New Guinea and Australia (from data in Read et al. 2010)

Discussion

There is still much to be learned in relation to distributional changes of key taxa such as *Nothofagus*, and the factors eliciting these changes and contributing to the evolution of the traits that today distinguish the temperate and tropical taxa. We note that other rainforest taxa, by having other ancestral traits and different phylogenetic constraints, may have been influenced by environmental changes in different ways from *Nothofagus*. In addition, species adapted to warmer conditions (e.g. tropical lowlands) or to different microenvironments, such as the rainforest understorey or lower canopy, may differ in a broad range of traits from *Nothofagus* species, which today typically dominate the uppermost canopy of montane or temperate lowland forests. Hence, the history and comparative biology of *Nothofagus* taxa may illustrate only part of the ecological history of rainforest in the region.

For *Nothofagus*, the palaeoecological history of this period appears to be as follows. During the wet and mild Palaeocene and warmer early Eocene, *Nothofagus* may have been present only in localised highland areas (Truswell 1993; McGowran et al. 2000). This is consistent with the current occurrence of evergreen *Nothofagus* in generally cool to mild conditions, even in the tropics (Read et al. 2005), with relatively low photosynthetic temperature optima

(Read 1990). Given the ever-wet conditions thought to occur at this time, upland areas of even low elevation may have been exposed to frequent cloud-lie, as commonly experienced in New Guinea, and often associated with suboptimal conditions for plant growth (Grubb 1977; Read et al. 2005). In New Guinea, *Nothofagus* typically extends to lower altitudes (c. 600-1200 m asl) only on suboptimal sites, although current distribution patterns are likely to be moderated by the highly diverse flora of this region, particularly in the warmer lowland climates (Read et al. 2005). It is possible that 'megathermal' *Nothofagus* species existed in the Palaeogene that were lost in later cooling episodes, but there is little evidence to date. It is not clear to what extent declining CO_2 levels contributed to the increasing abundance of *Nothofagus* in the late Eocene and the loss of some megathermal taxa.

Conditions were still cool in the Oligocene (but with more intense climatic gradients and so more variation in vegetation: McGowran et al. 2000) when *Nothofagus* diversity, including subgenus *Brassospora*, appeared to reach its maximum. At this time, 'tropical' (subgenus *Brassospora*) and 'temperate' (other subgenera) species were commonly contemporaneous, even within single fossil deposits (Hill 2001). There may have been some altitudinal separation of species, with subgenus *Brassospora* growing at lower elevations (Specht 1981), but macrofossil evidence suggests that diverse *Nothofagus* taxa coexisted at a finer scale (Hill 2001). This may have been possible because of the extremely wet and mild climate of the time, allowing coexistence of groups that may not be able to coexist in current climate regimes of this region (Hill 2001). Differential changes in distribution since that time suggest that differentiation in ecophysiology (and possibly in responses to disturbance) occurred among these groups of species. If that is the case, habitat segregation may have allowed landscape-level (or finer-scale) coexistence – for example, with 'tropical' taxa occurring in habitats that experienced less variability and extremes of climate than 'temperate' taxa (Read et al. 1990a; Hill 2001).

By the early-middle Miocene, *Nothofagus* was less abundant in inland than coastal regions (Kershaw et al. 1994), and by the late Miocene to Pliocene, rainforest had declined in abundance across Australia and subgenus *Brassospora* had generally declined in abundance relative to subgenus *Lophozonia* (McGowran et al. 2000). The apparent decrease in suitable habitat, together with the decrease in the extent of rainforest *per se* (given that rainforest provides ameliorated microenvironments), is likely to have reduced refugial opportunities in both extent and environmental heterogeneity. Hence, there is likely to have been reduced capacity for effective plant migration through periods of environmental change, particularly for rainforest species that were more sensitive to changing conditions and/or had less-dispersible seeds – i.e. loss of some species may have been caused more by failure to keep pace with environmental change, than to regional loss of suitable environments. The reduction in the extent of rainforest is likely to have been strongly related to the decrease in annual precipitation (e.g. Brodribb and Hill 1998). However, conditions were also becoming cooler, with increased seasonality of both temperature and rainfall. Moreover, during the Plio-Pleistocene, there was a shift to winter-high rainfall in southeast Australia (Bowler 1982; Macphail et al. 1995). Hence, the contraction in rainforest may have been influenced by a broader suite of factors that influenced growth, reproduction and survival of the component species. By contrast, the Pliocene saw an expansion in montane habitat for *Nothofagus* in New Guinea and it expanded to maximum extent during glaciations in the Pleistocene (Hope 1996).

The apparent loss of diversity of *Nothofagus* in southeast Australia by the mid-ate Miocene, and the differential changes in distribution shown by the various taxa, were probably strongly influenced by cumulative effects of declining rainfall and temperature combined with increasing seasonality of both temperature and rainfall. Many of these species appear to have been lost from southern Australia before the severe deterioration associated with the glaciations of the late Neogene, judging from the reduced diversity of some Miocene fossil deposits, but timing of losses is very uncertain. In particular, the pairing of a seasonal temperature regime with drying summers may have contributed most to the loss of some evergreen *Nothofagus* taxa and the evolution of *N. cunninghamii*. The later shift to a winter-high rainfall regime would further contribute to the seasonal contrasts in conditions. To be successful in the altered climate regime of the southern rainforests, it appears to have been important (based on the biology of extant species) to optimise photosynthesis towards the growing season, part of which (summer) became drier than in the past. This could be achieved by (a) a high acclimatory capacity of photosynthesis, allowing high rates of photosynthesis through the wetter but cooler parts of spring and autumn, plus (b) strategies that enhance water uptake so that photosynthesis is not heavily limited during the warmer but drier months. These are the characteristics of *N. cunninghamii*, whose leaf type does not appear until relatively late in the fossil record (Jordan 1997), or at least is infrequent until that time. *Nothofagus moorei*, which also has the high acclimatory capacity of photosynthesis in relation to temperature of *N. cunninghamii*, appears to have higher stomatal sensitivity than its southern relative, consistent with the warmest time of the year being wet through its current range. This strategy reduces the requirement for alternative and less appropriate (in this climate) expenditure on means of increasing water uptake, e.g. increased allocation to root biomass, and is consistent with the loss of large-leafed species similar to *N. moorei* from southernmost Australia during the period when summers were becoming drier. That is, a *N. moorei*-like species may have had sufficiently broad temperature tolerance to persist for longer in the cooler and more seasonal climate, but would be predicted to have reduced growth rates (leading to reduced competitive ability) in climates with drier summers. It would be predicted to survive longest in climates with wetter summers, such as western Tasmania, and in oceanic rather than continental climates.

The climate deterioration in southeast Australia would particularly constrain growth and competitiveness of fossil species of subgenus *Brassospora,* if similar in biology to their extant relatives. Climatic deterioration may have at first just restricted species to wetter sites, since rainfall seasonality *per se* is not the problem, just its potential consequence of low-rainfall periods, particularly when these are associated with the summer of a seasonal temperature regime. This is consistent with the apparent loss of subgenus *Brassospora* and some other taxa with tropical affinities from all but the wettest areas during the Pliocene (Macphail et al. 1995; McGowran et al. 2000). At the same time, a drying climate before the switch to winter-high rainfall may have exacerbated the effects of climatic cooling and increased the likelihood of frosts. Some species of subgenus *Brassospora* may have persisted in southeast Australia longer than others, depending on differences in sensitivities to particular environmental factors. The loss of subgenus *Brassospora* from the wet highlands of northeast Australia during the Plio-Pleistocene (Kershaw and Sluiter 1982) might also have been influenced by increasing temperature seasonality and drier conditions, in addition to constraints imposed by the reduced areal extent of rainforest in response to drier conditions (Kershaw and Sluiter 1982).

It is not clear what limited the evolution of traits that enhanced tolerance of seasonality in the *Brassospora* group. A better knowledge of the biology of the New Caledonian *Nothofagus* species may help in this regard, since they occur in a more seasonal climate. At this stage, we can only note that evolution of tolerance of this climate regime in *N. cunninghamii* appears to have involved a coordinated suite of traits (even just for tolerance of summer-low rainfall), and that if its ancestor was more ecophysiologically similar to *N. moorei*, it was probably pre-adapted by some traits to the onset of the more severe conditions of the Plio-Pleistocene. A *N. moorei*-like ancestor has been suggested previously for *N. cunninghamii* (Hill 2001), but recent molecular evidence (Worth et al. 2009) casts some doubt on this hypothesis. There is also some recent evidence from fossil beetle remains that high annual rainfall and a summer-high rainfall climate persisted until c. 1.5 Ma in southern Australia (Sniderman et al. 2009). This suggests a lesser role for the switch to winter-high rainfall for those rainforest taxa that had already declined in abundance in the region, and may provide further evidence of multiple and interacting factors in the loss of taxa that are restricted today to low latitudes, with probable variation among species in their sensitivity and responses to particular environmental changes.

One final trait of note in *N. cunninghamii* is its phenotypic variability, with considerable variation in leaf form and some physiological traits across its range. Data from juveniles grown in common environments suggest considerable genetic variation among populations in traits ranging from leaf form (Hovenden and Vander Schoor 2004) to ecophysiology (Read 1990; Hovenden and Brodribb 2000), possibly enhanced by the suggested multiple glacial refugia occupied during the Pleistocene (Worth et al. 2009). Genetic variability may be a key to the success of this species, and possibly of other geographically widespread species of *Nothofagus* (e.g. Premoli 1997). This, with the apparently superior acclimatory capacity of *N. cunninghamii*, may have engendered a greater vagility than some of its congeners, allowing it and its ancestors to keep pace with the climatic fluctuations (environmental 'steps and reversals': McGowran et al. 2000) seen through the Cenozoic in southeast Australia.

Concluding remarks

Study of living relatives of fossil *Nothofagus* species shows the potential responsiveness of vegetation and its component species to changes in seasonality, or at least to changes in seasonal conditions. Interpretation of climate based on nearest living relatives needs to take this into account – i.e. that species occurrence may reflect seasonal conditions and variability or even periodic extremes more than mean annual conditions. However, it is unclear to what degree the comparative biology of living species of *Nothofagus* indicates general mechanisms of change in diversity and species distributions in the history of southern rainforests. We also note the potential difficulty in unravelling effects of different features of the changing environment on species distribution, especially since environmental factors (e.g. climate, CO_2 levels, soil and disturbance regimes) can influence species survival by both cumulative and interactive effects on plants that affect their competitiveness. One aspect not explored here is the contribution of changing climates to potentially complex changes in vegetation through factors such as altered fire regimes.

It is also becoming apparent that the recent evolutionary history of some species may be more complex than previously has been understood. In particular, a study of haplotype diversity suggests nesting of a *N. moorei* haplotype within the *N. cunninghamii* lineage (Worth et al. 2009), and so a more intricate and dynamic history than previously imagined. Clearly, further investigations of the fossil record, phylogenetic relationships and the biology of nearest living relatives are needed to assist in unravelling the history of past responses to environmental change in order to better understand vegetation responses of the future.

Acknowledgements

The great contribution by Geoff Hope to the vegetation history and ecology of the region is warmly acknowledged by his co-authors. We have been inspired variously by his skill, insights, dedication and enthusiasm as a teacher, research colleague, field guide and friend.

References

Ash, J. 1982. The *Nothofagus* Blume (Fagaceae) of New Guinea. In J.L. Gressitt (ed.), *Biogeography and ecology of New Guinea*, Monographiae Biologicae Vol. 42 pp355-380. The Hague: Dr W. Junk Publishers.

Axelrod, D.I. 1984. An interpretation of Cretaceous and Tertiary biota in polar regions. *Palaeogeography, Palaeoclimatology, Palaeoecology* 45:105-147.

Bannister, P. 1976. *Introduction to Physiological Plant Ecology*. Oxford: Blackwell Scientific Publications.

Beerling, D.J. and C.K. Kelly 1996. Evolutionary comparative analyses of the relationship between leaf structure and function. *New Phytologist* 134:35-51.

Beerling, D.J., B.H. Lomax, D.L. Royer, G.R. Upchurch Jr and L.R. Kump 2002. An atmospheric pCO_2 reconstruction across the Cretaceous-Tertiary boundary from leaf megafossils. *Proceedings of the National Academy of Sciences* 99:7836-7840.

Bowler, J.M. 1982. Aridity in the late Tertiary and Quaternary of Australia. In W.R. Barker and P.J.M. Greenslade (eds), *Evolution of the flora and fauna of arid Australia*, pp35-45. Frewville, South Australia: Peacock Publications, in association with the Australian Systematic Botany Society and ANZAAS, South Australian Division, Inc.

Brodribb, T. and R.S. Hill 1998. The photosynthetic drought physiology of a diverse group of southern hemisphere conifer species is correlated with minimum seasonal rainfall. *Functional Ecology* 12:465-471.

Carpenter, R.J., G.J. Jordan and R.S. Hill 2007. A toothed Lauraceae leaf from the Early Eocene of Tasmania. *International Journal of Plant Sciences* 168:1191-1198.

Cunningham, S.C. and J. Read 2002. Comparison of temperate and tropical rainforest tree species: photosynthetic responses to growth temperature. *Oecologia* 133:112-119.

Cunningham, S.C. and J. Read 2003a. Do temperate trees have a greater capacity to acclimate to changing temperatures than tropical rainforest trees? *New Phytologist* 157:55-64.

Cunningham, S.C. and J. Read 2003b. Comparison of temperate and tropical rainforest tree species: Growth responses to temperature. *Journal of Biogeography* 30:143-153.

Donoso, C. 1996. Ecology of *Nothofagus* forests in Central Chile. In T.T. Veblen, R.S. Hill and J. Read (eds), *The ecology and biogeography of* Nothofagus *forests*, pp271-292. New Haven: University of Yale Press.

Givnish, T.J. 1979. On the adaptive significance of leaf form. In O.T. Solbrig, P.H. Raven, S. Jain and G.B. Johnson (eds), *Topics in plant population biology*, pp375-407. New York: Columbia University Press.

Givnish, T.J. 2002. Adaptive significance of evergreen vs. deciduous leaves: solving the triple paradox. *Silva Fennica* 36:703-743.

Grime, J.P. 1965. Comparative experiments as a key to the ecology of flowering plants. *Ecology* 46:513-515.

Grubb, P.J. 1977. Control of forest growth and distribution on wet tropical mountains: with special reference to mineral nutrition. *Annual Review of Ecology and Systematics* 8:83-107.

Hayes, P.A., J.E. Francis, D.J. Cantrill and J.A. Crame 2006. Palaeoclimate analysis of Late Cretaceous angiosperm leaf floras, James Ross Island, Antarctica. In J.E. Francis, D. Pirrie, and J.A. Crame (eds) *Cretaceous-Tertiary high-latitude palaeoenvironments, James Ross Basin, Antarctica*. Geological Society, London, Special Publications 258:49-62.

Hill, R.S. 1983. Evolution of *Nothofagus cunninghamii* and its relationship to *N. moorei* as inferred from Tasmanian macrofossils. *Australian Journal of Botany* 31:453-465.

Hill, R.S. 1991. Tertiary *Nothofagus* (Fagaceae) macrofossils from Tasmania and Antarctica and their bearing on the evolution of the genus. *Botanical Journal of the Linnean Society* 105:73-112.

Hill, R.S. 1994. The history of selected Australian taxa. In R.S. Hill (ed.) *History of the Australian Vegetation: Cretaceous to Recent,* pp390-419. Cambridge: Cambridge University Press.
Hill, R.S. 2001. Biogeography, evolution and palaeoecology of *Nothofagus* (Nothofagaceae): the contribution of the fossil record. *Australian Journal of Botany* 49:321-332.
Hill, R.S. and M.E. Dettmann 1996. Origin and diversification of the genus *Nothofagus.* In T.T. Veblen, R.S. Hill and J. Read (eds) *The ecology and biogeography of* Nothofagus *forests,* pp11-24. New Haven: University of Yale Press.
Hill, R.S. and J. Read 1991. A revised infrageneric classification of *Nothofagus* (Fagaceae). *Botanical Journal of the Linnean Society* 105:37-72.
Hill, R.S. and L.J. Scriven 1997. Palaeoclimate across an altitudinal gradient in the Oligocene-Miocene of northern Tasmania: an investigation of nearest living relative analysis. *Australian Journal of Botany* 45:493-505.
Hope, G.S. 1996. History of *Nothofagus* in New Guinea and New Caledonia. In T.T. Veblen, R.S. Hill and J. Read (eds) *The ecology and biogeography of* Nothofagus *forests,* pp257-270. New Haven: University of Yale Press.
Hope, G. 2009. Environmental change and fire in the Owen Stanley Ranges, Papua New Guinea. *Quaternary Science Reviews* 28:2261-2276.
Hovenden, M.J. and T. Brodribb 2000. Altitude of origin influences stomatal conductance and therefore maximum assimilation rate in Southern Beech, *Nothofagus cunninghamii. Australian Journal of Plant Physiology* 27:451-456.
Hovenden, M.J. and J.K. Vander Schoor 2004. Nature vs nurture in the leaf morphology of Southern beech, *Nothofagus cunninghamii* (Nothofagaceae). *New Phytologist* 161:585-594.
Howard, T.M. 1973. Studies in the ecology of *Nothofagus cunninghamii* Oerst. II. Phenology. *Australian Journal of Botany* 21:79-92.
Jordan, G.J. 1997. Evidence of Pleistocene plant extinction and diversity from Regatta Point, western Tasmania, Australia. *Botanical Journal of the Linnean Society* 123:45-71.
Jordan, G.J. and R.S. Hill. 1999. The phylogenetic affinities of *Nothofagus* (Nothofagaceae) leaf fossils based on combined molecular and morphological data. *International Journal of Plant Sciences* 160:1177-1188.
Jordan, G.J., R.A. Dillon and P.H. Weston 2005. Solar radiation as a factor in the evolution of scleromorphic leaf anatomy in Proteaceae. *American Journal of Botany* 92:789-796.
Kershaw, A.P. and I.R. Sluiter 1982. Late Cenozoic pollen spectra from the Atherton Tableland, north-eastern Australia. *Australian Journal of Botany* 30:279-295.
Kershaw, A.P., J.S. Clark, A.M. Gill and D.M. D'Costa 2002. A history of fire in Australia. In R.A. Bradstock, J.E. Williams and M.A. Gill (eds) *Flammable Australia: the fire regimes and biodiversity of a continent,* pp3-25. Cambridge: Cambridge University Press.
Kershaw, A.P., H.A. Martin and J.R.C. McEwan Mason 1994. The Neogene: a period of transition. In R.S. Hill (ed.), *History of the Australian vegetation: Cretaceous to recent,* pp299-327. Cambridge: Cambridge University Press.
Larcher, W. 2003. *Physiological plant ecology: ecophysiology and stress physiology of functional groups.* 4th ed. New York: Springer.
Larcher, W. 2005. Climatic constraints drive the evolution of low temperature resistance in woody plants. *Journal of Agricultural Meteorology* 61:189-202.
Lowman, M.D. 1992. Leaf growth dynamics and herbivory in five species of Australian rain-forest canopy trees. *Journal of Ecology* 80:433-447.
Macphail, M.K., E.A. Colhoun and S.J. Fitzsimons 1995. Key periods in the evolution of the Cenozoic vegetation and flora in western Tasmania: the Late Pliocene. *Australian Journal of Botany* 43:505-526.
McElwain, J.C. and W.G. Chaloner 1996. The fossil cuticle as a skeletal record of environmental change. *Palaios* 11:376-388.

McGowran, B., M. Archer, P. Bock, T.A. Darragh, H. Godthelp, S. Hageman, S.J. Hand, R. Hill, Q. Li, P.A. Maxwell, K.J. McNamara, M. Macphail, D. Mildenhall, A.D. Partridge, J. Richardson, S. Shafik, E.M. Truswell and M. Warne 2000. Australasian palaeobiogeography: the Palaeogene and Neogene record. *Memoir of the Association of Australasian Palaeontologists* 23:405-470.

McKenzie, G.M. 1995. Shifting patterns of *Nothofagus cunninghamii* (Hook.) Oerst rainforest in the late Quaternary of southeastern Australia: evidence from pollen. *Historical Biology* 9:257-268.

McKenzie, G.M. 2002. The late Quaternary vegetation history of the south-central highlands of Victoria, Australia. II. Sites below 900 m. *Austral Ecology* 27:32-54.

Nix, H.A. 1986. A biogeographic analysis of elapid snakes. In R. Longmore (ed.), *Atlas of elapid snakes of Australia*, pp4-15. Canberra: Australian Flora and Fauna Series No. 7, Australian Government Publishing Service.

Ogden, J., G.H. Stewart and R.B. Allan 1996. Ecology of New Zealand *Nothofagus* forests. In T.T. Veblen, R.S. Hill and J. Read (eds) *The ecology and biogeography of* Nothofagus *forests*, pp25-82. New Haven: University of Yale Press.

Otto-Bliesner, B.L. and G.R. Upchurch Jr 1997. Vegetation-induced warming of high-latitude regions during the Cretaceous period. *Nature* 385:804-807.

Pagani, M., J.C. Zachos, K.H. Freeman, B. Tipple and S. Bohaty 2005. Marked decline in atmospheric carbon dioxide concentrations during the Paleogene. *Science* 309:600-603.

Poole, I. and D.J. Cantrill 2006. Cretaceous and Cenozoic vegetation of Antarctica integrating the fossil wood record. In J.E. Francis, D. Pirrie, and J.A. Crame (eds) *Cretaceous-Tertiary high-latitude palaeoenvironments, James Ross Basin, Antarctica.* Geological Society, London, Special Publications 258:63-81.

Pole, M.S. and M.K. Macphail 1996. Eocene *Nypa* from Regatta Point, Tasmania. *Review of Palaeobotany and Palynology* 92:55-67.

Premoli, A.C. 1997. Genetic variation in a geographically restricted and two widespread species of South American *Nothofagus*. *Journal of Biogeography* 24:883-892.

Read, J. 1990. Some effects of acclimation temperature on net photosynthesis in some tropical and extra-tropical Australasian *Nothofagus* species. *Journal of Ecology* 78:100-112.

Read, J. and M.J. Brown 1996. Ecology of Australian *Nothofagus* forests. In T.T. Veblen, R.S. Hill and J. Read (eds) *The ecology and biogeography of* Nothofagus *forests*, pp131-181. New Haven: University of Yale Press.

Read, J. and G. Farquhar 1991. Comparative studies in *Nothofagus* (Fagaceae). I. Leaf carbon isotope discrimination. *Functional Ecology* 5:684-695.

Read, J. and J.E. Francis 1992. Responses of some Southern Hemisphere tree species to a prolonged dark period and their phytogeographic and palaeoecological implications for high-latitude Cretaceous and Tertiary floras. *Palaeogeography, Palaeoclimatology, Palaeoecology* 99:271-290.

Read, J. and R.S. Hill 1989. The response of some Australian temperate rainforest tree species to freezing temperatures and its biogeographical significance. *Journal of Biogeography* 16:21-27.

Read, J., R.S. Hill and G.S. Hope 2010. Contrasting responses to water deficits of *Nothofagus* species from tropical New Guinea and southern temperate forests: can rainfall regimes constrain latitudinal range? *Journal of Biogeography.*

Read, J. and G.S. Hope 1989. Foliar frost resistance of some evergreen tropical and extra-tropical Australasian *Nothofagus* species. *Australian Journal of Botany* 37:361-373.

Read, J. and G.S. Hope 1996. Ecology of *Nothofagus* forests of New Guinea and New Caledonia. In T.T. Veblen, R.S. Hill and J. Read (eds) *The ecology and biogeography of* Nothofagus *forests*, pp200-256. New Haven: University of Yale Press.

Read, J., G.S. Hope and R.S. Hill 1990a. Integrating historical and ecophysiological studies in *Nothofagus* to examine the factors shaping the development of cool rainforest in southeastern Australia. In J.G. Douglas and D.C. Christophel (eds) Proceedings of the 3rd International Organisation of Palaeobotany Symposium, 1988. pp97-106.

Read, J., G.S. Hope and R.S. Hill 1990b. The dynamics of some *Nothofagus*-dominated rainforests in Papua New Guinea. *Journal of Biogeography* 17:185-204.

Read, J., G.S. Hope and R.S. Hill 2005. Phytogeography and climate analysis of *Nothofagus* subgenus *Brassospora* in New Guinea and New Caledonia. *Australian Journal of Botany* 53:297-312.

Royer D.L., C.P. Osborne and D.J. Beerling 2003. Carbon loss by deciduous trees in a CO_2-rich ancient polar environment. *Nature* 424:60-62.

Royer, D.L. and P. Wilf 2006. Why do toothed leaves correlate with cold climates? Gas exchange at leaf margins provides new insights into a classic paleotemperature proxy. *International Journal of Plant Sciences* 167:11-18.

Sauquet, H., P.H. Weston, C.L. Anderson, N.P. Barker, D.J. Cantrill, A.R. Mast and V. Savolainen 2009. Contrasted patterns of hyperdiversification in Mediterranean hotspots. *Proceedings of the National Academy of Sciences* 106:221-225.

Scriven, L.J., S. McLoughlin and R.S. Hill 1995. *Nothofagus plicata* (Nothofagaceae), a new deciduous Eocene macrofossil species, from southern continental Australia. *Review of Palaeobotany and Palynology* 86:199-209.

Sniderman, J.M.K., N. Porch and A.P. Kershaw 2009. Quantitative reconstruction of Early Pleistocene climate in south-eastern Australia and implications for atmospheric circulation. *Quaternary Science Reviews* 28:3185-3196.

Specht, R.L. 1981. Evolution of the Australian flora: some generalizations. In A. Keast (ed.) *Ecological Biogeography of Australia*, pp783-805. The Hague: Dr W Junk Publishers.

Specht, R.L., M.E. Dettmann and D.M. Jarzen 1992. Community associations and structure in the Late Cretaceous vegetation of southeast Australasia and Antarctica. *Palaeogeography, Palaeoclimatology and Palaeoecology* 94:283-309.

Spicer, R.A. 1987. The significance of the Cretaceous flora of northern Alaska for the reconstruction of the climate of the Cretaceous. *Geologisches Jahrbuch A* 96:265–291.

Turner, I.M. 1994. Sclerophylly: primarily protective? *Functional Ecology* 8:669-675.

Truswell, E.M. 1993. Vegetation changes in the Australian Tertiary in response to climatic and phytogeographic forcing factors. *Australian Systematic Botany* 6:533-557.

Van Steenis, C.G.G.J. 1971. *Nothofagus*, key genus of plant geography, in time and space, living and fossil, ecology and phylogeny. *Blumea* 19:5-98.

Veblen, T.T., C. Donoso, T. Kitzberger and A.J. Rebertus 1996. Ecology of southern Chilean and Argentinean *Nothofagus* forests. In T.T. Veblen, R.S. Hill and J. Read (eds) *The ecology and biogeography of* Nothofagus *forests*, pp293-353. New Haven: University of Yale Press.

Walker, D. and G.S. Hope 1982. Late Quaternary vegetation history. In J.L. Gressitt (ed.) *Biogeography and ecology of New Guinea*, pp263-285. The Hague: Dr W. Junk Publishers.

Wilf, P., S.L. Wing, D.R Greenwood and C.L. Greenwood 1998. Using fossil leaves as paleoprecipitation indicators: An Eocene example. *Geology* 26:203-206.

Woodward, F.I. 1987. *Climate and plant distribution*. Cambridge: Cambridge University Press.

Worth, J.R.P., G.J. Jordan, G.E. McKinnon and R.E. Vaillancourt 2009. The major Australian cool temperate rainforest tree *Nothofagus cunninghamii* withstood Pleistocene glacial aridity within multiple regions: evidence from the chloroplast. *New Phytologist* 182:519-532.

3

Beneath the peat: A refined pollen record from an interstadial at Caledonia Fen, highland eastern Victoria, Australia

A. Peter Kershaw

School of Geography and Environmental Science, Monash University, Australia
peter.kershaw@arts.monash.edu.au

G. Merna McKenzie

School of Geography and Environmental Science, Monash University, Australia

Jonathan Brown

School of Geography and Environmental Science, Monash University, Australia

Richard G. Roberts

GeoQuEST Research Centre, School of Earth and Environmental Sciences, University of Wollongong, Australia

Sander van der Kaars

School of Geography and Environmental Science, Monash University, Australia, and Department of Palynology and Climate Dynamics, Albrecht-von-Haller-Institute for Plant Sciences, University of Göttingen, Germany

Introduction

The reconstruction of past vegetation has traditionally been based on peat deposits because of their accessibility and abundance in previously glaciated landscapes of northwestern Europe where the interest originated, the fact that pollen analysis can be combined with identification of macrofossils and peat stratigraphy to maximise knowledge of past vegetation and its controls, particularly climate, and the ease with which sediment cores for analysis can be extracted. The interest was extended to the Australasian region largely by researchers from, or trained at, European institutions, especially the Godwin Institute for Quaternary Research (GIQR) at the University of Cambridge. Peat studies caught on well in New Zealand and have been practised for many years, originally by Lucy Cranwell and subsequently by Neville Moar, Matt McGlone and others. But the predominantly unglaciated and arid landscape of Australia did not initially seem to have great appeal or promise, although GIQR-trained David Churchill and Sue Duigan undertook initial pollen studies on peat deposits. There was general disappointment when Donald Walker, on his arrival at the Australian National University to establish a palynological laboratory, declared the general unsuitability of Australia for such research and commenced work on the peatlands of previously glaciated highland New Guinea.

As a student of Walker's, Geoff Hope was sent to New Guinea, where he developed a lifetime passion for peatland investigation in exotic places and, when geographically constrained, for investigation of the rare bogs of southeastern Australia. Hope's approach of accumulating sediments with a small peat sampler, often hidden away in 'tourist' baggage, or even borrowing a spade if no sampler was available (see Figure 1) may appear to be hit-and-run tactics, but the conservation value of these acquisitions has become progressively apparent as this natural archive has been increasingly destroyed by land-use activities, peatland and bushfires and, recently, greenhouse-induced climate change.

Figure 1. Geoff has always been into peat stratigraphy (from Hope 1968)

Although what was often a one-man operation can be sufficient to extract sediment accumulated since the last glacial period, especially in shallow glaciated depressions, it may fail to reveal any older material. The construction of substantially older records was traditionally considered of limited and dubious value because problems in dating and correlation often led to endless chronological and palaeoecological debate and speculation. However, improvements in the applicability of radiocarbon dating, the development of techniques capable of producing ages beyond those available from radiocarbon dating, and an increasing interest in pre-Holocene events that may have bearing on the understanding of the present landscape and prediction of future environmental change, has elevated the status of longer records.

Caledonia Fen, a shallow peat deposit in the southeastern highlands of Victoria, after initially being dismissed as too condensed to provide a good Holocene record (Phil Ladd pers comm.), was later found, surprisingly, to extend into polleniferous 'glacial' sediments that could not be penetrated far with hand-operated corers, nor dated using available conventional radiocarbon methods. A pollen diagram was produced and published in summary form (Kershaw et al. 1983, 1991), before research was suspended for many years. The impetus for further study was provided by the interest of technicians Jack Misson and David Tooth in the development of appropriate coring devices, a passion for highland vegetation history and the persuasive powers of one of us (GMM), 'willing' student power, and the development of accelerator mass spectrometry (AMS) radiocarbon dating. The presence of a narrow band of organic sediment near the base of an 8 m core led to speculation that this represented climatic amelioration during the Last Interglacial – until a subsequent broader organic layer, that might be better related to the Last Interglacial, was recorded towards the base of an 18 m core. Pollen analysis seemed to demonstrate, from forest expansion, that there was a similar increase in temperature within both layers, generally equivalent to that of the Holocene. A combination of AMS radiocarbon ages on plant and insect macrofossils and optically stimulated luminescence (OSL) ages on quartz grains strongly suggested that the upper organic band occurred during the last glacial period and the lower layer most likely represented the Last Interglacial. Selected features of the full palynological record, analysed at the high resolution of one sample per 4 cm, are shown in Figure 2.

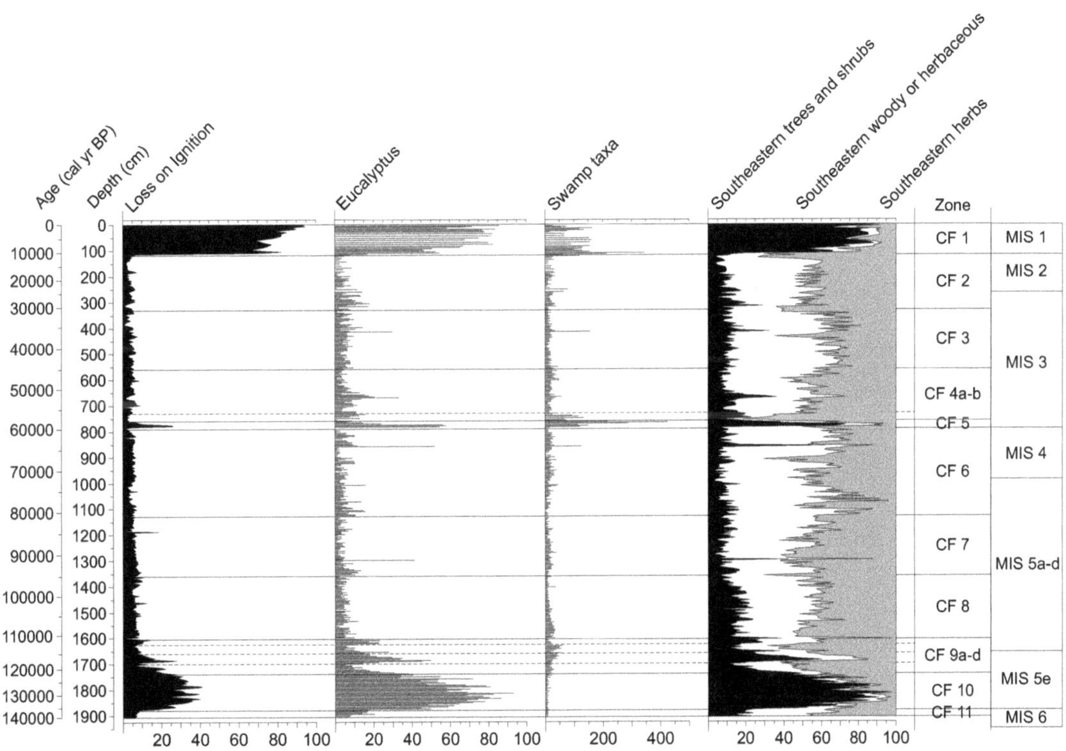

Figure 2. Selected features of the palynological record from Caledonia Fen (Kershaw et al. 2007) in relation to the Vostok ice-core chronology of Pettit et al. (1999)

This paper is focused on the age and duration of the upper organic layer and its significance for understanding the dynamics of high mountain vegetation and the nature and causes of glacial climate variation. The layer can be defined as an interstadial. This term was originally introduced to define a period of warming during a glacial period that was too short or of insufficient magnitude to lead to the full expansion of deciduous forest in northern Europe, as occurs during interglacials (West 1961). However, it has been applied to other areas where forest migration is not a problem and, as at Caledonia Fen, forest development occurred for only a short period. In some regions, especially those adjacent to, or greatly influenced by, the North Atlantic climate, phases of short-term forest development within the last glacial period have been linked to semi-regular Dansgaard-Oeschger (D-O) warming events, as defined in Greenland ice cores (Dansgaard et al. 1984), and currently termed Greenland Interstadials (GIs), with intervening cold phases termed Greenland Stadials (GSs) (Svensson et al. 2006). The vegetation response to D-O forcing appears to be variable within as well as between records (e.g. Sanchez Goñi et al. 2008), and it would be valuable to determine whether the single prominent interstadial at Caledonia Fen can be accommodated by any D-O GI, or whether there is any other explanatory mechanism.

For special investigation of this event, the sampling resolution, where material had not been completely lost during sediment removal for OSL dating, has been increased from every 4 cm to contiguous 1 cm intervals, while two additional OSL ages have been determined. These numerical ages provide the age model for the core section centred on the event. The section in question was extended temporally to allow examination of any short-term stadial variability and, in particular, to capture and examine the significance of occasional, single sample spikes in *Eucalyptus* pollen, sometimes accompanied by aquatic pollen peaks, in the glacial period (see Figure 2). Palynological analysis, sediment organic matter determination and OSL dating techniques follow those detailed in Kershaw et al. (2007).

The site and its regional context

Caledonia Fen (37° 20' S, 146° 44' E) is a small depression, about 300 m by 160 m, perched on a valley side at an altitude of 1280 m between the Caledonia River (c. 1150 m) and the Snowy Range (c. 1650 m) within the mountainous southeastern Australian Alpine National Park (Figure 3). The site was formed as part of a stream-eroded valley that was subsequently blocked off by mass-movement deposits and then infilled with fluviolacustrine and swamp sediments (Brown In prep.; Kershaw et al. 2007). Currently, the fen receives water from a number of small inlet streams and is drained by a single intermittent outlet stream. The basin is covered with a mix of bog and fen plant communities, most notably aquatic fen dominated by the emergent sedges *Eleocharis sphacelata* and *Scirpus crassiusculus* and submerged species *Myriophyllum pedunculatum* and *M. propinquum* (Haloragaceae), fen dominated by the sedge *Carex gaudichaudiana* and peat moss *Sphagnum cristatum*, and raised bog dominated by *Sphagnum cristatum*, the restiad *Empodisma minus* (Restionaceae), *Carex gaudichaudiana* and the heath *Epacris paludosa*. The most likely explanation for this vegetation mix is precipitation, as the current estimated 1530 mm/year (Houlder et al. 1999) is close to the threshold between dominance of wetter raised bog and drier fen within highland southeastern Australia (Kershaw et al. 2007).

The region experiences a temperate climate, with more than one-third of mean annual rainfall falling in the winter quarter (Bureau of Meteorology 1989). Although the westerly wind system brings moist and cloudy weather through much of the year, the region also experiences cold air masses emanating from the Southern Ocean close to Antarctica and also moist and warm tropical maritime air from the southern Pacific Ocean (Sturman and Tapper 2005). Mean annual temperature is estimated to be 7.3°C, mean maximum temperature 19.1°C, and mean minimum temperature -1.2°C. In cooler months, blizzard conditions may bring sub-zero maximum daily temperatures for several days at a time and frosts are common even in the warmest months. At altitudes above 1400 m, winter precipitation is predominantly as snow, which accumulates from June until the spring thaw begins in September, providing the major source of water to Caledonia Fen. Generally speaking,

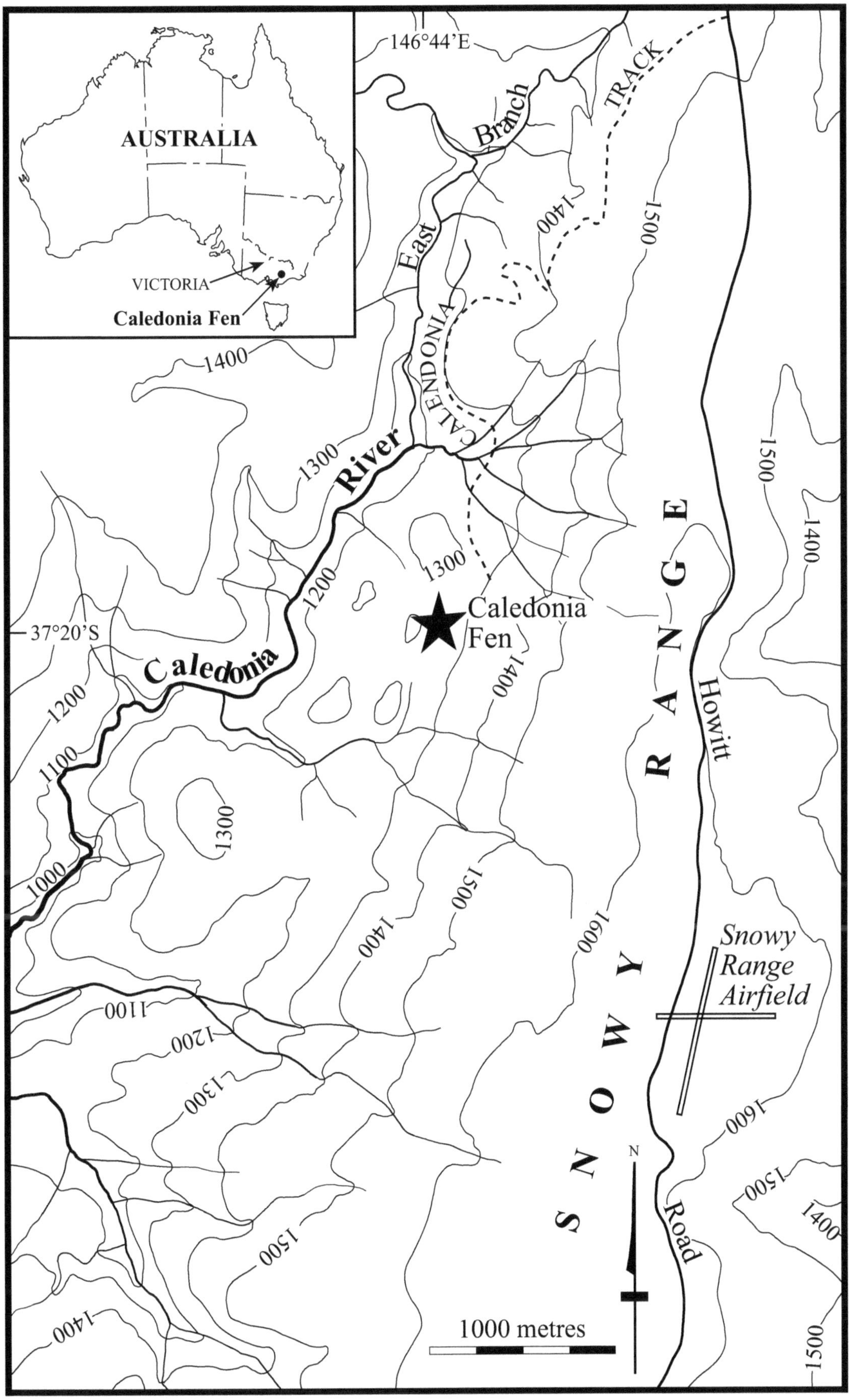

Figure 3. Location of Caledonia Fen. Based on the Tamboritha-Moroka 8223-S, 1:50,000 Topographic Map, Survey and Mapping Victoria, 1992 (modified from Kershaw et al. 2007)

precipitation in the alpine and subalpine regions is often erratic and may lead to water stress in summer (Farrell and Ashton 1973).

Variations in regional vegetation can be broadly related to altitude (Figure 4). Caledonia Fen lies towards the upper limit of montane sclerophyll forest, dominated by *Eucalyptus delegatensis* on moister slopes and with a variety of other eucalypts, including *E. dalrympleana* and *E. rubida*, on drier and steeper aspects. These forests have an understorey composed of small trees, woody shrubs, herbs and ferns. They become more open with increased altitude and give way to sub-alpine woodland composed of *E. pauciflora, E. stellulata* and *E. rubida*, with the myrtaceous shrub *Leptospermum grandifolium* and a variable understorey of grasses, scattered shrubs and herbs around 1400 m. A mosaic of treeless communities, including short and tall herbfield dominated by the Asteraceae, moist and wet heathland, grassland, bog and fen, occurs within sub-alpine woodland on the Snowy Range plateau, due to cold air drainage. Dry heathland, which forms a floristically and ecologically well-defined closed community, exists on rocky exposed ledges or steep gullies. It is characterised by the sole representative of the fire-sensitive, temperate rainforest conifers on mainland southeastern Australia, *Podocarpus lawrencei* (Podocarpaceae). The altitudinal treeline, estimated from that on Mount Kosciuzko to be around 1700 m (Figure 4), is not reached within the region.

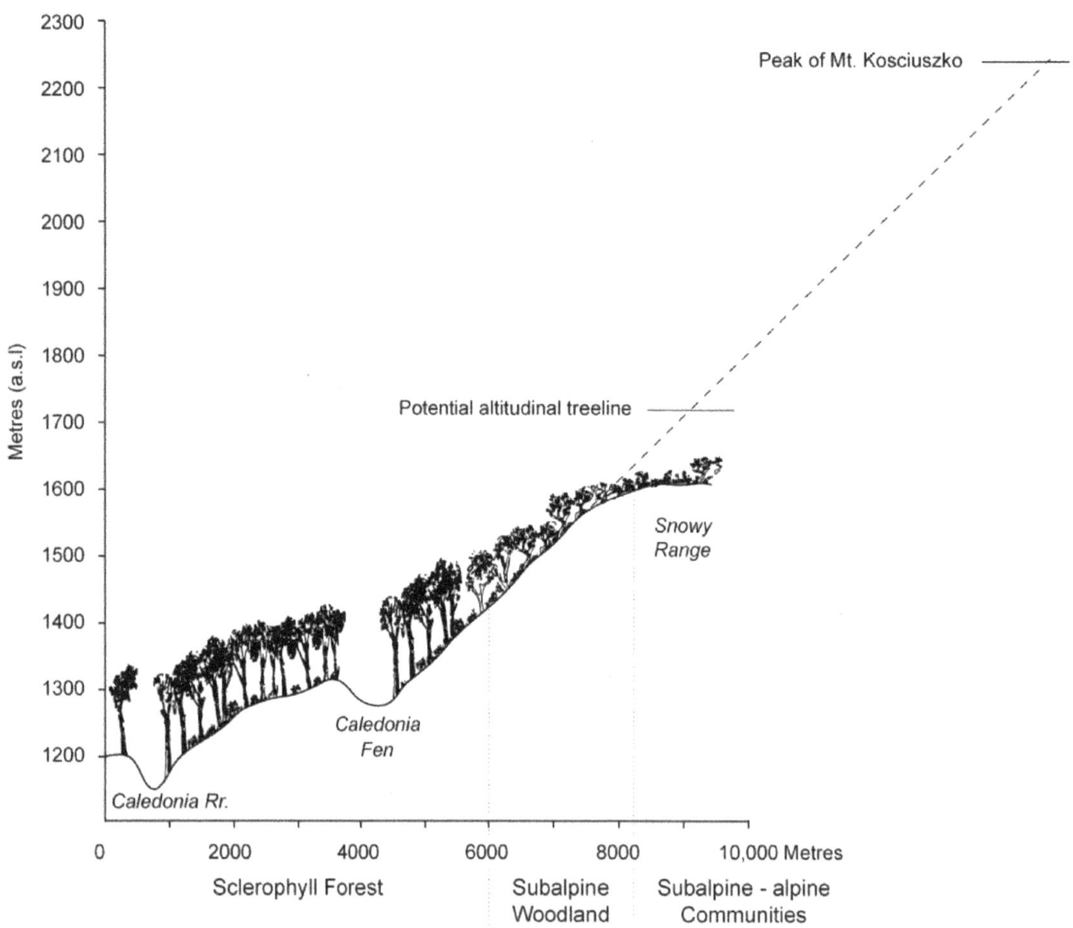

Figure 4. Vegetation transect from the Caledonia River to the Snowy Range in relation to the potential altitudinal vegetation zonation, including the estimated location of the tree line

Chronological considerations

The chronology established for the whole core was based on its comparison with the Vostok temperature curve of Pettit et al. (1999), taking into account available AMS and OSL dates from the top 8 m of sediment. Actual ages were established for the zone and sub-zone boundaries, shown in Figure 2, and the age-depth curve is shown in Figure 5. For the section selected for study in this paper, it was considered appropriate to establish an independent timescale based on two original and two new OSL ages from within or bracketing the section (Figure 5), especially as the ages increase with depth (Table 1). The experimental uncertainties of several millennia associated with the ages mean that the inferred changes in sedimentation rate should be viewed with caution, but, overall, the ages probably provide a good indication of the length of time represented by the section. In relation to actual age, all dates are systematically older than predicted from correlation with the Vostok record and, although this may be the case, it could result from some incorrect assumptions about past changes in water content, as discussed in Kershaw et al. (2007). However, error bars just or nearly bisect the age-depth curve of the Vostok fitted chronology (Figure 5).

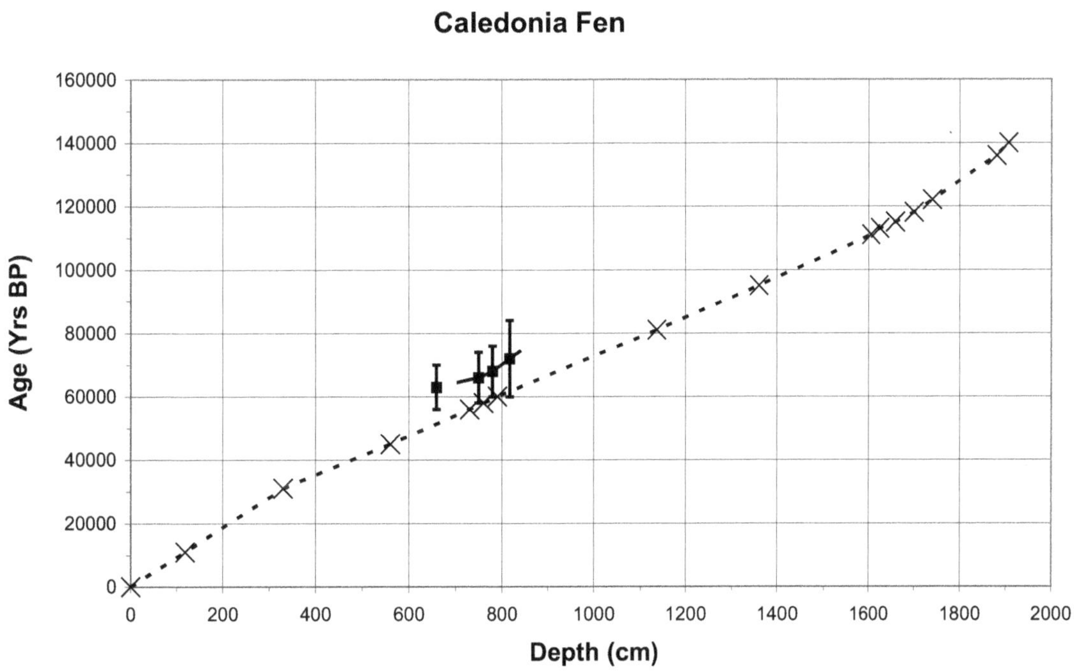

Figure 5. Age-depth curve for the Caledonia Fen record of Kershaw et al. (2007), determined from comparison of pollen zone and subzone boundaries (marked by Xs) with the Vostok ice-core chronology of Pettit et al. (1999). The offset curve (solid line) relates to the refined component of the record aged from the mean points of OSL age ranges within and bracketing this period. See Table 1 for details of OSL ages

Sample code	Depth in core (cm)	Optical age (yr BP)	Source
CF9	659	63,000 ± 7000	This paper
CF4	748-755	66,000 ± 8000	Kershaw et al. 2007
CF3	776-783	68,000 ± 8000	Kershaw et al. 2007
CF10	817	72,000 ± 12,000	This paper

Table 1. Luminescence results associated with the interstadial at Caledonia Fen

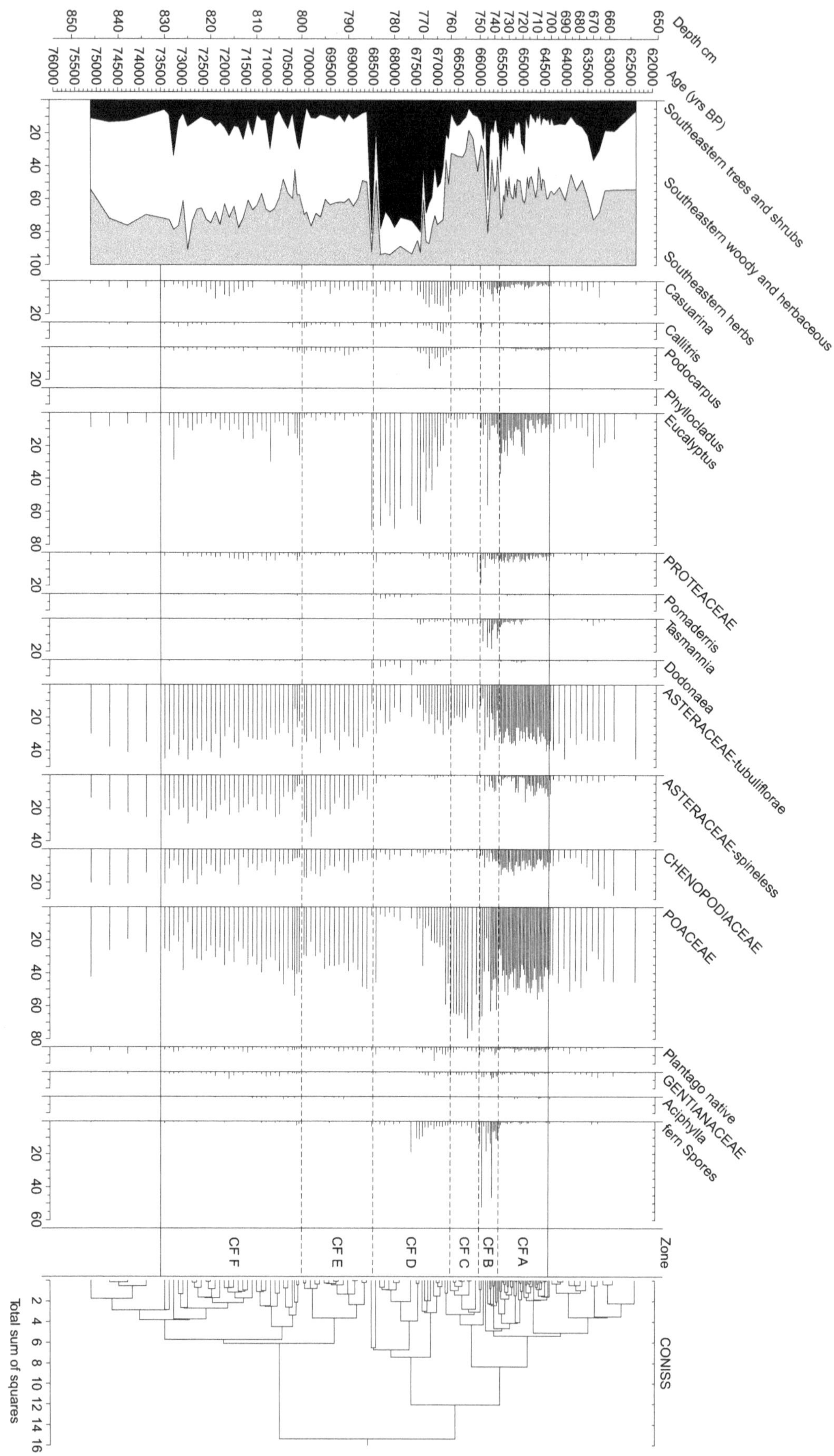

Figure 6. Selected dryland attributes of the refined component of the palynological record from Caledonia Fen embracing the major interstadial period. Taxon values are expressed as percentages of the dryland southeast Australian pollen sum (D'Costa and Kershaw 1997) for the appropriate sample

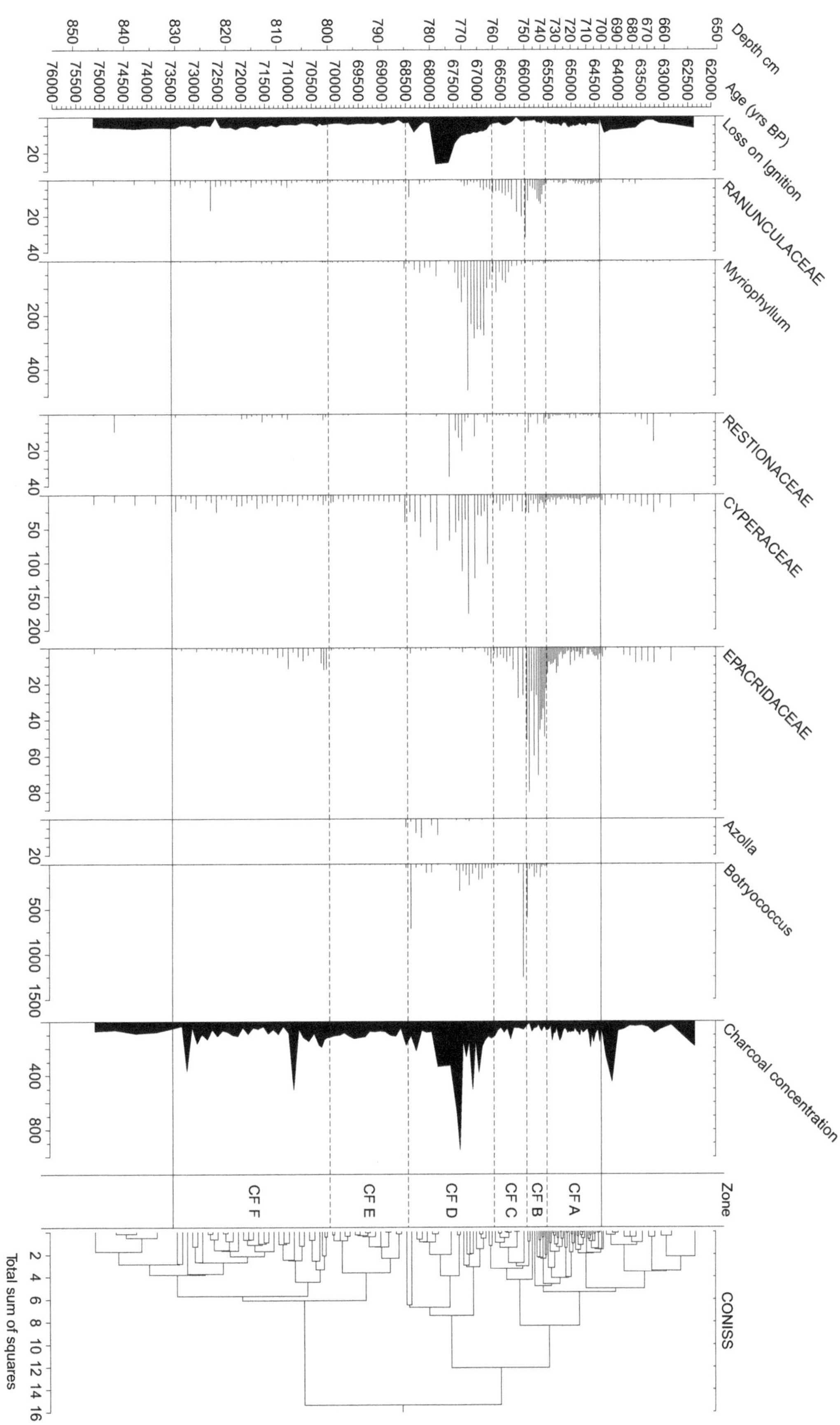

Figure 7. Selected aquatic attributes of the refined component of the palynological record from Caledonia Fen embracing the major interstadial period. Taxon values are expressed as percentages of the dryland southeast Australian pollen sum (D'Costa and Kershaw 1997) for the appropriate sample

The palynological record

Selected data are presented in Figures 6 and 7. They include major dryland (Figure 6) and aquatic taxa (Figure 7), all expressed as a percentage of the southeastern Australian pollen sum of regionally dispersed dryland taxa, which include the illustrated taxa *Casuarina, Podocarpus, Phyllocladus, Eucalyptus, Pomaderris, Dodonaea,* Asteraceae-tubuliflorae, Asteraceae-spineless, Poaceae and *Plantago*. Pollen sum taxa have been used also to construct the ratio diagram of trees and shrubs, woody and herbaceous taxa, and herbs. Also included are fern spore (predominantly tree fern) percentages in Figure 6, and charcoal concentration values in particles/cm^3, and loss on ignition (or organic content) values in Figure 7. The stratigraphically constrained classification, CONISS, based on pollen sum taxon values, has been employed to help determine a zonation scheme, although potentially slightly different response times to forcing influences at this resolution necessitate some flexibility in zone description. The refined component of the diagram extends from a depth of 830 cm, with an estimated mean age of 73,400 years, to 700 cm, with an estimated mean age of 64,300 years.

Zone CFF (830-800 cm; c. 73,400-70,200 yr BP)

High values for Poacaeae and Asteraceae pollen indicate the regional dominance of grassland and herbfield, while significant representation of Proteaceae, Gentianaceae and *Aciphylla* together indicate that these communities were alpine in character (McKenzie 1997; Martin 1999). High values of Chenopodiaceae suggest a drier facies of alpine vegetation than today, and vegetation that was probably steppic in character with no modern analogue (Martin 1999). Assuming the present-day lapse rate of 1.1°C lowering of temperature with every 100 m rise in altitude, mean annual temperatures would have been at least 4.5°C lower than today. Significant eucalypt pollen percentages could have been derived mainly from a much lower altitude treeline, but the likely absence of a marked treeline during the last glacial period, as indicated by lowland and other highland sites in the region (Kershaw 1998), suggests pollen transport to the site from distant to perhaps proximal forest, or woodland patches existing under locally moist and sheltered conditions. Consistently low loss-on-ignition values, combined with low values for aquatic taxa, suggest that the site was a low-productivity lake, probably because low temperatures resulted in ice cover for much of each year. However, fairly consistent or frequent representation of *Ranunculus,* Restionaceae, Cyperaceae and Epacridaceae may suggest that limited fen and bog vegetation grew on the site margin.

The zone shows significant variation in relative proportions of major dryland summary groups and, although this is perhaps exaggerated by the higher sample resolution than for the basal samples, it is notable in comparison with most other zones. It incorporates three sharp peaks in *Eucalyptus* pollen. Charcoal values are consistently low, apart from two sharp peaks that may indicate fires in the catchment.

Zone CFE (800-785 cm; c. 70,200-68,600 yr BP)

This zone is marked by overall lowest *Eucalyptus* percentages for the record, but generally higher values for other major woody taxa, Casuarinaceae, *Callitris* and *Podocarpus.* The consistency in percentages of *Eucalyptus* and Casuarinaceae strongly suggests regional to long-distance pollen dispersal, with the absence of a local source. As in the previous zone, the dominant woody or herbaceous taxa, Asteraceae and Chenopodiaceae, generally decrease through the zone relative to Poaceae. Aquatic taxon percentages are lowest for the record, suggesting minimal swamp representation in the basin. In fact, the extreme similarity in sample values for Cyperaceae is more consistent with a regional than proximal source. The overall impression from the pollen assemblages is that conditions were the most adverse for both woody and aquatic plants in the record. Background values for charcoal suggest also that the catchment vegetation did not carry fire.

Zone CFD (785-764 cm; c. 68,600-66,800 yr BP)

This zone is characterised by high values of tree pollen and reductions in Poaceae and the woody or herbaceous plants Asteraceae-spineless and Chenopodiaceae. Tree-pollen percentages are dominated by *Eucalyptus*, which rise sharply at the base, from less than 5% to near 70%, then plateau, before falling gradually but erratically to 20% at the top. The extent of expansion of forest, altitudinally as well as areally, is demonstrated by the near confinement of the pollen of lower-altitude wet and dry sclerophyll trees, *Pomaderris* and *Dodonaea* respectively, to this zone. The forest-pollen assemblage changes towards the top of the zone, as these taxa have reduced values, while *Tasmannia* and *Podocarpus*, as well as tree ferns, become prominent. Temperatures were clearly higher in the period represented by the zone as a whole and, from the overall assemblage composition, probably achieved levels similar to those of the Holocene. The changes towards the top of the zone suggest the temperature may have been lower to some degree and effective rainfall higher. The fact that there are also peaks for the diagram in *Casuarina* and *Callitris* within the upper part of the zone appears counter-intuitive, as they are best represented regionally under drier and warmer conditions. Either these trees were represented by taxa now absent from the region (that could include Cupressaceae pollen morphologically indistinguishable from *Callitris* and now confined to wetter highlands in Tasmania) or they indicate a high, long-distance component from greatly expanded lowland forests. Charcoal concentrations achieve much higher levels and are more frequent than elsewhere in the record, indicating a positive relationship between tree proximity and density and fire activity.

Within the basin, climatic amelioration clearly led to reduction in ice-cover duration, as aquatics were able to become well established. Cyperaceae has high values through the zone. Accompanying *Azolla* and *Botryoccocus* demonstrate the presence of open water in the early part of the zone. However, productivity within the lake, leading to significant organic matter accumulation, was delayed by about 500 years. The major peaks for the record in *Myriophyllum* and Restionaceae may indicate successional advancement, although the mix of open water to bog plants suggests a variable fen surface similar to that of today.

Zone CFC (764-750 cm; c. 66,800-66,000 yr BP)

This zone shows a major reversal from forest to open assemblages, similar to zone CFE except that Poaceae is dominant, Asteraceae-spineless is still largely absent and Chenopodiaceae is reduced to lowest percentages for the diagram. The achievement of highest values for Ranunculaceae and partially maintained percentages of *Myriophyllum* together suggest that temperatures were sufficiently high to maintain relatively ice-free conditions on site, while the ephemeral fen nature of this assemblage may explain, through oxidation, reduced organic content of the sediment. Drier conditions are also suggested by the near absence of bog Restionaceae, and possibly also by the absence of a clear charcoal peak, if this is interpreted as the inability of the vegetation to carry fire due to a sparse coverage. The postulation of a combination of dry and relatively warm conditions is difficult to reconcile with the low values of *Eucalyptus* and particularly Chenopodiaceae pollen. With *Eucalyptus*, it is possible perhaps that parent species in the area were not adapted to dryness or that they were poor pollen producers, as noted by Martin (1999) for *Eucalyptus pauciflora*, the present dominant of sub-alpine woodland. With Chenopodiaceae, it could have been that a change in the predominant wind direction or a substantial reduction in wind velocity had resulted in the loss of a long-distance component of salt-bush pollen from interior semi-arid shrublands. The nearest analogue in the extended

Caledonia Fen record is a short Poaceae-dominated period that just precedes the Holocene (Figure 2), a time when temperatures were close to those of today (Williams et al. 2009).

Zone CFB (750-735 cm; c. 66,000-65,500 yr BP)

Shows sharp peaks in three major taxa (Asteraceae tubuliflorae, Poaceae and *Eucalyptus*) and consistently higher values of Asteraceae-spineless and Chenopodiaceae than in the previous zone. It is characterised as having highest values for *Tasmannia,* fern spores, and Epacridaceae for the diagram. The high values for Epacridaceae, together with relatively high percentages of Restionaceae, suggest the establishment of a bog within the basin, although high values for *Botryococcus* indicate the presence also of open water. The absence of *Myriophyllum* may relate to a higher level of acidity. Deposited sediments are surprisingly inorganic, suggesting that open water was not consistently present, a feature perhaps consistent with variable representation in dryland pollen components, or that oxidation of organics took place after deposition of these sediments. Although variable and apparently transitional in nature, conditions were sufficiently mild to allow survival of ferns and aquatic vegetation. The fact that *Tasmannia* is a frequent understorey taxon of, and palynologically can be indicative of, *Eucalyptus pauciflora* sub-alpine woodland (Martin 1999) may provide a more certain explanation for the relatively low eucalypt pollen values.

Zone CFA (735-700 cm; c. 65,500-64,300 yr BP)

Stadial conditions very similar to those of zone CFF are indicated, apart from generally lower values of Asteraceae-spineless – a feature of unknown significance due to the inability, thus far, to confidently identify source plants in the present landscape (Macphail and Martin 1991). In the lower part of the zone, even the variability in ratios of the summary dryland diagram are apparent, and there is an isolated peak in *Eucalyptus* towards the top of the zone.

Discussion

The nature of the interstadial

The interstadial is more extensive than originally envisaged, in that it began before the switch to a relatively high organic sediment component. It could be considered to embrace zone CFD, defined on continuous high values of *Eucalyptus*, which is estimated to have lasted about 1800 years. However, on proposed temperature grounds inferred largely from the representation of tree-fern spores and site evidence for colonisation by aquatic plants and some successional development within aquatic vegetation, it could have extended to the end of zone CFB, covering some 3000 years. In either case, the beginning is shown to be more abrupt than the ending. Regardless of definition, this period in the history of the fen is shown to be complex, with relatively frequent climate-driven changes.

Affiliation of the interstadial

So-called interstadials during the last glacial cycle in the Australian region have generally been relatable to the orbital- and ice-forced periods MIS 5c, MIS 5a and early MIS 3 emanating from the North Atlantic region. However, the duration of the Caledonia Fen interstadial appears to be much shorter than these interstadials (that typically relate to semi-precessional cycles of about 10,000 years duration) and, therefore, the period is more likely to be driven or influenced also by a shorter-term millennial process. There are three major suggested causes of millennial-scale variability within the last glacial period: Dansgaard-Oeschger (D-O) cycles, Antarctic climatic reversals, and high El Niño-Southern Oscillation activity. All are suggested to have some expression in the few records from the Australian region that

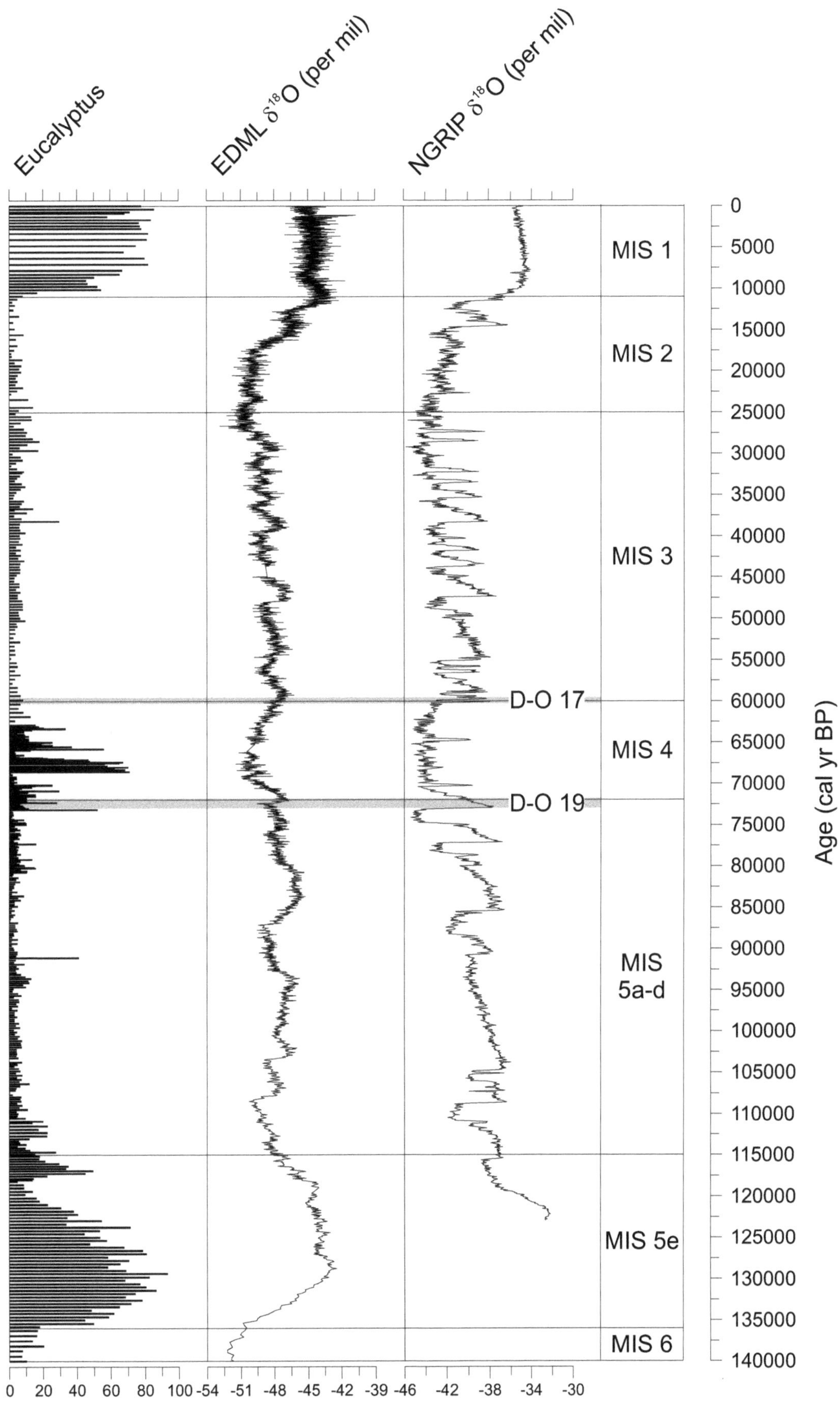

Figure 8. Comparison of the *Eucalyptus* record from Caledonia Fen with the oxygen isotope records from ice cores showing millennial scale variability during the last glacial period; the EDML (EPICA) core displaying Antarctic warm reversals and the NGrip Greenland core showing Dansgaard-Oeschger (D-O) warm events (EPICA Community Members 2006). D-O events 17 and 19 are marked

are sufficiently long and refined to capture them, but their signals are relatively weak and inconsistent between records. Sea-surface temperature variations from marine records in the Southern Ocean have been related to Antarctic warm events (Barrows et al. 2007), while those from IMAGES Core MD01-2378, off northwestern Australia, show close affinity with Northern Hemisphere propagated D-O warm events (Dürkop et al. 2008). At Lynch's Crater, in northeast Queensland, Turney et al. (2004) recorded variations of a similar frequency and phasing to D-O events within the organic components of the sediments; these were suggested to have been the result of ENSO oscillations, whose signals were atmospherically transmitted to the North Atlantic, triggering the ice meltings that gave rise to the D-O cycles. However, variations were related to rainfall rather than temperature, had insufficient impact to notably alter adjacent dryland pollen assemblages, and the record only extends back to about 50,000 years ago.

In terms of duration and pattern, the interstadial does resemble a well-developed D-O cycle, rather than an Antarctic warm event (Figure 8), with an apparent sharp increase in temperature from stadial levels and then a more gradual but oscillating reduction, regardless of whether the longer or shorter period is considered. However, the dating falls into a period, between about 60,000 and 70,000 yr BP, when no such major event has been recorded. As this corresponds largely with MIS 4, which experienced cold conditions, including within the southern Australasian region (Barrows et al. 2007), warm phases are not to be expected. If the dates are systematically too old, as could be suggested from a comparison with the Kershaw et al. (2007) chronology, a correspondence with D-O 17, centered on 59,000 years ago (at the beginning of MIS 3), could be possible, although, from pattern-matching, a fit with D-O 19 (on the MIS 5a-4 boundary), with an estimated age of about 72,000 yr BP, is much more likely (EPICA Community Members 2006). This event had a duration of about 2000 years, similar to that estimated for zone CFD, and has a following short, but D-O unrecognised, warm event that could be correlated with zone CTB. Although the age difference appears to be about 6000 years, this is well within OSL dating uncertainty for Caledonia Fen. An older age may be plausible also on the grounds that OSL ages are systematically younger than those of radiocarbon at this site (Kershaw et al. 2007).

The status of other abrupt events

One of the major problems with the acceptance of the Caledonia Fen interstadial as a D-O event is that it is singular. Certainly, there is a great deal of variation in the expression of D-O events at different sites and in different locations: for example, Core MD01-2378 records only four identifiable events out of the most recent 17, but apparent representation of only one is extreme. However, it has been mentioned that there were other events, marked by sharp peaks in *Eucalyptus* and generally minor aquatic pollen representation, that were considered as potential D-O candidates in the full record in Caledonia Fen by Kershaw et al. (2007), on the basis that their frequency appeared comparable. It was hoped that a more certain impression of their duration as well as frequency could be gained from this temporally precise study. Unfortunately, the analysed sequence is beyond the main period of expression of D-O events and few clear minor events could be determined. Those that do exist suggest that eucalypt peaks extended over no more than one sample and hardly qualify as interstadials. It could well be that most can be accounted for by occasional over-representation through vagaries of pollen transport or flowering, but, considering that the strong eucalypt pollen rise at the beginning of zone CFD occurred within one sample, probably representing fewer than 100 years, there is reason to believe that marked sub-centennial forest or woodland responses to short-lived climatic events can be identified in the palaeo-record.

Conclusions

The analysis of a contiguously sampled section of a sediment core from Caledonia Fen, centred on the only marked interstadial within the last glacial period, has revealed details of vegetation and climate at a centennial scale. The interstadial is estimated to cover a maximum of about 3000 years. It shows an initial abrupt change from alpine steppe vegetation to montane forest, perhaps within a century, with temperatures and precipitation close to those achieved within the Holocene. This expansion lasted for little more than 1000 years before there was a less abrupt decline in forest and its replacement by a predominantly grassland vegetation, considered to indicate much drier, but still relatively warm, conditions. A short reversal to moister conditions is suggested by a predominantly wet scrub vegetation without a eucalypt canopy, before a return to alpine steppe.

The best age estimate for the interstadial, based on OSL dating, is 66,800- 65,500 yr BP, which falls within the globally cool period of MIS 4. However, from comparison with various identified major global events and taking into consideration the uncertainties associated with the OSL ages, the best match is with D-O 19 that spans the MIS 5-4 boundary around 73,500 yr BP. Acceptance of this correlation, though, is constrained by the lack of any mechanism that will allow a signal, generally considered to have emanated from the North Atlantic, to be so prominent, and by the absence of any other marked D-O interstadials. Other events are recognised in the sequence, but are considered too short and frequent to qualify as interstadials. Based on existing knowledge, any explanation of the cause of the interglacial has to be considered speculative. It is fortunate, though, that, unlike the superficial Holocene peats, this archive will remain in the landscape to await the tools and knowledge necessary for explanation in the future.

Acknowledgements

We thank Kara Rasmanis and Stuart Clift for assistance with drafting illustrations.

References

Barrows, T.T., S. Juggins, P. De Deckker, E. Calvo and C. Pelejero 2007. Long-term sea surface temperature and climate change in the Australian-New Zealand region. *Paleoceanography* 22:1-17.

Bureau of Meteorology 1989. *Climate of Australia,* Australian Government Publishing Service: Canberra.

Dansgaard, W., S. Johnsen, H.B. Clausen, D. Dahl-Jensen, N. Gundestrup, C.U. Hammer and H. Oeschger 1984. North Atlantic climatic oscillations revealed by deep Greenland ice cores. In J. E. Hansen, and T. Takahashi (eds) *Climate Processes and Climate Sensitivity,* pp288-298. Washington: American Geophysical Union.

D'Costa, D.M. and A.P. Kershaw 1997. An expanded recent pollen database from southeastern Australia and its potential for refinement of palaeoclimatic estimates. *Australian Journal of Botany* 45:583-605.

Dürkop, A., A. Holbourn, W. Kuhnt, R. Zuraida, N. Anderson and P.M. Grootes 2008. Centennial-scale climate variability in the Timor Sea during Marine Isotope Stage 3. *Marine Micropalaeontology* 66:208-221.

EPICA Community Members 2006. One-to-one coupling of glacial climate variability in Greenland and Antarctica. *Nature* 444:195-198.

Farrell, T.P. and D.H. Ashton 1973. Ecological studies on the Bennison High Plains. *Victorian Naturalist* 90:286-298.

Hope, G.S. 1968. Pollen studies at Wilsons Promontory. M.Sc. Thesis, Australia: Melbourne University.

Houlder, D.J., M.F. Hutchinson, H.A. Nix and J.P. McMahon 1999. *ANUCLIM User Guide, Version 5.0.* Canberra: Centre for Resource and Environmental Studies, Australian National University.

Kershaw, A.P. 1998. Estimates of regional climatic variation within southeastern Australia since the Last Glacial Maximum from pollen data. *Palaeoclimates - Data and Modelling* 3:107-134.

Kershaw, A.P., D.M. D'Costa, J. McEwen Mason and B.E. Wagstaff 1991. Quaternary vegetation of mainland southeastern Australia. *Quaternary Science Reviews* 10:391-404.

Kershaw, A.P., G.M. McKenzie, N. Porch, R.G. Roberts, J. Brown, H. Heijnis, L.M. Orr, G. Jacobsen and P.R. Newall 2007. A high resolution record of vegetation and climate through the last glacial cycle from Caledonia Fen, south-eastern highlands of Australia. *Journal of Quaternary Science* 22:481-500.

Kershaw, A.P., W. Southern, J.M. Williams and L.J. Joyce 1983. The vegetation record from the southeastern highlands of mainland Australia 40,000-3000 B.P. In J.M.A. Chappell and A. Grindrod (eds) *Proceedings, First CLIMANZ Conference, Howman's Gap, Victoria, February 1981,* pp16, 37-8, 62, 80-1, 101. Canberra: Dept. of Biogeography and Geomorphology, Australian National University.

Macphail, M. and A.R.H. Martin 1991. 'Spineless' Asteraceae II. *Palynological and Palaeobotanical Association of Australasia Newsletter* 23:1-2.

Martin, A.R.H. 1999. Pollen analysis of Digger's Creek Bog, Kosciuszko National Park: vegetation history and tree-line change. *Australian Journal of Botany* 47:725-744.

McKenzie, G.M. 1997. The late Quaternary vegetation history of the south-central highlands of Victoria, Australia: sites above 900m. *Australian Journal of Ecology* 22:19-36.

Pettit, J.R., J. Jouzel, D. Raynaud, N.I. Barkov, J.-M. Barnola, I. Basile, J. Bender, J. Chappellaz, M. Davis, G. Delaygue, M. Delmotte, V.M. Kitlyakov, M. Legrand, V.Y. Lipenkov, C. Lorius, L. Pepin, C. Ritz, E. Saltzman and M. Stievenard 1999. Climate and atmospheric history of the past 420,000 years from the Vostok ice core, Antarctica. *Nature* 399:429-436.

Sanchez Goñi, M.F., A. Landais, W.J. Fletcher, F. Naughton, S. Desprat and J. Duprat 2008. Contrasting impacts of Dansgaard-Oeschger events over a western European latitudinal transect modulated by orbital parameters. *Quaternary Science Reviews* 27:1136-1151.

Sturman, A. and N. Tapper 2005. *The Weather and Climate of Australia and New Zealand.* Second Edition. Oxford: Oxford University Press.

Svensson, A., K.K. Andersen, B. Bigler, H.B. Clausen, D. Dahl-Jensen, S.M. Davies, S.J. Johnsen, R. Muscheler, S.O. Rasmussen, R. Röthlisberger, J. P. Steffensen and B. M. Vinther 2006. The Greenland Ice Core Chronology 2005, 15-42 ka. Part 2: comparison to other records. *Quaternary Science Reviews* 25:3258-3267.

Turney, C.S.M., A.P. Kershaw, S. Clemens, N. Branch, P.T. Moss and L.K. Fifield 2004. Millennial and orbital variations in El Niño/Southern Oscillation and high latitude climate in the last glacial period. *Nature* 428:306-310.

West, R.G. 1961. Interglacial and interstadial vegetation in England. *Proceedings of the Linnean Society of London* 172:81-89.

Williams, M., E. Cook, S. van der Kaars, J. Shulmeister, T. Barrows and A.P. Kershaw 2009. Glacial and deglacial climatic patterns in Australia and surrounding regions from 35 000 to 10 000 years ago reconstructed from terrestrial and near-shore proxy data. *Quaternary Science Reviews.*

4

The vegetation cover of New Zealand during the Last Glacial Maximum: Do pollen records under-represent woody vegetation?

Matt S. McGlone

Landcare Research, Lincoln, New Zealand
mcglonem@landcareresearch.co.nz

Rewi M. Newnham

School of Geography, Environment & Earth Sciences, Victoria University of Wellington, Wellington, New Zealand

Neville T. Moar

Research Associate, Landcare Research, Lincoln, New Zealand

Introduction

The Last Glacial Maximum (LGM) was a global event characterised by cool, often dry climates, low atmospheric CO_2 concentrations, low sea level and high ice volume, with polar and mountain glaciers and ice sheets at their maximum extent in most regions. The most widely used definition of its duration is 18 ka to 24 ka (ka = 1000 calendar years before present), suggested by Mix et al. (2001). Newnham et al. (2007) have demonstrated that this is not appropriate for New Zealand (and almost certainly globally) and have proposed an extended LGM of c. 29 ka to c. 19 ka, which definition we will use here. The LGM is of key importance in understanding how the global climate system functions and a considerable effort over recent decades has been put into generating and collating datasets that document how components of the ocean-land-atmosphere system reacted during this period. One of these components is vegetative cover, important because it integrates diverse climatic drivers over wide areas. In this paper, 'trees' refers to woody plants that regularly achieve ≥6 m in height, and 'shrubs' to woody plants <6 m.

The vegetative cover of New Zealand during the LGM has been assessed largely from pollen evidence. These records have been interpreted as showing extensive non-forest communities (scablands, grassland, shrublands) in the southern North Island and the South Island, with continuous and extensive forest tracts mostly confined to the North Island north of 38° S

(Newnham et al. 1999). Nevertheless, on the basis of pollen, macrofossil and biogeographic evidence, it has been hypothesised that most forest trees and shrubs typical of a given region survived in situ, albeit mainly in small stands (McGlone 1985).

Recently, this interpretation of the LGM landscape has been challenged. Beetle evidence, it is claimed, shows that closed, woody vegetation was more extensive than indicated by pollen records at low to mid-altitudes in the south of New Zealand (Burge and Shulmeister 2007) and that at times the LGM environment was much warmer and wetter, and very much more woody than hitherto suspected (Marra et al. 2006; Woodward and Shulmeister 2007). Burge and Shulmeister (2007:130) suggest the time has come to reassess the situation:

> 'More important than any temperature inference is the observation that the current interpretation of glacial-age vegetation from pollen data appears not to be robust. There are now a number of studies ... that demonstrate that local site indicators from around New Zealand suggest woody vegetation persisting through glacial times. A variety of mechanisms have been proposed by the authors so that these differing records can be reconciled to the pollen-based interpretation of glacial vegetation. Our site clearly indicates that there is at least as good a case to reinterpret the pollen information.'

To summarise, they argue that interpretations of pollen records have consistently underestimated the extent of LGM forest or closed woody vegetation. Thus, the issue is not whether forest trees and shrubs survived, but the extent of forest or closed woody vegetation cover during the LGM and the consequent implications for vegetation and climate reconstruction.

This review critically examines this claim that LGM pollen records are systematically under-representing forest and woody vegetation, as suggested from the beetle evidence. We also consider plant macrofossils, as they provide independent evidence. This reassessment of pollen and macrofossil evidence indicates that there is at least as good a case to reinterpret the beetle conclusions.

Pollen representation of LGM vegetation

Here, we present two sorts of pollen evidence. First, we assess the general pattern of pollen distribution during the LGM by using selected, single pollen samples from a wide geographic spread of sites; and second, we look in more detail at several representative sites which cover all, or significant proportions, of the LGM chron. The widespread Kawakawa Tephra (c. 27 ka) from a large ignimbrite eruption of the central North Island is an important marker horizon for this study (Pillans et al. 1991), as is the Rerewhakaaitu Tephra (c. 17.6 ka) (Newnham et al. 2003). In all the pollen diagrams presented, pollen percentages have been calculated from a pollen sum which includes all terrestrial pollen types, but excludes ferns, wetland graminoids and forbs and the wetland shrub *Leptospermum*. The term *Fuscospora* refers to *Nothofagus* subgenus *Fuscospora* beeches. Locations mentioned in the text are shown in Figure 1.

In Figure 2, we present the pollen and spore spectra from 60 pollen-analysed sites dating to the LGM. Where the site pollen record contains more than one pollen sample dated to the LGM, the sample with least arboreal pollen was chosen so that the *minimum* extent of forest pollen presence is depicted. We therefore refer to this as the *minimum forest LGM* dataset. As the climates of the LGM period were variable (Newnham et al. 2007), and contamination of ^{14}C samples with younger carbon a greater problem in that time range than later, a pollen 'snap-

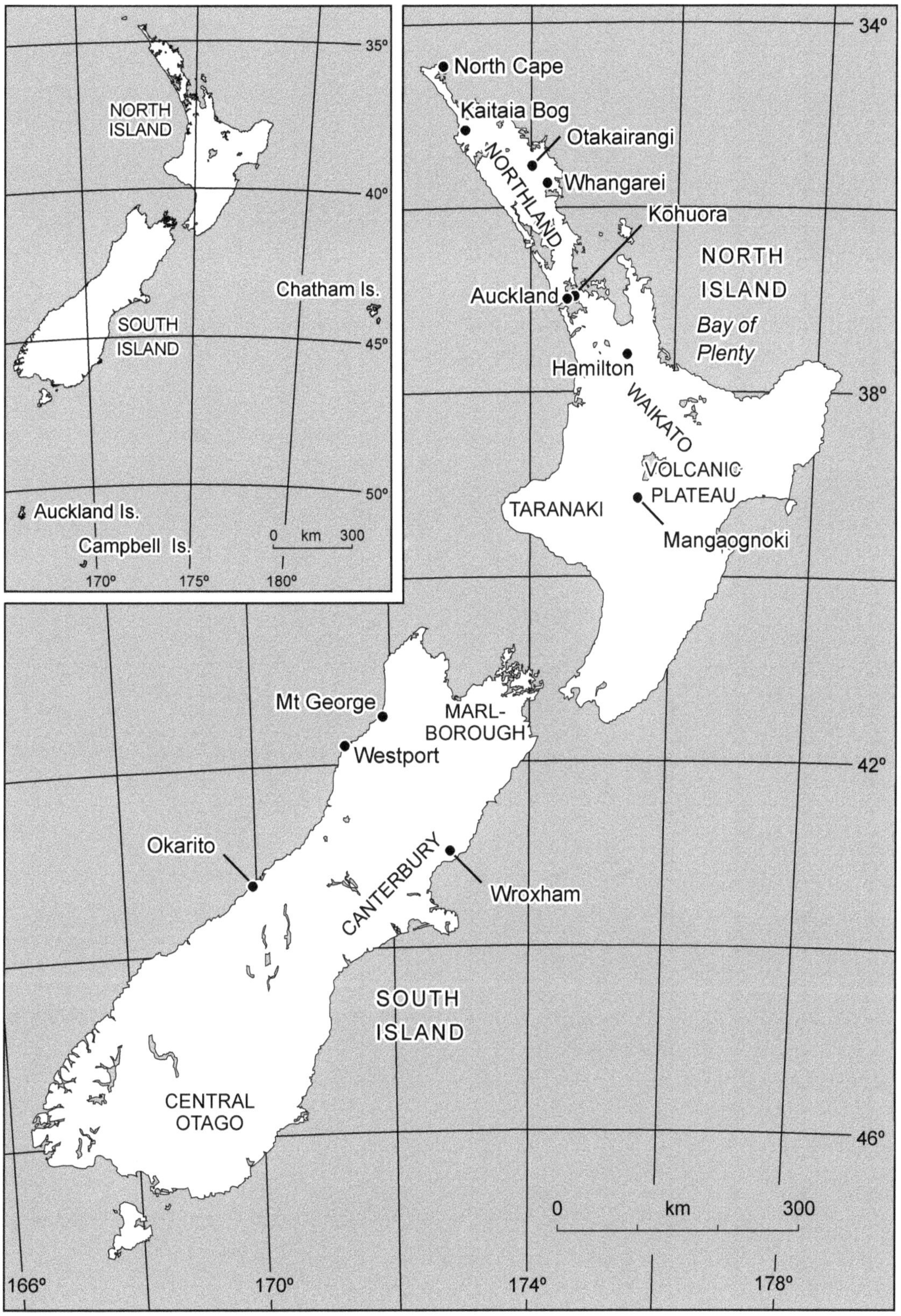

Figure 1. New Zealand region, showing locations mentioned in the text

shot' approach introduces uncertainty. At most of the shorter sequence sites, we are uncertain whether LGM interstadial or stadial conditions are represented. Examples of this variation within the LGM can be seen in the representative pollen diagrams shown in Figure 3. It should also be noted that temporal variation in forest cover within the LGM may explain some of the

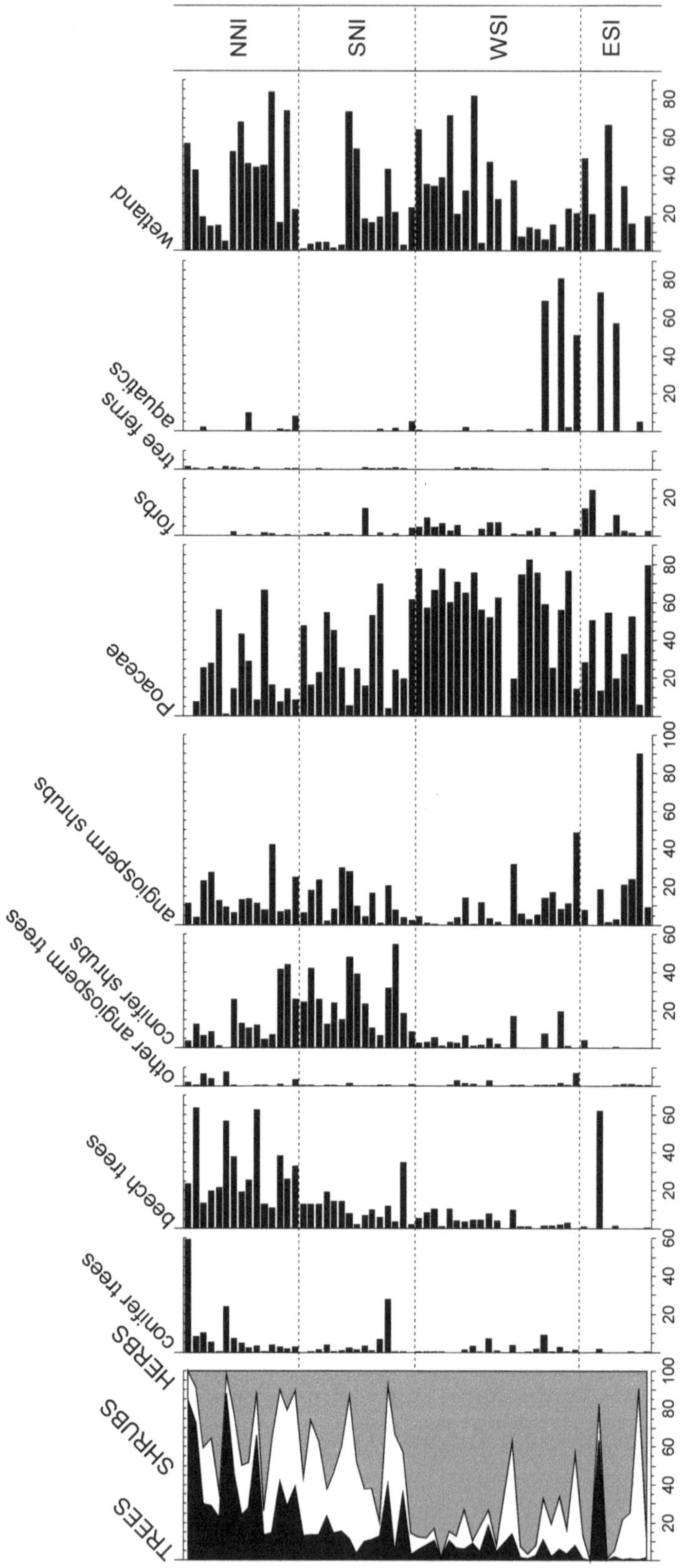

Figure 2. Pollen and spore results for selected forest minimum samples, LGM. Arranged in order of increasing latitude within broad zones: northern North Island (NNI – north of latitude 37° S); southern North Island (SNI – south of 37° S but including the Nelson district); western South Island (WSI – south of Nelson, west of the Main Divide); eastern South Island (ESI – south of Nelson, east of the Main Divide). Pollen sum: all terrestrial types excluding ferns, lycopods and wetland forbs, rushes and sedges

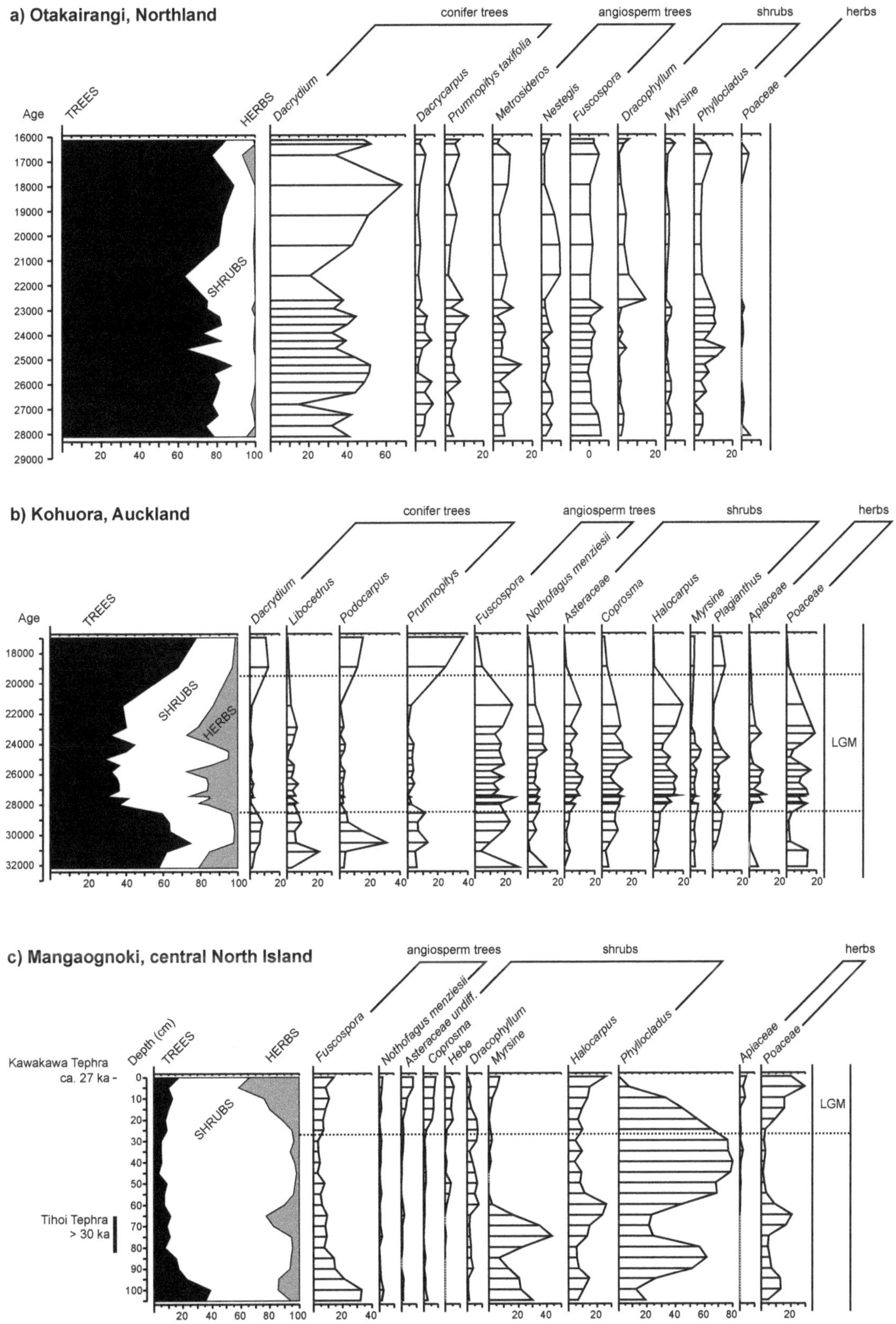

Figure 3. Representative pollen diagrams covering part or most of the LGM chron for: Northland (a) northern North Island (b), central North Island (c), western South Island (Figure 3 continues on page 54)

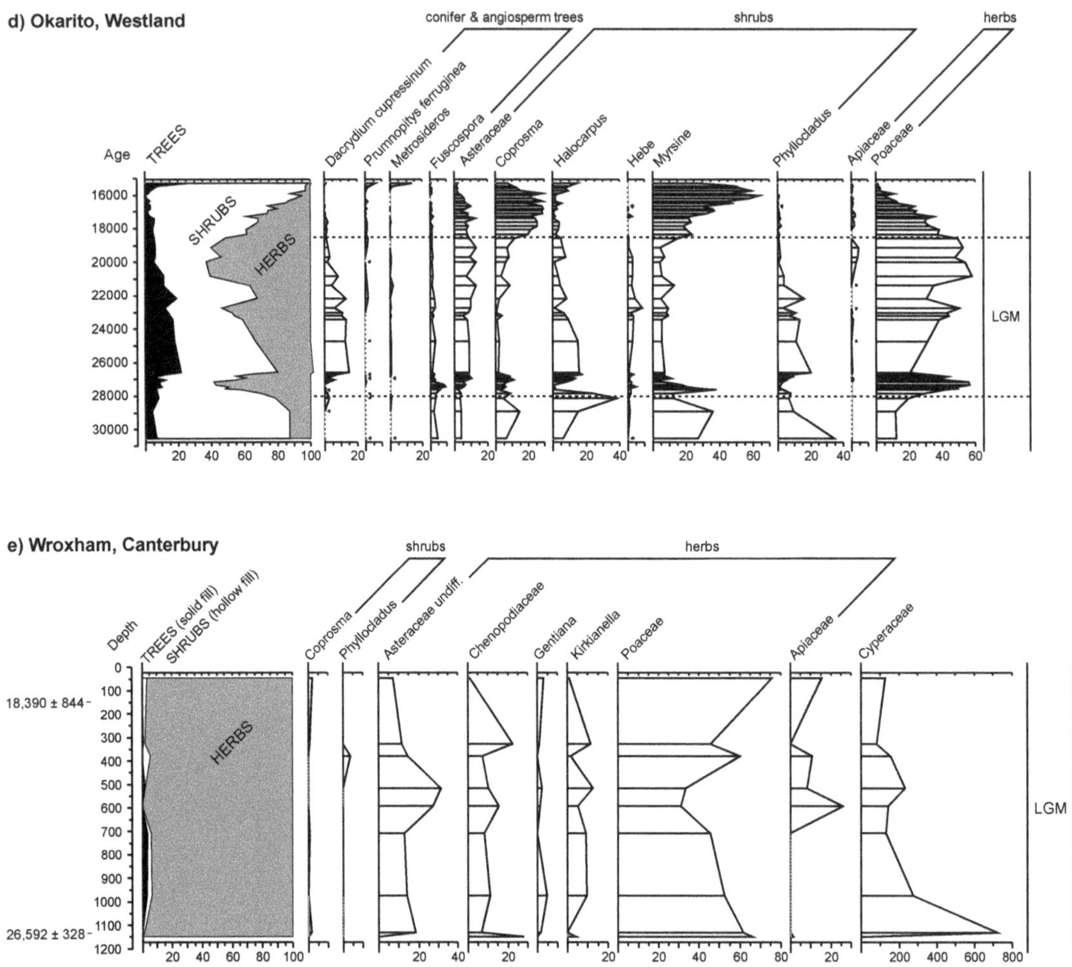

Figure 3. (cont.) Representative pollen diagrams covering part or most of the LGM chron for: Northland (d), and eastern South Island (e). Pollen sum as for Figure 1. All ages are in cal yr BP

differences reported between palaeoecological sequences, particularly when different sites and different proxies are compared (Newnham et al. 2007).

It is not possible to be too choosy about site characteristics for LGM pollen spectra if a broad coverage is wanted. The LGM pollen sample sites form a highly heterogeneous set, including lake and pond deposits, peats, organic-rich soils, and thin organic silts enclosed in gravel or loess matrices. Most are likely to have been derived from soils or relatively shortlived ponds or peaty backwaters in aggrading fluvial channels. Even those from a relatively small number of lakes or peat bogs tend to have been deposited during times of inorganic sedimentation within the basin.

The composition and source area of a given pollen sample is strongly affected by local site characteristics. Peats and soils usually are dominated by pollen and spores of vegetation growing within a few metres of the collection point. The degree to which they also sample the regional pollen rain depends on the local vegetation. Soils beneath forest stands generally record little other than the trees immediately in the vicinity; peats or soils with a low or herbaceous vegetation cover will adequately record pollen from surrounding dryland vegetation. Fens, lakes and ponds not only receive pollen rain from aerial dispersal of pollen from the wider surrounding region, but obtain a variable amount of water-transported pollen and spores. The relative source area of ponds and lakes depends on the size of their basin: the smaller they are, the larger the amount of pollen derived from immediately adjacent vegetation; the larger, the more regional and extra-regional pollen sources are represented (Sugita 1994). Marine sediments (not included in this terrestrial dataset) receive some pollen from direct deposition on the sea surface, but much must come from fluvial transport. Inevitably, some of the pollen is reworked from non-contemporaneous deposits. Transport along waterways and settling over weeks through the water column enriches the spectra in some pollen and spore types (e.g. fern spores, winged pollen such as that of many conifers, and reworked fossil types; McGlone 2001a). Marine-core pollen is good at recording long-term fluctuations in pollen and spore input, but poor at sensing details of the onshore vegetation, and thus we use it only as an adjunct.

Geographic pattern of pollen and spore representation during the LGM

The main feature of the minimum forest LGM pollen and spore assemblages is the low percentages of tree pollen and the high levels of non-woody pollen types (Poaceae, Asteraceae, forbs) (Figure 2). The mean for forest tree pollen over all sites is 15.1 ± 18% and only 14 sites (23%) have more than 25% forest tree pollen. Tree ferns form a particularly low component of the total (0.7 ± 1.1%) and conifer trees make up 2.8 ± 7.2%, while 88% of all the forest pollen recorded is of *Nothofagus*. This is a strong contrast with modern late-Holocene pre-deforestation assemblages (135 sites distributed throughout New Zealand; Wilmshurst et al. 2007), which average c. 80% tree pollen from conifer (c. 44%) or angiosperm (c. 35%) sources, c. 7% from tree ferns, and low amounts from herbaceous (c. 4%) or shrub (c. 17%) sources.

Turning to regional distinctions, some broad patterns are evident. Four northern sites suggest a nearly complete forest cover north of Auckland at the LGM. Dodson et al. (1988) report a podocarp-hardwood forest similar to that of the present at North Cape (34.5° S); *Fuscospora*-conifer-hardwood forests dominated at Kaitaia Bog (35° S; Elliot 1998), near Whangarei (35.5° S; Newnham 1992) and coastal Waikato (37.5° S; Lees et al. 1998). Analysis of the pore number of the *Fuscospora* pollen type suggests *Nothofagus truncata* was the major pollen contributor at these northern sites (Lees et al. 1998; Newnham 1992), consistent with its status as the major *Fuscospora* species of the northern North Island at present. A solitary South Island site (Mount George; Moar et al. 2008) is also dominated by forest, largely *Fuscospora* and *Nothofagus menziesii*. The Mount George site is distinctive in that it is close to the LGM shoreline on the flanks of a steep coastal range, exactly the type of site long predicted as being the only place where extensive forest tracts could persist in the south.

The northern LGM pollen spectra can be easily matched with those from extant forests far to the south or at higher altitudes. In general, similar pollen spectra characterise the lowland western South Island at present.

We now exclude the five largely forested sites in order to characterise the extensive partially forested landscapes (Figure 4). Leaving aside Northland, LGM New Zealand can usefully be split into four regions: Auckland, Bay of Plenty and the Waikato Basin (NNI); the Volcanic Plateau southwards to northern Marlborough (SNI); the South Island west of the glaciers (WSI); and the eastern South Island (ESI). Tree pollen shows a decline from north to south and west to east: NNI 41%; SNI 15%; WSI 7%; and ESI 4%. Its variability is high in NNI, but much lower further south. Shrub conifers are abundant in NNI and, in particular, SNI, making up c. 17% and 26% respectively of the pollen rain. Grass, other shrubs and forbs increase from north to south, collectively averaging 88% in WSI and 98% in ESI.

LGM pollen records from the Auckland Isthmus (Figure 3b; Newnham et al. 2007) depict a landscape in which forest, largely dominated by *Fuscospora*, *Nothofagus menziesii*, *Libocedrus* and *Phyllocladus*, forms a mosaic with shrubland and grassland. Although there are some traces of typically northern species (e.g. *Agathis australis*), arboreal species which later become common in the postglacial and Holocene are at low levels or absent (e.g. *Metrosideros*, *Cyathea*, *Nestegis*) (Lancashire et al. 2002; Sandiford et al. 2003). Similar pollen spectra occur further south in the North Island and a cluster of sites in the Hamilton Basin are close in composition to those from the Auckland Isthmus (Newnham et al. 1989). We do not have LGM records

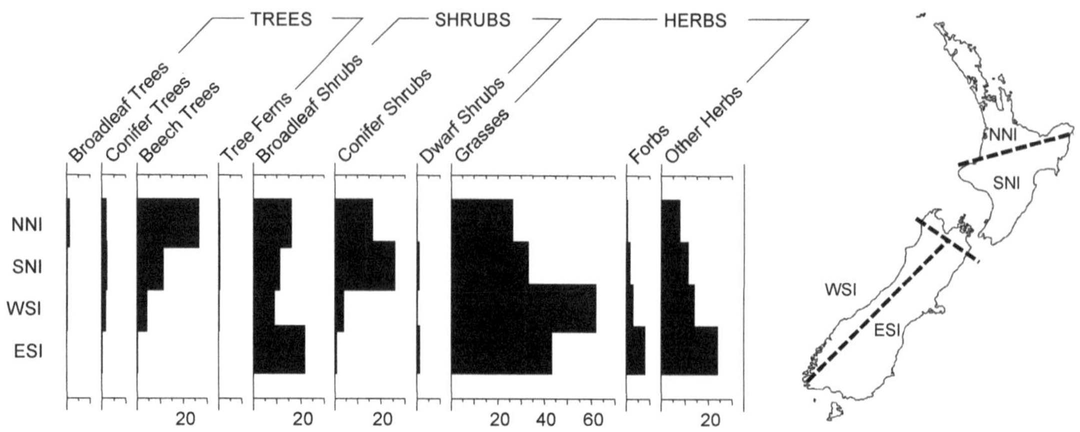

Figure 4. Summary pollen averaged for NNI, SNI, WSI and ESI LGM vegetation zones

from coastal Bay of Plenty in the east, but pollen records for the interstadial immediately preceding indicate *Nothofagus*-conifer forest, conifer shrubland and grassland mosaics with more forest but otherwise similar to those prevailing in the Auckland Isthmus (McGlone et al. 1984). *Nothofagus*, conifer shrubland and grass-dominated pollen assemblages are recorded at sites up to 800 m on the Volcanic Plateau (Figure 3c; McGlone and Topping 1983). The *Fuscospora* percentages are relatively consistent across the altitudinal gradient, suggesting derivation from scattered stands, while high, variable percentages of grass, conifer shrubs and other shrub types are consistent with local shrubland/grassland mosaics. Lowland Taranaki sites at current sea level have repeatedly experienced shrub and grass-dominated intervals, with low tree-pollen percentages during the last glaciation (McGlone et al. 1984), and the two LGM sites at current sea level are dominated by conifer shrubs, other shrubs and grassland (McGlone 1980).

Although the Auckland Isthmus, Waikato Basin, Volcanic Plateau, Bay of Plenty and southwestern North Island pollen sites range over 800 m of altitude and 4° of latitude, they all record much the same sort of open shrub and grassland-dominated vegetation with a greater or lesser forest component (Figure 2). Tree percentages are generally less than 50% and dominated by *Fuscospora*, *Nothofagus menziesii* and *Libocedrus*. While forest types are more abundant and tall conifer and hardwood elements more commonly encountered in lower altitude and latitude sites, there is no clear zonation with successive dominance by lowland forest, montane, shrubland and grassland with increasing altitude, as might be expected from comparison with present-day zonations. Moreover, there is a great deal of variability in the non-arboreal representation: grass pollen has a mean of 40.8% ± 26.4% and ranges from 7.5% to 67.4%. The general pollen and spore composition of a number of these sites (particularly those of the Waikato Basin) is close to that of modern and Holocene sites from the inland eastern South Island, which was partially forested with tall conifer shrubs before deforestation (McGlone and Moar 1998).

It has been long noted that in the central North Island, areas above 600 m to 800 m lack tephra older than the early postglacial, and this is consistent with a lowering of the treeline by about 800 m ± 100 m, as predicted from LGM glaciers on the Tongariro Mountains (Pillans et al. 1991). The presence of extensive grassland and shrubland at present-day sea level must have resulted from other factors than depression of mean annual temperatures (McGlone 1985).

WSI pollen spectra differ most noticeably from SNI spectra in their lower amounts of shrub conifer (4% versus 26%) and much greater amounts of grass and forb pollen (71% versus 44%). However, despite this indication of lower stature and more open vegetation, there are sites in this region with high tree-pollen levels (c. 10%; mainly *Dacrydium cupressinum*, *Prumnopitys*, *Fuscospora* and *Nothofagus menziesii*. *Metrosideros* (likely to be *M. umbellata*) is more common in this region than anywhere else). Inland sites close to the LGM ice margin tend to have very high levels of grass pollen and low levels of shrubs (Moar 1980).

Only a few LGM sites have been analysed from the eastern South Island (Figures 2 and 3e). Their most distinctive feature is the paucity of tree or shrub-conifer pollen (1.4 ± 1.5%; minimum 0%, maximum 4.9%). This is almost exactly the same percentage as recorded by modern pollen sites on subantarctic Campbell Island, some 600 km south of the New Zealand mainland (McGlone and Meurk 2000). The LGM pollen spectra are dominated by Asteraceae and grasses and have a range of low-growing forbs. Common herbaceous elements such as the Lactuceae (dandelion group), *Chenopodium*, Brassicaceae and *Rumex* indicate open herb and dwarf shrub communities similar to those found today in riverbeds or saline Central Otago terraces (Moar 1980).

In summary, LGM pollen spectra indicate forest survival throughout New Zealand, even at times of 'minimum forest', with a general trend of greater forest reduction towards the south and east.

Do interpretations of LGM pollen data underestimate the extent of woody vegetation?

The argument that pollen data under-represents the abundance of woody and arboreal vegetation during the New Zealand LGM is not a replay of the debates that have raged in Europe over pollen interpretations of the LGM. In Europe, LGM pollen records have been interpreted by some as supporting a treeless landscape north of the Alps, with comparatively moderate percentages of tree pollen attributed to reworking or long-distance dispersal, whereas more recently, the same pollen records are being aligned with macrofossil and genetic evidence suggesting forest survival in northern 'cryptic refugia'. By contrast, New Zealand pollen analysts have been consistent in claiming forest and shrubland was present throughout most of New Zealand during the LGM, despite the dominance of southern pollen sites by non-arboreal pollen and spores and very low representation of tree pollen in some sites (Figure 2). Indeed, it is relevant to note that a recent review of the glacial refugia debate in Europe acknowledged that New Zealand pollen analysts were the first to propose micro-refugia for trees during LGM times (Bennett and Provan 2008). In particular, McGlone (1985:731) demonstrated that evidence for postglacial spread from discrete northern LGM refugia was weak and that universal survival of all plants in the regions they currently occupied was a viable hypothesis:

> 'The conclusion which can be drawn from the suddenness with which forest reoccupied the landscape, and the lack of migratory waves or extended successions, is that podocarp-hardwood forest was close to nearly every site yet investigated. Survival of some forest throughout New Zealand has been proposed on biogeographic grounds, particularly as regards the *Nothofagus* species which are reputedly slow movers.... If *Podocarpus spicatus* and *Ascarina lucida* could survive the glacial maximum in the southern regions of New Zealand, there is every reason to suppose that nearly all other southern trees and shrubs could also. It therefore seems that this model of widespread survival of forest trees can be expanded to cover nearly every plant at present resident within a region.'

Thus, New Zealand pollen records have long been interpreted as showing forest-shrub-grassland mosaic with local survival of core regional species in small patches.

This interpretation is corroborated by plant macrofossil evidence. Wood, stems, leaves and seeds of forest trees are uncommon in LGM sediments south of the Auckland Isthmus – that is, from areas where pollen evidence shows trees were likely to have been scarce or patchy.

Nothofagus menziesii and *Nothofagus cliffortioides* leaves and twigs have been reported from sites in north Westland (Moar and Suggate 1996; Turney et al. 2002; Marra and Leschen 2004), but there are no other reports. Many kilometres of exposures of alluvial silts and gravels dating to the LGM period are known, and in the south, they are barren of wood. In the central North Island, three major ignimbrite eruptions, the Kawakawa Tephra (27,097 ± 957 cal yr BP), the Okareka Tephra (21,800 ± 500 cal yr BP) and the Rerewhaakaaitu Tephra (17,625 ± 425 cal yr BP; all ages from Lowe et al. 2008) covered many square kilometres of landscape, but tree charcoal and wood associated with these deposits is uncommon. This relative absence of tree macrofossils cannot be attributed to preservational artefacts. Wood, charcoal and leaves of forest trees are common in late glacial and Holocene alluvial sediments (Molloy et al. 1963; Pullar and Patel 1972) and post-LGM ignimbrites close to source tend to have charcoal and charred wood associated with them. Moreover, macrofossils of shrubs (including conifer shrubs) and herbs are often encountered in LGM sediments (for instance, McGlone et al. 1978; Soons and Burrows 1978; Neall 1979; McGlone and Topping 1983), thus confirming the presence of suitable sites for preservation. There have been a number of isolated reports of macrofossils from LGM pollen sites, but detailed studies are rare. Soons and Burrows (1978) report on a rich macrofossil LGM site in inland Canterbury at an altitude of 780 m, which was dominated by herbaceous or prostrate woody vegetation, with a complete absence of shrub or tree macrofossils. Together with information from pollen in the same sediments, it is clear that sparse, herb-dominated vegetation occupied the site.

As pollen evidence clearly points to the presence of forest, albeit in limited amounts, throughout non-glaciated regions, and recent beetle evidence (Marra and Leschen 2004; Burge and Shulmeister 2007) shows that trees were growing in inland eastern South Island locations, the near absence of woody plant macrofossils is significant. Absence of tree wood or other macrofossils can probably be explained by two factors. Wood of forest trees normally becomes preserved in alluvial deposits because of undercutting of river banks or slips in steep catchments and subsequent transport by water and then rapid, deep burial by aggregating silt and gravel deposits. This is a common depositional context for LGM deposits containing macrofossils. For example, in a recent investigation of plant macrofossils from sites extending across Eurasia, Binney et al. (2009) report that all 45 samples of trees dated to the LGM (25-17 kyr cal BP) are from sites located in or near a river floodplain, whereas for samples younger than 15 ka, all types of landscape position are represented in the samples. In New Zealand, if LGM forest tree stands were both much rarer than now, and mainly growing on protected, stable sites, they would have had a much smaller chance of being incorporated in the alluvial deposits from which most of our information comes. Such stable sites are also unlikely to preserve soil charcoal, as charcoal is normally preserved by soil movement after burning (Wardle 2001).

In Figure 5, we summarise the vegetation cover of New Zealand during the coolest phases of the LGM based on our interpretation of biological and physical evidence.

Can we reconcile pollen and beetle evidence?

Burge and Shulmeister (2007) claim that beetle fossils indicate much more closed woody vegetation than pollen evidence would suggest. This claim relies primarily on their beetle fossil sites in the Westport area (coastal northwestern South Island), which they say demonstrate the close-by presence of closed canopy woodland, while the pollen interpretation (Moar and Suggate 1979) is of an open grassland with a woody element. Clearly, this is a critical site for resolving the problem.

We note that in the beetle faunas from this site classic wet-forest saproxylic taxa like *Pycnomerus* and other zopherids, for example, are completely absent from this sequence,

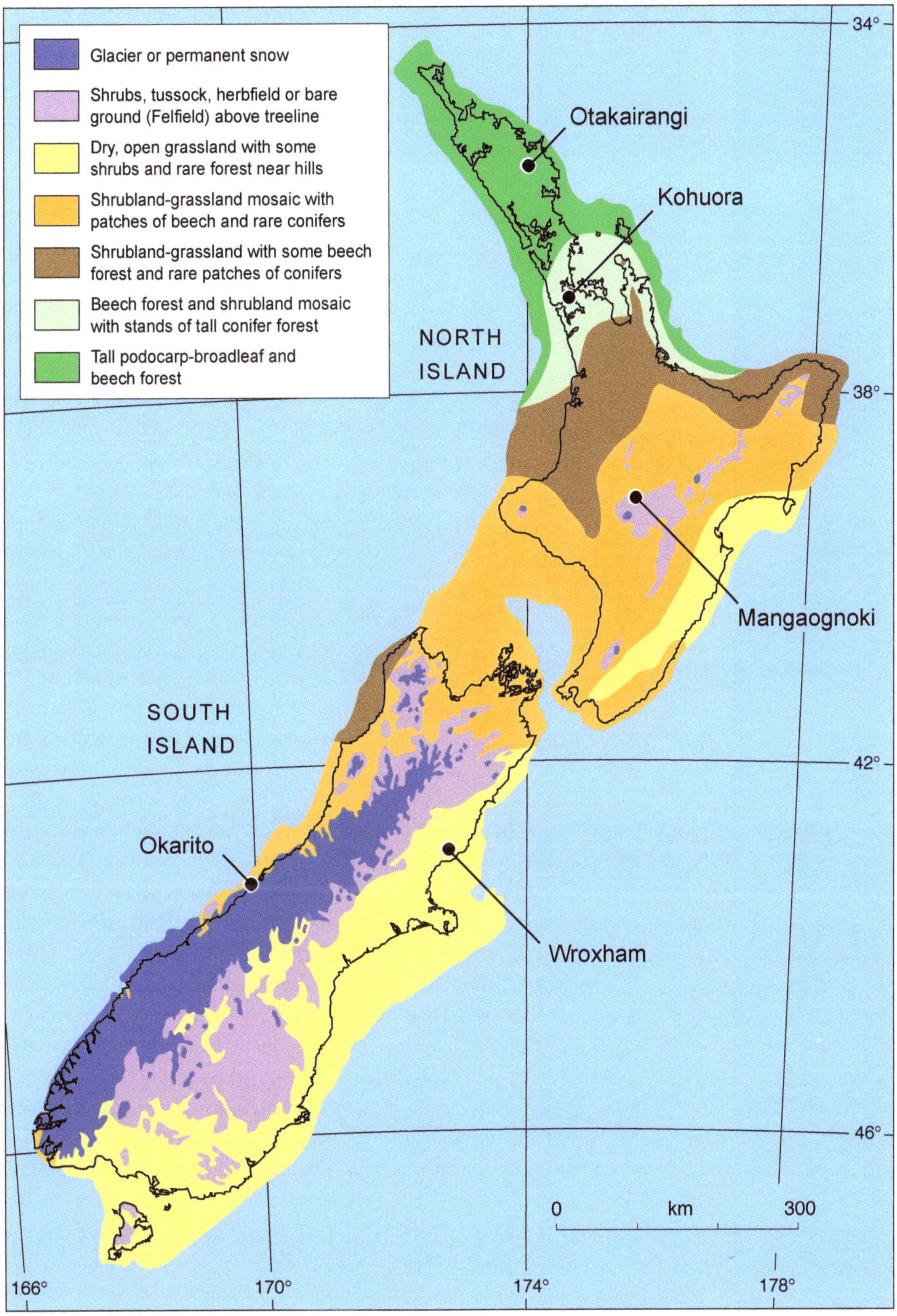

Figure 5. New Zealand vegetation cover at the height of the LGM as reconstructed from pollen, macrofossil, beetle and geomorphic evidence

except for a single sample towards the base (and potentially pre-LGM). The remaining taxa are not necessarily obligates of closed canopies, but almost certainly would be well suited to woody vegetation of many structures (Nick Porch pers comm. 2009). We note that the pollen percentages at this site typically have c. 30% tree and shrub pollen, thus indicating near-local presence of some woody vegetation. Instead, Burge and Shulmeister (2007) propose 'The contrast between the pollen and beetle reconstructions may thus simply represent a combination of an enlarged pollen-collection area and low pollen-producing taxa (including arboreal taxa) being overshadowed by abundant input from increased areas of grassland and local wetland taxa such as *Myriophyllum* and Cyperaceae' (p 129). At this particular site, *Myriophyllum* and Cyperaceae did not affect the pollen sum in this way. Aquatics are by convention not included inside the pollen sum (that is, are not part of the percentage calculation) and when Cyperaceae, which were included by Moar and Suggate (1979), are excluded, there is still a substantial pollen sum of c. 150 grains. For the interval dated to the LGM (18,650 ± 250 ^{14}C yr BP; 162 cm), 6% of the pollen is from trees, 26% from shrubs and 67% from grass, a little higher than, but similar to, the original publication figures. Excluded wetland types make up only about a third of the total terrestrial and wetland pollen sum, a similar level to that found at modern pollen sites surrounded by forest, grassland or shrub (McGlone 1982; McGlone and Moar 1997). There are traces of under-represented pollen types such as *Hebe* and *Dracophyllum*. These pollen results therefore suggest a grassland-shrubland in the vicinity of the site, or a grassland-dominant site with a woody element, as per the interpretation in Moar and Suggate (1979). The simplest explanation for the apparent discrepancy between beetles and pollen is that there isn't one.

At a more general level, there is little reason to believe that dominant shrubland and forest could remain essentially invisible in terms of plant fossils across the LGM landscape. Small stands of wind-pollinated trees or shrubs within grassland communities are not under-represented in the present landscape. About 20% of pollen from modern sites from grassland in central Canterbury inter-montane basins derives from stands of *Nothofagus*, essentially in the same proportion as its regional coverage (Randall 1990). On the other hand, trees and shrubs with poorly dispersed pollen are less likely to be represented if they are even a few tens of metres from a site (e.g. *Leptospermum*, *Weinmannia racemosa*). If the woody vegetation of the LGM consisted almost entirely of plants with poor pollen dispersal, it is possible that it could be massively under-represented in the pollen rain. However, woody vegetation dominated by such plants is rather rare in the current vegetation (an exception being the subantarctic islands, which we discuss below). Shrubland taxa, including wind-pollinated types such as *Coprosma* and insect-pollinated types such as *Muehlenbeckia* and *Myrsine*, are very well represented in late-glacial transitions (e.g. McGlone and Neall 1994; Newnham, Vandergoes et al. 2007). Unless there was a major shift in the woody taxa that characterised shrubland during the LGM, for instance from *Hebe* or *Dracophyllum*, to better-represented shrub types, the late-glacial transition experienced a rapid increase in woody vegetation cover. This, in turn, implies that during the LGM, New Zealand had a much larger representation of herbaceous vegetation. However, if we make the assumption the LGM woody cover was largely of these more cryptic taxa, what would we expect the pollen representation to have been? Fortunately, we have a close analogy in subantarctic Campbell Island, a small island of some 113 km^2, 600 km south of the New Zealand mainland (Figure 1).

In general terms, the LGM pollen rain in the western North and South Islands, is similar to that of subantantarctic Campbell Island today. About two-thirds of Campbell Island is in herbaceous vegetation and the major shrubland dominants both in size and abundance are *Dracophyllum* spp. Low *Dracophyllum* forest-tall shrubland, which dominates the sheltered eastern side of the island, gives way with altitude to *Dracophyllum*/*Myrsine*/*Coprosma* shrubland, tussockland and tundra herbfield. A total of 46 pollen sites have been used to characterise the modern pollen rain (McGlone and Meurk 2000). Grass pollen makes up

37% ± 26% of the terrestrial pollen rain, and shrubs 27% ± 26%. As we estimate Campbell Island grassland and woody cover to be approximately a third each, these results suggest that pollen fairly represents the vegetation. Although the dominant woody element, *Dracophyllum* spp, is under-represented in mainland sites (McGlone 1982), this is not so apparent on Campbell Island (McGlone and Meurk 2000), largely due to the complete absence of tall, wind-pollinated trees and shrubs. Short-statured vegetation does not contribute much to the regional pollen rain, as pollen released near ground level is less likely to be lofted into the air stream and typically is transported only a short distance. Therefore, poorly dispersed local pollen sources will not be diluted (as percentages) by extra-local well-dispersed pollen sources. The 13 Campbell Island sites classified as grassland or *megaphyllous* forb/grassland had a mean herbaceous pollen percentage of 69% ± 20%, of which grass pollen comprises 65% ± 23%, with a shrub contribution of 6% ± 4%. Grass percentages ranged from 27% to 96%, similar to those of southern South Island sites in the minimum forest LGM dataset.

The analogy between subantarctic Campbell Island and LGM southern New Zealand is not exact: the subantarctic islands have extensive peatlands which were rare during the LGM, and the pollen rain has considerable amounts of ground fern spores (62% ± 55%; outside pollen sum), much higher than southern LGM New Zealand (10% ± 54%). Nevertheless, the LGM pollen match is much closer to Campbell Island than to present-day or Holocene pollen mainland assemblages. For instance, the Ahukawakawa Swamp, an extensive wetland-grassland complex on the flanks of Mount Taranaki, although recording much grass and wetland pollen, generally had 50% or more woody pollen types (McGlone et al. 1988). In fact, extensive modern-day pollen sampling in New Zealand rarely has identified indigenous vegetation sites that yield more than 50% grass pollen, the exceptions being induced indigenous or exotic lowland grasslands (Randall 1990; McGlone 2001b). The minimum forest LGM sites regularly exceed this value, and average roughly this amount in WSI and ESI.

Our preferred explanation for dominance by herbaceous pollen during the LGM, therefore, is that the pollen spectra more or less proportionally represent broad vegetation cover classes. However, for completeness, we will consider two possible alternative explanations.

1) *Low atmospheric concentrations of CO_2 at the LGM may have led to comparatively less pollen production by woody plants which must invest in carbon-rich support structures.* There is some evidence that pollen production diminishes with low CO_2 (Ladeau and Clark 2006; Rogers et al. 2006). However, as described above, forest persisted in the north throughout the LGM and forest pollen increased sharply in North Island sites around 18,000 years ago (Newnham et al. 2003), several thousand years before there was any significant increase in atmospheric CO_2 concentration. Similarly, a mid-LGM warming and period of forest expansion evident in some NZ pollen records (Vandergoes et al. 2005; Alloway et al. 2007; Newnham, Vandergoes et al. 2007) coincides with minimum atmospheric CO_2 concentration. If there was a CO_2 effect on flowering, it certainly was insufficient to prevent adequate forest pollen representation.

2) *Sites that preserve LGM pollen spectra may have been more likely to be in grassland.* LGM peatlands are rare (McGlone 2009), and many LGM pollen sites are temporary ponds and sloughs, often associated with valley bottoms or in terrace deposits associated with gravels or inorganic sediments. These low-lying situations, often subject to frequent sediment deposition, flooding and erosion, and hard winter frosts, thus may have favoured grassland. Nevertheless, modern pollen evidence strongly suggests that small pockets of grassland in stable shrub or forest-covered landscapes are not dominated by grassland pollen. Moreover, many sites are stable and not in topographic depressions. The Westland site where the beetle and pollen evidence is apparently in

conflict accumulated organic silts over more than 10,000 years, and thus was not unstable, although dominated by grassland pollen types (Moar and Suggate 1979), and nor were the Auckland Isthmus or Volcanic Plateau sites (McGlone and Topping 1983; Newnham, Lowe et al. 2007). Certainly, much of the variability in the pollen results from the LGM derives from the topographic position of the sites but, rather than distorting the picture, is probably a fairly faithful representation of the actual distribution of unstable terrain.

It seems clear that during the LGM, woody vegetation, although common, ceded dominance to herbaceous communities and bare ground over much of New Zealand south of Northland. In the inland and eastern South Island, woody vegetation of any sort was quite limited. Forest trees, although present throughout, made up such a low proportion of the vegetative cover that they contributed little to the pollen rain. Note that this does not imply large regions without trees, as our preferred model is scattered stands of trees or patches of forest throughout. As tree pollen is the dominant component of long-distance pollen transport, absence of this pollen source automatically increases the percentage representation of local vegetation cover relative to the situation during the Holocene. This percentage effect is not necessarily misleading if the local vegetation fairly represents the broader landscape. With regard to this issue, there is some possibility of under-representation of low-growing shrubs interspersed with, or forming extensive patches within, grassland. For instance, the Rotoaira Basin site on the Volcanic Plateau, central North Island, has more grass and less shrub pollen represented than the contemporaneous nearby sites on rolling hill country (McGlone and Topping 1983). However, the same interpretation would be taken from both sets of sites: extensive grassland and shrub cover.

Although the attribution of beetle fossils to closed woody vegetation or litter under closed woody vegetation (Marra and Leschen 2004; Burge and Shulmeister 2007) seems well founded, there is little specificity in the attributions and no basis for conclusions about plant abundance. According to the brief descriptions given, most of the beetles could persist under any sort of woody vegetation. Of the tall trees, only *Nothofagus* has been identified from its fossil beetle fauna (Marra and Leschen 2004). We note that there is little published ecology on New Zealand beetles, and the basis for most of the conclusions about fossil assemblages is sparse notes attached to specimens or appended to morphological descriptions in faunas. No modern survey of present-day beetle remains in relation to their current environment has been undertaken in New Zealand, and thus basic facts (for instance, how well the local vegetation is reflected in the fossil fauna; from what distance the beetles may have come) are simply unknown. If there were a more diverse and abundant beetle fauna on woody vegetation relative to grassland plants, or if beetles inhabiting woody vegetation were more likely to become fossils, the inferences made on the basis of beetles would be misleading. We contrast this situation with the well-documented and quantified information on modern pollen-vegetation relationships in New Zealand.

Causation of the pattern of LGM vegetation in New Zealand

The essential problem in interpreting the LGM vegetation cover of New Zealand is that the depression in mean annual temperatures does not in itself explain the substantial reduction in forest cover indicated by pollen and macrofossil records. On the basis of glacial features and reconstructions of sea-surface temperatures derived from marine cores off southern New Zealand, mean annual temperatures fell during the LGM by only 4-5°C (Pelejero et al. 2006; Barrows et al. 2007). Note that no land temperature adjustment for lower sea level is necessary (as erroneously suggested by many), as the atmospheric mass above any terrestrial point was unaffected by falling sea levels due to compensatory storage of water as ice on land, while observed sea-surface temperatures should be adjusted down by c. 1°C when compared with contemporaneous terrestrial temperatures to account for increased atmospheric mass above the ocean surface (Osmaston 2006). A fall of 5°C equates to a lowering of vegetation zones in much of southern New Zealand by c. 800 m, which leaves some 300-400 m in altitude above present sea level for montane forest even during the coldest phases (McGlone and Bathgate 1983). Further north, forest should have been able to dominate up to 700 m in altitude above present sea level. As we have seen, the pattern of disturbed ground in the central North Island, as indicated by absence of pre-LGM tephras, supports the inference drawn from the pollen that continuous vegetation extended to about this altitude. However, pollen assemblages and plant macrofossils strongly support the concept of low shrubland-grassland being the dominant cover. The question thus becomes one of why forest and tall woody vegetation generally was not more extensive.

As discussed earlier, it is unlikely that forest absence was a direct effect of low atmospheric CO_2 concentrations, as forest persisted in the north, and in small patches throughout. On the current landscape, forest and tall woody vegetation is generally excluded below the regional tree line by (1) frequent disturbance including by river channel activity, landslides, storms and fire; (2) low precipitation (under 450 $mm.a^{-1}$); (3) intense frost, which can exclude *Nothofagus*, the main montane tree species (Wardle 1985); and (4) persistent wind. While there was undoubtedly a great deal of ground disturbance during the LGM, as evidenced by enhanced aggradation recorded as gravel deposits, loess sheets and high levels of silt in offshore marine cores, it seems improbable that it would have been frequent or widespread enough outside the river and stream flood channels to have permanently repressed tall woody vegetation. Although precipitation is likely to have been lower (Alloway et al. 1992) and to have constrained forest in eastern districts (McGlone 2002a), westerly winds continued to prevail, making it highly unlikely that precipitation could have dropped to levels low enough to restrict forest in western South Island and most North Island districts outside of the southeast. Persistent wind keeps the windward flanks of subantarctic Campbell and Auckland Islands nearly entirely clear of forest and most shrubs (McGlone 2002b), but on mainland New Zealand such strong effects are highly localised on upland ridges and exposed coastal cliffs. If the extraordinarily strong, persistent winds of the subantarctics fail to keep forest and tall woody vegetation from the sheltered sides of those islands, LGM winds were unlikely to have done so on the mainland.

Although all of these factors must have played a role, we are left essentially with outbreaks of very cold, dry air off cold oceans and expanded sea ice to the south as the most probable single explanation for the widespread elimination of forest and tall woody vegetation (McGlone et al. 1993). A similar explanation has been used regarding LGM pollen assemblages in southern Australia (Harle et al. 2004). Tree dormancy is broken and cold resistance is lost in late spring, and freezing events during this period are likely to have been particularly effective in retarding tree growth. Southern Hemisphere plants are particularly susceptible because of their low freezing tolerance (Bannister 2007). The fact that grass-dominated sites occurred from current sea level to 600 m, and from the far south of the South Island to the Auckland Isthmus, provides a strong support for frequent, brief-duration cold events differentially affecting sites because of their topography and latitude. A present-day analogue is the July 1996 Southland frost, during which lowland temperatures fell as low as -15°C and remained well below freezing for more than a week, leading to widespread mortality of indigenous plants, including podocarp trees (Bannister 2003). Considering that the Southland frost occurred during one of the warmest decades of the past century, it seems at least possible that during the LGM, with surrounding seas much cooler, extended Antarctic sea ice closer to the mainland, outbreaks of icebergs to as far north as the Chatham Rise (Carter et al. 2002; Crundwell et al. 2008; Fraser et al. 2009) and extensive glaciers along the axial ranges, similar or longer or more severe outbreaks would be much more common. A recent modelling simulation of New Zealand LGM climate showed an increase in the frequency and strength of the southerlies, and showed they were capable of bringing very cold polar air over most of the country, with the east of the South Island especially affected (Drost et al. 2007). As the severity of these southerly incursions would have lessened towards the north, this explanation is consistent with the generally higher levels of forest pollen observed at northern sites in the minimum forest LGM dataset.

Conclusions

The traditional accounts of an LGM southern landscape in which forest and tall woody vegetation was attenuated towards the south and nearly absent in the southeast of the South Island, and grassland vegetation prominent from the Auckland Isthmus southwards are supported by our analysis. However, the exact proportion of the landscape in either forest or closed woody vegetation is difficult to determine because of the relatively few sites dated to the LGM, and the possibility of bias towards LGM sites either frequently disturbed or subject to cold-air drainage and hence greater-than-normal frost stress. A straightforward reading of the LGM pollen results would suggest that as much as 65% of the NNI, 55% of the SNI, 20% of the WSI and 25% of the ESI was 'woody' – that is, in either forest or shrubland. It is difficult to see how the current beetle studies can challenge this conclusion, especially given their lack of quantification and modern validation. Instead, they give firm evidence for the presence of shrubs, closed woody vegetation and *Nothofagus* in the few sites where they have been analysed. If the *woody vegetation* indicated by beetles equates primarily to montane-subalpine woody vegetation, shown by pollen to be extensive throughout southern New Zealand during the LGM, then the two proxies are easily reconciled. We thus contend that the beetle evidence, rather than being in conflict, is strongly supportive of the broad pollen and macrofossil conclusions of:

- widespread survival of forest, albeit in small and scattered patches with larger remnants close to the LGM coastline on coastal hills;
- abundant tall shrubland in the North Island and relatively less in the south;
- grassland throughout, patchy in the far north, but becoming more extensive in the south;
- very open herbaceous and low-growing shrub communities in many parts of the eastern South Island.

Acknowledgements

We thank Simon Haberle (editor), Nick Porch and an anonymous referee for their helpful comments on a draft of this contribution.

References

Alloway, B.V., D.J. Lowe, D.J.A. Barrell, R.M. Newnham, P.C. Almond, P.C. Augustinus, N.A.N. Bertler, L. Carter, N.J. Litchfield, M.S. McGlone, J. Shulmeister, M.J. Vandergoes, P.W. Williams and N.I. Members 2007. Towards a climate event stratigraphy for New Zealand over the past 30 000 years (NZ-INTIMATE project). *Journal of Quaternary Science* 22(1):9-35.

Alloway, B.V., R.B. Stewart, V.E. Neall and C.G. Vucetich 1992. Climate of the Last Glaciation in New Zealand, based on aerosolic quartz influx in an andesitic terrain. *Quaternary Research* 38(2):170-179.

Bannister, P. 2003. Are frost hardiness ratings useful predictors of frost damage in the field? A test using damage records from the severe frost in South Otago and Southland, New Zealand, July 1996. *New Zealand Journal of Botany* 41(3):555-569.

Bannister, P. 2007. A touch of frost? Cold hardiness of plants in the Southern Hemisphere. *New Zealand Journal of Botany* 45(1):1-33.

Barrows, T.T., S. Juggins, P. De Deckker, E. Calvo and C. Pelejero 2007. Long-term sea surface temperature and climate change in the Australian-New Zealand region. *Paleoceanography* 22(2).

Bennett, K. D. and J. Provan 2008. What do we mean by 'refugia'? *Quaternary Science Reviews* 27(27-28):2449-2455.

Burge, P.I. and J. Shulmeister 2007. Re-envisioning the structure of last glacial vegetation in New Zealand using beetle fossils. *Quaternary Research* 68(1):121-132.

Carter, L., H.L. Neil and L. Northcote 2002. Late Quaternary ice-rafting events in the SW Pacific Ocean, off eastern New Zealand. *Marine Geology* 191(1-2):19-35.

Crundwell, M., G. Scott, T. Naish and L. Carter 2008. Glacial-interglacial ocean climate variability from planktonic foraminifera during the Mid-Pleistocene transition in the temperate Southwest Pacific, ODP Site 1123. *Palaeogeography Palaeoclimatology Palaeoecology* 260(1-2):202-229.

Dodson, J.R., N.J. Enright and R.F. McLean 1988. A late Quaternary vegetation history for far northern New Zealand. *Journal of Biogeography* 15(4):647-656.

Drost, F., J. Renwick, B. Bhaskaran, H. Oliver and J. McGregor 2007. Simulation of New Zealand's climate using a high-resolution nested regional climate model. *International Journal of Climatology* 27(9):1153-1169.

Elliot, M.B. 1998. Late Quaternary pollen records of vegetation and climate change from Kaitaia Bog, far northern New Zealand. *Review of Palaeobotany and Palynology* 99(2):189-202.

Fraser, C.I., R. Nikula, H.G. Spencer and J.M. Waters 2009. Kelp genes reveal effects of subantarctic sea ice during the Last Glacial Maximum. *Proceedings of the National Academy of Sciences of the United States of America* 106(9):3249-3253.

Harle, K.J., A.P. Kershaw and E. Clayton 2004. Patterns of vegetation change in southwest Victoria (Australia) over the last two glacial/interglacial cycles: evidence from Lake Wangoom. *Proceedings of the Royal Society of Victoria* 116:105-137.

Ladeau, S.L. and J.S. Clark 2006. Pollen production by *Pinus taeda* growing in elevated atmospheric CO2. *Functional Ecology* 20(3):541-547.

Lancashire, A.K., J.R. Flenley and M.A. Harper 2002. Late glacial beech forest: an 18,000-5000-BP pollen record from Auckland, New Zealand. *Global and Planetary Change* 33:315-327.

Lees, C.M., V.E. Neall and A.S. Palmer 1998. Forest persistence at coastal Waikato, 24000 years BP to present. *Journal of the Royal Society of New Zealand* 28(1):55-81.

Lowe, D.J., P.A.R. Shane, B.V. Alloway and R.M. Newnham 2008. Fingerprints and age models for widespread New Zealand tephra marker beds erupted since 30,000 years ago: a framework for NZ-INTIMATE. *Quaternary Science Reviews* 27(1-2):95-126.

Marra, M. and R.A.B. Leschen 2004. Late Quaternary paleoecology from fossil beetle communities in the Awatere Valley, South Island, New zealand. *Journal of Biogeography* 31(4):571-586.

Marra, M.J., B.V. Alloway and R.M. Newnham 2006. Paleoenvironmental reconstruction of a well-preserved Stage 7 forest sequence catastrophically buried by basaltic eruptive deposits, northern New Zealand. *Quaternary Science Reviews* 25(17-18):2143-2161.

McGlone, M.S. 1980. *Late Quaternary Vegetation History of central North Island, New Zealand*, University of Canterbury.

McGlone, M.S. 1982. Modern pollen rain, Egmont National Park, New Zealand. *New Zealand Journal of Botany* 20(3):253-262.

McGlone, M.S. 1985. Plant biogeography and the late Cenozoic history of New Zealand. *New Zealand Journal of Botany* 23(4):723-749.

McGlone, M.S. 2001a. A late Quaternary pollen record from marine core P69, southeastern North Island, New Zealand. *New Zealand Journal of Geology and Geophysics* 44(1):69-77.

McGlone, M.S. 2001b. The origin of the indigenous grasslands of southeastern South Island in relation to pre-human woody ecosystems. *New Zealand Journal of Ecology* 25(1):1-15.

McGlone, M.S. 2002a. A Holocene and latest Pleistocene pollen record from Lake Poukawa, Hawke's Bay, New Zealand. *Global and Planetary Change* 33(3-4):283-299.

McGlone, M.S. 2002b. The late quaternary peat, vegetation and climate history of the southern oceanic islands of New Zealand. *Quaternary Science Reviews* 21(4-6):683-707.

McGlone, M.S. 2009. Postglacial history of New Zealand wetlands and implications for their conservation. *New Zealand Journal of Ecology* 33(1):1-23.

McGlone, M.S. and J.L. Bathgate 1983. Vegetation and climate history of the Longwood Range, South Island, New Zealand, 12,000 BP to the present. *New Zealand Journal of Botany* 21(3):293-315.

McGlone, M.S. and C.D. Meurk 2000. Modern pollen rain, subantarctic Campbell Island, New Zealand. *New Zealand Journal of Ecology* 24(2):181-194.

McGlone, M.S. and N.T. Moar 1997. Pollen-vegetation relationships on the subantarctic Auckland Islands, New Zealand. *Review of Palaeobotany and Palynology* 96(3-4):317-338.

McGlone, M.S. and N.T. Moar 1998. Dryland Holocene vegetation history, Central Otago and the Mackenzie Basin, South Island, New Zealand. *New Zealand Journal of Botany* 36(1):91-111.

McGlone, M.S. and V.E. Neall 1994. The late Pleistocene and Holocene vegetation history of Taranaki, North Island, New Zealand. *New Zealand Journal of Botany* 32(3):251-269.

McGlone, M.S., V.E. Neall and B.D. Clarkson 1988. The effect of recent volcanic events and climatic changes on the vegetation of Mt Egmont (Mt Taranaki), New Zealand. *New Zealand Journal of Botany* 26(1):123-144.

McGlone, M.S., V.E. Neall and B.J. Pillans 1984. Inaha terrace deposits - a late Quaternary terrestrial record in south Taranaki, New Zealand. *New Zealand Journal of Geology and Geophysics* 27(1):35-49.

McGlone, M.S., C.S. Nelson and T.M. Hume 1978. Palynology, age and environmental significance of some peat beds in the Upper Pleistocene Hinuera Formation, south Auckland, New Zealand. *Journal of the Royal Society of New Zealand* 8(4):385-393.

McGlone, M.S., C.S. Nelson and A.J. Todd 1984. Vegetation history and environmental significance of pre-peat and surficial peat deposits at Ohinewai, lower Waikato lowland. *Journal of the Royal Society of New Zealand* 14(3):233-244.

McGlone, M.S., M.J. Salinger and N.T. Moar 1993. Palaeovegetation studies of New Zealand's climate since the Last Glacial Maximum. In H. E. Wright, J. E. Kutzbach, T. Webb III, W. F. Ruddiman, F. A. Street-Perrrott and P.J. Bartlein (eds) *Global Climates since the Last Glacial Maximum*, pp 294-317. Minneapolis: University of Minnnesota Press.

McGlone, M.S. and W.W. Topping 1983. Late Quaternary vegetation, Tongariro region, central North Island, New Zealand. *New Zealand Journal of Botany* 21(1):53-76.

Mix, A.C., E. Bard and R. Schneider 2001. Environmental processes of the ice age: land, oceans, glaciers (EPILOG). *Quaternary Science Reviews* 20(4):627-657.

Moar, N., R.P. Suggate and C. Burrows 2008. Environments during the Kaihinu Interglacial and Otira Glaciation, coastal north Westland, New Zealand. *New Zealand Journal of Botany* 46(1):49-63.

Moar, N.T. 1980. Late Otiran and early Aranuian grassland in central South Island. *New Zealand Journal of Ecology* 3:4-12.

Moar, N.T. and R.P. Suggate 1979. Contributions to the Quaternary history of the New Zealand flora 8. Interglacial and glacial vegetation in the Westport District, South Island. *New Zealand Journal of Botany* 17(3):361-387.

Moar, N.T. and R.P. Suggate 1996. Vegetation history from the Kaihinu (last) interglacial to the present, west coast, South Island, New Zealand. *Quaternary Science Reviews* 15(5-6):521-547.

Molloy, B.P. J., C.J. Burrows, J.E. Cox, J.A. Johnson and P. Wardle 1963. Distribution of subfossil remains in eastern South Island, New Zealand. *New Zealand Journal of Botany* 1:68-77.

Neall, V.E. 1979. Sheets P19, P20, & P21, New Plymouth, Egmont, and Manaia. In *Geological Map of New Zealand 1:50 000*. 1 ed. Wellington: New Zealand Geological Survey, DSIR.

Newnham, R.M. 1992. 30,000 year pollen, vegetation and climate record from Otakairangi (Hikurangi), Northland, New Zealand *Journal of Biogeography* 19(5):541-554.

Newnham, R.M., D.N. Eden, D.J. Lowe and C.H. Hendy 2003. Rerewhakaaitu Tephra, a land-sea marker for the Last Termination in New Zealand, with implications for global climate change. *Quaternary Science Reviews* 22(2-4):289-308.

Newnham, R.M., D.J. Lowe, T. Giles and B.V. Alloway 2007. Vegetation and climate of Auckland, New Zealand, since ca. 32 000 cal. yr ago: support for an extended LGM. *Journal of Quaternary Science* 22(5):517-534.

Newnham, R.M., D.J. Lowe and J.D. Green 1989. Palynology, vegetation and climate of the Waikato lowlands, North Island, New Zealand, since c. 18,000 years ago. *Journal of the Royal Society of New Zealand* 19(2):127-150.

Newnham, R.M., D.J. Lowe and P.W. Williams 1999. Quaternary environmental change in New Zealand: a review. *Progress in Physical Geography* 23(4):567-610.

Newnham, R.M., M.J. Vandergoes, C.H. Hendy, D.J. Lowe and F. Preusser 2007. A terrestrial palynological record for the last two glacial cycles from southwestern New Zealand. *Quaternary Science Reviews* 26(3-4):517-535.

Osmaston, H.A. 2006. Should Quaternary sea-level changes be used to correct glacier ELAs, vegetation belt altitudes and sea level temperatures for inferring climate changes? *Quaternary Research* 65(2):244-251.

Pelejero, C., E. Calvo, T.T. Barrows, G.A. Logan and P. De Deckker 2006. South Tasman Sea alkenone palaeothermometry over the last four glacial/interglacial cycles. *Marine Geology* 230(1-2):73-86.

Pillans, B., M. McGlone, A. Palmer, D. Mildenhall, B. Alloway and G. Berger 1991. The Last Glacial Maximum in central and southernmost North Island, New Zealand – a paleoenvironmental reconstruction using the Kawakawa Tephra Formation as a chronostratigraphic marker. *Palaeogeography, Paleoclimatology, Palaeoecology* 101:283-304.

Pullar, W.A. and R.N. Patel 1972. Identification of tree stumps and driftwood associated with tephra layers in alluvium, peat, and dune sands. *New Zealand Journal of Botany* 10:605-614.

Randall, P.M. 1990. A Study of modern pollen deposition, Southern Alps, South Island, New Zealand. *Review of Palaeobotany and Palynology* 64:263-272.

Rogers, C.A., P.M. Wayne, E.A. Macklin, M.L. Muilenberg, C.J. Wagner, P.R. Epstein and F.A. Bazzaz 2006. Interaction of the onset of spring and elevated atmospheric CO2 on ragweed (*Ambrosia artemisiifolia* L.) pollen production. *Environmental Health Perspectives* 114(6):865-869.

Sandiford, A., R. Newnham, B. Alloway and J. Ogden 2003. A 28,000-7600 cal yr BP pollen record of vegetation and climate change from Pukaki Crater, northern New Zealand. *Palaeogeography Palaeoclimatology Palaeoecology* 201(3-4):235-247.

Soons, J.M. and C.J. Burrows 1978. Dates for Otiran deposits, including plant microfossils and macrofossils, from Rakaia Valley. *New Zealand Journal of Geology and Geophysics* 21:607-615.

Sugita, S. 1994. Pollen Representation of Vegetation in Quaternary Sediments: Theory and Method in Patchy Vegetation. *Journal of Ecology* 82(4):881-897.

Turney, C.S.M., J.E. Hunt and C. Burrows 2002. Deriving a consistent delta C-13 signature from tree canopy leaf material for palaeoclimatic reconstruction. *New Phytologist* 155(2):301-311.

Vandergoes, M.J., R.M. Newnham, F. Preusser, C.H. Hendy, T.V. Lowell, S.J. Fitzsimons, A.G. Hogg, H.U. Kasper and C. Schluchter 2005. Regional insolationforcing of late Quaternary climate change in the Southern Hemisphere. *Nature* 436(7048):242-245.

Wardle, P. 1985. New Zealand timberlines. 3. A synthesis. *New Zealand Journal of Botany* 23:263-271.

Wardle, P. 2001. Holocene forest fires in the upper Clutha district, Otago, New Zealand. *New Zealand Journal of Botany* 39(3):523-542.

Wilmshurst, J.M., M.S. McGlone, J.R. Leathwick and R.M. Newnham 2007. A pre-deforestation pollen-climate calibration model for New Zealand and quantitative temperature reconstructions for the past 18 000 years BP. *Journal of Quaternary Science* 22(5):535-547.

Woodward, C.A. and J. Shulmeister 2007. Chironomid-based reconstructions of summer air temperature from lake deposits in Lyndon Stream, New Zealand spanning the MIS 3/2 transition. *Quaternary Science Reviews* 26(1-2):142-154.

5

Holocene vegetation history of a high-elevation (1200 m) site in the Lake Heron Basin, inland Canterbury, New Zealand

J. M. Pugh

Department of Geological Sciences, University of Canterbury, Christchurch, New Zealand

J. Shulmeister

School of Geography, Planning and Environmental Management, University of Queensland, Australia
james.shulmeister@uq.edu.au

Introduction

The Canterbury high country is a favourable location to examine climate-change histories because it lies in the lee of the Southern Alps. This causes the area to be a rain-shadow region and it is sensitive to changes in the strength and persistence of the regional westerly flow. Strong westerly flow is associated with droughts and high summer temperatures. In contrast, weakened westerly flow allows moisture from the east to penetrate these upland basins. As a consequence, this is an important area to study changes in the Southern Hemisphere westerly winds in this sector of the Southern Ocean. This record is unusual because it comes from near the natural tree line and as a consequence should be particularly sensitive to climate change and other environmental forcing. There are a number of significant palaeoecological questions that relate to this setting, including: (1) the persistence of montane podocarp woodland dominated by *Phyllocladus* and *Halocarpus* into the Holocene and the timing and cause of its subsequent replacement by beech forest; (2) the role played by fire in controlling vegetation structure and species composition; and (3) human impacts in the high country, especially with the transfer of high-country land into the conservation estate and consequential issues of ecological and landscape management (Armstrong et al. 2005).

Site description

Geology and physiography

The Lake Heron basin is a 30 km long, 7-8 km wide inter-montane basin in the mid-Canterbury region between the Arrowsmith Range to the west and the Mount Somers Range to the east. It has a mean floor elevation of about 700 m amsl. The basin formed between the Canterbury front ranges and the main ranges of the Southern Alps and is the product of reverse and back thrusting (Pettinga et al. 2001) in the Quaternary. Several active thrust faults occur in the basin (Oliver and Keene 1990). The regional bedrock is Triassic Torlesse greywackes, but there are localised outcrops of both Jurassic volcanics (andesites) and Tertiary sedimentary rocks, the latter mainly in fault-angle depressions (Oliver and Keane 1990).

Staces Tarn (NZMS 260 J35/581516) is a shallow (<1 m deep) mesotrophic lake situated in a trough on a ridgeline 1.5 km south of Staces Hill (1479 amsl) at the confluence of the Cameron Valley and Lake Heron Basin (Figure 1). Staces Tarn is approximately 130 m long and 30 m wide and occurs at 1200 m (amsl). The tarn has no outlet, as a moraine impounds the lake along its southwest margin.

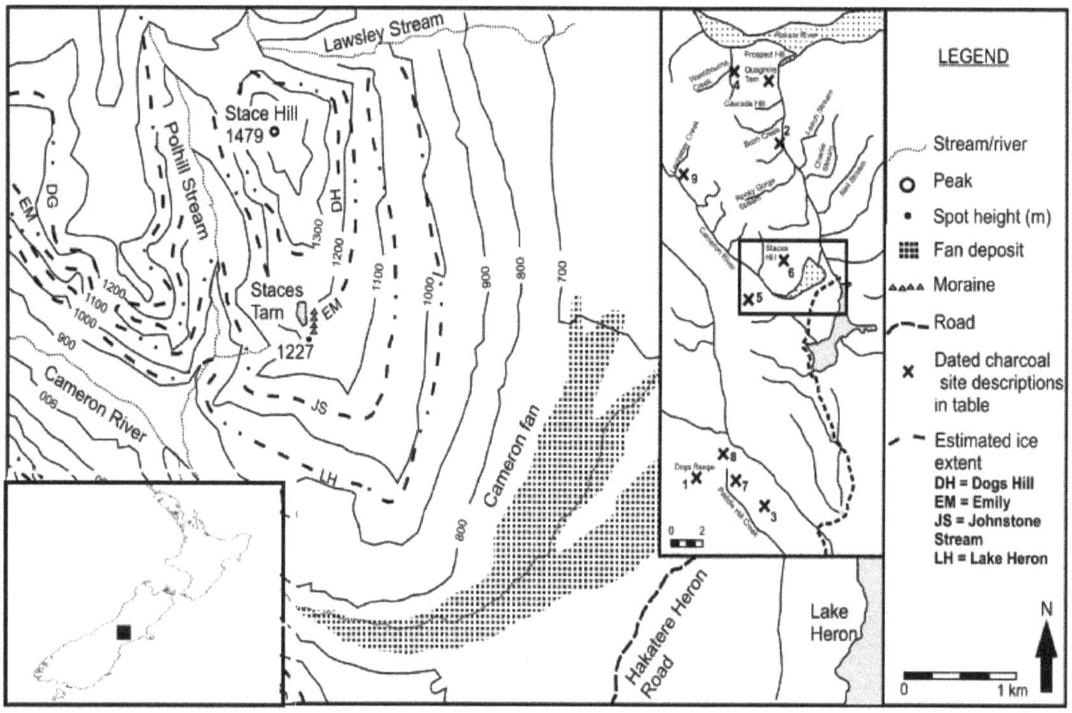

Figure 1. Location map for Staces Tarn. The inset map of New Zealand shows general location in the central South Island. The regional map marks on sites mentioned in the text and the other regional localities for dated charcoal occurrences. The main map details the area around Staces Tarn. The dashed lines relate to former glacial limits as defined by Mabin (1980). All advances except Dogs Hill are presumed to be last glacial cycle in age (<75 ka). Dogs Hill is attributed to oxygen isotope stage 6

The study area has experienced multiple glaciations (Mabin 1980; Pugh 2008) inferred to extend through at least three glacial-interglacial cycles, and at its maximum extent during the last glaciation ice was thick enough to overrun Staces Tarn. The tarn formed following the retreat of a small offshoot of ice from the main Lake Heron ice lobe about 18,000 years ago, dammed behind the end moraine as the glacier stagnated (Pugh 2008).

Climate

The area has mild summers and cool winters and frosts are common. The mean annual rainfall at Upper Lake Heron Station is 1047 mm (1969-1981), with the mean monthly rainfall varying from 46 mm in February to 119 mm in August (Bowden 1983). Mean monthly temperatures are not available for the study area. They would be comparable to those of the adjacent Rakaia valley to the east, which range between 3.5°C (July) and 15.3°C (January) at 364 m amsl (some 836 m below Staces Tarn) (New Zealand Meteorological Service 1983). Using an environmental lapse rate of 6°C/km, mean January temperature at Staces Tarn would be approximately 10.3°C.

Vegetation

Forest cover would have been extensive below the tree line, but natural fires, Polynesian burning and European clearing have removed most forest patches except where they have survived in protected gullies. The present vegetation cover in the Lake Heron basin is predominantly tall tussock grassland, dominated by *Chionochloa* sp. cf. *rigida* (Burrows and Russell 1990). Snow tussock grassland is extensively developed on hill summits and ridges (Bowden 1983), including around Staces Tarn. There are few native trees in the study area, other than isolated stands of *N. menziesii* (silver beech), *Hoheria lyallii* and *Sophora microphylla* (Burrows and Russell 1990). Burrows and Russell also noted patches of *Kunzea ericoides* and *Leptospermum scoparium* occurring near where the Cameron River valley enters the Lake Heron basin.

Previous pollen studies and historical accounts from the Canterbury foothill country indicate that podocarp forest, typically dominated by *P. totara, P. halli* and *Prumnypitys taxifolia,* was widespread on the valley floors and lower elevation slopes (e.g. Burrows 1977). *Nothofagus* (southern beech) forest, mostly comprising mountain and silver beech (*N. solandri* var *cliffortioides* and *N. menziesii*), cloaked upper slopes below the tree line, and tussock grassland and sub-alpine herbfeld dominated above the tree line (Moar 1971; McGlone 1988).

The arrival of *Nothofagus* forest is a relatively recent phenomenon in the Canterbury hill country (e.g. Burrows and Lord 1993), with montane podocarp forest dominated by *Phyllocladus* spp. and *Halocarpus bidwilli* occupying the near-tree-line ecotone in the earlier part of the Holocene. The time of arrival of *Nothofagus* forest varies from area to area. In this area, Burrows et al. (1990) dated the burial of a *Phyllocladus alpinus* woodland by outwash in the Cameron valley to 9520 ± 95 yr BP (NZ 688). Based on dated macrofossils, *Phyllocladus* was still present in the Cameron valley below approximately 1100 m until at least 2840 ± 70 yr BP (NZ 1880; Burrows and Lord 1993).

Isolated stands of *N. menziesii* currently occur in tributary valleys in the north of the basin, the closest being located in Rocky Gorge Stream, 5 km north of Staces Tarn. Other stands to the north occur in Bush Creek, in an unnamed valley below Cascade Hill and in Downs Hut Stream. They also occur in tributaries in the east of the basin in Leach Stream, Charlie Stream and Neil Stream (Burrows and Russell 1990).

Staces Tarn has not been investigated previously and it is an excellent site for palaeoclimate studies because it is located just below the regional tree line and hence should be very sensitive to environmental changes. In this paper, we present results from a detailed pollen analytical study of the site supported by limited AMS radiocarbon and luminescence dating, to provide a Holocene vegetation and fire history for this site.

Methods

The core was recovered from the margin of a delta at the northern end of the tarn (see Figure 2). A hand-operated D-section corer with a 50 cm x 8 cm chamber was used for core extraction. The samples were extracted alternately from two holes approximately 50 cm apart to allow for complete core recovery and to minimise down-hole contamination. The material changed from an organic gyttja at the top to a compacted grey clay with increasing depth. Sampling ceased at a depth of 193 cm as the corer was unable to penetrate further. Locations and elevations were determined using a handheld Garmin GPS Etrex H and a 1:50,000 scale topographic map.

Figure 2. Panoramic view of Staces Tarn. The core came from the margin of the delta at the north end of the lake

Description of the core follows Troels-Smith (1955). Sixteen 1 cm^3 sub-samples were removed at 10 cm intervals for the top 1 m of core, and at 20 cm intervals from 110 cm to 190 cm, for pollen analysis. The remainder of the cores was wrapped in plastic film and archived in a refrigerator at the Department of Geological Sciences, University of Canterbury. Pollen slides were prepared following the standard methods outlined in Moore et al. (1991). A lycopodium tablet added to each sample allowed the calculation of pollen concentrations. Counting was done at 400x magnification using a Leitz Diaplan comparison microscope. Palynomorphs were identified using the pollen atlas of Moar (1993) and publications by Pocknall (1981) and Large and Braggins (1991). A target of 250 dryland pollen grains for each sample was attempted but the grey silts of these cores are not pollen rich. Individual counts ranged from 52 to 306 dryland grains, with low values in the top and bottom three to four samples. Total palynomorph counts ranged from 260 to 712 per sample.

Charcoal abundance and volume are estimated through the counting of charcoal particles in a predetermined area using a fixed and random number of points, as described by Clark (1982). Charcoal values are presented in pollen diagrams as surface area per unit volume.

A pollen diagram was constructed using the PSIMPOLL program (Bennett 2002).

Chronology

A sample block of inorganic lake muds was removed at 160-170 cm core depth, wrapped in tin-foil and tape, then submitted for luminescence dating to the Victoria University, Wellington, luminescence laboratory. The VUW laboratory used the multiple aliquot-additive dose technique (MAAD) for Infra-Red Stimulated Luminescence (IRSL) on the 4-8 micron polymineralic fraction in these muds. A 14.8 g charcoal sample was removed from a significant charcoal band at 76 cm depth and sent to the Rafter Radiocarbon Laboratory at GNS Science, Wellington, New Zealand, for AMS radiocarbon dating.

Results

Chronology and sedimentation rates

The luminescence sample (ST 160-170) yielded an age of 7.65 ± 0.55 ka, while the radiocarbon sample (NZA 29676) at 76 cm yielded an uncalibrated age of 4644 ± 40 yr BP (5475-5298 cal yr BP [p>.95] (using Calib 5.0: Stuiver and Reimer 1993)).

With the exception of the top 15 cm of the sample, which is rich in organic material and poorly compacted, the sediments are uniform lake muds, and sedimentation rates can be constructed from the dated charcoal layer at 67 cm depth, yielding a date of 5475-5298 cal. yr BP and the luminescence age of 7.65 ± 0.5 ka from 160-170 cm depth. This yields rates of about 0.4 mm/yr in the lower part of the core and about 0.1 mm/yr above the charcoal layer at 67 cm. These rates are extremely slow by New Zealand standards. The unconsolidated material in top 15 cm is clearly historical (based on palynolomorphs – see below) and there is a possibility of loss or disturbance of material at this level, due to animal trampling and/or episodic desiccation. The arrival of beech in the pollen record at 20 cm suggests that this is unlikely to be significant.

Pollen results

Figure 3 displays the concentration of dryland taxa without swamp elements and fern spores.

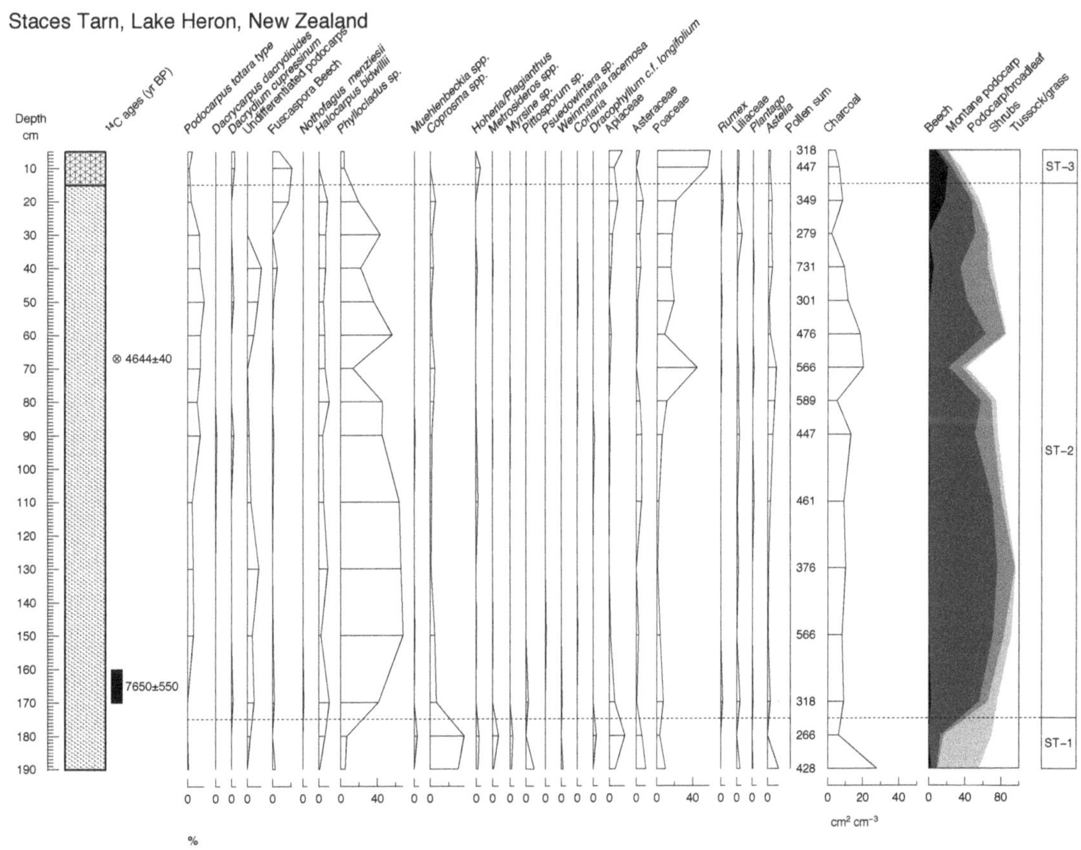

Figure 3. Summary pollen diagram for Staces Tarn. Note that the older age is a luminescence age, while the younger age is a radiocarbon result. The pollen sums are the total counts; dryland totals are lower

The dryland pollen results are presented in the diagrams in five summary groups.

1. Beech forest, consisting of *Nothofagus* fuscaspora type, a composite taxon of five indistinguishable beech species, and *Nothofagus menziesii.* We include *Peraxilla* sp. with the beeches as this mistletoe is usually a parasite on beech (Allan 1961).
2. Montane podocarp forest, which consists of *Halocarpus bidwillii, Phyllocladus* spp. and *Podocarpus nivalis.* These taxa can occur as forest trees, as heathland elements, or even as prostrates.
3. Podocarp/broadleaf forest, consisting mainly of *Podocarpus/prumnopitys* type, *P. totara* type (represented by both *P. hallii* and *P. traversii),* and small counts of *Dacrycarpus dacrydioides, Dacrydium cupressinum,* and other lowland forest types. This forest type occurs mainly below 500 m and is relatively diverse botanically, especially in the underrepresented broadleaf elements. Podocarp pollen grains that were unidentifiable due to damaged sacchi or corpus are included in the total count as 'undifferentiated podocarps'.
4. Shrubland: small trees and tall shrubs dominated in the pollen records by *Coprosma,* with *Muehelenbeckia* spp., *Hoheria* sp., *Myrsine* spp. and *Pseudowintera* sp. and the forest margin/disturbance indicator *Coriaria.*
5. Tussock grasses, heaths and herbfeld, comprised mainly of Poaceae (grasses), with smaller contributions from Apiaceae, *Gentiana, Dracophyllum,* Asteraceae, Liliaceae, and *Astelia.*

We also counted but do not present wetland taxa consisting of Cyperaceae with Restionaceae, *Phormium* (flax), *Myriophyllum,* Chenopodiaceae and *Potamogeton.* Trilete and Monolete fern spores, the most common palynomorphs, were included in this group, as ferns were concentrated around the lake margin.

European disturbance indicators and pasture weeds, including *Rumex* and *Plantago,* are noted, and *Rumex* is common enough to warrant presentation as a curve. Other than *Rumex,* European weeds were present in only trace amounts.

Zone ST-1: 190-175 cm. *Coprosma* with *Gentiana,* Apiaceae, Asteraceae and *Astelia* zone

Coprosma varies between 30% and 35%, with *Gentiana* decreasing from 10% to 5% at the top of the zone. Of the tussocks and grasses, Apiaceae, Asteraceae and *Astelia* fluctuate between 5% and 15%. Poaceae comprises 10% of the dryland total. At the base of the zone, almost 0.03 cm^2/cm^3 of charcoal content is recorded, the largest count in the core.

Zone interpretation

Zone ST-1 is characterised by the dominance of shrubs and herbs (*Coprosma, Gentiana,* Apiaceae, Asteraceae, *Astelia*). The pollen assemblage of ST-1 represents a heathland environment, comprising shrubs and herbs that are typically slow growing or stunted. Heathland occurs in the South Island in mountainous areas and on glacial outwash surfaces, including wet areas above the tree line. Coprosmas occur in a range of environments, including sub-alpine landscapes, and are tolerant of both poorly drained and exposed habitats (Macphail and McQueen 1983). While the shrubs range from lowland to montane/sub-alpine, they tend to be common in records from cooler and higher-elevation sites. *Dracophyllum* is a member of the Ericaceae and most taxa display typical heath morphologies, except the tree-lily-like *D. traversii* (mountain neinei). Most are montane to sub-alpine in distribution and *D. longifolium* is very common in the modern vegetation around the core site. It is severely underrepresented in pollen diagrams (Macphail and McQueen 1983). Gentians, Apiaceae and the dreaded Astelias (Spaniard) are typical elements of sub-alpine grassland and herbfield vegetation in New Zealand (e.g. Wardle 1986). They normally indicate either above-tree-line or frost-basin-type environments. All of these taxa are indicative of a near-tree-line setting, with the site inferred to sit above the tree line at this time.

Two anomalous pollen grains occur in ST-1. The first is *Peraxilla* sp. of the genus Lorantheaceae, one of the three New Zealand beech mistletoes. *P. tetrapetala* is an arboreal xylem parasite dependent on its host for water and nutrients (Ullmann et al 1985). The parasite is commonly found on *Nothofagus* spp., but it has been recorded also on *Quintinia* (Allan 1961). These mistletoes are usually dependent on birds for pollination, although insects may play a role for some species (Kuijt 1969). The presence of mistletoe indicates that beech trees may have been growing near Staces Tarn at this time or the grain may have washed off a bird. However, the second anomalous grain is that of a *Brassospora* beech. Given that this pollen group has been extinct in New Zealand since the mid-Pleistocene (Mildenhall and Biryami 2003), it is likely that it is derived from Tertiary sediments that outcrop at lower elevations in the Cameron valley (Warren 1967) and which may have outcropped upstream of the tarn.

Zone ST-2: 175-15 cm. *Phyllocladus* and lowland podocarps zone

Phyllocladus dominates the zone, varying between 25% and 70% of the total dryland taxa. *Podocarpus/Prumnopitys* type increases from 1% at the bottom of the zone to 17% at the top, while undifferentiated *Podocarpus* pollen fluctuates between 2% and 15%. Low counts of *Dacrycarpus dacrydioides* (1%) and *D. cupressinum* (3%) occur. Shrubland taxa have very low counts; *Coprosma* decreases from 7% to 2% upwards through the zone. The grasses Apiaceae and Poaceae fluctuate between 1% and 5%. Asteraceae and *Astelia* increase from 1% to approximately 7% at the top of the zone. The charcoal content fluctuates between 2 and 6 cm^2/cm^3.

Within Zone ST-2 at 70 cm depth, Poaceae increases briefly to 47%, with *Phyllocladus* declining to 10%. *Podocarpus* types remain much the same, at 12%. *Astelia* increases slightly to 9%. Microscopic charcoal content at this depth peaks at approximately 0.02 cm^2/cm^3 and remains elevated at 65 cm depth. At 67 cm depth, a 2 mm thick band of macroscopic charcoal occurs.

Zone ST-2 is characterised by the dominance of celery pine *Phyllocladus* spp. It is interpreted as a montane podocarp forest. Now, *Phyllocladus* forest occurs in regions with an annual precipitation of at least 1250 mm (Wardle 1969), suggesting that the region was wetter than at the present day. As the floor of the Lake Heron basin currently receives around 1000 mm rain per year, it probably indicates relatively similar conditions to today, though it may have been marginally wetter.

The lowland podocarps are abundant pollen producers and have long-distance dispersal capabilities. They do not currently occur much above 600 m elevation and therefore are unlikely to have ever occurred close to Staces Tarn, so they provide a measure of the regional pollen rain. *D. cupressinum* is a widespread taxon but is more common in the wetter western regions. *D. dacrydioides* is widely distributed on the West Coast of the South Island, and is not currently found in inland Canterbury (Lintott and Burrows 1973). It currently occurs in lowland forests, dominating fertile, free-draining floodplains and the wet margins of the lowland swamps and bogs (Metcalf 2002). *Casuarina* is another taxon that is extinct in New Zealand. Unlike Brassospora beech, it is a persistent but rare component of New Zealand pollen diagrams and is interpreted as being derived from Australia. It confirms the strong westerly sourcing of long-distance-derived pollen in Staces Tarn.

There is a gradual decline in *Phyllocladus* in the top half of this zone, matched by a gradual increase in grass (Poaceae). There are also generally higher levels of charcoal in the upper half of the zone and there is a macroscopic charcoal layer at 67 cm depth.

This spike in grass is short lived and tree and shrub values recover to near pre-fire values by 60 cm, even though this is the peak microscopic charcoal level. We infer the relationship between *Phyllocladus* and grass to represent changes in either the structure of the woodland (becoming more open) or proximity to a grassland woodland boundary.

Zone ST-3: 15-0 cm. Fuscaspora type and Poaceae zone

Nothofagus fusca type increases to 15% before dropping away to 7% in the uppermost zone. Poaceae is the other dominant taxa and increases from 20% to almost 60% at the top of the zone. *Hoheria* peaks at 5% and Apiaceae fluctuates between 6% and 15%. Asteraceae, Liliaceae and *Astelia* decrease slightly to 2%. *Phyllocladus* decreases from 15% to 2%, while *P. totara* type fluctuates between 1% and 5% and *D. cupressinum* is 4% of the dryland taxa. Charcoal values decrease upwards through the zone from 0.006 to 0.004 cm^2/cm^3.

Zone ST-3 is characterised by the emergence of Fuscaspora beech tree pollen, which we attribute to mountain beech *Nothofagus solandri* var *cliffortioides* based on site elevation and proximity to the tree line. Silver beech (*N. menziesii*) and mountain beech are the most common taxa of near-tree-line assemblages in this region and *N. menziesii* pollen is easily distinguishable from the other beech pollens. In addition, *Phyllocladus* is replaced by grass and Apiaceae. We infer two changes, firstly the invasion of beech into already declining *Phyllocladus* woodland, and secondly a general opening up of vegetation, with grassland becoming widespread. The former is natural, with similar expansions of *Nothofagus* forest recorded in many pollen diagrams in Canterbury (Moar 1971; Lintott and Burrows 1973; Russell 1980; Burrows and Russell 1990) and elsewhere in New Zealand. The latter is evidence for human impact, and specifically, European settlement.

Discussion and conclusions

Holocene vegetation history of Staces Tarn and Lake Heron area

Vegetation in the lowest zone (ST-1) is characterised by the dominance of shrubs and herbs and is typical of a late-glacial-period pollen record from inland Canterbury (Burrows and Russell 1990). We have no age control in this zone and we cannot discount the alternative possibility that it could be early Holocene in age. We note that near-Holocene temperatures were reached in many parts of eastern New Zealand by about 14,000 years ago (Turney et al. 2003). An early-Holocene thermal optimum, between 1.5°C and 3°C warmer than present, is suggested for the New Zealand region from about 9000 to 7000 yr BP (Weaver et al. 1998; Wilmshurst et al. 2007).

At Prospect Hill (13 km north and 470 m lower), low-montane forest replaced shrubland at 10,000 yr BP. However, forest arrival at Staces Tarn challenges the presumed early-Holocene warmer temperatures, as trees are not dominant at the site until 7500 yr BP. Podocarps disperse and colonise sites quickly if conditions are favourable (Macphail and McQueen 1983) and it would have been expected that podocarps would have colonised the site earlier in the thermal optimum. Hence, some other factor or factors must have been inhibiting forest spread. The upper limit of tree growth has long been associated with the summer 10°C isotherm (e.g. Koeppen 1923), and West (1977) noted that at least three months with mean temperatures greater than 10°C are required for forests to dominate. Small changes in the mean monthly temperatures could result in noticeable changes in vegetation structure. It is proposed that low seasonality during the early Holocene resulted in cooler summers, and delayed the forest colonisation of the high-elevation Staces Tarn area. This is despite the fact that mean *annual* temperatures were at least as warm as they are today.

As climates became more seasonal and summer temperatures rose, we infer that montane podocarps spread uphill, reaching the site about 7700 radiocarbon years ago. They remained dominant until about 5000 yr BP. After this time, they steadily declined coeval with a gradual increase in grasses. McGlone and Wilmshurst (1999) suggested that as the subtropical high retreated northwards from its early-Holocene position, temperatures gradually declined. The westerlies also increased through the Holocene (Shulmeister et al. 2004) and the gradual decline of podocarp forests may reflect both a gradual cooling, and more significantly, an increase in the occurrence of droughts as eastern areas dried out under the enhanced westerly flow.

Based on sedimentation rates, beech pollen rose from trace values after about 3000 yr BP at Staces Tarn, but it was not until about 1500 yr BP that a pronounced pollen change occurred as beech replaced the montane podocarp forest as the dominant forest cover. The beech history at Staces Tarn is similar to adjacent areas. Beech pollen was present at Prospect Hill in the Rakaia Valley to the north at approximately 4500 yr BP. However, it was not until after 2000 yr BP that *N. fusca*/*N. menziesii* became the dominant forest type (Burrows and Russell 1990) at that site. In the southeastern South Island, *Nothofagus* forest expansion occurred around 2500 yr BP (Pocknall 1981).

Beech was eliminated from extensive parts of its South Island range during the glacial maximum, or earlier. Unlike podocarp forest, which is vagile, beech responded slowly to post-glacial climate change. The slow beech expansion was and still is the result of a range of factors, including slow dispersal rates, specialist soil mycorrhizal requirements (Baylis 1980), and poor competitiveness with podocarps on nutrient-rich soils (Rogers and McGlone 1989). Climate change also played a role. Cool, moist early-Holocene summer climates favoured montane podocarps over beech forest. Warmer and drier summers later in the Holocene favoured beech (e.g. Lintott and Burrows 1973). Increased disturbances caused by fire to the already weakened podocarp forests will have encouraged their replacement by beech. In short, the arrival of beech is an ecological change, but is set in the context of ongoing climate changes that favoured this change.

The very top level of the core displays a drop in beech pollen and a dramatic increase in grass pollen. This is almost certainly an anthropogenic signal relating to the opening up of the basin to grazing, which was associated with burning.

Fire history

The role of fire in New Zealand vegetation history has been an issue of some debate in the New Zealand ecological literature. Lightning strikes are the only source of natural fires in the South Island and McGlone (2001) shows that the total area burned under modern conditions is very minor. Whether pre-deforestation (and pre-human settlement) trees and shrubs were more susceptible to fire is not known, but McGlone (2001) suggests that this is at least possible. Natural fires are inferred to have played a role in South Island vegetation changes in the late Holocene, and if any area in New Zealand is likely to display ecological responses to fire, the dry (rain-shadow) inland basins of Canterbury and Otago are the best candidates.

Previous research has indicated that the Arrowsmith Range and upper Rakaia Valley region has been subject to Holocene fire activity (Harvey 1974; Burrows 1983, 1988, 1996; Rodbell 1986), but there are difficulties in making correlations between events due to the variety of settings and the likelihood that only a few of many tens of fires are recorded. At Staces Tarn, a macroscopic charcoal layer in the sediment core is dated at 4644 ± 40 BP (NZA 29676) at 67 cm. Other macroscopic charcoal layers are recognised at 30 cm, 37 cm, 47 cm and 91 cm, and throughout the core there appears to be a stable concentration of microscopic charcoal fragments, indicating persistence of fire through the Holocene, albeit at low levels. Applying a steady rate of sedimentation between the macroscopic charcoal layers suggests fires occurred about 6900, 3750, 2950 and 2400 cal yr BP (rounded to nearest 50 years). For comparisons with previous studies, uncalibrated ages are more useful, and when interpolated from the uncalibrated radiocarbon age, these fires occurred at 6300, 3250, 2100 and 2100 yr BP. Sedimentation rates were previously used to estimate the ages of charcoal layers in a sediment core from Prospect Hill, in the adjacent Rakaia catchment, and yielded ages of c. 5800, 3800, 3500, 2600 and 860 yr BP (Burrows and Russell 1990). They confirm increased fire occurrence in the second half of the Holocene and are consistent with the climatological inferences.

Of significance to this study are dated charcoal fragments from Mount Pyramid (Rodbell 1986), only 3 km northeast of Staces Tarn. Two fire events at this location have been dated, at 5240 ± 110 and 2180 ± 100 yr BP (see Table 1). While the first event is not identified in the sediments at Staces Tarn, the event at ~2180 yr BP may correlate with the charcoal layer estimated to have occurred at 2100 yr BP at Staces Tarn. If this is the case, the fire must have been regionally significant given that it occurred on both sides of the Cameron River and reached high elevations (Figure 1).

Table 1. Buried soil horizons with charcoal from the Lake Heron area (* indicates wood sample, inferred to date a fire)

Site number	Laboratory number	Radiocarbon date (half life 5568 yr)	Location	Grid reference, altitude	Reference
1	NZ 1684	>40,900	Dogs Range, Paddle Hill Creek	J36/525374 1340 m	Harvey 1974
2	Wk 2637*	8880 ± 60	Bush Creek, Lake Stream Valley	J35/587596 660 m	Burrows 1996
3	NZ 6810	6940 ± 150	Dogs Range, Paddle Hill Creek	J36/574358 1045 m	Rodbell 1986
2	Wk 3451*	5910 ± 60	Bush Creek, Lake Stream Valley	J35/587596 660 m	Burrows 1996
4	NZ 3942	5830 ± 130	Washbourne Stream, Prospect Hill	S73/637833 800 m	Burrows and Russell 1990
5	NZ 6803	5240 ± 110	Mount Pyramid, Cameron Valley	J35/559491 1580 m	Rodbell 1986
6	NZA-29676	4644 ± 40	Staces Tarn, Staces Hill	J35/581516 1210 m	This study
5	NZ 6808	2180 ± 100	Mount Pyramid, Cameron Valley	J35/559491 1580 m	Rodbell 1986
7	NZ 1686	1860 ± 70	Upper Paddle Hill Creek	J36/548371 609 m	Harvey 1974
8	NZ 1687	1820 ± 70	Upper Paddle Hill Creek	J36/544392 731 m	Harvey 1974
4	NZ 3941	860 ± 50	Washbourne Stream, Prospect Hill	S73/637833 800 m	Burrows and Russell 1990
9	NZ 3943	620 ± 60	Lochaber Creek, Cameron Valley	J35/511574 1200 m	Burrows et al. 1990

In summary, fire is persistent in these inland basins through much of the Holocene but at a relatively low base level. There is evidence for an increase in the frequency of fire events from about 5000 yr BP and a significant increase about 3000 yr BP. McGlone et al. (1995) reported a similar time for the occurrence of fires in Central Otago. Fires in east Otago are dated to have started slightly later (McGlone and Wilmshurst 1999). Why should this pattern be apparent? As already noted, throughout the Holocene, summer insolation had gradually strengthened. This resulted in increasing summer insolation and enhanced westerly circulation, with thresholds apparently exceeded at about 5000 yr BP when the West Coast glaciers reactivated (e.g. Shulmeister 1999) and about 3000 yr BP when wet podocarp forest occupied Southland (e.g. Markgraf et al. 1992). The stronger insolation and more frequent westerly winds, especially in late summer and autumn, would have dried out the basins and made them susceptible to lightning-strike fires.

Did these fires have an ecological consequence? From the Staces Tarn record, the suggestion is yes. Unlike many areas where grass pollen becomes widespread only in post-Maori times, grass begins to increase steadily from about 5000 years ago. This is unlikely to be a lowering of the tree line for the climatological reasons outlined earlier. Instead, it reflects the gradual opening up and ultimate replacement of the *Phyllocladus* woodland. It is unlikely that this would have occurred without fire to maintain the ecological pressure on the woodlands. The arrival of beech forest may also be a direct effect of the intensified burning regime.

Acknowledgements

Jeremy Pugh would like to thank the landowners, Phillip Todhunter of Lake Heron Station and Eric and Sally Watson of Mount Arrowsmith Station, for giving him permission to carry out this work on their land. Jamie Shulmeister would like to acknowledge the inspirational role of Geoff Hope in the development of both his scholarship and teaching. I am particularly grateful for Geoff showing me, by example, how to run field courses – a source of much enjoyment and numerous graduate students over the years. Age-control and field-work costs were covered by a Mason Trust grant to Jeremy Pugh. We thank Uwe Rieser (VUW) for the luminescence result.

References

Allan, H.H. 1961. *Flora of New Zealand. Vol. 1.* Wellington, Government Printer.

Armstrong, D.L., R.L. Engelbrecht and R.L. Jefferies 2005. *A review of pastoral lease rental and tenure review outcomes associated with pastoral lands throughout the South Island of New Zealand.* Land Information New Zealand.

Baylis, G.T.S. 1980. Mycorrhizas and the spread of beech. *New Zealand Journal of Ecology 3*:151-153.

Bennett K.D. 2002. *Psimpoll 3.10: C Programs for Plotting Pollen Diagrams and Analysing Pollen Data.* Uppsala University, Sweden, 117pp.

Bowden, M.J. 1983. *The Rakaia River and catchment – a resource survey.* Christchurch, New Zealand: North Canterbury Catchment Board and Regional Water Board, (4 vols.). 593pp.

Burrows, C.J. 1977. Forest and scrub flora of the upper Rangitata, Rakaia and Wilberforce Valleys. *Canterbury Botanical Society Journal* 10:1-8.

Burrows, C.J. 1983. Radiocarbon dates from Late Quaternary deposits in the Cass District, Canterbury, New Zealand. *New Zealand Journal of Botany* 21:443-454.

Burrows, C.J. 1988. Late Otiran and early Aranuian radiocarbon dates from South Island localities. *New Zealand Natural Sciences* 15:25-36.

Burrows, C.J. 1996. Radiocarbon dates for Holocene fires and associated events, Canterbury, New Zealand. *New Zealand Journal of Botany* 34:111-121.

Burrows, C.J. and J.B. Russell 1990. Aranuian vegetation history of the Arrowsmith Range, Canterbury. 1. Pollen diagrams, plant macrofossils and buried soils from Prospect Hill. *New Zealand Journal of Botany* 28:323-345.

Burrows, C.J., K.W. Duncan and J.R. Spence 1990. Aranuian vegetation history of the Arrowsmith Range, Canterbury II. Revised chronology for moraines of the Cameron Glacier. *New Zealand Journal of Botany* 28:455-66.

Burrows, C.J. and J.M. Lord 1993. Recent colonisation of Nothofagus Fusca at Cass, Canterbury. *New Zealand Journal of Botany* 31:139-146.

Clark R. 1982. Point count estimation of charcoal in pollen preparations and thin sections. *Pollen et Spores* 24:523-525.

Harvey, M.D. 1974. Soil studies in a high country catchment, Paddle Creek, South Canterbury. Unpublished MAgSc thesis, Lincoln College, University of Canterbury, New Zealand.

Koeppen, W. 1923. Die klimate der Erde Grundriss der Klimakunde. Berlin. Walter de Gruyter (Summarised, in 1937, by the Committee of the Geophysical Association). *Geography* 22(4):252-282.

Kuijt, J. 1969. *The biology of flowering parasitic plants.* Berkeley: University of California Press.

Large, M.F. and J.E. Braggins 1991. *Spore Atlas of New Zealand Ferns and Fern Allies.* Wellington, New Zealand: SIR Publishing 167 pp.

Lintott, W.H. and C.J. Burrows 1973. A pollen diagram and macrofossils from Kettlehole Bog, Cass, South Island, New Zealand. *New Zealand Journal of Botany* 11:269-282.

Mabin, M.C.G. 1980. *Late Pleistocene glacial sequences in the Rangitata and Ashburton Valleys, South Island, New Zealand.* Unpublished Ph.D. thesis, University of Canterbury, Christchurch.

Macphail, M.K. and D.R. McQueen 1983. The value of New Zealand pollen and spores as indicators of Cenozoic vegetation and climate. *Tuatara* 26:37-59.

Markgraf, V., J.R. Dodson, A.P. Kershaw, M.S. McGlone and N. Nicholls 1992. Evolution of late Pleistocene and Holocene climates in the circum-South Pacific land areas. *Climate Dynamics* 6:193-211.

McGlone, M.S. 1988. *New Zealand.* In B.Huntley and T. Webb III (eds) *Vegetation History,* pp557-599. London: Kluwer Academic publishers.

McGlone, M.S. 2001. The origin of the indigenous grasslands of southeastern South Island in relation to pre-human woody ecosystems. *New Zealand Journal of Ecology* 25:1-15.

McGlone, M.S. and J. Wilmshurst 1999. A Holocene record of climate, vegetation change and peat bog development, east Otago, South Island, New Zealand, *Journal of Quaternary Science* 14:239-254.

McGlone, M.S., A.F. Mark and D. Bell 1995. Late Pleistocene and Holocene vegetation history, Central Otago, South Island, New Zealand. *Journal of the Royal Society of New Zealand* 25:1-22.

McQueen, D.R., D.C. Mildenhall and C.J.E. Bell 1968. Paleobotanical evidence for changes in the Tertiary climates of New Zealand. *Tuatara* 16:49-56.

Metcalf, L. 2002. *Trees of New Zealand.* Auckland: New Holland.

Mildenhall, D.C. and M.L. Biryami, 2003. A redescription of *Podosporites parvus* (Couper) Mildenhall emend. Mildenhall and Byrami from the Early Pleistocene, and late extinction of plant taxa in northern New Zealand. *New Zealand Journal of Botany* 41:147-160.

Moar, N.T. 1971. Contributions to the Quaternary history of the New Zealand flora. 6. Aranuian pollen diagrams from Canterbury, Nelson and North Westland, South Island. *New Zealand Journal of Botany* 9:80-145.

Moar N.T. 1993. *Pollen Grains of New Zealand Dicotyledonous Plants.* Lincoln, New Zealand: Manaaki Whenua Press, 200 pp.

Moore P.D., J.A. Webb and M.E. Collinson 1991. *Pollen Analysis.* Oxford, England: Blackwell Scientific Publications, 216 pp.

New Zealand Meteorological Service. 1983. New Zealand Meteorological Service, Temperature Normals 1951-80. *New Zealand Meteorological Service Miscellaneous Publication* 183. Wellington: New Zealand Meteorological Service.

Oliver, P.J. and H.W. Keene 1990. *Sheet J36 BD and part sheet J35 – Clearwater.* Geological map of New Zealand 1:50 000. Wellington: Department of Scientific and Industrial Research.

Pettinga, J., M.D. Yetton, R.J. Van Dissen and G. Downes 2001. Earthquake Source Identification and Characterisation for the Canterbury Region, South Island, New Zealand. *Bulletin of the New Zealand Society for Earthquake Engineering* 34:282-317.

Pocknall, D.T. 1981. Pollen morphology of the New Zealand species of *Dacrydium* Solander, *Podocarpus* L'Heritier, and *Dacrycarpus* Endlicher (Podocarpaceae). *New Zealand Journal of Botany* 19:67-95.

Pugh, J.M. 2008. *The late Quaternary history of the Lake Heron Basin, mid-Canterbury, New Zealand.* Unpublished MSc thesis, University of Canterbury, Christchurch.

Rodbell, D.T. 1986. *The use of lichenometry, rock weathering and soil development to estimate ages of moraines and fluvial terraces in the upper South Branch, 142 Ashburton Valley, South Island, New Zealand.* Unpublished M.Sc. thesis, University of Colorado, Boulder.

Rogers, G.M. and M.S. McGlone 1989. A postglacial vegetation history of the southern central uplands of North Island, New Zealand. *Journal of the Royal Society of New Zealand* 19:229-248.

Russell, J.B. 1980. *Aranuian pollen diagrams from montane Canterbury, New Zealand.* Unpublished Ph.D. thesis, University of Canterbury, Christchurch.

Shulmeister, J.P 1999. Australasian evidence for mid-Holocene climate change implies precessional control of Walker Circulation in the Pacific. *Quaternary International* 57/58:81-91.

Shulmeister, J.P., I. Goodwin, K. Renwick, L. Armand, M.S. McGlone, E. Cook, J. Dodson, P.P. Hesse, P. Mayewski and M. Curran 2004. The Southern Hemisphere westerlies in the Australasian sector over the last glacial cycle: A synthesis. *Quaternary International* 118-119:23-53.

Stuiver, M. and P.J. Reimer 1993. Extended 14C data base and revised CALIB 3.0 14C age calibration program. *Radiocarbon* 35:215-230.

Troels-Smith, J. 1955. Characterization of unconsolidated sediments. *Geological Survey of Denmark* 10:1-73.

Turney, C.S.M., M.S. McGlone and J.M. Wilmshurst 2003. Asynchronous climate change between New Zealand and the North Atlantic during the last deglaciation. *Geology* 31:223-226.

Ullmann, I., O.L. Lange, H. Ziegler, J. Ehleringer, E.-D. Schulze and I.R. Cowan 1985. Diurnal courses of leaf conductance and transpiration of mistletoes and their hosts in central Australia. *Oecologia* 67:577-587.

Wardle, P. 1969. Biological flora of New Zealand. 4. *Phyllocladus alpinus* Hook f. (Podocarpaceae) Mountain Toatoa, Celery Pine. *New Zealand Journal of Botany* 7:76-95.

Wardle, P. 1986. Alpine vegetation of New Zealand: An Ecological Review. In B.A. Barlow (ed) *Flora and Fauna of Alpine Australasia*, pp45-62. Melbourne: CSIRO.

Warren, G. 1967. *Sheet 17-Hokitika. Geological map of New Zealand 1:250,000.* Wellington, New Zealand: Department of Scientific and Industrial Research.

Weaver, A.J, M. Eby, A.F. Fanning and E.C. Wiebe 1998. Simulated influence of carbon dioxide, orbital forcing and ice sheets on the climate of the Last Glacial Maximum. *Nature* 394:847-853.

West, R.G. 1977. *Pleistocene Geology and Biology.* London: Longman Group Limited.

Wilmshurst, J.M., M.S. McGlone, J.R. Leathwick and R.M. Newnham 2007. A pre-deforestation pollen-climate calibration model for New Zealand and quantitative temperature reconstructions for the past 18,000 years BP. *Journal of Quaternary Science* 22:535-547.

6

Last Glacial Maximum habitat change and its effects on the grey-headed flying fox (*Pteropus poliocephalus* Temminck 1825)

J. G. Luly

School of Earth and Environmental Sciences, James Cook University, Townsville, Queensland
jonathan.luly@jcu.edu.au

David Blair

School of Marine and Tropical Biology, James Cook University, Townsville, Queensland

Jennifer G. Parsons

School of Marine and Tropical Biology, James Cook University, Townsville, Queensland

Samantha Fox

School of Marine and Tropical Biology, James Cook University, Townsville, Queensland; Department of Primary Industries and Water, Hobart, Tasmania

Jeremy VanDerWal

School of Marine and Tropical Biology, James Cook University, Townsville, Queensland; Centre for Tropical Biodiversity and Climate Change, James Cook University, Townsville, Queensland

Introduction

The grey-headed flying fox (*Pteropus poliocephalus*) is a large phytophagous bat found in coastal and near-coastal eastern Australia, from Mackay in the north to Geelong in the south (Parsons et al. 2008; Roberts et al. 2008). It is among the best known of the Australian flying foxes. Eby (1991), Parry-Jones and Augee (1991, 1992, 2001), Tidemann and Nelson (2004), McDonald-Madden et al. (2005), Parris and Hazell (2005), and Williams et al. (2006) documented aspects of *P. poliocephalus* ecology. Eby (1991) and Parry-Jones and Augee (2001) focused on movements between colony sites and feeding areas, while Tidemann and Nelson (2004) demonstrated long-distance movements between colonies. Breeding biology was dealt with by Martin et al. (1985), O'Brien (1993), and Fox (2006), and conservation status was summarised by Dickman and Fleming (2002).

The application of molecular tools to flying-fox phylogeny and phylogeography is in its infancy but the approach has obvious potential to solve many of the intractable problems posed to traditional methodologies by this elusive nocturnal species. Among other uses, analyses of genetic diversity can indicate whether a species has undergone changes in population size. In this paper, we place the molecular evidence for population change into a palaeoenvironmental context to gauge the antiquity of any such change and assess its implications for understanding long-term flying-fox responses to contemporary habitat fragmentation and habitat change.

The grey-headed flying fox

The grey-headed flying fox is a species that inhabits coastal forests and woodlands. The northern limit of its range is near Mackay in northern Queensland, and until recently, it was found as far south as Mallacoota in far eastern Victoria (Figure 1). Although a sporadic visitor to Victoria for at least 100 years, a small permanent population became established in Melbourne in 1986. This southwards extension of range was attributed to the development of a warm, humid urban microclimate associated with the growth of the city (Parris and Hazell 2005). Williams et al. (2006) emphasised the capacity of the rich and reliable food sources provided in urban landscapes to sustain a permanent population of flying foxes.

Inland limits for the species are not clearly defined, and *P. poliocephalus* is found episodically from the coast to the western slopes of the Great Dividing Range, coinciding approximately with the 800 mm isohyet (Churchill 2008). The altitudinal range is similarly ill-defined, though migrants pass through the Stanthorpe district of southern Queensland (elevation approximately 870 m) and the flying foxes are occasional summer visitors to Canberra at an elevation of approximately 600 m (D. Hansman pers comm.; J. Luly pers obs.). At present, the core range of the species lies in the tall eucalypt forests of coastal and northern New South Wales (Law et al. 2002).

The grey-headed flying fox is a canopy feeder, consuming pollen, nectar, flowers, fleshy fruit and occasionally leaves (Eby 1998; Williams et al. 2006). Nectar and pollen are most often from sclerophyllous forests (dietary studies indicate a preference for *Eucalyptus* and *Melaleuca*), while the fruit component of the diet tends to be provided by rainforest-related taxa, such as figs, and in the modern setting, urban plantings and orchards. Reliance on nectar and pollen imposes significant constraints on *P. poliocephalus*. Spatial and temporal heterogeneity of flowering in eucalypt-dominated sclerophyllous forests is the norm (Law et al. 2000, 2002), obliging flying foxes to forage widely. Nocturnal foraging may involve round trips of up to 50 km, though most foraging activity occurs within a 20 km radius (or less) of the daytime roost (Eby 1991; Tidemann 1999). Individual bats have been shown to make periodic movements of at least 2000 km, covering more than 4 degrees of latitude over the course of several months (Tidemann and Nelson 2004). Despite a capacity to forage over large distances, the phenology of key food trees imposes annual periods of significant food shortage, especially in late winter or early spring (Law et al. 2002). Species such as *Eucalyptus tereticornis, E. robusta, E. siderophloia* and *Melaleuca quinquenervia* (northern New South Wales) and *E. melliodora* and *E. albens* (western slopes) are key food sources that sustain *P. poliocephalus* through the winter-spring famine and the exigencies of the birthing season when females produce their single young (Law et al. 2002). In poor years, particularly in times of drought, food shortages lead to reduced reproductive success and increased mortality (Eby 1999; Eby et al. 1999). Conversely, *P. poliocephalus*, like other pteropodid bats, is unable to

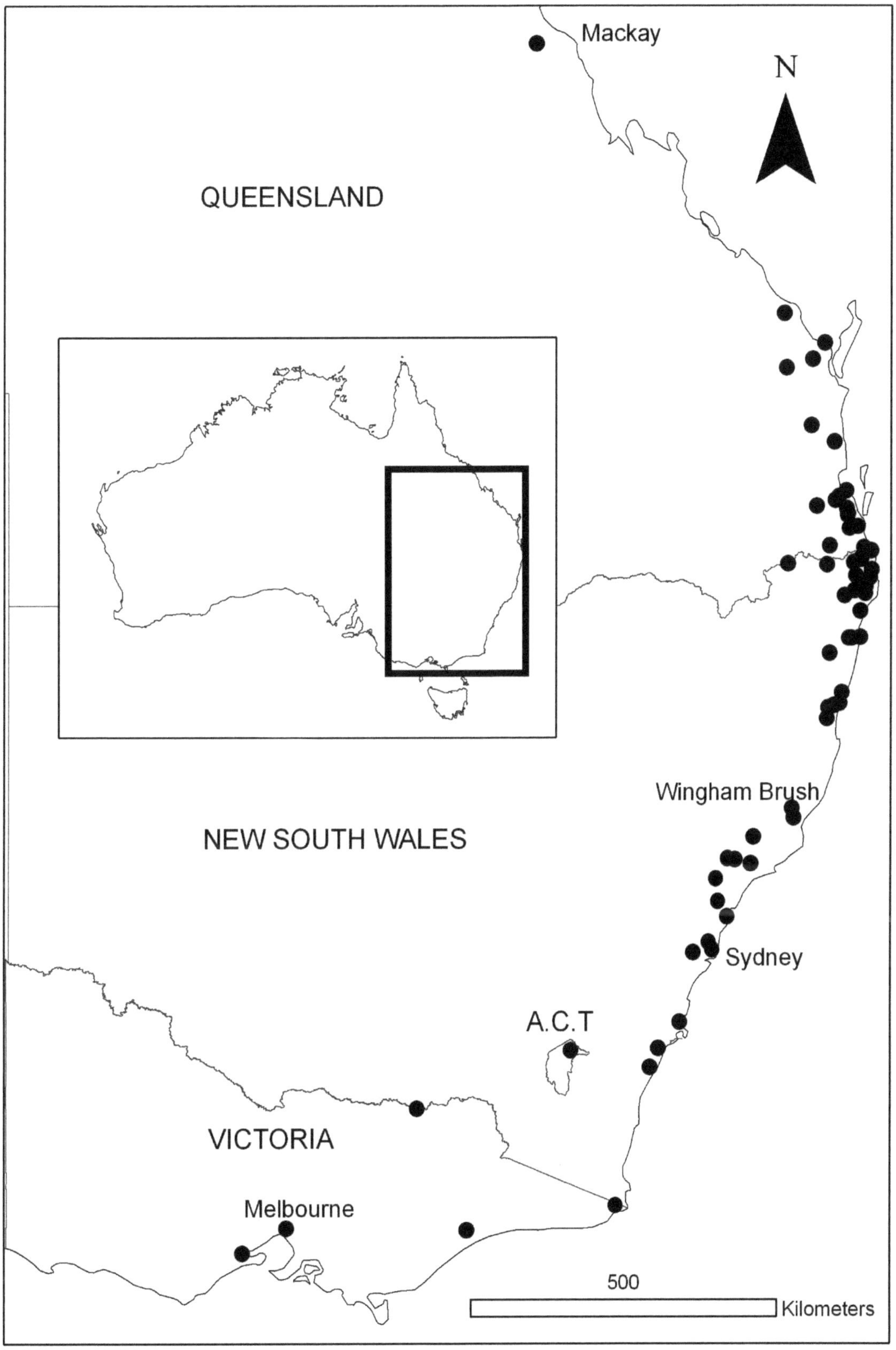

Figure 1. Current distribution of the grey-headed flying fox (*Pteropus poliocephalus*). Sampled localities are marked

increase its reproductive output in times of plenty, though a greater proportion of juveniles may survive their first year. Accounts of opportunistic recruitment (Tidemann 1999), which would allow rapid reproductive response to good seasons or routinely replace young lost early in the breeding season, are mistaken (Fox et al. 2008).

Local populations of *P. poliocephalus*, and other flying foxes are sensitive to fluctuations in environmental conditions. Mass mortality events attributed to excessive heat (Ratcliffe 1932; Welbergen et al. 2008), cold (J. Maclean pers comm. 2008) and starvation (Ratcliffe 1932) have been noted, as well as a number of instances of spontaneous mass abortion of near-term young (Parsons pers obs.; J. Maclean pers comm. 2007). Grey-headed flying-fox populations have a low intrinsic capacity to increase (Martin and McIlwee 2002; McIlwee and Martin 2002) and numbers are presently in decline. Writing in 1931, Ratcliffe suggested flying-fox numbers had decreased by approximately 50% in eastern Australia since European settlement as a result of habitat clearing and, possibly, competition with the black flying fox (*Pteropus alecto*). More recent censuses suggest there has been a 30% decline in *P. poliocephalus* populations in the 10 to 15 years before 2000 (Richards 2000), as a result of clearing of habitat, culling by orchardists, losses to powerlines and barbed-wire fencing, as well as climatic extremes. The decline has almost certainly continued since then. The most recent reliable estimate based on colony census (Eby 2004) suggests the total grey-headed flying-fox population to be approximately 425,000 animals.

These data were sufficient to convince the IUCN, the Commonwealth and the states of New South Wales and Victoria to list *P. poliocephalus* as vulnerable on the IUCN Red List (2008), under the Environment Protection and Biodiversity Act 1999 (Commonwealth), the Threatened Species Conservation Act 1995 (NSW) and the Fauna and Flora Guarantee Act 1988 (Victoria).

In summary, *P. poliocephalus* is an extremely mobile species with a capacity to exploit spatially and temporally variable resources but which experiences significant resource bottlenecks on inter-seasonal and intra-seasonal timescales. The species has a demonstrated sensitivity to clearance of habitat and a low intrinsic capacity to recover from population declines. As a consequence, it is likely to be severely affected by the climatically modulated changes in vegetation predicted for the near-term future. It may well also have suffered severely from habitat change accompanying climatic changes in the past.

Quaternary climate and habitat change

Palaeoclimate

Predicted climatic change and instability is considered to pose a significant threat to a panoply of species in Australia, including flying foxes (Welbergen et al. 2008). Changes anticipated over the next century are relatively small in magnitude and are overlaid on a 10,000-year Holocene history remarkable for its stability and predictability. Before 10,000 BP, stability and predictability were far from the norm and severe climatic stress influenced individual species and the habitats they occupied. The most recent period of extreme climatic stress is at and around the Last Glacial Maximum between 24,000 BP and 19,000 BP (Barrows et al. 2002). This period was the culmination of a slide from warm, interglacial conditions centering on 125,000 BP, to cold, dry climates of a glacial period. The changing climate drove radical alterations in continental area, hydrology, landscape function and biota not experienced in the Holocene.

The magnitude of climate changes during the Pleistocene is estimated through analyses of proxy data and application of numerical models, and the timing estimated with reference to radiometric dating of climate-sensitive geological and geochemical sequences on land and in the ocean. For the most part, there is convergence between the findings of proxy-based and model-based reconstructions of palaeoclimate. Estimates of temperatures at the Last Glacial Maximum range between 6°C and 9°C below current values in inland Australia (Miller et al. 1997), 8°C to 9°C in the Snowy Mountains (Galloway 1965) and about 5°C in the tropical north (Kershaw 1995, though rather equivocally). Proximity to the ocean is likely to have moderated the underlying temperature depression in coastal areas, however sea-surface temperatures were also lower in extra-tropical latitudes than they are today, and models that incorporate the influence of the ocean indicate that reductions of the order of 3°C might have been typical in coastal localities (Hope 2005).

Evidence from lake levels (e.g. Bowler and Hamada 1971; Chen et al. 1993; Bowler et al. 2001; Magee et al. 2004,), speleothems (Ayliffe et al. 1998) sand dunes and dust (Ash and Wasson 1983; Wasson 1983; Hesse 1994; Hesse et al. 2003), and vegetation (Hope 1994; Kershaw 1995) all indicates that cold conditions at the Last Glacial Maximum were accompanied by a drier climate. Quantitative estimates of rainfall change are less amenable to precise delineation than palaeotemperature, as analyses of palaeo-water balance are complicated by interactions between precipitation, temperature, wind velocity and vegetation cover. Wide bands of uncertainty are attached to figures presented in all studies. Estimates of Last Glacial Maximum rainfall based on vegetation suggest effective rainfall in the tropics was lower than modern values by about 40% (Kershaw 1986; Thomas et al. 2001) and 50% or more in southern Australia. Estimates based on closed lake systems, the so-called rain-gauge lakes, tend to be rather higher, 50% reductions being not unusual (Luly 1993; Thomas et al. 2001).

Additional moisture stress was applied to vegetation during Last Glacial Maximum times by reduced levels of atmospheric carbon dioxide (Barnola et al. 1987), which acted to markedly reduce water-use efficiency in C3 plants, notably trees. Farquhar (1997) noted that, with the observed Last Glacial Maximum CO_2 concentration of about 180 ppm, half the pre-industrial concentration of recent times, water-use efficiency would also be half that expected in the latest part of the Holocene. In essence, this means that any given quantum of photosynthetic activity at the Last Glacial Maximum would require twice the water it would at CO_2 levels seen in pre-industrial modern times, adding a layer of physiological drought to the climatic drought imposed at the Last Glacial Maximum.

Last Glacial Maximum vegetation

The fossil record of vegetation at the Last Glacial Maximum is sparse but still presents a clear picture of vegetative responses to the prevailing climate. Most sites preserving a record of Last Glacial Maximum vegetation indicate widespread occurrence of treeless, semi-steppe vegetation, sparse grassland or extremely open cool-climate savanna.

Key sites in, or near, the southern portions of the modern range of the grey-headed flying fox include Lake George (Singh and Geissler 1985), sites in western Victoria (e.g. McKenzie and Kershaw 2000; Cook 2009), Bass Strait (Hope 1978; D'Costa et al. 1993; Harle 1997) and the Pilliga Scrub (Dodson and Wright 1989). In the northern portion of the grey-headed flying-fox range, sites on the Bunya Mountains (Moravek unpubl.) and Fraser Island (Longmore 1997; Longmore and Heijnis 1999) also demonstrate radical reductions in tree cover at and immediately following the Last Glacial Maximum. Severe thinning (cover reduced to 20-40% of modern values) of vegetation around that time can also be inferred from formation or re-activation of linear dunes in places, such as the Blue Mountains, East Gippsland and Tasmania, that are currently stabilised by forest (Ash and Wasson 1983; Wasson 1983; Hesse et al. 2003). The sparsity of eucalypts in Last Glacial Maximum landscapes is well illustrated by mean percentages of eucalypt pollen in all samples from southeastern Australian sites at selected time slices (Kershaw 1995). Eucalypt pollen comprises about 42% in samples from pre-European modern-day vegetation and a mere 7% in the Last Glacial Maximum flora. These data, in the sense of indicating sensitivity of eucalypts to climatic change, are consistent with the results of bioclimatic modelling presented by Hughes et al. (1996), who argued that eucalypts will be significantly affected by the relatively minor temperature increases posited for an anthropogenically warmed greenhouse climate.

It is self-evident that areas of refugial woody vegetation persisted throughout the last glacial cycle, otherwise woody vegetation and the organisms depending on it would now be extinct. Sites which appear to have been close to arboreal refugia, such as Lake Baraba (Black et al. 2006) or Boulder Flat (Kenyon 1989), rarely exhibit an abundance of pollen from eucalypts or other food plants used by *P. poliocephalus*. That said, the sporadic occurrence of low-altitude populations of montane species such as *Eucalyptus pauciflora* in contemporary Australia, and the mixed ecological affinities of vertebrate fossils found in Last Glacial Maximum sites lead Hope (1994), with support from Kershaw (1995) and Hope et al. (2005), to suggest that Australia then exhibited a complex mosaic of vegetation occupying locally protected sites on a generally low-relief landscape, rather than a wholesale downslope movement of altitudinally zoned vegetation such as in Papua New Guinea, Europe and the United States. This messy series of patches could be exploited by a highly mobile animal like the grey-headed flying fox, allowing survival but not prosperity.

The continental shelf

The Last Glacial Maximum coincided with a fall in sea level of 120-130 m (Yokoyama et al. 2001), exposing most of the continental shelf. Extensive areas were exposed on the Queensland Plateau in the north of the modern range of *P. poliocephalus* and in what is now Bass Strait in the south. Between these areas, the shelf to the 100 m isobath is rarely wider than 20 km, and Last Glacial Maximum low-sea-level stands would not provide the space needed for extensive forests capable of sustaining a large number of flying foxes. The vegetation of the Bass Strait sector of the shelf was open, comprising grassland, shrub-heath and communities of halophytes (Hope 1978; D'Costa et al. 1993; Harle 1997), none being of much value for the grey-headed flying fox. There is less evidence to reconstruct the vegetation of the Queensland Plateau at low-sea-level times, however long pollen records from Fraser Island (Longmore 1997; Longmore and Heijnis 1999) indicate that although edaphic and hydrological conditions peculiar to sand mass sites influence vegetation responses to climate, tree-dominated vegetation contracted markedly at Last Glacial Maximum times, to be replaced by the shrub-dominated 'wallum' vegetation characteristic of older dune landscapes of the modern island. Eastwards from there, information is scarce, but the trend noted from Fraser Island and places to the north (Torgersen et al. 1988; Moss and Kershaw 2000; Chivas et al. 2001) suggests the prospect is remote that extensive forested habitat suitable for use by the grey-headed flying fox existed on the exposed shelf.

Molecular evidence of population history

The application of molecular genetic methods to population biology in modern and ancient environments is a recent and rapidly developing field (e.g. Hofreiter and Stewart 2009). Past changes in population size leave signatures in the genome that can be detected by analysis of, for example, alignments of mitochondrial d-loop DNA sequences from many individuals. Here we analyse mitochondrial sequences to investigate past population sizes in *P. poliocephalus* and seek to correlate this with habitat and environmental histories.

Methods

DNA used in this investigation was extracted from 61 grey-headed flying foxes killed by hail and heat stress at Wingham Brush about 250 km north of Sydney and 55 live animals captured at the Royal Botanic Gardens in Sydney (Divljan 2008). DNA was extracted from tissue cut from the plagiopatagium (Worthington Wilmer and Barratt 1996), using the EDNA HiSpEx tissue extraction kit, following manufacturer's instructions. A portion of the mitochondrial d-loop was amplified using PCR primers developed for *Pteropus vampyrus* by Kevin Olival (Columbia University, New York). Sequencing protocols are presented in Fox et al. (In prep.).

Panmixia (essentially, unrestricted gene flow throughout the range of the species) in the grey-headed flying fox was inferred by Webb and Tidemann (1996) and Fox et al. (In prep.), and as no evidence of restricted gene flow is evident in mitochondrial sequence data (Fox et al. In prep.), samples from Wingham Brush and Sydney are considered to come from a single population. Accordingly, genetic data from both sites were combined in our analyses.

Signatures of changes in population size can be detected in the pattern of pairwise differences among mitochondrial sequences (Rogers and Harpending 1992). A plot of the frequencies of all pairwise differences (mismatch distribution) forms a smooth unimodal curve in populations that have undergone recent growth but not in populations that have been static for long periods or have declined. Parameters were estimated in Arlequin (Excoffier et al. 2005) under the sudden expansion model (null model). Arlequin is a software package implementing numerous methods for population genetic data analysis. The observed pairwise mismatch distribution was plotted, along with its simulated value under this scenario, and the upper and lower 95% bounds (20,000 replicates) (Figure 2). The sum of squared deviation (deviation of simulated from observed – SSD) was calculated and tested against that expected under the sudden expansion model. A related statistic, the raggedness index (Harpending 1994), was also calculated. Its significance is tested in a manner similar to that of the SSD. Values for tau (τ), can be used to estimate the time at which a population started to expand using the equation $\tau = 2ut$, where t is the number of generations since the crest of the expansion and u is the mutation rate per generation per sequence (Rogers and Harpending 1992).

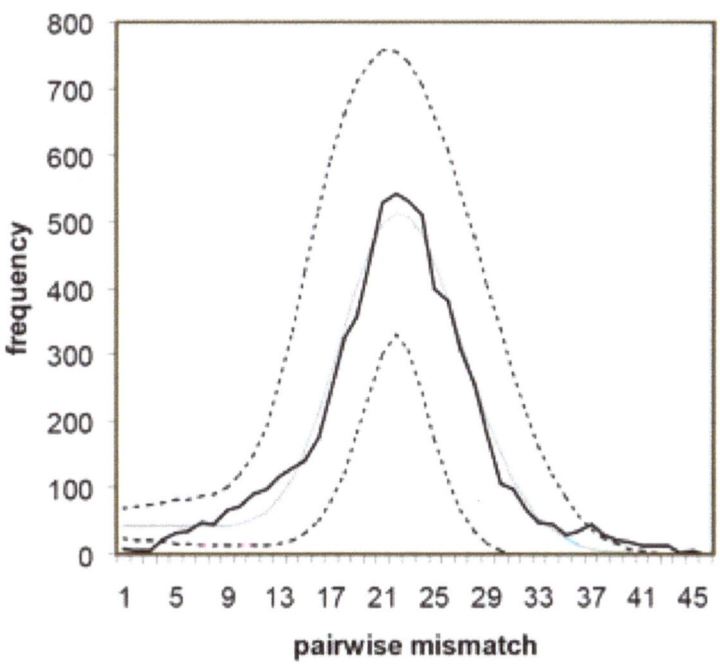

Figure 2. Pairwise mismatch distributions calculated from all available GHFF sequences (solid black line). Upper and lower dashed lines represent the upper and lower 95% limits based on 2000 replicates in Arlequin. The model frequency is the grey line close to the line for the observed frequencies

Demographic histories can also be estimated from genealogies (phylogenies) in a Bayesian statistical framework using the program BEAST v1.4.6 (Drummond and Rambaut 2007: http:// beast.bio.ed.ac.uk). Given a set of aligned sequences and a model for sequence evolution, this program samples phylogenetic trees from which simultaneous estimates of a number of demographic and evolutionary parameters are calculated. Runs of MODELTEST (Posada and Crandall 1998) indicated that the HKY+G+I model was the appropriate description of nucleotide substitution to use with BEAST. This model, first developed by Hasegawa et al. (1985), accommodates variation in substitution rate along the length of a sequence and permits a proportion of sites to be invariant. BEAST runs were of sufficient length (typically 30 million or more) to ensure effective sample sizes (ESSs) were always over 100, and usually substantially over this value. Bayesian skyline plots (Drummond et al. 2005) generated by BEAST show changes in effective population size over time, along with credibility intervals. This approach avoids the need to choose a single demographic scenario such as 'constant population size' or 'exponential growth' from the options available in BEAST, and minimises the role of subjective judgement in setting analytical parameters.

The alignment of the 116 sequences was 535 nucleotides in length. Results of all molecular analyses are consistent with expansion in populations of *P. poliocephalus*. For the pairwise-mismatch data, the sum of squared deviation (SSD) value was 0.000584 ($P = 0.768$), indicating no departure from the unimodal curve that is consistent with population expansion. The analogous Harpending's raggedness index was 0.0014 ($P = 0.9023$), also consistent with molecular response to population expansion.

Solution of the equation $\tau = 2ut$ requires that one knows either the time since expansion or the mutation rate. Since we know neither with certainty, we explored several possibilities. In light of the palaeoenvironmental data summarised above, we take one very likely starting point for a population expansion to be soon after the culmination of the Last Glacial Maximum at about 18,000 BP. The value for τ estimated in ARELQUIN under the sudden expansion model was 21.887 (confidence interval when $\alpha = 0.05$, 16.568 – 24.664). The estimated mutation rate (μ) under this scenario is within the range 86%-128% substitutions per site per million years. Other estimates of mutation rates for the d-loop region of bats are few. Chen et al. (2006) proposed a rate of 20% per site per million years for the d-loop of *Rhinolophus monoceros* (Microchiroptera: Rhinolophidae). Applying this to the mitochondrial data from *P. poliocephalus* gives a time since expansion between 75,562 and 113,974 years BP, a period early in the most recent glacial cycle. Expansion from these dates would imply, quite unreasonably, that Last Glacial Maximum declines in habitat extent and quality had no effect on the bat population.

A Bayesian skyline plot was constructed to show the changes in population size over time (Figure 3). The plot shows effective population size (multiplied by generation time in years) (the Y-axis) and the associated upper and lower 95% confidence values (highest posterior densities). Plots of mean and median values are very similar in appearance: the median is shown. The time scale on the X-axis in Figure 3 is inferred using the mutation rate estimated assuming population expansion after the Last Glacial Maximum. In this scenario, a marked expansion began at about 10,000 years BP. If a different mutation rate were to be specified, the time scale on the X-axis would be scaled accordingly. For example, if the rate estimated by Chen et al. (2006) of 20% per site per million years is used, the X-axis would cover about five times the time span shown in Figure 3 and the marked expansion would be pushed back to more than 50,000 years ago. No anthropogenic reduction in *P. poliocephalus* populations can be detected in Figure 3, as the Bayesian skyline plot is derived from phylogenies which require sequences to have mutated during the time scale in question. Mutations presumably have been too few to be detectable during the ~200 years since European settlement.

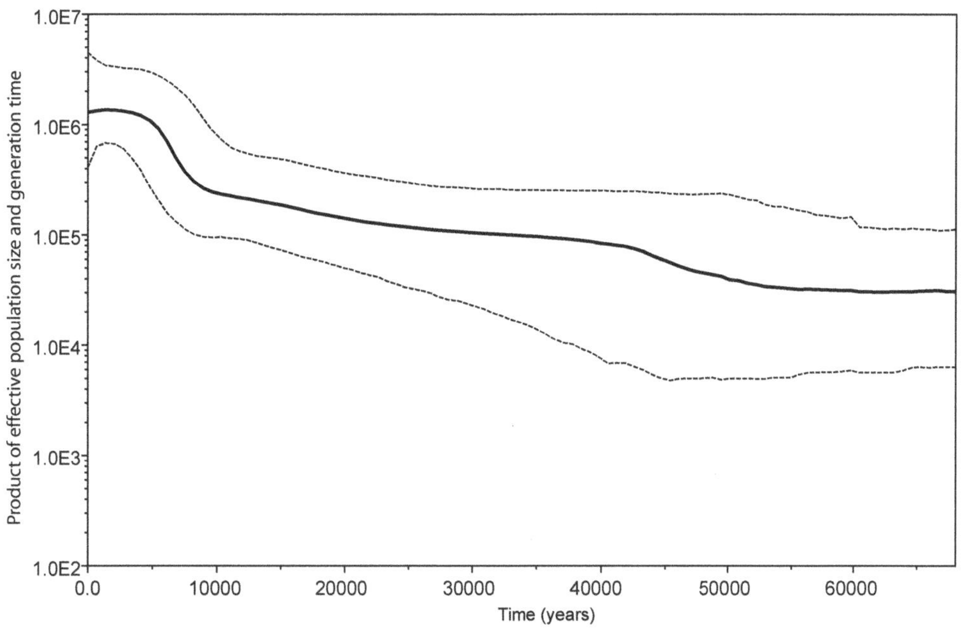

Figure 3. Bayesian skyline plot: Population size change over time is indicated (median value is solid thick line), with 95% HPD intervals (dashed lines). The X-axis shows time in years back to a specified cut-off (in this case the mean of the root height). Time scales are dependent on mutation rate: the scale shown assumes a mutation rate of 1.06e-6. The Y-axis (note logarithmic scale) indicates effective population size estimates multiplied by generation time

Bioclimatic modelling

Climate models developed for modern-day grey-headed flying-fox camps in Australia were projected on to global climate models for the Last Glacial Maximum using the maximum entropy algorithm Maxent (Phillips et al. 2006), to predict the extent of habitat capable of supporting colonies of *P. poliocephalus* at the Last Glacial Maximum. Maxent is a presence-only species-distribution modelling approach, for which locations of *P. poliocephalus* roosts were gleaned from the literature (Parris and Hazell 2005) and unpublished data (Parsons et al. In prep.). The environmental correlates used in the models included annual mean temperature, temperature seasonality, maximum temperature of warmest period, minimum temperature of coldest period, temperature annual range, annual precipitation, precipitation seasonality, precipitation of wettest quarter, and precipitation of driest quarter. The surfaces for modern conditions were derived using Anuclim 5.1 software (McMahon et al. 1995) and a nine second digital elevation model (Version 2; Geoscience Australia http://www.ga.gov.au/). We assumed the glacial maximum sea level to be 100 m below the present level, and maps (Figures 4a, 4b) show the area of continental shelf landward of the 100 m isobath.

Maxent builds the models, contrasting where a species occurs with the 'background' or pseudo-absences, i.e. presence is definite, whereas 'absence' may reflect true absence or gaps in the data. In this study, 10,000 background points were drawn at random from localities within 1000 km of known colonies to minimise regional biases in the models (Van Der Wal et al. 2009). Further description of model parameterisation can be found in Parsons et al. (In prep.). Once built, the models were projected on to two global climate models (CCSM and MIROC) for the Last Glacial Maximum (~21,000 BP) at 5km resolution. These surfaces were downloaded from Worldclim-1 (www.worldclim.org) and are to this extent independent of the Last Glacial Maximum climatic changes summarised from the Australian palaeoenvironmental literature. As differences between predictions yielded by the two Last Glacial Maximum GCM models were minimal, we chose only to present the results from CCSM, as this gives the more conservative estimate of past distribution (Figure 4a). Maxent thresholds were set at 0.01 for all analyses. Estimates of abundance or carrying capacity relative to current abundance were made by summing the suitability values of the past predictions and contrasting these with the present modelled suitability, as per Van Der Wal et al. (2009).

Perhaps unsurprisingly, given that the models were constructed using the modern distribution of *P. poliocephalus* overlaid on a modern climate surface, they describe the current

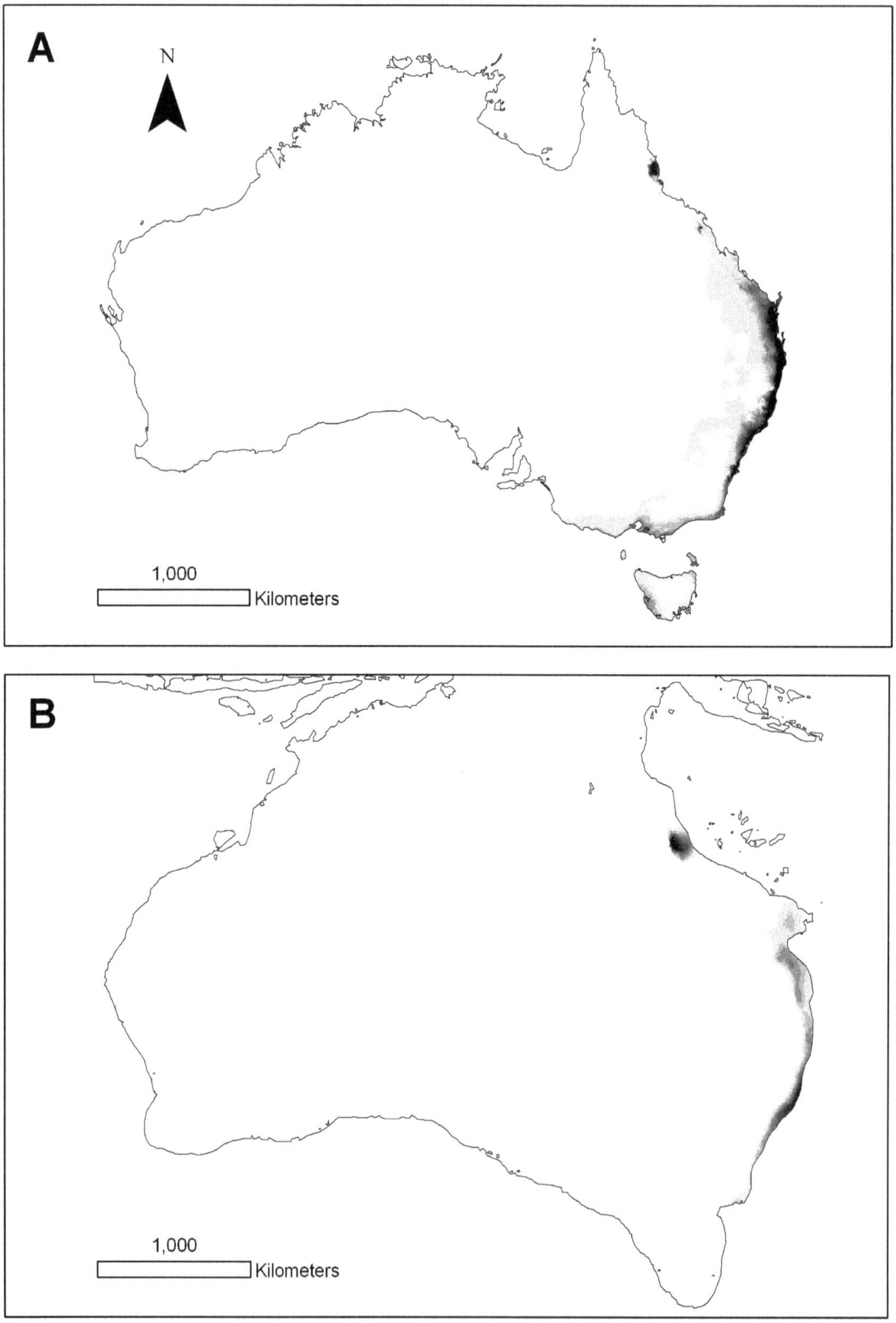

Figure 4. Maxent model of suitable environmental space for grey-headed flying foxes in the current climate (A) using BIOCLIM and at the Last Glacial Maximum (B) using CCSM (Worldclim). Darker areas indicate more suitable environmental conditions

climatic circumstances of the bat well. There is an outlier of climatically suitable habitat identified on the Atherton Tablelands, approximately 700 km north of the most northerly known *P. poliocephalus* colony at Finch Hatton near Mackay. Though anecdotal accounts suggest that very occasional individual *P. poliocephalus* have been sighted in that region (Parsons pers comm.), the species is not resident there. The most parsimonious explanation for this discrepancy – always assuming that the model is generally applicable – lies in the interactions between *P. poliocephalus* and the three other species of flying fox (*Pteropus conspicillatus*, *P. alecto* and *P. scapulatus*), which occupy a common competitive space. All three species use the same resource base as *P. poliocephalus*, though *P. conspicillatus* and *P. alecto* are more generalist feeders than the nectar-dependent *P. poliocephalus*. *Pteropus scapulatus* is also a nomadic nectarivore but ranges much further inland and north than the other species (Hall and Richards 2000), and it is likely that the most direct competition is between *P. poliocephalus* and *P. alecto*. Some authors argue that the apparent southwards range extension of *P. alecto* in the 20th century has been at the expense of *P. poliocephalus* (Eby et al. 1999), suggesting that *P. poliocephalus* would find it difficult to colonise an area if *P. alecto* is present. Under these circumstances, successful establishment of a disjunct population of *P. poliocephalus* so distant from the core range of the species seems to be vanishingly unlikely, even if the habitat is suitable.

The CCSM climate models centred on 21,000 BP suggest that habitat for *P. poliocephalus* decreased radically, to some 33% of its modern extent, and the carrying capacity of the habitat declined even more spectacularly, to 16% of the modern-day figure (Figure 4b). The core glacial maximum distribution of *P. poliocephalus* remains focused in central New South Wales, with the envelope of usable bioclimate shifting eastwards on to the continental shelf. In southern Australia, this opens little additional habitat as the shelf is narrow. In southern Queensland, where the shelf widens substantially, modelled habitat exists but is not of high ecological value for the species. The patch of potential habitat available on the Atherton Tablelands remains. However, the gap to be surmounted by any putative colonisation by *P. poliocephalus* is even wider than at present.

Discussion

Each of the three avenues employed to investigate the likely population history of *Pteropus poliocephalus* has its own set of assumptions and foibles. Simple conclusions drawn from interpretations of the palaeoenvironmental record are vulnerable to uncertainties related to the biology of the creature in question. This difficulty is well illustrated by the debates that lie behind extinction of the Australian megafauna and the small mammal fauna extirpated during the tenure of European dominance of the continent. In each instance, incomplete appreciation of ecology, breeding biology and habitat requirements hinders efforts to discriminate between the many alternative explanations advanced for their demise.

Similarly, too-literal adherence to conclusions drawn from bioclimatic modelling may yield internally consistent results that are not consonant with what is known of the palaeoenvironmental record for the species or region of interest. A splendid example is provided by Graham et al. (2006), who improved the explanatory power of models seeking to explain the distribution of biodiversity in the wet tropics of Queensland by incorporating Bioclim-generated layers of historical vegetation to their analyses. The modelled extent of historical vegetation at the Last Glacial Maximum is considerably greater than that reconstructed by palaeoenvironmental methods, in part because focus on climatic controls on vegetation do not capture the effects flowing from pre-European human influence on vegetation and landscape. In our study, modelled habitat potentially suitable for *P. poliocephalus* occurs in the wet tropics region, but there is absolutely no reason to believe that the purported habitat is of use to the species now or has been in the past.

Finally, the application of molecular methods to historical questions is an innovation adopted with great enthusiasm across a range of disciplines and for a diversity of purposes. The analytical methods are rigorous but interpretation of results retains elements of the subjectivity which characterises analyses of palaeoenvironments by more traditional means. Our analysis of the timing of expansion in populations of *P. poliocephalus* is based on a mutation rate that is high by accepted standards in the field (although comparable with recent estimates for some other vertebrates, e.g. Hay et al. 2008), but which yields results that are consistent with conclusions drawn from modelling and evidence of habitat change drawn from the fossil record. Application of more conservative mutation rates yields ages of expansion that are believable on molecular grounds but cannot be sustained in light of other lines of evidence.

The population history of *P. poliocephalus* is unlikely to be unique to this species. Hints of a Last Glacial Maximum genetic bottleneck can be detected in the spectacled flying fox (*Pteropus conspicillatus*; Fox 2006) and it would be most surprising if other species were not similarly affected. Flying foxes and other volant species are well equipped to traverse habitat discontinuities and appear to be able to persist in small numbers for significant periods of time by exploiting habitat patches. The density-dependent ecological roles played by flying foxes (McConkey and Drake 2006) may have been significantly disrupted during periods of population contraction and provide an additional, hitherto unsuspected, stressor on community function during glacial episodes.

Conclusions

Climatic events that culminated at the Last Glacial Maximum produced catastrophic declines in the area of habitat available to the grey-headed flying fox and other species dependent on wooded landscapes, but the effects of these changes on flying-fox populations are conjectural, as the fossil record of all but a very select few species is entirely absent. Molecular explorations of the population history of *Pteropus poliocephalus* suggest that it has undergone a major expansion; the timing of the expansion cannot be determined with confidence on molecular grounds alone, but in combination with palaeoenvironmental reconstructions and bioclimatic modelling, the timing of likely changes in population can be refined and the extent of population changes quantified. During the Last Glacial Maximum, *P. poliocephalus* numbers may have fallen to as few as 16% of the modern population, or approximately 68,000 individuals. The recovery following the Last Glacial Maximum is likely to have coincided with the recovery of forested habitat from about 10,000 BP.

In modern times, *P. poliocephalus* is undergoing a rapid decline in population as a result of human actions. The limited intrinsic capacity to recover from population declines, coupled with the complexity of the causes of contemporary declines and the unpopularity of bats in general and flying foxes in particular, makes *P. poliocephalus* a likely victim of the application of 'ecological triage' (*sensu* Walker 1992) when it comes to allocation of scarce conservation resources to stem the decline. Our results show that the species has a capacity for long-term survival in fragmented landscapes and give some cause to hope that *P. poliocephalus* may have the genetic resilience to persist at low numbers and to recover if given sufficient time.

Acknowledgements

J.P. thanks Townsville Airport Pty Ltd, Qantas and Virgin Blue for financial and logistical support.

References

Ash, J. E. and R.J. Wasson 1983. Vegetation and sand mobility in the Australian dunefield. *Zeitschrift für Geomorphologie N.F., Supplement Band* 45:7-25.

Ayliffe, L.K., P.C. Marianelli, K.C. Moriarty, R.T. Wells, M.T. McCulloch, G.E. Mortimer and J.C. Hellstrom 1998. 500 ka precipitation record from southeastern Australia: Evidence for interglacial relative aridity. *Geology* 26:147-150.

Barnola, J.M., D. Raynaud, Y.S. Koretkevich and C. Lorius 1987. Vostok ice core provides 160,000 year record of atmospheric CO_2. *Nature* 329:408-414.

Barrows, T.T., J.O. Stone, L.K. Fifield and R.G. Cresswell 2001. Late Pleistocene glaciation of the Kosciuszko massif, Snowy Mountains, Australia. *Quaternary Research* 55:179-189.

Barrows, T.T., J.O. Stone, L.K. Fifield and R.G. Cresswell 2002. The timing of the last glacial maximum in Australia. *Quaternary Science Reviews* 21:159-173.

Black, M.P., S.D. Mooney and H.A. Martin 2006. A >43,000-year vegetation and fire history from Lake Baraba, New South Wales, Australia. *Quaternary Science Reviews* 25:3003-3016

Bowler, J.M. and T. Hamada. 1971. Late Quaternary Stratigraphy and Radiocarbon Chronology of Water Level Fluctuations in Lake Keilambete, Victoria. *Nature* 232: 330-332.

Bowler, J.M., K-H. Wyrwoll and L. Yanchuou 2001. Variations of the northwest Australian summer monsoon over the last 300,000 years: the paleohydrological record of the Gregory (Mulan) Lakes System. *Quaternary International* 83-85:63-80.

Brereton, R., S. Bennett and I. Mansergh 1995. Enhanced greenhouse climate change and effects on selected fauna of south-eastern Australia: A trend analysis. *Biological Conservation* 72:339-354.

Catterall, C. and M. Kingsford 1993. *Remnant bushland of south-east Queensland in the 1990s: its distribution, loss, ecological consequences and future prospects.* Brisbane: Institute of Applied Environmental Research, Griffith University and Brisbane City Council.

Chen, X.Y., J.M. Bowler and J.W. Magee 1993. Late Cenozoic stratigraphy and hydrologic history of Lake Amadeus, a central Australian playa *Australian Journal of Earth Sciences* 40:1-14.

Chen, S.F., S.J. Rossiter, C.G. Faulkes and J. Jones 2006. Population genetic structure and demographic history of the endemic Formosan lesser horseshoe bat (*Rhinolophus monoceros*). *Molecular Ecology* 15:1643-1656.

Chivas, A.R., A. García, S. van der Kaars, M.J.J. Couapel, S. Holt, M.R. Reeves, D.J. Wheeler, A.D. Switzer, C.V. Murray-Wallace, D. Banerjee, D.M. Price, S.X. Wang, G. Pearson, N.T. Edgar, L. Beaufort, P. De Deckker, E. Lawson and C.B. Cecil 2001. Sea-level and environmental changes since the last interglacial in the Gulf of Carpentaria, Australia: an overview. *Quaternary International* 83-85:19-46.

Churchill, S.K. 2008. *Australian Bats*. Second edition. Sydney: Jacana Books, 255pp.

Collins, L. 1999. Wildlife caring and conservation: the effect of drought on the flying-foxes of the New South Wales north coast. *Australasian Bat Society Newsletter* 12:31-32.

Cook, E. 2009. A record of late Quaternary environments at lunette-lakes Bolac and Turangmoroke, Western Victoria, Australia, based on pollen and a range of non-pollen palynomorphs. *Review of Palaeobotany and Palynology* 153:185-224.

Crucifix, M., R.A. Betts and C.D. Hewitt 2005. Pre-industrial-potential and Last glacial maximum global vegetation simulated with a coupled climate-biosphere model: diagnosis of bioclimatic relationships. *Global and Planetary Change* 45:295-312.

D'Costa, D., J. Grindrod and R. Ogden 1993. Preliminary environmental reconstructions from Late Quaternary pollen and mollusc assemblages at Egg Lagoon, King Island, Bass Strait. *Australian Journal of Ecology* 18:351-366.

Dickman, C. and M. Fleming 2002. Pest or passenger pigeon? The New South Wales Scientific Committee's assessment of the status of the Grey-headed flying-fox. In P. Eby and D. Lunney (eds) *Managing the Grey-headed Flying-fox as a Threatened Species*, pp20-28. Sydney: Royal Zoological Society of New South Wales.

Divljan, A. 2008. Population ecology of the grey-headed flying-fox, *Pteropus poliocephalus*: a study of the age structure and the effects of mortality on a vulnerable species. Unpublished PhD thesis, Australia: University of Sydney.

Dodson, J.R. and R.V.S. Wright 1989. Humid to arid to subhumid vegetation shift on Pilliga Sandstone, Ulungra Springs, New South Wales. *Quaternary Research* 32:182-192.

Drummond, A.J. and A. Rambaut 2007. BEAST: Bayesian evolutionary analysis by sampling trees. *BMC Evolutionary Biology* 7:214.

Drummond, A.J., A. Rambaut, B. Shapiro and O.G. Pybus 2005. Bayesian coalescent inference of past population dynamics from molecular sequences. *Molecular Biology and Evolution* 22:1185-1192.

Duncan, A., G.B. Baker and N. Montgomery 1999. Action Plan for Australian Bats. *http://www.environment.gov.au/biodiversity/threatened/publications/action/bats/index.html*

Eby, P. 1991. Seasonal movements of grey-headed flying-foxes, *Pteropus poliocephalus* (Chiroptera: Pteropodidae), from two maternity camps in northern New South Wales. *Wildlife Research* 18:547-559.

Eby, P. 1991. Low reproductive output in grey-headed flying-fox associated with a short period of food scarcity. *Australasian Bat Society Newsletter* 14:17-20.

Eby, P. 1998. An analysis of diet specialization in frugivorous *Pteropus poliocephalus* (Megachiroptera) in Australian subtropical rainforest. *Australian Journal of Ecology* 23:443-456.

Eby, P., G. Richards, L. Collins and K. Parry-Jones 1999. The distribution, abundance and vulnerability to population reduction of a nomadic nectarivore, the grey- headed flying-fox *Pteropus poliocephalus* in New South Wales during a period of resource concentration. *Australian Zoologist* 31:240-243.

Eby, P. 2004. *National count of grey-headed flying foxes April 3 & 4, 2004.* A report to the Department of Environment and Heritage, Canberra.

Excoffier, L., G. Laval and S. Schneider 2005. ARLEQUIN (version 3.0): an integrated software package for population genetics data analysis. *Evolutionary Infomatics Online* 1:47-50.

Farquhar, G.D. 1997. Carbon dioxide and vegetation. *Science* 278:1411-1413.

Field, J.H., J.R. Dodson and I.R. Prosser 2002. A Late Pleistocene vegetation history from the Australian semi-arid zone. *Quaternary Science Reviews* 21:1023-1037.

Fox, S.J. 2006. *Population structure in the spectacled flying-fox,* Pteropus conspicillatus*: a study of genetic and demographic factors.* Unpublished PhD thesis, School of Marine and Tropical Biology, James Cook University, Queensland, Australia.

Fox, S.J., K. Parry-Jones, D. Blair, J.G. Luly, A. Divjlan and G. Wardle, In prep. Panmixia in the grey-headed flying-fox in New South Wales inferred using mitochondrial and nuclear markers.

Fox, S.J., J.G. Luly, C.B. Mitchell, J. Maclean and D.A. Westcott 2008. Demographic indications of decline in the spectacled flying-fox (*Pteropus conspicillatus*) on the Atherton Tablelands of northern Queensland. *Wildlife Research* 35:417-424.

Fox, S., H. Spencer and G. O'Brien 2008. Analysis of twinning in flying-foxes (Megachiroptera) reveals superfoetation and multiple paternity. *Acta Chiropterologica* 10:271-278.

Galloway, R.W. 1965. Late Quaternary climates in Australia. *Journal of Geology* 73:603-618.

Graham, C.H., C. Moritz and S.E. Williams 2006. Habitat history improves prediction of biodiversity in rainforest fauna. *Proceedings of the National Academy of Sciences USA* 103:632636.

Hall, L. and G. Richards 2000. *Flying-foxes. Fruit and blossom bats of Australia*. Sydney: University of New South Wales Press. 135pp.

Harle, K.J. 1997. Late Quaternary vegetation and climate change in southeastern Australia: palynological evidence from marine core E55-6. *Palaeogeography, Palaeoclimatology, Palaeoecology* 131:465-483.

Harpending, H.C. 1994. Signature of ancient population growth in a low-resolution mitochondrial mismatch distribution. *Human Biology* 66:591-600.

Hasegawa, M., H. Kishino and Y. Yano 1985. Dating of the human-ape splitting by a molecular clock of mitochondrial DNA. *Journal of Molecular Evolution* 22:160-174.

Hay, J.M., S. Subramanian, C.D. Millar, E. Mohandesan and D.M. Lambert 2008. Rapid molecular evolution in a living fossil. *Trends in Genetics* 24:106-109.

Hesse, P., G.S. Humphreys, P.M. Selkirk, D.A. Adamson, D.B. Gore, D.C. Nobes, D.M. Price, J.-L. Schwenninger, B. Smith, M. Tulau and F. Hemmings 2003. Late Quaternary aeolian dunes on the presently humid Blue Mountains, Eastern Australia. *Quaternary International* 108:13-32.

Hesse, P. P. 1994. The record of continental dust from Australia in Tasman Sea sediments. *Quaternary Science Reviews* 13:257-272.

Hofreiter, M. and J. Stewart 2009. Ecological change, range fluctuations and population dynamics during the Pleistocene. *Current Biology* 19:R584-R594.

Hope, G.S. 1978. The Late Pleistocene and Holocene vegetational history of Hunter Island, northwestern Tasmania. *Australian Journal of Botany* 26:490-514.

Hope, G.S. 1994. Quaternary vegetation. In R. Hill (ed.) *Australian Vegetation: Cretaceous to Recent* pp368-389. Cambridge UK: Cambridge University Press.

Hope, G.S., A.P. Kershaw, S. van der Kaars, X-J. Sun, P.-M. Liew, L.E. Heusser, H. Tahahara, M. McGlone, N. Miyoshi, and P.T. Moss 2005. History of vegetation and habitat change in the Austral-Asian region. *Quaternary International* 118-119:103-126.

Hope, P. 2005. *The Weather and Climate of Australia at the Last glacial maximum*. Unpublished PhD Thesis, School of Earth Sciences, University of Melbourne, 266pp.

Hughes, L., E.M. Causey and M. Westoby 1996. Climatic range of *Eucalyptus* species in relation to future climate change. *Global Ecology and Biogeography Letters* 5:23-29.

Kenyon, C.E. 1989. *A late Pleistocene and Holocene pollen diagram from Boulder Flat, east Gippsland.* Unpublished BSc (Hons.) thesis, Department of Botany, Monash University.

Kershaw, A.P. 1986. Climatic-change and aboriginal burning in Northeast Australia during the last two glacial interglacial cycles. Nature 322:47-49.

Kershaw, A.P. 1995. Environmental change in greater Australia. *Antiquity* 69:656-675.

Law, B., L. Mackowski, L Schoer and T. Tweedie 2000. Flowering phenology of myrtaceous trees and their relation to climatic, environmental and disturbance variables in northern New South Wales. *Austral Ecology* 25:165-178.

Law, B., P. Eby and D. Somerville 2002. Tree planting to conserve flying-foxes and reduce orchard damage. In P. Eby and D. Lunney (eds) *Managing the Grey-headed Flying-fox as a Threatened Species*, pp84-94. Sydney: Royal Zoological Society of New South Wales.

Longmore, M.E. 1997. Quaternary palynological records from perched lake sediments, Fraser Island, Queensland, Australia: rainforest, forest history and climatic control. *Australian Journal of Botany* 45:507-526.

Longmore, M.E. and H. Heijnis 1999. Aridity in Australia: Pleistocene records of palaeohydrological and palaeoecological change from the perched lake sediments of Fraser Island, Queensland, Australia. *Quaternary International* 57-58:35-47.

Luly, J.G. 1993. Holocene palaeoenvironments near Lake Tyrrell, semi-arid northwestern Victoria, Australia. *Journal of Biogeography* 20:587-596.

Magee, J.W., G.W. Miller, N.A. Spooner and D. Questiaux 2004. Continuous 150 k.y. monsoon record from Lake Eyre, Australia: Insolation-forcing implications and unexpected Holocene failure. *Geology* 32:885-888.

Martin, L., P.A. Towers, M.A. McGuckin, L. Little, H. Luckhoff and A.W. Blackshaw 1985. Reproductive biology of flying-foxes (Chiroptera: Pteropodidae). *Australian Mammalogy* 10:115-118.

Martin, L. and A.P. McIlwee 2002. The reproductive biology and intrinsic capacity for increase of the Grey-headed Flying-fox *Pteropus poliocephalus* (Megachiroptera), and the implications of culling. In P. Eby and D. Lunney (eds) *Managing the Grey-headed Flying-fox as a Threatened Species*, pp91-108. Sydney: Royal Zoological Society of New South Wales.

McConkey, K.R. and D. R. Drake 2006. Flying-foxes cease to function as seed dispersers long before they become rare. *Ecology* 87:271-276.

McDonald-Madden, E., G.S.G. Schreiber, D.M. Forsyth, D. Choquenot and T.F. Clancy 2005. Factors affecting grey-headed flying-fox (*Pteropus poliocephalus*: Pteropodidae) foraging in the Melbourne metropolitan area, Australia. *Austral Ecology* 30:600-608.

McIlwee, A. and L. Martin 2002. On the intrinsic capacity for increase of Australian flying-foxes (*Pteropus* spp. Megachiroptera). *Australian Zoologist* 32:76-100.

McKenzie, G.M. and A.P. Kershaw 2000. The last glacial cycle from Wyelangta, the Otway region of Victoria, Australia. *Palaeogeography, Palaeoclimatology, Palaeoecology* 155:177-193.

McMahon, J.P., M.F. Hutchinson, H.A. Nix and K.D. Ord 1995. ANUCLIM user's guide, version 1. Canberra: Centre for Resource and Environmental Studies, Australian National University.

Miller, G.H., J.W. Magee and A.J.T. Jull 1997. Low-latitude glacial cooling in the Southern Hemisphere from amino-acid racemisation in emu egg shells. *Nature* 385:241-244.

Moravek, S., J.G. Luly, J.F. Grindrod, R.J. Fairfax, R.J. Fensham and S.S. Smithers In prep. Origin and development of the grassy bald at Dandabah, Bunya Mountains, southeastern Queensland, Australia.

Moss, P.T. and A.P. Kershaw 2000. The last glacial cycle from the humid tropics of northeastern Australia: comparison of a terrestrial and marine record. *Palaeogeography, Palaeoclimatology, Palaeoecology* 155:155-176.

O'Brien, G.M. 1993. Seasonal reproduction in flying-foxes, reviewed in the context of other tropical mammals. *Reproduction, Fertility, Development* 5:499-521.

Parris, K.M. and D.L. Hazell 2005. Biotic effects of climate change in urban environments: The case of the grey-headed flying-fox (*Pteropus poliocephalus*) in Melbourne, Australia. *Biological Conservation* 124:267-276.

Parry-Jones, K. and M.L. Augee 1991. Food selection by grey-headed flying-foxes (*Pteropus poliocephalus*) occupying a summer colony site near Gosford, New South Wales. *Wildlife Research* 18:111-124.

Parry-Jones, K.A. and M.L. Augee 1992. Movements of Grey-headed flying-foxes (*Pteropus poliocephalus*) to and from a colony site on the central coast of New South Wales. *Wildlife Research* 19:331-340.

Parry-Jones, K.A. and M.L. Augee 2001. Factors affecting the occupation of a colony site in Sydney, New South Wales by the grey-headed flying-fox *Pteropus poliocephalus* (Pteropodidae). *Austral Ecology* 26:47-55.

Parsons, J.G., S.K.A. Robson and L.A. Shilton 2008. One for all and all four one! First report of all four *Pteropus* species co-roosting at one site in Australia. *Australasian Bat Society Newsletter* 31:10.

Parsons, J.G., S.K.A. Robson, L.A. Shilton and J. VanDerWal. 2010. The distribution of flying fox camps in Australia: new perspectives in a changing climate. In preparation.

Phillips, S. J., M. Dudik and R. Schapire 2006. MaxEnt, version 2. AT&T Labs-Research, Princeton University, and the Centre for Biodiversity and Conservation, American Museum of Natural History.

Posada, D. and K.A. Crandall. 1998. MODELTEST: testing the model of DNA substitution Bioinformatics 14: 817-818.

Ratcliffe, F.N. 1931. *The Flying-fox* (Pteropus) *in Australia.* CSIR Bulletin No. 53 Melbourne.

Ratcliffe, F.N. 1932. Notes on the fruit bats (*Pteropus* spp.) of Australia. *The Journal of Animal Ecology* 1:32-57.

Roberts, B.J., C.P. Catterall, J. Kanowski and P. Eby 2008. A re-evaluation of the northern distributional limit of the Grey-headed Flying-fox, *Pteropus poliocephalus. Australasian Bat Society Newsletter* 31:16-19.

Rogers, A.R. and H. Harpending 1992. Population growth makes waves in the distribution of pairwise genetic differences. *Molecular Biology and Evolution* 9:552-569.

Singh, G. and E.A. Geissler 1985. Late Cainozoic history of vegetation, fire, lake levels and climate at Lake George, New South Wales, Australia. *Philosophical Transactions of the Royal Society of London B* 311:379-477.

Tidemann, C.R. 1999. The biology and management of the Grey-headed Flying-fox. *Acta Chiropterologica* 1:151-164.

Tidemann, C.R. and J.C. Nelson 2004. Long-distance movements of the grey-headed flying-fox (*Pteropus poliocephalus*). *Journal of the Zoological Society of London* 263:141-146.

Thomas, I. et al 2001.

Torgersen, T., J.G. Luly, P. De Deckker, M.R. Jones, D.E. Searle and W.J. Ullman 1988. Late Quaternary environments of the Carpentaria Basin, Australia. *Palaeogeography, Palaeoclimatology, Palaeoecology* 67:245-261.

Van Der Wal, J., L. Shoo and S.E. Williams 2009. New approaches to understanding late Quaternary climate fluctuations and refugial dynamics in Australian wet tropical rainforests. *Journal of Biogeography* 36:291-301.

Walker, B.H. 1992. Biodiversity and ecological redundancy. *Conservation Biology* 6:18-23.

Wasson, R.J. 1983. The Cainozic history of the Strzelecki and Simpson dunefields. *Zeitschrift für Geomorphologie N.F., Supplement Band* 45:85-115.

Webb, L.J. 1964. An historical interpretation of the grass balds of the Bunya Mountains, south Queensland. *Ecology* 45:159-62.

Webb, N.J. and C.R. Tidemann 1996. Mobility of Australian flying-foxes, *Pteropus* spp. (Megachiroptera): evidence from genetic variation. *Proceedings of the Royal Society - Biological Sciences* 263:497-502.

Welbergen, J., S.M. Klose, N. Markus and P. Eby 2008. Climate change and the effects of temperature extremes on Australian flying-foxes. *Proceedings of the Royal Society B* 275:419-425.

Williams, S. and J. Middleton 2008. Climatic seasonality, resource bottlenecks and abundance of rainforest birds: implications for global climate change. *Diversity and Distributions* 14:69-77.

Williams, S.E., E.E. Bolitho and S. Fox 2003. Climate change in Australian tropical rainforests: an impending environmental catastrophe. *Proceedings of the Royal Society of London Series B-Biological Sciences* 270:1887-1892.

Williams, N.S.G., M.J. McDonnell, G.K. Phelan, L.D. Keim and R. van der Ree 2006. Range expansion due to urbanisation: Increased food resources attract, Grey-headed Flying-foxes (*Pteropus poliocephalus*) to Melbourne. *Austral Ecology* 31:190-198.

Worthington Wilmer, J. and E.M. Barratt 1996. A non-lethal method of tissue sampling for genetic studies of Chiropterans. *Bat Research News* 37:1-3.

Yokoyama, Y., A. Purcell, K. Lambeck and P. Johnson 2001. Shore-line reconstructions around Australia during the Last glacial maximum and Late Glacial Stage. *Quaternary International* 83-85:9-18.

7

Observations on feeding frequencies among native and exotic birds and fruit bats at *Erythrina variegata* and *Dysoxylum* trees on American Samoa

Julie A. Sherman

School of Geographical Sciences, Arizona State University, Tempe, United States
julie.sherman@asu.edu

Patricia L. Fall

School of Geographical Sciences, Arizona State University, Tempe, United States

Introduction

Like most oceanic islands, Polynesian islands have experienced environmental degradation and native species losses, due in part to deforestation and predation by humans and human-introduced species (Dahl 1984a, b; Steadman et al. 1999; Cocklin and Keen 2000; Steadman and Martin 2003), as well as natural disturbances such as hurricanes (Craig et al. 1994a; Elmqvist et al. 1994; Pierson et al. 1996; Hjerpe et al. 2001; Whistler 2002; Franklin et al. 2004; McConkey et al. 2004a). Fossil evidence indicates the majority of Polynesian bird and bat species became extinct or were extirpated in the wake of human colonisation (Steadman 1993, 1995, 2006; Koopman and Steadman 1995). Indigenous Polynesian island fruit bats and birds are key pollinators and seed dispersers of native rainforest flora (Fujita and Tuttle 1991; Cox et al. 1991; Rainey et al. 1995; McConkey and Drake 2002; Meehan et al. 2002; Fall et al. 2007). Koopman and Steadman (1995) suggested that a loss of vertebrate frugivores in Polynesia will lead to a decline in plant diversity, and indeed fossil pollen data from Tonga indicate that diminished rainforest plant diversity accompanied the loss of native dispersers and pollinators (Fall 2005). Ancient Polynesian and subsequent European colonists introduced many exotic species now naturalised in the Pacific Islands (Rinke 1986; Wiles et al. 1997; Richardson et al. 2000; Steadman 2006). Introduced fauna prey on or harass native species (Thorsen et al. 2000; Thibault et al. 2002; Blanvillain et al. 2003; Yap and Sodhi 2004), and compete with endemic animals for food (Banko et al. 2002; Hansen et al. 2002).

American Samoa offers an excellent locale in which to study the ecological implications of exotic wildlife and human habitat modification on Polynesian islands. The human population of American Samoa's main island, Tutuila, is growing, while human habitation on the smaller island of Ofu is decreasing (United States Census Bureau 2001). Human-introduced exotic birds are well established on Tutuila, but have not yet spread to outlying Ofu, Olosega and Ta'u (Watling 2004). Freifeld (1999, 2004) has investigated the habitat relationships of avifauna on Tutuila, and there has been considerable research on the feeding behaviour of American Samoa's fruit bats (Banack 1996, 1998; Nelson et al. 2000a, b, 2005; Brooke 2001; Banack and Grant 2002). Few data are available that describe the interactions between exotic and native species at key indigenous nectar sources in American Samoa. We present observations of native and exotic avifauna and fruit bats at indigenous *Erythrina variegata* and *Dysoxylum maoto* and *Dysoxylum samoensis* trees during their peak flowering and fruiting period on Ofu and Tutuila islands, American Samoa. *Erythrina* nectar and *Dysoxylum* fruit are important foods for native birds and fruit bats. Limited data are available on feeding frequencies and inter-specific interactions among native and exotic species at *Eyrthrina variegata* on American Samoa. We observed visitation and feeding by native birds, introduced (exotic) birds and native fruit bats at *Erythrina* and *Dysoxylum* trees in native rainforest, rural plantation and native-forest-edge habitats.

American Samoa

Study sites were located on Tutuila and Ofu islands in American Samoa, part of the Western Polynesian biogeographic subregion (Mueller-Dombois and Fosberg 1998). American Samoa is comprised of five main volcanic islands and two coral atolls located 13-15° S and 168-173° W in the South Pacific Ocean (Figure 1). The tropical maritime climate here has average daily temperatures between 24°C and 30°C, and rainfall from about 300 cm to more than 500 cm annually, with year-round high relative humidity (Western Regional Climate Center 2005). Ofu (5 km^2) is characterised by a narrow strip of littoral and lowland rainforest habitats, some of which are used for agriculture. Areas of native forest are most common on steeper slopes and peaks up to 494 m (Whistler 1994). Ofu's single town has a population of 289 people (United States Census Bureau 2001), a decrease from 353 in 1990 (World Gazetteer 2006). Tutuila has a larger landmass (142 km^2) and higher elevations (up to 653 m) than Ofu. Tutuila's 55,876 people (United States Census Bureau 2001) and its agricultural lands are concentrated within the littoral and lowland rainforest zones.

Birds and bats of American Samoa

America Samoa's only native mammals are bats (Banack and Grant 2002). The islands are home to two indigenous species of Pteropodidae (fruit bats): the Tongan fruit bat (*Pteropus tonganus*) and the Samoan fruit bat (*Pteropus samoensis*). While both species are vulnerable to human impacts, the Samoan fruit bat has been extirpated on the nearby archipelago of Tonga (Koopman and Steadman 1995), and is endangered in Samoa (Mickleburgh et al. 1992). In 1992, the government of American Samoa banned hunting of all fruit bats to help their populations recover from over-exploitation, as well as from decimation by tropical cyclones in 1987, 1990 and 1991 (Banack 1996).

Native land birds in American Samoa include two honeyeaters: the cardinal honeyeater (*Myzomela cardinalis*) and the wattled honeyeater (*Foulehaio carunculata*). Cardinal honeyeaters occur only on Tutuila; they are absent from Ofu (Watling 2004). The sole land bird endemic to Samoa and American Samoa is the Samoan starling (*Aplonis atrifusca*) (Freifeld 2004). Four pigeons are native to American Samoa: Pacific pigeon (*Ducula pacifica*), many-coloured fruit

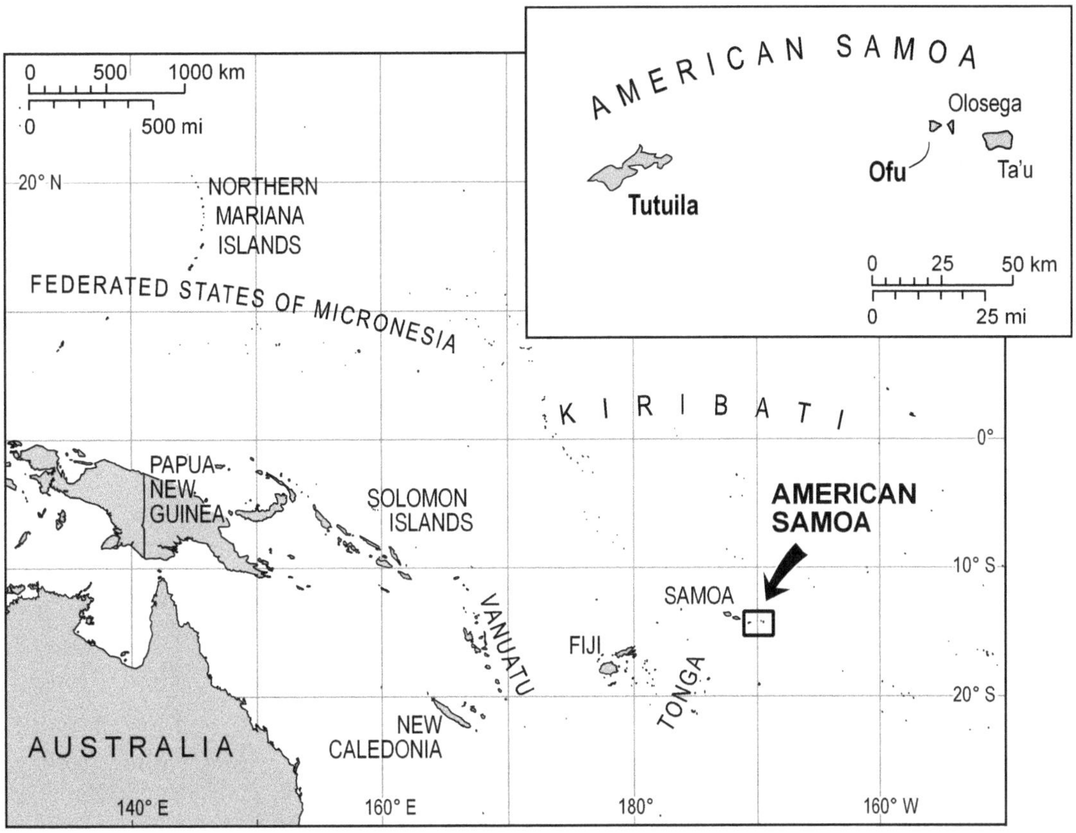

Figure 1. Location of Tutuila and Ofu islands, American Samoa, in the South Pacific. Created by Barbara Trapido-Lurie

dove (*Ptilinopus perousii*), purple-capped fruit dove (*Ptilinopus porphyraceus*), and shy ground dove (*Gallicolumba stairi*) (Steadman and Friefeld 1999). American Samoa banned hunting of all pigeon species in 1992 (Freifeld 2004).

In addition to its native avifauna, American Samoa has been colonised by three exotic, human-introduced bird species: the red-vented bulbul (*Pycnonotus cafer bengalensis*), the common myna (*Acridotheres tristis*) and the jungle myna (*Acridotheres fuscus*) (Freifeld 1999). The red-vented bulbul was introduced into American Samoa in the late 1950s and became established quickly (Freifeld 1999). Bulbuls and mynas are found only on Tutuila. Researchers indicate a need for data regarding behaviour and competition among native birds and exotic bulbuls for native food sources (Freifeld 1999; Watling 2004), and detailing the feeding behaviour of fruit bats at nectar sources (Banack 1998). This study presents data on these matters.

Methods

We made observations at 16 flowering *Erythina variegata* and five fruit-bearing *Dysoxylum* focal trees on Ofu and Tutuila over a 22-day period between 19 July and 9 August 2001 (Table 1). We chose individual and small groups of overlapping trees that provided visibly comparable canopy sizes and food resource availability. In some cases, smaller adjacent trees with overlapping canopies were observed simultaneously, as if they constituted a single tree. Our data were collected at focal trees located in native-forest stands, rural plantations and a transitional native-forest edge. We observed additional, non-focal trees in village and native-forest habitats for limited periods. Data gathered on feeding or visitations at non-focal trees are included in the discussion.

Table 1. *Erythrina variegata, Dysoxylum maoto* and *Dysoxylum samoensis* focal trees on Ofu and Tutuila islands, American Samoa

Habitat	Tree	Time range of observations	Total no. of 2-min. intervals
***Erythrina variegata* trees**			
Ofu Island			
Native forest			
	E1	05:17-07:31	158
	E2	05:20-07:02	131
	E3	06:04-07:04	31
	E4	05:17-07:27	119
	E5	05:24-06:30	41
	E6	05:18-07:31	101
Tutuila Island			
Native forest			
	E7	06:30-07:30	43
	E8	06:34-07:44	47
Native-forest edge			
	E9	06:30-08:42	57
Rural plantation			
	E10	05:50-07:30	115
	E11	05:50-07:30	115
	E12	05:54-07:30	112
***Dysoxylum maoto* and *Dysoxylum samoensis* trees**			
Ofu Island			
Native forest			
	D1	06:37-11:42	106
Habitat	**Tree**	**Time range of observations**	**Total no. of 2-min. intervals**
Tutuila Island			
Native forest			
	D2	06:30-07:30	61
	D3	06:34-07:44	123
Rural plantation			
	D4	05:50-07:30	93

Trees E6 and E8 each denote two adjacent trees with overlapping canopies; tree E12 is three adjacent trees with overlapping canopies. Tree D1 denotes two adjacent trees with overlapping canopies.

Erythrina variegata nectar is an important food for American Samoa's native fruit bats and honeyeaters (Banack 1998; Freifeld 1999; Webb et al. 1999). *Dysoxylum maoto* and *Dysoxylum samoensis* provide foods favoured by the native Pacific pigeon (Steadman and Freifeld 1999). *Erythrina variegata* generally is considered native to American Samoa (e.g. Freifeld 1999), although Whistler (1992) suggests ancient Polynesians may have introduced it to some Pacific Islands, where it has been naturalised for millennia. *Erythrina variegata* trees flower between July and September (Whistler 1994), when few other trees are flowering (Banack 1998; Webb et al. 1999). *Dysoxylum maoto* and *Dysoxylum samoensis*, both indigenous to Samoa (Whistler 1994), are in peak fruit from January through September (Webb et al. 1999).

We recorded visitation and feeding based on scan sampling (Altmann 1974), a method used commonly to assess behaviour of both birds and fruit bats (Pitter and Christiansen 1997; Sun and Moermond 1997; Banack 1998; Ochoa-Acuna and Kunz 1999; Munoz-Romo 2006), as well as to investigate visitation and feeding at focal trees (Boinski and Scott 1988; Tirado Herrera et al. 2003; Kirika et al. 2008). Observations at *Erythrina* focal trees were made between 05:00 and 09:00 to capture peak feeding times for most avian and fruit-bat species. Although Samoan fruit bats have been observed feeding during all hours of the day, their peak daylight foraging period is between 05:00 and 08:00 (Banack 1996). Daylight activity of the largely nocturnal Tongan fruit bats also peaks between 05:00 and 08:00, and again between 16:00 and 17:59 (Morrell and Craig 1995; Banack and Grant 2002). Our initial observations showed intensive bird feeding at *Erythrina* during the early morning (06:00-08:00), but no evening feeding (17:40-19:00). Thus, we chose to make observations between 05:00 and 09:00, the optimal interval for studying feeding by birds and both species of bats. We collected quantitative data on feeding frequencies over an average period of three days for each tree. Night-vision binoculars were used to observe focal trees before dawn and at dusk. Sunrise during the study period varied between 06:41 and 06:48, with dawn between 06:19 and 06:21. Sunset occurred between 18:07 and 18:16, and dusk continued between 18:30 and 18:38 (United States Naval Observatory 2007). Observations at *Dysoxylum* trees were made between 05:50 and 12:00, a time range which captures peak activity periods for birds and fruit bats. Data for *Dysoxylum samoensis* and *D. maoto* were combined, since these species share a habitat range encompassing primary and secondary coastal and lowland forests (Whistler 1994). Both trees are characterised by alternate, pinnately compound leaves and brown sub-globose fruits (Whistler 1994).

Data were collected at two-minute intervals. At each interval, we recorded the number and species of birds and bats visible in a tree, as well as their behaviour. Aggressive inter-specific interactions (e.g. vocalisation and rapid flight in pursuit of an approaching animal) were noted when observed. Feeding was recorded when visible; animals were recorded as perching when engaged in some non-feeding posture in the tree, such as perching, hanging, or preening. Animals flying in or around the focal trees were not recorded. Bird species were identified, while data for Tongan and Samoan fruit bats were combined since often they were indistinguishable in pre-dawn, low-light conditions. We analysed the mean recorded visitation (number) of animals and the mean recorded number of feeding animals at each focal tree. Mean recorded abundance is a common measure to analyse avian habitat relationships (see, for example, Freifeld 1999); we use mean recorded frequencies to present our quantified observations at individual focal trees.

Results

We observed cardinal and wattled honeyeaters, red-vented bulbuls and the Tongan and Samoan fruit bats feeding on *Erythrina variegata* trees. *Dysoxylum samoensis* and *D. maoto* trees were visited by the Samoan starlings, Pacific pigeons, wattled honeyeaters, collared kingfishers, common and jungle mynas and the red-vented bulbul (Table 2).

Table 2. Birds and fruit bats observed in *Erythrina variegata, Dysoxylum maota* and *Dysoxylum samoensis* trees on Ofu and Tutuila islands, American Samoa. Species origin (N = native; E = exotic, human-introduced); tree usage shows activity while in tree (F = feeding; P = perching); if feeding, the floral or fruit consumed is noted (fr = fruit, sds = seeds, flwrs = flowers, nec = nectar)

***Erythrina variegata* trees**				
Birds				
Scientific name	Common name	Samoan name	Origin	Tree usage
Foulehaio carnunculata	Wattled honeyeater	'Iao	N	F, nec
Myzomela cardinalis	Cardinal honeyeater	Segasegamau'u	N	F, nec
Pycnonotus cafer	Red-vented bulbul	Manu palagi	E	F, flwrs, nec
Fruit bats				
Scientific name	Common name	Samoan name	Origin	Tree usage
Pteropus samoensis	Samoan fruit bat	Pe'a vao	N	F, flwrs, nec
Pteropus tonganus	Tongan fruit bat	Pe'a fanua	N	F, nec
***Dysoxylum maoto* and *Dysoxylum samoensis* trees**				
Birds[1]				
Scientific name	Common name	Samoan name	Origin	Tree usage
Aplonis atrifusca	Samoan starling	Fuia	N	F, fr, sds
Ducula pacficia	Pacific pigeon	Lupe	N	F, fr, sds
Foulehaio carnunculata	Wattled honeyeater	'Iao	N	P
Halcyon chloris	Collared kingfisher	Tiotala	N	P
Acridotheres tristis	Common myna	Maina fanua	E	P
Acridotheres fuscus	Jungle myna	Maina vao	E	P
Pycnonotus cafer	Red-vented bulbul	Manu palagi	E	P, F, fr

[1]An indigenous columbrid, the many-coloured fruit dove (*Ptilinopus perousii* Peale), was seen flying by or possibly from the Tutuila *Dysoxylum* trees, but was not recorded perching or feeding in the trees. Nomenclature, authority and Samoan names follow Utzurrum 2002; Watling 2004; Freifeld 1999; and Steadman 2006.

Erythrina trees

On both Ofu and Tutuila, all observed birds and bats fed most intensely on *Erythrina* nectar between 06:30 and 07:30 (Table 3). Cardinal honeyeaters, red-vented bulbuls and fruit bats were not recorded perching in focal trees, although all frequently flew out of trees or to other inflorescences. Honeyeaters arrived and began feeding between 06:07 and 06:44. Red-vented bulbuls followed a similar pattern, with the earliest recorded arrival and feeding at 06:24. Fruit bats were recorded already feeding in *Erythrina* trees when we arrived before 05:20. By 09:00 there was only very limited bird activity, and bats were generally absent from the trees. Additional observations between 17:40 and 19:00 revealed very limited visitation by any bird or bat species. All observed visitations in this time period involved only perching, and no evening feeding.

We recorded wattled honeyeaters and both Tongan and Samoan fruit bats feeding at *Erythrina* in Ofu's native forest (Figure 2). Tutuila's wattled and cardinal honeyeaters fed at the highest frequencies at trees in the native forest, where exotic red-vented bulbuls were not observed feeding. Although red-vented bulbuls were not observed feeding in native forests, three individuals were recorded in trees adjacent to the road in native forest (near focal trees E7 and E8).

Table 3. Feeding frequencies at *Erythrina variegata* focal trees on Ofu and Tutuila islands, American Samoa, between 05:00 and 09:00. Data are presented as means (se), indicating the average number of feeding animals per two-minute interval; HE: honeyeater; se: standard error

	Wattled HE	Cardinal HE	Fruit bats	Bulbuls
Island, Habitat and Tree				
Ofu Island				
Native forest				
E1	2.47 (0.20)	0	0.38 (0.04)	0
E2	2.19 (0.18)	0	0.63 (0.05)	0
E3	2.45 (0.29)	0	0.03 (0.03)	0
E4	3.52 (0.33)	0	0.60 (0.08)	0
E5	1.02 (0.25)	0	0.61 (0.07)	0
E6	3.55 (0.35)	0	0.09 (0.03)	0
Sum	2.72 (0.12)	0	0.43 (0.03)	0
Tutuila Island				
Native forest				
E7	4.74 (0.26)	0.86 (0.16)	0.58 (0.13)	0
E8	5.30 (0.20)	1.47 (0.14)	0.57 (0.12)	0
Sum	5.03 (0.16)	1.18 (0.11)	0.58 (0.09)	0
Native-forest edge				
E9	2.63 (0.14)	0.11 (0.05)	0	0.32 (0.12)
Rural plantation				
E10	0.04 (0.02)	0.08 (0.04)	1.54 (0.06)	5.20 (0.41)
E11	0.03 (0.02)	0	0.82 (0.07)	3.90 (0.34)
E12	0.04 (0.02)	0.18 (0.04)	1.70 (0.10)	6.61 (0.55)
Sum	0.04 (0.01)	0.08 (0.02)	1.35 (0.05)	5.23 (0.26)

Trees E6 and E8 each denote two adjacent trees with overlapping canopies; tree E12 is three adjacent trees with overlapping canopies.

On Tutuila's rural plantations, we observed a stark contrast between high feeding frequencies by red-vented bulbuls and very low frequencies for wattled and cardinal honeyeaters (Figure 3). Red-vented bulbuls actively defended focal *Erythrina* trees and inflorescences from both species of honeyeaters. Many honeyeaters were quickly chased off by bulbuls, which would fly directly at them, vocalising continuously, and pursuing them out of the tree. We observed 43 instances of bulbuls chasing honeyeaters. On only one occasion was a cardinal honeyeater observed chasing a bulbul to another branch. At tree E12 we noted cardinal honeyeaters feeding only at the outer edges of the canopy, away from the central canopy where the bulbuls were concentrated. A focal tree at a native-forest edge in Tutuila (E9) showed substantial feeding by wattled honeyeaters, only modest feeding instances by cardinal honeyeaters and red-vented bulbuls, and no feeding by fruit bats (although Tongan fruit bats were frequently noted feeding at trees in similar habitats). Overall fruit-bat feeding frequency on Tutuila was highest at rural plantation trees. Samoan fruit bats were recorded feeding at native-forest trees, but were not definitively identified at rural plantation trees. Both fruit-bat species were seen feeding at native-forest trees on Tutuila and Ofu.

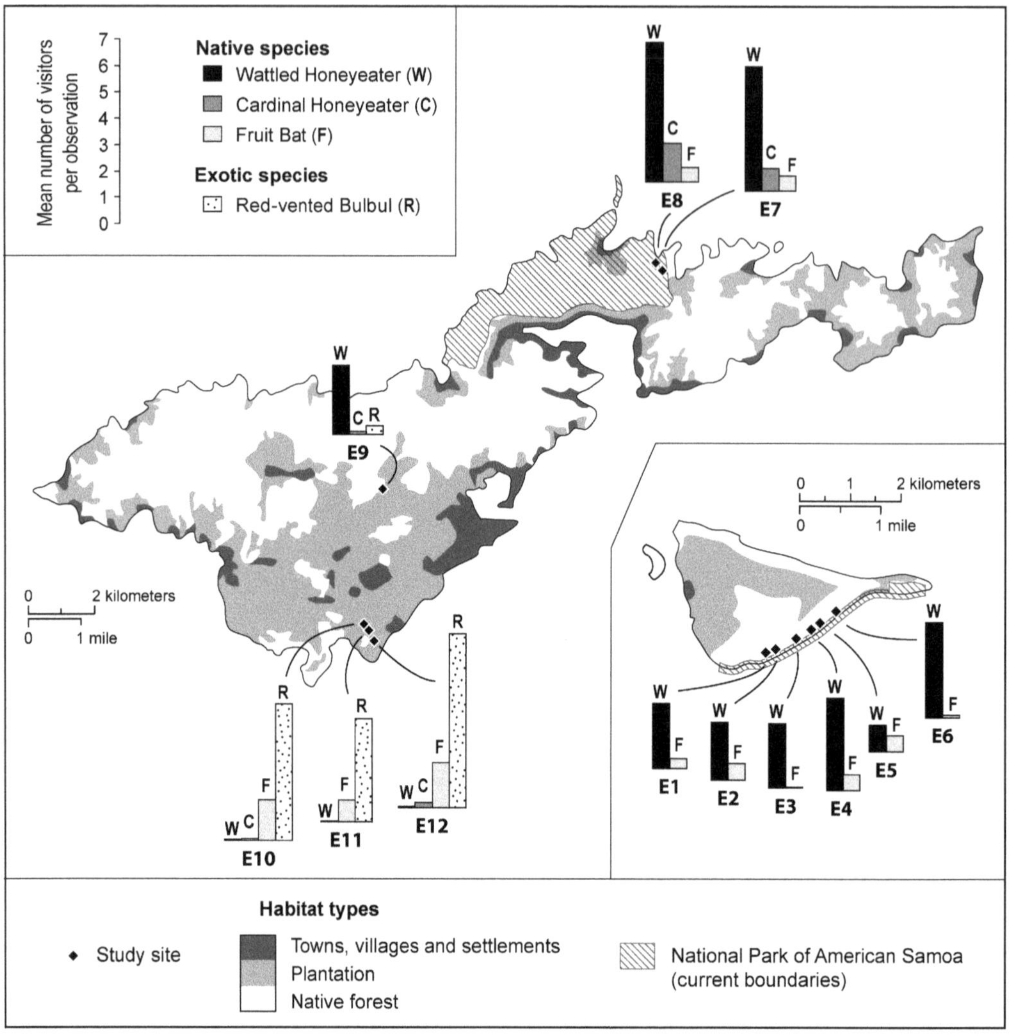

Figure 2. Mean feeding frequencies of birds and bats observed at *Erythrina variegata* trees on Tutuila and Ofu (inset) islands; habitat types after Whistler (1994). Created by Barbara Trapido-Lurie

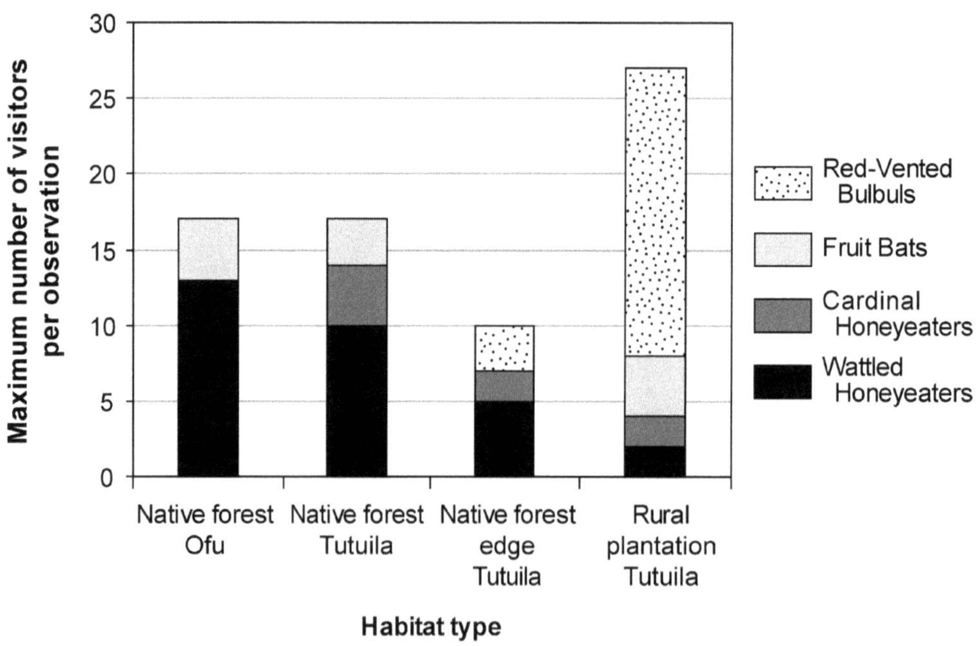

Figure 3. Maximum numbers of birds and bats observed feeding simultaneously at a two-minute interval at *Erythrina variegata* trees on Tutuila and Ofu islands by focal-tree habitat type. Created by Barbara Trapido-Lurie

Dysoxylum trees

We observed Samoan starlings in *Dysoxylum* trees in Ofu and Tutuila's native forests, and in a rural plantation on Tutuila (Figure 4; Table 4). Pacific pigeons were recorded only in native forest focal trees on Ofu and Tutuila. During most of our observations, these species were perching, not feeding. Similarly, wattled honeyeaters, collared kingfishers (*Halcyon chloris*) and mynas were observed perching, but not feeding, in *Dysoxylum* trees on Ofu and Tutuila. Red-vented bulbuls fed at focal *Dysoxylum* in rural plantations on Tutuila. Samoan starlings were the most frequent visitors to *Dysoxylum* on both islands. On Tutuila, starling visitation frequency was highest in a rural plantation.

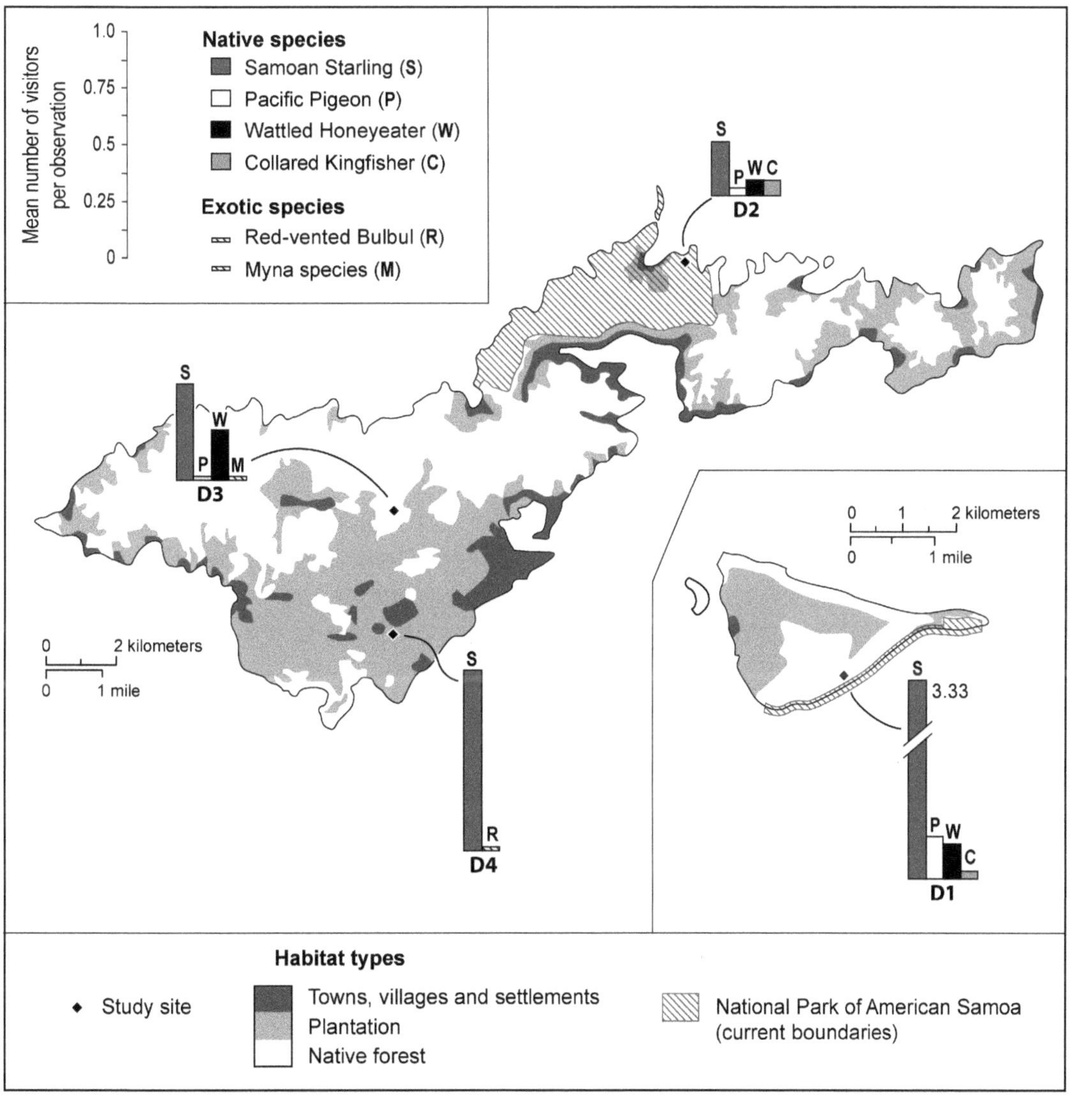

Figure 4. Mean total visitation frequency (feeding and perching) of birds observed at *Dysoxylum maoto* and *Dysoxylum samoensis* trees on Tutuila and Ofu islands; habitat types after Whistler (1994). Created by Barbara Trapido-Lurie

Table 4. Feeding and visitation frequencies at *Dysoxylum maoto* and *Dysoxylum samoensis* focal trees on Ofu and Tutuila islands, American Samoa. Data are presented as mean (se), indicating average number of visitors and feeders per two-minute interval; HE: honeyeater; se: standard error

	Native feeders				**Exotic feeders**		**Other visitors (no feeding observed)**		
	Samoan starlings		Pacific pigeons		Bulbuls		Wattled HE	Kingfishers	Mynas
	Feeding	Perching	Feeding	Perching	Feeding	Perching	Perching	Perching	Perching
Island, Habitat and Tree									
Ofu Island									
Native forest									
D1	0.38 (0.11)	3.33 (0.13)	0.02 (0.01)	0.19 (0.01)	0	0	0.15 (0.04)	0.04 (0.02)	0
Tutuila Island									
Native forest									
D2	0.03 (0.02)	0.23 (0.06)	0	0.03 (0.02)	0	0	0.07 (0.04)	0.07 (0.04)	0
D3[1]	0.07 (0.03)	0.42 (0.06)	0[1]	0.01 (0.01)	0	0	0.22 (0.01)	0	0.02 (0.01)
Rural plantation									
D4	0.13 (0.05)	0.79 (0.11)	0	0	0.02 (0.02)	0.05 (0.02)	0.01 (0.01)	0	0

[1]Fruit and fruit peels were observed dropping beneath Pacific pigeons in tree D3, although feeding behaviour was not visible. Two unidentified birds also were observed perching in tree D3. Total mean visitation frequency including feeding and perching is displayed in Figure 4. Tree D1 denotes two adjacent trees with overlapping canopies.

Discussion

The highest feeding frequency of wattled honeyeaters occurred at trees in relatively undisturbed native forests on both Ofu and Tutuila, in accordance with the reported preference of this species for feeding in native forests (Freifeld 1999; Steadman et al. 1999). In contrast, we recorded honeyeaters feeding rarely and for very brief periods in rural plantations. Large flocks of wattled honeyeaters fed simultaneously for extended periods at focal trees not visited by bulbuls. Watling (2004) notes that wattled honeyeaters attempt to protect preferred nectar sources from bulbuls, sometimes in vain. Our observations show that in rural plantation habitats, red-vented bulbuls fed in and successfully defended *Erythrina* trees from most feeding attempts by wattled honeyeaters. The wattled honeyeater currently is the most abundant land bird on Tutuila (Freifeld 1999), but continued destruction of native forests may increasingly drive it to forest-edge or plantation habitats where it finds itself at a competitive disadvantage against introduced birds at favoured food sources, such as *Erythrina* trees.

The highest feeding frequency of cardinal honeyeaters was at *Erythrina* trees in Tutuila's native forest. Freifeld (1999) suggests that wattled honeyeaters may attempt to exclude cardinal honeyeaters from *Erythrina* in native-forest habitats. We did not observe inter-specific chases among wattled and cardinal honeyeaters, although an intraspecific chase between two cardinal honeyeaters was recorded at a rural plantation tree. Our data show both cardinal honeyeater and wattled honeyeater feeding frequency increased at trees in native forests, and decreased dramatically in rural plantations, where large numbers of red-vented bulbuls were feeding. As with wattled honeyeaters, we observed red-vented bulbuls chasing cardinal honeyeaters from the trees. Cardinal honeyeaters were recorded feeding at these trees for brief periods between chases. Cardinal honeyeaters were surprisingly uncommon at the forest-edge tree, where we anticipated them in abundance due to the relatively low feeding frequency of red-vented bulbuls, and because studies suggest cardinal honeyeaters prefer edge and village habitats to native forests (Freifeld 1999). However, our forest-edge tree may have offered insufficient food resources, as its flowers were dropping during the last day of observations. Feeding frequencies of both birds and fruit bats at *Erythrina variegata* appeared to vary with changes in the focal trees' flowers. Visitation frequencies decreased noticeably as flowers dropped petals. This occurred over a period of a few days to a week. The process happened more quickly at trees with higher visitation frequencies, suggesting that intensive nectar feeding contributes to floral demise.

Red-vented bulbuls were prolific feeders at trees on rural plantations where they aggressively defended *Erythrina* flowers from both species of honeyeaters, through aerial chases and vocalisation. Bulbuls were less common visitors to the tree in forest-edge habitat, and they were not observed feeding at any trees in native rainforest. In addition to inserting their beaks into *Erythrina* flowers, we also observed them eating flowers. Bulbuls outside their native range often are described as limited to human-modified habitats (Freifeld 1999; Eguchi and Amano 2004; Watling 2004; Yap and Sodhi 2004; Peh et al. 2006), however, on Oahu, Hawai'i, they have spread into undeveloped areas once used exclusively by native birds (McAllan and Hobcroft 2005), tripling their range in only six years (van Riper et al. 1979; Williams and van Giddings 1984). We observed the greatest frequency of bulbuls in trees on a rural plantation 8 km from the nearest village. McAllan and Hobcroft (2005) suggest that roads may be one vector for the dispersal of red-vented bulbuls into non-village habitats in Samoa. Perhaps reflecting this tendency, three bulbuls were observed in native forest alongside a road near the small town of Afono. The growth and geographical spread of human population in American Samoa may facilitate expansion of the bulbuls' range, and increase their conflict or competition with native birds.

Fruit bats of both species fed in native-forest trees on Ofu and Tutuila, hanging on branches to feed at flowers, also inserting their mouths into blossoms to lick nectar, and occasionally consuming petals. Fruit bats fed most frequently at *Erythrina* in rural plantation habitats on Tutuila. Nelson and others (2000a) found that the Samoan fruit bats prefer native foods to agricultural fruits, while the Tongan fruit bats eat both native and agricultural fruits (Banack 1996). The visitation and feeding frequency of fruit bats at *Erythrina* trees did not change notably with the presence or abundance of exotic or native birds. We observed fruit bats having frequent vocal interactions, and displacing one another from trees or inflorescences. In contrast to birds, bats were not seen pursuing other species. However, we noted very high concentrations of fruit bats and no visible birds at a cluster of *Erythrina* trees on the edge of a hotel development, suggesting the possibility of inter-species exclusion. Our results may underestimate fruit-bat feeding at *Erythrina* flowers, as both species also feed at night and, in the case of the Samoan bat, during other daylight hours (Banack 1996; Brooke 2001). Proximity to diverse food sources may affect fruit-bat feeding frequencies at individual *Erythrina* trees, as both bat species tend to visit numerous different trees for flowers or fruit (Banack 1996). Availability of roost trees also may influence Samoan fruit-bat feeding choices, since they use long-term roosts and small core habitat areas (Brooke 2001). Researchers found Samoan fruit bats roost most frequently in areas with the highest percentage of food-producing trees (Brooke et al. 2000). Brooke (2001) radio-tracked two Samoan fruit bats and found that they foraged over areas between 1.2 km^2 and 8.18 km^2.

We observed the highest feeding frequency of Samoan starlings at *Dysoxylum* trees in native forest on Ofu. On Tutuila, our data show most frequent feeding by starlings at a rural plantation tree. Freifeld (1999) suggests that Samoan starlings may be the only native birds that feed in agricultural habitats. Many native foods preferred by starlings, including *Dysoxylum,* are common in disturbed landscapes and along habitat edges (Whistler 1994; Freifeld 1999). Samoan starlings are key contributors to recolonisation of disturbed areas by native plants, due to their dispersal of a variety of native seeds.

Pacific pigeons were recorded only at *Dysoxylum* trees in native rainforest. This supports research indicating that Pacific pigeons avoid human-modified habitats, and are dependent on relatively undisturbed native forests, where they feed exclusively (Freifeld 1999; Franklin and Steadman 2001; Freifeld and Steadman 2001; McConkey et al. 2004b). Pacific pigeon visitation frequency was low for all observed locations. Our data may underestimate feeding and visitation by Pacific pigeons, as the species is notably cryptic, especially near human populations (Freifeld 1999). Current populations of Pacific pigeons have not rebounded from hurricanes and subsequent hunting pressures (Craig et al. 1994a, b; Freifeld 2004), suggesting recovery to past population levels may be unlikely, as more of the native lowland rainforest on Tutuila is lost.

Conclusions

Our data show feeding frequency differences at individual trees and between trees in broad habitat categories. Native birds fed frequently at focal trees in native forests, but seldom in rural plantations. Native honeyeaters were largely excluded from *Erythrina* trees where large numbers of exotic red-vented bulbuls were feeding. Native Pacific pigeons were observed feeding only at trees in native forest. The presence of red-vented bulbuls may be an important determinant in feeding patterns of American Samoan native birds at *Erythrina variegata* trees. Bulbuls aggressively defended *Erythrina* trees on rural plantations and largely excluded native honeyeaters from feeding, whereas fruit bats fed at *Eyrthrina* in both native forest and rural plantations, even when large numbers of red-vented bulbuls were present. Similarly, human habitat modifications may affect feeding of native birds at fruiting *Dysoxylum* trees. Pacific pigeons were not recorded in disturbed habitats, and were observed feeding at *Dysoxylum* only in native forests. Both Samoan starlings and Pacific pigeons visited and fed at the greatest frequency at *Dysoxylum* trees in Ofu's native forest. This study underscores the need for more studies on the impacts of America Samoa's invasive birds on native avian and fruit-bat habitat and feeding preferences across a broader range of seasons, food sources and habitat types, as well as the critical role of human land-use and wildlife management policies for the conservation of American Samoa's native fauna.

Acknowledgements

We are grateful to Suzanne Nelson and Darin Masters for their invaluable help in American Samoa. We thank Peter Craig and the staff of the National Park of American Samoa, Vaoto Lodge, and our families for assistance in American Samoa. The research was funded by two Matthew G. Bailey Scholarships from the School of Geographical Sciences, Arizona State University. We thank Steve Falconer, Taly Drezner, Anthony Brazel and David E. Brown for comments on an early draft of this manuscript, Jon Luly and an anonymous reviewer for helpful suggestions and Barbara Trapido-Lurie for creating the figures.

References

Altmann, J. 1974. Observational study of behavior: sampling methods. *Behavior* 69:227-263.

Banack, S.A. 1996. Flying foxes, genus *Pteropus*, in the Samoan Islands: Interactions with forest communities. Unpublished PhD thesis. Berkeley: University of California.

Banack, S.A. 1998. Diet selection and resource use by flying foxes (genus *Pteropus*). *Ecology* 79:1949-1967.

Banack, S.A. and G.S. Grant 2002. Spatial and temporal movement patterns of the flying fox, *Pteropus tonganus*, in American Samoa. *Journal of Wildlife Management* 66:1154-1163.

Banko, P.C., P.T. Oboyski, J.W. Slotterback, S.J. Dougill, D.M. Goltz, L. Johnson, M.E. Laut and T.C. Murray 2002. Availability of food resources, distribution of invasive species, and conservation of a Hawaiian bird along a gradient of elevation. *Journal of Biogeography* 29:789-808.

Blanvillain, C., J.M. Salducci, G. Tutururai and M. Maeura 2003. Impact of introduced birds on the recovery of the Tahiti flycatcher (*Pomarea nigra*), a critically endangered forest bird of Tahiti. *Biological Conservation* 109:197-205.

Boinski S. and P.E. Scott 1988. Association of birds with monkeys in Costa Rica. *Biotropica* 20:136-143.

Brooke, A.P. 2001. Population status and behaviours of the Samoan flying fox (*Pteropus samoensis*) on Tutuila Island, American Samoa. *Journal of Zoology* 254:309-319.

Brooke, A.P., C. Solek and A. Tualaulelei 2000. Roosting behavior of colonial and solitary flying foxes in American Samoa (Chiroptera: Pteripodidea). *Biotropica* 32:338-350.

Cocklin, C. and M. Keen 2000. Urbanization in the Pacific: Environmental change, vulnerability and human security. *Environmental Conservation* 27:392-403.

Cox, P.A., T. Elmqvist, E.D. Pierson and W.E. Rainey 1991. Flying foxes as strong interactors in South Pacific island ecosystems: A conservation hypothesis. *Conservation Biology* 5:448-454.

Craig, P., P. Trail and T.E. Morrell 1994a. The decline of fruit bats in American Samoa due to hurricanes and overhunting. *Biological Conservation* 69:261-266.

Craig, P., T.E. Morrell and K. So'oto 1994b. Subsistence harvest of birds, fruit bats and other game in American Samoa, 1990-1991. *Pacific Science* 48:344-352.

Dahl, A.L. 1984a. Biogeographical aspects of isolation in the Pacific. *Ambio* 13:302-305.

Dahl, A.L. 1984b. Oceania's most pressing environmental problems. *Ambio* 13:296-301.

Eguchi, K. and H.E. Amano 2004. Spread of exotic birds in Japan. *Ornithological Science* 3:3-11.

Elmqvist, T., W.E. Rainey, E.D. Pierson and P.A. Cox 1994. Effects of tropical cyclones Ofa and Val on the structure of a Samoan lowland rain forest. *Biotropica* 26:384-391.

Fall, P.L. 2005. Vegetation change in the coastal-lowland rainforest at Avai'o'vuna Swamp, Vava'u, Kingdom of Tonga. *Quaternary Research* 64:451-459.

Fall, P.L., T.D. Drezner and J. Franklin 2007. Dispersal ecology of the lowland rain forest in the Vava'u island group, Kingdom of Tonga. *New Zealand Journal of Botany* 45:393-417.

Franklin, J., D.R. Drake, K.R. McConkey, F. Tonga and L.B. Smith 2004. The effects of Cyclone Waka on the structure of lowland tropical rain forest in Vava'u, Tonga. *Journal of Tropical Ecology* 20:409-420.

Franklin, J. and D.W. Steadman 2001. The potential for conservation of Polynesian birds through habitat mapping and species translocation. *Conservation Biology* 5:506-521.

Freifeld, H.B. 1999. Habitat relationships of Samoan forest birds on Tutuila Island, American Samoa. *Journal of Biogeography* 26:1191-1213.

Freifeld, H.B. 2004. Temporal variation in forest bird survey data from Tutuila Island, American Samoa. *Pacific Science* 58:99-117.

Freifeld, H.B. and D.W. Steadman 2001. Landbirds on offshore islands in Samoa. *Journal of Field Ornithology* 72:72-85.

Fujita, M.S. and M.D. Tuttle 1991. Flying foxes (Chiroptera: Pteropodidae): Threatened animals of key ecological and economic importance. *Conservation Biology* 5:455-463.

Hansen, D.M., J.M. Olesen and C.G. Jones 2002. Trees, birds and bees in Mauritius: Exploitative competition between introduced honey bees and endemic nectarivorous birds. *Journal of Biogeography* 29:721-734.

Hjerpe, J., H. Hedenås and T. Elmqvist 2001. Tropical rain forest recovery from cyclone damage and fire in Samoa. *Biotropica* 33:249-259.

Kirika, J.M., N. Farwig and K. Bohning-Gaese 2008. Effects of local disturbance of tropical forests on frugivores and seed removal of a small-seeded Afrotropical tree. *Conservation Biology* 22:318-328.

Koopman, K.R. and D.W. Steadman 1995. Extinction and biogeography of bats on 'Eua, Kingdom of Tonga. *American Museum of Novitates* 3125:1-13.

McAllan, I.A.W. and D. Hobcroft 2005. The further spread of introduced birds in Samoa. *Notornis* 52:16-20.

McConkey, K.R. and D.R. Drake 2002. Extinct pigeons and declining bat populations: are large seeds still being dispersed in the tropical Pacific. In D. Levey, W. Silva, and M. Galetti (eds) *Frugivory and seed dispersal: evolutionary and conservation perspectives*, pp381-395. Wallingford: CAB International.

McConkey, K.R., D.R. Drake, J. Franklin and F. Tonga 2004a. Effects of Cyclone Waka on flying foxes (*Pteropus tonganus*) in the Vava'u Islands of Tonga. *Journal of Tropical Ecology* 20:555-561.

McConkey, K.R., H.J. Meehan and D.R. Drake 2004b. Seed dispersal by Pacific pigeons (*Ducula pacifica*) in Tonga, Western Polynesia. *Emu* 104:369-376.

Meehan, H.J., K.R. McConkey and D.R. Drake 2002. Potential disruptions to seed dispersal mutualisms in Tonga, Western Polynesia. *Journal of Biogeography* 29:695-712.

Mickleburgh, S.P., A.M. Hutson and P.A. Racey 1992. *Old World fruit bats: An action plan for their conservation*. Gland, Switzerland: Chiroptera Specialist Group, Species Survival Commission, IUCN.

Morrell, T.E. and P. Craig 1995. Temporal variation in fruit bats observed during daytime surveys in American Samoa. *Wildlife Society Bulletin* 23:36-40.

Mueller-Dombois, D. and F.R. Fosberg 1998. *Vegetation of the tropical Pacific Islands*. New York: Springer-Verlag.

Munoz-Romo, M. 2006. Ethogram and diurnal activities of a colony of *Artibeus lituratus* (Phyllostomidae: Stenodermatinae). *Acta Chiropterologica* 8:231-238.

Nelson, S.L., M.A. Miller, E.J. Heske and G.C. Fahey, Jr. 2000a. Nutritional consequences of a change in diet from native to agricultural fruits for the Samoan fruit bat. *Ecography* 23:393-401.

Nelson, S.L., M.A. Miller, E.J. Heske and G. C. Fahey, Jr. 2000b. Nutritional quality of leaves and unripe fruit consumed as famine foods by the flying foxes of Samoa. *Pacific Science* 54:301-312.

Nelson, S.L., T.H. Kunz and S.R. Humphrey 2005. Folivory in fruit bats: Leaves provide a natural source of calcium. *Journal of Chemical Ecology* 31:1683-1691.

Ochoa-Acuña, H. and T. H. Kunz 1999. Thermoregulatory behavior in the small island flying fox, *Pteropus hypomelanus* (Chiroptera:Pteropodidae). *Journal of Thermal Biology* 24:15-20.

Peh, K.S-H., N.S. Sodhi, J. de Jong, C.H. Sekercioglu, C.A.-M.Yap and S.L.-H. Lim 2006. Conservation value of degraded habitats for forest birds in southern Peninsular Malaysia. *Diversity and Distributions* 12:572-581.

Pierson, E.D, T. Elmqvist, W.E. Rainey and P.A. Cox 1996. Effects of tropical cyclonic storms on flying fox populations on the South Pacific islands of Samoa. *Conservation Biology* 10:438-451.

Pitter, E. and M.B. Christiansen 1997. Behavior of individuals and social interactions of the red-fronted macaw *Ara rubrogenys* in the wild during the midday rest. *Ornitologia Neotropical* 8:133-143.

Rainey, W.E., E.D. Pierson, T. Elmqvist and P.A. Cox 1995. The role of flying foxes (Pteropodidae) in oceanic island ecosystems of the Pacific. *Symposia of the Zoological Society of London* 67:47-62.

Richardson, D.M., P. Pysek, M. Rejmanek, M.G. Barbour, D. Panetta and C.J. West 2000. Naturalization and invasion of alien plants: concepts and definitions. *Diversity and Distributions* 6:93-107.

Rinke, D.R. 1986. The status of wildlife in Tonga. *Oryx* 20:146-151.

Steadman, D.W. 1993. Biogeography of Tongan birds before and after human impact. *Proceedings of the National Academy of Sciences, U.S.A.* 90:818-822.

Steadman, D.W. 1995. Prehistoric extinctions of Pacific island birds. *Science* 267:1123-1131.

Steadman, D.W. 2006. *Extinction and biogeography of tropical Pacific birds*. Chicago: University of Chicago Press.

Steadman, D.W. and H.B. Freifeld 1999. The food habits of Polynesian pigeons and doves: A systematic and biogeographic review. *Ecotropica* 5:13-33.

Steadman, D.W., J. Franklin, D.R. Drake, H.B. Freifeld, L.A. Bolick, D.S. Smith and T.J. Motley 1999. Conservation status of forests and vertebrate communities in the Vava'u Island Group, Tonga. *Pacific Conservation Biology* 5:191-207.

Steadman, D.W. and P.A. Martin 2003. The late Quaternary extinction and future resurrection of birds on Pacific islands. *Earth Science Reviews* 61:133-147.

Sun, C. and T.C. Moermond 1997. Foraging ecology of three sympatric turacos in a montane forest in Rwanda. *Auk* 114:396-404.

Thibault, J.C., J.L. Martin, A. Penloup and J.Y. Meyer 2002. Understanding the decline and extinction of monarchs (Aves) in Polynesian Islands. *Biological Conservation* 108:161-174.

Thorsen, M., R. Shorten, R. Lucking and V. Lucking 2000. Norway rats *(Rattus norvegicus)* on Fregate Island, Seychelles: The invasion; subsequent eradication attempts and implications for the island's fauna. *Biological Conservation* 96:133-138.

Tirado Herrera, E.R., T. Franke, C. Knogge, J. Skrabal and E.W. Heymann 2003. Flower and Fruit Visitors of *Marcgravia longifolia* in Amazonian Peru. *Plant Biology* 5:210-214.

United States Census Bureau 2001. Census 2000 Population Counts for American Samoa. United States Census Bureau Public Information Office. http://www.census.gov/census2000/americansamoa.html.

United States Naval Observatory 2007. Complete Sun and Moon Data for One Day. USNO website. http://aa.usno.navy.mil/data/docs/RS_OneDay.php.

Utzurrum, R.C.B. 2002. Flying foxes (fruit bats). In P. Craig (ed) *Natural history guide to American Samoa: A collection of articles*, pp34-36. Pago Pago: National Park of American Samoa and Department of Marine and Wildlife Resources.

van Riper C. III, S.G. van Riper and A.J. Berger 1979. The red-whiskered bulbul in Hawaii. *Wilson Bulletin* 91:323-328.

Watling, D. 2004. *A Guide to the Birds of Fiji and Western Polynesia.* Suva: Environmental Consultants Ltd.

Webb, E.L., B.J. Stanfield and M.L. Jensen 1999. Effects of topography on rainforest tree community structure and diversity in American Samoa, and its implications for frugivore and nectivore populations. *Journal of Biogeography* 26:887-897.

Western Regional Climate Center 2005. *Pago Pago, American Samoa: Normals, means, and extremes.* Western Regional Climate Center. http://www.wrcc.dri.edu/cgi-bin/clilcd.pl?pi61705.

Whistler, W.A. 1992. *Flowers of the Pacific Island seashore: a guide to the littoral plants of Hawai'i, Tahiti, Samoa, Tonga, Cook Islands, Fiji, and Micronesia*. Honolulu: Isle Botanica.

Whistler, W.A. 1994. *Botanical inventory of the proposed Tutuila and Ofu units of the National Park of American Samoa.* Cooperative National Park Resources Studies Unit Technical Report 87. Honolulu: University of Hawaii at Manoa, Department of Botany.

Whistler, W.A. 2002. *The Samoan Rainforest: a guide to the vegetation of the Samoan archipelago*. Honolulu: Isla Botanica.

Wiles, G.J., J. Engbring and D. Otobed 1997. Abundance, biology, and human exploitation of bats in the Palau Islands. *Journal of Zoology* 241:203-227.

Williams, R.N. and L. van Giddings 1984. Differential range expansion and populations growth of bulbuls in Hawaii. *Wilson Bulletin* 96:647-655.

World Gazetteer 2006. American Samoa: largest cities and towns and statistics of their population. World Gazetteer online. http://www.world-gazetteer.com

Yap, C.A.M. and N.S. Sodhi 2004. Southeast Asian invasive birds: Ecology, impact and management. *Ornithological Science* 3:57-67.

8

Megafaunal extinctions and their consequences in the tropical Indo-Pacific

Richard T. Corlett

Department of Biological Sciences, National University of Singapore, Singapore
corlett@nus.edu.sg

Introduction

The global Quaternary Megafauna Extinction (QME) event eliminated two-thirds of all mammal genera and half (c. 178) of all species of body mass >44 kg, with most well-dated extinctions occurring between c. 50,000 and 3000 years ago (Barnosky 2008). The QME differed from other extinction events in the fossil record by the absence of replacements, the fact that similar episodes occurred in widely separated sites at different times, and the fact that the timing can often be loosely linked with the arrival of modern humans. The causes and consequences of this event have been debated at length in the literature for Australia, Madagascar, northern Eurasia, and both North and South America, with Africa often used as a 'control' in comparisons (Koch and Barnosky 2006; Barnosky 2008; Johnson 2009), but tropical Asia, Wallacea (the Indonesian Islands separated by deep water from the Asian and Australian continental shelves) and New Guinea have generally been omitted from these debates. This partly reflects uncertainties in the fossil and archaeological records of the region – in particular, about the timing of large vertebrate extinctions in relation to the arrival first of early *Homo* and then of modern humans (Bird et al. 2004; Louys et al. 2007; Corlett 2009a). However, the Asian mainland, all large islands and many smaller ones supported more large vertebrate species and much more large vertebrate biomass in the Pleistocene than they do now (Long et al. 2002; Louys et al. 2007; Louys 2008) and an ongoing extinction episode threatens all the survivors (Corlett 2007). Here, I ask what has been lost, when and why, and what the ecological consequences have been.

Who was where when?

The earliest human remains in the region are assigned to *Homo erectus*. Most skeletal remains are from Java, 1.0-1.8 million years ago, but the earliest securely dated stone tools are from the island of Flores (Morwood et al. 1998) and South China (Hou et al. 2000), only 800,000 years ago. Most of the faunal remains associated with *H. erectus* finds suggest open woodland or savanna and it is not clear whether the species could live in closed forest. Other unknowns are its ability to manipulate fire and to make voluntary sea crossings, such as those needed to reach Flores. Later Asian populations of *H. erectus* appear to have diverged increasingly from their African ancestors, and mainland East Asia was occupied between about 500,000 and 70,000 years ago by hominids who may have represented evolutionary developments from local *H. erectus* populations and/or archaic forms of *H. sapiens* and/or additional *Homo* species (Bacon et al. 2006; Louys et al. 2007). A remarkable dwarf hominin, *H. floresiensis*, inhabited a limestone cave on Flores from 95 ka to 16.6 ka (Moore et al. 2009).

There is no clear archaeological evidence for anatomically and culturally modern humans in Eurasia before about 45,000-50,000 years ago. This fits with the growing amount of archaeological, genetic, pollen and charcoal evidence in support of a single dispersal event that brought modern humans from Africa, where they had emerged 150,000-200,000 years ago, along the coastlines of South and Southeast Asia, to New Guinea and Australia, within a period of a few thousand years (Mellars 2006; Pope and Terrell 2008). This 'coastal express train' model of modern human dispersal assumes that the pioneer populations lived initially on coastal resources, moving on as those resources became depleted, and only later moved inland.

When was what where?

There is no clear functional justification for a 44 kg cut-off in the definition of megafauna (Martin and Klein 1984; Hansen and Galetti 2009), but this limit has been widely used in the literature and is followed here for convenience. Owen-Smith's 1000 kg minimum has much stronger functional justification (Owen-Smith 1988), but includes few Pleistocene taxa in the region between tropical China and New Guinea covered here. Holocene extinctions of smaller (<44 kg) mammals (only island rats and bats in this region) are covered by Turvey (2009b). The only known terrestrial megafaunal or near-megafaunal species on the tropical islands east of New Guinea were the land turtles, terrestrial crocodiles and giant flightless birds that persisted into the Holocene on New Caledonia and Fiji (Turvey 2009a).

It would be interesting to examine the large vertebrate fauna of the region over the full period of hominin occupation, but the problems of interpretation of the fossil record increase rapidly as one goes back in time, so this review is limited to the past million years, with the main focus on the late Pleistocene, from c. 130 ka, and the Holocene (Table 1). This summary builds on those by Long et al. (2002), Louys et al. (2007) and Louys (2008), updated where necessary. I have checked the primary sources wherever possible and taken a critical stance when examining the Asian literature, excluding some doubtful records. Scientific names of all taxa are given in Table 1. This table underestimates the regional loss of megafaunal species because late Pleistocene fossils have usually been assigned to the nearest living relative, or to a single extinct taxon, while recent molecular studies have often divided extant taxa (e.g. orangutans, clouded leopards) into geographically separated species. It also gives a very misleading impression of the extent of local megafaunal losses, since many large vertebrates still persist somewhere in the region, despite occupying <10% of their maximum Holocene ranges and an even smaller percentage of their maximum Pleistocene ranges.

Table 1. Terrestrial vertebrates with body mass >44 kg recorded from the Indo-Pacific region (tropical China to New Guinea) from the Late Pleistocene to present day, showing the maximum geographical range, most recent time period when the species was extant within the region, and estimated range of body masses. The table *underestimates* extinctions, since Late Pleistocene fossils have usually been assigned to the nearest living relative or to a single extinct taxon, while recent molecular studies have tended to divide extant taxa into geographically separated species. Almost all extant species occupy <10% of their maximum Holocene range and an even smaller percentage of their Pleistocene one (Table 1 continues on page 120)

Taxon		Maximum range	Last known extant	Mass (kg)	Common name
SQUAMATA					
Varanidae	*Varanus* spp.	Wallacean islands	late Pleistocene	??	monitor lizards
	Varanus komodoensis	Flores, Komodo etc.	extant	<165	Komodo dragon
	V. salvadorii	New Guinea	extant	<50?	crocodile monitor
Pythonidae	*Morelia* spp.	New Guinea	extant	??	pythons
	Python molurus	S. China-Sulawesi	extant	<140	Burmese python
	P. reticulatus	SE Asia-Wallacea	extant	<160	reticulated python
TESTUDINATA					
Testudinidae	*Geochelone* spp.	Wallacean islands	Pleistocene	??	giant tortoises
CASUARIIFORMES					
Casuariidae	*Casuarius casuarius*	New Guinea	extant	30-60	southern cassowary
	C. unappendiculatus	New Guinea	extant	40-60	northern cassowary
DIPROTODONTIA					
Diprotodontidae	*Hulitherium*	New Guinea	after 40,000 BP	150	
	Maokopia	New Guinea	after 40,000 BP	100	
	Zygomaturus	New Guinea	after 40,000 BP	300	
Macropodidae	*Protemnodon* spp.	New Guinea	after 40,000 BP	45-100	kangaroo
PROBOSCIDEA					
Elephantidae	*Elephas* spp.	Sulawesi, Luzon	Pleistocene	??	dwarf elephants
	Elephas maximus	China to Java	extant	2500-4000	Asian elephant
	Palaeoloxodon naumanni	China, Vietnam	late Pleistocene	??	straight-tusked elephant
	Stegodon spp.	China to Timor	mid P. to Holocene	300-3000?	stegodons
PRIMATES					
Hominidae	*Pongo* spp.	China to Java	late P. to Holocene	>50	orangutans
	P. abelii	Sumatra	extant	40-90	Sumatran orangutan
	P. pygmaeus	Borneo	extant	30-90	Bornean orangutan
PHOLIDOTA					
Manidae	*Manis* sp.	Borneo	40,000 BP	50?	giant pangolin
CARNIVORA					
Felidae	*Panthera pardus*	China to Java	extant	30-70	leopard
	P. tigris	China to Java	extant	75-250	tiger
Hyaenidae	*Crocuta crocuta*	China to Thailand	extant (Africa)	45-80	spotted hyena
	Pachycrocuta brevirostris	China to Java	late Pleistocene?	>100	giant hyena
Ursidae	*Ailuropoda melanoleuca*	China to Thailand	extant	70-115	giant panda

Taxon		Maximum range	Last known extant	Mass (kg)	Common name
	Helarctos malayanus	China to Java	extant	25-65	sun bear
	Ursus thibetanus	China to Thailand	extant	50-180	Asiatic black bear
PERISSODACTYLA					
Rhinocerotidae	*Dicerorhinus sumatrensis*	China to Borneo	extant	500-1000	Sumatran rhinoceros
	Rhinoceros philippinensis	Luzon	Pleistocene	<800	
	R. sinensis	China, Vietnam	after 40,000 BP	>1000	
	R. sondaicus	China to Java	extant	1500-2000	Javan rhinoceros
	R. unicornis	China to Java	extant (S. Asia)	1600-3000	Indian rhinoceros
Tapiridae	*Megatapirus augustus*	China to Laos	Holocene	>500	giant tapir
	Tapirus indicus	China to Java	extant	200-400	Malayan tapir
ARTIODACTLYA					
Suidae	*Babyrousa* spp.	Sulawesi & islands	extant	50-100	babirusa
	Sus spp.	China to Sulawesi	extant	10-150	pigs
Cervidae	*Axis calamianensis*	Calamian Islands	extant	30-90	Calamian deer
	A. kuhlii	Bawean	extant	36-50	Bawean deer
	A .porcinus	China to Thailand	extant	36-50	hog deer
	Elaphodus davidianus	China	recent	150-200	Père David's deer
	Cervus eldii	China to Thailand	extant	<150	Eld's deer
	Rucervus schomburgki	Thailand	1932	100-120	Schomburgk's deer
	Rusa alfredi	Visayan Islands	extant	25-80	Visayan spotted deer
	R. marianna	Philippines	extant	40-60	Philippine deer
	R. timorensis	Java &Bali	extant	75-160	Javan rusa
	R. unicolor	China to Borneo	extant	100-300	sambar
Bovidae	*Bos frontalis*	China to Malaysia	extant	700-1500	gaur
	B. javanicus	China to Java	extant	600-800	banteng
	B. sauveli	Vietnam to Thailand	recent	<900	kouprey
	Bubalus spp.	Philippines	late Pleistocene		
	B. arnee	Myanmar-Java	extant	800-1200	wild water buffalo
	B. depressicornis	Sulawesi	extant	150-300	lowland anoa
	B. mephistopheles	China	mid Holocene	<1000	short-horned water buffalo
	B. mindorensis	Mindoro	extant	180-300	tamaraw
	B. quarlesi	Sulawesi	extant	150-300	mountain anoa
	Pseudoryx nghetinhensis	Laos, Vietnam	extant	100	saola
	Capricornis milneedwardsii	China to Thailand	extant	50-140	Chinese serow
	C. sumatraensis	Thailand to Java	extant	50-140	Sumatran serow

New Guinea

The largest extant varanid in New Guinea (*Varanus salvadorii*) and the largest pythons (*Morelia* spp.) may occasionally reach megafaunal size, and New Guinea supports two megafaunal cassowary species (Wright 2005). The late Pleistocene fauna of the New Guinea highlands also included several now-extinct marsupials of megafaunal size, including large browsing kangaroos (*Protemnodon* spp.) and diprotodontids (*Hulitherium*, *Maokopia*, *Zygomaturus*) (Long et al. 2002; Fairbairn et al. 2006).

Wallacea (excluding Sulawesi)

Potentially megafaunal reptiles on the oceanic islands of Wallacea (Figure 1) included varanids, pythons and giant tortoises. Megafaunal varanids (*Varanus komodoensis*) are now confined to Komodo, western Flores and a couple of nearby islands, and all giant tortoises are extinct, but pythons (*Python reticulatus*) occur on most islands, apparently as natives. However, although this species reaches megafaunal sizes on the mainland and continental islands, with a 6-7 m snake weighing >50 kg (Fredriksson 2005), at least some of the Wallacean forms are dwarfed (Auliya et al. 2002). Proboscids were apparently the only mammals able to swim the sea straits east of Bali. Pleistocene fossil dwarf stegodons are known from several localities on Timor, along with a giant tortoise and a large varanid, but all these seem to have vanished before the first evidence for modern humans at 40,000 BP (O'Connor 2007). Fossil dwarf stegodons of uncertain age have also been reported from Sumba (Sartono 1979) and from the tiny island of Sangihe, between Sulawesi and Mindanao (van den Bergh et al. 1996). On Flores, a dwarf (c. 300 kg) stegodon species from 900 ka was replaced by 850 ka by a larger species, which in turn became dwarfed, before disappearing c. 12 ka (van den Bergh et al. 2008). The Komodo dragon persisted throughout this sequence and is the only extant native megafaunal species, but a giant tortoise was last recorded at 900 ka.

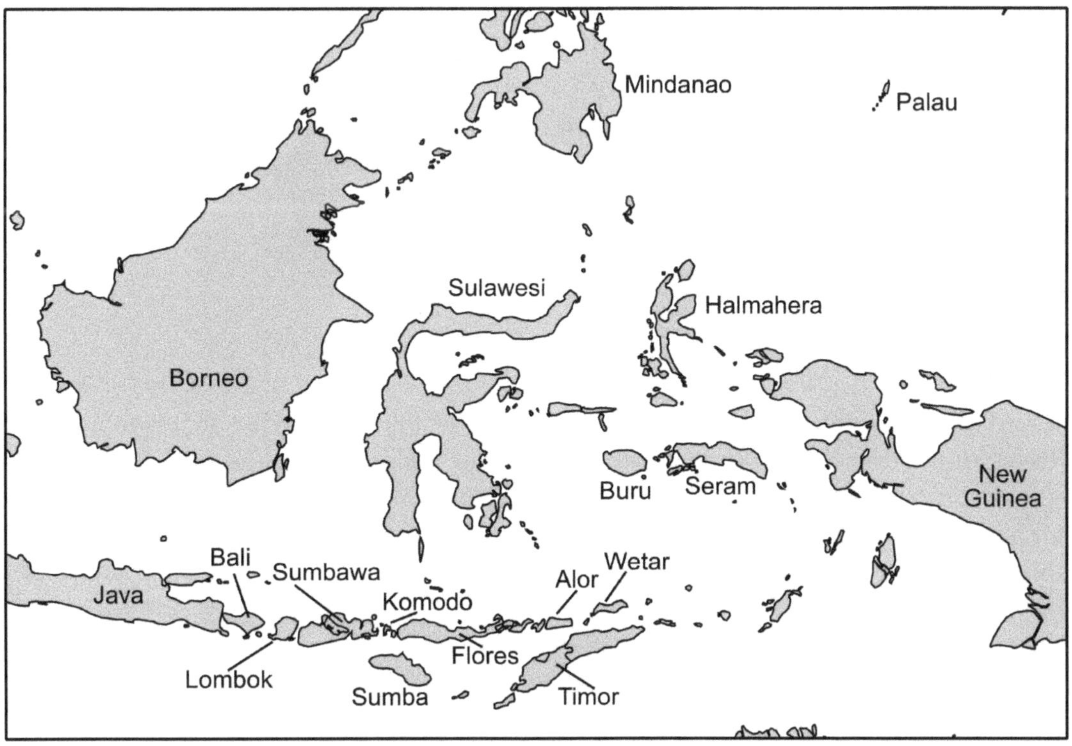

Figure 1. Map of Wallacea and the surrounding region

Sulawesi

At the Last Glacial Maximum, the Makassar Straits separating Sulawesi from Borneo were only 45 km wide at the narrowest point. The Pleistocene fossil fauna had a number of extinct genera, including giant tortoises, stegodonts, elephants and a large suid, *Celebochoerus heekereni*, but it is not clear whether any of these survived into the late Pleistocene (van den Bergh et al. 2001). The extant megafauna includes the babirusas, a pig (*Sus celebensis*), two species of anoa, and the reticulated python.

Philippines

At Pleistocene low sea levels, most of the land area of the Philippines was in six major islands, separated by narrow straits. Only Palawan and the adjacent small islands may ever have been connected to Borneo and it is not clear whether or when this last happened. Palawan supports a subset of the Bornean fauna, including species not found elsewhere in the Philippines, but it also has several endemic species and subspecies. Tigers were present until at least the late Pleistocene and a deer became extinct in the mid Holocene (Piper et al. 2008). The extant Calamian deer is endemic to islands in the Palawan Group. Excluding the Palawan Group, the extant megafauna of the Philippines consists of endemic pigs, deer and the tamaraw, plus the widespread reticulated python (which is possibly a human introduction on some islands). There are also fossils of extinct *Bubalus* species on Luzon and Cebu, suggesting this genus may once have been present throughout the Philippines (Croft et al. 2006). There are poorly dated Pleistocene fossils of stegodonts on Luzon and Mindanao, and fossils of both elephants and rhinoceroses on Luzon (Bautista 1991; van den Bergh et al. 1996).

Java

The megafauna coexisting with *H. erectus* on middle Pleistocene Java included stegodons, hippopotamuses and hyenas (Louys et al. 2007), but Java appears to have had a fully modern fauna by 120 ka (Westaway et al. 2007), and – assuming that the fossils are correctly attributed to modern species – no species recorded since then has become globally extinct. The late Pleistocene megafauna consisted of the Asian elephant, orangutan, tiger, leopard, sun bear, Malayan tapir, Javan rhinoceros, Javan rusa, banteng, wild water buffalo, Sumatran serow, Javan warty pig, Eurasian wild pig and two species of python. The elephant, orangutan, tiger, sun bear, tapir and serow disappeared in prehistoric to early colonial times and the Javan rhinoceros has been reduced to a single population of 60 individuals. An extant endemic species of deer occurs on Bawean Island, 150 km north of Java.

Borneo and Sumatra

The only fossil evidence of proboscideans from the huge continental island of Borneo is teeth of uncertain provenance (Cranbrook and Piper 2007), while the current population of wild elephants in the northeast seems best explained by the traditional story of introduction from Java via Sulu (Cranbrook et al. 2008). A giant (probably megafaunal) species of pangolin is known from parts of a single individual dated at c. 40,000 ka (Piper et al. 2007a). The known Holocene megafauna included three taxa: Javan rhinoceros (Cranbrook and Piper 2007), Malayan tapir (Cranbrook and Piper 2009) and tiger (Piper et al. 2007b), which became extinct before – possibly just before – modern times. The surviving megafauna consists of the Bornean orangutan, sun bear, Sumatran rhinoceros, bearded pig, sambar deer, banteng and reticulated python. Sumatra has a poorer fossil record than Borneo and the only known megafaunal extinction to occur between the late Pleistocene and historical times is the leopard (Louys et al. 2007), while the Javan rhinoceros is a recent extinction. In contrast to Borneo, elephants, tigers and tapirs still survive, along with the Sumatran orangutan, sun bear,

Sumatran rhinoceros, bearded pig, sambar deer and reticulated python. All but pigs, deer, tapirs and pythons are currently severely threatened by hunting (Corlett 2007).

Continental Southeast Asia and South China

During the middle Pleistocene, hominins in South China and Vietnam coexisted with a diverse megafauna that included a giant (estimated 540 kg) ape, *Gigantopithecus*, which became extinct around 300 ka (Rink et al. 2008). Some elements of this fauna persisted into the late Pleistocene and several taxa that are now globally extinct survived into the early Holocene. These late survivors include at least one species of stegodon, a giant tapir and the spotted hyena (Louys et al. 2007). Note, however, that these Holocene dates are considered questionable by Turvey (2009a). In the middle to late Pleistocene, orangutans and giant pandas occurred at many sites from southern China to Thailand (Louys et al. 2007; Zhang et al. 2007). The Malayan tapir is also recorded from late Pleistocene sites in China.

Megafaunal range declines and extinctions have been better documented in China than elsewhere, with more archaeological sites and longer historical records. Stegodons, Malayan and giant tapirs, hyenas and orangutans disappeared before the start of historical records. The short-horned water buffalo (*Bubalus mephistopheles*) became globally extinct in early historical times (Yang et al. 2008). The decline of the giant pandas started in the late Pleistocene and accelerated in historical times (Loucks et al. 2001; Zhang et al. 2007), while the retreat of the elephants and rhinoceroses (one to two species) has occurred largely over the past 3000 years, with a progressive withdrawal of the northern boundaries of their distributions in the face of growing population pressures (Corlett 2007). Historical data suggest a threshold value of about four people per square kilometres for rhinos, and 20 people for elephants, above which they did not persist (Liu 1998). Rhinoceroses became extinct within the past 30 years in China and elephants have been reduced to tiny populations near the southwestern borders. The declines started later in continental Southeast Asia, but only a tiny population of Javan rhinoceroses and a few scattered Sumatran rhinoceroses still survive, and the Asian elephant occupies a fraction of its 19th century range (Corlett 2007). Tigers have been eliminated from southern China and are in rapid decline everywhere else in the region, while leopards and the two Southeast Asian bear species have been eliminated from most of the more accessible parts of their continental ranges.

All tropical Asian bovids are currently threatened by hunting (Corlett 2007) and the kouprey may be extinct. Deer are often considered relatively resilient to hunting, but all species have declined severely in recent decades, including the widespread sambar, and this group includes the only two global extinctions (in the wild) in the region within the past 150 years. Schomburgk's deer inhabited marshy grasslands in the central plains of Thailand, but has been extinct since 1938, while further north, Père David's deer was eliminated from riverine marshland in the lower reaches of the Yangtze River more than a century ago and recently has been reintroduced. Eld's deer, which occurs in open forests, has also had its range drastically reduced over the past century. Pigs are the 'last large mammal standing' in many areas, but pig populations have shown massive recent declines as a result of overhunting.

Why?

Many early and middle Pleistocene extinctions appear to represent a natural process of species turnover, although there are some conspicuous exceptions, such as the giant tortoises and hippopotamuses (Louys et al. 2007), but by the late Pleistocene, extinctions without replacement predominate, matching the pattern seen in the Quaternary Megafauna Extinction event elsewhere on the planet (Barnosky 2008). Unlike other well-documented examples of the QME, however, there is no evidence to suggest a single, sharp extinction event, and no obvious peak with the presumed arrival of modern humans around 50,000 years ago. Instead, there has been a steady trickle of regional then global extinctions, mostly within the past 40,000 years and many within the Holocene. The inadequacies of the fossil record and its dating have inevitably blurred the true history of megafaunal extinctions in the region, but surely not enough to account for the broad spread of apparent extinction dates.

The megafaunal losses described above have been attributed by different authors to various combinations of direct (i.e. hunting) and/or indirect (habitat modification) human impacts and natural environmental change. If the region had been completely uninhabited for the past million years, it would be possible to build a plausible case for attributing most of the pre-recent losses to environmental change, in particular the development of dense, closed-canopy forest over all but the driest parts of the region at the start of the Holocene. Ground-dwelling herbivores are denied access to the majority of the leaf production in a closed-canopy forest, since this takes place in the canopy. A decline in the density of terrestrial herbivores in turn threatens the largest carnivore, the tiger, which does not have access to arboreal prey, as well as putting pressure on the leopard. The extinction of the giant pangolin may also reflect the loss of open habitats that have a high density of accessible ant and termite nests at the start of the Holocene (Medway 1972). A problem with such environmental explanations, however, is they require that late Pleistocene environments and/or the transition to the Holocene were uniquely difficult periods for megafaunal species. Each glacial and interglacial period is different, but there is no independent evidence that the past 100,000 years has been significantly more difficult for large vertebrates than previous glacial cycles (Barnosky et al. 2004). Volcanoes have also been invoked as agents of extinction, particularly on Flores, where faunal turnovers at 900 ka and 12 ka both coincided with volcanic eruptions. However, the largest eruption on Earth during the late Pleistocene, that of Toba on Sumatra 74,000 years ago, did not coincide with any known extinctions in the region (Louys 2007).

Environmental explanations that may be plausible in the absence of people become less plausible in their presence. The capabilities of *Homo erectus* are largely unknown, but it is hard to imagine a large, relatively intelligent, omnivorous, social primate coexisting for long with something as vulnerable as a giant tortoise. Giant tortoises are known from Sulawesi, Flores and other islands in the early to middle Pleistocene, and disappeared on Flores around the time hominins, presumably *H. erectus*, arrived (van den Bergh et al. 2001). *H. erectus* was associated at various times and places with other large vertebrates that no longer survive in the region. A dwarf stegodon was lost from Flores at the same time as the giant tortoise, and the bones of its dwarfed successor are associated with those of *H. floresiensis* in late Pleistocene cave deposits (van den Bergh et al. 2008). Extinct megafauna, including elephants, stegodons and rhinoceroses, are associated with early stone tools in the Philippines (Bautista 1991). Human fossils and/or stone tools from middle and late Pleistocene cave sites in southern China and Indochina are also often associated with the remains of stegodons and the giant tapir (Ciochon and Olsen 1991; Bekken et al. 2004; Schepartz et al. 2005). Early *Homo* was also associated at some sites with *Gigantopithecus* (Ciochon et al. 1996; Harrison et al. 2002). These associations suggest some form of interaction, but there is no direct evidence for hunting, and scavenging is equally plausible in most cases.

Although the evidence for an impact of *H. erectus* on megafaunal survival is weak, it is surely significant that the clearest association between human arrival and megafaunal loss is in the New Guinea highlands (Fairbairn et al. 2006; Field et al. 2008), which *H. erectus* did not reach. The lack of a single, clear extinction event in tropical Asia could thus simply reflect the complex chronology of hominin habitation. In contrast to most other regions of the world outside Africa, where naive megafauna were faced by expanding populations of technologically sophisticated, habitat-generalist, modern humans, tropical Asia's fauna had a relatively gradual introduction to the perils of humanity. If, as seems likely, *H. erectus* avoided closed-canopy forest and, at least initially, lacked projectile weapons and mastery of fire, impacts may have been slow and selective in comparison with those of modern humans, and may have allowed time for some adaptive evolution.

With modern *Homo sapiens*, there is no need to speculate on its hunting ability, but it is reasonable to question whether human population densities before the second half of the Holocene were high enough to lead to faunal extinctions. The absence of a well-defined mass-extinction event in tropical Asia in the 60,000-40,000 BP period during which modern humans probably arrived is consistent with the 'coastal express train' model, with coastal populations moving on as they depleted resources, and only later moving inland. The extinction of the giant pangolin soon after the presumed arrival date was attributed above to the loss of open habitats, but a slow-moving, 2 m long mammal whose only defence was to roll up must also have been exceptionally vulnerable to hunting. A similar argument can be made for the slow-moving, slow-breeding orangutans, which ranged widely over tropical East Asia in the late Pleistocene, from southern China to Java, but were confined to the sparsely populated ever-wet rainforests of Borneo and Sumatra by historical times (Delgado and van Schaik 2000). The giant panda and the New Guinea megafaunal marsupials may also have been particularly vulnerable to hunters for similar reasons, but there seems to be no obvious pattern to most other pre-recent extinctions.

Many authors have combined the two major hypotheses and argued that a deteriorating environment combined with pressure from hunting or anthropogenic habitat-modification to push species over the edge into extinction. This is in many ways an unsatisfactory compromise, and is hard to test or refute, but it is equally difficult to argue that either factor – drastic changes in climate and vegetation, or the arrival and expansion of successive human species – had no significant impact on megafaunal populations. A predominance of human impacts over environmental change becomes increasingly clear over the past few thousand years and the role of human exploitation in the drastic range reductions shown by all surviving megafaunal species in the past century is undeniable (Corlett 2007).

Consequences

Whatever the causes of the large reductions in megafaunal diversity and biomass over the past 130,000 years, the consequences are likely to have been, and continue to be, significant. Indeed, if megafaunal impacts were significant, the vegetation and associated fauna of much of the region may now be in a state of long-term relaxation from these impacts (Johnson 2009). Very large herbivores may have kept vegetation in a more open, patchy condition than exists today. More generally, they may have acted as 'ecosystem engineers', modifying the physical environment in a way that affects other species (Pringle 2008). The largest surviving megaherbivores, the forest elephants and rhinoceroses, are browsers, with the strength to push down, break off, or uproot shrubs, saplings and small trees (Corlett 2007). Elephant movements can lead to the creation of extensive networks of trails. Loss of megafauna-maintained open habitats would have impacted shade-intolerant plants and smaller terrestrial herbivores that would have suffered reduced access to plant biomass. Large deer and cattle do less incidental damage to vegetation, but are more selective as browsers and grazers. Plants that have coevolved with these large herbivores may invest in physical or chemical defences that are ineffective against smaller herbivores, and may thus lose out in competition with plants that lack these defences when large browsers are removed (Johnson 2009). Dung piles from large herbivores have multiple impacts, creating patchiness and providing habitat for other species (Campos-Arceiz 2009).

In central Africa, forest elephants at natural densities probably disperse more seeds than any other vertebrate species and disperse them over much larger distances (Blake et al. 2009), but dispersal by megaherbivores has received little attention in tropical Asia. Many ripe, fleshy fruits reach the ground uneaten, where many are consumed by large terrestrial herbivores, these herbivores including probably all extant species of pigs, deer, cattle, tapirs, elephants and rhinoceroses (Corlett 1998). The indehiscent pods produced by some legumes are consumed by the same animals. A 'megafaunal syndrome' of very large fleshy fruits (>10 cm diameter with numerous small seeds, or 4-10 cm diameter with a few large seeds) dispersed largely by mammals with >1000 kg body mass has been identified in Africa and Brazil (Guimaraes et al. 2008), but the distribution of such fruits in the tropical Indo-Pacific has not yet been documented. Asian elephants have been reported to prefer large, yellow, sweet-smelling fruits with large, hard seeds (Kitamura et al. 2007). Some studies suggest that they consume less fruit and fewer species than their African relatives, but fruits from 29 species were recorded by mahouts in Myanmar as eaten by work elephants (Campos-Arceiz et al. 2008a), and potential seed dispersal distances are very large (<6 km, Campos-Arceiz et al. 2008b). Rhinoceroses also eat large fruits and can potentially move >10 km within plausible gut-passage times (Corlett 2009b). Tapirs have a similar potential.

The 44 kg cut-off used in this study brings in a wider range of fruit types than the classic megafaunal syndrome described from studies on African elephants. Large deer disperse seeds both by regurgitation of large, hard seeds from fleshy fruits (e.g. Prasad et al. 2006) and by defecation of small ones, with some of the latter swallowed incidentally during consumption of foliage (Myers et al. 2004; Yamashiro and Yamashiro 2006). Pigs and bovids disperse small seeds in the same way as deer, but their role, if any, in dispersing larger seeds is not known. The extinct megafaunal marsupials of New Guinea probably dispersed some seeds in the same way as extant grazing and browsing megaherbivores (Webb 2008), but any role they had in the dispersal of large seeds is probably covered by the extant cassowaries, which still disperse seeds in large fruits (<6 cm) for long distances (Wright 2005; Bradford et al. 2008). Orangutans are the largest arboreal frugivores and eat fruits of many different types, including the biggest species available. Their ability to move large seeds over long distances (>1 km) is shared with only a few other large-bodied frugivores (Corlett 2009b).

Megafaunal introductions

Many megafaunal species have been deliberately introduced by people outside their natural ranges and a proportion of these have established wild populations. These successful invasions provide evidence for the existence of 'empty niches' in communities throughout the region, particularly on islands, although in some cases these niches have been created or expanded recently by human modification of the natural vegetation. Pigs (*Sus scrofa*, *S. celebensis*, Larson et al. 2007) and deer (mostly *Rusa timorensis*) were spread throughout the region during the Holocene. The archaeological record for Flores, for example, shows that the only megafaunal animals at the start of the Holocene were people and Komodo dragons, but that people subsequently introduced first the Sulawesi warty pig (*Sus celebensis*), c. 7000 years ago, and subsequently the Eurasian pig (*S. scrofa*) and several smaller mammal species (van den Bergh et al. 2008). The archeological record of south Sulawesi shows that deer (*Cervus timorensis*) first appeared there about 4000 years ago (Simons and Bulbeck 2004). The elephant population on Borneo is apparently also a recent introduction (Cranbrook et al. 2008).

Megafaunal reintroductions?

If the extinct megafauna played a unique role in the ecosystems they inhabited, it makes sense to consider their reintroduction or replacement. Where megafaunal species have been lost from an area in recent times, but persist elsewhere, then reintroduction is an option. Indeed, many megafaunal species are likely to be favoured by the leaf and shoot biomass available near ground level in disturbed and fragmented habitats. Piper and Cranbrook (2007) propose the reintroduction of the Malayan tapir to the Planted Forest Zone in Sarawak, which consists of 490,000 ha of primary, secondary and industrial plantation forests, while several projects are attempting to re-establish orangutans in areas from which they have been extirpated (Corlett 2009a). With an estimated 16,000 captive elephants in Asia and many of these unwanted as their use in logging declines, the major problem with the Asian elephant is a lack of suitable habitat, rather than a shortage of animals for release (Leimgruber et al. 2008). In contrast, both species of forest rhinoceros are critically endangered and neither has a captive population.

Where the extinction was many tree generations ago, it is likely that the current vegetation represents a new equilibrium in which previously suppressed plant species are more abundant and others less abundant than when the megafauna was present at natural densities (Johnson 2009). While restoring the natural situation may seem a laudable aim, it is also possible that the disruption of fragmented forest communities will simply promote invasive alien species, since these are likely to exert a much greater 'propagule pressure' than any rare native species.

Where the missing taxon is globally extinct, but an ecologically similar relative persists, then the introduction of this as a substitute could be considered. A major problem, however, lies in assessing 'ecological equivalence'. For proboscidians, the Asian elephant is the only Asian survivor of a once diverse group. The fact that some of these other taxa coexisted with Asian elephants in the past suggests that they are *not* complete equivalents, but how much this matters is hard to determine.

Conclusions

The relatively small number of global extinctions in the Indo-Pacific megafauna in comparison with the rest of the world masks a catastrophic collapse in local species diversity and biomass since the middle Pleistocene. The evidence from the region is frustratingly incomplete, but I consider it most likely that hominin impacts have been the major factor behind most large vertebrate extinctions and range restrictions in the past 130,000 years and probably some earlier ones. Large vertebrates have evolved in or invaded every accessible land mass on Earth and appear to have been a major component of all vertebrate communities since soon after vertebrates evolved, apart from the period immediately after the K/T extinctions 65 million years ago. For their disappearance during the brief period following the expansion of the genus *Homo* out of Africa to be a mere coincidence would require much stronger evidence than any that is currently available for uniquely extreme environmental pressures during this time. Any uncertainties about the role of human populations disappear by the mid Holocene, and the devastating human impacts of the past century are well documented.

If an anthropogenic explanation for the megafaunal collapse is accepted as most likely, this then raises the question of why there is no clear evidence of rapid overkill soon after the arrival of humans. I have suggested a two-part explanation for this: that *Homo erectus* lacked (or, at least, initially lacked) the technical skills and forest adaptations needed to cause rapid population collapse, except in the most vulnerable of non-forest taxa, while *H. sapiens* spread initially along the coast and invaded the forested interior only gradually and unevenly. Indeed, much of the region had extremely low human population densities into historical times (Corlett 2009a). Although the megafauna is equally absent whether it was wiped out by climate change or by humans, the former would be an interesting palaeoecological phenomenon, while the latter, if true, is a historical tragedy and one which we would reverse if we could. Yet the surviving megafauna is everywhere in retreat and more extinctions are likely (Corlett 2007). Conservation successes in the Indo-Pacific tend to be local and small scale, while the larger-scale efforts needed to save viable populations of the largest species appear to be beyond our current capabilities. Some species, such as tigers, will survive in captivity whatever their fate in the wild, but others, including the two critically endangered forest rhinoceroses, lack viable captive populations. More than a billion people live in the region covered by this review and this population will continue to rise until at least 2050 (Corlett 2009a). Conservation problems are many and urgent as a result, but the largest vertebrates deserve special attention.

Acknowledgements

Geoff Hope was hugely influential in my early career, so he is the first person I need to acknowledge. I have also benefited greatly from discussions with Ahimsa Campos-Arceiz.

References

Auliya, M., P. Mausfeld, A. Schmitz and W. Boehme 2002. Review of the reticulated python (*Python reticulatus* Schneider, 1801) with the description of new subspecies from Indonesia. *Naturwissenschaften* 89:201-213.

Bacon, A.-M., F. Demeter, S. Rousse, V.T. Long, P. Duringer, P.-O. Antoine, N.K. Thuy, B. T. Mai, N. T. M. Huong, Y. Dodo, H. Matsumura, M. Schuster and T. Anezaki 2006. New palaeontological assemblage, sedimentological and chronological data from the Pleistocene Ma U'Oi cave (northern Vietnam). *Palaeogeography Palaeoclimatology Palaeoecology* 230:280-298.

Barnosky, A.D., P.L. Koch, R.S. Feranec, S.L. Wing and A.B. Shabel. 2004. Assessing the causes of Late Pleistocene extinctions on the continents *Science* 306:70-75.

Barnosky, A.D. 2008. Megafauna biomass tradeoff as a driver of Quaternary and future extinctions. *Proceedings of the National Academy of Sciences of the United States of America* 105:11543-11548.

Bautista, A.P. 1991. Recent zooarchaeological researches in the Philippines. *Jurnal Arkeologi Malaysia* 4:45-58.

Bekken D., L.A. Schepartz, S. Miller-Antonio, H. Yamei and H. Weiwen 2004. Taxonomic abundance at Panxian Dadong, a Middle Pleistocene Cave in South China. *Asian Perspectives* 43: 333-359

Bird, M.I., G. Hope and D. Taylor 2004. Populating PEP II: The dispersal of humans and agriculture through Austral-Asia and Oceania. *Quaternary International* 118-119:145-163.

Blake, S., S.L. Deem, E. Mossimbo, F. Maisels and P. Walsh 2009. Forest elephants: tree planters of the Congo. *Biotropica* 41:459-468.

Bradford, M.G., A.J. Dennis and D.A. Westcott 2008. Diet and dietary preferences of the southern cassowary (*Casuarius casuarius*) in North Queensland, Australia. *Biotropica* 40:338-343.

Campos-Arceiz, A. 2009. Shit happens (to be useful)! Use of elephant dung as habitat by amphibians. *Biotropica* 41:406-407.

Campos-Arceiz, A., T.Z. Lin, W. Htun, S. Takatsuki and P. Leimgruber 2008a. Working with mahouts to explore the diet of work elephants in Myanmar (Burma). *Ecological Research* 23:1057-1064.

Campos-Arceiz, A., A.R. Larrinaga, U.R. Weerasinghe, S. Takatsuki, J. Pastorini, P. Leimgruber, P. Fernando and L. Santamaria 2008b. Behavior rather than diet mediates seasonal differences in seed dispersal by Asian elephants. *Ecology* 89:2684-2691.

Ciochon, R. and J.W. Olsen 1991. Paleoanthropological and archaeological discoveries from Lang Trang caves: a new Middle Pleistocene hominid site from northern Vietnam. *Indo-Pacific Prehistory Association Bulletin* 10:59-73.

Ciochon, R., V. T. Long, R. Larick, L. Gonzalez, R. Grun, J. De Vos, C. Yonge, L. Taylor, H. Yoshida and M. Reagan 1996. Dated co-occurrence of *Homo erectus* and *Gigantopithecus* from Tham Khuyen Cave, Vietnam. *Proceedings of the National Academy of Sciences of the United States of America* 93:3016-3020.

Corlett, R.T. 1998. Frugivory and seed dispersal by vertebrates in the oriental (Indomalayan) region. *Biological Reviews* 73:413-448.

Corlett, R.T. 2007. The impact of hunting on the mammalian fauna of tropical Asian forests. *Biotropica* 39:292-303.

Corlett, R.T. 2009a. *The ecology of tropical East Asia*. Oxford: Oxford University Press.

Corlett, R.T. 2009b. Seed dispersal distances and plant migration potential in tropical East Asia. *Biotropica* 41:592-598.

Cranbrook, E.O., J. Payne and C.M.U. Leh 2008. Origin of the elephants *Elephas maximus* L. of Borneo. *Sarawak Museum Journal* 63(84):95–125.

Cranbrook, E.O. and P.J. Piper. 2007. The Javan rhinoceros *Rhinoceros sondaicus* in Borneo. *Raffles Bulletin of Zoology* 55:217-220.

Cranbrook, E.O. and P.J. Piper 2009. Borneo records of Malay tapir, *Tapirus indicus* Desmarest: a zooarchaeological and historical review. *International Journal of Osteoarchaeology* 19:491-507

Croft, D.A., L.R. Heaney, J.J. Flynn and A.P. Bautista 2006. Fossil remains of a new, diminutive *Bubalus* (Artiodactyla : Bovidae : Bovini) from Cebu Island, Philippines. *Journal of Mammalogy* 87:1037-1051.

Delgado, R.A., Jr. and C.P. van Schaik 2000. The behavioral ecology and conservation of the orangutan (*Pongo pygmaeus*): A tale of two islands. *Evolutionary Anthropology* 9:201-218.

Fairbairn, A.S., G.S. Hope and G.R. Summerhayes 2006. Pleistocene occupation of New Guinea's highland and subalpine environments. *World Archaeology* 38:371-386.

Field, J., M. Fillios and S. Wroe 2008. Chronological overlap between humans and megafauna in Sahul (Pleistocene Australia-New Guinea): A review of the evidence. *Earth Science Reviews* 89:97-115.

Fredriksson, G.M. 2005. Predation on sun bears by reticulated python in East Kalimantan, Indonesian Borneo. *Raffles Bulletin of Zoology* 53:165-168.

Guimaraes, P.R., Jr., M. Galetti and P. Jordano 2008. Seed dispersal anachronisms: rethinking the fruits extinct megafauna ate. *PLoS One* 3: Article No. e1745.

Hansen, D.M. and M. Galetti 2009. The forgotten megafauna. *Science* 324:42-43.

Harrison, T., J. Xueping and D. Su 2002. On the systematic status of the late Neogene hominoids from Yunnan Province, China. *Journal of Human Evolution* 43:207-227.

Hou, Y.M., R. Potts and B.Y. Yuan 2000. Mid-Pleistocene Acheulean-like stone technology of the Bose basin, South China. *Science* 287:1622-1626.

IUCN. 2008. *2008 IUCN Red List of threatened species*. United Kingdom: International Union for Conservation of Nature and Natural Resources.

Johnson, A., R. Bino and P. Igag 2004. A preliminary evaluation of the sustainability of cassowary (Aves: Casuariidae) capture and trade in Papua New Guinea. *Animal Conservation* 7:129-137.

Johnson, C.N. 2009. Ecological consequences of Late Quaternary extinctions of megafauna. *Proceedings of the Royal Society B* 276: 2509-2519.

Kitamura, S., T. Yumoto, P. Poonswad and P. Wohandee 2007. Frugivory and seed dispersal by Asian elephants, *Elephas maximus*, in a moist evergreen forest of Thailand. *Journal of Tropical Ecology* 23:373-376.

Koch, P.L. and A.D. Barnosky 2006. Late Quaternary extinctions: State of the debate. *Annual Review of Ecology, Evolution and Systematics* 37: 215-250

Larson, G., T. Cucchi, M. Fujita et al. 2007. Phylogeny and ancient DNA of *Sus* provides insights into neolithic expansion in island southeast Asia and Oceania. *Proceedings of the National Academy of Sciences of the United States of America* 104:4834-4839.

Leimgruber, P., B. Senior, Uga, M. Aung, M.A. Songer, T. Mueller, C. Wemmer and J.D. Ballou 2008. Modeling population viability of captive elephants in Myanmar (Burma): implications for wild populations. *Animal Conservation* 11:198-205.

Liu, H. 1998. The change of geographical distribution of two Asian species of rhinoceros in Holocene. *Journal of Chinese Geography* 8:83-88.

Long, J., M. Archer, T. Flannery and S. Hand 2002. *Prehistoric mammals of Australia and New Guinea: one hundred million years of evolution.* Sydney: University of New South Wales Press.

Loucks, C.J., Z. Lu, E. Dinerstein, H. Wang, D.M. Olson, C. Zhu and D. Wang 2001. Ecology: Giant pandas in a changing landscape. *Science* 294:1465.

Louys, J. 2007. Limited effect of the Quaternary's largest super-eruption (Toba) on land mammals from Southeast Asia. *Quaternary Science Reviews* 26:3108-3117.

Louys, J. 2008. Quaternary extinctions in southeast Asia. In A.M.T. Elewa (ed) *Mass extinction*, pp159-190 Berlin: Springer.

Louys, J., D. Curnoe and H. Tong 2007. Characteristics of Pleistocene megafauna extinctions in Southeast Asia. *Palaeogeography Palaeoclimatology Palaeoecology* 243:152-173.

Martin, P.S. and R.G. Klein 1984. *Quaternary extinctions: a prehistoric revolution.* Tucson: University of Arizona Press.

Medway, L. 1972. The Quaternary mammals of Malesia: a review. In P. S. Ashton and H. M. Ashton (eds) *The Quaternary era in Malesia*, pp 63-98. Hull: University of Hull.

Mellars, P. 2006. Going east: New genetic and archaeological perspectives on the modern human colonization of Eurasia. *Science* 313:796-800.

Moore, M. W., T. Sutikna, Jatmiko, M. Morwood and A. Brumm 2009. Continuities in stone flaking technology at Liang Bua, Flores, Indonesia. *Journal of Human Evolution* 57:503-526.

Morwood, M.J., P.B. O'Sullivan, F. Aziz and A. Raza 1998. Fission-track ages of stone tools and fossils on the east Indonesian island of Flores. *Nature* 392:173-176.

Myers, J.A., M. Vellend and S. Gardescu 2004. Seed dispersal by white-tailed deer: Implications for long-distance dispersal, invasion and migration of plants in eastern North America. *Oecologia* 139:35-44.

O'Connor, S. 2007. New evidence from East Timor contributes to our understanding of earliest modern human colonisation east of the Sunda Shelf. *Antiquity* 81:523-535.

Owen-Smith, R.N. 1988. *Megaherbivores: the influence of very large body size on ecology.* New York: Cambridge University Press.

Piper, P.J. and E.O. Cranbrook 2007. The potential of large protected plantation areas for the secure re-introduction of Borneo's lost 'megafauna': a case for the Malay tapir *Tapirus indicus*. In R. Stuebing, J. Unggang, J. Ferner, J. Ferner, B. Giman and K.K. Ping, (eds) *Proceedings of the Regional Conference of Biodiversity Conservation in Tropical Planted Forests in Southeast Asia*, pp184-191. Kuching: Forest Department.

Piper, P.J., R.J. Rabett and E.O. Cranbrook 2007a. New discoveries of an extinct giant pangolin (*Manis* cf. *palaeojavanica* Dubois) at Niah Cave, Sarawak, Borneo: biogeography, palaeoecology and taxonomic relationships. *Sarawak Museum Journal* 84:207-226.

Piper, P.J., E.O. Cranbrook and R.J. Rabett 2007b. Confirmation of the presence of the tiger *Panthera tigris* (L.) in late Pleistocene and Holocene Borneo. *Malayan Nature Journal* 59:257-265.

Piper, P.J., J. Ochoa, H. Lewis, V. Paz and W.P. Ronquillo 2008. The first evidence for the past presence of the tiger *Panthera tigris* (L.) on the island of Palawan, Philippines: Extinction in an island population. *Palaeogeography Palaeoclimatology Palaeoecology* 264:123-127.

Pope, K.O. and J.E. Terrell 2008. Environmental setting of human migrations in the circum-Pacific region. *Journal of Biogeography* 35:1-21.

Prasad, S., J. Krishnaswamy, R. Chellam and S.P. Goyal 2006. Ruminant-mediated seed dispersal of an economically valuable tree in indian dry forests. *Biotropica* 38:679-682.

Pringle, R.M. 2008. Elephants as agents of habitat creation for small vertebrates at the patch scale. *Ecology* 89:26-33.

Rink, W. J., W. Wei, D. Bekken and H.L. Jones 2008. Geochronology of *Ailuropoda-Stegodon* fauna and *Gigantopithecus* in Guangxi Province, southern China. *Quaternary Research* 69:377-387.

Sartono, S. 1979. The discovery of a pygmy stegodon from Sumba, East Indonesia: an announcement. *Modern Quaternary Research in Southeast Asia* 5:57-63.

Schepartz, L.A., S. Stoutamire and D.A. Bekken 2005. *Stegodon orientalis* from Panxian Dadong, a Middle Pleistocene archaeological site in Guizhou, South China: Taphonomy, population structure and evidence for human interactions. *Quaternary International* 126-128:271-282.

Simons, A. and D. Bulbeck 2004. Late Quaternary faunal successions in South Sulawesi, Indonesia. *Modern Quaternary Research in Southeast Asia* 18:167-190.

Turvey, S.T. 2009a. In the shadow of the megafauna: prehistoric mammal and bird extinctions across the Holocene. In S.T. Turvey (ed.) *Holocene extinctions*, pp17-39. Oxford: Oxford University Press.

Turvey, S.T. 2009b. Holocene mammal extinctions. In S.T. Turvey (ed.) *Holocene extinctions*, pp41-61. Oxford: Oxford University Press.

van den Bergh, G., P.Y. Sondaar, J. De Vos and F. Aziz 1996. The proboscideans of the South-East Asian islands. In J. Shoshani and P. Tassy (eds) *The Proboscidea: evolution and palaeoecology of elephants and their relatives*, pp240-248. Oxford: Oxford University Press.

van den Bergh, G.D., R.D. Awe, M.J. Morwood, T. Sutikna, Jatmiko and E. Wahyu Saptomo 2008. The youngest stegodon remains in Southeast Asia from the Late Pleistocene archaeological site Liang Bua, Flores, Indonesia. *Quaternary International* 182:16-48.

van den Bergh, G.D., J. de Vos and P.Y. Sondaar 2001. The Late Quaternary palaeogeography of mammal evolution in the Indonesian Archipelago. *Palaeogeography Palaeoclimatology Palaeoecology* 171:385-408.

Webb, S. 2008. Megafauna demography and late Quaternary climatic change in Australia: A predisposition to extinction. *Boreas* 37:329-345.

Westaway, K.E., M.J. Morwood, R.G. Roberts, A.D. Rokus, J.X. Zhao, P. Storm, F. Aziz, G. van den Bergh, P. Hadi, Jatmiko and J. de Vos 2007. Age and biostratigraphic significance of the Punung Rainforest Fauna, East Java, Indonesia and implications for *Pongo* and *Homo*. *Journal of Human Evolution* 53:709-717.

Wright, D.D. 2005. Diet, keystone resources and altitudinal movement of dwarf cassowaries in relation to fruiting phenology in a Papua New Guinean rainforest. In J. L. Dew and J. P. Boubli (eds) *Tropical fruits and frugivores: the search for strong interactors*, pp205-236. Dordrecht: Springer.

Yamashiro, A. and T. Yamashiro 2006. Seed dispersal by kerama deer (*Cervus nippon keramae*) on Aka Island, the Ryukyu Archipelago, Japan. *Biotropica* 38:405-413.

Yang, D.Y., L. Liu, X. Chen and C.F. Speller 2008. Wild or domesticated: DNA analysis of ancient water buffalo remains from North China. *Journal of Archaeological Science* 35:2778-2785.

Zhang, B., M. Li, Z. Zhang, B. Goossens, L. Zhu, S. Zhang, J. Hu, M. W. Bruford and F. Wei 2007. Genetic viability and population history of the giant panda, putting an end to the "Evolutionary dead end"? *Molecular Biology and Evolution* 24:1801-1810.

9

Marsupials as introduced species: Long-term anthropogenic expansion of the marsupial frontier and its implications for zoogeographic interpretation

Thomas E. Heinsohn

Invasive Animals Co-operative Research Centre and Institute for Applied Ecology, University of Canberra, Australia; Visiting Fellow, Archaeology and Natural History, Research School of Pacific and Asian Studies, Australian National University, Australia

Thomas.Heinsohn@canberra.edu.au

Introduction: Those 'dull and inferior ... Marsupials'

Traditionally, Southern mammals such as the marsupials and monotremes of Australasia have been viewed as further down the evolutionary tree and thus inferior to the placental mammals of Afro-Eurasia and the Americas. This Eurocentrism or Northern-Hemisphere-centrism is typified in the writings of turn-of-the-century British naturalist McCabe (1910), who wrote a paper entitled 'Australia – A Museum of Living Antiquities' in which he characterised the continent as a last bastion for inferior species:

> '...circumstances point to an invasion of Australia by the land-bridge from Asia ... Under the stress of the new carnivores of the northern world the dull inferior Monotremes and Marsupials escaped over the land-bridge to the southern continent.' (McCabe 1910:44)

Renowned late 19th century German naturalist Haeckel (1893) took Eurocentrism a step further by suggesting that the minute one crosses the Wallace Line, which marks the boundary between the Oriental faunal region of Eurasia and the Australasian (or Australian) faunal region, one is, in effect, stepping back in time to a world of primitive species:

> 'Crossing the narrow but deep Lombok Strait [Wallace's Line] we go with a single step from the Present Era to the Mesozoicum.' (Haeckel 1893:5)

We now know that there was never a Cenozoic land bridge between the Eurasian continent and Australia, and that marsupials and monotremes reached Australasia via southern Gondwanan connections, but the McCabian and Haeckelian-type views of general Northern evolutionary ascendancy and biological chauvinism lingered on well into the second half of the 20th century (as discussed in Calaby 1984 and Heinsohn 2006a). We now also appreciate that the Australasian marsupial fauna contains a full spectrum of forms adapted to various niches, some of which proved to be quite vulnerable to foreign invasive species such as dingos, foxes, cats, rabbits and humans, and some of which proved to be quite resilient and potentially invasive themselves – for example, Australia's recent challenges in managing the ecological impacts of over-abundant eastern grey kangaroo *Macropus giganteus* populations in peri-urban parks and reserves. Indeed, kangaroos *Macropus* spp. along with many other marsupials have proved highly adapted to the Australian environment, with macropodids discovered to have one of the most energy-efficient modes of locomotion known for a terrestrial mammal (Dawson 1995; Flannery 1997; Tyndale-Biscoe 2005; Van Dyck and Strahan 2008). Similarly, another marsupial, the common brushtail possum *Trichosurus vulpecula*, is renowned for its ability to adapt to both the urban and agricultural environment and actively invades or re-invades these newly created highly anthropogenic landscapes and niches. Ironically, among the vulnerable Australasian native mammals that are now either extinct or threatened are suites of native placental murid rats derived from waif-dispersing Asian ancestors, and this parallels the status of many recently extinct or threatened marsupial species with Gondwanan origins (Flannery et al. 1990; Van Dyck and Strahan 2008). Thus in the biogeography of continents and islands, it is probably evolution in isolation more than phylogenetic history that determines the vulnerability of a species. In modern Australasia with its extensive human impacts, there are both 'losers' and 'winners' among the extant marsupials, just as there are 'losers' and 'winners' among the native placental murid rats (Flannery et al. 1990; Low 2002; Van Dyck and Strahan 2008).

As an inverse test of their adaptability, this paper aims to review the success of Australasian (Australidelphian) marsupials as introduced species on foreign landmasses. This is assessed through literature review combined with some of the author's own fieldwork data on anthropogenic biological invasions by translocated marsupials in the Indo-Australian Archipelago. A significant epistemological problem encountered in such a study is that while marsupial introductions in regional archipelagos during recent historic times for the most part appear to have been historically recorded, those that occurred in the prehistoric past were not. Thus prehistoric marsupial translocations have had to be reconstructed from a combination of palaeontological and archaeological data, combined with other lines of evidence such as molecular systematics and broader ethnozoological and biogeographic contextual analysis. Such an analysis enables the uncovering of prehistorically introduced 'camouflaged exotics' (Heinsohn 1997a) or 'cryptogenic' (Carlton 1996) populations of introduced marsupials on islands, which might otherwise go unnoticed by superficially appearing to be natural. This is referred to as uncovering the 'crypto-anthropogenic factor' of hidden long-term human influences on zoogeography (Heinsohn 1997a, 1998a, 2003, 2006a).

Some of the lines of evidence considered for distinguishing translocated from naturally occurring island vertebrate populations, but here applied to marsupials, are listed below (from Heinsohn 2006a):

1. Patchy, improbable or strangely disjunct distributions;
2. Limited distributions on a landmass, where a species appears to have only arrived comparatively recently and the biological invasion is still in progress;
3. Occurrences on islands that follow a commensal or semi-commensal pattern where species are only found in and around human settlements or other disturbed and highly anthropogenic habitats;
4. Molecular systematics evidence, such as genetic similarity over wide inter-island distributions, that may indicate recent introduction;
5. Sudden appearances of species in archaeological sequences;
6. A lack of fossil or subfossil material in palaeontological or archaeological sequences that may indicate recent introduction;
7. Ethnozoological evidence of long associations between humans and certain wild species that are known to be carried around in watercraft as pets, or for food, utilitarian, trade or ceremonial purposes;
8. Island assemblages of vertebrates that favour human-associated species over natural dispersal ability;
9. Directly documented accounts of species introduction, such as written historical records or oral historical records of translocation and/or subsequent biological invasions by a successfully introduced species;
10. Analogy from biogeographic patterns created by historically documented biological invasions by introduced species; and
11. Big-picture analysis of the global distribution of invasive species and their current invasion frontier, with back-tracing of their probable historic/prehistoric anthropogenic invasion pathway from an original natural source area, so as to distinguish their 'natural' versus introduced/anthropogenic distributions.

In light of the potential ambiguity of single strands of evidence, it is often a combination of some of these indicators that tips the scales of probability in favour of considering translocation by human agency.

Given the problem of camouflaged exotics or cryptogenic populations of marsupials in the Indo-Australian Archipelago, a phenomenon which for the most part has been recognised only recently in relation to marsupials (as reviewed by Heinsohn 2006a), a further aim of this paper is to assess the degree to which the classic zoogeographers and later observers were aware of the aforementioned crypto-anthropogenic factor in marsupial zoogeography. Did varying degrees of naivety or awareness influence their respective interpretations of island marsupial faunas and the placement of various zoogeographic lines (Figure 1)? Thus this interdisciplinary review combines biogeographic analysis with a history of zoogeographic ideas. In particular, this article is juxtaposed with an earlier paper by Gilmore (1977), 'The success of marsupials as introduced species', to show how much our collective knowledge of marsupial biogeography has progressed in the past few decades.

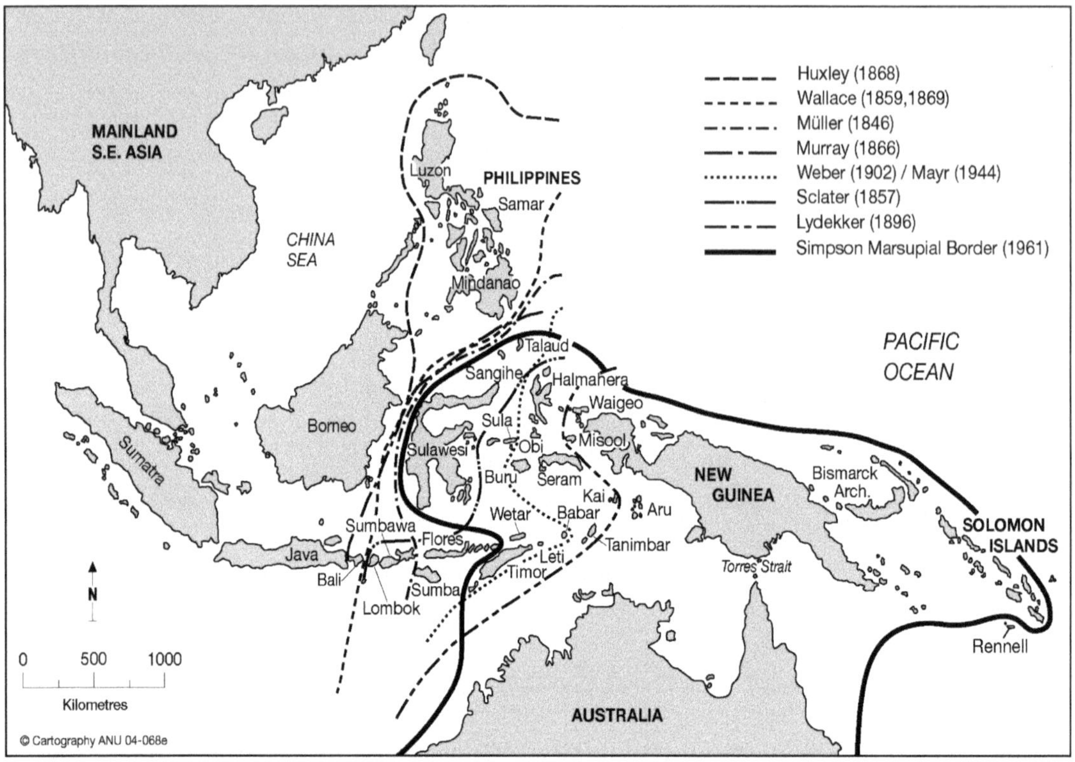

Figure 1. Some traditional zoogeographic boundary lines in the Indo-Australian Archipelago most of which attempt to define the demarcation between Asian and Australian faunas. Simpson's Marsupial Line (1961) is marked in bold. Simpson naively treats this boundary as the natural distributional limit of marsupials in the tropics (after Heinsohn 2006a)

Study area: The realm of the Australidelphian (Australasian) marsupial

The study area includes all of the regions of the world in which Australidelphian (Australasian) marsupials occur as natural or introduced species (not to be confused with the Ameridelphian marsupials of the Americas). For Australidelphian marsupials this includes: (1) Australia and its continental shelf islands such as Tasmania, which together have approximately 158 marsupial species; (2) New Guinea and its fringing Papuan subregion continental shelf islands, which have about 80 species; (3) the oceanic Wallacean islands of Eastern Indonesia, comprising the Lesser Sundas, Sulawesi subregion and the Moluccas, which collectively have about 12 species (seven of which are Wallacean endemics); (4) the oceanic northern Melanesian islands (the Bismarck Archipelago and Solomon Islands) of Near Oceania, which have about seven species (many, or possibly all, of which may be introduced); (5) New Zealand and its satellite islands in southern Remote Oceania, which have six recently introduced species; (6) the Hawaiian Islands in northern Remote Oceania, which have one recently introduced species on Oahu Island; and (7) the British Isles in the European Palaearctic region, which has a single recently introduced species (Tomich 1986; Whitten et al. 1987; Corbet and Hill 1992; Flannery 1994; Flannery 1995a, b; Monk et al. 1997; Heinsohn 2001a, 2002a, b, c, 2003, 2004a, b, 2005a, b, c, 2006a; Ziegler 2002; Long 2003; Helgen and Flannery 2004; King 2005; Nowak 2005; Heinsohn and Hope 2006; Harris and Yalden 2008; Van Dyck and Strahan 2008).

A particular focus of this study, however, is the tropical Australasian portion of the Indo-Australian Archipelago that stretches to the east of the Oriental Region. The term 'Indo-Australian Archipelago' can be confusing in that it is generally used as a synonym for 'Malay Archipelago' (islands stretching from the Malay Peninsula to New Guinea), but some authors use it in a broader sense to also include Australia and the islands of Near Oceania (etymology reviewed in Heinsohn 2006a). Some significant recognised boundaries include the Wallace/Huxley Line, which marks the eastern edge of the Sundaic continental shelf and the eastern limit of complex continental Oriental Region faunas (Figure 2). This is paralleled further east by Lydekker's Line, which marks the western edge of the Meganesian continental shelf and the western limit of complex continental Meganesian (also called 'Greater Australian', 'Sahul' or 'Australo-Papuan') faunas. Between the continental Sundaic region and continental Meganesia lie the oceanic Wallacean islands with depauperate and unbalanced oceanic faunas that exhibit limited Oriental influences from the west and limited Meganesian (Australo-Papuan) influences from the east. To the east of Meganesia lie the oceanic northern Melanesian islands (the Bismarck Archipelago and Solomon Islands) of Near Oceania, which also have depauperate and unbalanced oceanic faunas that exhibit salient New Guinean influences on their biotas (Heinsohn 2006a; Heinsohn and Hope 2006; Morwood and Van Oosterzee 2007). For a delineation of specific tropical Australasian zoogeographic subregions to the immediate east of the Wallace Line, see Figure 3.

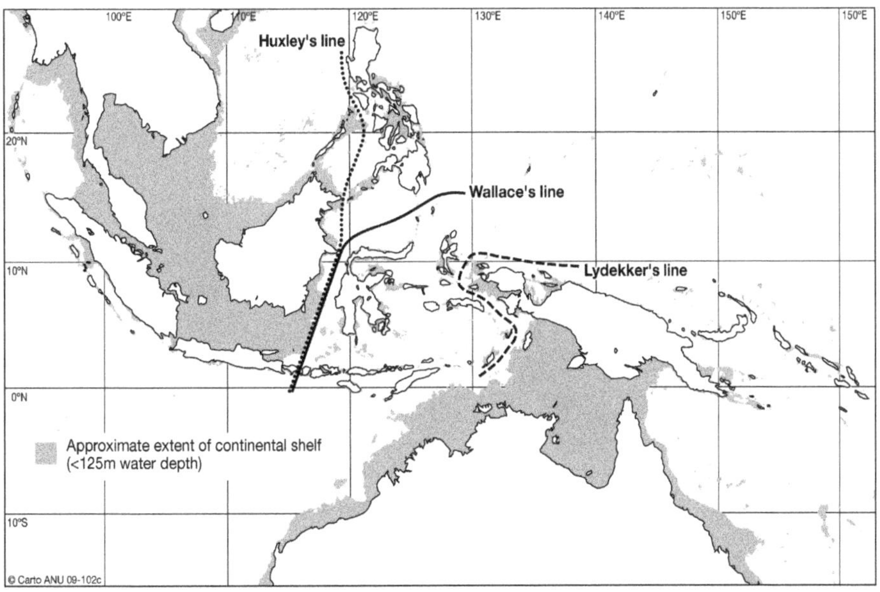

Figure 2. Indo-Australian Archipelago showing land and sea during the Last Glacial Maximum at *c.* 19,000 BP when sea levels may have fallen by 120 m to 130 m. The Wallace/Huxley Line essentially marks the eastern edge of the Sunda continental shelf and eastern limit of complex Oriental continental faunas; while Lydekker's Line essentially marks the western edge of the Sahul/Greater Australian continental shelf and the western limit of complex Australo-Papuan continental faunas. Between these two lines lie the oceanic Wallacean Islands (after Heinsohn 2006a)

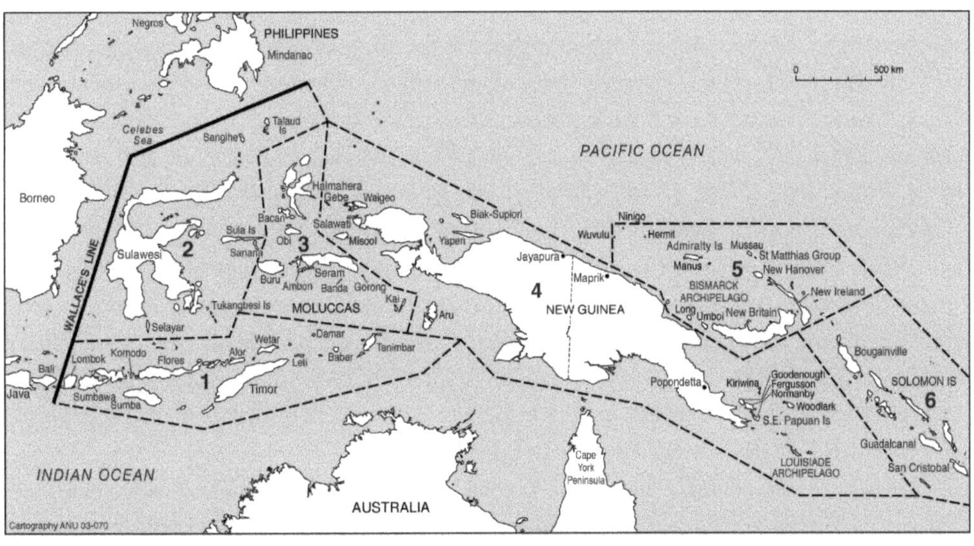

Figure 3. Map of the zoogeographic subregions to the immediate east of the Wallace Line. (1) Lesser Sundas; (2) Sulawesi Subregion; (3) Moluccas; (4) Papuan Subregion; (5) Bismarck Archipelago; and (6) Solomon Islands. Wallacea is comprised of subregions 1, 2 and 3; while Near Oceania is comprised of subregions 5 and 6 (after Heinsohn 2003)

During times of significantly lowered sea levels, such as the Last Glacial Maximum (LGM) at about 19,000 BP, when sea levels fell by 120 m to 130 m, large areas of both the Asian Sunda shelf and the Greater Australian Sahul shelf became exposed (Figure 2). In the west, this caused the Greater Sunda islands of Sumatra, Java, Bali and Borneo to coalesce with each other and the Asian mainland to form a continuous land bridge stretching from the Malay Peninsula to Bali and Borneo-Palawan. In the east, this caused New Guinea and its continental satellites such as Aru to coalesce with Australia and Tasmania to form a single Meganesian or Greater Australian continental landmass. Despite some limited internal coalescences, however, the intervening Wallacean islands (Lesser Sundas, Sulawesi subregion and Moluccas), separated by deep channels and seas, remained insular and this prevented the formation of a land bridge from Asia to Australia (Figure 2). Similarly, due to a series of deep intervening channels and seas, the northern Melanesian islands (the Bismarck Archipelago and Solomon Islands) of Near Oceania, despite internal coalescence within some island groups, also remained insular and oceanic and never formed land bridges with New Guinea or between principal island subregions. Furthermore, the Philippines (other than Palawan with its Bornean connections), despite some internal coalescences, also remained significantly insular and oceanic and thus could almost be viewed as a northern extension of oceanic Wallacea (Chappell and Shackleton 1986; Chappell 1987; Haq et al. 1987; Fairbanks 1989; Hope 1996; Hall 2001; Inger and Voris 2001; Heinsohn 2006a; Morwood and Van Oosterzee 2007).

The dominant natural vegetation of the tropical Indo-Australian Archipelago (excluding Australia) is Malesian and Papuasian rainforest and monsoon forest, but with sizeable areas of either natural or anthropogenic savannah and savannah woodland, particularly in drier parts of the archipelago such as across much of the Lesser Sundas and southern New Guinea. The highest mountains in the region occur in the central cordillera of New Guinea, which has peaks approaching 5000 m in altitude, with the island supporting a spectrum of lowland to montane, subalpine and alpine habitats, including sizeable areas of montane anthropogenic grasslands. In other stretches of the archipelago, such as in parts of Wallacea and the northern Melanesian islands, there are island mountain ranges that reach between 2000 m and 3000 m in altitude, many of which support significant areas of montane forest, and upland anthropogenic grasslands (Whitten et al. 1987; Hope 1996; Monk et al. 1997; Heinsohn and Hope 2006).

The earliest recorded human occupation in the Indo-Australian Archipelago is that of *Homo erectus* in Java to the west of the Wallace Line at about 1.2 million years ago, where both fossil remains and stone tools have been found (Bellwood 1997). However, recent discoveries of ancient stone tools, possible early anthropogenic faunal extinction (turnover) horizons, and the fossil remains of *Homo floresiensis* on the Lesser Sunda island of Flores, variously dating between 880,000 and 18,000 years ago, have indicated the presence of pre-modern hominids to the east of the Wallace Line in Wallacea (Morwood and van Oosterzee 2007). Indeed, the *H. floresiensis* remains recently have been interpreted as possibly belonging to a relictual pre-*Homo erectus* hominid population that was probably once widespread in Asia and was able to hang on in Flores until about 18,000 BP. Indeed, suspected possible Middle Pleistocene stone-tool assemblages and/or possible Middle Pleistocene anthropogenic faunal extinction (turnover) horizons from places such as Timor, Sulawesi and the Philippines may be indicative that early pre-*H. erectus* hominids akin to *H. floresiensis* reached other parts of Wallacea and more northerly oceanic areas such as the Philippines (Morwood and Van Oosterzee 2007; Morwood pers comm. 2009, Mulvaney Lecture, Australian National University, 13 May 2009). If this is the case, then Sulawesi, with its ancient endemic marsupials, may be the first place where hominids and marsupials came into contact.

In contrast to the above early hominid record, the initial invasion of Wallacea and Meganesia (Greater Australia or Sahul) by modern *Homo sapiens* is generally thought to have occurred between 40,000 and 55,000 years ago in the late Pleistocene, with the islands of the Bismarck Archipelago in Near Oceania also being colonised within a broadly similar timeframe

(c. 40,000 BP), but slightly later for the Solomon Islands at about 32,000 BP (Bellwood 1997; Thorne et al. 1999; Wickler 2001; Leavesley and Chappell 2004; Morwood and van Oosterzee 2007; Summerhayes 2007). Following the relative stabilisation of sea levels in the lower Holocene, a further major phase of human colonisation was Austronesian-speaking agricultural and maritime peoples' expansion into Wallacea, around coastal New Guinea, into Near Oceania and the more distant Pacific islands of Remote Oceania in the past 4000 years, probably reaching New Zealand in the last millennium (Bellwood 1997; Kirch 1997; Spriggs 1997; Summerhayes 2007). This was a period that saw the rapid spread of various plant cultivars and the domesticated triumvirate of dogs, pigs and chickens into Oceania, along with anthropogenic spread of commensal murids such as the Pacific rat *Rattus exulans* (Kirch 1997; Spriggs 1997; Summerhayes 2007).

In the Indo-Australian Archipelago as far as western New Guinea, the aforementioned expansion was followed by ongoing economic and cultural influences from South and Southeast Asia, including the early spice and produce trade, and finally by the impact of the 'modern' spice trade in the past millennium, which involved the mercantile empires of the Chinese, Arabs, Portuguese, Spanish, Dutch and British (Swadling 1996; Black 1999). In Australasia, Australia and New Zealand were progressively colonised by British settlers from the late 18th and early 19th centuries respectively (Derbyshire 1993; Nile and Clerk 1996). Most of New Guinea and northern Melanesia remained as scattered indigenous realms until the 19th century when colonial rivalries and the lure of copra and gold led the Dutch, Germans and British to establish territories and colonial outposts (Souter 1963; Derbyshire 1993; Black 1999). The activities of the colonial acclimatisation societies in the 19th and early 20th centuries led to the deliberate introduction of a whole range of Eurasian, American and Australian vertebrate species to New Zealand, and a range of mostly Eurasian vertebrates to Australia, with significant impacts on both landmasses (Lever 1992; McDowall 1994; Young 2004; Heinsohn 2006a). In the tropical Indo-Australian Archipelago and Oceania, Dutch, German, British, Australian and French colonists contributed to the further spread of Oriental vertebrates such as the Javan rusa deer *Cervus timorensis*, adding to a traditional long-term process of translocation of multiple Oriental vertebrate species eastwards into Wallacea and Oceania by the native population, which had also carried some New Guinean/Australo-Papuan vertebrate species, including some marsupials, westwards into Wallacea (Glover 1971, 1986; Groves 1984; Heinsohn 1997a, 1998a, b, 2001a, 2002a, 2003, 2004a, 2005a, b, 2006a; Heinsohn and Hope 2006). This is the deeper historical and cultural backdrop to the anthropogenic translocation of Australasia's Australidelphian marsupials.

Results and discussion: A survey of records and postulations of long-term Australidelphian marsupial translocation

Based on literature review and some of the author's own field observations, Table 1 summarises records and postulations of historic and prehistoric Australidelphian marsupial translocations, while Table 2 summarises the postulated impact of translocation on the composition and status of island marsupial faunas. Analysis reveals that there are three principal putative source areas for translocated Australidelphian marsupials: (1) continental Australia (including continental shelf islands such as Tasmania and the Bass Strait islands); (2) continental New Guinea and the broader Papuan subregion (including continental shelf satellites such as Aru, Misool, Salawati and the southeast Papuan islands); and (3) oceanic Wallacea (specifically Sulawesi). These are discussed separately below.

Table 1. Records and postulations of anthropogenic translocation of Australasian (Australidelphian) marsupials during historic and prehistoric times (Table 1 continues on pages 141-148)

Taxa/species and value/ reason for translocation	Presumed natural distribution	Where thought to be introduced	Historical context	References
New Guinea / Papuan exports				
Dasyuridae				
New Guinea quoll (*Dasyurus albopunctatus*) [? Pet, trade animal, bilas fur, food and small game species, rat-catcher]	New Guinea	(??I) Malai Island in the Siassi group	Lilley (1986) reports finding the bones of a small game animal, the New Guinea quoll (normally restricted to the New Guinea mainland) from Holocene deposits on Malai Island in the oceanic Siassi group (near Umboi) in the Vitiaz Strait. This may merely represent a captive animal or preserved salted or smoked carcass/carcass portion transported in from the New Guinea mainland. Or it may represent a now-extinct naturally occurring population due to waif dispersal on a natural raft, or possibly a now-extinct ephemeral population of animals derived from anthropogenic translocation.	Lilley 1986; Heinsohn 2006a
Peroryctidae				
Common echymipera or common spiny bandicoot (*Echymipera kalubu*) [? Food and small game species, trade animal]	New Guinea and some adjacent islands	(?I) Manus Is.; (?I) Some other oceanic New Guinea satellites	Slightly ambiguous archaeological evidence indicates that the population on Manus Is. in Northern Melanesia may have been introduced in the Terminal Pleistocene at around 13,000 BP (Spriggs 1997; Summerhayes 2007). Alternatively, Manus Is. and other oceanic populations may represent unaided dispersal on natural rafts washed out to sea during floods, storms, king tides or tsunamis.	Flannery 1995b; Spriggs 1997; Heinsohn 2003; Long 2003; Heinsohn 2006a; Summerhayes 2007
Rufous echymipera or rufous spiny bandicoot (*Echymipera rufescens*) [? Food and small game species, trade animal]	New Guinea; Aru Is.	(?I) Kai Is.	Possibly introduced to the oceanic Kai Is. through captives carried for food, trade or stock purposes from Aru. Alternatively, the Kai population may represent unaided dispersal on natural rafts washed out to sea during floods and storms.	Flannery 1995b; Heinsohn 1997a; Heinsohn 2003, 2006a
Petauridae				
Sugar glider (*Petaurus breviceps papuanus*) [Papuan region subspecies] [? Pet, trade animal, bilas fur, delicacy]	New Guinea & some Papuan Region satellite islands	(?I)Halmahera; (?I) Bacan; (?I)Gebe; (?I) Some other New Guinea satellites	Populations on Halmahera and Bacan, and on some other oceanic New Guinea satellites may possibly be due to long-term human agency through captives carried as pets or as a small mammal food delicacy; or possibly through stowaways in traditional cargos such as large bunches of bananas.	Flannery 1994b; Flannery 1995b; Flannery et al. 1995; Monk et al. 1997; Heinsohn 2003; Long 2003; Heinsohn 2004a, 2006a

Phalangeridae				
Common spotted cuscus (*Spilocuscus maculatus*) [Pet, trade animal, bilas fur and fibre, food and game animal]	New Guinea	Selayar; Buru; Ambon; Seram and satellites; Banda; Kai; St. Matthias group.; New Ireland; ?New Britain (adventive populations)	Probably introduced to much, or all, of its oceanic distribution through long-term human agency, through captives carried as pets or for food, trade and stock purposes. Archaeological evidence from the St. Matthias group indicates that it was probably introduced there within the past few thousand years (Kirch 1988, 1997). Populations in New Ireland appear to have been introduced sometime in the decade leading up to the commencement of the Second World War (Flannery and White 1991; Heinsohn 1998a, b, 2004a), and the species is still spreading down the island from its suspected original release point at Kavieng at the far northwestern tip of the long narrow island (main invasion frontier reaching as far south as Lemakot area), but with secondary introduced populations derived from cuscuses carried further on in motor vehicles or boats now starting to appear in other more southern locations (eg. Pangeifua area to the southwest of the main invasion frontier and the Weiten River Valley in far southern New Ireland). These separate populations will eventually coalesce and speed up the rate of complete colonisation of New Ireland by *S. maculatus* (Heinsohn 2006a). A recent development is that adventive populations of *S.maculatus*, probably carried across the St. George Straight from New Ireland in boats in the late 20th century and early 21st century, have begun to appear in adjacent northern New Britain (Sebastian Högberg pers comm. 2009) and these will probably go on to establish invasion beachheads on that island.	Musser 1987; Whitten et al. 1987; Kirch 1988; Flannery 1990a; Flannery and White 1991; Flannery 1994b; Flannery 1995a, b; Monk et al. 1997; Spriggs 1997; Emmons and Kinbag 2001; Long 2003; Heinsohn 1997a, 1998a, b, 2002a, b, 2003, 2004a, b, 2006a
Admiralty cuscus (*Spilocuscus kraemeri*) [Pet, trade animal, bilas fur and fibre, food and game animal]	Ancestors may be from New Guinea or fringing Papuan subregion islands	(?I) Admiralty Is.; Hermit Is.; Ninigo group.; Wuvulu [All of the above are a part of the Manus Province Archipelago] ?(I) Bali Is., Witu Is., West New Britain Province?	May have been introduced to the aforementioned islands of the Manus Province Archipelago through long-term human agency through captives carried as pets or for food, trade and stock purposes (Flannery 1994, 1995b; Heinsohn 2004a). Spriggs (1997) and Summerhayes (2007) interpret archaeological evidence from the site of Pamwak on Manus Island to conclude that *S.kraemeri* may have been introduced there in the terminal Pleistocene at around 13,000 years ago, presumably from unknown sources in the Papuan subregion/northern New Guinea. Author considers it possible that the apparent uniqueness of *S. kraemeri* may be due to anthropogenic admixture/introduction of a mixed *Spilocuscus* spp. founding stock leading to a differentiated hybrid population. Alternatively, *S. kraemeri* may have naturally rafted to part of the Admiralty Islands, such as Greater Manus Island where it differentiated, and was then subsequently spread by human agency to other islands in the vast Manus Province Archipelago, such as the Hermit Is., Ninigo group and Wuvulu. On the basis of a single specimen, an adventive population derived from Manus Province trade animals may now also occur on Bali Is., Witu Is. of West New Britain Province to the southeast of Manus Province	Flannery 1994b; Flannery 1995b; Spriggs 1997; Anthony 2001; Long 2003; Heinsohn 1997a, 1998a, b, 2003, 2004a, 2006a; Helgen and Flannery 2004; Summerhayes 2007

Taxa/species and value/ reason for translocation	Presumed natural distribution	Where thought to be introduced	Historical context	References
Northern common cuscus, grey cuscus or brown cuscus (*Phalanger orientalis*) [Pet, trade animal, bilas fur and fibre, food and game animal]	New Guinea	Timor; Leti; Wetar; Sanana; Buru; Ambon; Seram and satellites; Banda; Kai; Bismarck Archipelago; Solomon Is.; some other oceanic New Guinea satellite islands	Probably introduced to much, or all, of its oceanic distribution through long-term human agency through captives carried as pets or for food, trade and stock purposes. Archaeologists of the Lapita Homeland Project consider, on the basis of material from a range of cave sites in New Ireland in the Bismarck Archipelago, that *P. orientalis* may have been introduced there as early as 23,500 to 20,000 BP in the Late Pleistocene (Flannery and White 1991; Spriggs 1997; Summerhayes 2007), while archaeological evidence from Timor (Glover 1971, 1986) and the North Solomons (Wickler 2001; Summerhayes 2007) indicates that the species may have been introduced to those areas in the middle to upper Holocene respectively. Long Island off the north of New Guinea is reported to have been deliberately restocked with northern common cuscus *P. orientalis* following a cataclysmic volcanic eruption three centuries ago that wiped out virtually all animal life. Diamond (as cited by Koopman 1979) and Mayr and Diamond (2001), citing oral historical evidence from the inhabitants of Long Island, report that this happened in the 19th century through the efforts of a man called Ailimai, apparently from sources on Tolokiwa Island to the east.	Glover 1971; Koopman 1979; Ball and Hughes 1982; Groves 1984a; Glover 1986; Menzies and Pernetta 1986; Flannery et al. 1988; Flannery and White 1991; Flannery 1993b; Flannery 1994b; Flannery 1995ab; Monk et al. 1997; Spriggs 1997; Mayr and Diamond 2001; Wickler 2001; Heinsohn 2002c; Heinsohn 2003; Long 2003; Heinsohn 2004a, 2005ab, 2006a; Summerhayes 2007
Woodlark cuscus (*Phalanger lullulae*) [Pet, trade animal, bilas fur and fibre, food and game species]	Woodlark Is.	(?I) Alcester Is.	Possibly introduced to Alcester Is. through captives from Woodlark Is. carried as pets, or for food, trade, ceremonial or stock purposes	Flannery 1994b; Flannery 1995b; Heinsohn 2003; Long 2003; Heinsohn 2004a, 2006a
Macropodidae				
Huon tree-kangaroo (*Dendrolagus matschiei*) [Pet, trade animal, food and game species, bilas fur and fibre; bones, teeth and other animal products]	New Guinea	(?I) Umboi; (?I) ?New Britain	Probably introduced through long-term human agency through captives carried as pets or for food or trade purposes. Juveniles captured during hunting are often kept for a while as pets, with some subsequently escaping.	Koopman 1979; Honacki et al. 1982; Maynes 1989; Flannery 1995a, b; Flannery et al. 1996; Heinsohn 2003; Long 2003; Martin 2005; Heinsohn 2006a
Brown dorcopsis				
(*Dorcopsis muelleri*) [Pet, trade animal, food and game species, bones, teeth, pelt and other animal products]	New Guinea; Misool	(?I)Halmahera; (?I) Gebe. Now extinct on both islands	May have been introduced from an adjacent Papuan region continental shelf satellite island such as Misool or Salawati, or the New Guinea mainland in the early to middle Holocene through captives carried as pets, or for food, trade or stock purposes, but now extinct on Halmahera and Gebe.	Flannery et al. 1995; Flannery et al. 1998; Heinsohn 2003, 2006a; Summerhayes 2007

Northern pademelon (*Thylogale browni*) [Pet, trade animal, food and game species, bones, teeth, pelt and other animal products]	New Guinea	Bagabag; Umboi; New Britain; New Ireland; New Hanover Now extinct introduced populations appear to have occurred within recent historic times on a string of New Ireland satellites, including Tabar Is., Lihir group, Tanga Is. and Feni Is. (Heinsohn 2005c)	Probably introduced through long-term human agency through captives carried as pets, or for food, trade, ceremonial or stock purposes. Archaeological evidence from cave sites in New Ireland indicates that it may have been introduced there around 8400 to 7000 BP (Flannery and White 1991; Spriggs 1997; Summerhayes 2007). Egloff (1975, 1979) and Flannery et al. (1988) discuss archaeological evidence of macropodid translocation in the Northern Melanesian Islands during the Holocene.	Egloff 1975; Koopman 1979; Flannery et al. 1988; Maynes 1989; Flannery and White 1991; Flannery 1992; Flannery 1995a, b; Spriggs 1997; Heinsohn 2003; Long 2003; Heinsohn 2005c, 2006a
Dusky pademelon (*Thylogale brunii*) [Pet, trade animal, food and game species, bones, teeth, pelt and other animal products]	Aru Is.; New Guinea	(?I) Kai Is.	Possibly introduced to the oceanic Kai Is. through captives carried from Aru as pets, or for food, trade or stock purposes.	Maynes 1989; Heinsohn 1997a; Heinsohn 2003; Long 2003; Heinsohn 2006a
Agile wallaby (*Macropus agilis*) [New Guinea form of this Australo-Papuan macropodid is sometimes described as the subspecies *M. a. papuanus*] [Pet, trade animal, food and game species; bones, teeth, pelt and other animal products]	Northern and northeastern Australia; Southern New Guinea	(?I) Goodenough; (?I) Fergusson; (?I) Normanby; (?I) Kiriwina; (?I) Savannah woodland areas in mainland southeast Papua to the north of New Guinea's Central Range	May have been prehistorically introduced to the aforementioned islands of Southeast Papua from the New Guinea mainland through captives carried as pets, or for food, trade, ceremonial or stock purposes. Possibly also introduced to the isolated savannah/savannah-woodland patches of the Southeast Papuan mainland to the north of New Guinea's Central Cordillera. Egloff (1975, 1979) discusses archaeological evidence of macropodid translocation in the Southeast Papuan Islands during the Holocene.	Brass 1956; Brass 1959; van Deusen 1972a, b; Schodde and Calaby 1972; Liem 1977; Ziegler 1977; Egloff 1979; Maynes 1989; Flannery 1995b; Heinsohn 2003; Long 2003; Heinsohn 2006a
Sulawesi exports				
Sulawesi bear cuscus (*Ailurops ursinus*) [Pet, trade animal, food and game species, pelt, fur, fibre, bones, teeth and other animal products]	Sulawesi and some close satellite islands	(?I) Salebabu Island in the remote Talaud group	The remote and patchy occurrence on Salebabu Island in the oceanic Talaud group may possibly be due to long-term human agency through captives carried as pets, or for food, trade, ceremonial or stock purposes. Alternatively, the patchy and disjunct Talaud population may possibly be relict and due to natural dispersal such as on rafts washed out to sea by floods, storms and surges.	Flannery 1994b; Flannery 1995b; Heinsohn 1997a, 1998a, b, 2001a, b, 2003, 2004a, 2006a

Taxa/species and value/ reason for translocation	Presumed natural distribution	Where thought to be introduced	Historical context	References
Australian exports to Remote Oceania				
Phalangeridae				
Common brush-tail possum (*Trichosurus vulpecula*) [Pelt for fur industry]	Tasmania and mainland Australia	New Zealand (North Is.; South Is.; Stewart Is.; Chatham Is. and other small NZ satellite islands)	Deliberately introduced to New Zealand from Tasmania and mainland Australia starting in the mid 19th century to establish a fur industry. Due to selective over-browsing on New Zealand's mammalian-folivore-naive forests leading to the death of certain palatable tree species from exhaustion following repeated defoliation, *T. vulpecula* is reported to have caused possum-induced die-back and changes in forest compostion in many areas of the North and South islands. Other impacts include predation on native invertebrates and small vertebrates such as nestling birds, and competition with indigenous species such as birds for tree-hollows and other habitat space.	Thomson 1922; Donne 1924; Wodzicki 1950; Murphy 1951; De Vos et al. 1956; Pracy 1962; Davidson 1965; Poole 1970; Gibb and Flux 1973; Tyndale-Biscoe 1973; Bull and Whitaker 1975; Gilmore 1977; King 1984; Lever 1985; Cowan 1990; King 1990; Holmes 1993; Flannery 1994b; McDowall 1994; Montague 2000; Isern 2002; Long 2003; Heinsohn 2004a; Cowan 2005; Heinsohn 2006a; Heinsohn and Barker 2006
Brush-tailed rock wallaby or black-tailed rock wallaby (*Petrogale penicillata*) [Acclimatisation movement and menagerie escapees]	Eastern Australia	Kawau Is., Rangitoto Is. and Motutapu Is. off the coast of the North Island of New Zealand; Oahu Is. in the Hawaiian Is.	Deliberately introduced to small satellite islands in New Zealand commencing in the early 1870s as a part of acclimatisation efforts. Also found on Oahu Island in Hawaii, as a result of a male and female that escaped from a private menagerie in 1916.	Tinker 1941; Wodzicki 1950; De Vos et al. 1956; Gilmore 1968; Tomich 1969; Poole 1970; Kramer 1971; Gibb and Flux 1973; Tyndale-Biscoe 1973; Bull and Whitaker 1975; Gilmore 1977; Lazell 1980; Tinker 1980; Lazell 1981; Lazell et al. 1984; Archer et al. 1985; Lever 1985; Moulton and Pimm 1986; Tomich 1986; Maynes 1989; King 1990; Flannery 1995b; Ziegler 2002; Long 2003; King 2005; Warburton 2005e; Heinsohn 2006a
Swamp wallaby or black wallaby (*Wallabia bicolor*) [Acclimatisation movement]	Eastern Australia	Kawau Is., off the coast of New Zealand's North Island	Probably introduced to Kawau Is. by Sir George Grey in circa 1870 as a part of acclimatisation efforts.	Wodzicki 1950; De Vos et al. 1956; Gilmore 1968; Poole 1970; Gibb and Flux 1973; Bull and Whitaker 1975; Gilmore 1977; Lever 1985; Maynes 1989; King 1990; Flannery 1995b; Long 2003; King 2005; Warburton 2005f; Heinsohn 2006a
Bennett's wallaby, red-necked wallaby, brush wallaby or scrub wallaby (*Macropus rufogriseus*) [Acclimatisation movement]	Tasmania and eastern Australian mainland	New Zealand's South Island (Waimate and Hawea districts)	Deliberately introduced to the Hunters Hills area of New Zealand's South Island in 1874 as a part of acclimatisation efforts, with the population growing and spreading to occupy a large area of the central South Island to the east of the Southern Alps.	Murphy 1951; Davidson 1965; Gilmore 1968; Poole 1970; Gibb and Flux 1973; Tyndale-Biscoe 1973; Bull and Whitaker 1975; Gilmore 1977; Lever 1985; Maynes 1989; King 1990; McDowall 1994; Flannery 1995b; Long 2003; King 2005; Warburton 2005b; Heinsohn 2006a

Tammar wallaby, dama wallaby, Kangaroo Island wallaby or silver-grey wallaby (*Macropus eugenii*) [Acclimatisation movement]	Southwestern Western Australia and southern South Australia	Kawau Is. and Rotorua district of New Zealand's North Island	Introduced to Kawau Island by Sir George Grey in circa 1870 as a part of acclimatisation efforts; and subsequently introduced to the Rotorua area of the North Island in circa 1912 from whence it continues to expand its range.	Wodzicki 1950; De Vos et al. 1956; Gilmore 1968; Poole 1970; Gibb and Flux 1973; Bull and Whitaker 1975; Gilmore 1977; Lever 1985; Maynes 1989; King 1990; McDowall 1994; Flannery 1995b; Long 2003; King 2005; Warburton 2005a; Heinsohn 2006a
Parma wallaby, white-throated wallaby or small brown wallaby (*Macropus parma*) [Acclimatisation movement]	Northeastern and central eastern New South Wales, Australia	Kawau Is. off the coast of New Zealand's North Island	Probably introduced circa 1870 as a part of acclimatisation efforts.	Gilmore 1968; Poole 1970; Wodzicki and Flux 1971; Gibb and Flux 1973; Tyndale-Biscoe 1973; Bull and Whitaker 1975; Gilmore 1977; Honecki et al. 1982; Archer et al. 1985; Lever 1985; Maynes 1989; King 1990; McDowall 1994; Flannery 1995b; Long 2003; King 2005; Warburton 2005c; Heinsohn 2006a
Black-striped wallaby or scrub wallaby (*Macropus dorsalis*) [Acclimatisation movement]	Eastern Australia	?Kawau Is. off the coast of the North Island of New Zealand	This species is presumed to have been liberated on Kawau Is. by Sir George Grey circa 1870 and may have persisted there for more than 80 years with the last apparently authenticated record in 1954, but the species may now be either extinct or extremely rare on Kawau. There is, however, ongoing debate as to whether *M. dorsalis* ever occurred on Kawau, with possible misidentification of other introduced macropodid species on the island leading to a false record of its occurrence there.	Wodzicki 1950; De Vos et al. 1956; Gilmore 1968; Poole 1970; Gibb and Flux 1973; Bull and Whitaker 1975; Gilmore 1977; Lever 1985; Maynes 1989; King 1990; Flannery 1995b; Long 2003; King 2005; Warburton 2005d; Heinsohn 2006a
Australian exports to Europe				
Bennett's wallaby, red-necked wallaby, brush wallaby or scrub wallaby (*Macropus rufogriseus*) [Zoo and menagerie escapees, and acclimatisation movement]	Tasmania and eastern Australian mainland	Great Britain (virtually extinct); Germany (extinct)	(UK) A series of escapes and deliberate liberations from the latter half of the 19th century to the first half of the 20th century led to a number of ephemeral populations in Great Britain. A small and dwindling extant population in the Peak District is derived from an escape of five animals from an enclosure in Staffordshire in 1939-40. (Germany) A series of acclimatised ephemeral populations also existed at scattered locations in Germany from the late 19th century until the 1940s.	Boettger 1943; Grzimek 1967; Mallon 1970; Yalden and Hosey 1971; Tyndale-Biscoe 1973; Corbet 1974; Gilmore 1977; Lever 1977; Yalden 1982; Lever 1985; Maynes 1989; Baker 1990; Corbet and Harris 1991; Yalden 1999; Long 2003; Heinsohn 2006a; Harris and Yalden 2008
Australian exports to local Continental Shelf islands				
Dasyuridae				
Southern dibbler (*Parantechinus apicalis*) [Conservation]	Southwestern Western Australia	Escape Is. off the west coast of Western Australia	Deliberately introduced between 1998 and 2001 as a part of conservation efforts, with the species appearing to become established.	Long 2003; Heinsohn 2006a

Taxa/species and value/ reason for translocation	Presumed natural distribution	Where thought to be introduced	Historical context	References
Peramelidae				
Long-nosed bandicoot (*Perameles nasuta*) [? Reason]	Eastern Australia	(?I) Badu Is. (Mulgrave Is.) in the Torres Strait, Australia	Long-nosed bandicoots may have been introduced to Badu Is. 'within living memory'.	Long 2003
Eastern barred bandicoot (*Perameles gunnii*) [Conservation]	Southeastern Australia, including Tasmania	Maria Is. off the coast of Tasmania.	Fifty-five individuals were introduced in 1971 as a part of wildlife management efforts.	Weidenhofer 1977; Rounsevell et al. 1991; Long 2003; Heinsohn 2006a
Southern brown bandicoot (*Isoodon obesulus*) [Conservation]	Australia, including Tasmania	Maria Is. off the coast of Tasmania	Apparently introduced to Maria Is. off the Tasmanian coast in 1971 as a part of wildlife management efforts.	Weidenhofer 1977; Rounsevell et al. 1991; Long 2003; Heinsohn 2006a
Vombatidae				
Common wombat (*Vombatus ursinus*) [Conservation]	Southeastern Australia, including Flinders Is. and Tasmania	Maria Is. off the coast of Tasmania	Between 1969 and 1971 28 individuals of the subspecies *V. ursinus ursinus* from Flinders Is. were introduced to Maria Is. off Tasmania as a part of wildlife management efforts.	Weidenhofer 1977; Long 2003; Heinsohn 2006a
Southern hairy-nosed wombat (*Lasiorhinus latifrons*) [Conservation]	Central southern Australia	Wedge Is. off the coast of South Australia.	Introduced in 1971 as a part of wildlife management efforts.	Copley 1995; Long 2003; Heinsohn 2006a
Phascolarctidae				
Koala (*Phascolarctos cinereus*) [Conservation]	Eastern Australia	Kangaroo Is.; Phillip Is.; French Is.; some other small continental islands	Introduced to Phillip Is. and French Is. in 1920 in Victoria, and to Kangaroo Is. in South Australia between 1923 and 1958, becoming successfully established. Over-population on islands such as Kangaroo Is. has caused habitat damage such as high mortality rate for rough-barked manna gum *Eucalyptus viminalis* through over-browsing.	Condon 1967; Inns et al. 1979; Kemper 1985; Archer et al. 1987; Lee and Martin 1988; Phillips 1990; Backhouse and Crouch 1991; Copley 1995; Inns 2002; Long 2003; Heinsohn 2006a

Petauridae				
Sugar glider (*Petaurus breviceps breviceps*) [Southern and eastern Australian subspecies of the Australo-Papuan sugar glider] [Escaped pets]	Australia	Tasmania	Probably introduced into Tasmania in the middle years of the 19th century through escaped pets imported from Victoria.	Gunn 1851; Lord 1919; Lord and Scott 1924; Smith 1973; Honecki et al. 1982; Rounsevell et al. 1991; Lever 1985; Flannery 1994b; Long 2003; Heinsohn 2004a, 2006a
Pseudocheiridae				
Common ringtail possum (*Pseudocheirus peregrinus*) [Acclimatisation movement and conservation]	Eastern Australia, including Tasmania	Kangaroo Is. off the coast of South Australia in 1926; Maria Is. off the coast of Tasmania in 1971	Deliberately introduced as a part of acclimatisation and conservation efforts.	Waite and Jones 1927; Condon 1967; Inns et al. 1979; Kemper 1985; Rounsevell et al. 1991; Copley 1995; Inns 2002; Long 2003; Heinsohn 2004a, 2006a
Potoroidae				
Tasmanian bettong (*Bettongia gaimardi*) [Conservation]	Southeastern Australia (extinct); and Tasmania (extant)	Maria Is. off the coast of Tasmania	Between 1969 and 1971, 123 individuals were released on to Maria Is. as part of wildlife management efforts, with the species becoming successfully established there.	Weidenhofer 1977; Long 2003; Heinsohn 2006a
Brush-tailed bettong (*Bettongia penicillata*) [Conservation]	Australia	Venus Bay Island 'A'; Baird Bay Is.; St. Peter Is.; Wedge Is. off the coast of South Australia	Introduced to islands off South Australia in the last quarter of the 20th century as a part of conservation efforts.	Long 2003; Nelson et al. 1992; Copley 1995; Short et al. 1992; Delroy et al. 1986; Heinsohn 2006a
Macropodidae				
Rufous hare-wallaby or Mala (*Lagorchestes hirsutus*) [Conservation]	Western and central Australia	Trimouille Is., Montebello group off the coast of Western Australia	Thirty individuals were successfully introduced to Trimouille Is. in the Montebello group in 1998 as a part of conservation efforts.	Long 2003; Heinsohn 2006a
Tasmanian pademelon (*Thylogale billardierii*) [Conservation]	Southeastern Australia (extinct); Tasmania and larger islands of the Bass Strait (extant)	Maria Is. off the coast of Tasmania	Thirteen individuals were released on Maria Is. in 1971 as part of wildlife management efforts.	Weidenhofer 1977; Long 2003; Heinsohn 2006a

Taxa/species and value/ reason for translocation	Presumed natural distribution	Where thought to be introduced	Historical context	References
Black-footed rock wallaby (*Petrogale lateralis*) [Conservation]	Western and central Australia; some islands off South Australia.	Central Pearson Is., South Pearson Is., Thistle Is., Wedge Is., off the coast of South Australia.	Subspecies *Petrogale lateralis pearsoni* is endemic to North Pearson Is., South Australia, but was translocated to other islands off South Australia in the latter part of the 20th century as a part of conservation efforts.	Long 2003; Copley and Alexander 1997; Copley 1995; Short et al. 1992; Heinsohn 2006a
Rothschild's rock wallaby (*Petrogale rothschildi*) [Conservation]	Northwestern Australia in the Hamersley Range area, including some rocky offshore islands of the Dampier Archipelago	West Lewis Is. off the coast of northwestern Western Australia	Deliberately re-introduced to West Lewis Is. in 1981 as a part of conservation efforts.	Strahan 1995; Pearson and Kinnear 1997; Long 2003; Heinsohn 2006a
Bennett's wallaby, red-necked wallaby, brush wallaby or scrub wallaby (*Macropus rufogriseus*) [Conservation]	Tasmania and eastern Australia	Maria Is. off the coast of Tasmania	Between 1969 and 1970, 127 Tasmanian Bennett's wallabies (*Macropus rufogriseus rufogriseus*) were successfully introduced to Maria Is. off Tasmania as a part of wildlife management efforts.	Weidenhofer 1977; Long 2003; Heinsohn 2006a
Tammar wallaby, dama wallaby, Kangaroo Island wallaby or silver-grey wallaby (*Macropus eugenii*) [Food animal and conservation]	Southwestern Western Australia and southern South Australia	Greenly Is. off the coast of South Australia; North Island in the Abrolhos Archipelago of Western Australia	*M. eugenii* was introduced to Greenly Island off the coast of South Australia in about 1905, apparently to provide an emergency food supply for shipwreck survivors; and introduced to North Island in the Abrolhos Archipelago of Western Australia in the latter half of the 20th century.	Inns et al. 1979; Short et al. 1992; Copley 1995; Long 2003; Heinsohn 2006a
Eastern grey kangaroo (*Macropus giganteus*) [Tasmanian subspecies *M.giganteus tasmaniensis*] [Conservation]	Eastern Australia, including northeastern Tasmania	Maria Is. off the coast of Tasmania	Between 1969 and 1971, 45 individuals of the Tasmanian subspecies *Macropus giganteus tasmaniensis* subspecies were successfully introduced to Maria Is. as a part of conservation efforts.	Wapstra 1976; Weidenhofer 1977; Rounsevell et al. 1991; Long 2003; Heinsohn 2006a
Western grey kangaroo (*Macropus fuliginosus*) [? Reason]	Southwestern and southern Australia west of the Great Dividing Range	Woody Is. in the Recherche Archipelago of Western Australia	Successfully introduced to Woody Is. in 1948.	Short et al. 1992; Long 2003; Heinsohn 2006a

(?I) = May be introduced. ? = May occur on that landmass.

Australian exports

The translocation of Australian marsupials to new landmasses can be broken down into three categories: (1) species introduced to Europe in recent historic times; (2) species introduced to Pacific islands in Remote Oceania in historically recent times; and (3) species introduced locally to Australia's own continental shelf islands in historically recent times. Currently, there is very little information on possible prehistoric translocation of strictly Australian (as opposed to Australo-Papuan) marsupial species, other than, for example, postulations by McNiven and Hitchcock (2004) based on archaeological and ethnohistorical evidence that Torres Strait Island peoples may have enhanced macropodid populations on some of the Torres Strait islands during insular post-land-bridge Holocene times through introductions or re-introductions of macropodid species, including the Australian common wallaroo *Macropus robustus*. While this is quite plausible, such postulation currently runs into the problem of clearly distinguishing between natural relictual macropodid populations left over from when the islands were part of the Australo-Papuan Torresian land bridge, and possible subsequently introduced/re-introduced populations. This could potentially be resolved through more extensive archaeological sampling of the area, and molecular systematics analysis of DNA extracted from subfossil bone/teeth and from tissue samples from extant regional macropodid populations.

Only one Australian marsupial species has been successfully introduced to Europe – the red-necked wallaby *Macropus rufogriseus* from eastern and southeastern Australia and Tasmania, with Tasmanian stock principally used. Several small populations were successfully established in Germany and the United Kingdom in the latter half of the 19th century and the first half of the 20th century, through deliberate acclimatisation efforts and escapees from zoos and private menageries. With successful reproductive recruitment, these populations appeared to be naturally viable and able to withstand Palaearctic predators such as foxes and dogs, but for the most part, apart from the impacts of some severe winters, they eventually dwindled because of human predation (especially during the hungry war years in Germany) and the indirect impacts of high human population densities. Today, out of all of these former introduced populations, there are probably only a few animals left, such as in the Peak District of England in the United Kingdom (Long 2003; Heinsohn 2006a; Harris and Yalden 2008).

Seven species of marsupials have been successfully translocated to Pacific islands in Remote Oceania. These include one phalangerid, the common brushtail possum *Trichosurus vulpecula*, and six macropodids including the swamp wallaby *Wallabia bicolor*, red-necked wallaby *Macropus rufogriseus*, tammar wallaby *M. eugenii*, parma wallaby *M. parma*, and black-striped wallaby *M. dorsalis* to New Zealand; and the brushtailed rock wallaby *Petrogale penicillata* to both New Zealand and Hawaii. Of these, *T. vulpecula* is now almost ubiquitous in New Zealand, having been introduced to the North Island, South Island, Stewart Island and many satellite islands, including the remote eastern Chatham Islands (pers obs. 2009). Introduction of this folivorous possum began in the mid 19th century to establish a fur industry, with both Tasmanian and mainland Australian stock being imported before use of New Zealand-bred stock for later liberations (Thompson 1922; Donne 1924; Wodzicki 1950; Pracy 1962; Montague 2000; Heinsohn 2004a; Cowan 2005; Tyndale-Biscoe 2005). In New Zealand, the possum has become an invasive over-abundant pest species with a significant impact on ecosystems and biodiversity. The principal impact is selective over-browsing on New Zealand's mammalian-folivore-naive forests (there were no native mammalian folivores in New Zealand), leading to the death of certain palatable tree species from exhaustion following repeated defoliation. This is reported to have caused possum-induced die-back and changes in forest composition in many areas of the North and South Islands. For example, species such as rata *Metrosideros* spp., fuchsia *Fuchsia excorticata*, titoki *Alectryon excelsa* and kamahi *Weinmannia racemosa* have declined in many areas. Possums can also affect plant recruitment in a number of species through consumption of seedlings on the ground or through eating the palatable bark on favoured saplings, effectively ring-barking young trees, and this, too, can contribute to changes in plant community composition. Other possum impacts include predation on native invertebrates such as snails and insects and on small vertebrates such as nestling birds, and competition with indigenous species such as birds for tree hollows and other habitat space (King 1984; Montague 2000; Heinsohn 2004a; Cowan 2005; Tyndale-Biscoe 2005; Amanda Baird pers comm. 2009).

In comparison, New Zealand's six macropodids were all introduced as a part of 19th century or early 20th century acclimatisation efforts, and have much more restricted ranges, and limited impacts. Indeed, three species, including *W. bicolor*, *M. parma* and *M. dorsalis,* have been restricted to Kawau Island in the Hauraki Gulf of the North Island since being liberated there by Sir George Grey in c. 1870, and one, *M. dorsalis*, appears to have died out in the mid 20th century after persisting there for 80 years. A further two species, *P. penicillata* and *M. eugenii*, were also introduced to Kawau Island by Sir George Grey in c. 1870, where they still persist, but were additionally introduced to other parts of New Zealand. Specifically, in c 1870, *P. penicillata* was also introduced to Rangitoto and Motutapu Islands in the Hauraki Gulf, where it persisted for about 120 years until being exterminated in the 1990s; while *M. eugenii* was additionally introduced to the Rotorua area of the North Island in c. 1912, where it is now firmly established and from where it continues to expand its range. In contrast, New Zealand's sixth wallaby, *M. rufogriseus*, was introduced to the Hunters Hills area of the South Island in 1874, and is now firmly established in a considerable area of the central South Island to the east of the Southern Alps. It is the only macropodid in the south, and was never established in northern New Zealand (Wodzicki 1950; Lever 1985; Maynes 1989; Long 2003; King 2005; Warburton 2005a, b, c, d, e, f).

Of the aforementioned Australian macropodids introduced to temperate New Zealand, the brushtailed rock wallaby *Petrogale penicillata* is now also established on the Hawaiian island of Oahu in tropical Remote Oceania. This occurred as a result of a single male and female pair that escaped from a private menagerie in 1916. A small, apparently viable population persists to this day in steep rocky terrain on Ewa Kalihi in the lower Ko'olau Range on Oahu Island, where it is somewhat contained by geography and habitat (Kramer 1971; Tinker 1980; Lazell 1981; Lazell et al. 1984; Lever 1985; Moulton and Pimm 1986; Tomich 1986; Maynes 1989; Ziegler 2002; Long 2003; Heinsohn 2006a).

Of the approximately 20 Australian marsupial species translocated locally to surrounding continental shelf islands in historically recent times, most were introduced or re-introduced to island reserves and refuges in the 20th century for species management and conservation purposes. These mostly fairly minor translocations include: a carnivorous dasyurid; three species of bandicoot; two wombat species; koalas; sugar gliders; common ringtail possums; two bettong species; a hare wallaby; Tasmanian pademelons; two species of rock wallaby; and four wallabies/kangaroos in genus *Macropus* (Long 2003; Heinsohn 2004a, 2006a; Van Dyck and Strahan 2008) (see Table 1). Of the above, probably the most significant introduction in terms of the scale of the subsequent biological invasion is the mainland Australian sugar glider *Petaurus breviceps* to Tasmania in the mid 19th century through escaped pets brought from Victoria. After initial escapes in northern Tasmania, perhaps with some further human-assisted dispersal, it is recorded to have spread across the entire length of the island by the early 20th century, and today *P. breviceps* occupies virtually all areas, other than non-congenial habitats such as parts of the damp southwest region (Gunn 1851; Lord 1919; Lord and Scott 1924; Lever 1985; Rounsevell et al. 1991; Flannery 1994; Long 2003; Heinsohn 2004a; Van Dyck and Strahan 2008).

Curiously, the introduction of the koala *Phascolarctos cinereus* to Kangaroo Island in South Australia in the 1920s has had some parallels to the impact of the brushtail possum *T. vulpecula* in New Zealand, in that this fairly large arboreal folivore has become over-abundant on the island and is responsible for the death of certain palatable tree species such as rough-barked manna gum *Eucalyptus viminalis* through over-browsing. Repeated defoliation through excessive browsing by over-abundant koalas eventually causes a significant proportion of palatable trees to die from exhaustion, thus bringing about changes in forest composition (Inns 2002; Heinsohn pers obs. 2008). Kangaroo Island has a second introduced arboreal folivore, the common ringtail possum *Pseudocheirus peregrinus*, also introduced in the 1920s (Waite and Jones 1927; Copley 1995; Inns 2002; Long 2003; Heinsohn 2004a), and this species may possibly contribute to cumulative browsing pressures on palatable plant species.

New Guinea and Papuan subregion exports

With an estimated 80 marsupial species, New Guinea, including its fringing Papuan subregion satellite islands, is the world's second major centre for Australidelphian marsupial radiation (after Australia, which has nearly 160 species) (Table 2). Despite having half the number of marsupials, the Papuan subregion is, however, the world's greatest exporter of translocated marsupials to surrounding oceanic archipelagos, with 10 or more of its species suspected to have been exported through long-term human agency (Table 1). Furthermore, nearly all of these postulated translocations, including the export of two peroryctid bandicoots, a petaurid glider, several phalangerid possums and several macropodids, are thought to have occurred in prehistoric times. Several of these postulations regarding prehistoric translocations are based principally on archaeological evidence of sudden appearances of skeletal material in archaeological sequences, while other postulations, often regarding areas where little archaeological sampling has occurred, are based on broader ethnozoological and biogeographic contextual analysis, sometimes also including available evidence from the emerging field of molecular systematics.

The oldest postulated marsupial translocation in the region, based on archaeological evidence, is that of the New Guinean northern common cuscus *Phalanger orientalis* to New Ireland in the outer northwestern arm of the Bismarck Archipelago in Near Oceania at 23,500 BP to 20,000 BP in the late Pleistocene (Flannery and White 1991; Flannery 1993; Spriggs 1997; Leavesley 2005; Summerhayes 2007). This is also the world's oldest recorded anthropogenic animal translocation event, making *P. orientalis* the world's oldest known 'ethnotramp', an animal species dispersed through human agency (as defined in Heinsohn 1997a, 1998a, b, 2001a, 2003, 2004a, 2006a). Furthermore, as *P. orientalis* may have been translocated through New Britain to New Ireland, populations on the former, if also introduced, may have been translocated at an even earlier date. Other Pleistocene marsupial translocations postulated on the basis of archaeological evidence include the introduction of the New Guinean common spiny bandicoot *Echymipera kalubu* and the ancestors of the Admiralty cuscus *Spilocuscus kraemeri* to Manus Island in the northeastern Bismarck Archipelago, possibly at around 13,000 BP in the terminal Pleistocene (Spriggs 1997; Summerhayes 2007). This compares with archaeological evidence for prehistoric Holocene marsupial translocations in Near Oceania which supports postulated introduction of: the New Guinean northern pademelon *Thylogale browni* to New Ireland between 8400 and 7000 BP; the New Guinean common spotted cuscus *Spilocuscus maculatus* to the St. Matthias group in the far northern Bismarck Archipelago in the late Holocene; and secondary introductions of *P. orientalis* to Nissan and the Solomon Islands in the middle to upper Holocene (Kirch 1988; Flannery 1995b; Kirch 1997; Spriggs 1997; Summerhayes 2007).

There is also historical and archaeological evidence of further ephemeral (now extinct) introduced *T. browni* macropodid populations on some of the smaller New Ireland satellites, including Tabar Island, the Lihir group, Tanga Island and Feni Island, and archaeological evidence of possible ephemeral (now extinct) introduced populations of this macropodid on Buka in the North Solomons, and Eloaue Island in the St. Matthias group in the upper

Holocene (Egloff 1975; Flannery et al. 1988; Flannery 1995b; Spriggs 1997; Heinsohn 2005c; Summerhayes 2007). Combined archaeological, ethnozoological and biogeographic evidence also indicates that another macropodid, the Australo-Papuan agile wallaby *Macropus agilis*, may have been introduced to its contemporary island range in the continental southeast Papuan islands during mid to late Holocene times (Brass 1956, 1959; Egloff 1979; Maynes 1989; Flannery 1995b; Heinsohn 2003; Long 2003; Heinsohn 2006a).

To the immediate west of New Guinea in the oceanic Moluccas of Wallacea, there has been considerably less archaeological investigation, but limited excavations from places such as Halmahera in the north Moluccas and Gebe, an adjacent oceanic Papuan subregion outlier, indicate that a now extinct, probably introduced, macropodid population of the Papuan brown dorcopsis *Dorcopis muelleri* occurred on the aforementioned islands in the early and middle Holocene respectively (Flannery et al. 1995; Flannery et al. 1998; Summerhayes 2007). Further archaeological evidence of marsupial translocation comes from Timor in the eastern Lesser Sundas, where more extensive archaeological excavations indicate that the New Guinean northern common cuscus *Phalanger orientalis* was probably introduced there in the

Table 2. Proportions of introduced marsupials on landmasses in the Indo-Australian Archipelago, Oceania and Europe

	Marsupial species	Number introduced	Percentage introduced
Continental Meganesia			
Australia	158	0	0%
New Guinea	80	0	0%
Oceanic Wallacea			
Halmahera	2	? 1	? 50%
Seram	3	2	66%
Buru	2	2	100%
Kai Is.	5	? 3 to 5	? 60% to 100%
Sulawesi subregion			
(Sulawesi and satellites including Selayar and Peleng)	4	1	25%
Timor			
(and Lesser Sundas as a whole)	1	1	100%
Near Oceania			
Admiralty Islands	2	? 0 to 2	? 0% to 100%
New Ireland	3	3	100%
New Britain	? 6	? 3 to 6	? 50% to 100%
Umboi	4	? 3 to 4	? 75% to 100%
Solomon Is.	1	1	100%
Remote Oceania			
New Zealand	6	6	100%
Hawaiian Is.	1	1	100%
European Continental islands			
British Isles	1	1	100%

References: Tomich 1986; Whitten et al. 1987; Corbet and Hill 1992; Flannery 1994b; Flannery 1995a, b; Monk et al. 1997; Heinsohn 2001a, 2002a, b, c, 2003, 2004a, b, 2005a, b, c, 2006a; Ziegler 2002; Long 2003; Helgen and Flannery 2004; King 2005; Nowak 2005; Heinsohn and Hope 2006; Harris and Yalden 2008; Van Dyck and Strahan 2008

middle Holocene (Glover 1971, 1986; Flannery 1995b; Bellwood 1997; Monk et al. 1997; Heinsohn 2003; Summerhayes 2007).

Other than palaeontology and archaeology, a further line of evidence for anthropogenic translocation of New Guinea region marsupials is provided by ethnohistorical and contemporary records of live animals being carried around in watercraft, as pets, living-larder food animals, trade commodities and sources of animal products such as decorative pelts, fibre, bone and teeth. While diminutive marsupials such as young sugar gliders *Petaurus breviceps* are principally kept as pocket pets (pers obs.), other larger marsupials such as bandicoots, cuscuses, and macropodids can provide significant parcels of bush meat, while in some areas spotted cuscuses *Spilocuscus* spp. in particular are highly valued for their often colourful pelts, which are worn as esteemed body adornments ('bilas' in New Guinea pidgin), with some tree kangaroo *Dendrolagus* spp. pelts also prized as bilas (pers obs.). Marsupial fur is also valued as a source of traditional fibre used to make products such as woven bilum bags and hats, with cuscuses and some other possums being plucked for this purpose. Young cuscuses including *Spilocuscus* spp. and *Phalanger* spp. are also popular pets, although virtually any young marsupial, including wallaby and tree-kangaroo joeys captured during hunting, may be kept for a while as ephemeral pets or live-animal trade commodities and carried about. Any carriage of pets, trade stock, or living-larder captives increases the chance of escapees establishing new populations beyond their natural range, particularly given the flimsy or less-than-secure state of various cages, baskets, sacks, woven bags and bindings used to restrain captive marsupials (Heinsohn 1997a, 1998a, b, 2001a, 2003, 2004a, b, 2006a, pers obs.).

There are also contemporary and ethnohistorical records of deliberate attempts to introduce various marsupials such as cuscuses and wallabies to faunally depauperate islands as part of a game-enhancement strategy to provide a future source of food and other animal products. This parallels the well-recorded tradition of European sailors, fishermen and whalers who are historically known to have left breeding stock of goats, pigs and fowls on stop-over islands to ensure a future food supply. The author, for example, recorded details of late 20th century attempts by local Melanesians to introduce the common spotted cuscus *S. maculatus* to Djaul Island off New Ireland, and the Admiralty cuscus *S. kraemeri* to small satellites such as Bipi Island off Manus, while there is also ethnohistorical evidence that local people traditionally kept some of New Ireland's satellite islands deliberately stocked with northern pademelons *Thylogale browni*, restocking from larger source populations on the adjacent New Ireland mainland (Heinsohn 1998b, 2005c).

A further recorded case of probable deliberate translocation of New Guinea region marsupials during the historical period is that of the northern common cuscus (*Phalanger orientalis*) being introduced (or re-introduced) to Long Island off the northeast coast of New Guinea some time in the past 300 years. The oral traditions of local people indicate that the island was deliberately restocked with *P. orientalis*, a valued food animal, following a cataclysmic volcanic eruption of Krakatoan magnitude that occurred three centuries ago and destroyed virtually all life (Koopman 1979; Ball and Hughes 1982; Ian Hughes pers comm. 1990). Jared Diamond (as cited by Koopman 1979:3), reporting on the testimonies of the island's inhabitants, considers that this restocking may have happened in the 19th century, from phalangerid sources on Tolokiwa Island to the east. This is subsequently elaborated on in Mayr and Diamond (2001:18): 'Even the name of the person who introduced *Phalanger orientalis* to Long Island in the 19th century is known: a man called Ailimai.'

There are also historically recorded cases of secondary marsupial introductions in the Bismarck Archipelago that apparently happened as a result of accidental escape of captive marsupials. This underscores the diversity of likely scenarios for the introduction of New Guinea region marsupials to surrounding archipelagos. Thus, while a great many introductions in the prehistoric past may have been deliberate, as a part of game enhancement strategies to add to the faunal resources on islands, a proportion of introductions may have occurred as a result of pets, living-larder food animals, captive trade stock, or simply a hunter's live catch (caught in traps, snares or by hand) that escaped captivity on arrival at a destination (Heinsohn 1997a, 1998a, b, 2001a, 2003, 2004a, b, 2006a).

Contemporary anthropogenic spread of Spilocuscus maculatus in the Bismarck Archipelago

The first recorded translocation of the New Guinean common spotted cuscus *S. maculatus* into the Bismarck Archipelago is its prehistoric introduction to the remote St. Matthias group in the late Holocene, where the animal remains strongly established to this day (Kirch 1988, 1997; Flannery and White 1991; Heinsohn 1998b). A salient historical case of marsupials apparently established on an island through escaped captives is provided by the story of how *S. maculatus* subsequently made it to New Ireland from sources in the St. Matthias group. Flannery and White (1991:111) were the first to report in any detail on this case:

> 'Sanila Talevat, a senior man of Madina, related the facts of the introduction of this species as they were known to him: several *S. maculatus* were brought to Kavieng, northern New Ireland, in 1929 from the St. Matthias Group by Australian Government officials (or possibly their Melanesian employees). [The population on the St. Matthias Group itself results from human introduction.] The animals subsequently escaped. They are now found over a large area of northern New Ireland ... Sanila suggests that they had not spread much farther south than the narrowest part of New Ireland, ~60km south of Kavieng.' (Flannery and White, 1991:111, reporting on data collected in 1988)

The contemporary and ongoing human-induced spread of *S. maculatus* provides an illuminating analogy for anthropogenic biological invasions in the past. The author's own spotlighting survey and oral historical investigation of the distribution of *S. maculatus* in New Ireland revealed that by 1990, it had invaded about 40 km to 50 km down the long narrow island (Heinsohn 1998b, 2004a, 2006a) (Figure 4). However, even then, there were reports of isolated adventive populations, apparently mostly derived from escaped or liberated pets carried in by boat or motor vehicle, that were beginning to appear at scattered localities beyond the main invasion front, such as one near Pangeifua some 20 km further south (Heinsohn 1998b, 2004a). By the mid 1990s, an adventive population of *S. maculatus* had been recorded in lower elevations of the Weiten River Valley of far southern New Ireland (Emmons and Kinbag 2001). This far southern record is more than 200 km to the south of the main invasion frontier, and probably also represents an adventive population established fairly recently through secondary boat or vehicle transport of pets and captives (author's assessment), and no doubt will go on to form a further invasion beachhead, with such anthropogenic secondary invasion beachheads eventually coalescing to speed up the complete colonisation of New Ireland by *S. maculatus* (Heinsohn 2006a). Indeed, rapid development of oil palm plantations and the spread of logging activities to far corners of New Ireland has probably hastened the spread of *S. maculatus* through greatly increased vehicular and boat traffic and movement of people, including cashed-up workers who buy the animals at markets for use as food or popular status-enhancing pets.

The latest development is that recently translocated adventive populations of *S. maculatus*, probably from introduced sources in adjacent New Ireland, have begun to appear around villages and plantations in northern New Britain (Sebastian Högberg, pers comm. 2009). Thus, the predicted anthropogenic spread of *S. maculatus* from New Ireland to New Britain has already begun. Adventive populations may also be starting to appear on other New Ireland satellites, such as New Hanover (Lavongai). For a reconstruction of the historical invasion of New Ireland and some adjacent islands by *S. maculatus*, see Figure 4 (Heinsohn 1998b, 2004a, 2006a).

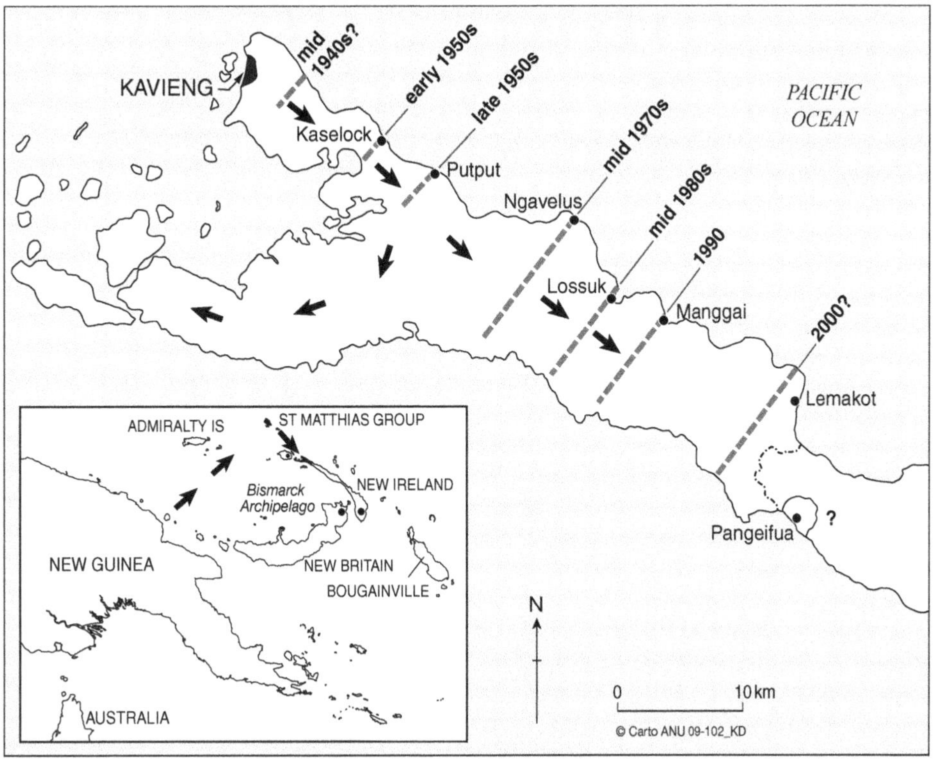

Figure 4. Ongoing biological invasion of the Bismarck Archipelago by the introduced northern New Guinean subspecies of the common spotted cuscus *Spilocuscus maculatus maculatus*. Inset – This species was prehistorically introduced to the St. Matthias group in the Bismarck Archipelago in the late Holocene, from whence it was subsequently historically introduced to Kavieng at the far northwestern tip of New Ireland in the decade leading up to the commencement of the Second World War. Main map – Reconstruction of the subsequent biological invasion of New Ireland in the latter half of the 20th century. Dot at Pangeifua represents an adventive population beyond the main invasion front. Inset – Dot at the far southern end of New Ireland shows an isolated secondary adventive population of *S. maculatus* recorded at lower elevations in the Weiten River Valley of far southern New Ireland in the mid 1990s more than 200 km beyond the main invasion front. This is probably due to a secondary introduction of animals brought in from the north by boat or motor vehicle. Dot in northern New Britain represents secondary adventive populations of *S. maculatus* that have recently appeared in and around villages and plantations on that part of the island, probably through boat traffic from adjacent New Ireland. Adventive populations may now also occur on other New Ireland satellites such as New Hanover (Lavongai) marked by a dot (after Heinsohn 2006a)

In New Ireland, it was found that recent biological invasion by a slightly larger and more generalist *S. maculatus* had caused a slight niche contraction in pre-existing prehistorically introduced populations of *P. orientalis* (see Figures 5 and 6). Principally, in broad sympatry with *S. maculatus* within the northwestern New Ireland invasion zone, *P. orientalis* tended to retreat to core habitats within closed forest, whereas in allopatry beyond the invasion zone, where it was the sole cuscus species, it exhibited a more relaxed niche that also included plantations and some open forest areas. In contrast, the invading *S. maculatus* tended to occupy the full spectrum of closed forest, open forest and plantation, displacing *P. orientalis* from the more marginal plantation and open-forest habitats (Heinsohn 1998a, b, 2004 a, b). It is likely that this pattern will be repeated as *S. maculatus* invades New Britain and other adjacent islands that were once the sole reserve of *P. orientalis*.

The big picture of postulated anthropogenic export of Papuan region marsupials is summarised below.

Summary of postulated anthropogenic Papuan marsupial exports to surrounding oceanic islands

Taking all of the above contextual information into account, broad ethnozoological and biogeographic analysis indicates that:

1. Possibly all of the oceanic distribution of the New Guinean northern common cuscus *Phalanger orientalis* in the Bismarck Archipelago and Solomon Islands in Near Oceania; on Seram, Buru and satellites in the Central Moluccas; on the Kai Islands in the South Moluccas; on Sanana Island in the Sulawesi subregion; and on Timor and its satellites in the Lesser Sundas is due to long-term human agency;
2. Possibly all of the oceanic distribution of the New Guinean common spotted cuscus *Spilocuscus maculatus* in the St. Matthias group, New Ireland and New Britain in the Bismarck Archipelago; on Seram, Buru and satellites in the Central Moluccas; in the Kai Islands in the South Moluccas; and on far-flung Selayar Island in the Sulawesi subregion is due to long-term human agency;
3. All of the oceanic distribution of the Admiralty cuscus *Spilocuscus kraemeri* in the Admiralty Islands and other Manus Province Archipelago islands in the Bismarck Archipelago may be due to long-term human agency, with the distinctive specific status possibly being due to human-induced hybridisation and founder effect caused by anthropogenic admixture or introduction of a mixed *Spilocuscus* spp. founding stock from New Guinea or Papuan subregion sources;
4. The oceanic distribution of the Australo-Papuan sugar glider *Petaurus breviceps* in the north Moluccas and on some other oceanic New Guinea satellites such as New Britain may be due to long-term human agency;
5. The oceanic distribution of the New Guinean common spiny bandicoot *Echymipera kalubu* on Manus Island in the Bismarck Archipelago, and on some other oceanic New Guinea satellites such as New Britain, may be due to long-term human agency;
6. The oceanic distribution of the New Guinean rufous spiny bandicoot *Echymipera rufescens* in the Kai Islands of the south Moluccas may be due to long-term human agency;
7. The oceanic distribution of the New Guinean Huon tree kangaroo *Dendrolagus matschiei* on Umboi Island, and possible occurrence on New Britain, in the Bismarck Archipelago, is probably due to human agency;
8. The oceanic distribution of the New Guinean northern pademelon *Thylogale browni* in the Bismarck Archipelago is probably due to long-term human agency;
9. The oceanic distribution of the New Guinean dusky pademelon *Thylogale brunii* in the Kai Islands of the south Moluccas is probably due to long-term human agency;
10. The oceanic distribution of past populations of the New Guinean/Papuan subregion brown dorcopsis *Dorcopsis muelleri* in the north Moluccas can probably be attributed to long-term human agency;
11. The distribution of the Australo-Papuan agile wallaby *Macropus agilis* in the continental southeast Papuan islands is probably due to long-term human agency; and
12. The secondary occurrence of the endemic Woodlark Island cuscus *Phalanger lullulae* on Alcester Island may be due to long-term human agency (See Table 1).

Figure 5. Introduced New Guinean common spotted cuscus *Spilocuscus maculatus maculatus* invading New Ireland. Photograph by T. E. Heinsohn

Figure 6. Northern common cuscus *Phalanger orientalis* in New Ireland, where archaeological evidence indicates that it was probably first introduced between 23,500 and 20,000 years ago in the late Pleistocene. This is the oldest known case of anthropogenic translocation of an animal species making *P. orientalis* the world's oldest known ethnotramp. Photograph by T. E. Heinsohn

Wallacean exports

After Australia and New Guinea, including fringing islands of the Papuan subregion, the third largest centre for Australidelphian marsupial radiations is the oceanic Wallacean islands, which currently have a dozen recorded marsupial species, seven of which are endemic. These seven Wallacean endemics are: (1) three phalangerid possums endemic to the Sulawesi subregion – the Sulawesi bear cuscus *Ailurops ursinus*, the small Sulawesi cuscus *Strigocuscus celebensis*, and the Peleng cuscus *Strigocuscus pelengensis*; (2) three phalangerids endemic to the north Moluccan area, the ornate cuscus *Phalanger ornatus*, the Obi cuscus *Phalanger rothschildi*, and the Gebe cuscus *Phalanger alexandrae*; and (3) one bandicoot, the Ceram bandicoot *Rhynchomeles prattorum*, endemic to Seram in the central Moluccas. Gebe is a relatively small oceanic island lying near the boundary between the Moluccan and Papuan subregions. Principally because of its avifaunal composition, it is traditionally placed in the latter subregion and treated as an oceanic Papuan outlier (Rand and Gilliard 1967; Coates and Bishop 1997). However, because the Gebe cuscus *P. alexandrae* is closely related to the ornate cuscus *P. ornatus* of adjacent Halmahera in the north Moluccas, it is treated here as a part of the oceanic Wallacean marsupial fauna.

Apart from the Wallacean endemics, as discussed in greater detail above, the other non-endemic marsupials in Wallacea may all have been introduced from New Guinea and its fringing Papuan subregion islands through long-term human agency. These include two probably introduced New Guinean phalangerid possums *Phalanger orientalis* and *Spilocuscus maculatus*, one probably introduced Australo-Papuan petaurid glider *Petaurus breviceps*, one possibly introduced New Guinean bandicoot *Echymipera rufescens*, and one probably introduced New Guinean macropodid *Thylogale brunii* (Table 1).

Out of the aforementioned endemic Wallacean marsupials, only one, the Sulawesi bear cuscus *Ailurops ursinus*, is postulated to have possibly been subjected to human-assisted dispersal beyond its natural range on Sulawesi and some close satellite islands, to Salebabu Island in the remote oceanic Talaud (or 'Talaut') group:

> '*A. u. melanotis* from the Talaut Group is small, and the fur is more profusely tipped with yellow than in other forms. It is provisionally recognised here as a distinct subspecies. However, it is apparently restricted to a single island in the distant Talaut Group, north of Sulawesi, which suggests that it may have been introduced there prehistorically by humans' (Flannery 1994:160)

Indeed, its apparent distinctiveness as a subspecies on Salebabu (as with *Spilocuscus kraemeri* and *Phalanger orientalis breviceps* in the northern Melanesian islands of Near Oceania) could possibly be an artefact of long-term human-induced founder effect or anthropogenic allopatric differentiation resulting from thousands of years of isolation on oceanic Salebabu Island following long-term prehistoric introduction there. However, further research, including more extensive zoological surveys, molecular systematics and greater archaeological/palaeontological sampling of regional fossil/subfossil sites, is required to determine whether the outlying Salebabu Island, Talaud group, *A. ursinus* population is relict and due to ancient natural dispersals or an artefact of Quaternary human agency (Heinsohn 2004a, 2006a). Regarding its ethnozoological status, in Sulawesi *A. ursinus* is used for bush meat and sometimes carried about live as a living-larder trade commodity or pet (pers obs.), thus making it a candidate for possible deliberate or accidental introduction to adjacent islands via liberated or escaped captives.

Diagnosing the natural marsupial frontier: The problem of ambiguous oceanic marsupial faunas

In attempting to define the natural marsupial frontier, a line that marks the natural (as opposed to anthropogenic) distributional limits of Australidelphian marsupials, researchers encounter the problem of ambiguous marsupial faunas on certain fringing oceanic islands where there is currently insufficient evidence to determine whether all or some of the marsupial fauna is due to introduction by human agency, or indigenous and due to natural processes such as sweepstakes dispersal on natural rafts. This is because in the absence of historical records or adequate archaeological sampling, prehistorically introduced species, in particular, can mimic recently dispersed naturally occurring species (and vice versa), thus creating cryptogenic populations of unknown status or camouflaged exotic species which appear natural. In the Indo-Australian Archipelago, principally due to insufficient archaeological sampling, cryptogenic marsupial populations occur on New Britain and the Admiralty Islands in the Bismarck Archipelago of Near Oceania, and on Salebabu Island in the Talaud group in the Sulawesi subregion of Wallacea. The ambiguous zoogeographic status of various marsupials on these islands is described in Table 3.

Table 3. Ambiguous marsupial faunas on oceanic islands at the marsupial frontier (Table 3 continues on pages 160 and 161)

Species	Putative origins and zoogeographic context	Zoogeographic and ethnozoological status
Near Oceania		
New Britain	Non-volant mammal fauna is oceanic and unbalanced. Several endemic murid rats recorded, but depauperate compared to New Guinea mainland (4).	
Marsupials		
Common echymipera *Echymipera kalubu* (New Britain)	(a) Possibly arrived from New Guinea by over-water dispersal on a natural raft, or (b) possibly introduced through long-term human agency.	Cryptogenic – origins unclear on current evidence. Utilised food and game animal. Source of bush meat. (1, 2, 4, 8)
Sugar glider *Petaurus breviceps* (New Britain)	(a) Possibly arrived from New Guinea by over-water dispersal on a natural raft, or (b) possibly introduced through long-term human agency.	Cryptogenic – origins unclear on current evidence. Occasional pet and eaten as small game animal delicacy. (1, 2, 4, 8, 9)
Northern common cuscus *Phalanger orientalis* (New Britain)	(a) Possibly arrived from New Guinea by over-water dispersal on a natural raft, or (b) possibly introduced through long-term human agency.	Cryptogenic – origins unclear on current evidence. Utilised ethnotramp. Pet, food, game and trade animal. Source of animal products such as bush meat and bilum fibre. (1, 2, 4, 8, 9)
Common spotted cuscus *Spilocuscus maculatus maculatus* (New Britain)	This species appears to have been recently introduced to northern New Britain in the very late 20th century and/or very early 21st century, probably from adjacent sources in New Ireland.	Recently introduced species with adventive populations established. Utilised ethnotramp. Pet, food, game and trade animal. Source of animal products including bush meat, decorative fur pelts, and bilum fibre. (11)

Species	Putative origins and zoogeographic context	Zoogeographic and ethnozoological status
?Admiralty cuscus ? *Spilocuscus kraemeri* (New Britain) Probably not a part of the New Britain fauna.	If pets or small adventive populations of this Admiralty Islands species do occur anywhere on New Britain or on any of its small satellite islands such as Bali Island, Witu group, then these are likely to be due to trade animals carried in from the Admiralty Islands where live animals are extensively traded for bush meat, other animal products and pet-keeping purposes (pers obs.). The author disagrees with the theory put forward by Helgen and Flannery (2005) that *S. kraemeri* may be endemic to New Britain, the original source, but is cryptic there and thus not recorded (13). This seems extremely unlikely as *S. kraemeri* is an adaptable and prominent species, which, like *S. maculatus*, tends to be easily located during spotlighting surveys (pers obs.). As such, it is unlikey to have gone undetected on New Britain until now.	?Introduced species if present in New Britain. Utilised ethnotramp. Pet, food, game and trade animal. Source of animal products including bush meat, decorative fur pelts, and bilum fibre.
Northern pademelon *Thylogale browni* (New Britain)	Probably introduced from New Guinea through long-term human agency (as this species is not a likely candidate for rafting, floating or swimming).	Prehistorically introduced species. Utilised ethnotramp. Food, game and trade animal, and occasional pet. Source of bush meat. (1, 2, 4, 8, 10)
Tree-kangaroo *Dendrolagus matschiei* (New Britain)	Probably introduced from New Guinea through human agency, possibly from secondary introduced sources on Umboi Island (as this relatively large marsupial is not a likely candidate for rafting, floating or swimming).	?Introduced species. Traditional source of bushm eat and pelts. Juveniles kept as occasional pets. (5, 7)
Admiralty Islands	Non-volant mammal fauna is oceanic and depauperate. One endemic murid rat recorded (4).	
Marsupials		
Admiralty cuscus *Spilocuscus kraemeri* (Admiralty Islands)	(a) Ancestral stock possibly arrived in the Admiralty Islands from New Guinea by over-water dispersal on a natural raft, or (b) ancestral stock was possibly introduced through long-term human agency. Distinctiveness may possibly be due to human-induced founder effect such as anthropogenic hybridisation in a mixed founding stock. Helgin and Flannery (2004) theory of introduction from undiscovered populations in New Britain (13) seems unlikely as this is a highly visible and adaptable species (pers obs.) that is unlikely to have remained undiscovered on New Britain.	Cryptogenic – origins unclear on current evidence. Apparently endemic to Manus Province Archipelago, but paradoxically also possibly introduced. Utilised ethnotramp. Pet, food, game and trade animal. Source of animal products including bush meat, decorative fur pelts and bilum fibre. (1, 2, 4, 6, 8, 9). In Seventh Day Adventist parts of the Admiralty Islands where there is reduced hunting pressure, the species is reported to become overabundant and become a minor pest in food gardens. (12)
Common echymipera *Echymipera kalubu* (Admiralty Islands)	(a) Possibly arrived from New Guinea by over-water dispersal on a natural raft, or (b) species was possibly introduced through long-term human agency.	Cryptogenic – origins unclear on current evidence. Utilised food and game animal. Source of bush meat. (1, 2, 4, 8)

Species	Putative origins and zoogeographic context	Zoogeographic and ethnozoological status
Sulawesi Subregion		
Talaud Islands	Non-volant mammal fauna is oceanic and depauperate. Two endemic murid rats recorded. (4)	
Marsupials		
Sulawesi bear cuscus *Ailurops ursinus melanotis* (Salebabu Island, Talaud Group)	(a) Ancestral stock may have arrived from Sulawesi or one of its near satellites by over-water dispersal on a natural raft, or (b) possibly introduced to Talaud Islands through long-term human agency.	Cryptogenic – origins unclear on current evidence. Utilised food and game animal. Occasional pet and sometimes traded. (1, 2, 3, 4, 8, 9)

Sources: (1) Tate et al. 1945; (2) Laurie and Hill 1954; (3) Corbet and Hill 1992; (4) Flannery 1995b; (5) Flannery et al. 1996; (6) Helgen and Flannery 2004; (7) Martin 2005; (8) Heinsohn 1998a, 2003, 2006a; (9) Heinsohn 2004a, 2005b; (10) Heinsohn 2005c; (11) Sebastian Högberg, pers comm. 2009; (12) Heinsohn 1998b, 2004a; (13) Helgen and Flannery 2004.

Thus the marsupial fauna of the Talaud group, consisting of a single species *A. ursinus*, may be wholly introduced or wholly natural, while the marsupial fauna of the Admiralty Islands, comprised of two species *S. kraemeri* and *E. kalubu*, may be wholly introduced, 50% introduced, or wholly natural. New Britain, with up to seven, but probably six recorded species, has a marsupial fauna that may be wholly introduced or partially introduced, but not wholly natural, as the most recent introduction, that of *S. maculatus*, is historically recorded. In contrast to ambiguous New Britain, where there has been a lack of zooarchaeological sampling at appropriate sites, the marsupial fauna of neighbouring New Ireland, due to thorough archaeological sampling and historical records, is known to be wholly introduced (Flannery and White 1991; Flannery 1995b; Spriggs 1997; Heinsohn 1998b; Summerhayes 2007).

However, given that all of the marsupial species occurring in the Bismarck Archipelago are utilised or traded ethnotramp species (Heinsohn 2003, 2004a), with even the diminutive sugar glider *Petaurus breviceps* kept as a pocket pet, and given the lack of stochastic arrival of other non-ethnotramp marsupials such as pseudocheirids and dasyurids by sweepstakes dispersal during the late Cenozoic, the author considers it quite possible that the entire marsupial population in the archipelago was introduced through long-term human agency. Indeed, given all of the known precedents of marsupial translocation in the Bismarck Archipelago and the Solomon Islands, an application of the Occam's razor principle of looking for the simplest explanation could support attribution of the entire marsupial population of Near Oceania to long-term human agency, with the same principle also applied to the outlying *A. ursinus* population in the remote and oceanic Talaud Group. This leads to the author's proposed line of actual natural marsupial distributional limits marked by the bold broken line in Figure 7. This contrasts with Figure 8 which depicts the contemporary line of Australasian marsupial distributional limits when one takes into account all natural and introduced marsupial populations in the region. This represents a considerable anthropogenic expansion of the marsupial frontier into Near Oceania, Remote Oceania and southern Wallacea.

On the basis of the Occam's razor principle of parsimonious explanation and some ethnozoological and ecological considerations, the plausibility of the recent Helgen and Flannery (2004) radical hypothesis regarding possible New Britain origins for the Admiralty cuscus *Spilocuscus kraemeri* in the Bismarck Archipelago is questioned below. *S. kraemeri* is recorded from the Admiralty Islands and some outlying satellites such as the Hermit Islands, the Ninigo group and Wuvulu in the vast oceanic Manus Province Archipelago of Papua New Guinea (Laurie and Hill 1954; Flannery 1994, 1995b; Heinsohn 2004a). The species has the curious status of being apparently endemic to the Manus Province Archipelago, but paradoxically also considered to have been possibly introduced there from unknown sources in the New Guinea region (Flannery 1994, 1995b). On the basis of limited archaeological sampling from the Pamwak site on Manus Island, it is considered that this may have happened in the terminal Pleistocene (Spriggs 1997; Summerhayes 2007). The author has postulated that the apparent distinctiveness of *S. kraemeri* may be due to anthropogenic founder effect and genetic drift which have exaggerated allopatric differentiation since introduction thousands of years ago, and that this founder effect possibly may have involved anthropogenic hybridisation or admixture following introduction of a mixed *Spilocuscus* spp. founding stock from sources in the New Guinea region (Heinsohn 1997a, 1998a, b, 2003, 2004a, 2006a). A regional precedent for this model is provided by the introduced pigs of New Guinea which may be derived from hybrids of *Sus scrofa vittatus* and *Sus celebensis* carried in from sources in the Indonesian Archipelago during the Holocene (Groves 1981; Flannery 1995a). If humans can inadvertently almost create a new taxon with ethnotramp pigs, then why not with ethnotramp phalangerids? A further possibility, however, regarding the origins of *S. kraemeri* in the Admiralty Islands is that it is due to natural waif dispersal of ancestral founding stock from New Guinea or its fringing Papuan subregion islands in the prehistoric past via animals washed out to sea on natural rafts as a result of floods, king tides, storms or tsunamis. However, even if *S. kraemeri* did make it to the Admiralty Islands by natural means, at least part of the animal's distribution within the vast Manus Province Archipelago, such as that on the outlying Hermit Islands, the Ninigo group and Wuvulu, is probably due to human agency, as the species is extensively traded at markets in that province, and to this day continues to be carried about in traditional vine baskets and other cages (pers obs.; Heinsohn 1998a, b, 2003, 2004a, 2006a).

A radical new hypothesis regarding the origins of *S. kraemeri* in the Admiralty Islands was presented by Helgen and Flannery (2004), who on the basis of two specimens, one bought at a New Britain market and reputed to come from Bali Island in the Witu group of West New Britain and a further museum specimen labelled 'New Britain', proposed that the species differentiated in New Britain becoming an island endemic, and was subsequently introduced to the Admiralty Islands by prehistoric human agency. Furthermore, they proposed that the species may still be in New Britain, but is cryptic there, thus remaining undiscovered and unrecorded on that island or its near satellites (Helgen and Flannery 2004). The author's own view is that the above hypothesis is improbable as it is extremely unlikely that New Britain would have been home to a prominent species such as *S. kraemeri* without it being firmly recorded by science over recent centuries. This is underscored by the fact that in the Admiralty Islands, *S. kraemeri* (as with *S. maculatus* in New Ireland) is a highly adaptable animal found in primary and secondary forest habitats as well as spreading into plantations and gardens, with its presence being well known by local villagers and hunters, and other than where locally over-hunted, is also relatively easily detected by biologists with spotlights and head torches (pers obs.; Heinsohn 1998b). Thus, it is extremely unlikely not to have been discovered on New Britain if present there.

It is much more likely that the few specimens of *S. kraemeri* with a New Britain provenance represent animals traded in from the Admiralty Islands. After all, if *S. kraemeri* can be traded as far as the Wuvulu, Ninigo and Hermit groups far to the west of the main Admiralty Islands, as is well recorded, then one would also consider it quite likely for animals to be traded to the southeast as far as New Britain and satellites such as Bali Island in the Witu group of West New Britain Province, the latter of which is about the same distance from the Admiralty Islands as Wuvulu. Indeed, confusion caused by the uncertain provenance of trade animals or naive acquisition-point labelling of specimens collected from traders is a well-recorded epistemological problem in the interpretation of historical museum systematics collections (as discussed in Heinsohn 2006a).

Helgen and Flannery (2004) do offer a possible explanation for the cryptic nature of putative undiscovered New Britain *S. kraemeri* populations, such as competitive suppression by local populations of *Phalanger orientalis breviceps*, making them rare. However, in the

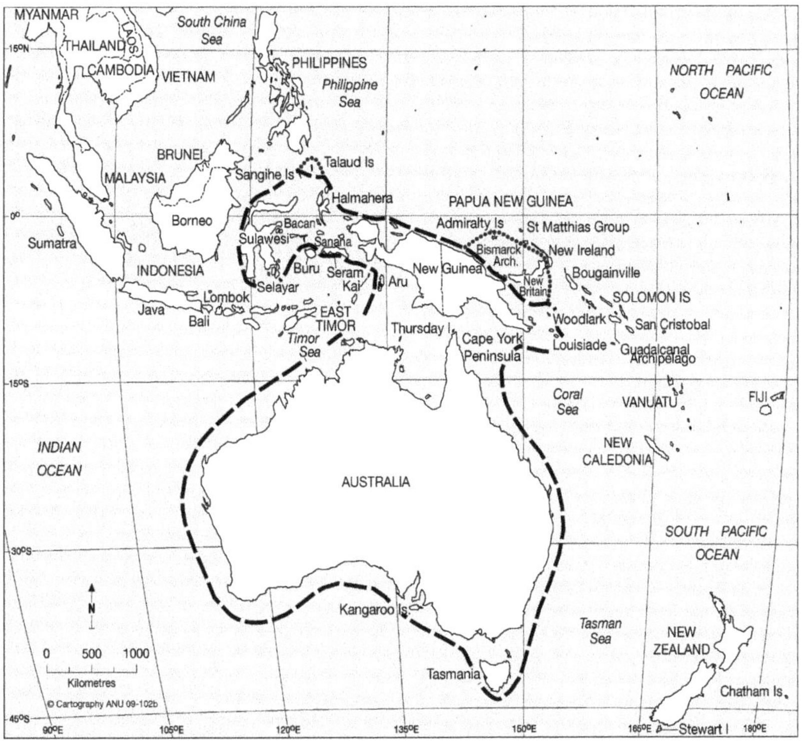

Figure 7. Author's postulated actual natural marsupial distribution when one subtracts probably or possibly introduced marsupial populations (bold dashed line). This represents an older natural marsupial frontier prior to prehistoric and historic human influences on zoogeography. The bold broken line represents the maximum degree to which the natural marsupial frontier retreats if ambiguous marsupial faunas on New Britain, Admiralty Islands and the Talaud Group turn out to be introduced. Fainter dotted line extensions represent the degree to which the natural marsupial frontier expands if ambiguous elements of the New Britain, Admiralty Islands and Talaud marsupial faunas turn out to be naturally occurring through waif dispersal such as on natural rafts (after Heinsohn 2006a)

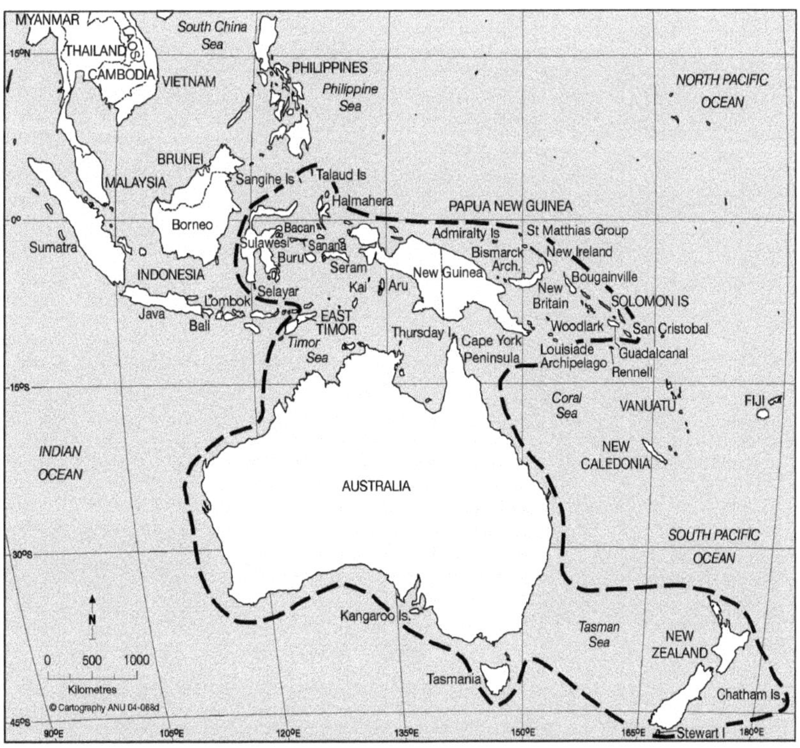

Figure 8. Line of actual marsupial distributions in Australasia taking into account all natural and introduced marsupial populations. This represents the contemporary human-influenced marsupial frontier in Australasia (after Heinsohn 2006a)

author's own observations of interspecific interactions between sympatric populations of slightly dwarfed island populations of *Spilocuscus maculatus maculatus* and *Phalanger orientalis breviceps* in northwestern New Ireland during a six-month field season, it was the larger *Spilocuscus* that had a competitive suppressing effect on *Phalanger*, rather than the other way around, with *Spilocuscus* being generally easier to detect in northwestern New Ireland than *Phalanger* (Heinsohn 1998b, 2004a, b). As *S. kraemeri* is only slightly smaller than *S. maculatus maculatus*, and probably a little larger than island populations of *P. orientalis breviceps*, it is unlikely for *P. orientalis breviceps* to cause competitive suppression of *S. kraemeri* should the two come into sympatric contact. Thus, the Helgen and Flannery (2004) theory is regarded as implausible on ecological and ethnozoological grounds.

Thus two apparent records of *S. kraemeri* on Bali Island, the Witu group, West New Britain Province, and on the New Britain mainland respectively are most likely due to trade animals carried in from Manus Province. However, this does raise the possibility that adventive populations of *S. kraemeri* derived from escaped or liberated trade animals from Manus Province may have become established in places such as the Witu islands of West New Britain Province. Furthermore, if recently introduced adventive populations of both *S. maculatus* and *S. kraemeri* are beginning to appear on New Britain or some of its satellite islands, then this may eventually lead to interbreeding and the creation of a new hybrid *Spilocuscus* population, thus setting off a new evolutionary concatenation in the anthropogenic island biogeography of phalangerids in Near Oceania. Thus, as has probably happened in the past, humans may again be on the verge of creating new taxa through anthropogenic hybridisation.

Discussion and conclusions: Implications of recent recognition of long-term marsupial translocation in the Indo-Australian Archipelago

As discussed above, in the Indo-Australian Archipelago, up to five out of 12 extant oceanic Wallacean marsupials (~42%) may have been introduced from New Guinea or fringing Papuan subregion islands through long-term human agency (the other seven ~58% being Wallacean endemics). This compares with the oceanic Bismarck Archipelago and Solomon Islands of Near Oceania, where possibly all of the marsupials may have been introduced from New Guinea or Papuan subregion sources. However, this strong long-term anthropogenic effect on Indo-Australian oceanic marsupial faunas has only recently been uncovered. This started with the 1960s archaeological work of Glover (1971, 1986) in Timor, who revealed that the island's single marsupial, the New Guinean northern common cuscus *Phalanger orientalis*, had probably been introduced in the mid Holocene. Following this revelation, came the 1980s and 1990s work of the Lapita Homeland Archaeological Project, which revealed that a whole range of marsupials in Near Oceania (the Bismarck Archipelago and Solomon Islands) had also been prehistorically introduced, some as early as the late Pleistocene (Allen and Gosden 1991; Flannery and White 1991; Kirch 1997; Spriggs 1997; Summerhayes 2007). This exposed the naivety of early marsupial biologists and most of the classic zoogeographers who had failed to imagine that tropical Indo-Australian marsupials, as economically valued commodities (in common with various placental species such as rusa deer *Cervus timorensis*), might also have been subject to long-term anthropogenic translocation and range expansion (Heinsohn 1997a, 1998a, b, 2001a, 2003, 2006a).

As late as 1977, marsupial biologist Gilmore, for example, in an essay entitled 'The success of marsupials as introduced species', while showing great awareness of historically recent Australidelphian marsupial introductions to the United Kingdom, New Zealand and Hawaii, automatically attributed the occurrence of all other far-flung oceanic populations, such as those in the Solomon Islands to 'natural spread' (Gilmore 1977:169). Furthermore, as late as the 1960s, world-renowned zoogeographer Simpson, in an essay entitled 'Historical Zoogeography of Australian Mammals' (Simpson 1961) and in a subsequent book 'The Geography of Evolution' (Simpson 1965), defined his marsupial line, which purported to show the natural distributional limits of Australasian marsupials (see Figure 1). Other than excluding the obviously introduced marsupials in New Zealand and Hawaii, the Simpson Marsupial Line treated all other oceanic marsupial distributions in the Indo-Australian Archipelago and Australasia as natural.

Going further back in time, while classic 19th century zoogeographer Alfred Russel Wallace demonstrated a precocious awareness that various Oriental placental mammals such as Javan rusa deer *Cervus timorensis*, long-tailed macaque *Macaca fascicularis* and Malay civet *Viverra tangalunga* had been translocated into various parts of Wallacea, he demonstrated an almost complete blind spot regarding the possibility of marsupial translocation, assuming all island marsupial distributions to be natural (Wallace 1869; Heinsohn 1997a, 1998b, 2006a). One cannot help but wonder whether, had Wallace recognised that much of the Wallacean marsupial biota was introduced, he would, in addition to recognising the Wallace Line as the eastern limit of complex Oriental continental faunas, have also perceived the current Lydekker's Line as the western limit of complex Australo-Papuan continental faunas. Instead, however, while getting the Wallace Line substantially right, he erroneously viewed the Wallacean islands as vicariant remnants of a western extension of a sunken 'great Pacific continent', thus mistakenly treating oceanic Wallacea as continental in origin (Wallace 1860, 1869, 1876). Furthermore, virtually all of the other classic 19th century zoogeographers, such as Salomon Müller, Philip Lutley Sclater, Thomas Henry Huxley, Andrew Dickson Murray, Max Wilhelm Weber and Richard Lydekker, who proposed various faunal boundary lines demarcating the Asian and Australian faunas (Figure 1) in their various writings, treated all marsupial faunas in the Malay Archipelago as natural (Heinsohn 2006a). Indeed, even by the mid 20th century, the only classic zoogeographer known to have demonstrated faint early awareness that some marsupial populations in Wallacea may have been introduced through long-term human agency was Philip Jackson Darlington, who in his classic tome *Zoogeography: The Geographical Distribution of Animals* (1957:468) said, 'Whether some of these marsupials may be introduced on some islands I do not know.'

Latter-day recognition of the status of some oceanic marsupials as introduced species also changes society's ecological perceptions of them. They go from being automatically treated as a benevolent natural part of the environment to being viewed as perhaps out-of-place translocated exotics which may have caused an anthropogenic impact. For example, just as the introduced folivorous common brushtail possum *Trichosurus vulpecula* is recorded to have caused a considerable impact on New Zealand's indigenous forests and wildlife, so to some degree may have another phalangerid possum, the northern common cuscus *Phalanger orientalis*, when it first arrived in New Ireland 23,500 to 20,000 years ago, or when it and other phalangerids were first introduced to other tropical oceanic Indo-Australian islands through long-term human agency. For example, during ecological field work in New Ireland in 1990, the author did notice some indications of possum impacts on some trees due to combined browsing pressure from introduced populations of both *P. orientalis* and the common spotted cuscus *S. maculatus*, with cuscuses actually observed in the process of defoliating the branches of favoured food trees such as *Intsia bijuga* (Heinsohn 1997a, 1998b, 2004a, b).

Indeed, both *P. orientalis* and *S. maculatus*, as adaptable and widely introduced ethnotramp possums in Wallacea and Near Oceania, could be viewed as invasive species, which, among other impacts, could potentially threaten the region's island-endemic phalangerids such as the Papuan subregion's restricted Woodlark Island cuscus *Phalanger lullulae*, or the restricted north Moluccan and Sulawesi subregion phalangerids through interspecific competition. In other instances, the threat may arise from both competition and possible hybridisation, should, for example, the closely related *S. maculatus* be introduced to the island realm of the endemic Waigeo Island cuscus *Spilocuscus papuensis*. Whatever the case, adaptable ethnotramp species such as *P. orientalis* and *S. maculatus*, and possibly also *S. kraemeri*, particularly with the advent of modern rapid transportation in the region such as motor boats and air services and in the absence of effective biosecurity measures, are likely to continue to expand their ranges. *S. maculatus*, for example, will rapidly invade all of New Ireland, having been subject recently to secondary translocation into the far southern Weiten River Valley (Emmons and Kinbag 2001; Heinsohn 2006a), far beyond the original northern New Ireland invasion frontier (Heinsohn 1998b, 2004a), with new secondary invasion beachheads eventually coalescing until all suitable habitats on New Ireland are occupied and animals spread further afield to other New Ireland satellites such as New Hanover (Lavongai). From New Ireland, *S. maculatus* trade animals and pets will continue to be carried across the St. George's Channel to neighbouring New Britain, where further adventive populations will become established, and these will coalesce eventually into new invasion fronts from whence locally sourced animals will be traded and carried to other parts of New Britain to repeat the whole process of complete invasion of a new landmass. Then New Ireland and New Britain may become launching platforms for the future spread of *S. maculatus* to Bougainville Province and the Solomon Islands, with the species eventually invading much of Near Oceania. Furthermore, a widely traded *S. kraemeri* from Manus Province may increase its range through anthropogenic translocation to other parts of the Bismarck Archipelago, perhaps even coming into contact with *S. maculatus* in some locations to form new hybrid populations.

In the Lesser Sundas, introduced populations of *P. orientalis* on Timor and some of its satellites, where it is used as a valued game animal sold at markets and kept as a pet (pers obs.), may gradually spread to other adjacent islands, perhaps eventually making their way westwards along the Lesser Sundas chain; while in the Solomon Islands, the species may eventually be spread to some of the last uninvaded islands such as those in far southern Rennell and Bellona Province, or those of far eastern Temotu (Santa Cruz) Province, or perhaps even beyond to Vanuatu. Thus, owing to some highly adaptable ethnotramp species and anthropogenic translocation, the marsupial frontier is likely to expand rather than contract and contribute to the general homogenisation of biotas during this recent era of accelerated anthropogenic impacts that some have dubbed the Homogenocene (Samways 1999) or Anthropocene (Crutzen and Stoermer 2000; Ruddiman 2003).

However, such translocations are not necessarily always deleterious to all aspects of an island's biota. Introduction of *P. orientalis* to the Solomon Islands in the mid to upper Holocene, just as many of that archipelago's large endemic murid rats were in decline or going extinct under human impacts (Flannery and Wickler 1990; Flannery 1995b; Spriggs 1997; Flannery and Roberts 1999), may have allowed a timely prey-switch for the endemic Sanford's eagle *Haliaeetus sanfordi* and the fearful owl *Nesasio solomonensis*, both of which are now recorded to habitually prey on the archipelago's introduced cuscuses (Hadden 1981, 2004; Coates 1990; Doughty et al. 1999). Furthermore, even in places such as New Ireland that lacked giant endemic murids, local populations of the white-breasted sea eagle *Haliaeetus leucogaster* have learned to prey regularly on both *P. orientalis* and *S. maculatus*, which now provide a locally abundant additional source of prey (Heinsohn 2000; Beehler et al. 2001). A similar prey-shift by various raptors may have occurred also in Timor, with the mid-Holocene introduction of *P. orientalis* and recorded upper-Holocene extinction (Glover 1971, 1986) of a whole suite of giant endemic murid rats.

Another observation that highlights the adaptability and potential invasiveness of some marsupials is the fact that some successfully established populations of both phalangerids and macropodids have arisen from very small founding populations. The entire population of brushtailed rock wallabies *Petrogale penicillata* on Oahu Island, Hawaii, for example, was established from a single pair that escaped in 1916; while the population of red-necked wallabies *Macropus rufogriseus* that established in the Hunters Hills area of New Zealand's

South Island is attributed to a single male and two females released in 1874 (Maynes 1989; Warburton 2005b). Similarly, the entire population of New Guinean common spotted cuscus *Spilocuscus maculatus* in New Ireland, which now occupies a large swath of the 300 km long island and probably numbers in the tens of thousands, was probably established from only a few individuals, possibly as few as a single pair, or even just a single gravid female or mother with pouch young (Heinsohn 1998b, 2004a, b). Furthermore, in its most far-flung introduced population on Selayar Island in the Sulawesi subregion, *S. maculatus* even manages to coexist with introduced Oriental placental mammals, including scansorial viverids such as the common palm civet *Paradoxurus hermaphroditus*, arboreal sciurids such as the plantain squirrel *Callosciurus notatus* and moderately dense human populations in a highly anthropogenic island environment (Heinsohn 2002a, 2004a). Similar adaptability is demonstrated by *P. orientalis* in Timor, where that species shares that island's significantly anthropogenic environment with the introduced viverid *P. hermaphroditus* and a scansorial cercopithecid monkey, the long-tailed macaque *Macaca fascicularis* (Heinsohn 2004a, 2005a, b).

On high oceanic islands such as Seram, Buru, Timor, New Britain, New Ireland, Bougainville and Guadalcanal, introduced populations of *P. orientalis* in relative competitive release have also demonstrated considerable ecophysiological adaptability by exhibiting a much greater altitudinal range than occurs in their natural continental context on the New Guinea mainland. For example, whereas *P. orientalis* is found from sea level to 1500 m in New Guinea where its altitudinal range appears to be restricted by competition with specialised montane cuscuses, on Seram it is recorded to occur right up to the limits of forest, just below the 3020 m rocky summit of Mount Binaiya, with these higher altitude populations exhibiting thicker coats than their lowland counterparts (Heinsohn 2005a, b; Heinsohn and Hope 2006).

Just as some marsupials have proved highly vulnerable to threatening processes and extinction, others have proved in the right circumstances to be highly adaptable invasive species that as translocated ethnotramps are bringing about an overall expansion of the marsupial frontier. See the contrast between Figure 7 and Figure 8, the first showing the author's postulated line of natural marsupial distributions, compared with the latter depiction of the contemporary greatly expanded anthropogenic marsupial frontier. Marsupials also push some ethnozoological temporal 'frontiers', in that the world's oldest recorded anthropogenic translocation of an animal is that of the New Guinean cuscus *P. orientalis* to New Ireland in Near Oceania at 23,500 to 20,000 BP. Thus, the world's oldest known ethnotramp species is a humble marsupial that may have been deliberately harnessed for use on oceanic islands as part of an ancient late-Pleistocene game-enhancement production system. Removal of a traditional Eurocentric theoretical blind spot in relation to Australasian mammals reveals that there may be much more to the humble marsupial than meets the eye. They carry a long human history in their pouches.

Acknowledgements

The author would like to thank local people across the region and the governments of Indonesia, Timor-Leste, Papua New Guinea, the Solomon Islands and New Zealand for their hospitality and for allowing study visits to various parts of their realms. Regional biologists, Quaternary researchers and archaeologists, especially those associated with the pioneering Lapita Homeland Archaeological Project, are collectively thanked for building much of the knowledge upon which this analysis is based. Special thanks goes to the Department of Archaeology and Natural History, RSPAS, ANU, for support of my studies over the years. Geoffrey Hope, Simon Haberle, Matthew Prebble, Janelle Stevenson, Sue O'Connor, Atholl Anderson, Glen Summerhayes, Matthew Spriggs, Peter Bellwood, Colin Groves, Ken Aplin, J. Peter White, Ian Glover, Tim Flannery, Lester Seri, Robin Hide, Chris Tidemann,

Chris Ballard, Matthew Leavesley, Gerry Maynes, Kay Dancey and Sebastian Högberg are particular names that come to mind for acknowledgement. Special thanks goes to Professor Geoffrey Hope for being a supervisor, mentor and friend over the years. Graffiti in the ANU Geography Department once read, 'Without Geoff this department would be Hopeless.' You have certainly added heart, soul, vision, and a touch of playful wickedness to the departments you have inhabited.

References

Allen J. and C. Gosden (eds) 1991. *Report of the Lapita Homeland Project*. Occasional Papers in Prehistory 20. Canberra: Department of Prehistory, Research School of Pacific Studies, the Australian National University.

Allen J., C. Gosden and J.P. White 1989. Human Pleistocene adaptations in the tropical island Pacific: recent evidence from New Ireland, a Greater Australian outlier. *Antiquity* 63:584-61.

Anthony N. 2001. Small Mammals: An Annotated Checklist. In G. Allez (ed) *New Britain Biological Survey, West New Britain Province, Papua New Guinea*, pp34-45. Middleton: University of Wisconsin New Britain Biological Survey.

Archer M., T.F. Flannery and G.C. Grigg 1985. *The Kangaroo*. Sydney: Kevin Weldon.

Archer M., S. Cork, S. Hand, S. Phillips and M. Smith 1987. *Koala: Australia's Endearing Marsupial*. Sydney: Leonard Cronin/Reed Books.

Backhouse G. and A. Crouch 1991. Koala management in the Western Port region Victoria. In A.K. Lee, K.A. Handasyde and G.D. Sanson (eds) *Biology of the Koala*, pp313-321. Chipping Norton: Surrey Beatty and Sons..

Baker S.J. 1990. Escaped exotic mammals in Britain. *Mammal Review* 20(2/3):75-96.

Ball E.E. and I.M. Hughes 1982. Long Island, Papua New Guinea – People, Resources and Culture. *Records of the Australian Museum* 34(10):463-525.

Beehler B.M., J.P. Angle, D. Gibbs, M. Hedemark and D. Kuro 2001. A Field Survey of Resident Birds of Southern New Ireland. In B.M. Beehler & L.E. Alonso (eds) *Southern New Ireland, Papua New Guinea: A Biodiversity Assessment*. RAP (Rapid Assessment Program) Bulletin of Biological Assessment 21, pp61-66. Washington DC: Conservation International.

Bellwood P. 1997. *Prehistory of the Indo-Malaysian Archipelago, Revised Edition*. Honolulu: University of Hawaii Press.

Black J. (ed.) 1999. *Atlas of World History*. London: Dorling Kindersley.

Boettger C.R. 1943. Känguruhs in Deutschland in freier Wildbahn. *Natur und Volk* 73:331-336.

Brass L.J. 1956. Results of the Archbold Expeditions. No. 75. Summary of the Fourth Archbold Expedition to New Guinea (1953). *Bulletin of the American Museum of Natural History* 111(2):77-152.

Brass L.J. 1959. Results of the Archbold Expeditions. No. 79. Summary of the Fifth Archbold Expedition to New Guinea (1956-1957). *Bulletin of the American Museum of Natural History* 118(1):1-70.

Bull P.C. and A.H. Whitaker 1975. The Amphibians, Reptiles, Birds and Mammals. In G. Kuschel (ed) *Biogeography and Ecology in New Zealand*, pp231-276. The Hague: W. Junk.

Calaby J. 1984. Forward. In M. Archer and G. Clayton (eds) *Vertebrate Zoogeography and Evolution in Australasia: Animals in Space and Time*, p. v. Sydney: Hespernian Press.

Carlton J.T., 1996. Biological invasions and cryptogenic species. *Ecology* 77(6):1653-1655.

Carter T.D., J.E. Hill and G.H.H. Tate 1945. *Mammals of the Pacific World*. New York: Macmillan.

Chappell J. 1987. Late Quaternary sea-level changes in the Australian region. In M.J. Tooley and I. Shennan (eds) *Sea-level Changes*, pp296-331. Oxford: Basil Blackwell.

Chappell J. and N.J. Shackleton 1986. Oxygen isoptopes and sea level. *Nature* 324:137-140.

Coates B.J. 1990. *The Birds of Papua New Guinea, Including the Bismarck Archipelago and* Bougainville. *Volumes 1 & 2*. Alderley: Dove Publications.

Coates B.J. and K.D. Bishop 1997. *A Guide to the Birds of Wallacea: Sulawesi, The Moluccas and Lesser Sunda Islands, Indonesia*. Alderley: Dove Publications.

Coates B.J. and W.S. Peckover 2001. *Birds of New Guinea and the Bismarck Archipelago: A Photographic Guide*. Alderley: Dove Publications.

Condon H.T. 1967. Kangaroo Island and its Vertebrate Land Fauna. *Australian Natural History* 15(12):409-412.

Copley P.B. 1995. Translocations of native vertebrates in South Australia: a review. In M. Serena (ed) *Reintroduction Biology of Australian and New Zealand Fauna*, pp35-42. Chipping Norton: Surrey Beatty and Sons.

Copley P.B. and P.J. Alexander 1997. Overview of the status of rock-wallabies in South Australia. *Australian Mammalogy* 19(2):153-162.

Corbet G.B. 1974. The Distribution of Mammals in Historic Time. In D.L. Hawksworth (ed) *The Changing Flora and Fauna of Britain*, pp179-202. London: Academic Press.

Corbet G.B. and S. Harris (eds) 1991. *The Handbook of British Mammals. Third Edition*. Oxford: Blackwell Scientific Publications.

Corbet G.B. and J.E. Hill 1992. *The Mammals of the Indomalayan Region: A Systematic Review*. Oxford: Natural History Museum Publications / Oxford University Press.

Cowan P.E. 1990. Brushtail Possum. In C.M. King (ed) *The Handbook of New Zealand Mammals*, pp68-98. Auckland: Oxford University Press..

Cowan P.E. 2005. Brushtail Possum. In C.M. King (ed) *The Handbook of New Zealand Mammals. Second Edition*, pp56-80. Melbourne: Oxford University Press.

Cowan P.E. and A. Moeed 1987. Invertebrates in the diet of brushtail possums, *Trichosurus vulpecula*, in lowland podocarp/broadleaf forest, Orongorongo Valley, Wellington, New Zealand. *New Zealand Journal of Zoology* 14:163-177.

Crutzen P.J. and E.F. Stoermer 2000. The Anthropocene. *Global Change Newsletter* 41:17-18.

Darlington P.J. 1957. *Zoogeography: The Geographical Distribution of Animals*. New York: John Wiley and Sons.

Davidson M.M. 1965. *Changes in Wild Mammal Populations in Cantebury*. Wellington: New Zealand Forest Service.

Davies M., C.R. Twidale and M.J. Tyler (eds) 2002. *Natural History of Kangaroo Island*. Adelaide: Royal Society of South Australia.

Dawson T.J. 1995. *Kangaroos: Biology of the largest marsupials*. Sydney: UNSW Press.

Delroy L.B., J. Earl, I. Radbone, A.C. Robinson and M. Hewett 1986. The Breeding and Re-establishment of the Brush-tailed Bettong, *Bettongia penicillata*, in South Australia. *Australian Wildlife Research* 13:387-396.

Derbyshire I.D. (ed) 1993. *The Hutchinson Dictionary of World History*. Oxford: Helicon.

De Vos A., R.H. Manville and R.G. van Gelder 1956. Introduced Mammals and Their Influence on Native Biota. *Zoologica: Scientific Contributions of the New York Zoological Society* 41(19):163-194.

Dickman C. 2005. Marsupials of the World: An Introduction. In R.M. Nowak (ed) *Walker's Marsupials of the World*, pp1-67. Baltimore: The John Hopkins University Press.

Dickman C. and R. Woodford Ganf 2007. A fragile balance: The extraordinary story of Australian Marsupials. Fishermans Bend, Vic.: Craftsman House.

Donne T.E. 1924. *The Game Animals of New Zealand*. London: John Murray.

Doughty C., N. Day and A. Plant 1999. *Birds of The Solomons, Vanuatu and New Caledonia.* London: Christopher Helm / A. and C. Black.

Egloff B.J. 1975. Archaeological investigations in the coastal Madang area and on Eloaue Island of the St. Matthias Group. *Records of the Papua New Guinea Public Museum and Art Gallery* 5:1-43.

Egloff B.J. 1979. *Recent Prehistory in Southeast Papua.* Terra Australis 4. Canberra: Department of Prehistory, Research School of Pacific Studies, Australian National University.

Emmons L.H. and F. Kinbag 2001. Survey of Mammals of Southern New Ireland. In B.M. Beehler and L.E. Alonso (eds) *Southern New Ireland, Papua New Guinea: A Biodiversity Assessment,* RAP (Rapid Assessment Program) Bulletin of Biological Assessment 21, pp67-69. Washington DC: Conservation International.

Fairbanks R.G. 1989. A 17,000-year glacio-eustatic sea-level record: influence of glacial melting rates on the Younger Dryas event and deep ocean circulation. *Nature* 342:637-642.

Flannery T.F. 1990a. *Mammals of New Guinea. First Edition.* Carina: Robert Brown and Associates.

Flannery T.F. 1992. Taxonomic revision of the *Thylogale brunii* complex (Macropodidae: Marsupialia) in Melanesia, with description of a new species. *Australian Mammalogy* 15:7-23.

Flannery T.F. 1993. Moving animals from place to place. In G. Burenhult (ed) *The Illustrated History of Humankind Volume 1,* p175. San Francisco: Harper.

Flannery T.F. 1994. *Possums of the World: A Monograph of the Phalangeroidea.* Sydney: Australian Museum/Geo Productions.

Flannery T.F. 1995a. *Mammals of New Guinea. Revised and updated edition.* Sydney: Australian Museum/Reed Books.

Flannery T.F. 1995b. *Mammals of the South-West Pacific and Moluccan Islands.* Sydney: Australian Museum/Reed Books.

Flannery T.F. 1997. The fate of empire in low- and high-energy ecosystems. In T. Griffiths and L. Robin (eds) *Ecology and Empire: Environmental History of Settler Societies,* pp46-59. Melbourne: Melbourne University Press.

Flannery T.F., P. Bellwood, J.P. White, T. Ennis, G. Irwin, K. Schubert and S. Balasubramaniam 1998. Mammals from Holocene Archaeological Deposits on Gebe and Morotai Islands, Northern Moluccas, Indonesia. *Australian Mammalogy* 20(3):391-400.

Flannery T.F., P. Bellwood, P. White, A. Moore, Boeadi and G. Nitihaminoto 1995. Fossil marsupials (Macropodidae, Peroryctidae) and other mammals of Holocene age from Halmahera, North Moluccas, Indonesia. *Alcheringa* 19:17-25.

Flannery T.F., P. Kendall and K. Wynn-Moylan 1990. *Australia's Vanishing Mammals: Endangered and Extinct Native Species.* Sydney: Readers Digest.

Flannery T.F., P.V. Kirch, J. Specht and M. Spriggs 1988. Holocene mammal faunas from archaeological sites in island Melanesia. *Archaeology in Oceania* 23(3):89-94.

Flannery T.F., R. Martin and A. Szalay 1996. *Tree Kangaroos: A Curious Natural History.* Melbourne: Reed.

Flannery T.F. and R.G. Roberts 1999. Late Quaternary Extinctions in Australasia: An Overview. In R.D.E. MacPhee (ed) *Extinctions in Near Time: Causes, Contexts, and Consequences,* pp239-255. New York: Kluwer Academic / Plenum Publishers.

Flannery T.F. and J.P. White 1991. Animal Translocation: Zoogeography of New Ireland Mammals. *National Geographic Research and Exploration* 7(1):96-113.

Flannery T.F. and S. Wickler 1990. Quaternary Murids (Rodentia: Muridae) from Buka Island, Papua New Guinea, with Descriptions of Two New Species. *Australian Mammalogy* 13:127-139.

Gibb J.A. and J.E.C. Flux 1973. Mammals. In G.R. Williams (ed) *The Natural History of New Zealand: An Ecological Survey,* pp334-371. Wellington: Reed.

Gilmore D.P. 1968. Wallabies in New Zealand. *Animals* June 1968:62-66.

Gilmore D.P. 1977. The success of marsupials as introduced species. In B. Stonehouse and D. Gilmore (eds) *The Biology of Marsupials*, pp169-178. Baltimore: University Park Press.

Glover I.C. 1971. Prehistoric Research in Timor. In D.J. Mulvaney and J. Golson (eds) *Aboriginal Man and Environment in Australia*, pp158-181. Canberra: Australian National University Press.

Glover I.C. 1986. *Archaeology in Eastern Timor, 1966-67*. Canberra: Department of Prehistory, Research School of Pacific Studies, The Australian National University.

Groves C.P. 1981. *Ancestors for the pigs: taxonomy and phylogeny of the genus Sus*. Technical Bulletin No.3. Canberra: Department of Prehistory, Research School of Pacific Studies, Australian National University.

Groves C.P. 1984. Of mice and men and pigs in the Indo-Australian Archipelago. *Canberra Anthropology* 7(1 and 2):1-19.

Grzimek B. 1967. *Four-legged Australians: Adventures with the Animals and Men in Australia*. London: Collins.

Gunn R.C. 1851. On the Introduction and Naturalization of *Petaurus sciureus* in Tasmania. *Papers and Proceedings of the Royal Society of Van Diemen's Land* 1:253-255.

Hadden D. 1981. *Birds of the North Solomons. Wau Ecology Handbook No. 8*. Wau: Wau Ecology Institute.

Hadden D. 2004. *Birds and Bird Lore of Bougainville and the North Solomons*. Alderley: Dove Publications.

Haeckel E. 1893. Zur Phylogenie der Australischen Fauna. *Denkschriften der Medizinisch-Naturwissenschaftlichen Gessellschaft zu Jena* 4:5.

Hall R. 2001. Cenozoic reconstructions of SE Asia and the SW Pacific: changing patterns of land and sea. In I. Metcalfe, J.M.B. Smith, M. Morwood and I. Davidson (eds) *Faunal and Floral Migrations and Evolution in SE Asia-Australasia*, pp35-56. Lisse: A.A. Balkema.

Harris S. and D.W. Yalden 2008. *Mammals of the British Isles: Handbook, 4th Edition*. Southampton: The Mammal Society.

Haq B.U., J. Hardenbol and P.R. Vail 1987. Chronology of Fluctuating Sea Levels Since the Triassic. *Science* 235:1156-1167.

Heinsohn T.E. 1997a. Phantoms in the foliage: human influences on the rainforests of the New Guinea Archipelago. In J. Dargavel (ed) *Australia's Ever-Changing Forests 111: Proceedings of the Third National Conference on Australian Forest History*, pp278-295. Canberra: Australian Forest History Society/Centre for Resource and Environmental Studies, Australian National University.

Heinsohn T. E. 1997b. *Taz-Hatz: Tasmanian Native Mammal Fur Industry Collection Assessment Report* (unpublished research report on the Tasmanian fur industry). Canberra: People and the Environment Section, National Museum of Australia.

Heinsohn T.E. 1998a. Captive Ecology. *Nature Australia* 26(2):36-43.

Heinsohn T.E. 1998b. *The Realm of the Cuscus: Animal Translocation and Biological Invasions East of Wallace's Line*. Canberra: Unpublished Master of Science Thesis, Geography Department, School of Resource and Environmental Management, Australian National University.

Heinsohn T.E. 2000. Predation by the White-breasted Sea Eagle *Haliaeetus leucogaster* on Phalangerid Possums in New Ireland, Papua New Guinea. *Emu* 100:245-6.

Heinsohn T.E. 2001a. Human influences on vertebrate zoogeography: Animal translocation and biological invasions across and to the east of Wallace's Line. In I. Metcalfe, J.M.B. Smith, M. Morwood and I. Davidson (eds) *Faunal and Floral Migrations and Evolution in SE Asia-Australasia*, pp153-170.Lisse: A.A. Balkema.

Heinsohn T.E. 2001b. A Giant Among Possums: Sulawesi Bear Cuscus. *Nature Australia* 26(12):24-31.

Heinsohn T.E. 2002a. Status of the common spotted cuscus *Spilocuscus maculatus* and other wild mammals on Selayar Island, Indonesia, with notes on Quaternary faunal turnover. *Australian Mammalogy* 24(2):199-207.

Heinsohn T.E. 2002b. Observations of probable camouflaging behaviour in a semi-commensal common spotted cuscus *Spilocuscus maculatus maculatus* (Marsupialia: Phalangeridae) in New Ireland, Papua New Guinea. *Australian Mammalogy* 24(2):243-245.

Heinsohn T.E. 2002c. Possum extinctions at the marsupial frontier: the status of the northern common cuscus *Phalanger orientalis* on Santa Ana Island, Makira Province, Solomon Islands. *Australian Mammalogy* 24(2):247-248.

Heinsohn T.E. 2003. Animal translocation: long-term human influences on the vertebrate zoogeography of Australasia (natural dispersal versus ethnophoresy). *Australian Zoologist* 32(3):351-376.

Heinsohn T.E. 2004a. Phalangeroids as ethnotramps: A brief history of possums and gliders as introduced species. In R. Goldingay and S.M. Jackson (eds) *The Biology of Australian Possums*, pp506-526. Sydney: Surrey Beatty and Sons.

Heinsohn T.E. 2004b. Ecological variability in the common spotted cuscus *Spilocuscus maculatus* in the Australasian Archipelago - a review. iIn R. Goldingay and S.M. Jackson (eds) *The Biology of Australian Possums*, pp527-538. Sydney, Surrey Beatty and Sons.

Heinsohn T.E. 2005a. Den sites and habitats utilised by the northern common cuscus *Phalanger orientalis* (Marsupialia: Phalangeridae) in East Timor. *Australian Mammalogy* 27(1):99-101.

Heinsohn T.E. 2005b. The Cuscus That Fooled Science [Northern common cuscus *Phalanger orientalis*]. *Nature Australia* 28(5):26-33.

Heinsohn T.E. 2005c. Wallaby extinctions at the macropodid frontier: Changing status of the northern pademelon *Thylogale browni* in New Ireland Province, Papua New Guinea. *Australian Mammalogy* 27(2):175-183.

Heinsohn T.E. 2006a. *Secret life of the cuscus and the cassowary: The crypto-anthropogenic factor and zoogeographic interpretation in the Indo-Australian Archipelago and Oceania 1846-2006 (With a guide to introduced terrestrial vertebrates in the region).* Canberra: Unpublished PhD thesis, Department of Archaeology and Natural History, Research School of Pacific and Asian Studies, Australian National University.

Heinsohn T.E. 2006b. Spread of the cane toad *Bufo marinus* to San Cristobal (Makira) Island, Solomon Islands. *Australian Zoologist* 33(4):474-475.

Heinsohn T.E. 2009. *The marsupial frontier: a history of Australasian marsupials as introduced species*. Monograph in prep.

Heinsohn T.E. and R.G. Barker 2006. Observations of scavenging carnivory by the common brushtail possum *Trichosurus vulpecula* on macropodid carcasses in Namadgi National Park, montane Southeastern Australia. *Australian Zoologist* 33(3):295-296.

Heinsohn T.E. and G.S. Hope 2006. The Torresian Connections: Zoogeography of New Guinea. In J.R. Merrick, M. Archer, G.M. Hickey and M.S.Y. Lee (eds) *Evolution and Biogeography of Australasian Vertebrates*, pp71-93. Sydney: Auscipub.

Helgen K.M. and T.F. Flannery 2004. Notes on the phalangerid marsupial genus *Spilocuscus* with description of a new species from Papua. *Journal of Mammalogy* 85:825-833.

Holmes D. 1993. *My Seventy Years on the Chatham Islands.* Christchurch: Shoal Bay Press.

Honacki J.H., K.E. Kinman and J.W. Koeppl (eds) 1982. *Mammal Species of the World: A Taxonomic and Geographic Reference.* Lawrence: The Association of Systematics Collections/Allen Press.

Hope G.S. 1996. Quaternary change and the historical biogeography of Pacific islands. In A. Keast and S.E. Miller (eds) *The Origin and Evolution of Pacific Island Biotas, New Guinea to Eastern Polynesia: Patterns and Processes*, pp165-190.. Amsterdam: SPB Academic Publishing.

Inger R.F. and H.K. Voris 2001. The biogeographical relations of the frogs and snakes of Sundaland. *Journal of Biogeography* 28:863-891.

Inns R.W., P.F. Aitken and J.K. Ling 1979. Mammals. In M.J. Tyler, C.R. Twidale and J.K. Ling (eds) *Natural History of Kangaroo Island*, pp91-102. Adelaide: Royal Society of South Australia.

Inns R.W. 2002. Terrestrial Mammals. In M. Davies, C.R. Twidale and M.J. Tyler (eds) *Natural History of Kangaroo Island*, pp74-79. Adelaide: Royal Society of South Australia.

Isern T.D. 2002. Companions, stowaways, imperialists, invaders: Pests and weeds in New Zealand. In E. Pawson and T. Brooking (eds) *Environmental Histories of New Zealand*, pp233-245. Melbourne: Oxford University Press.

Kemper K. 1985. Mammals. In H.J. Aslin (ed) *A List of the Vertebrates of South Australia*, pp7-19. Adelaide: South Australian Department of Environment and Planning.

King C.M. 1984. *Immigrant Killers: Introduced predators and the conservation of birds in New Zealand.* Auckland: Oxford University Press.

King C.M. (ed.) 1990. *The Handbook of New Zealand Mammals.* Auckland: Oxford University Press.

King C.M. (ed.) 2005. *The Handbook of New Zealand Mammals. Second Edition.* Oxford: Oxford University Press.

Kirch P.V. 1988. The Talepakemalai Lapita Site and Oceanic Prehistory. *National Geographic Research* 4(3):328-242.

Kirch P.V. 1997. *The Lapita Peoples: Ancestors of the Oceanic World.* Oxford: Blackwell.

Koopman K.F. 1979. Zoogeography of Mammals from Islands Off the Northeastern Coast of New Guinea. *American Museum Novitates* 2690(2):1-17.

Kramer R.J. 1971. *Hawaiian Land Mammals.* Rutland: Charles E. Tuttle.

Laurie E.M.O. and J.E. Hill 1954. *List of Land Mammals of New Guinea, Celebes and Adjacent Islands 1758-1952.* London: British Museum of Natural History.

Lazell J.D. 1980. Kalihi Rock Wallaby of Hawaii. *Tiger Paper* 7(2):31-32.

Lazell J.D. 1981. Strange Rock Wallabies of Oahu. *Explorers Journal* 59(2):66-67.

Lazell J.D., T.W. Sutterfield and W.D. Giezentanner 1984. The Population of Rock Wallabies (Genus *Petrogale*) on Oahu, Hawaii. *Biological Conservation* 30:99-108.

Leavesley M.G. 2001. Thirty-five thousand years in New Ireland. *Paradise* 145:45-47.

Leavesley M.G. 2005. Prehistoric Hunting Strategies in New Ireland, Papua New Guinea: The Evidence of the Cuscus (*Phalanger orientalis*) Remains from Buang Merabak Cave. *Asian Perspectives* 44(1):207-218.

Leavesley M.G. and J. Chappell 2004. Buang Merabak: additional early radiocarbon evidence of the colonisation of the Bismarck Archipelago, Papua New Guinea. *Antiquity* 78(301):1-5 online.

Lee A. and R. Martin 1988. *The Koala: A Natural History.* Sydney: New South Wales University Press.

Lever C. 1977. *The Naturalized Animals of the British Isles.* London: Hutchinson.

Lever C. 1985. *Naturalized mammals of the world.* London: Longman.

Lever C. 1992. *They Dined on Eland: The Story of the Acclimatisation Societies.* London: Quiller Press.

Liem D.S. 1977. A Preliminary report on the ecosystem assessment and the possible re-establishment of agile wallaby (*Macropus agilis*) on Goodenough Island, Milne Bay Province. *Wildlife in Papua New Guinea* 77/22:1-32.

Lilley I. 1986. *Prehistoric Exchange in the Vitiaz Strait, Papua New Guinea.* Canberra: Unpublished PhD Thesis, Prehistory Department, Australian National University.

Long J.L. 2003. *Introduced Mammals of the World: Their History, Distribution and Influence.* Melbourne, CSIRO Publishing.

Lord C.E. 1919. Notes on the Mammals of Tasmania. *Papers and Proceedings of the Royal Society of Tasmania for the Year 1918*:16-52.

Lord C.E. and H.H. Scott 1924. *A Synopsis of the Vertebrate Animals of Tasmania*. Hobart: Oldham, Beddome and Meredith.

Low T. 2002. *The New Nature: Winners and Losers in Wild Australia*. Melbourne: Viking.

Mallon D. 1970. Britain's Wild Wallabies. *Animals* 13(6):256-257.

Martin R. 2005. *Tree-kangaroos of Australia and New Guinea*. Melbourne: CSIRO Publishing.

Maynes G.M. 1989. Zoogeography of the Macropodoidea. In G. Grigg, P. Jarman and I. Hume (eds) *Kangaroos, Wallabies and Rat-kangaroos*, 2 Volumes, pp47-66. Chipping Norton: Surrey Beatty & Sons.

Mayr E. and J. Diamond 2001. *The Birds of Northern Melanesia: Speciation, Ecology & Biogeography*. Oxford: Oxford University Press.

McCabe J. 1910. Australia - A Museum of Living Antiquities. *The Lone Hand* November 1, 1910:38-46.

McDowall R.M. 1994. *Gamekeepers for the Nation: The story of New Zealand's acclimatisation societies 1861-1990*. Christchurch: Canterbury University Press.

McNiven I.J. and G. Hitchcock 2004. Torres Strait Islander Marine Subsistence Specialisation and Terrestrial Animal Translocation. *Memoirs of the Queensland Museum Cultural Heritage Series* 3(1):105-162.

Menzies J.I. and J.C. Pernetta 1986. A taxonomic revision of cuscuses allied to *Phalanger orientalis* (Marsupialia: Phalangeridae). *Journal of Zoology, London. Series B*. 1(3):551-618.

Monk K.A., Y. de Fretes and G. Reksodiharjo-Lilley 1997. *The Ecology of Nusa Tenggara and Maluku*. Singapore: Periplus Editions.

Montague T.L. (ed.) 2000. *The Brushtail Possum: Biology, Impact and Management of an Introduced Marsupial*. Christchurch: Manaaki Whenua Press.

Morwood M. 2009. Hobbits in context: hominin biogeography in island Southeast Asia. 2009 Mulvaney Lecture, Australian National University, 13 May 2009.

Morwood M. and P. Van Oosterzee 2007. *The Discovery of the Hobbit: The Scientific Breakthrough that Changed the Face of Human History*. Sydney: Random House.

Moulton M.P. and S.L. Pimm 1986. Species Introductions to Hawaii. In H.A. Mooney and J.A. Drake (eds) *Ecology of Biological Invasions of North America and Hawaii*, pp231-249. New York: Springer-Verlag.

Murphy R.C. 1951. The Impact of Man Upon Nature in New Zealand. *Proceedings of the American Philosophical Society* 95(6):569-582.

Musser G.G. 1987. The mammals of Sulawesi. pp73-93 in *Biogeographical Evolution in the Malay Archipelago*, edited by T.C. Whitmore. Oxford: Clarendon Press.

Nelson L.S., R.F. Storr and A.C. Robinson 1992. *Plan of Management for the Brush-tailed Bettong Bettongia penicillata Gray, 1837 (Marsupialia, Potoroidae) in South Australia*. Adelaide: National Parks and Wildlife Service, Department of Environment and Planning.

Nile R. and C. Clerk 1996. *Cultural Atlas of Australia, New Zealand and the South Pacific*. Sydney: Readers Digest Press.

Nowak R.M. 2005. *Walker's Marsupials of the World*. Baltimore: The John Hopkins University Press.

Pearson D.J. and J.E. Kinnear 1997. A review of the distribution, status and conservation of rock-wallabies in Western Australia. *Australian Mammalogy* 19(2):137-152.

Phillips B. 1990. *Koalas: The Little Australians we'd all hate to lose*. Canberra: Australian Government Publishing Service.

Poole A.L. 1970. *Wild Animals in New Zealand*. Wellington: A.H. and A.W. Reed.

Pracy L.T. 1962. *Introduction and Liberation of the Opossum (Trichosurus vulpecula) into New Zealand*. Wellington: New Zealand Forest Service Information Series No. 45.

Rand A.L. and E.T. Gilliard 1967. *Handbook of New Guinea Birds*. London: Weidenfeld and Nicolson.

Rounsevell D.E., R.J. Taylor and G.J. Hocking 1991. Distribution Records of Native Terrestrial Mammals in Tasmania. *Wildlife Research* 18:699-717.

Ruddiman W.F. 2003. The anthropogenic greenhouse era began thousands of years ago. *Climate Change* 61(3):261-293.

Samways M. 1999. Translocating fauna to foreign lands: here comes the Homogenocene. *Journal of Insect Conservation* 3:65-66.

Schodde R. and J.H. Calaby 1972. The Biogeography of the Australo-Papuan Bird and Mammal Faunas in Relation to the Torres Strait. In D. Walker (ed) *Bridge and Barrier: The Natural and Cultural History of Torres Strait*, pp257-300.. Canberra: Department of Biogeography and Geomorphology, Research School of Pacific and Asian Studies, Australian National University.

Short J., S.D. Bradshaw J. Giles, R.I.T. Prince and G.R. Wilson 1992. Reintroduction of macropods (Marsupialia: Macropodoidea) in Australia – A review. *Biological Conservation* 62:189-204.

Simpson, G.G. 1961. Historical zoogeography of Australian Mammals. *Evolution* 15:431-46.

Simpson, G.G. 1965. *The Geography of Evolution: Collected Essays*. Philadelphia: Chilton.

Smith M.J. 1973. *Petaurus breviceps. Mammalian Species* 30:1-5.

Souter G. 1963. *New Guinea: The Last Unknown*. Sydney: Angus and Robertson.

Spriggs M. 1997. *The Island Melanesians*. Oxford: Blackwell.

Strahan R. (ed.) 1995. *The Mammals of Australia*. Sydney: Australian Museum/Reed.

Summerhayes G.R. 2007. Island Melanesian Pasts: A View from Archaeology. In J.S. Friedlaender (ed) *Genes, Language and Culture History in the Southwest Pacific*, pp10-35. Oxford: Oxford University Press.

Swadling P. 1996. *Plumes from Paradise: Trade cycles in outer Southeast Asia and their impact on New Guinea and nearby islands until 1920*. Coorparoo: Papua New Guinea National Museum/Robert Brown and Associates.

Thomson G.M. 1922. *The Naturalisation of Animals and Plants in New Zealand*. Cambridge: Cambridge University Press.

Thorne, A., R. Grun, G. Mortimer, N.A. Spooner, J.J. Simpson, M. McCulloch, L. Taylor and D. Curnoe 1999. Australia's oldest human remains: age of the Lake Mungo 3 skeleton. *Journal of Human Evolution* 36:591-612.

Tinker S.W. 1941. *Animals of Hawaii: A Natural History of the Amphibians, Reptiles, and Mammals Living in the Hawaiian Islands*. Honolulu: Tongg Publishing Co.

Tinker S.W. 1980. *A List of the Amphibians, Reptiles, and Mammals of the Hawaiian Islands (Exclusive of the Whales)*. Honolulu: Spencer Wilkie Tinker.

Tomich P.Q. 1969. *Mammals in Hawaii: A Synopsis and Notational Bibliography*. Bernice P. Bishop Museum Special Publication 57. Honolulu: Bishop Museum Press.

Tomich P.Q. 1986. *Mammals in Hawaii: A Synopsis and Notational Bibliography, Second Edition*. Bernice P. Bishop Museum Special Publication 76. Honolulu: Bishop Museum Press.

Tyndale-Biscoe H. 1973. *Life of Marsupials*. Melbourne: Edward Arnold.

Tyndale-Biscoe H. 2005. *Life of Marsupials*. Melbourne: CSIRO Publishing.

Van Deusen H.M. 1972a. Mammals. In P. Ryan (ed) *Encyclopaedia of Papua and New Guinea, 3 volumes*, pp688-694. Melbourne: Melbourne University Press / University of Papua and New Guinea.

Van Deusen H.M. 1972b. Marsupials. In P. Ryan (ed) *Encyclopaedia of Papua and New Guinea, 3 volumes*, pp711-714.. Melbourne: Melbourne University Press / University of Papua and New Guinea.

Van Dyck S. and R. Strahan 2008. *The Mammals of Australia. Third Edition*. Sydney: Reed New Holland.

Waite E.R. and F.W. Jones 1927. The Fauna of Kangaroo Island, South Australia No.2 – The Mammals. *Transactions and Proceedings of the Royal Society of South Australia* 51:322-325.

Wallace A.R.1860. On the Zoological Geography of the Malay Archipelago. *Zoological Journal of the Proceedings of the Linnean Society* 4:172-184.

Wallace A.R. 1869. *The Malay Archipelago: The Land of the Orang-utan, and the Bird of Paradise.* New York: Harper and Brothers.

Wallace A.R. 1876. *The Geographical Distribution of Animals: With a study of the relations of living and extinct faunas as elucidating the past changes of the Earth's surface. 2 Volumes.* London: Macmillan and Co..

Wapstra J.E. 1976. *Forester Kangaroo in Tasmania: Status and Management 1976. Wildlife Division Technical Report 76/1.* Hobart: Tasmanian National Parks and Wildlife Service.

Warburton B. 2005a. Dama wallaby. In C.M. King (ed) *The Handbook of New Zealand Mammals. Second Edition*, pp32-39. Melbourne: Oxford University Press.

Warburton B. 2005b. Bennett's wallaby. In C.M. King (ed) *The Handbook of New Zealand Mammals. Second Edition*, pp39-45. Melbourne: Oxford University Press.

Warburton B. 2005c. Parma wallaby. In C.M. King (ed) *The Handbook of New Zealand Mammals. Second Edition*, pp45-49. Melbourne, Oxford University Press.

Warburton B. 2005d. Black-striped wallaby. In C.M. King (ed) *The Handbook of New Zealand Mammals. Second Edition*, pp49. Melbourne: Oxford University Press.

Warburton B. 2005e. Brushtailed rock wallaby. In C.M. King (ed) *The Handbook of New Zealand Mammals. Second Edition*, pp50-53. Melbourne: Oxford University Press.

Warburton B. 2005f. Swamp wallaby. In C.M. King (ed) *The Handbook of New Zealand Mammals. Second Edition*, pp53-55. Melbourne: Oxford University Press.

Weidenhofer M. 1977. *Maria Island, a Tasmanian Eden.* Hobart: Darlington Press.

Whitten T., G.S. Henderson and M. Mustafa 1987. *The Ecology of Sulawesi.* Singapore, Periplus Editions.

Wickler S. 2001. *The Prehistory of Buka: A Stepping Stone Island in the Northern Solomons. Terra Australis 16.* Canberra: Department of Archaeology and Natural History, and Centre for Archaeological Research, Australian National University.

Wodzicki K.A. 1950. *Introduced Mammals of New Zealand: An Ecological and Economic Survey.* Bulletin No. 98. Wellington: New Zealand Department of Scientific and Industrial Research.

Wodzicki K.A. and J.E.C. Flux 1971. The Parma Wallaby and its Future. *Oryx* 11(1):40-47.

Yalden D. 1982. When did the mammal fauna of the British Isles arrive? *Mammal Review* 12(1):1-57.

Yalden D. 1991. History of the Fauna. In G.B. Corbet and S. Harris (eds) *The Handbook of British Mammals, Third Edition*, pp7-18.Oxford: Blackwell Scientific Publications.

Yalden D. 1999. *The History of British Mammals.* London: T. and A.D. Poyser.

Yalden D. and G.R. Hosey 1971. Feral wallabies in the Peak District. *Journal of Zoology, London* 165:513-520.

Young D. 2004. *Our Islands, Our Selves: A History of Conservation in New Zealand.* Dunedin: University of Otago Press.

Ziegler A.C. 1977. Evolution of New Guinea's marsupial fauna in response to a forested environment. In B. Stonehouse and D. Gilmore (eds) *The Biology of Marsupials*, pp117-138. Baltimore: University Park Press.

Ziegler A.C. 1982. An ecological check-list of New Guinea Recent mammals. In J.L. Gressitt (ed) *Biogeography and Ecology of New Guinea*, pp863-894.The Hague: W. Junk.

Ziegler A.C. 2002. *Hawaiian Natural History, Ecology, and Evolution.* Honolulu: University of Hawaii Press.

10

The empty coast: Conditions for human occupation in southeast Australia during the late Pleistocene

Sandra Bowdler

Archaeology M405, School of Social and Cultural Studies, University of Western Australia, Perth, Australia
sandra.bowdler@uwa.edu.au

Introduction

Archaeology, particularly Australian archaeology, and particularly the archaeology of the pre-European Australian past, is significantly dependent on other disciplines. One of our main interests has been understanding the relationship of Aboriginal people to their environments, and to this end, we have collaborated, with differing degrees of closeness, with geomorphologists, palaeontologists and, of course, palynologists. Geoff Hope is an outstanding example of the last, and has contributed greatly to our knowledge of past environments and their significance for human occupation. I have personally been privileged to work with Geoff, and his contribution to my research, and to the wider questions raised by that research, has been inestimable.

Geoff Hope and the Pleistocene occupation of Tasmania

Some 40 years ago, Rhys Jones published a paper entitled *The geographical background to the arrival of man in Australia and Tasmania* (Jones 1968). This was in itself an important example of the need for input from other disciplines to our understanding of the human past and its environmental context. In that paper, Jones was concerned to explain the (then) lack of evidence for humans in Tasmania during the Pleistocene. The oldest human occupation sites known at the time were all well within the Holocene, with the oldest being Rocky Cape South, dated to around 8000 BP. The lowest levels of Rocky Cape South contained evidence for a well-developed coastal economy, consisting of dense shell midden material containing marine shell and fish remains, as well as land mammals and birds. Because of its steep offshore profile, 8000 BP was close to the time when the sea reached its present position at Rocky Cape. Jones argued that the inhabitants of the cave had probably been living on the Pleistocene coast at times of lower sea level, and arrived at Rocky Cape with the rising sea.

This left the problem of where people were during the Pleistocene, and why they were not occupying what were at the time inland parts of Tasmania. Jones (1968:200) argued that, given the evidence for a firmly established coastal economy and lifestyle, Tasmanian Aborigines during the Pleistocene were tightly constrained to the coast by dense, cool, temperate rainforest that, based on the evidence of the immediate pre-European past, was not conducive to human occupation. Those parts of Tasmania which were not clothed with rainforest were covered with glaciers and permafrost: 'most of Tasmania was inhospitable and probably uninhabited' (Jones 1968:200).

My doctoral research on Hunter Island, northwest Tasmania, showed human occupation dating to c. 22,000 years ago (Bowdler 1984). This in itself did nothing to dispel Jones's theory. Hunter Island during the Pleistocene was a small hill on the Bassian Plain, probably about 100 km from the coast. Human occupation in Cave Bay Cave, on the east side of the island, was intermittent: between 22,000 and 18,000 BP people lit fires, hunted land mammals and left behind stone and bone artefacts. A single hearth represented a visitation about 15,000 years ago, but more intensive occupation did not recur until c. 7000 years ago, when the sea reached its present level. The evidence for that period resembles that from the basal layers of Rocky Cape South, consisting of shell midden material containing some fish bones, sea-bird bones and small mammals.

What was crucial for a new interpretation of Pleistocene human occupation of Tasmania was the pollen evidence which recurred throughout the Cave Bay Cave sediments, and which was the subject of study by Geoff Hope (Hope 1978). Hope was able to show that, far from being covered by dense rainforest, the environment of northwest Tasmania, at least, including the exposed Bassian Plain, was an expanse of sweeping grassland. This was supported by the analysis of faunal remains from Cave Bay Cave, but the pollen evidence was the important factor, since bolstered by other studies.

If there was no environmental barrier, why were people apparently restricted to the Pleistocene coast? In the first instance, I argued (Bowdler 1977) that the Aboriginal colonists of Australia were coastally adapted, and rarely made forays away from the immediate coast. Further evidence, however, showed that Pleistocene Tasmanians were not, in fact, so restricted. Starting with Kutikina Cave on the Franklin River in the early 1980s, the discovery of inland sites in southwest Tasmania opened a new window on Pleistocene Tasmania (e.g. Jones 1995).

In the historic period, southwest Tasmania comprised dense, cool, temperate rainforest with a narrow high-energy coastal fringe. All the ethnohistorical records, and the archaeological evidence such as it was to that time, suggested that while dense shell middens showed the coastal strip was well-utilised by Aboriginal people, the rainforest interior was practically unoccupied. It had been pointed out by Jones (1968) that Aborigines had used fire to convert some of the edges of the rainforest to more productive sedgeland, and also to maintain tracks through it, but essentially the rainforest interior was inimical to human occupation.

Once it was realised, however, that the rainforest had not expanded during the Pleistocene, but on the contrary, had shrunk to small pockets, it was evident that it did not form a barrier to human occupation or at least utilisation of the areas it had covered in the recent past. A number of sites dating from about 13,000 to 35,000 years ago show exploitation of cave sites on river systems in southern Tasmania up to 100 km from the coast, with a concentration on the exploitation of wallabies (Allen 1996).

Further evidence of Pleistocene occupation of Tasmania has been found in the Furneaux Islands, on what would have been the exposed Bassian Plain. Mannalargenna Cave on Prime Seal Island, off the west coast of Flinders Island, was occupied from c. 22,000 to 8000 BP. Human remains were found in Cliff Cave on the west side of King Island, dated to around 14,000 years ago. Other Bass Strait island sites were occupied mostly between about 10,000 and 4000 years ago, and were, apart from the Hunter Group, not occupied or visited by humans in the ethnographic present (Sim 1994).

It is still the case that no Pleistocene sites are known from the north, northeast or southeast coasts of Tasmania. Apart from the sites located on the river systems of the southern forest region (see Allen 1996), there are no interior Pleistocene sites. The site of Warragarra, some 80 km from the sea in northern Tasmania, was expected to contain such deposits, but its basal occupation levels are dated to c. 11,000 BP (Allen and Porch 1996).

Eastern Tasmania is thus devoid of evidence for Pleistocene occupation. When the sea reached its present level, however, there is evidence that people began exploiting littoral resources, just as in the northwest (contra Cane 2001:150). In northwest Tasmania, as we have seen, Rocky Cape South and Cave Bay Cave both showed evidence of coastal economies dating to 8000-7000 BP and in the southeast, the sites at Carlton Bluff and Apollo Bay on Bruny Island demonstrate something similar (Dunnett 1993). The main difference is that in the northwest, people were exploiting a high-energy rocky shoreline, and in the southeast, the sites are in sheltered embayments and contain shellfish species from low-energy, estuarine habitats. For both situations, however, the question remains: where were these people during the earlier time of lower sea level? It seems unlikely that this lack of evidence is due to differential preservation, in that rock shelters capable of preserving Pleistocene evidence occur in these areas, particularly the Rocky Cape caves and others in that area. Were people for some reason tethered to the Pleistocene coastline or did they become archaeologically instantaneous coastal exploiters just as the sea reached its present level? Since the latter seems unlikely (just as Jones argued in 1968), the former is a better hypothesis, but at this stage, it is difficult to establish whether it is culturally or environmentally determined. It might also be wondered whether the wallaby hunters of the southern Tasmanian forests were striking inland from coastal bases, or represent a quite different adaptation from the coastal dwellers of the northwest and southeast. Recent research by Pike-Tay et al. (2008) on dental growth-increment analysis of wallaby teeth from relevant sites seems to support year-round occupation of the interior region, which does suggest a different non-coastal adaptation.

Eastern mainland Australia during the Pleistocene

Moving north to what is now the Australian mainland, we find even less evidence for Pleistocene occupation on the eastern Australian coastal strip, which is here considered to extend from the shoreline to the bottom of the eastern mountains, and from Mornington Peninsula to Cooktown. Within this extensive area, after many years of concentrated archaeological attention, only three generally accepted sites of Pleistocene age are known. They are Burrill Lake, Bass Point and Wallen Wallen Creek: all are problematic with respect to their dating (Bowdler 1999). The last is a sandy site with dating (c. 21,000 BP) based on dispersed charcoal; it has not been fully reported (Neal and Stock 1986; Ulm et al. 1995). The lowest layer of the Bass Point site is also a sandy deposit, for which the dating (c. 17,000 BP) is also based on scattered charcoal which could have blown into the site from anywhere (Bowdler 1970, 1976). The dating of the lowest level of the Burrill Lake rock-shelter site might equally be called into question. It consists of a 'clay wedge', which is thought to have been deposited with great rapidity, possibly from local soil instability due to Aboriginal firing (Hughes 1981). This idea is supported by the two dates for this layer, one from the top (20,801±810 BP) and one from the bottom (20,760±800 BP), which are statistically identical. The source of the dates was dispersed charcoal (Lampert 1971). It can be argued that the material dated was charcoal swept into the rock shelter with clay, entirely unconnected with the human occupation. If we dismiss these three dated sites, we are left with little convincing evidence of human occupation during the Pleistocene on the eastern Australian coastal strip.

A small number of sites in the Sydney region are said to be of Pleistocene age, but all consist of artefacts in open sand contexts, and none have been reported in detail. Only one appears to have a published date. A sand body on the Paramatta River has produced a date of 30,735±407 BP (Wk-17435) (McDonald 2008:36). Any information on the dating material, collection method, archaeological association and so on for this site, and the others, would seem to be buried in unpublished reports. Until such time as these dates can be validated by published details, I choose to discount them.

The oldest well-established dates on the coastal strip generally date to the mid Holocene. Right on the coast, there are some dates for human occupation, which, as in Tasmania, coincide approximately with the sea reaching its present level. Examples are Nara Inlet 1, on Hook Island in the Whitsundays on the Queensland coast (c. 8000 BP, Barker 2004), Border Island near Proserpine also on the Queensland coast (c. 6500 BP, Barker 2004), Swansea Inlet near Newcastle (c. 7900 BP, Gillespie and Temple 1977; Dyall 1982), Curracurrang near Sydney (c. 7500 BP, Megaw 1968), Currarong at Jervis Bay (c. 6000 BP, Lampert 1971), and Wilsons Promontory (c. 6600 BP, Coutts 1967). The dating of all of these is not absolutely secure, it must be said, but the dates from Nara Inlet, Curracurrang and Currarong are well-published and convincing. Only Nara Inlet seems to be obviously associated with evidence of a coastal economy, such as we find in Tasmania and parts of the southern Australian coastline west of Wilson's Promontory (Cane 2001). This phenomenon of an apparently late development of coastal economies has been considered by Beaton (1985), as well as Cane (2001). From the wider perspective with which I am concerned here, the question is, are these sites evidence of people pushed back from narrowly inhabited Pleistocene shoreline by post-glacial sea-level rise, or have the inhabitants come from elsewhere?

When we look at more interior sites, which are still on the coastal strip but not right on the coast, we find a plethora of sites dating to between 12,000 and 8000 BP (e.g. Boot 1993, 1994; Stockton 1993; Attenbrow 2004). Moving into the more elevated areas, evidence for human occupation is considerably older. In the Blue Mountains, convincing dates for human occupation are of the order of 12,000-14,000 years (Stockton 1993). A date of c. 22,000 BP is less convincing in view of its depositional situation and its being associated only with a single stone flake (Bowdler 1981). For similar reasons, a date of c. 21,000 BP for human occupation at the site of Birrigai in the ACT is not as credible as a somewhat younger one of c. 16,000 BP (Flood et al. 1987). Also somewhat problematic is a site in the rugged areas near the Hunter Valley north of Sydney, located on Fal Brook. Charcoal associated with stone artefacts in an open soil matrix produced three dates, one of c. 35,000 BP, one of >20,000 BP and one of c. 13,000 BP. The charcoal associated with the second date is said to have come from a hearth (Attenbrow 2004:350). No published information is available about the stratigraphy and archaeological associations of these dates.

In more northerly parts of the eastern highlands, in south-central Queensland, human occupation has been dated to c. 19,000 at the Kenniff Cave site (Mulvaney and Joyce 1965). Further north again, we have one of the oldest sites in Australia, Ngarrabullgan Cave on Mount Mulligan in the interior of Cape York, some 100 km from the coast and dated to about 40,000 BP (David et al. 1997). The nearby site of Nonda Rock has occupation from about the same age (David et al. 2007). Further west of the Great Dividing Range, the semi-arid region of the Willandra Lakes has produced evidence for human occupation of similar antiquity (e.g. Bowler et al. 2003). These sites have all been published in considerable detail, and the associations are clear.

This takes us back to the observation made by Jones in 1968, before Burrill Lake had been dated, that perhaps there was a west-to-east gradient in the prehistoric colonisation of

Australia. It is, of course, possible that this pattern is completely illusory, due to the vagaries of research, or the result of differential site preservation over time, with old sites being destroyed by natural processes (cf. Rowland 1996:195). This is not, however, the case along the western and southern margins of the continent, where Pleistocene sites occur not far from the coast, many showing evidence that the inhabitants either visited the coast themselves, or had contact with people who lived there (Bowdler 1999; Morse 1999; O'Connor 1999; Cane 2001). It might also be pointed out that archaeological investigation along the eastern coastal strip has been highly intensive, particularly compared with the southern coastline of Australia.

If we accept that the eastern coast of Australia was indeed empty during the Pleistocene, and this is not an artefact of research or erosion, how might we account for it? One line of inquiry might be to consider environmental factors which might have influenced human occupation during the Pleistocene.

Environmental considerations

Callaghan (1980) pointed out nearly 30 years ago that the east coast of Australia might not have been all that desirable a place to live during the Pleistocene. What we now see as a well-watered, fertile area abounding in resources for human occupation was very different during times of lowered sea level. Lowered temperatures and increased aridity due particularly to decreased effective precipitation may have contributed to a more difficult environment for human occupation. Fluctuating sea levels may also have led to a less productive shoreline (see also Lampert and Hughes 1974).

It has often been suggested (e.g. Morse 1999) that where the continental shelf is at its narrowest, we are most likely to find Pleistocene occupation sites if people are coastally oriented, economically and/or culturally, and this does appear to be the case in Western Australia. On the east coast of Australia, however, as Callaghan (1980) pointed out, the continental shelf is generally equally narrow, but this narrowness at times of lower sea level would have created a most unproductive shoreline for human exploitation. In the central-south coast area of New South Wales, the mid-to-late Holocene shoreline was a productive combination of rocky headlands ringed with rock platforms and short sandy beaches, interspersed with river and lagoonal estuarine situations. At times of lower sea level, it may have consisted of a far less productive stretch of long sandy beaches, fewer rocky expanses, and smaller estuaries (Attenbrow 2004:212). Further north, lowered sea level would have left the coastal strip similarly more challenging for human exploitation. Rather than a gentle tropical shore protected by the Great Barrier Reef, the Pleistocene coastline would have consisted of cliffed limestone ridges, rising 100 m or more above sea level, following the outer edge of the present reef (Border 1999:130).

If the coasts were not as benign as in more recent times, we may wonder why there is not more evidence for occupation of the hinterland earlier than the terminal Pleistocene. There is little direct data for the nature of the eastern Australian strip during the Pleistocene, with most pollen studies being restricted to the Holocene (e.g. Dodson and Thom 1992). Research by Hesse et al. (2004) indicates extreme conditions during the Last Glacial Maximum (LGM), with dune formation taking place on the Blue Mountains, suggesting high aridity and sparse vegetation cover. Dune formation was also active on the coast itself; at Kurnell, strong westerlies blew up dunes from the exposed bed of Botany Bay (Lees 2006). In northeast Queensland, evidence from Lynch Crater on the Atherton Tableland suggests decreased effective precipitation during this period (Kershaw 1995). Kershaw (1995), extrapolating mainly from research in Tasmania (including that of Hope) and western Victoria, suggested that much of southeastern Australia was covered with a fairly homogeneous Poaceae-Asteraceae steppe during the LGM, but there is no *direct* evidence that this applied to the eastern coastal strip.

McGowan et al. (2008) discuss atmospheric dust loads during the Quaternary, based on research on Stradbroke Island off southeast Queensland, and conclude that increased dust load in eastern Australia during the LGM was due to severe and prolonged drought. They calculate that the total aeolian sedimentation flux during the LGM reached a maximum of more than 80 times that calculated for the early Holocene (McGowan et al. 2008:177). They infer more frequent dry southwesterly winds, less frequent maritime easterly winds, reduced rainfall, lower temperatures, increased continentality as a result of lower sea levels (-120 m), leading to a marked change in the floristic composition of local vegetation, with thinner vegetation cover and an increased percentage of bare ground. While it might be argued that the period of the LGM was but a passing and relatively short-term (at least in archaeological terms) event, further research on Stradbroke Island shows there were two periods of severe cold, at c. 30,000 BP and c. 21,700 BP, and that the entire period c. 33,000 to 18,000 BP was 'punctuated by abrupt periods of increasing aridity' (Petherick et al. 2008:800; see also Petherick et al. 2009). During the LGM, and in the period leading up to it, it would seem that much of the coastal strip of eastern Australia was a distinctly rebarbative environment – cold, dry, sparsely vegetated and subject to widespread dust storms.

Conclusions

It might be concluded, therefore, that the eastern Australian coastal strip during the Pleistocene was not a haven for human occupation, certainly not in the period 30,000-18,000 BP. It is, of course, possible that it was occupied before that, in the period now generally accepted as the timeframe for the human colonisation of Australia, 60,000-40,000 BP. It is further possible that that occupation was too sparse to leave any archaeological traces. On the other hand, perhaps we can suggest that the coastal colonisation of Australia (cf. Bowdler 1977) did not, in fact, embrace the east coast, which a small exploring population found distinctly unattractive and unnecessary. We know that people were on the western side of the Great Dividing Range by 40,000 years ago, and began to penetrate the western slopes of the eastern highlands not long after. Perhaps after the retreat of the glaciers, the east coast began to look like a more attractive proposition, luring travellers from the west to filter down its precipitous eastern cliffs and gullies to explore the newly emerging well-watered, resource-rich coasts and rivers of an eastern Eden.

Acknowledgements

Thanks to the Thylacine Club and ace member Geoff Hope all those years ago. Thanks also to Jane Balme for running an eagle eye over this, but of course she is not responsible for the contentious bits.

References

Allen, J. (ed) 1996. *Report of the southern forests archaeological project. Volume 1. Site descriptions, stratigraphies and chronologies*. Bundoora: School of Archaeology, La Trobe University.

Allen, J. and N. Porch 1996. Warragarra rockshelter. In J. Allen (ed) *Report of the southern forests archaeological project. Volume 1. Site descriptions, stratigraphies and chronologies*, pp195-217. Bundoora: School of Archaeology, La Trobe University.

Attenbrow, V. 2004. *What's changing: population size or land-use patterns? The archaeology of the Upper Mangrove Creek, Sydney Basin*. Terra Australia 21. Canberra: Pandanus Books.

Barker, B. 2004. *The Sea People: Late Holocene maritime specialisation in the Whitsunday Islands, central Queensland.* Terra Australis 20. Canberra: Pandanus Books.

Beaton, J.M. 1985. Evidence for a coastal occupation time-lag at Princess Charlotte Bay (North Queensland) and implications for coastal colonisation and population growth theories for Aboriginal Australia. *Archaeology in Oceania* 20:1-20.

Boot, P. 1993. Pleistocene date from archaeological excavations in the hinterlands of the New South Wales south coast. *Australian Archaeology* 37:59.

Boot, P. 1994. Recent research into the prehistory of the hinterland of the south coast of New South Wales. In M. Sullivan, S. Brockwell and A. Webb (eds) *Archaeology in the North: Proceedings of the 1993 Australian Archaeological Association Conference,* pp319-340. Darwin: North Australia Research Unit (ANU).

Border, A. 1999. Aboriginal settlement of offshore islands in the Southern Great Barrier Reef Province, central Queensland. In J. Hall and I. J. McNiven (eds) *Australian Coastal Economies,* pp129-139. Canberra: ANH Publications, Department of Archaeology and Natural History, RSPAS, ANU.

Bowdler, S. 1970. Bass Point: the excavation of a southeast Australia shell midden showing cultural and economic change. Unpublished BA (Hons) thesis, Sydney, University of Sydney.

Bowdler, S. 1976. Hook, line and dillybag: an interpretation of an Australian coastal shell midden. *Mankind* 10:248-258.

Bowdler, S. 1977. The coastal colonisation of Australia. In J. Allen, J. Golson and R. Jones (eds) *Sunda and Sahul: Prehistoric Studies in Southeast Asia, Melanesia and Australia,* pp205-246. London: Academic Press.

Bowdler, S. 1981. Hunters in the highlands: Aboriginal adaptations in the eastern Australian uplands. *Archaeology in Oceania* 16:99-111.

Bowdler, S. 1984. *Hunter Hill, Hunter Island* Terra Australis 8. Canberra: Dept. of Prehistory, Research School of Pacific Studies, Australian National University.

Bowdler, S. 1999. Research at Shark Bay, WA, and the nature of coastal adaptations in Australia. In J. Hall and I. J. McNiven (eds) *Australian Coastal Economies,* pp79-84. Canberra: ANH Publications, Department of Archaeology and Natural History, RSPAS, ANU.

Bowler, J.M., H. Johnston, J.M. Olley, J.R. Prescott, R.G. Roberts, W. Shawcross and N.A. Spooner 2003. New ages for human occupation and climatic change at Lake Mungo, Australia. *Nature* 421: 837-840.

Callaghan, M. 1980. Some previously unconsidered environmental factors of relevance to south coast prehistory. *Australian Archaeology* 11:43-49.

Cane, S. 2001. The Great Flood: eustatic change and cultural change in Australia during the Late Pleistocene and Holocene. In A. Anderson, I. Lilley and S. O'Connor (eds) *Histories of Old Ages: essays in honour of Rhys Jones,* pp141-165. Canberra: Pandanus Books.

Coutts, P.J.F. 1967. Coastal dunes and field archaeology in S. E. Australia. *Archaeology and Physical Anthropology in Oceania* 2:28-34.

David, B., R. Roberts, C. Tuniz, R. Jones and J. Head 1997. New optical and radiocarbon dates from Ngarrabullgan Cave, a Pleistocene archaeological site in Australia: implications for the comparability of time clocks and for the human colonization of Australia. *Antiquity* 71:183-188.

David, B., R.G. Roberts, J. Magee, J. Mialanes, C. Turney, M. Bird, C. White, K. Fifield and J. Tibby 2007. Sediment mixing at Nonda Rock: investigations of stratigraphic integrity at an early archaeological site in northern Australia and implications for the human colonisation of the continent. *Journal of Quaternary Science* 22:449-479.

Dodson, J.R. and B.G. Thom 1992. Holocene vegetation history from the Hawkesbury Valley, New South Wales. *Proceedings of the Linnean Society of New South Wales* 113:121-134.

Dunnett, G. 1993. Diving for dinner: some implications from Holocene middens for the role of coasts in the late Pleistocene of Tasmania. In M. A. Smith, M. Spriggs and B. Fankhauser (eds) *Sahul in Review: Pleistocene Archaeology in Australia, New Guinea and Island Melanesia*, pp247-257. Occasional Papers in Prehistory No. 24. Canberra: Department of Prehistory, Research School of Pacific Studies.

Dyall, L.K. 1982. Aboriginal fishing stations on the Newcastle coastline, New South Wales. In S. Bowdler (ed) *Coastal Archaeology in Eastern Australia*, pp52-62. Canberra: Department of Prehistory, Research School of Pacific Studies, Australian National University.

Flood, J., B. David, J. Magee and B. English 1987. Birrigai: a Pleistocene site in the south-eastern highlands. *Archaeology in Oceania* 22:9-26.

Gillespie, R. and R.B. Temple 1977. Radiocarbon dating of shell middens. *Archaeology and Physical Anthropology in Oceania* 12:26-37.

Hesse, P.P., G.S. Humphreys, P.M. Selkirk, D.A. Adamson, D.B. Gore, D.C. Nobes, D.M. Price, J. -L. Schwenninger, B. Smith, M. Tulau and F. Hemmings 2004. Late Quaternary aeolian dunes on the presently humid Blue Mountains, eastern Australia. *Quaternary International* 108:13-32.

Hope, G.S. 1978. The late Pleistocene and Holocene vegetational history of Hunter Island, north-western Tasmania. *Australian Journal of Botany* 24:493-514.

Hughes, P.J. 1981. The impact of bushfires on soils and soil erosion. In P. Stanbury (ed) *Bushfires: Their Effect on Australian Life and Landscape*, pp33-37. Sydney: The Macleay Museum, University of Sydney.

Jones, R. 1968. The geographical background to the arrival of man in Australia and Tasmania. *Archaeology and Physical Anthropology in Oceania* 3:186-215.

Jones, R. 1995. Tasmanian archaeology: establishing the sequences. *Annual Review of Anthropology* 24:423-446.

Kershaw, A.P. 1995. Environmental change in Greater Australia. *Antiquity* 69:656-675.

Lampert, R.J. 1971. *Burrill Lake and Currarong*. Terra Australis 1. Canberra: Department of Prehistory, Research School of Pacific Studies, Australian National University.

Lampert, R.J. and P.J. Hughes 1974. Sea level change and Aboriginal coastal adaptations in southern New South Wales. *Archaeology and Physical Anthropology in Oceania* 9:226-235.

Lees, B. 2006. Timing and formation of coastal dunes in northern and eastern Australia. *Journal of Coastal Research* 22:78-79.

McDonald, J. 2008. *Dreamtime Superhighway: Sydney Basin Rock Art and Prehistoric Information Exchange*. Terra Australia 27. Canberra: Pandanus Books, Australian National University.

McGowan, H.A., L.M. Petherick and B.S. Kamber 2008. Aeolian sedimentation and climate variability during the late Quaternary in southeast Queensland, Australia. *Palaeogeography, Palaeoclimatology, Palaeoecology* 265:171–181

Megaw, J.V.S. 1968. A dated cultural sequence for the south Sydney region of New South Wales. *Current Anthropology* 9:325-329.

Morse, K. 1999. Coastwatch: Pleistocene resource use on the Cape Range Peninsula. In J. Hall and I. McNiven (eds) *Australian Coastal Archaeology*, pp73-78. Canberra: ANH Publications, R.S.P.A.S., Australian National University.

Mulvaney, D.J. and E.B. Joyce 1965. Archaeological and geomorphological investigations on Mt Moffatt Station, Queensland, Australia. *Proceedings of the Prehistoric Society* 31:147-212.

Neal, R. and E. Stock 1986. Pleistocene occupation in the south-east Queensland region. *Nature* 323:618-621.

Petherick, L.M., H.A. McGowan and B.S. Kamber 2009. Reconstructing transport pathways for late Quaternary dust from eastern Australia using the composition of trace elements of long traveled dusts. *Geomorphology* 105:67-79.

Petherick, L., H. McGowan and P. Moss 2008. Climate variability during the Last Glacial Maximum in eastern Australia: evidence of two stadials? *Journal of Quaternary Science* 23:787-802.

O'Connor, S. 1999. *30,000 years of Aboriginal occupation: Kimberley, north west Australia.* Terra Australis 14. Canberra: ANH Publications and the Centre for Archaeological Research, Australian National University.

Pike-Tay, A., R. Cosgrove and J. Garvey 2008. Systematic seasonal land use by late Pleistocene Tasmanian Aborigines. *Journal of Archaeological Science* 35:2532–2544.

Rowland, M.J. 1996. Prehistoric archaeology of the Great Barrier Reef province - retrospect and prospect. In P. Veth and P. Hiscock (eds) *Archaeology of Northern Australia: Regional Perspectives,* pp190-212. Tempus 4. St Lucia: Anthropology Museum, University of Queensland.

Sim, R. 1994. Prehistoric human occupation in the King and Furneaux Island regions, Bass Strait. In M. Sullivan, S. Brockwell and A. Webb (eds) *Archaeology in the North: Proceedings of the 1993 Australian Archaeological Association Conference*, pp358-374. Darwin: North Australia Research Unit (ANU).

Stockton, E.D. 1993. Archaeology of the Blue Mountains. In Eugene Stockton (ed) *Blue Mountains Dreaming the Aboriginal Heritage*, pp23-52. Winmalee (NSW): Three Sisters.

Ulm, S., B. Barker, A. Border, J. Hall, I. Lilley, I. McNiven, R. Neal and M. Rowland 1995 Pre-European coastal settlement and use of the sea: a view from Queensland. *Australian Archaeology* 41:24-26.

11

Early Holocene human occupation and environment of the southeast Australian Alps: New evidence from the Yarrangobilly Plateau, New South Wales

Ken Aplin

Australian National Wildlife Collection, CSIRO Division of Sustainable Ecosystems, Canberra, Australia
Ken.Aplin@csiro.au

Fred Ford

Australian National Wildlife Collection, CSIRO Division of Sustainable Ecosystems, Canberra, Australia

Peter Hiscock

School of Archaeology and Anthropology, Australian National University, ACT, Australia

Introduction

The sciences of Quaternary studies and archaeology have developed hand in hand in Australasia so that, as a rule, there is a close match for any particular biome between knowledge of late-Quaternary environments and both the time depth and richness of regional archaeological records. The southeast Australian Alps represent a significant exception to this rule. For while the environmental history of this biome is relatively well known, at least since the termination of the last glaciation at c. 16,000 BP (e.g. Kershaw and Strickland 1989; Martin 1999; Barrows et al. 2001; Hope 2003; Kershaw et al. 2007), evidence of human activity above 1000 m elevation currently extends back no further than 4000 years (Flood 1980; Kamminga 1992, 1995; Lourandos 1997; Mulvaney and Kamminga 1999). Even more remarkably, the regional palaeoenvironmental record does not contain a strong signature of prehistoric human activity. For example, Geoffrey Hope's extensive palynological and sedimentological studies of mountain mires in New South Wales and the ACT found no obvious signature of prehistoric burning that might be linked to human activity (Macphail and Hope 1985; Hope 2003), a result that is replicated elsewhere in the region (Ladd 1979; Green et al. 1988; Kershaw and Strickland 1989; Dodson et al. 1994; McKenzie 1997, 2002; Mooney et al. 1997). In contrast, many sites show a marked increase in charcoal associated with the appearance of *Pinus* pollen, marking the onset of European land management practices (Hope 2003). Indeed, as Hope (2003:74) remarked, 'The impact of European pastoralism … is the greatest alteration to the environment that can be seen in the Holocene.'

Extensive archaeological survey of the southeast Australian Alps by Josephine Flood in the 1970s failed to locate any sites older than 4000 years in the wider region, and only very sparse lithic scatters at elevations above 1200m asl (Flood 1980). Further survey work in the 1980s and '90s (Kamminga 1992, 1995) failed to alter this situation; at present, the earliest evidence of occupation above 1000 m asl is dated to c. 4000 cal BP at Nursery Swamp (1140 m asl) in the ACT (Flood et al. 1987). In contrast, there is evidence of considerably earlier occupation from various localities around the margin of the southeastern Australian Alps, including Cloggs Cave at 76 m asl in East Gippsland, where evidence of human occupation extends back to c. 21,000 cal BP (Flood 1980), and Birrigai in the ACT at 730 m asl, with a basal occupation date of c. 25,000 cal BP (Flood et al. 1987). A question of considerable interest in this regional context concerns the antiquity of the ethnographically documented seasonal exploitation of bogong moths, a food resource rich in protein and fat that is available during the summer months at high-elevation aestivation sites (Flood 1980).

The enigma of the apparently late aboriginal occupation of the southeast Australian Alps, and its seemingly gentle footprint, is given additional emphasis by the documentation over the past two decades (Kiernan et al. 1983; Jones 1995; Cosgrove 1999) of human occupation of river valleys on the western margin of the central plateau of Tasmania throughout the period of the Last Glacial Maximum (LGM; c. 26,000-16,000 BP; Barrows et al. 2001), involving exploitation of a subalpine landscape featuring a mosaic of low tree heaths, sedgelands, grasslands and herbfields (Colhoun and van de Geer 1986, 1994; Cosgrove et al. 1990). This occupation started around 31,000 BP and continued until termination of glacial conditions, when the region became densely forested. If people were so successfully exploiting subalpine and even alpine environments in Tasmania through the LGM, it seems reasonable to expect that they were at least technologically equipped to exploit a large area of broadly similar biomes in the southeast Australian Alps, located a relatively short distance north across the exposed Bassian Plain.

Here we report new evidence for early Holocene human activity in the southeast Australian Alps, derived from a shallow test excavation in cave Y259, located at 1100 m asl near the Yarrangobilly Caves, New South Wales (Figure 1). We also present our preliminary observations of a rich, associated non-cultural mammalian fauna that allows inferences to be drawn regarding local environmental conditions during and after the period of human occupation of the site.

Site Y259

Y259 in the Australian Speleological Federation Karst Index Database (http://www.caves.org.au/kid/) is located in a doline on what is informally known as the plateau area of the Yarrangobilly karst. The doline is around 1100 m asl and lies at the intersection of two bioregions (Thackway and Creswell 1995): the Australian Alps and the southeast Highlands (the latter represented by the Bago subregion) (see Figure 1). The doline has a maximum cliff height of around 20 m, but more subdued relief at the northern end allows straightforward pedestrian access. The vegetation in the doline and immediate surrounds is open snowgum (*Eucalyptus pauciflora* subsp. *pauciflora*) and black sallee (*E. stelullata*) woodland with a snowgrass (*Poa sieberi*) dominated understorey. This open vegetation continues along a vague drainage line trending westwards across the top of the gently sloping plateau of the Yarrangobilly Limestone until reaching the steep gorge of the Yarrangobilly River, which has very open vegetation on rocky slopes. Steeper slopes of sediments of the Ravine Beds and Gooborrangandra Porphyry

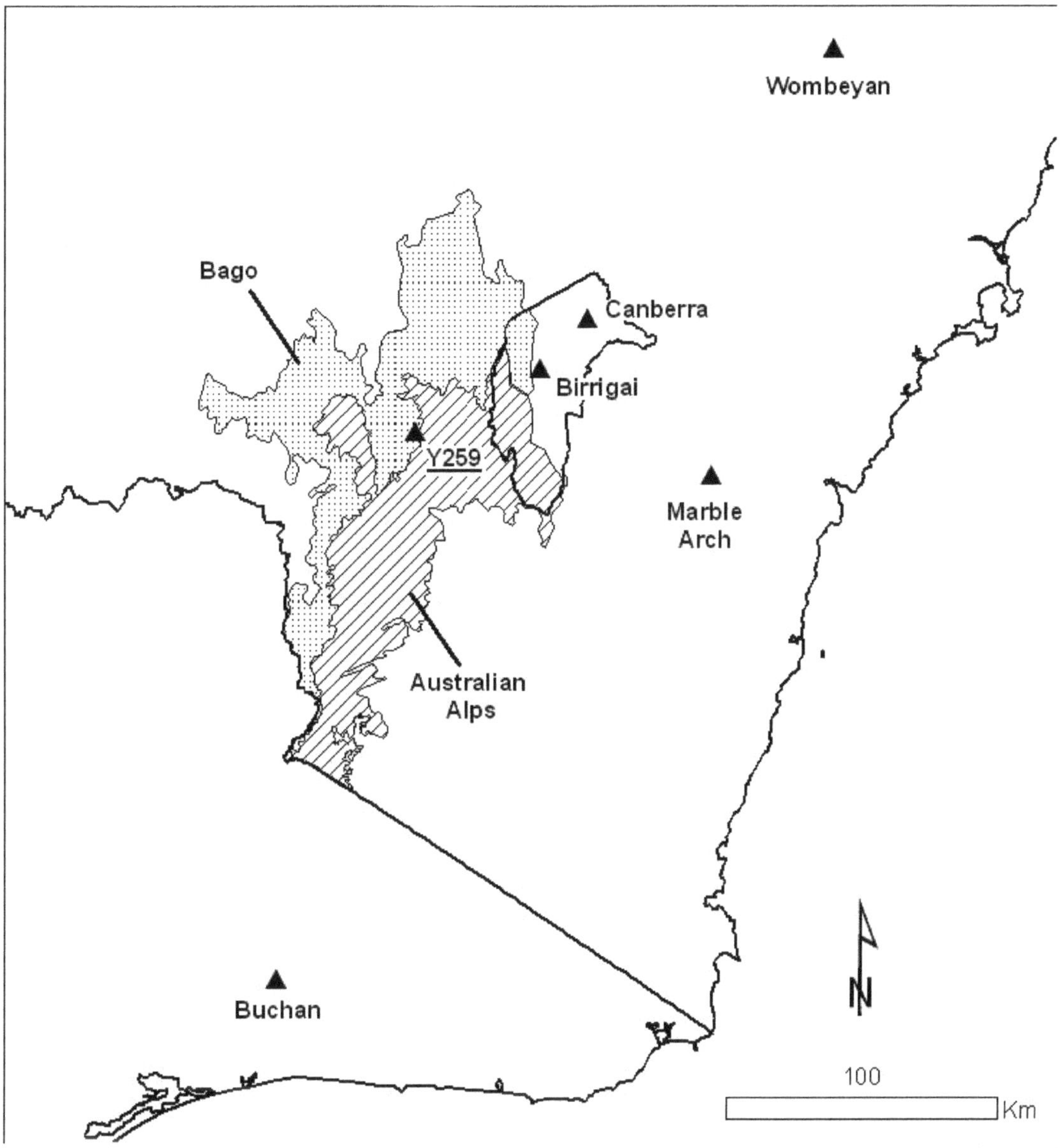

Figure 1. Map of southeastern Australia showing the boundaries (for New South Wales only) of the Bago sub-bioregion (Southeast Highlands bioregion) and the Australian Alps bioregion of the Interim Biogeographic Regionalisation of Australia (IBRA; Thackway and Creswell 1995), and the location of various archaeological and palaeontological localities mentioned in the text. Site Y259 is located almost exactly on the boundary between the two IBRA regions. The location of Buchan includes both Cloggs Cave and Pyramids Cave

immediately to the east and north of the doline possess denser montane open forests dominated by mountain gum (*E. dalrympleana*) and narrow-leafed peppermint (*E. robertsonii*), with ribbon gum (*E. viminalis*) along creek lines, and scattered individuals and small stands of alpine ash (*E. delegatensis*). An open midstorey of blackwood (*Acacia melanoxylon*) and silver wattle (*A. dealbata*) and an open shrubby understorey of mat rush (*Lomandra* spp.), bracken (*Pteridium esculentum*) and beard heath (*Leucopogon hookeri*) covers most of these slopes.

Y259 is essentially a cleft situated between the doline wall (which forms its western wall) and several massive, detached blocks that form both its eastern wall and a partial roof. The floor area is approximately 15 m by 5 m, with its long axis orientated approximately north-south (Figure 2). The northern end of the cleft is entered via a short descent from the floor of the doline; this area is well-lit and largely open to the elements. The southern end of the cleft is more confined and sheltered, with a low roof along the eastern side and a high but narrow cleft leading to an elevated opening on the western side. Active cave formation is present on the sloping western wall and on the lower roof at the southern end.

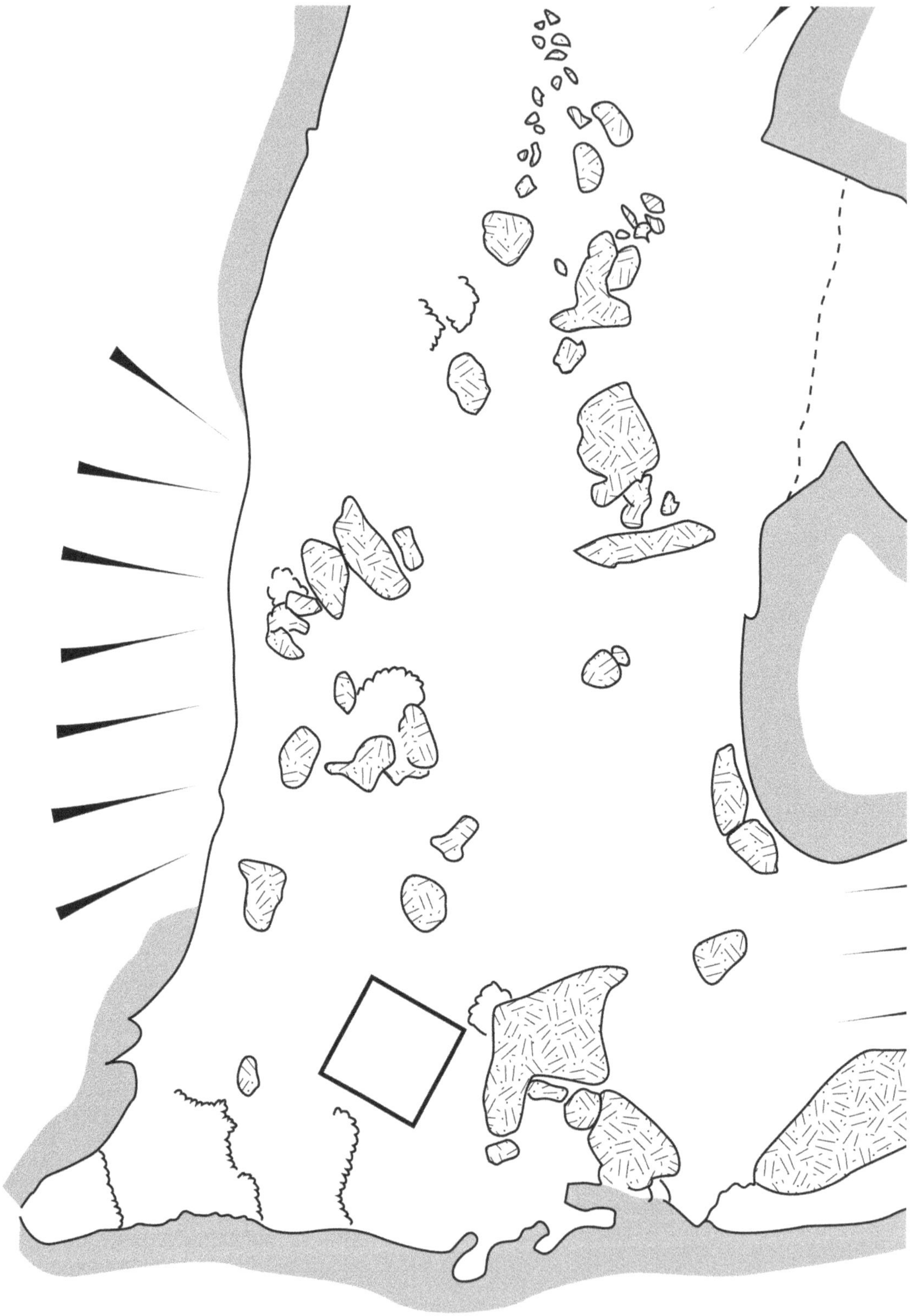

Figure 2. Floor plan of Y259. The position of the test excavation is indicated. Arrows indicate the slope down from the three external openings of the cave

The cave deposit comprises a sub-horizontal accumulation of fine, in-washed sediments bedded around scattered pieces of roof fall. Sediment probably enters at both ends of the cave and the surface slopes down to a low point located in the centre on the eastern side; this area of the cave floor is prone to inundation. Elsewhere, the surface is damp and supports a thin layer of mosses wherever light penetrates to the cave floor. In the darker reaches, the surface bears a powdery white 'frosting' of calcium carbonate. Dissociated small vertebrate bone is scattered over much of the floor surface but is concentrated at both ends of the cleft, where shelves suitable for roosting owls are present several metres above the floor.

Numerous remains of bogong moths were noted in the cave at the time of excavation and on other visits, and these identify Y259 as a minor aestivation site. The site is much lower than most known aestivation sites, which tend to be found higher than 1400 m asl (Mansergh et al. 1990; Broome 2001), but these are generally boulder-strewn sites that lack the depth or coolness of limestone caves. Y259 is probably subject to more significant diurnal temperature variations than many other local caverns that lack moths due to its open roof and small size. Cold overnight temperatures might benefit aestivating moths when compared with more stable climates in other caves, while daytime temperatures are still cool due to the massive limestone roof.

Drummond (1963) performed an excavation in Y259 (reported as Y50, which occupies the same doline; Andy Spate of the Canberra Speleological Society indicated the location of Drummond's excavation to us) and reported that 'many thousands of bones' were collected (p. 35). He recorded 14 species of mammal in two samples: 375 identifiable bones from the 'Surface' and 136 bones from 'Layer 3 Down' (Drummond 1963:40) and made some remarks about the significance of differences in faunal composition between the two reported layers. The whereabouts of Drummond's collection is not known.

The test excavation

Our excavation in Y259 was undertaken as part of a regional survey of the 'original' (i.e. late prehistoric to early historic) mammalian fauna of the southeast Australian Alps and adjacent Southern Tableland regions. This work is primarily focussed on bone deposits accumulated by non-human predators, many of which are in small caverns unsuitable for human occupation and with little or no sediment accumulation. However, a few larger cave sites, including Y259, have prey remains incorporated into deeper sediment accumulations. These provide opportunities for recovering faunal samples from stratified and potentially dateable contexts.

A 1 m^2 test pit was positioned at the southern end of the floor area of Y259 (shown on Figure 2), in an area that contained the highest density of surface bone, the minimal amount of surface limestone rubble, and an absence of obvious recent surface wash. Loose material to a depth of approximately 1 cm was brushed up as a 'Surface' sample, hereafter termed 'Spit S'. Subsequent excavation was carried out by visible stratification, or approximately 2 cm spits, when sediment appeared homogenous. On completion of Spit 6, a flowstone-capping brecciated sediment was encountered. The flowstone was breached only over the eastern half of the square and excavation continued for three more spits. Charcoal samples for C14 dating were collected from each spit.

Post-excavation treatment

All excavated sediment was bagged and transported to the laboratory without sieving or other treatment. Before the main treatment, a subsample of the brecciated deposit from Spit 9 was washed, dried and examined. The presence of burnt bone and stone artefacts, not observed during excavation, alerted us to the archaeological nature of the deposit. Charcoal samples from Spits 7 and 9, spanning the depth of the brecciated unit, were submitted to the Waikato Radiocarbon Dating Laboratory for age determination using Accelerator Mass Spectrometry (AMS).

Material recovered from Spits 7-9 is the primary focus of this paper. For comparisons, we also present data for the Surface, Spit 1 and Spit 5 samples, the latter representing the first of the two flowstone-sealed sedimentary units.

The samples from Spits S, 1 and 5 were soaked in water, washed gently through a fine mesh (approximately 0.5 mm), then dried. The samples from Spits 7-9 were subjected to the same treatment except that chunks of brecciated sediment were removed and placed in a weak acetic acid solution (approximately 2-5%, depending on degree of consolidation) for dissolution of calcium carbonate. The acid-treated residue was put through three or more changes of water to flush residual acid, then air-dried. Any material that was still brecciated was subjected to a second cycle of acid treatment.

The samples from Spits 2-4 and 6, as well as one bag of deposit from Spit 7, remain unprocessed at the time of writing, as does a significant volume of Spit 1.

Sorting

All tooth-bearing elements (dentaries, maxillae and premaxillae) and isolated teeth were separated from the remaining bone. Also picked out were cranial and selected postcranial elements (all major limb bones and pedal elements) of all taxa except murids, including birds and reptiles. All burnt bone was recovered, irrespective of taxon.

The washed and dried samples from Spits S, 5 and 7-9 were sorted in entirety, while those from Spit 1 were picked over less thoroughly due to time constraints.

Taxonomic determinations

Taxonomic determinations were checked against reference specimens held by the Australian National Wildlife Collection (ANWC) and were made in the context of taxonomic reviews of several key groups, most notably the murid rodents. Whenever possible, specimens derived from localities in the southeastern Highlands were consulted in preference to material from more remote locations. We have not yet attempted to identify the small number of reptile and bird specimens.

Several marsupials and murid represented in the Y259 assemblage fall well outside their contemporary geographic ranges as currently documented. The identification of several of these taxa warrants some brief comment.

Two species of *Sminthopsis* are clearly represented in all samples except Spit S. Both lack lower molar entoconids but are clearly distinguishable on tooth size and the configuration of the premolar series (less crowded in the larger-toothed form). These differences correspond with those documented by Archer (1981) between southeastern Australian samples of *S. murina* (the smaller-toothed form) and *S. leucopus*.

The taxon listed as *Antechinus* sp. cf. *A. flavipes* is a member of what might be called the *A. flavipes* group; this broadly distributed assemblage is in need of taxonomic revision.

Dentaries and maxillae attributed to Leadbeater's possum (*Gymnobelideus leadbeateri*) are very similar in overall size and shape to those of the sugar glider (*Petaurus breviceps*), but the two are distinguishable on the basis of molar size and corresponding alveolar dimensions (Wakefield 1972b). To further confirm the species identification, additional cranial elements (squamosal, petrosal, alisphenoid and frontal bones) of small petaurids were retrieved from the bulk bone samples from Spits 5 and 7, and compared with reference specimens of both species. Cranially, *G. leadbeateri* differs from *P. breviceps* in having a less pneumatised glenoid and auditory region and a narrower interorbital region. Cranial elements of both taxa are readily distinguished in the Y259 sample.

Drummond (1963) reported *Pseudomys australis* from his excavation. His specimens probably belonged to the morphologically similar *P. higginsi*, which is represented by numerous well-preserved specimens in Surface and Spits 1, 5 and 7. This taxon is known as a living animal only from Tasmania, but is recorded in various subfossil and fossil assemblages from mainland southeastern Australian sites (Wakefield 1972a, b; Hope 1976; Flood 1980). The Y259 material is consistent in all respects with the Pyramids Cave sample used by Wakefield

(1972b) in his description of the mainland subspecies *Pseudomys higginsi australiensis*. Ancient DNA techniques are being applied to Yarrangobilly specimens in an effort to confirm the taxonomic affinities of mainland populations of *P. higginsi*.

Pseudomys oralis is not known as a living animal south of Mount Royal, in the New South Wales Central Coast hinterland, but it is recorded from recent cave deposits as far south as East Gippsland (Wakefield 1972a, b). During sorting of the Y259 and other regional bone samples, we observed that specimens with the distinctive morphology of *P. oralis* fall into two distinct size classes, with other anatomical differences observable in more complete cranial fragments. We suspect that these represent two different species, probably siblings, and herein designate them as *P. oralis* A (smaller-toothed) and *P. oralis* B (larger-toothed). Attempts are currently underway to extract ancient DNA from some of the better-preserved specimens of each form to explore the taxonomic implications of this morphological dichotomy and the relationship of these taxa to mtDNA lineage diversity described from extant *P. oralis* populations (Jerry et al. 1998).

Faunal quantification and analysis

For small mammals, only teeth and tooth-bearing specimens were selected out for identification. For comparison of taxonomic composition between excavated samples, we use a total count of allocated specimens for each taxon, expressed as the Number of Individual Specimens (NISP). We used NISP values in preference to Minimum Number of Individuals (MNI) because in such a small excavated area and with highly dissociated remains, there is no reason to believe that MNI values are any more representative of original relative prey abundances. Moreover, their use would certainly deflate (most to unity) the relative contribution of all of the less-abundant taxa. For the larger-bodied taxa, other cranial and postcranial elements were also identified to an appropriate taxonomic level (often only to family; e.g. as a medium-sized macropodid) and these elements are included in NISP values for these taxa. NISP values for each sample are presented in Table 1.

Table 1. Taxonomic composition of the Y259 vertebrate fauna recovered from each of Spits S and Spit 1 (Unit I), Spit 5 (Unit II), and Spits 7-9 (Unit III). All values are NISP. Taxa listed as 'Indet' are identified only to the indicated taxonomic level, e.g. 'Indet *Sminthopsis* sp.' are determinable as *Sminthopsis* but could not be identified to species level (Table 1 continues on page 194)

Taxon	Surface	Spit 1	Spit 5	Spit 7	Spit 8	Spit 9
Rattus fuscipes	707	462	393	106	141	29
Rattus lutreolus	34	40	11	7	12	5
Indet. *Rattus* sp.	47	63	19	33	10	8
Mastacomys fuscus	99	64	39	14	26	10
Pseudomys fumeus	103	92	37	6	13	5
Pseudomys higginsi	30	16	2	1		
Pseudomys novaehollandiae	1	8	3	5	1	
Pseudomys oralis A	14	5	1	1	29	
Pseudomys oralis B	2	11	5	4		
Total murids	977	753	515	171	240	58
Acrobates pygmaeus	14	9	10	3		
Cercartetus nanus	22	33	33	5	1	
Total small possums	36	42	43	8	1	
Gymnobelideus leadbeateri	8	7	3			
Petaurus breviceps	1	1	15	4	4	

Taxon	Surface	Spit 1	Spit 5	Spit 7	Spit 8	Spit 9
Total medium possums	1	1	23	11	7	
Petaurus australis	2					
Pseudocheirus pergerinus	7	3	22	4	3	
Trichosurus sp.	1					
Total large possums	7	3	24	5	3	
Potorous spp.	2	1	2	5		
Bettongia sp.	1	2	2	2		
Total potoroids	3	2	1	2	7	2
Petrogale penicillata	4	3	4			
Indet. small macropodid	1	3				
Total small macropodids	4	1	6	4		
Macropus rufogriseus	2					
Indet. medium-sized macropodid	2					
Total medium-sized macropodids	2	2				
Perameles nasuta	3	5	1			
Isoodon obesulus	7	3	5			
Indet. peramelid	4					
Total peramelids	7	3	8	9	1	
Sminthopsis murina	10	11	6	2	2	1
Sminthopsis leucopus	2	4	1	5	2	
Indet. *Sminthopsis* sp.	1					
Antechinus agilis	86	64	94	15	32	6
Antechinus swainsonii	6	7	18	6	2	2
Antechinus sp. cf. *A. flavipes*	2	1	2			
Indet. small dasyurid	8	7	2			
Phascogale tapoatafa	2	1				
Total small dasyurids	100	80	124	26	44	13
Dasyurus viverrinus	3	1	1			
Dasyurus maculates	2					
Nyctophilus geoffroyi	1	1	1			
Nyctophilus gouldii	1	1	1			
Falsistrellus tasmaniensis	1					
Scotorepens sp.	1					
Total microchiropterans	1	2	2	2	1	
Oryctolagus cuniculus	1					
Total mammals	1132	884	745	215	329	83
Scincidae	8	20	11	2		
Elapidae	1					
Birds	1	2	4	4	1	
Total all vertebrates	1141	906	760	215	336	84

Stratigraphy and dating

The surface sediments (Unit I) of the excavation are unconsolidated, organic, rich, brown, loamy sediment with a small quantity of limestone rubble and organic litter, including visible wood-charcoal fragments. The density of bone and charcoal was noted to increase between Spit S and Spit 1. Spit 2 maintained the high density of bone and charcoal; it was removed to expose a distinctly greyer unit across the entire square. Spits 3 and 4 removed this sub-unit, which was also found to contain abundant bone and charcoal. An undulating flowstone horizon was exposed locally in Spit 3 and across the entire square on completion of Spit 4, with a vertical fall of 3-4 cm from north to south. The flowstone consists of a 1-2 mm thick layer of relatively clean calcite, sealing partially brecciated sediments; it was broken through and the underlying bone and charcoal-rich unit removed in two spits (5 and 6). The flowstone and Spits 5-6 are grouped as Unit II. Spit 6 exposed a second and more substantial flowstone across the entirety of the square, this time showing a vertical fall of 5 cm from north to south. This lower flowstone was also broken through but only over the northern half of the square. Excavation proceeded in this reduced area for three spits, by breaking up the brecciated sediments with a small crowbar. The sediment contained a much larger quantity of limestone rubble than the upper units, as well as rounded fragments of a hard, red claystone of uncertain origin. While this superficially resembles burnt sediments from hearths, final identification is reserved pending microscopic examination. The lower flowstone and underlying sediments are designated as Unit III. Section drawings illustrate the relative thickness of the recognised stratigraphic units (Figure 3).

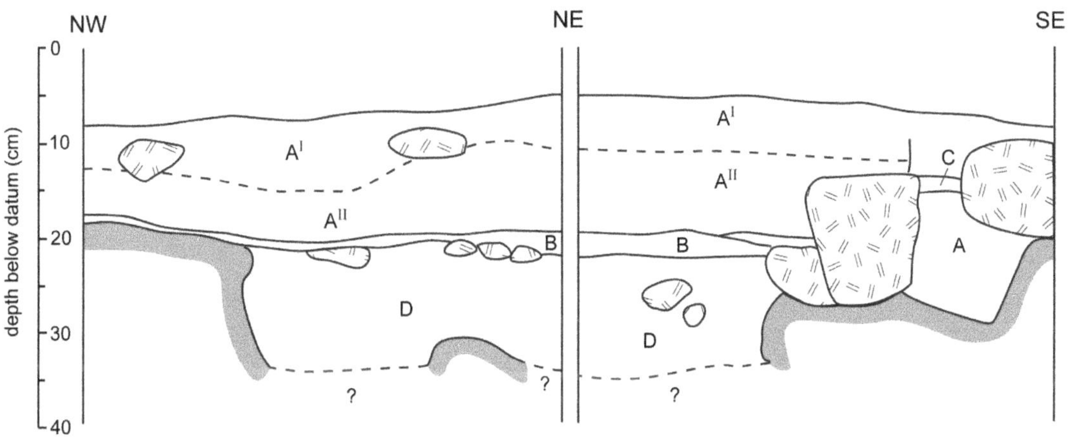

Figure 3. Section drawings for the NE and NW walls of the excavation in Y259. Unexcavated deposit of unknown depth is indicated at the bottom of each section. Limestone blocks are shown with crosshatch. The fine layer of flowstone that separates Units I and II was not obvious in the section; its approximate position, based on depth measurements at the time of excavation, is indicated by the dashed line

The presence of the two unbroken flowstone units gives us great confidence that our excavation was placed away from any disturbance caused by Drummond (1963). Furthermore, the sharp nature of the upper surface of both flowstones, compared with their more diffuse lower boundaries, together with the north-to-south dip on both surfaces, suggests to us that each flowstone was deposited during a brief hiatus in sedimentation, and that they represent primary features in the stratigraphy. The point is of interest in so far as it impacts on our perception of the level of stratigraphic integrity of the deposit and contained faunal and cultural sequence, which would clearly be poorer if formation of the flowstones substantially post-dated the main period of accumulation (and potential disturbance) of the deposit. Moreover, if the features are primary, then the consistent dip on both flowstone-sealed surfaces indicates that the bulk of the sediments entered the deposit from the northern side, i.e. the main doline, rather than through the more elevated southern entrance. The origin of the flowstones might be tested by direct dating of the calcites relative to charcoal in the underlying and overlying sedimentary units.

Two AMS dates were obtained on discrete chunks of wood charcoal removed from weakly brecciated sediments of Spits 7 and 9, respectively (Table 2). The results indicate a relatively rapid accumulation of Unit III during the early Holocene, in the interval between 9700 and 9120 cal. yr BP.

Table 2. Radiocarbon dates from Y259. Dates were produced by the University of Waikato Radiocarbon Dating Laboratory using Accelerator Mass Spectrometry. Calibrated ages are based on OxCal v3.10 (Bronk Ramsey 2001)

Spit	Lab code	Material	d13C (‰)	Radiocarbon age	Southern Hemisphere Calibration 2 sigma interval
7	Wk-18838 (AMS)	Charcoal	-25.6 ± 0.2	8343 ± 43 BP	9440-9120 cal yr BP
9	Wk-18839 (AMS)	Charcoal	-25.3 ± 0.2	8668 ± 43 BP	9700-9500 cal yr BP

The faunal assemblage

Substantial quantities of bone were recovered from all excavated levels. The assemblages from Spits S and 1 are comprised entirely of large quantities of unburnt and mostly unbroken bones derived from small to medium-sized mammals, with occasional remains of birds and reptiles (Table 1 and Figure 4). The samples from Spits 5 (Unit II) and Spits 7-9 (Unit III) are similar except that the assemblages also contain small numbers of mostly fragmented bones derived from larger-bodied mammals. In Unit III, a significant proportion of these larger bone fragments is also burnt to varying degrees.

Taphonomy

We interpret the Y259 assemblage as a composite of two faunal components with distinct origins. The first component, comprising the bulk of the remains from each level, is interpreted as the dissociated remains of regurgitated pellets or 'casts' deposited by owls on to the cave floor from their daytime roosts. Signature characteristics of a bone deposit accumulated by owls include a preponderance of nocturnally active small mammals over larger-bodied mammals and diurnally active vertebrates (e.g. birds or reptiles), a large proportion of intact skeletal elements, and an absence or low incidence of burning (Wakefield 1972b; Kusmer 1990). The largest mammals in Spits S and 1 are possums, rat kangaroos and bandicoots, but these are represented primarily by juveniles.

Three species of owls found in southeastern Australia are known to roost in caves and to accumulate large quantities of prey remains – sooty owl (*Tyto tenebricosa*), masked owl (*Tyto novaehollandiae*) and barn owl (*Tyto alba*). All have been recorded in the region of the deposit (Table 3), although sooty and masked owls are currently rare or vagrant. Clear criteria for distinguishing the prey accumulations of each species have not been established. However, the larger-bodied sooty and masked owls are understood to be capable of taking a wider size range of prey than the smaller-bodied barn owl, while sooty owls are thought to forage more often in wetter forests than masked owls, and to make greater use of arboreal mammals (Morton et al. 1977; Lundie-Jenkins 1993; Kavanagh and Murray 1996; Bilney et al. 2006). The relatively poor representation of arboreal mammals in the Y259 assemblage suggests that masked owls and/or barn owls may have been responsible for the bulk of the bone accumulation.

The second component of the assemblage, confined to Units II and III, is interpreted as prey remains deposited by occasional human visitors to the site. This component is most clearly indicated by the representation of larger mammals, such as several species of wallabies, that are considerably beyond the predatory capacity of owls, and by the physical condition of the remains of these larger animals, which are typically fragmented and, at least in Unit III, often show signs of heat modification, whether from cooking or post-discard burning. Careful examination of the samples of burnt bone from each of Spits 7 to 9 shows they are primarily derived from medium-sized to large mammals, and only rarely from small mammals.

Some caution must be exercised in interpreting burnt bone as an indicator of human predation or activity (Nicholson 1993; Asmussen 2009). Bone can burn in natural bushfires and it might conceivably have washed into the cave from outside. In the present case, we regard this as unlikely, for two reasons. Firstly, none of the larger mammal bones, including the various burnt specimens, show any rounding of fracture edges of the kind that typifies bone subject to overland transport. And secondly, many of the burnt bones show extreme charring and even calcination, due to heating to temperatures well above those usually attained under a natural fire (Nicholson 1993; Stiner et al. 1995). We conclude that the bones were burnt in or below hearths, probably situated within the cave itself. But does this imply human predation? Again, not necessarily so, as construction of a hearth on a cave deposit rich in non-cultural bone presumably can result in intense burning of buried bones. While this might explain some of the burnt bone in Y259, we believe that most of the burnt bones are the remains of animals that were either hunted or scavenged by people, and then cooked, eaten and their

Table 3. Faunal comparison of Y259 to surrounding faunal records based on records within the *New South Wales Wildlife Atlas* (Table 3 continues on page 198)

Taxa recorded in Y259	10km	Alps	Bago
Marsupials			
Dasyurids			
Antechinus sp. cf. *A. flavipes*			+
Antechinus agilis	+	+	+
Antechinus swainsonii		+	+
Dasyurus maculates	+	+	+
Dasyurus viverrinus		+	
Phascogale tapoatafa		+	+
Sminthopsis leucopus			
Sminthopsis murina			
Bandicoots			
Perameles nasuta		+	+
Isoodon obesulus			+
Macropodoids			
Bettongia sp.			
Potorous sp.			+
Macropus rufogriseus	+	+	+
Petrogale penicillata		+	
Possums and gliders			
Acrobates pygmaeus		+	+
Cercartetus nanus	+	+	+
Gymnobelideus leadbeateri			
Petaurus australis		+	+
Petaurus breviceps	+	+	+
Pseudocheirus peregrinus	+	+	+
Trichosurus sp.	+	+	+
Murid rodents			
Mastacomys fuscus	+	+	+

Taxa recorded in Y259	10km	Alps	Bago
Pseudomys fumeus	+	+	+
Pseudomys higginsi			
Pseudomys novaehollandiae			
Pseudomys oralis A			
Pseudomys oralis B			
Rattus fuscipes	+	+	+
Rattus lutreolus			
Introduced mammals			
Oryctolagus cuniculus	+	+	+
Potential owl species contributing to deposit			
Tyto alba	+	+	+
Tyto novaehollandiae		+	
Tyto tenebricosa		+	
Potential owl prey not recorded in Y259			
Burramys parvus		+	
Hydromys chrysogaster		+	+
Petauroides volans		+	+
Petaurus norfolcensis		+	+
Introduced mammals			
Mus musculus	+	+	+
Rattus rattus	+	+	+
Potential human prey not definitively recorded in Y259			
Macropus giganteus	+	+	+
Macropus robustus		+	+
Phascolarctos cinereus		+	+
Pteropus scapulatus			+
Tachyglossus aculeatus	+	+	+
Vombatus ursinus	+	+	+
Wallabia bicolour	+	+	+

remains discarded into hearths. To argue that the larger mammal bones might have entered the site naturally and been burnt coincidently beneath a hearth, it would be necessary to explain the absence of burning of more than just a few of the very large number of associated small mammal bones that littered the cave floor during accumulation of Unit III.

Taxonomic composition

The faunal assemblages from each of the analysed samples are summarised in Table 1 and Figure 4. A minimum of 33 species of mammals is represented, along with at least one species of skink and several different but currently undetermined bird species.

The various samples differ in both gross composition and detailed taxonomic representation. Looking first at gross composition (Figure 4a), a clear difference is seen between Unit III, where wallabies make up from 20% to 40% of total NISP, and Unit I, where these taxa are not represented at all. As argued above, the wallaby remains almost certainly represent a distinct archaeological (i.e. human-derived) component, present within the assemblage from Unit III, but absent from the two uppermost samples. The sample from Unit II, in which wallabies make up 4% of total NISP, is intermediate in this regard and may include a small archaeological (i.e. human-derived) component.

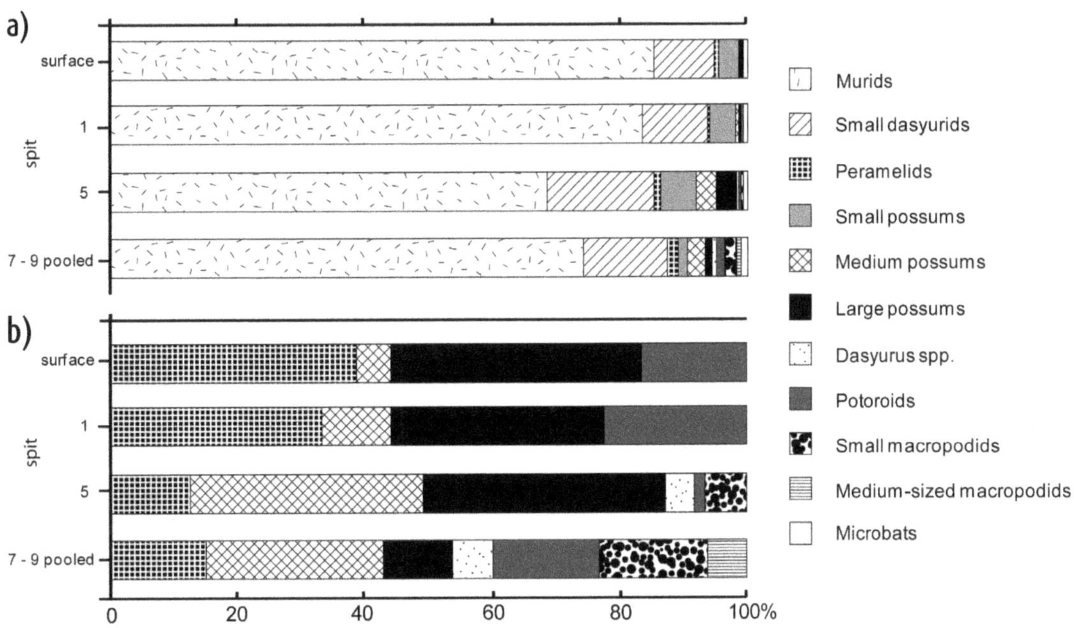

Figure 4. Summary taxonomic composition of the Y259 vertebrate fauna recovered from each of Spits S and Spit 1 (Unit I), Spit 5 (Unit II), and combined Spits 7-9 (Unit III). Plotted values are based on NISP. Figure 4a includes all taxa; 4b excludes the smallest mammals

If wallabies are excluded from these calculations, the samples from Units II and III are still distinguished in having a higher proportion of medium-sized mammals, especially possums, a slightly higher proportion of small dasyurids, and correspondingly fewer murids (Figure 4b). The presence of possum-sized bones among the fragmented and burnt component of the Unit III samples suggests that some of the medium-sized mammals are a product of human predation. However, it is also possible that the change reflects a decreased emphasis on medium-sized mammals by the avian predators, perhaps due to a greater contribution by barn owls during more recent times. Both factors may be partly responsible for the observed changes.

A notable feature of the finer-scale taxonomic composition is the dominance throughout of three murids, *Rattus fuscipes, Mastacomys fuscus* and *Pseudomys fumeus*, which together contribute 53% (Spit 9) and 78% (Spit S) of the total NISP for mammals. The following temporal changes are apparent among the numerous sub-dominant taxa:

1. Recent loss of *Sminthopsis leucopus* (present in Spit 1 and below but absent in the large sample from Spit S);
2. Earlier loss of *Gymnobelideus leadbeateri* (present in Spit 5 and below; absent in Spits 1 and S) and coincident reduction in each of *Pseudomys novaehollandiae*, *Antechinus swainsonii* (both still present in Spits 1 and S but more abundant in Spit 5 and below) and *Sminthopsis* spp. overall;
3. Apparent absence of *Perameles nasuta* above Spit 5 and coincident appearance of *Isoodon obesulus*; and
4. Considerable increase in abundance in Spit 5 and above of *Cercartetus nanus*.

Notable absences from all samples are the greater glider (*Petauroides volans*), squirrel glider (*Petaurus norfolcensis*), mountain pygmy possum (*Burramys parvus*), and rabbit rat (*Conilurus albipes*). The mountain pygmy possum and rabbit rat are likely to be precluded due to habitat preferences (see Discussion). However, the absence of the larger gliders from deposits is enigmatic, particularly given the abundance of *P. volans* in nearby forests today.

The introduced rabbit (*Oryctolagus cuniculus*) is represented only in Spit S. Several rabbit skeletons were found in the nearby entrance of Y50, and it seems likely that rabbit material may not be derived from avian predators, but from incidental deaths in the cave, or from dog or fox activity. This interpretation is supported by the absence of *Mus musculus* and *Rattus rattus*. Drummond (1963) was not confident in stating that non-native species were absent from the cave deposit because he was unable to differentiate *Rattus* species; however, it seems that all species in the owl-derived fauna are native, and the deposition of such material appears to have ceased in Y259 before the local arrival of introduced murids.

Stone artefacts

Artefacts were recovered only from Spits 7, 8 and 9 (Table 4). A total of 16 specimens display unambiguous signatures of human manufacture, such as positive hertzian fractures, with clear platforms and, in most cases, negative dorsal scars from the same platform, indicating that they were struck from uni-platform cores. The clustering of artefacts in Spits 7-9 may indicate one relatively intensive phase of human use of the site. However, multiple visits to the site are indicated by the different raw materials in each of the spits: all artefacts in Spit 9 are volcanic, whereas all artefacts in Spit 7 are quartz, while chert is present only in Spit 8 (Table 4). These patterns are unlikely to result from taphonomic processes and are best seen as the product of multiple, discrete periods of human visitation.

Table 4. Description of the stone artefacts from Y259

Spit	*Raw material*	*Flake dimensions in mm*			*Description*
		Length	Width	Thickness	
7	Quartz	11.4	5.6	1.8	Complete tertiary flake with feather termination
7	Quartz	3.1	4.7	2.4	Medial flake fragment
7	Quartz	4.2	5.5	2.5	Proximal fragment of tertiary flake
8	Sedimentary (chert)	7.0	4.0	1.1	Proximal fragment of tertiary flake
8	Quartz	5.9	8.4	2.5	Medial flake fragment
8	Quartz	7.7	6.6	1.8	Complete tertiary flake
8	Volcanic (tuff)	7.0	6.0	2.8	Distal fragment with feather termination
8	Volcanic (tuff)	10.9	6.3	2.1	Flake fragment
8	Quartz	10.9	10.0	3.1	Longitudinally split flake with feather termination
8	Sedimentary (chert?)	8.8	10.6	1.6	Complete tertiary flake with feather termination
8	Volcanic (tuff)	13.3	15.6	2.6	Distal fragment with step termination
8	Volcanic (tuff)	13.2	20.9	2.6	Complete flake with cortex on dorsal face
8	Volcanic	30.8	17.5	4.8	Complete tertiary flake with shattered platform
9	Volcanic	10.4	9.1	1.5	Complete tertiary flake with step termination
9	Volcanic	8.3	4.2	1.6	Flake fragment
9	Volcanic	7.7	20.1	6.2	Complete tertiary flake with feather termination

The stone artefacts are all unretouched flakes, hence are not diagnostic of particular functions or activities. As shown in Figure 5, none are fewer than 3 mm in length or 4 mm in width. Since the sieve size used in excavation was smaller than this and the samples were carefully sorted, the lack of small artefacts between 1 mm and 3 mm is significant. Knapping of stone artefacts typically produces large quantities of small fragments in the 1-3 mm range and these are a reliable indicator of artefact manufacture in a site. In the case of Y259, the absence of artefacts of that size class reveals that artefacts were typically made elsewhere, either in a different site or a different part of this site, and that limited artefact reduction occurred here. This pattern is consistent with short-term visits to the cave by mobile groups which were transporting retouched flakes or cores made elsewhere and which undertook only very limited amounts of stone working when occupying Y259.

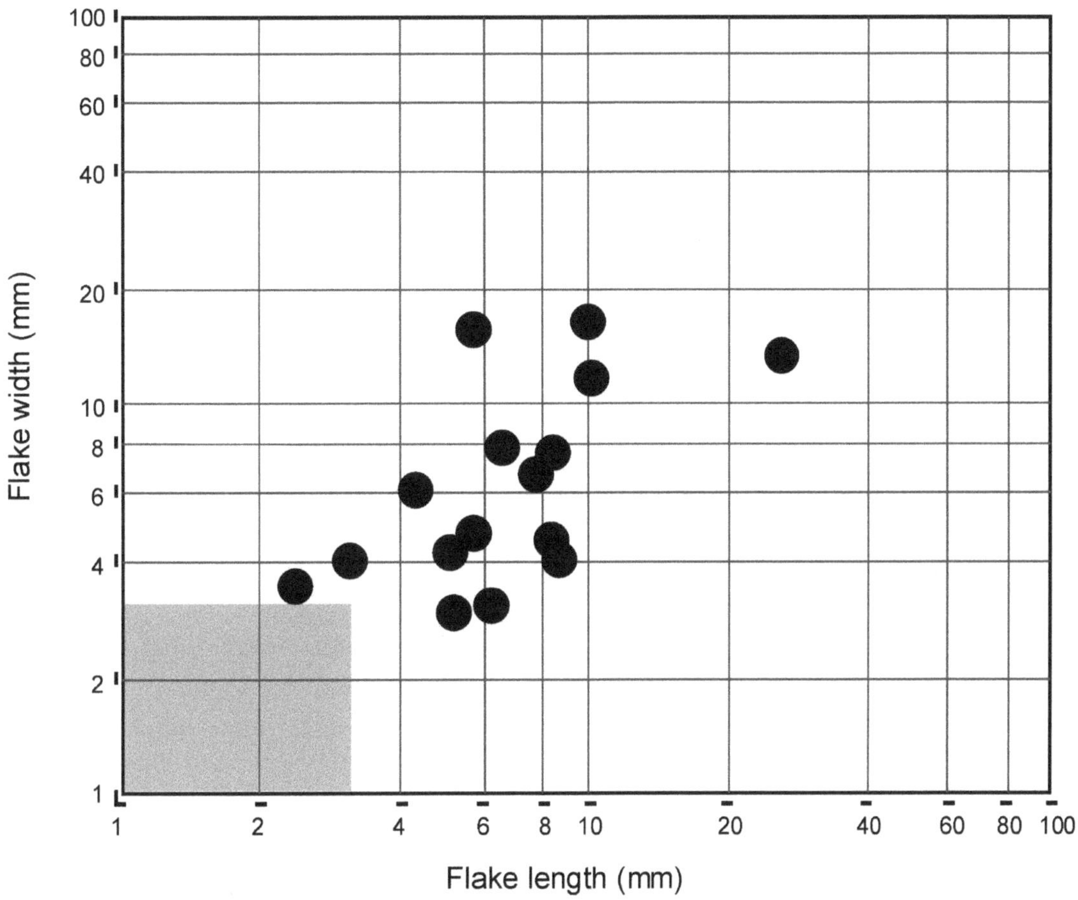

Figure 5. Scatterplot of the dimensions of stone artefacts in Y259. Grey area shows dimensions between 1 mm and 3 mm where on-site stone working debris is expected to fall

Discussion

The cave-floor deposit in Y259 provides important new information on three separate topics of interest for the natural and human history of the Southeastern Highlands. These are: (1) The nature of the 'original', pre-European mammal fauna; (2) Environmental conditions during the early Holocene; and (3) The history of aboriginal occupation.

The 'original' pre-European mammal fauna

This topic is the subject of a wider investigation involving many more sites and will be treated in full elsewhere. However, a number of important observations warrant mention in this context. As contended by Drummond (1963), the 'original' fauna (i.e. very late prehistoric into early historical times) of the Yarrangobilly Plateau recorded in Y259 was substantially more diverse than the fauna recorded by modern records from the surrounding bioregions (Table 3). Drummond's (1963) list of 14 mammal species from Y259 is a subset of those reported herein, with the exception of *Pseudomys australis* (see taxonomic comments above).

Hope (1976) listed 18 terrestrial mammals and three bats from five caves at Yarrangobilly, based on surface collections made by M. and J. Mendrum and R. Curtis in 1970-72. This also represents a subset of the more extensive list obtained from Y259, but with the addition of two larger marsupials (*Phascolarctos cinereus* and *Vombatus ursinus*) and one bat (*Chalinolobus morio*). Seebeck and Johnson (1980) noted a partial subfossil skull of *Potorous longipes* from a nearby cave at Yarrangobilly, and we have also collected remains of this species (as well as *P. tridactylus*) from other nearby sites. Taxa recorded from the surface of Y259 and other sites at Yarrangobilly but never recorded as living animals in the surrounding region include: *Sminthopsis murina, Isoodon obesulus, Dasyurus viverrinus, Rattus lutreolus, Pseudomys higginsi, Pseudomys oralis, Pseudomys novaehollandiae, Potorous tridactylus, Potorous longipes* and *Bettongia gaimardi*. Several of these taxa (*D. viverrinus, B. gaimardi*) are presumed extinct on the Australian mainland, while others (*I. obesulus, Potorous tridactylus, Pseudomys oralis, R. lutreolus*) persist only on the eastern slopes of the Great Divide and along the coast. *Sminthopsis leucopus* is found in Spit 1 of Y259 and near-surface or surficial contexts in nearby caves.

Sminthopsis murina is broadly distributed across southeastern Australia but is most commonly found in open forest, woodland and heathland. The species has been recorded above 1200 m asl in the Blue Mountains region, but is unknown from the Australian Alps, although it occurs in adjacent lowland habitats near Canberra. In near-coastal habitats, *S. murina* attains its highest population densities at mid-successional stages after fire (Fox 2008). *Sminthopsis leucopus* is similar in its general ecology to *S. murina* (Wakefield 1972b; Lunney et al. 1989), and replaces that species in near-coastal habitats along the southern New South Wales and Victorian coasts and in Tasmania. It occurs at up to 900 m asl on the coastal scarp west of Bega.

Of the two bandicoots, *Perameles nasuta* is still regionally present in montane forests, although it is not recorded in Spit S. Although *P. nasuta* occupies a wide range of habitats, foraging activity tends to focus on open, grassy patches. Dickman and Stoddart (2008:189) state that it is 'most abundant at elevations of less than 1000 m that receive an annual rainfall of more than 750 mm'. In contrast, the regionally extinct *Isoodon obesulus* prefers to forage in dense understorey and avoids more open habitat (Claridge 1988). Surviving populations in eastern Australia tend to be associated with sandy substrate and heaths in near-coastal contexts.

The lack of contemporary regional records of *Dasyurus viverrinus, Bettongia gaimardi* and the two *Potorous* species is unsurprising given the scant representation of these taxa in historic collections and their widespread recent declines. Where taxa occur so infrequently in the cave deposits, direct dating of bones may be necessary to accurately document their history of decline.

The area around Y259 once possessed a rich rodent community that included *Pseudomys fumeus, P. higginsi, P. novaehollandiae, P. oralis* (A and B) and *Mastacomys fuscus*. In common with most of southeastern Australia, historical rodent records from the region are almost exclusively of *Rattus fuscipes*, which is also by far the most abundant taxon in the surficial unit of Y259 and also in all deeper levels. Happold (1989) recorded it from all elevations up to 2200+ m and in all habitats including forests, heathlands and boulder fields, although a general preference for wetter gullies and denser understorey typifies its preferred habitats (Bennett 1993). At Happold's main study site at c. 1600 m, near Smiggins Hole, *R. fuscipes* was captured in approximately equal numbers to *A. swainsonii* and *M. fuscus*. The extreme dominance of *R. fuscipes* in Y259 is difficult to explain purely in terms of habitat characteristics. This percentage dominance may be a result of the lack or very low abundance of other rodents such as *Pseudomys oralis* and *P. novaehollandiae*, rather than unusually high local abundance of *R. fuscipes*. The former species are much more abundant at lower-altitude sites in the region, contributing up to 46% of rodents at nearby sites (Ford et al. In prep).

Pseudomys fumeus is a very rare component of the modern Yarrangobilly fauna. It has never been captured live despite targeted surveys, and only three recently dead animals found in 1998 prove its continued existence (Ford 1998). Throughout the past 10,000 years, it was probably the second most abundant member of the forest rodent fauna, until it succumbed to the impacts of European colonisation. This species seems to prefer habitats with prominent heath and legume species, and thus tends to favour locally drier sites within the forest matrix; however, it has been recorded occupying wetter gullies (Ford et al. 2003).

Pseudomys novaehollandiae is found across a wide latitudinal range (Tasmania to south Queensland) and in a broad range of habitats from coastal heaths to moist open eucalypt forest (Lock and Wilson 1999). Most extant populations have a preference for sandy substrates which tend to develop a heath ground cover, and populations appear to peak six months to several years after fire (Fox and Fox 1978; Haering and Fox 1997). The species is only locally abundant in a few locations in northeastern NSW and coastal populations south of Sydney are small, isolated and extinction prone. The only known surviving inland population occurs in dry heaths of the Hervey Range, east of Parkes.

Pseudomys oralis has not been captured within 500 km of Yarrangobilly. It has been recorded as a subfossil from sites as far south as Buchan Caves (Wakefield 1972b), but living animals are known only from the Barrington Tops area northwards to the Lamington Plateau (Jerry et al. 1998). It seems to favour generally wetter forest mosaics, but will frequent open drier habitat components (Read and Tweedie 1996; Smith and Quin 1996). The specific ecological attributes of the two tentative taxa identified from Y259 are not known.

Mastacomys fuscus still occurs in the Yarrangobilly region, but is only patchily present, and seemingly rare compared with abundances in the Y259 deposit. It has disappeared from lower-altitude sites throughout the Southern Tablelands in the past 200 years (Ford et al. In prep), only persisting in the higher altitudes of Kosciuszko National Park and nearby state forests, usually inhabiting boulderfields, heaths and grasslands (Green and Osborne 2003). It still occurs at lower altitudes and in a broader range of habitats in Victoria (Wallis 1992).

Pseudomys higginsi was first identified on mainland Australia by Wakefield (1972a). Subsequent work by Wakefield (1972b) and Hope (1976) recorded the species in various deposits of confirmed or inferred late-Pleistocene age, usually found in association with *Burramys parvus*. Flood (1980) dated the apparent disappearance of *P. higginsi* from the Buchan area to the interval 14,000-17,000 BP, coincident with the last phase of continental deglaciation (Barrows et al. 2001; Turney et al. 2006), and noted its apparent replacement at that time by the similar-sized *P. oralis*. The present record of *P. higginsi* from a surface context in Y259 suggests that this species persisted in the Southeastern Highlands until quite recent times. This has been confirmed by direct dating by AMS of bones of this taxon to less than 500 years old from other sites in the Yarrangobilly region (Breed and Ford 2007; Ford et al. In prep).

Tasmanian populations of *P. higginsi* occupy a wide range of habitats, including rainforests, wet sclerophyll forest, open sclerophyll forests and boulder-strewn habitats (Stoddart and Challis 1991). The habitat preferences of mainland populations of *P. higginsi* are unknown. However, given its history of post-glacial reduction in range and apparent retreat to higher-elevation regions, it is reasonable to infer a preference for relatively cooler habitats. Direct dating of numerous referred bones is needed to establish the precise timing of distributional changes and the mainland disappearance of this species, however it almost certainly persisted at higher altitudes into historical times.

Early Holocene climate and environment

The last deglaciation of the Australian Alps was complete by 16,000 years ago (Barrows et al. 2001; Turney et al. 2006). Despite a return to cooler conditions during the Younger Dryas

(c. 13,000-11,500 years ago; Peteet 1995; Alley and Clark 1999; Andres et al. 2003), trees are registered locally at upland sites during this period, and by 10,000 years ago the tree line had attained its current position across the region (Kershaw and Strickland 1989; Martin 1999). Peat was forming in moist sites at 1000 m asl before the Younger Dryas, and in higher-elevation mires during the earliest Holocene (Kershaw and Strickland 1989; Hope 2003), suggesting widespread, early stabilisation of slope habitats.

In most parts of the world, climate proxies show a prolonged period of warm conditions during the early Holocene, beginning around 10,000 years ago and continuing for at least 4000 years (Grootes et al. 1993; Stager et al. 1997; Wang et al. 2001; Zhou et al. 2004). Mean annual temperatures probably peaked at c. 9500 years ago, with values 3°C above modern values at mid-latitudes in the northern hemisphere (Foley et al. 1994; Morgan et al. 2002). The 'Holocene Optimum' is registered in many Australian climate proxy records. In the southeast Australian Alps, regional palynological records show an increase in wet sclerophyll forest elements, most notably the understorey shrub *Pomaderris* (the '*Pomaderris* Interval'), commencing c. 9700 years ago and probably continuing until c. 4500 years ago (Kershaw and Strickland 1989; Martin 1999; Hope et al. 2004; Kershaw et al. 2007). Rapid growth of mires in the region is observed through the same time interval (Hope 2003). Lakes in southeastern Australia register their highest levels for the Holocene slightly later, between 7600 and 5500 years ago (Bowler 1981; D'Costa et al. 1989; Chivas et al. 1993). This is coincident with evidence for reduced regional aeolian activity (Stanley and De Deckker 2002).

The ^{14}C determinations for Unit III of Y259 correspond with the peak of the 'Holocene Optimum', and with the time of establishment of wetter forest communities throughout the surrounding region. The faunal assemblage from this period has several noteworthy characteristics, including the presence of *Gymnobelideus leadbeateri* and *Perameles nasuta*, the relatively high abundance of small dasyurids, especially *Antechinus swainsonii* and *S. leucopus*, and the relative scarcity of two species, *Pseudomys higginsi* and *Cercartetus nanus*. A crude method was used to aid interpretation of the potential climatic implications of the observed faunal changes. We compared climate attributes produced by BIOCLIM for the Y259 site (Table 5) with climate profiles for six mammal species and for the sooty owl, based on known living populations (Bennet et al. 1991). The current climate of Y259 falls within the climatic range currently occupied by five of the six mammal species and the sooty owl, despite the fact that two of the mammals, *Isoodon obesulus* and *Pseudomys novaehollandiae*, have suffered widespread local extinctions over the past 150 years and their modelled climatic preferences probably are not representative of their full natural range. In contrast, the mountain pigmy possum (*Burramys parvus*), a species today found only above 1300 m asl in the Australian Alps bioregion, has a climatic profile that is significantly wetter and colder than the Y259 locality. This species occurs regionally in deposits of late-Pleistocene age down to much lower elevations (Wakefield 1972b; Ride et al. 1989), and its absence from Y259 is consistent with the notion that the earliest part of the excavated sequence samples the wetter, warmer conditions of the early Holocene, rather than the preceding colder conditions of the Last Glacial Maximum.

Gymnobelideus leadbeateri is recorded as a living animal only in eastern Victoria. Previously published records of *G. leadbeateri* outside its current range include the Broom Breccia at Wombeyan, and Pyramids Cave, Buchan (primarily in the older 'Reddish Fraction'; minor representation in the younger 'Whitish Fraction') (Wakefield 1972b). In both of these assemblages, *G. leadbeateri* occurs with *Burramys parvus* in deposits that are presumed to be of late-Pleistocene antiquity. One other occurrence is from Marble Arch, near Braidwood (Hall 1975), where it occurs in association with an entirely modern fauna, though from an undated sub-surface context. The immediate local environment around Marble Arch is wet sclerophyll forest. Although the occurrence of *G. leadbeateri* in the Broom Breccia and Pyramids 'Reddish Fraction' faunas seems somewhat anomalous, the bulk of the evidence for this species suggests an association with wetter forest communities and warmer (lower-elevation) sites. The presence

Table 5. Comparison of the contemporary climate of Y259, representing the Yarrangobilly Plateau, with the current climatic profiles of six mammal species and one owl species

Climate variable	Y259	*I. ob*[1]	*G. le*[2]	*B. pa*[3]	*M. fu*[4]	*P. no*[1]	*P. fu*[4]	*T. te*[5]
Temperature (ºC)								
Annual mean	9.4	+4.5	+.01	-4.2	Same	+4.6	+0.3	+1
Min of coldest period	-1.6	+5.7	+2	-2.7	Same	+5.4	+2	+2.9
Max of warmest period	24.1	+0.5	Same	-6.5	Same	+0.7	Same	Same
Precipitation (mm)								
Annual	1212	-260	+281	+325	+82	-71	Same	Same
Wettest quarter	390	-79	+64	+74	Same	-15	Same	Same
Driest quarter	213	-43	+29	+74	+12	-2	Same	Same
Coldest quarter	387	-88	+40	+48	Same	-108	Same	Same
Warmest quarter	215	-41	+40	+84	+24	Same	Same	-3

Y259 values are based on BIOCLIM modelling performed according to the protocols of Bennett et al. (1991) from which species climate profiles were taken. 'Same' indicates that the Y259 value falls between the 25th and 75th percentiles of a species' climate profile. Numbers indicate the difference between a species' relevant percentile and the site value (e.g. +4.5 indicates that the 25% percentile (cooler end of the range) of annual temperature for *Isoodon obesulus* sites is 4.5ºC higher than the site value for Y259, while the 75th percentile (warmer end of range) for *Burramys parvus* sites is 4.2ºC lower than Y259). Values which fall outside the recorded range for all sites for a species are underlined. Note that profiles for recently extinct species are skewed by incomplete sampling of potential climatic range. I. ob = *Isoodon obesulus*; G. le = *Gymnobelideus leadbeateri*; B. pa = *Burramys parvus*; M. fu = *Mastacomys fuscus*; P. no = *Pseudomys novaehollandiae*; P. fu = *Pseudomys fumeus*; T. te = *Tyto tenebriosca.*

[1]Recent (post-European) local extinction. [2]Natural local extinction during mid? Holocene. [3]Not yet recorded in Y259. [4]Still locally present around Y259. [5]Some evidence of occasional local occurrence

of *G. leadbeateri* in Unit II demonstrates that it survived on the Yarrangobilly Plateau until sometime after 9000 years ago.

The peak abundance of small dasyurids during the early Holocene might reflect regionally higher productivity of forest communities, possibly associated with warmer and longer growing seasons. All members of this group are strictly animalivorous and feed primarily on terrestrial and arboreal invertebrates, variably supplemented by worms. Two species in particular are considerably more abundant in Unit III than in Unit I. *Antechinus swainsonii* has a broad altitudinal and latitudinal range in southeastern Australia but seems to be most abundant in alpine heath and in open forest communities with dense understorey of ferns or shrubs. In Kosciuszko National Park, Happold (1989:222) found this species to extend to 2000 m asl and to occupy 'stream valleys and moist heathlands … with abundant leaf-litter and soft soil'. *Sminthopsis leucopus* is known from numerous extant populations in Tasmania and around the coastal margin of the Southeastern Highlands. However, co-occurrence at relatively high abundance with *S. murina*, such as in Unit III of Y259, is unknown among modern populations, again suggesting a potentially high-productivity environment. However such environmental conditions are presumably mirrored at many modern sites, so why the two species co-occur in this site remains enigmatic.

It is difficult to interpret the apparent change in local dominance of *Perameles nasuta* in lower spits to *Isoodon obesulus* in higher levels because we have no knowledge of the ecology of inland populations of *I. obesulus*. Furthermore, although no *P. nasuta* was found in higher levels of Y259, the species is still present in the region, although uncommon. The absence of *P. nasuta* from higher spits is presumably partly an artefact of small sample size, but the relative dominance of *I. obesulus* is borne out at other Yarrangobilly sites and suggests an increased coverage of heath communities through the later Holocene. A similar indicator of climate may be the reduction in relative abundance of *Pseudomys oralis* from lower to higher spits and the relative increase in *P. fumeus*. All these species are capable of inhabiting a wide range of habitats, but *P. fumeus* and *I. obesulus* tend to be relatively more abundant in drier heath components of forest habitats than their counterparts (Claridge 1988; Ford et al. 2003).

Pseudomys higginsi is present but scarce in Unit III (and II), and considerably more abundant in Spits S and 1 of Unit I. As noted above, the mainland population of this species probably favoured relatively cooler conditions.

Cercartetus nanus is also significantly less abundant in Unit III than in Units I-II. This species occupies a wide range of habitats in southeastern Australia, from rainforest to coastal heath, but is most abundant in habitats that provide abundant nectar, such as coastal *Banksia* woodland (Bowen and Goldingay 2000). It was not listed by Happold (1989) as a member of the Southern Alps mammal fauna, but it has been recorded as a rare animal in all local bioregions, including the Australian Alps. The shrubby components, such as stands of *Banksia*, that tend to favour *C. nanus* are present, though scattered, in the area surrounding Y259.

In summary, the mammal assemblage from Unit III of Y259 suggests a local environment that included significant areas of wet sclerophyll forest and probably featured higher overall productivity than present-day communities. Indicators of understorey structure and composition suggest a mosaic including both moist, dense patches, supporting populations of such taxa as *Antechinus swainsonii*, and more open vegetation, supporting high local populations of *Perameles nasuta*, *Pseudomys novaehollandiae* and *Sminthopsis* spp.

Aboriginal occupation of the Southeastern Highlands

The small collection of stone artefacts and archaeological faunal remains from Unit III of the Y259 deposit, dating to the period 9700-9120 cal. BP, more than doubles the previously known 4500-year history of occupation above 1000 m asl in the southeast Australian Alps. The small archaeological assemblage recovered to date provides only scant clues as to what might have drawn people on to the Yarrangobilly Plateau during the early Holocene. The component of the faunal assemblage that can be attributed to human activity suggests fairly opportunistic hunting activity, without the strong focus seen in the LGM assemblages from southwest Tasmania. While the assemblage does include the remains of red-necked wallabies, the preferred prey species of late-Pleistocene Tasmanians, there are approximately equal numbers of other prey taxa, including rock wallabies, possums and a bandicoot. This broader range of prey items is probably due in part to the more diverse local fauna compared with that available to the Tasmanian hunters. Nevertheless, some prey selectivity is also apparent in the Y259 assemblage, with a clear preference for medium-sized to large mammals over the numerous small vertebrates that occurred locally around the site.

At present, it is unclear whether use of the site continued through the period of deposition of Unit II. No artefacts were recovered from Spit 5 of Unit II and only two bones show evidence of burning, and in neither case is it intense. However, the fauna from this period does contain a higher proportion of medium-sized mammals than the samples from Unit I. Further insight into this question may come from analysis of the sample from Spit 6.

There is no evidence whatsoever for human activity at the site during deposition of Unit I. The large faunal assemblage contains relatively few medium-sized mammals, displays no evidence of burning, and no stone artefacts were recovered within the excavated area or noted across the wider surface of the deposit. If people visited the site at all during the recent past, they left little or no record of their activities. Why this should be so is not clear. The cave provides an obvious shelter from both cold and hot temperatures and at least in recent years, it clearly harbours seasonal aggregations of bogong moths, a significant food resource for local aboriginal people during the early historic period.

The site of Birrigai, located at 730 m asl and approximately 55 km by air to the northeast of Yarrangobilly, provides evidence of a human presence in the wider region back at least to 25,000 years ago (Flood et al. 1987), at which time the world was firmly in the grip of the

Last Glacial Maximum (Barrows et al. 2002). Birrigai at that time was very likely above the local tree line and surrounded by grass-covered slopes affected by periglacial processes. People continued to visit Birrigai on an occasional basis through the remainder of the Pleistocene and into the Holocene, with no obvious change to indicate any major human response to the return of woodland to the valley or to the warmer conditions of the Holocene Optimum. The past 3000 years saw more frequent visitation to Birrigai. Unfortunately, the Birrigai deposit provides a poor environment for organic preservation and there is little direct evidence of economic activities through most of its long history of occupation.

Better organic preservation, including both animal and plant remains, is present in the inner chamber of Cloggs Cave, situated at 76 m asl on the southern flanks of the southeast Australian Alps. As in Y259, the faunal remains are clearly of mixed origin, including some derived from human activity and others introduced by owls and other predators. Although these components could not be distinguished with any certainty, Flood (1987:267) does note that 'the bone nearer the centre of the cave includes more large mammals such as macropodids, and is more likely to represent human dietary remains'. Judging from the relatively small quantities of discarded stone artefacts and larger mammal bone, Cloggs Cave was visited on an occasional basis only, between c. 21,000 cal. BP and c. 9800 cal. BP. After this date, no fires were lit in the inner chamber of Cloggs Cave and sedimentation slowed dramatically. Flood (1980:268) interpreted this as a possible response to the post-glacial amelioration of climate, perhaps encouraging a 'move out of the cave into the open'. Within the past 1000 years or so, people made limited use of the outer chamber of Cloggs Cave, but did not reoccupy its inner chamber.

It is interesting to note that the last definite evidence of human activity in Y259 coincides closely in time with abandonment of the Cloggs Cave inner chamber, and more generally, with the regional expansion of wet sclerophyll forest in response to the Holocene Optimum. On a regional scale, we also note with interest that the earliest evidence for reoccupation of the southeast Australian Alps, at around 4500 years ago, is broadly coincident with the termination of the *Pomaderris* Interval, presumably signalling the regional decline of wetter forest communities. If these observations gain further support, we might begin to ponder whether the history of human use of the southeast Australian Alps through the Holocene was driven in large part by the relative ease of access through vegetation communities changing in response to global climatic shifts.

Further excavation in Y259 clearly represents a high priority for archaeological research in the southeast Australian Alps. If the archaeological deposit in Y259 is deeper, continuing below the level of the limestone rubble, then it seems certain to extend back into the terminal Pleistocene. On the other hand, if the full sequence has already been documented for Y259, there is a strong case for recovering a larger sample of cultural remains representing the initial period of the Holocene Optimum and to clarify the timing and extent of any subsequent human activity.

Conclusion

Cave Y259 contains an important faunal sequence and significant cultural material that together provide a fascinating window into the time of the Holocene Optimum in the Southern Australian Highlands. People visited the cave on several occasions over the time interval 9700-9120 cal. BP, moving through a landscape that contained significant areas of wet sclerophyll forest, and left behind stone artefacts and the burnt and fragmented remains of their mammalian prey. After that time, the cave appears to have been unoccupied or visited very infrequently. However, many generations of owls continued to use the cave, leaving behind a rich record of local mammalian communities. More detailed analysis of the entire excavated sequence will shed light on the time of regional extinction of several species of key interest, including *Gymnobelideus leadbeateri* and *Pseudomys higginsi*.

Acknowledgements

We thank Rohan Bilney and Angela Frost for assistance with cave excavations at Yarrangobilly and cave managers Peter Bell and George Bradford for their interest and co-operation throughout the project. Angela Frost also assisted with sorting and identifications, and prepared the site diagrams. Aplin very belatedly thanks Geoff Hope for first suggesting that he study faunal remains as a way of keeping him out of even greater mischief. Excavation was carried out under a regional consent issued on September 26 2005 from NSW National Parks and Wildlife Service, South West Slopes Region, Tumut.

References

Alley, R.B. and P.U. Clark 1999. The deglaciation of the Northern Hemisphere: A global perspective. *Annual Review of Earth and Planetary Science* 27:149-182.

Andres, M.S., S.M. Bernasconi, J.A. McKenzie and U. Röhl 2003. Southern Ocean deglacial record supports global Younger Dryas. *Earth and Planetary Science Letters* 216:515-524.

Archer, M. 1981. Results of the Archbold Expeditions. No. 104. Systematic revision of the marsupial dasyurid genus *Sminthopsis* Thomas. *Bulletin of the American Museum of Natural History* 168:61-224.

Asmussen, B. 2009. Intention or incidental thermal alteration? Analysing site occupation through burnt bone. *Journal of Archaeological Science* 36:528-536.

Barrows, T.T., J. O. Stone, L.K. Fifield and R.G. Cresswell 2001. Late Pleistocene glaciation of the Kosciuszko Massif, Snowy Mountains, Australia. *Quaternary Research* 55:179–189.

Barrows, T.T., J.O. Stone, L.K. Fifield and R.G. Cresswell 2002. The timing of the Last Glacial Maximum in Australia. *Quaternary Science Reviews* 21:159-173.

Bennett, A.F. 1993. Microhabitat use by the long-nosed potoroo, *Potorous tridactylus*, and other small mammals in remnant forest vegetation of south-western Victoria. *Wildlife Research* 20:267-285.

Bennett, S., R. Brereton, I. Mansergh, S. Berwick, K. Sandiford and C. Wellington 1991. Enhanced greenhouse climate change and its potential effect on selected Victorian fauna. Arthur Rylah Institute Tech. Rep. No. 123. Department of Conservation and Environment, Victoria.

Bilney, R., R. Cooke and J. White 2006. Change in the diet of sooty owls (*Tyto tenebricosa*) since European settlement: from terrestrial to arboreal prey and increased overlap with powerful owls. *Wildlife Research* 33:17-24.

Bowen, M. and R. Goldingay 2000. Distribution and status of the eastern pygmy possum (*Cercartetus nanus*) in New South Wales. *Australian Mammalogy* 21:153-164.

Bowler, J.M. 1981. Australian Salt Lakes. *Hydrobiologia* 82:431-444.

Breed, W. and F. Ford 2007. *Native Mice and Rats*. Collingwood: CSIRO Publishing.

Bronk Ramsey, C. 2001. Development of the radiocarbon calibration program OxCal. *Radiocarbon* 43:355-363.

Broome, L.S., 2001. Intersite differences in population demography of Mountain Pygmy-possums *Burramys parvus* Broom (1987-1998): Implications for metapopulation conservation in Kosciuszko National Park. *Biological Conservation* 102:309-323.

Chivas, A.R., P. De Deckker, J. A. Cali, A. Chapman, E. Kiss and J.M.G. Shelley 1993. Coupled stable-isotope and trace-element measurements of lacustrine carbonates as palaeoclimatic indicators. American Geophysical Monographs 78:113-121.

Claridge, A.W. 1988. Diet and ecology of the Southern Brown and Long-nosed Bandicoots in South-eastern New South Wales. Honours Thesis, Department of Forestry, the Australian National University, Canberra.

Colhoun, E.A. and G. van de Geer 1986. Holocene to middle last glaciation vegetation history at Tullabardine Dam, western Tasmania. *Proceedings Royal Society of London* B229:177-207.

Colhoun, E.A., G. van de Geer, S.J. Fitzsimons and L.E. Heusser 1994. Terrestrial and marine records of the last glaciation from western Tasmania: Do they agree? *Quaternary Science Reviews* 13:293-300

Cosgrove, R. 1999. Forty-two degrees south: the archaeology of late Pleistocene Tasmania. *Journal of World Prehistory* 13:357-402.

Cosgrove, R., J. Allen and B. Marshall 1990. Palaeo-ecology and Pleistocene human occupation in south central Tasmania. *Antiquity* 64:59-78.

D'Costa D.M., P. Edney, A.P. Kershaw and P. De Deckker 1989. Late Quaternary palaeoecology of Tower Hill, Victoria, Australia. *Journal of Biogeography* 16:461-482.

Dickman, C. and M.D. Stoddart 2008. *Perameles nasuta*. In S. Van Dyke and R. Strahan (eds), *Mammals of Australia*, pp189-190. Sydney: Reed Books.

Dodson, J.R., T. De Salis, C.A. Myers and A.J. Sharp 1994. A Thousand Years of Environmental Change and Human Impact in the Alpine Zone at Mt. Kosciuszko, New South Wales. *Australian Geographer* 25:77-87.

Drummond, D.C. 1963. Bone deposits in caves. *Journal of the Sydney University Speleological Society* 6:29-42.

Flood, J. 1980. *The Moth Hunters. Aboriginal Prehistory of the Australian Alps.* Canberra: Australian Institute of Aboriginal Studies.

Flood, J., B. David, J. Magee and B. English 1987. Birrigai: a Pleistocene site in the southeastern highlands. *Archaeology in Oceania* 22:9-26.

Foley, J.A., J.E. Kutzbach, M.T. Coe and S. Levis 1994. Feedbacks between climate and boreal forests during the Holocene epoch. *Nature* 371:52-54.

Ford, F. 1998. *The smoky mouse in the Nullica region and Kosciuszko National Park: winter and spring 1998.* Report, New South Wales National Parks and Wildlife Service, Queanbeyan.

Ford, F., A. Cockburn and L. Broome 2003. Habitat preference, diet and demography of the smoky mouse, *Pseudomys fumeus* (Rodentia: Muridae), in south-eastern New South Wales. *Wildlife Research* 24:417-432.

Ford et al. In prep.

Fox, B.J. 2008. *Sminthopsis murina*. In S. Van Dyke and R. Strahan (eds), *Mammals of Australia*, pp152-154. Sydney: Reed Books.

Fox, B.J. and M.D. Fox. 1978. Recolonisation of coastal heath by *Pseudomys novaehollandiae*. *Australian Journal of Ecology* 3:447-465.

Fox, B.J., D.G. Read, E. Jeffereys and J. Luo 1994. Diet of the Hastings River mouse (*Pseudomys oralis*). *Wildlife Research* 21:491-505.

Green, K. and W.S. Osborne 2003. The distribution and status of the Broad-toothed Rat *Mastacomys fuscus* (Rodentia: Muridae) in NSW and the ACT. *Australian Zoologist* 32:229-237.

Green, D., G. Singh, H. Polach, D. Moss, J. Banks and E.A. Geissler 1988. A fine-resolution palaeoecology and palaeoclimatology from southeastern Australia, *Journal of Ecology* 76:790-806.

Grootes, P.M., M. Stuiver, J.W.C. White, S. Johnsen and J. Jouzel 1993. Comparison of oxygen isotope records from GISP2 and GRIP Greenland ice cores. *Nature* 366:552-554.

Haering, R. and B.J. Fox 1997. Habitat use by sympatric populations of *Pseudomys novaehollandiae* and *Mus domesticus* in coastal heathland. *Australian Journal of Ecology* 22:69-87.

Hall, L.S. 1975. A recent bone deposit at marble arch, N.S.W. *Proceedings of the 10th biennial conference of the Australian Speleological Federation*, pp35-44.

Happold, D.C.D. 1989. Small mammals of the Australian Alps. In R. Good (ed.), *The scientific significance of the Australian Alps*, pp221-239. Canberra: Australian Alps National Parks Liaison Committee.

Hope, G. 2003. The mountain mires of southern New South Wales and The Australian Capital Territory: their history and future. In J. Mackay & Associates (eds), *Celebrating Mountains. Proceedings of an International Year of Mountains Conference*, pp67-79. Canberra: Australian Alps Liaison Committee.

Hope, G., A.P. Kershaw, S. Van der Kaars, S. Xiangjun, P-M. Liew, L.E. Heusser, H. Takahara, M. McGlone, N. Miyoshi and P.T. Moss 2004. History of vegetation and habitat change in the Austral-Asian region. *Quaternary International* 118-119:103-26.

Hope, J. H. 1976. Quaternary fossil mammal localities in Southeastern New South Wales and Eastern Victoria. Unpublished manuscript.

Jerry, D.R., T.A. Dow, M.S. Elphinstone and P.R. Baverstock 1998. Historical and contemporary maternal population structuring in the endangered Hastings River mouse (*Pseudomys oralis*). *Conservation Biology* 12:1017-1022.

Jones, R. 1995. Tasmanian archaeology: Establishing the sequences. *Annual Review of Archaeology* 24:423-446.

Kamminga, J. 1992. Aboriginal settlement and prehistory of the Snowy Mountains. In B. Scougall (ed.), *Cultural Heritage of the Australian Alps*, pp101-124. Canberra: Australian Alps National Parks Liaison Committee.

Kamminga, J. 1995. Prehistory of the Snowy Mountains, southeastern Australia. In E. Johnson (ed.), *Ancient peoples and landscapes*, pp153-171. Lubbock: Museum of Texas Tech University.

Kavanagh, R.P. and M. Murray 1996. Home range, habitat and behaviour of the Masked Owl *Tyto novaehollandiae* near Newcastle, New South Wales. *Emu* 96:250-257.

Kershaw, A.P. and K.M. Strickland 1989. The development of alpine vegetation on the Australian mainland. In R. Good (ed.), *The scientific significance of the Australian Alps*, pp113-126. Canberra: Australian Alps National Parks Liaison Committee.

Kershaw, A.P., G.M. McKenzie, N. Porch, R.G. Roberts, J. Brown, H. Heijnis, M.L. Orr, G. Jacobsen and P.R. Newall 2007. A high-resolution record of vegetation and climate through the last glacial cycle from Caledonia Fen, southeastern highlands of Australia. *Journal of Quaternary Science* 22:481-500.

Kiernan, K., R. and D. Ranson 1983. New evidence from Fraser Cave for glacial age man in south-west Tasmania. *Nature* 301:28-32.

Kusmer, K.D. 1990. Taphonomy of owl pellet deposition. 64:629-637.

Ladd, P.G. 1979. Past and present vegetation on the Delegate River in the highlands of eastern Victoria. II. Vegetation and climatic history from 12,000 B.P. to present. *Australian Journal of Botany* 27:185-202.

Lock, M.L. and B.A. Wilson. 1999. The distribution of the New Holland mouse (*Pseudomys novaehollandiae*) with respect to vegetation near Anglesea, Victoria. *Wildlife Research* 26:565-577.

Lourandos, H. 1997. *Continent of hunter-gatherers: new perspectives in Australian prehistory.* Cambridge: Cambridge University Press.

Lundie-Jenkins, G. 1993. The diet of the Sooty owl *Tyto tenebricosa* in the Blue Mountains, N.S.W. *Emu* 93:124-127.

Lunney, D., M. O'Connell, J. Sanders and S. Forbes 1989. Habitat of the white-footed dunnart *Sminthopsis leucopus* (Gray) (Dasyuridae: Marsupialia) in a logged, burnt forest near Bega, New South Wales. *Austral Ecology* 14: 335-344.

McKenzie, G.M. 1997. The late Quaternary vegetation history of the south-central highlands of Victoria, Australia. I. Sites above 900 m. *Australian Journal of Ecology* 22:19-36.

McKenzie, G.M. 2002. The late Quaternary vegetation history of the south-central highlands of Victoria, Australia. II. Sites below 900 m. *Austral Ecology* 27:32-54.

Macphail, M.K. and G.S. Hope 1985. Late Holocene mire development in montane southeastern Australia: a sensitive climatic indicator. *Search* 15:344-349

Mansergh, I., B. Baxter, D. Scotts, T. Brady and D. Jolley 1990. Diet of *Burramys parvus* (Marsupialia: Burramyidae) and other small mammals in the alpine environment at Mt Higginbotham, Victoria. *Australian Mammalogy* 13:167-77.

Martin, A.R.H. 1999. Pollen analysis of Digger's Creek Bog, Kosciuszko National Park: vegetation history and tree-line change. *Australian Journal of Botany* 47:725-744.

Mooney, S.D., J.R. Watson and J.R. Dodson 1997. Late Holocene environmental change in an upper montane area of the Snowy Mountains, NSW. *Australian Geographer* 28:185-200.

Morgan, V., M. Delmotte, T. van Ommen, J. Jouze, J. Chappellaz, S. Woon, V. Masson-Delmotte and D. Raynaud 2002. Relative timing of deglacial events in Antarctica and Greenland. *Science* 297:1862-1864.

Morton, S.R., M. Happold, A.K. Lee and R.E. Macmillen 1977. The diet of the Barn owl, *Tyto alba*, in South-western Queensland. *Australian Wildlife Research* 4:91-97.

Mulvaney, J. and J. Kamminga 1999. *Prehistory of Australia.* Sydney: Allen and Unwin.

Nicholson, R.A. 1993. A morphological investigation of burnt animal bone and an evaluation of its utility in archaeology. *Journal of Archaeological Science* 20:411-428.

Peteet, D. 1995. Global Younger Dryas. *Quaternary International* 28:93-104.

Read, D.G. and T.D. Tweedie 1996. Floristics of habitats of *Pseudomys oralis* (Rodentia: Muridae). *Wildlife Research* 23:485-493.

Ride, W.D.L., G. Taylor, P.H. Walker and A.C. Davis 1989. Zoological history of the Australian Alps – the mammal fossil-bearing deposits of the Monaro. In R. Good (ed.), *The scientific significance of the Australian Alps*, pp79-110. Canberra: Australian Alps National Parks Liaison Committee.

Seebeck, J.H. and P.G. Johnston 1980. *Potorous longipes* (Marsupialia: Macropodidae); a new species from eastern Victoria. *Australian Journal of Zoology.* 28:119-34.

Stager, J.C., B. Cumming and L. Meeker 1997. A high-resolution 11,400-yr diatom record from Lake Victoria, East Africa. *Quaternary Research* 47:81-89.

Smith, A.P. and D.G. Quin 1996. Microhabitat requirements of the Hastings River mouse (*Pseudomys oralis*) (Rodentia: Muridae). New South Wales National Parks and Wildlife Service.

Stanley, S. and P. De Deckker 2002. A Holocene record of allochthonous, aeolian mineral grains in an Australian alpine lake; implications for the history of climate change in southeastern Australia. *Journal of Paleolimnology* 27:207-219.

Stiner, M.C., S.L. Kuhn, S. Weiner and O. Bar-Josef 1995. Differential burning, recrystallization, and fragmentation of archaeological bone. *Journal of Archaeological Science* 22:223-237.

Stoddart, D.M. and G. Challis 1991. The habitat and field biology of the long-tailed mouse (*Pseudomys higginsi*). Tasmanian Forest Research Council. Research Report No. 6.

Thackway R. and I.D. Creswell 1995. An Interim Biogeographic Regionalisation for Australia: a framework for establishing the national system of reserves, Version 4.0. Canberra: Australian Nature Conservation Agency.

Turney, C.S.M., S. Haberle, D. Fink, P. Kershaw, M. Barbetti, T.T. Barrows, M. Black, T.J. Cohen, T. Corrège, P.P. Hesse, Q. Hua, R. Johnston, V. Morgan, P. Moss, G. Nanson, T. Vanommen, S. Rule, N.J. Williams, J.-X. Zhao, D. D'Costa, Y.-X. Feng, M. Gagan, S. Mooney and Q. Xia 2006. Integration of ice-core, marine and terrestrial records for the Australian Last Glacial Maximum and Termination: a contribution from the OZ INTIMATE group. *Journal of Quaternary Science* 21:751-761.

Wakefield, N.A. 1972a. Studies in Australian Muridae: review of *Mastacomys fuscus*, and description of a new subspecies of *Pseudomys higginsi*. *Memoirs of the National Museum of Victoria* 33:15-31.

Wakefield, N.A. 1972b. Palaeoecology of fossil mammal assemblages from some Australian Caves. *Proceedings of the Royal Society of Victoria* 85:1-26.

Wallis, R.L. 1992. The broad-toothed rat (*Mastacomys fuscus*) in Dandenong Ranges National Park - a colony located in regenerating forest. *Victorian Naturalist* 109:177-178.

Wang, Y.J., H. Cheng, R.L. Edwards, Z.S. An, J.Y. Wu, C.-C. Shen and J.A. Dorale 2001. A high-resolution absolute-dated late Pleistocene monsoon record Hulu Cave, China. *Science* 294:2345-2348.

Zhou, W., W. Yu, A.J.T. Jull, G. Burr, J.Y. Xiao, X. Lu and F. Xian 2004. High-resolution evidence form southern China of an early Holocene optimum and a mid-Holocene dry event during the last 18,000 years. *Quaternary Research* 62:39-48.

12

Holocene lowland vegetation change and human ecology in Manus Province, Papua New Guinea

Matthew Prebble

Department of Archaeology and Natural History, College of Asia and the Pacific, Australian National University, Canberra, Australia

matthew.prebble@anu.edu.au

Jean Kennedy

Department of Archaeology and Natural History, College of Asia and the Pacific, Australian National University, Canberra, Australia

Wendy Southern

First Assistant Secretary, Department of the Prime Minister and Cabinet, Canberra, Australia

Introduction

This paper concerns evidence for past human impacts on the environment in the lowland tropical New Guinea region. Against a background regional overview, we consider two sequences, one archaeological, the other palaeoecological, from opposite ends of Manus Island, the largest island of the Admiralty Islands that now constitute Manus Province, Papua New Guinea. Contrasts in these local sequences prevent their easy alignment with grand narratives of regional prehistory. We show instead that closer examination of local contexts, especially the nature of agroecosystems, gives useful insights that help to disentangle natural processes of forest vegetation change and the effects of human activities. We consider aspects of the ecology of the tree genus *Calophyllum* L. (Clusiaceae), which occurs in both sequences, to assess the possibility of a human role in the dynamics of forest dominated by *Calophyllum euryphyllum* Lauterb. (Clusiaceae).

I. People and environment in the New Guinea region: An overview

Palaeoecological context

In the New Guinea region, the late-Quaternary environment since humans arrived 40,000-60,000 years ago was affected by altitudinal fluctuations of vegetation zonation in the highlands and sea-level changes around the coasts, both linked to the glacial cycle, with significant changes around the Last Glacial Maximum and less dramatic adjustments consequent on the high-sea-level stand in the mid Holocene.

Ecological mechanisms for vegetation change on tropical islands include competing species interactions, abrupt natural events (such as volcanic ash falls, tsunamis or cyclones), climate change, changing terrestrial insularity associated with geological activity and eustatic or isostatic sea-level fluctuations, and human interference. Palaeoecological data, particularly spanning the Holocene, have established partial chronological frameworks for these processes by recording ecological trends in different local contexts. Fine-scale vegetation changes can be measured from subfossil proxies (e.g. microfossil and macrobotanical analyses) and then used to infer conditions under which certain plant taxa have responded to disturbance. Similarly, palaeoclimatic patterns can be inferred from a number of subfossil, chemical and sedimentary proxies from the same archives. Surprisingly little evidence of this kind has been identified from the northern New Guinea lowlands and adjacent islands. Here we summarise the palaeoecological context for past vegetation changes in the region.

The Pleistocene

In New Guinea, investigating the role of humans in vegetation change is complicated by the lack of palaeoecological records that show changes before and after human colonisation, about 40,000-60,000 yr BP. Palynological records from mainland New Guinea that adequately represent these shifts have been reviewed recently by Hope (2007, 2009). The sites are mostly in basins in the high-elevation New Guinea ranges (Haberle 1998; Hope 2009). Archaeological evidence suggests people must have been in these areas since around 40,000 yr BP but sustained impacts on forest vegetation did not occur until around 30,000 yr BP at Kosipe and around 21,000 yr BP at Haeapugua.

Despite the growing evidence of Pleistocene occupation of the lowlands and coasts of New Guinea and its offshore islands (Allen 2003; O'Connor and Chappell 2003; O'Connell and Allen 2004), little palaeoecological evidence is available. Hope and Tulip (1994) summarised the main complicating factors that make the lowland tropics a challenge for palaeoecologists. Because rapid organic accumulation rates of between 10 cm and 80 cm in 100 years are not uncommon for both lakes and swamps, sections of peat or organic muds rarely exceed Holocene ages (Walker and Chen 1987; Hope and Tulip 1994; Osborne et al. 1996). Hope and Tulip (1994) point out that this problem is compounded by the low deposition rate of pollen in the humid tropics (after Kershaw and Hyland 1975). High local floristic diversity with a bias towards entomophilous and low pollen-producing taxa is a characteristic of the lowland tropical floras. Even when pollen preservation is high and deposition is continuous, many New Guinea palaeoecological records show marked transitions in forest vegetation during the Pleistocene but reveal no indication of ecological forcing factors, disturbance or anything else.

The pollen record from Hordorli, a swamp basin at 780 m altitude in the Cyclops Mountains near Lake Sentani, West Papua/Irian Jaya, Indonesia (Figure 1), is unique in that it provides a 50,000-year record of lowland vegetation change (Hope and Tulip 1994; Hope 1996). The suite of palaeoecological evidence that should mark initial human colonisation,

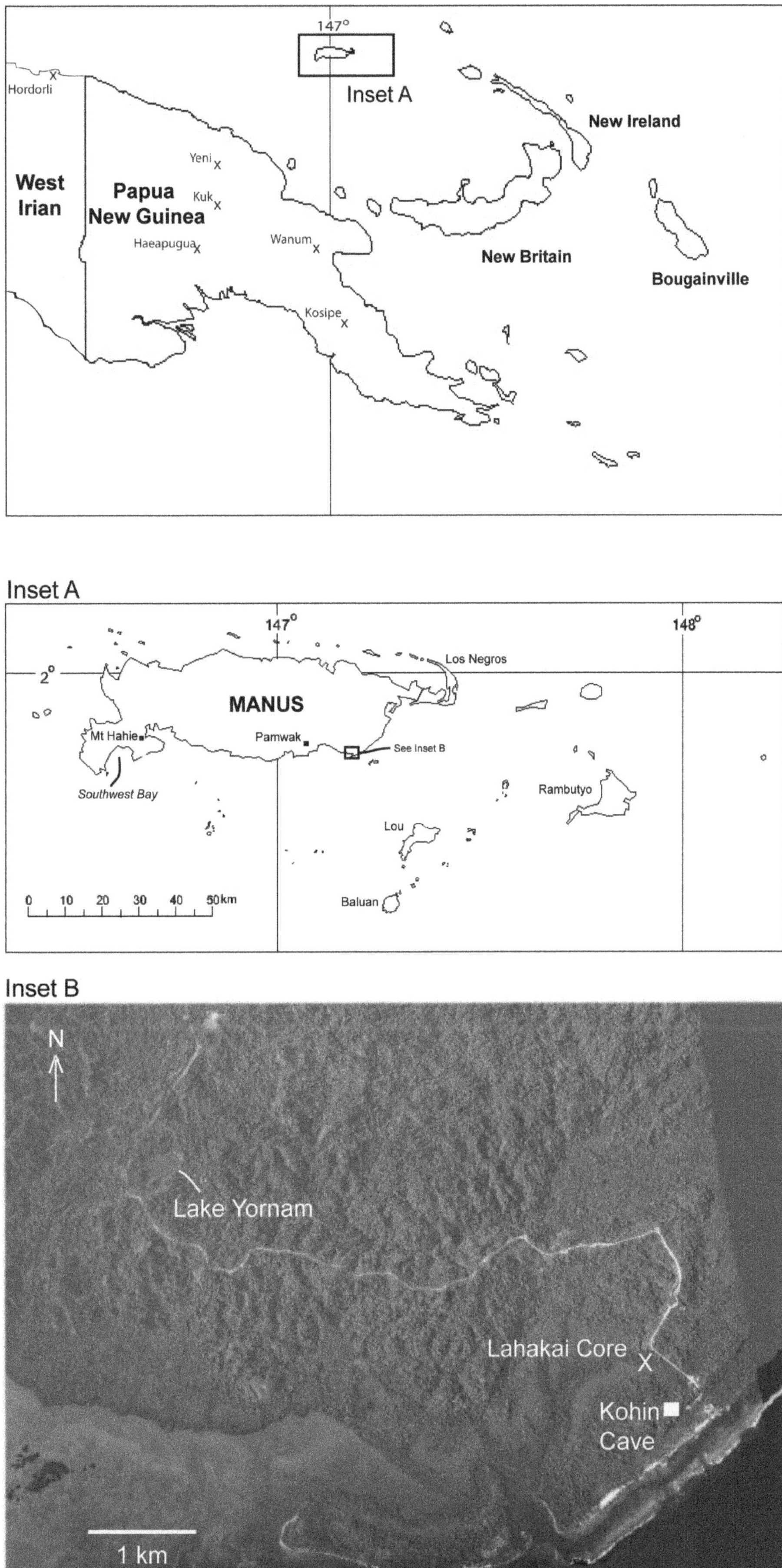

Figure 1. Map of Papua New Guinea, showing sites, the location of Manus Province and Lahakai Swamp in southeast Manus

such as increased fire frequency and increases in the abundance of secondary vegetation, is not conclusively represented at Hordorli until the Holocene. The Hordorli record also shows that closed tropical and lower montane forests have continuously occupied many areas despite human occupation. Hope and Haberle (2005) suggest that the lack of fire signals recorded at sites such as Hordorli may represent delayed intensive occupation of wetter sites.

There are no palaeoecological archives that span the Pleistocene from the tropical New Guinea islands. Given that the flora of these islands represents a subset of the mainland, there is no reason to presume that past vegetation change would be very different. Hope and Haberle (2005) suggest that the coastal lowland catchments of the Popondetta, Markham, upper Ramu and Sepik areas may have been drier during the Pleistocene than at present, dominated by *Nauclea* woodlands and possibly subject to natural fires, further complicating any palaeoecological signature of human presence. Such a model of vegetation change may also apply to the New Guinea islands.

The Holocene

Human influence on New Guinea vegetation change becomes more apparent in palaeoecological records of the Holocene, and can be correlated with archaeobotanical evidence for the beginning of plant domestication and agriculture (Yen 1996; Denham et al. 2004; Fairbairn 2005; Haberle 2007). Again, records are concentrated in the highlands of Papua New Guinea. Haberle and David (2004) have argued that increases in charcoal particles and *Casuarina* pollen, a secondary forest indicator, represented in a number of New Guinea highland palaeoecological records may signal the expansion of active and fallow cultivation areas during the mid Holocene.

In the coastal tropics of New Guinea, little palaeoecological research has concentrated on resolving questions of subsistence practice. Most pollen records have been retrieved in an effort to map the extent of mangrove sediments and the response of vegetation to coastal progradation and sea-level change. Mangrove sediments have been found hundreds of kilometres upriver in the Ramu and Sepik (Swadling and Hope 1992; Chappell 2005), indicating that a large estuary existed in the early mid Holocene before coastal progradation formed the present backswamp complex. Rapid siltation and coastal progradation has been demonstrated from mangrove sediments and geomorphological features found along the southern coast of New Guinea (Ellison 2005) and along the Fly-Digul platform, suggesting that the available land area for human occupation was limited during the early Holocene (Chappell 2005).

The pollen record from Lake Wanum, a lowland tropical lake in the Markham Valley, northern Morobe Province, Papua New Guinea (Figure 1) (Garrett-Jones 1979), provides the most continuous Holocene record so far of lowland vegetation change for mainland New Guinea. The lake has a rapid accumulation rate of organic-rich sediment during the Holocene. The record shows that an open sedge swamp surrounded by riparian forest/woodland persisted throughout the Holocene with an increase in dry-land grassland, secondary forest (including *Nauclea* and *Macaranga*) and burning after about 5500 yr BP.

In the Hordorli pollen record, charcoal particles first appear after 10,900 yr BP, and probably indicate a minor level of human activity. An increase in secondary forest species after 7000 yr BP may be a result of minimal human activity or, as Hope and Tulip (1994) suggest, may be the result of warmer conditions and shorter tree life in the lower montane forests.

The Wanum and Hordorli records both fail to provide any direct palaeoecological evidence for the initiation of root/tuber cultivation complexes and tree crops. Garrett-Jones (1979:295, 329) identified taro (*Colocasia esculenta*) pollen in the basal pollen zone of a 20 m deep core (core LWII), dated to around 9000 yr BP (identification confirmed by Haberle 1995). This represents the earliest subfossil record of *C. esculenta* from New Guinea, but the vegetation represented by the sample containing it appears to be a sedge swampland with riparian forest and grasslands present on the dryland slopes. Prebble et al. (In prep.) consider the early Holocene *C. esculenta* to represent most likely a wild form present as part of the natural distribution in marginal swampy habitats, or less likely, a feral population escaped from earlier human introductions.

Human prehistory of the northern New Guinea lowlands and adjacent islands

People reached New Britain and New Ireland around 35,000 yr BP, and Manus more than 20,000 yr BP. Although lowered Pleistocene sea levels never brought the islands of the Bismarck Archipelago significantly closer to each other or to the main Sahul landmass, New Britain and New Ireland could be reached by crossing between intervisible islands, distances shorter than those already accomplished through Wallacea. Manus, in contrast, has always been a very small target, far below the horizon from any direction.

Excavated sites are too few and scattered to provide a clear picture of the earliest human phase in the Bismarck region. The New Ireland Pleistocene sites suggest small groups, mobility, foraging for resources of land and sea, and a mixed diet (Gosden 1993, 1995; Allen 2000, 2003; Specht 2005).

In Manus, the basal occupation of the Pamwak rock shelter may be significantly older than the oldest ^{14}C date of 21,000 yr BP. Carbonised *Canarium* nut shell and faunal remains dating to the terminal Pleistocene have been interpreted to represent human introductions from mainland New Guinea (Fredericksen et al. 1993). Undated, but probably also of Pleistocene age, are flaked stone tools salvaged from southwest Manus open sites destroyed by logging (Kennedy et al. 1991; Kennedy 1992, 1997, 2002; Ambrose 2002a).

By the end of the Pleistocene, an emergent pattern of movement of resources across the Bismarcks suggests that people had begun modifying their environments and making use of inter-island connections to supplement locally available resources. Obsidian and animal translocations suggest people made interisland crossings between mainland New Guinea, New Britain, New Ireland and Manus (Flannery and White 1991; Gosden 1992; Allen and Gosden 1996; Heinsohn 2003; Specht 2005).

Archaeological sites of the early to middle Holocene in the Bismarcks have yielded material from wider catchments than before, and in larger quantities. There is a greater range of artefacts and manufacturing techniques, including grinding of both shell and stone axes (Spriggs 1997; Allen 2000; Specht 2005). Food remains at several sites where conditions favoured preservation include fruit, nuts and other useful trees, most of them familiar in gardens today.

Although the record for Manus in the early to middle Holocene is very thin, the general pattern of resource use is consistent with the rest of the Bismarcks. Whether tree crops were important in early to middle Holocene Manus is unknown, but it seems likely they were, given their earlier presence and prominence elsewhere in sites in the Bismarcks (Gosden 1995; Specht 2005).

From about 3500 years ago, sites of the Lapita cultural complex extend from the Bismarck Archipelago east to Tonga and Samoa. The nature of these sites and their relationship to earlier sites in the southwest Pacific region and further west has been the subject of extended debate (Spriggs 1997; Terrell and Welsch 1997; Kirch 2000; Green 2003; Terrell 2004). The standard interpretation sees Lapita sites as the settlements of immigrant agricultural people, who spread relatively rapidly from Southeast Asia through the Bismarck archipelago and further east, and whose descendants include the Polynesians. In extreme forms of this view, these new settlers have been distinguished from earlier-established inhabitants of the southwest Pacific by sharply defined cultural and technological contrasts, including in particular the introduction of agriculture. Current debate reflects more subtle positions about the relative contributions of putatively Southeast Asian and Melanesian subsistence practices and technological traditions (Green 2003; Golson 2005; Lilley 2006).

Manus does not fit neatly into a regional synthesis of intercommunicating Lapita peoples, for unequivocally Lapita material from Manus consists of only eight sherds from three sites, none of them in the beach locations typical of Lapita sites elsewhere. Thus, what part Manus played in the Lapita phenomenon is problematic (Kennedy 1981, 1982, 2002; Allen 1991; McEldowney and Ballard 1991), although obsidian from Lou Island occurs in Lapita sites throughout the Bismarks, and beyond. The Lapita-phase archaeological data from Manus are too tenuous to support arguments for local transformation of the patterns of settlement or subsistence (Kennedy 1983, 2002).

The end of the Lapita phase in the Bismarcks, North Solomons, Vanuatu and New Caledonia region has been characterised as a shift away from the regional interaction implied by the relative homogeneity of the Lapita cultural complex (Spriggs 1997; Bedford and Clark 2001, 2003), resulting in the present-day cultural diversity for which Melanesia is noted.

Again, the small number of excavated sites is inadequate to outline the post-Lapita period of Manus. Although the quantity and diversity of material and numbers of sites are greater than in preceding periods, much is from contexts which do not allow reliable dating. In general, the material suggests widening connections, throughout Manus and beyond.

The big picture: Changing subsistence and settlement patterns

Because people have been in the New Guinea islands region since the late Pleistocene, it is reasonable to expect that archaeological and palaeoecological sequences should register the shift from mobile foraging to more settled life based on agroecosystems. But because of the variability and flexibility of current subsistence practices, it is less clear what sort of changes might be expected. Despite the strongly asserted association of the shift to agroecosystems with the Lapita horizon, direct archaeobotanical or palaeoecological evidence is lacking, and change is inferred from proxies, such as the presence of pottery, shifts in settlement location, density and size, and linguistic reconstructions of plant and garden-related lexicons. Although this inferred subsistence change at present dominates the grand narrative of regional prehistory, the nature of emergent Holocene subsistence is open to question.

There is growing biogeographic and genetic evidence that important components of southwest Pacific agroecosystems are indigenous domesticates (e.g. bananas, breadfruit, *Pandanus* and sago; Yen 1991, 1996; Kennedy and Clarke 2004). The importance of the tree crops that form an integral part of the regional subsistence pattern, extending back at least to the middle Holocene, and probably to the late Pleistocene (Fairbairn 2005), has been underestimated. The significance of tree crops in mid-Holocene, pre-Lapita times is part of continuing debate about the Lapita cultural complex (Gosden 1992, 1995; Yen 1996; Kennedy and Clarke 2004).

The tree-crop-based systems of the present day, unlike the wide-ranging foraging of the earlier Pleistocene sites, involve significant alteration of natural forests, although this does not necessarily entail major disturbance. They show considerable regional variation in crop mixes and cultivation practices, and allow very rapid change in response to localised exigencies. As a result, local sequences may be quite disparate, and involve subtle changes difficult to detect in palaeoecological data.

II. Manus Island: Local cultural and environmental sequences

As sketched above, the archaeological record of Manus is meagre. Table 1 shows the excavated sites for which there is a reasonable assessment of radiocarbon ages or distinctive typology. There is a chronological bias towards the late Holocene, and a geographical one towards Lou Island and the eastern end of Manus. This largely reflects interest in the obsidian sources on Lou Island, shared by archaeologists and local sponsors of field work, as well as prehistoric inhabitants of the area. Historical records of European contact with Manus Province show a similar geographical bias (Kennedy et al. 1991).

Table 1. Archaeological sites of central/eastern Manus Province

Site	Pleistocene	Early-mid Holocene	Late Holocene		References
			Lapita	Post-Lapita	
Pamwak[a]	X	X		X	Fredericksen et al. 1993, Spriggs 1997, Ambrose 2002a
Peli Louson[a]		X		X	Kennedy 1983
Father's Water[a]		X		X	Kennedy 1983
Kohin[a]			X	X	Kennedy 1981
Mouk[b]			X	X	McEldowney and Ballard 1991
Sasi (Baun)[c]				X	Ambrose 1988, 1991, 2002a
Emsin[c]				X	Ambrose 1991, 2002a
Pisik[c]				X	Ambrose 1991, 2002b
Umleang[c]				X	Fullagar and Torrence 1991, Ambrose 2002a

a. Manus Island, b. near Baluan, c. Lou Island

The eastern half of Manus Island has villages and hamlets scattered throughout the interior, as well as along the coasts, with associated gardens and tree crops including planted sago stands. Tree crops are very important in village environs, and scattered throughout gardens and fallow. Garden clearance usually does not involve burning. The island's western half, in contrast, has neither settlements nor garden areas in the interior. The relatively few villages are coastal, and depend largely on sago (Hide et al. 1996). The contrast implies different local histories of settlement and subsistence, and requires a much finer focus of inquiry than the regional big picture.

Southwest Manus archaeological sequence

In southwest Manus, despite the absence of recent interior settlement, salvage archaeology during logging recorded more than 100 settlement sites. These are concentrated on high ground, especially ridge crests and their intersections, and marked by large quantities of pot sherds and worked obsidian (Kennedy et al. 1991; Wadra 1991; Kennedy 1992, 1997). The number and scale of these sites, exposed, recorded and then obliterated within a relatively short period, provide information difficult to integrate with that from the controlled conditions and close-up view of archaeological excavation, providing very detailed information about a tiny sample of the past. While archaeologists dissect a small part of a site, relatively modest logging and other industrial operations can simultaneously expose and destroy whole archaeological landscapes.

Despite destruction of the stratigraphic integrity of the southwest Manus sites, a few chronological and other inferences can be made from the salvaged artefacts. First, most of the sites are older than the forest, which can be no more than a few hundred years old: they are under, not in it. There is very little artefactual evidence referrable to the last few hundred years. Much of the distinctive pottery is not matched elsewhere in Manus or beyond. Since Lapita-style sherds are lacking, on conventional grounds the entire corpus may be characterised as post-Lapita, though quite what that means in Manus is unclear.

At least 25 sites contain distinctive, elaborately retouched obsidian points, technologically identical to those from a Lou Island site dated about 1720 yr BP (Kennedy 1997; Ambrose 2002a). Whether the southwest points are of imported obsidian is not clear, since there is a local southwest Manus obsidian source associated with Mount Hahie (Figure 1). But quite apart from intriguing questions of the geopolitics of resource acquisition and technological skills that these artefacts raise (Kennedy 1997), they imply a horizon during which settlement density was at least as high as the present-day pattern of eastern Manus. And this, in turn, suggests a similar subsistence base, with taro gardens and extensive tree crops.

At a few southwest sites, distinctive large retouched flakes similar to artefacts dated at Pamwak to the late Pleistocene/early Holocene suggest earlier occupation.

The archaeological evidence from under the forest of southwest Manus thus suggests people have been present there since the Pleistocene. At some point, they began to make and/or use pottery. About 1700 years ago, relatively intensive agriculture can be inferred. There is no direct evidence of what was grown, or when this phase ended. However, the lack of recent artefact types suggests that population decline and shift from inland to coastal settlement had occurred well before the period of European contact. Both the location and density of past settlements contrast with the present-day pattern. Since the growth of a forest dominated by *Calophyllum* spp. constitutes the most recent phase of the local sequence, understanding the dynamics of the forest is a necessary part of interpreting the human history of the southwest.

Logging in southwest Manus, begun in 1988, was prompted by the existence of a timber resource in a largely unpopulated area (Freyne and Bell 1982). In keeping with the aims of national forestry policy at the time, the Manus Provincial Government hoped the southwest Manus logging project would reduce the imbalance between the eastern and western halves of the island, bringing urban development and new technology to the west by construction of a township and veneer mill. Neither has eventuated. Appearances to the contrary, many of the trees are too old to yield useful timber, much less the peelable logs anticipated, and the project has been beset by long-running disputes (O'Collins and Lamothe 1989; Kennedy 1991; Ghai and Regan 1992:328; Anon. 2004). Logging proceeded despite recognition that the southwestern Manus *Calophyllum*-dominated forest is ecologically unusual, and that its dynamics are obscure (Kerenga and Croft 1984/85; Stevens 1995:63). While the archaeological record does not directly elucidate these dynamics, some relationship with preceding human activity is implied. Nevertheless, prominent NGO commentary on the timber resource insists it was in pristine condition before logging began (e.g. World Wildlife Fund 2001; Seacology 2008; National Geographic 2009). This is a familiar theme in critiques of forest resource use throughout the tropics (e.g. Bayliss-Smith et al. 2003; Willis et al. 2004).

Southeast Manus palaeoecological sequence: Lahakai Swamp

A preliminary palaeoecological investigation of Lahakai Swamp in southeastern Manus was undertaken with the aim of documenting Holocene vegetation change, in particular evaluating human impact and other disturbances on Manus (Southern 1987). After a broad survey of sites, Lahakai Swamp was chosen for its proximity to recent gardening and sago processing, and to Kohin Cave, an excavated archaeological site occupied intermittently from 3400 yr BP to the present (Kennedy 1981). It is worth noting that nearby Lake Yornam (Figure 1), formed probably within a karst sinkhole deposit, revealed alkaline conditions unsuitable for the preservation of pollen (Southern 1987).

A series of raised Plio-Pleistocene fringing reefs, now forming karst cliffs along part of the southeast coast of Manus, have blocked the drainage of many south-facing valleys. Diverted creeks are visible behind the cliffs (see Figure 1) and swamp deposits have accumulated large amounts of organic material suitable for palaeoecological analyses. In 1987, when the site was cored, this swamp was maintained as a sago plantation and was intensively exploited by local landowners. The vegetation was dominated by a spineless form of sago palm (*Metroxylon sagu*) which formed nearly monotypic stands with few other plant taxa present (Figure 2). The surrounding hill slopes behind the swamp were heavily gardened. Some areas were recently cleared for new gardens, and adjacent areas remained in fallow secondary forest regrowth that included many exploited fruit and nut trees.

Figure 2. Coring the Lahakai Swamp site. Jack Golson watching Wendy Southern and local assistants. Photograph by W. Ambrose, 1987

Methods

Lahakai Swamp was cored using a Russian D-Section corer, collected in 40 cm length sections to a depth of 9.6 m, at which point basal estuarine/marine-shell debris was encountered. The core was sampled at 10 cm intervals down to 630 cm and processed for palynomorphs using standard procedures (10% HCL, hot 10% KOH, 40% HF and acetolysis for all samples) as described by Faegri and Iversen (1975). Samples were then washed in alcohol followed by tertiary butyl alcohol to dehydrate them, before suspension in a recorded volume of silicon oil (2000 centistokes). This was used as an equivalent measure of pollen concentration, counter to more familiar techniques of adding a known number of *Lycopodium* marker spores to a sample. For microscopy, all counts were aimed at >150 pollen per sample. Fern and fern ally spores were also recorded but were not included in the total palynomorph sum. The palynomorphs were identified using the reference collection of what is now the Department of Archaeology and Natural History, ANU (see www.palaeoworks.anu.edu.au). All microscopic charcoal fragments were counted using the point-count method of Clark (1982), calculated as mm^2 cm^{-3}. The percentages of selected palynomorphs (major taxa >2% of total sum and a select range of indicator taxa), charcoal and total palynomorph concentration data were placed into stratigraphic diagrams using the program C2 Data Analysis version 1.5.2 (Juggins 2005).

Results and interpretation

The stratigraphy of the core is presented in Figures 3a, 3b and 3c, and divided into zones on the basis of the main vegetation signals and charcoal particle concentrations. The AMS radiocarbon dates from the bulk samples from the core are presented in Table 2, with the key features of each record and their palynological zones described in Table 3. Three phases of vegetation change are defined by the sequence from southeast Manus, demonstrating coastal edaphic changes in response to mid-Holocene sea-level fluctuations, followed by freshwater swamp forest development, and increasingly dryland forest dominance in the late Holocene. This later swamp forest-dryland phase is critical in that it allows some definition of land-use practices demonstrated by the presence of both horticultural (*Colocasia*) and tree-crop (e.g. *Metroxylon* and *Aleurites*) components.

Table 2. Radiocarbon ages from Lahakai core

Sample name	Depth cm	Radiocarbon age BP	Calibrated age 2σ
LKAI 1	210-220	340±70	151-500
LKAI 2	480-495	3654±185	3463-4418
LKAI 3	615-630	5280±185	5598-6394

All dates on bulk sediment. All ages calibrated using Calib v. 5.1.

Table 3. Palynological zones represented in the Lahakai Swamp sequence, Manus, and outlined on the basis of changes in dryland, wetland and human-impact signatures

Zone	Dryland	Wetland	Interpretation of human impact and other disturbances
5500-3000 cal. yr BP Zone I	Most common taxa recorded are Urticaceae/Moraceae, Arecaceae (which increases towards the top of the zone) and Myrtaceae.	High but declining values of mangrove taxa *Rhizophora* and *Brugiera*/*Ceriops*, with minimal representation of *Pandanus* and Centrolepidaceae/Restionaceae. Cyperaceae and cf *Hypolytrum* increase towards the top of the zone, as does *Lumnitzera*.	Micro-charcoal particles high at around 5000 cal. yr BP. Macro-charcoal band at around 4000 cal. yr BP particles.
3000-500 cal. yr BP Zone II	Characterised by the secondary taxa Ulmaceae, Poaceae and *Castanopsis*/*Lithocarpus*. *Macaranga*/*Mallotus* (also secondary forest taxa) and Urticaceae/Moraceae increase towards the top of the zone.	Mangroves largely absent, dominated by Cyperaceae.	Fern spores increase in this zone, with *Lygodium* showing a small peak. Two bands of macro-charcoal particles at around 1000 cal. yr BP.
500-present Zone III	Several taxa increase in this zone, including *Calophyllum*, *Elaeocarpus*, *Endospermum*, Myrtaceae, while the secondary forest taxa common in Zone II decrease.	Cyperaceae is represented in much lower proportions than in the previous zone. Most wetland taxa are poorly represented.	*Polypodium* increase and *Dicranopteris* appears consistently for the first time. Micro-charcoal particles high at around 500 cal. yr BP. Two macro-charcoal bands represented in this zone. Possible and probable plant cultigens including *Colocasia* and *Aleurites* are first represented in this zone.

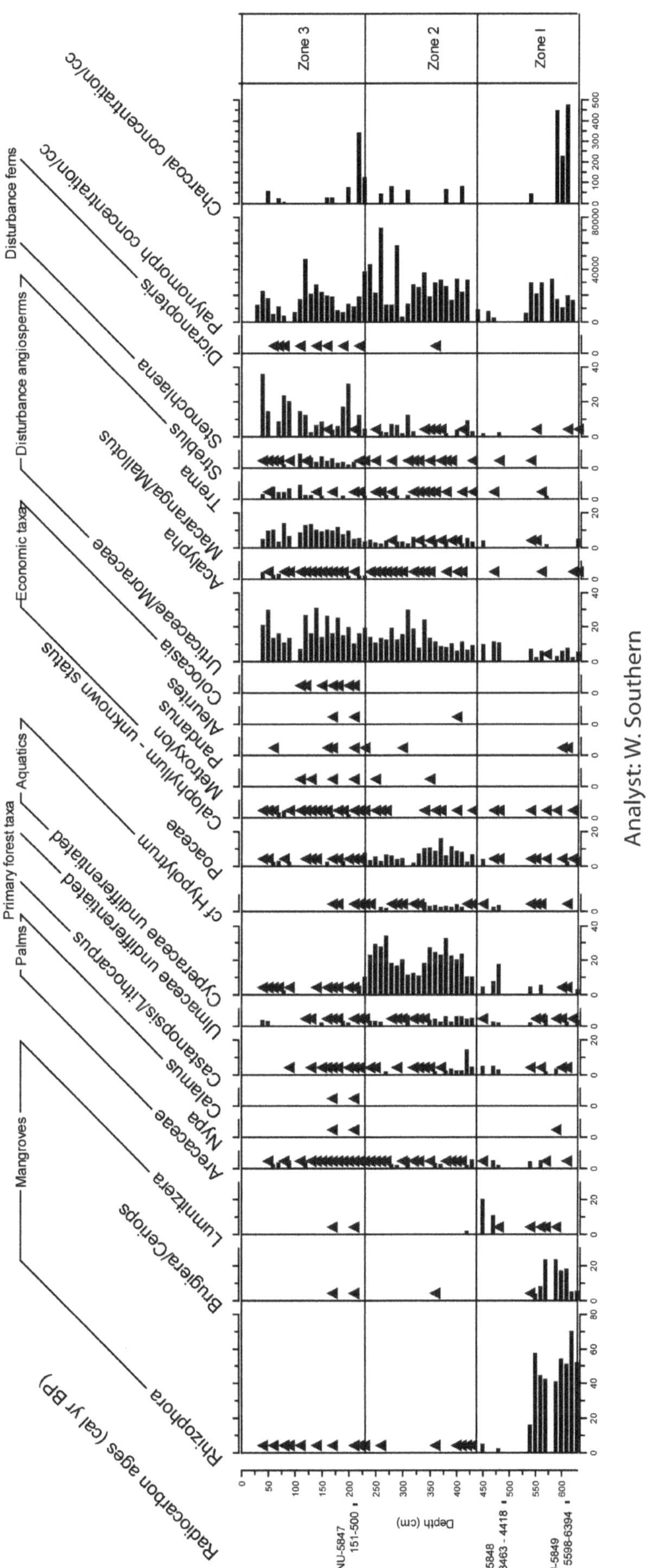

Figure 3a. Palynological diagram for the Lahakai core, Manus. The key abundant taxa are presented in bars (samples with taxa representing >2% of total sum) and triangles (samples with taxa representing <2% of total sum). Taxa are sorted within these groups in stratigraphic succession using the sort function in C2 Data Analysis. Taxa are grouped according to their respective vegetation associations as follows: Trees and shrubs; pteridophytes and bryophytes; sedges and other aquatics; grasses; economic taxa; unknown palynomorphs. Palynomorph concentrations are presented as grains/cm^{-3} x 10^{4}; charcoal concentrations are presented in mm^{2} cm^{-3} x 10^{2}

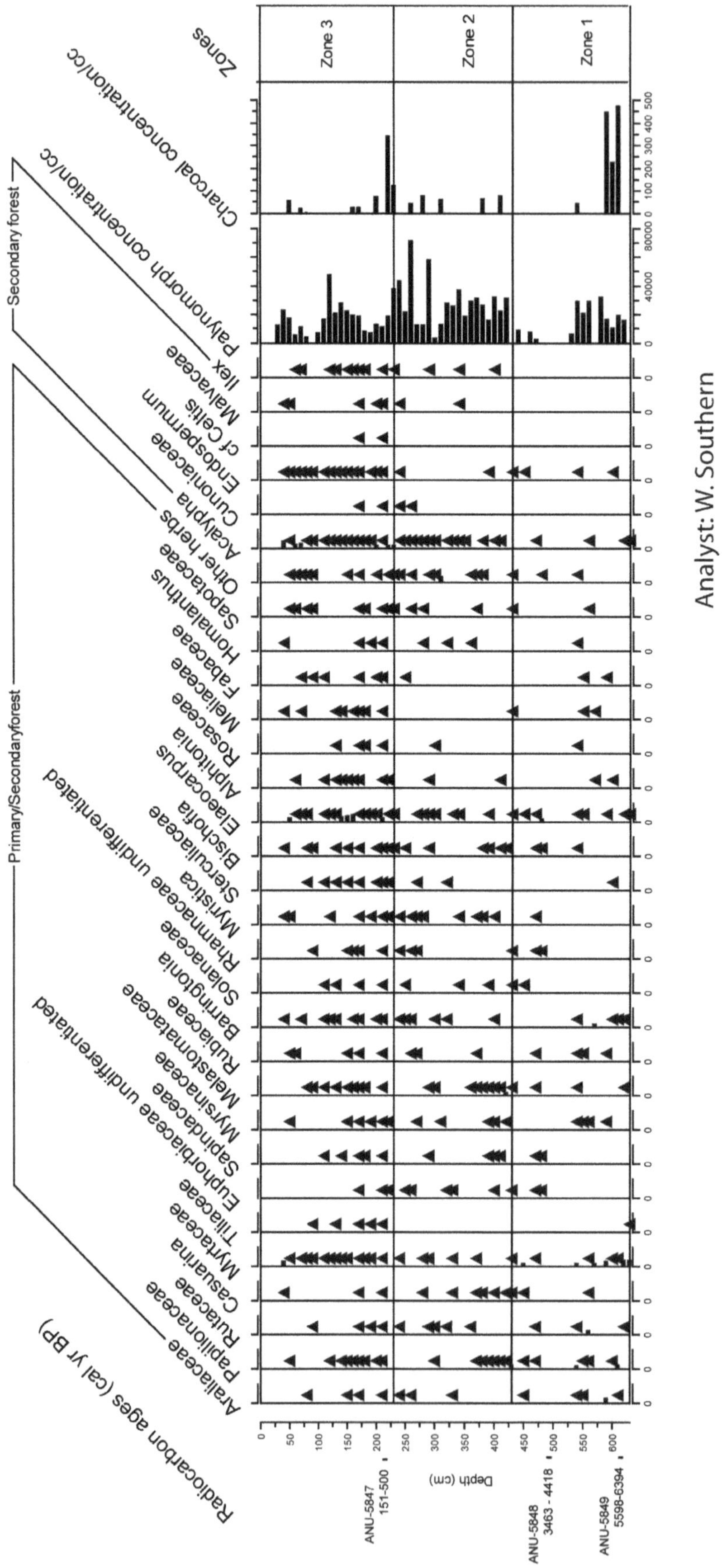

Figure 3b. Palynological diagram for the Lahakai core, Manus. More rare angiosperm tree, shrub and herb taxa are presented in bars (samples with taxa representing >2% of total sum) and triangles (samples with taxa representing <2% of total sum). Sorting, grouping and concentration as for Figure 3a

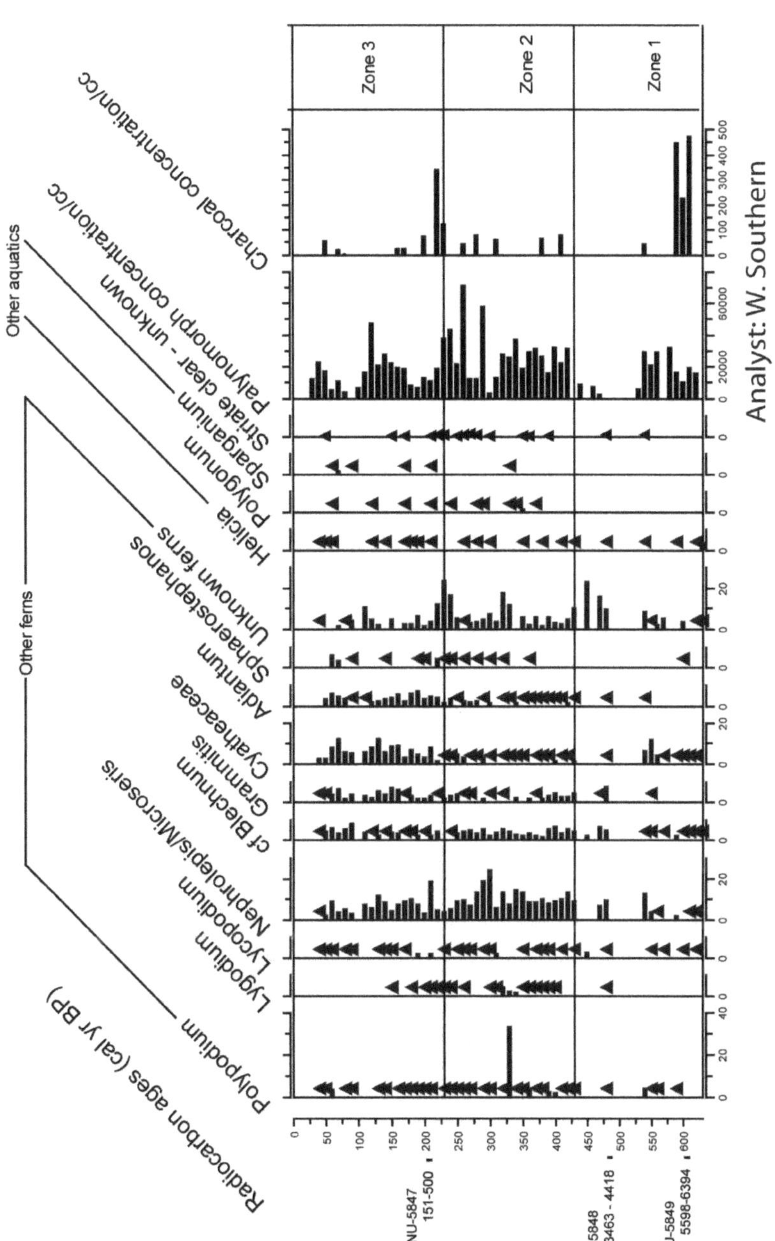

Figure 3c. Palynological diagram for the Lahakai core, Manus. More rare fern and aquatic taxa are presented in bars (samples with taxa representing >2% of total sum) and triangles (samples with taxa representing <2% of total sum). Sorting, grouping and concentration as for Figure 3a

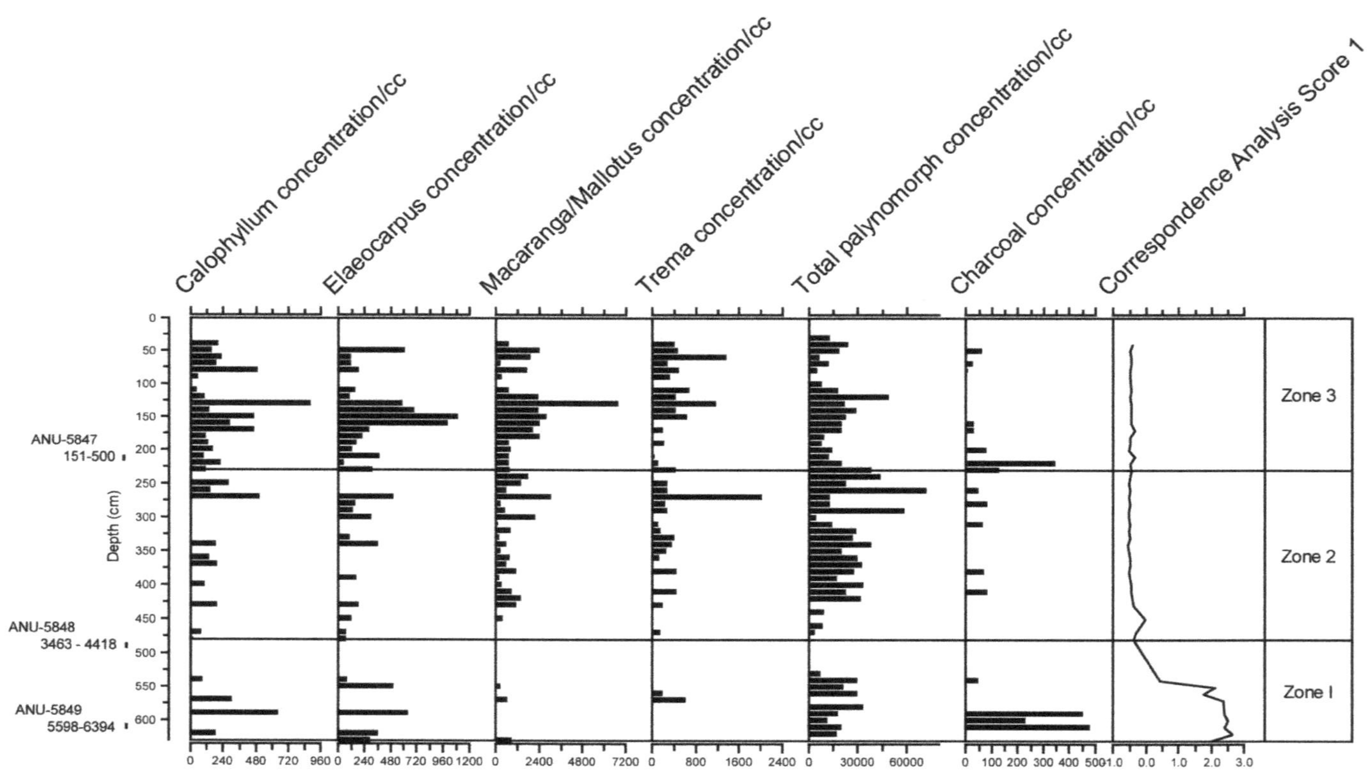

Figure 4. Palynomorph concentrations (grains/cm^{-3} x 10^4) of select taxa (including *Calophyllum*) and Axis 1 results of a Correspondence Analysis (CA) of the entire dataset from the Lahakai sequence are presented. Data were centred and standardised for the analysis and the Axis 1 results show 61% of the variance in the dataset. The plot and Correspondence Analysis were produced using the Program C2 Data Analysis version 1.4.3. (Juggins 2005)

One of the most striking features of this record is the lack of *Metroxylon* sago pollen in the upper part of the sequence, despite the core site being intensively managed for sago production in 1987. One explanation for the lack of sago pollen in the upper sequence of this core is that most trunks of this hapaxanthic (monocarpic) tree have been harvested before flowering. An alternative, that pollen production in sago is low, is ruled out by modern pollen assemblages from surface swamp soils from under sago stands in other parts of Papua New Guinea (M. Prebble unpublished data). This discrepancy in pollen representation highlights the complexity of the interpretation of fossil pollen data. Pollen representation of *Calophyllum*, important for the southwest Manus sequence, presents another complex case.

Calophyllum *at Lahakai*

Although few data are available on the pollination ecology of *Calophyllum*, some taxa in the genus are known to be entomophilous (Stevens 2007) and thus probably do not produce high concentrations of pollen. It is very difficult to gauge from the percentage data presented from the Lahakai record the magnitude of the response of *Calophyllum* to any ecological changes, including human impact. To address the problem of pollen representation, in Figure 4 we show pollen concentration data for *Elaeocarpus*, a primary forest tree, *Calophyllum*, and the two key disturbance taxa, *Macaranga/Mallotus* and *Trema*. These data mitigate the proportional bias imposed by anemophilous (wind dispersed) and locally abundant taxa.

In Zones 1 and 2, *Calophyllum* is represented in similar concentrations to *Elaeocarpus* but is not consistently represented across all of the samples, suggesting that both these taxa were ephemeral in the Lahakai environment at this time. These data contrast with *Macaranga/Mallotus* and *Trema* pollen, which are poorly represented in Zone 1, but are consistently represented across samples in Zone 2. All of these taxa are represented in all samples in Zone 3 and show similar changes in concentration across the zone. In Zone 2, the *Calophyllum* and *Elaeocarpus* concentrations are almost complementary, which suggests rapid changes in forest structure.

In order to assess whether these variations reflect changes exhibited across the entire record, correspondence analysis was performed on the percentage data of all pollen and spore taxa identified and all the samples analysed from the Lahakai core. Eigen values (Axis 1) generated for each sample are presented as proportions of the total variation in the dataset shown in Figure 4. These values account for 61% of the variation in the entire dataset, and are primarily controlled by the proportions of mangrove pollen in each sample. The greatest variation exhibited in this core occurs in Zone I. These data emphasise the point that variation in Zones 2 and 3 is minimal.

This raises the question of why *Calophyllum* behaves like a primary forest taxon in Zones 1 and 2 and a disturbance taxon in Zone 3.

III. *Calophyllum* forest in Manus and the Pacific

We have suggested that understanding the earlier settlement sequence of Manus requires understanding the dynamics of the forest. Since the growth of a forest dominated by *Calophyllum* spp. constitutes the most recent phase of the local sequence in southwest Manus, we consider what role this genus played in the southeast sequence represented at Lahakai. First, we review what is known of the ecology of *Calophyllum* in Manus and the wider region.

Information about the southwest Manus *Calophyllum* forest is limited but striking. Brief forestry and botanical surveys before logging began in 1988 highlighted its value for timber on the one hand, and ecological rarity on the other (Freyne and Bell 1982; Kerenga and Croft 1984/85; Stevens 1995:63). The area includes a very peculiar forest type, consisting of extensive plantation-like tracts in which large, even-sized *Calophyllum euryphyllum* Lauterb. (Clusiaceae) are almost the only tree, with abundant seedlings but virtually no saplings.

This monotypic forest type also occurs on Rambutyo Island in Manus Province (Figure 1), on Mussau and on small islands off New Ireland, growing on a range of different soils and substrates (Kerenga and Croft 1984/85:46). There are no detailed descriptions or maps of the extent of the monotypic *Calophyllum* stands in southwest Manus or elsewhere.

C. euryphyllum has a scattered distribution on the Aru Islands, the islands of Geelvink Bay and the Vogelkop peninsula of Irian Jaya, the north coast of New Guinea from Jayapura to the Sepik, the Bismarck Archipelago (except New Ireland) and in the Central and Milne Bay provinces of Papua New Guinea (Stevens 1995). *C. euryphyllum* is the dominant species of the monotypic forest, but it is not the only *Calophyllum* species on Manus. *C. soulattri* and *C. waliense*, the latter an endemic Manus species, are also recorded. Although detailed distributions are lacking, it is likely these three species grow together (Stevens 1974:355). *C. inophyllum*, a widespread Malesian strand species, is also indigenous to Manus, but is primarily adapted to the coastal fringe, though it can grow on wetland substrates (Stevens 1974, 1980, 1995, 2007).

Apart from the monotypic stands of *C. euryphyllum*, *Calophyllum* spp. are common trees in the rainforest canopy throughout Manus (Kerenga and Croft 1984/85). In central and south-central Manus, moderately rich mixed forest including *Calophyllum* spp. showed no distinct dominants (Kerenga and Croft 1984/85:44-46). In the survey of Manus forest resources that led to logging in western Manus, Freyne and Bell (1982:32-33) estimated that *Calophyllum* (species not identified, but probably mostly *C. euryphyllum*) comprised about 50% of the largest size class (dbh <70 cm) of trees in two out of three sampled rainforest blocks, and 84% in the third, the western Manus block that includes monotypic stands.

In a subsequent preliminary study in the southwest Manus logging area, O'Collins and Lamothe (1989) identified 11 forest associations including canopy, subcanopy and regenerating vegetation in logged, unlogged and burnt forests. Three forest types were also identified from an old coconut (*Cocos nucifera*) plantation. *Calophyllum* (cf. *C. euryphyllum*) dominated all forest types with the exception of logged sub-canopy forest and the canopy and sub-canopy forests of an old coconut plantation, where this tree forms a minor component of the vegetation.

In eastern Manus, *Calophyllum* is absent from forest regrowth in lowland areas heavily disturbed by human interference (Kerenga and Croft 1984/85:45). In interior eastern Manus, Rooney (1982) reported that *Calophyllum*-dominated rainforest forms a late stage of succession after clearance for taro (*Colocasia*) cultivation. He suggested that the conversion of earlier woody fallow to forest dominated by '*Calophyllum* and tropical hardwood species', which local inhabitants consider undesirable for taro gardens, resulted from management failure caused by labour shortage, following a modernisation movement in the 1940s that halved the local population (Rooney 1982:275). Though details of the species composition of this forest are unclear, Rooney's comment does not suggest monotypic *Calophyllum* stands.

Despite the scarcity of quantified descriptions, it seems that the frequency of *Calophyllum* spp. in Manus forest is variable, and that the monotypic stands of western Manus, Rambutyo and small islands off New Ireland represent a rare occurrence of hyper-dominance, the dynamics of which are unknown.

The survey of forest resources of Manus asserts that *Calophyllum* (species not stated) is a pioneer (Freyne and Bell 1982:29, 34). The basis for this statement seems to be the observation of seedlings in disturbed areas, but perhaps also the assumption that the even size of trees in the monotypic stands translates to even age. There is no comment about the absence of saplings in these stands noted by Kerenga and Croft (1984/85). Freyne and Bell's (1982:35) stand table showing regeneration of saplings, the only such data available for Manus, relates to more typical rainforest, in which large *Calophyllum* trees constitute about half the volume of usable timber, rather than the much higher figure for the western Manus block containing the monotypic stands. Freyne and Bell (1982:29) attribute this peculiar formation to catastrophic destruction, citing a Manus colleague's account of an oral tradition about a tidal wave and strong wind, thought to have occurred in the early 19th century (see also Johns 1986).

If the monotypic *C. euryphyllum* forest of southwest Manus was composed of a pioneer species in even-aged stands, the timing of forest initiation could be estimated, and suggest a terminal date for the shift from a pattern of inland villages most likely supported by taro gardening and tree crops to a smaller population living in exclusively coastal villages supported by sago and other tree crops.

This explanation for the southwest Manus monotypic forest lacks evidence. The argument from oral history fails on a number of grounds. There has been no comprehensive study of Manus oral history, available information being anecdotal at best. Attributing damage throughout the uplands of southwest Manus, or Rambutyo, to tsunamis lacks credibility, given the elevation of the terrain. Stories of catastrophes, including earthquakes, darkness and drought as well as high winds and tsunamis causing forest, garden and village destruction, are commonly recounted throughout Manus (J. Kennedy pers obs. 1977-1991). They are undoubtedly based on observation, but a widely repeated story does not account convincingly for peculiar effects of a localised catastrophe. In central Manus, where villagers tell a very elaborate account of how a tidal wave and wind long ago destroyed the local forest, leaving only one tree, a large *Ficus*, still standing, there are no dense *Calophyllum* stands (J. Kennedy pers obs. 1981). There is no reliable support from oral history for a local catastrophe in southwest Manus that might account for natural disturbance-mediated establishment of *Calophyllum* forest.

As to even age of the trees, the foresters' assumption that fairly uniform diameter translates to even age has not been borne out since logging began in southwest Manus. Many trees are too old to be any use (Anon. 2004). Age variation both within and between stands suggests more complex dynamics of establishment, as do other traits of the species.

Elsewhere on Manus, *C. euryphyllum* is a common component of rainforest. Its local value as timber, especially for canoe hulls, makes it likely that trees would once have been protected or conserved. Although there is little in the literature about its ecology, Rooney's (1982:275) comment about its role in late-stage fallows is suggestive. A recent forestry development program to propagate genetically improved seed for community tree-planting programs (Gunn 2007) shows that the seed has very high moisture content and is unlikely to retain viability for long. It is eaten by birds, bats and phalangers (Anon. n.d.).

Although *C. euryphyllum* is distributed from the Aru Islands and the Vogelkop, along the north coast of New Guinea to the Bismarck Archipelago, occurrences of monotypic stands are restricted to a few areas in Manus, Mussau and islands off New Ireland. Yet throughout the species' range, large-scale disturbance is common. This includes the eastern half of Manus, which lacks monotypic *Calophyllum* stands despite being closest to the volcanic eruptive centres of Lou Island and the St Andrew Strait that are the most potent source of disturbance in the region. In eastern Manus, fallow gardens support a familiar range of large-leaved, fast-growing pioneer species, in which *Calophyllum* is not prominent, if present at all (Rooney 1982; Kerenga and Croft 1984/85). While the reported distribution of monotypic stands of *C. euryphyllum* might well reflect incomplete knowledge, this is unlikely in the case of eastern Manus, or the case of New Britain, an island that has been thoroughly explored for its timber potential, and also has a rich recent history of natural catastrophes and widespread gardening.

Although there is little direct evidence of the ecology of *C. euryphyllum*, there is information to suggest its likely characteristics. All except one of the Papuasian species of *Calophyllum* surveyed in Stevens' (1980) systematic treatment of the genus are shade-tolerant rainforest trees (the exception is the common coastal/strand species *C. inophyllum*), with no evidence of pioneer habit. In Whitmore's (1989) expanded classification along a gradient from pioneers to progressively more shade-tolerant species, the Solomons *Calophyllum* species (*C. neo-ebudicum* and *C. peekelii*) both fall towards the shade-bearing end. Seeds have no dormancy and are poorly dispersed. Seedlings germinate readily in light or shade, but with high early mortality. Survivors grow up in high forest or small gaps (see also Greig-Smith et al. 1967; Whitmore 1984:89; Burslem and Whitmore 1996).

C. euryphyllum probably behaves in a similar fashion, with regeneration dependent on a population of seedlings rather than a seed bank. Thus, distribution of the trees may reflect conditions promoting the persistence and onward growth of seedlings into saplings, rather than the dispersal and germination of seeds (Bazzaz and Pickett 1980:293, 303). The monotypic *C. euryphyllum* forest is clearly a special case, requiring explanation of how other species are excluded. The absence of saplings noted by Kerenga and Croft (1984/85) suggests that the formation of monotypic stands involves peculiar dynamics in which on-growth of seedlings is suppressed, and that these may persist. It has been shown that in Bornean forests, most shaded seedlings of canopy tree species are at least 10 years old, and some may be much older (Delissio et al. 2002).

Ecological studies have examined forests where stands of *Calophyllum* spp. dominate, further east in the Pacific. *Calophyllum neo-ebudicum* occurs throughout Vanuatu in moist forest, but is most abundant in the southern islands of Erromango and Aneityum (Wheatley 1992; Mueller-Dombois and Fosberg 1998). On Aneityum, the natural vegetation is dense mixed lowland forest with a canopy of *C. neo-ebudicum* and *Agathis obtusa*, but much of this forest has been extensively logged, leaving much of the island in secondary scrub (Hope pers comm.). *C. neo-ebudicum* is the most common canopy associate in Erromango.

Drake et al. (1996) documented four types of forest on the raised limestone island of 'Eua, Kingdom of Tonga, two dominated by *C. neo-ebudicum*. The key feature of this study is the recognition that regeneration of disturbed areas in these forest types is not dominated by *C. neo-ebudicum*, but *Dendrocnide harveyi* (Urticaceae), *Bischofia javanica* (Euphorbiaceae) and *Rhus taitensis* (Anacardiaceae) in the mixed forest, and *Alphitonia zizyphoides* (Rhamnaceae) and *Elattostachys falcata* (Sapindaceae) in the upland *Calophyllum* forest. This suggests that disturbance has not maintained *C. neo-ebudicum* dominance, but instead supports a more diverse mixed forest assemblage.

Whitmore's long-running study of the regeneration of forests on Kolombangara, in the Solomon Islands, monitored all age classes of 12 canopy species on a range of sites and under disturbance conditions which included cyclone damage and gardening clearance, as well as small natural gaps (Whitmore 1989; Burslem and Whitmore 1996, 1999). Burslem et al. (2000) discuss forest regeneration after cyclone damage, to evaluate the suggestion that community composition of tropical forests may be influenced strongly by rare but large-scale disturbance events. They conclude that the relative abundance of common tree species on Kolombangara is only briefly disrupted by such events, and that a better explanation for spatial differences in the island's forest composition is provided by different histories of human disturbance. Bayliss-Smith et al. (2003) reach a similar conclusion for the Marovo lagoon area, and argue the value of interdisciplinary studies for understanding of rainforest dynamics.

As Whitmore (1989) showed for Kolombangara, the considerable spatial and temporal differences in population size and turnover of tree species following canopy disturbance make generalisations about forest dynamics unsafe. The functional and life-history characteristics of tree species vary independently across multiple parameters, which dichotomous typologies cannot reflect (Baker et al. 2003; Kraft et al. 2008; Zimmerman et al. 2008:111). Studies of relative species abundance and distribution in tropical forests, based on repeated censuses of large plots, have highlighted the difficulties of testing competing theories of the maintenance of the species diversity of tropical forests (Chave 2004, 2008; Carson and Schnitzer 2008). Notably, seedling-to-sapling stages of tropical forest dynamics have not yet been satisfactorily investigated (Zimmerman et al. 2008:108).

The high diversity of tropical forests is the central focus of studies of their ecology and dynamics. Recent approaches have shifted from relatively simple deterministic models to more sophisticated theoretical models, which integrate ideas from island biogeography and population genetics and have greater analytical power, especially capability to handle large datasets at multiple scales. A central issue generating much debate is whether stochastic dispersal-dependent sampling effects and demographic processes are sufficient to account for community structure, and the extent to which species-level differences in ecological strategy contribute to this structure (e.g. Connell 1978; Denslow 1987; Hubbell 2001; Potts et al. 2002; Chave 2004; Svenning et al. 2004; Wills et al. 2006; Carson and Schnitzer 2008; Kraft et al. 2008; Stokstad 2009). Recognition that tropical forests often have a few common species and a large number of rare ones has become central to the development of these insights in forest community ecology. Dealing quantitatively with spatial patterning of the rare species has required new approaches to both data collection and analysis, especially to measures of biodiversity (Chave 2004, 2008).

Monodominant tropical canopy species

High diversity notwithstanding, at the other extreme of the range of relative species abundance, dominance of the canopy by single species, ranging from 50% to 100% of individuals, is common throughout the tropics and has drawn extended commentary. Connell and Lowman (1989) classified these forests into two types, on the basis of whether or not the monodominant species persists beyond one generation. Stands of short-lived, light-demanding species with well-dispersed seeds capable of long dormancy do not regenerate in situ. Persistent dominance of the canopy by one species, they suggest, could arise in two ways: by colonisation of large gaps and subsequent regeneration, or by gradual replacement of other species. They proposed ectomycorrhizal association as one possible mechanism for such replacement and listed species which show persistent dominance from the families Dipterocarpaceae, Fagaceae, Lauraceae and Leguminosae (subfamily Caesalpinioideae), from Malesia and the New World.

Subsequent research has shown that monodominant canopy species do not share a simple set of causative traits or mechanisms. They may form at early or late successional stages after clearance, and this does not predict whether they regenerate or not. Those that form persistent stands of long-lived trees may or may not depend upon ectomycorrhizae or poor soils. Examples have expanded to include more species of the families Leguminosae, Guttiferae, and notably New Guinea and New Caledonian *Nothofagus* species (Connell and Lowman 1989; Hart et al. 1989; Hart 1990; Read et al. 1995, 2000; Torti et al. 2001; Potts et al. 2002; McGuire 2007).

C. euryphyllum on Manus provides another example of monodominance in a family (Clusiaceae) not included in recent discussions. Forest in which *C. euryphyllum* constitutes about 50% of canopy trees appears to be quite widespread throughout Manus. *Calophyllum* species further east in the Pacific seem to follow a similar pattern. The restricted patches in which *C. euryphyllum* reaches much higher proportions, in the west of Manus, on Rambutjo and elsewhere, we have singled out above as monotypic, emphasising not only the species' dominance but also the distinctive lack of growth stages between seedlings and mature trees.

Research on tropical canopy monodominance establishes the multiplicity of possible causal mechanisms, rendering generalisation unsafe. We need more evidence to explain both the widespread pattern of *Calophyllum* spp. dominance on Pacific islands, and especially the hyperdominance of the monotypic *C. euryphyllum* stands. On present evidence, it is impossible to decide whether the Lahakai pollen data might represent the more widespread, c. 50% dominance of *Calophyllum* spp. on Manus, or the rarer patches of monotypic stands.

Conclusion

The *Calophyllum*-rich forests of Manus are part of an island landscape that has supported a human population since the Pleistocene. We lack the detailed evidence necessary to show how people have impacted forest vegetation and vice versa. Nevertheless, *Calophyllum* spp are represented in both our local sequences, one based on a phase of human occupation on top of which has grown a distinctive monotypic forest, the other on palaeoecological signals from an area which archaeological evidence shows was well populated. In the most recent phase of both sequences, *Calophyllum* is somewhat anomalous, represented by a peculiar forest in the west, and a divergent pollen signal in the east.

Since we lack the observational data crucial for understanding the ecology of these trees and their community dynamics, we cannot make direct causal links between the former human settlement of the interior of southwest Manus and the succeeding monotypic forest, nor between the palynological record of Lahakai Swamp and the human activity represented by the nearby archaeological site. Nevertheless, the southwest Manus archaeological landscape makes it clear that the absence of human settlement in the recent forest of western Manus was not a permanent condition, and that the forest almost certainly post-dates agricultural disturbance. On the other hand, in eastern Manus, agricultural activities have continued to the present. This contrast in the history of human activity might be correlated with the distribution of monotypic *Calophyllum* forest. One possibility is that the monotypic stands formed after a pattern of moderate-scale, relatively frequent disturbance initiated by agriculture was interrupted by a gradual cessation of garden clearance, so that both frequency and scale of disturbance decreased. Tests of this hypothesis are offered by the unstudied archaeological potential and as yet undescribed ecological characteristics of other occurrences of monotypic *C. euryphyllum* forest, on Manus, Rambutyo and elsewhere. Further efforts to understand the phylo- and phytogeography of *Calophyllum* spp. in the southwest Pacific are warranted, including palaeoecological research focused on mapping the temporal and spatial distribution of *Calophyllum* from local sequences.

We conclude by noting that the benefits of interdisciplinary research include rephrasing and refocusing complex questions, such those posed by the ecological histories of islands, as much as answering them.

Acknowledgements

This paper draws on field work in Manus Province supported by the Manus Provincial Government and people of Mbunai village and Ndrahuke logging camp. Over its long and complex gestation, many people have helped with information and advice. They include Wal Ambrose, Tim Bayliss-Smith, Bill Clarke, Jim Croft, Brian Gunn, Ian Hughes, Bob Johns, Malcolm Levett and Peter Stevens. We thank them, and anonymous referees for constructive comments.

References

Allen, J. 1991. Introduction. In J. Allen and C. Gosden (eds), *Report of the Lapita Homeland Project. Occasional Papers in Prehistory* 20, pp1-8. Canberra: Department of Prehistory, Research School of Pacific Studies, Australian National University.

Allen, J. 2000. From beach to beach: The development of maritime economies in prehistoric Melanesia. In S. O'Connor and P. Veth (eds), *East of Wallace's Line: Studies of past and present maritime cultures of the Indo-Pacific region. Modern Quaternary Research in Southeast Asia* 16, pp139-175. Rotterdam: Balkema.

Allen, J. 2003. Discovering the Pleistocene in Melanesia. In C. Sand (ed.), *Pacific Archaeology: Assessments and prospects. Proceedings of the International conference for the 50th anniversary of the first Lapita excavation. Le Cahiers de l'Archéologie en Nouvelle Calédonie* 15, pp33-42. Nouméa: Départemente Archéologie, Service des Musées et du Patrimoine de Nouvelle-Calédonie.

Allen, J. and C. Gosden (eds) 1991. *Report of the Lapita Homeland Project. Occasional Papers in Prehistory* 20. Canberra: Department of Prehistory, Research School of Pacific Studies, Australian National University.

Allen, J. and C. Gosden 1996. Spheres of interaction: Modelling the culture history of the Bismarck Archipelago. In J. Davidson, G. Irwin, B.F. Leach, A. Pawley and D. Brown (eds), *Oceanic Culture History: Essays in honour of Roger Green. NZ Journal of Archaeology Special Publication*, pp183-197. Dunedin: NZ Journal of Archaeology.

Ambrose, W. 1988. An early bronze artefact from Papua New Guinea. *Antiquity* 62:483-491.

Ambrose, W. 1991. Lapita or not Lapita? The case of the Manus pots. In J. Allen and C. Gosden (eds), *Report of the Lapita Homeland Project. Occasional Papers in Prehistory* 20, pp103-112. Canberra: Department of Prehistory, Research School of Pacific Studies, Australian National University.

Ambrose, W. 2002a. From very old to new, obsidian artefacts in the Admiralty Islands. In C. Kaufmann, C. Kocher Schmid and S. Ohnemus (eds), *Admiralty Islands: Art from the South Seas*, pp67-72. Zürich: Museum Rietberg.

Ambrose, W. 2002b. Big pots on a small Lou Island. In S. Bedford, C. Sand and D. Burley (eds), *Fifty years in the field. Essays in honour and celebration of Richard Shutler Jr's archaeological career*, pp59-67. Auckland: New Zealand Archaeological Association.

Anonymous 2004. Review of current logging projects. Finalised individual project review report 2, Timber Permit 18-2 West Coast Manus. Papua New Guinea: Department of National Planning and Monitoring.

Anonymous n.d. *Nursery techniques and seed handling of PNG's tree species.* Waigani: PNG Forest Industries Association (Inc).

Baker, T.R., M.D. Swaine and D.F.R.P. Burslem 2003. Variation in tropical forest growth rates: Combined effects of functional group composition and resource availability. *Perspectives in Plant Ecology, Evolution and Systematics* 6:21-36.

Bazzaz, F.A. and S.T.A. Pickett 1980. Physiological ecology of tropical succession: A comparative review. *Annual Review of Ecology and Systematics* 11:287-310.

Bayliss-Smith, T., E. Hviding and T. Whitmore 2003. Rainforest composition and histories of human disturbance in Solomon Islands. *Ambio* 32:346-352.

Bedford, S. and G. Clark 2001. The rise and rise of the Incised and Applied Relief Tradition: A review and reassessment. In G. Clark, A. Anderson and S. Sorovi-Vunidilo (eds), *The Archaeology of Lapita Dispersal in Oceania. Terra Australis* 17, pp61-74. Canberra: Pandanus Books.

Burslem, F.R.P. and T.C. Whitmore 1996. *Silvics and wood properties of the common timber tree species on Kolombangara.* D.F.R.P. *Oxford Forestry Institute Tropical Forestry Papers* 34. *Solomon Islands Forest Records* 7. University of Oxford: Oxford Forestry Institute.

Burslem, F.R.P. and T.C. Whitmore 1999. Species diversity, susceptibility to disturbance and tree population dynamics in tropical rain forest. D.F.R.P. *Journal of Vegetation Science* 10:767-776.

Burslem, F.R. P., T.C. Whitmore and G.C. Brown 2000. Short-term effects of cyclone impact and long-term recovery of tropical rain forest on Kolombangara, Solomon Islands. D.F.R.P. *Journal of Ecology* 2000 88:1063-1078.

Carson, W.P. and S.A. Schnitzer (eds) 2008. *Tropical forest community ecology*. Chichester, UK: Wiley-Blackwell.

Chappell, J. 2005. Geographic changes of coastal lowlands in the Papuan past. In A. Pawley, R. Attenborough, J. Golson and R. Hide (eds), *Papuan Pasts: Cultural, linguistic and biological histories of Papuan-speaking peoples*, pp525-539. Canberra: Pacific Linguistics, RSPAS, ANU.

Chave, J. 2004. Neutral theory and community ecology. *Ecology Letters* 7:241-253.

Chave, J. 2008. Spatial variation in tree species composition across tropical forests: Pattern and process. In W.P. Carson and S.A. Schnitzer (eds), *Tropical forest community ecology*. pp11-30. Chichester, UK: Wiley-Blackwell.

Clark, R.L. 1982. Point count estimation of charcoal in pollen preparations and thin sections of sediments. *Pollen et Spores* 24:523-535.

Connell, J.H. and M.D. Lowman 1989. Low-diversity tropical rain forests; Some possible mechanisms for their existence. *American Naturalist* 134: 88-119.

Delissio, L.J., R.B. Primack, P. Hall and H.S. Lee 2002. A decade of canopy-tree seedling survival and growth in two Bornean rain forests: Persistence and recovery from suppression. *Journal of Tropical Ecology* 18:645-658.

Denham, T.P., Haberle, S.G. and Lentfer, C. 2004. New evidence and revised interpretations of early agriculture in Highland New Guinea. *Antiquity* 78:839-857.

Denslow, J.S., 1987. Tropical rainforest gaps and tree species diversity. *Annual Review of Ecology and Systematics* 18:431-451.

Drake, D.R., Whistler, W.A., Motley, T.J. and Imada, C.T. 1996. Rain forest vegetation of 'Eua, Kingdom of Tonga. *New Zealand Journal of Botany* 34:65-77.

Ellison, J. 2005. Holocene palynology and sea-level change in two estuaries in southern Irian Jaya. *Palaeogeography, Palaeoclimatology, Palaeoecology* 220:291-309.

Faegri, K. and J. Iversen 1975. *Textbook of pollen analysis*. Copenhagen: Munksgaard.

Fairbairn, A. 2005. An archaeobotanical perspective on Holocene plant-use practices in lowland northern New Guinea. *World Archaeology* 37:487-502.

Flannery, T and J.P. White 1991. Animal translocation. *National Geographic Research and Exploration* 7:96-113.

Fredericksen, C., M. Spriggs and W. Ambrose 1993. Pamwak rock shelter: A Pleistocene site on Manus Island, Papua New Guinea. In M.A. Smith, M. Spriggs and B. Fankhauser (eds), *Sahul in review: Pleistocene archaeology in Australia, New Guinea and Island Melanesia. Occasional Papers in Prehistory* 24, pp144-152. Canberra: Department of Prehistory, Research School of Pacific Studies, Australian National University.

Freyne, D.F. and G.S. Bell 1982. *Forest and land resources of Manus and Los Negros Islands. Research Bulletin* 31. Papua New Guinea: Department of Primary Industry.

Garret-Jones, S.E. 1979. Evidence for changes in Holocene vegetation and lake sedimentation in the Markham Valley, Papua New Guinea. Unpublished PhD thesis, Australian National University, Canberra.

Ghai, Y and A.J. Regan 1992. *The law, politics and administration of decentralisation in Papua New Guinea. Monograph* 30. Papua New Guinea: National Research Institute.

Golson, J. 2005. Introduction to the chapters on archaeology and ethnology. In A. Pawley, R. Attenborough, J. Golson and R. Hide (eds), *Papuan Pasts: Cultural, linguistic and biological histories of Papuan-speaking peoples*, pp221-233. Canberra: Pacific Linguistics, RSPAS, ANU.

Gosden, C. 1992. Production systems and the colonization of the western Pacific. *World Archaeology* 24:55-69.

Gosden, C. 1993. Understanding the settlement of Pacific Islands in the Pleistocene. In M. A. Smith, M. Spriggs, B. Fankhauser (eds), *Sahul in review: Pleistocene archaeology in Australia, New Guinea and Island Melanesia. Occasional Papers in Prehistory* 24, pp131-136. Canberra: Department of Prehistory, Research School of Pacific Studies. Australian National University.

Gosden, C. 1995. Arboriculture and agriculture in coastal Papua New Guinea. In J. Allen and J. O'Connell (eds), *Transitions: Pleistocene to Holocene in Australia and Papua New Guinea. Antiquity Special No.* 265, pp807–817. Oxford: Oxford University Press.

Green, R.C. 2003. The Lapita horizon and traditions signature for one set of Oceanic migrations. In C. Sand (ed.), *Pacific Archaeology: Assessments and prospects. Proceedings of the International conference for the 50th anniversary of the first Lapita excavation. Le Cahiers de l'Archéologie en Nouvelle Calédonie* 15, pp95-105. Nouméa: Départemente Archéologie, Service des Musées et du Patrimoine de Nouvelle-Calédonie.

Greig-Smith, P., M.P. Austin and T.C. Whitmore 1967. The application of quantitative methods to vegetation survey. I. Association analysis and principal component ordination of rain forest. *Journal of Ecology* 55:483–503.

Gunn, B. 2007. Facilitating the availability and use of improved germplasm for forestry and agroforestry in Papua New Guinea. *ACIAR Papua New Guinea and Solomon Islands Nius* 2(2):1-4.

Haberle, S.G. 1995. Identification of cultivated *Pandanus* and *Colocasia* in pollen records and the implications for the study of early agriculture in New Guinea. *Vegetation History and Archaeobotany* 4:195-210.

Haberle, S.G. 1998. Late Quaternary vegetation change in the Tari Basin, Papua New Guinea. *Palaeogeography, Palaeoclimatology, Palaeoecology* 137:1-24.

Haberle, S.G. and B. David 2004. Climates of change: Human dimensions of Holocene environmental change in low latitudes of the PEPII transect. *Quaternary International* 118–119:165-179.

Haberle, S.G. 2007. Prehistoric human impact on rainforest biodiversity in highland New Guinea. *Philosophical Transactions of the Royal Society B: Biological Sciences* 362:219-228.

Hart, T.B. 1990. Monospecific dominance in tropical rain forests. *Trends in Ecology and Evolution* 5:6-11.

Hart, T.B., J.A. Hart and P.G. Murphy 1989. Monodominant and species-rich forests of the humid tropics: Causes for their co-occurrence. *American Naturalist* 133:613-633.

Heinsohn, T. 2003. Animal translocation: Long-term human influences on the vertebrate zoogeography of Australasia (natural dispersal versus ethnophoresy). *Australian Zoologist* 32:351-376.

Hide, R.L., B.J. Allen, R.M. Bourke, D. Fritsch, R. Grau, J.L. Helepet, P. Hobsbawn, S. Lyon, M. Poienou, S. Pondrilei, K. Pouru, G. Sem and B. Tewi 1996. *Manus Province: Text summaries, maps, code lists and village identification. Agricultural systems of Papua New Guinea Working Paper* 18. Canberra: Department of Human Geography, Research School of Pacific and Asian Studies, Australian National University.

Hope, G.S. 1996. History of *Nothofagus* in New Guinea and New Caledonia. In T.T. Veblen, R.S. Hill and J. Read (eds), *The Ecology and biogeography of Nothofagus forests,* pp257-270. New Haven, Connecticut: Yale University Press.

Hope, G.S. 2007. Palaeoecology and palaeoenvironments of Papua. In A.J. Marshall and B.M. Beehler (eds), *The Ecology of Papua,* pp255–266. Singapore: Periplus Editions.

Hope, G.S. 2009. Environmental change and fire in the Owen Stanley Ranges, Papua New Guinea. *Quaternary Science Reviews* 28:2261-2276.

Hope, G.S. and S.G. Haberle 2005. The history of human landscapes of New Guinea. In A. Pawley, R. Attenborough, J. Golson and R. Hide (eds), *Papuan Pasts: Cultural, linguistic and biological histories of Papuan-speaking peoples*, pp541-554. Canberra: Pacific Linguistics, RSPAS, ANU.

Hope, G.S. and J. Tulip 1994. A long vegetation history from lowland Irian Jaya, Indonesia. *Palaeogeography, Palaeoclimatology, Palaeoecology* 109:385-398.

Hubbell, S.P. 2001. *The unified neutral theory of biodiversity and biogeography*. Princeton and Oxford: Princeton University Press.

Johns, R.H. 1986. The instability of the tropical ecosystem in New Guinea. *Blumea* 31:341-371.

Juggins, S., 2005. C2 Data Analysis Version 4.2. Newcastle: University of Newcastle.

Kennedy, J. 1981. Lapita colonisation of the Admiralty Islands? *Science* 213:757-759.

Kennedy, J. 1982. Archaeology in the Admiralty Islands: Some excursions. *Indo-Pacific Prehistory Association Bulletin* 3:22-35.

Kennedy, J. 1983. On the prehistory of Western Melanesia: The significance of new data from the Admiralties. *Australian Archaeology* 16:115-122.

Kennedy, J. 1991. Who should protect Papua New Guinea's archaeological heritage? in S.M. Eoe and P. Swadling (eds), *Museums and cultural centres in the Pacific*, pp148–151. Port Moresby: Papua New Guinea National Museum.

Kennedy, J. 1992. Two years of archaeological salvage in southwest Manus. *Research in Melanesia* 16:167-174.

Kennedy, J. 1997. The loneliness of an obsidian source in southwest Manus. *Archaeology in Oceania* 32:85-96.

Kennedy, J. 2002. Manus from the beginning: An archaeological overview. In C. Kaufmann, C. Kocher Schmid and S. Ohnemus (eds), *Admiralty Islands: Art from the South Seas*, pp17–28. Zürich: Museum Rietberg.

Kennedy, J. and W. Clarke 2004. *Cultivated landscapes of the southwest Pacific. RMAP Working Paper* 50. Canberra: Resource Management in Asia Pacific Program, Research School of Pacific and Asian Studies, Australian National University.

Kennedy, J., F. Wadra, U. Akon, R. Busasa, J. Papah and M. Piamnok 1991. Site survey of southwest Manus: A preliminary report. *Archaeology in Oceania* 26:114-118.

Kerenga, K. and J. Croft 1984/85. Notes on the vegetation at the collection sites of the Division of Botany expedition to Manus, 1981. *Science in New Guinea* 11:43-49.

Kershaw, A.P. and B.P.M. Hyland 1975. Pollen transfer and periodicity in a rainforest situation. *Review of Paleobotany and Palynology* 19:129-38.

Kirch, P.V. 2000. *On the road of the winds*. Berkeley and Los Angeles: University of California Press.

Kraft, N.J.B., R. Valencia, and D.D. Ackerly 2008. Functional traits and niche-based tree community assembly in an Amazonian forest. *Science* 322:580-882.

Lilley, I. 2006. Archaeology in Oceania: Themes and issues. In I. Lilley (ed.), *Archaeology of Oceania: Australia and the Pacific Islands*, pp1-28. Oxford, UK: Blackwell Publishing.

McEldowney, H. and C. Ballard 1991. The Mouk site: Manus as paradox or parable in reconstructions of the Lapita Cultural Complex. In J. Allen and C. Gosden (eds), *Report of the Lapita Homeland Project. Occasional Papers in Prehistory* 20, pp92-102. Canberra: Department of Prehistory, Research School of Pacific Studies, Australian National University.

McGuire, K.L. 2007. Common ectomycorrhizal networks may maintain monodominance in a tropical rain forest. *Ecology* 88:567-574.

Mueller-Dombois, D. and F.R. Fosberg 1998. *Vegetation of the tropical Pacific islands*. New York: Springer-Verlag.

National Geographic 2009. http://www.nationalgeographic.com/wildworld/profiles/terrestrial/aa/aa0101.html accessed 1 September 2009.

O'Collins, M. and L. Lamothe 1989. *The assessment of the impact of forest operations on the environment and village communities.* Unpublished report of field work undertaken on the west coast of Manus by students of the Forestry Department, Papua New Guinea University of Technology, Lae.

O'Connell, J.F. and J. Allen 2004. Dating the colonization of Sahul (Pleistocene Australia–New Guinea): A review of recent research. *Journal of Archaeological Science* 31:835-853.

O'Connor, S. and J. Chappell 2003. Colonisation and coastal subsistence in Australia and Papua New Guinea: Different timing, different modes? In C. Sand (ed.), *Pacific Archaeology: Assessments and prospects. Proceedings of the International conference for the 50th anniversary of the first Lapita excavation. Le Cahiers de l'Archéologie en Nouvelle Calédonie* 15, pp17-32. Nouméa: Départemente Archéologie, Service des Musées et du Patrimoine de Nouvelle-Calédonie.

Osborne, P.L., N.V.C. Polunin, and R.G. Totome 1996. Sediments as indicators of ecosystem function in four contrasting lakes in Papua New Guinea. In F. Schiemer and K. Boland (eds), *Perspectives in tropical limnology*, pp131–149. The Netherlands: SPB Academic Publishing.

Potts, M.D., P.S. Ashton, L.S. Kaufman and J.B. Plotkin 2002. Habitat patterns in tropical rain forests: A comparison of 105 plots in northwest Borneo. *Ecology* 83:2782–2797.

Prebble et al. In prep.

Read, J., P. Hallam and J.-F. Cherrier 1995. The anomaly of monodominant tropical rainforests: Some preliminary observations in the *Nothofagus*-dominated rainforests of New Caledonia. *Journal of Tropical Ecology* 11:359–389.

Read, J., T. Jaffré, E. Godrie, G.S. Hope and J.-M. Veillon 2000. Structural and floristic characteristics of some monodominant and adjacent mixed rainforests in New Caledonia. *Journal of Biogeography* 27:233–250.

Rooney, W.J. 1982. Changing food supply systems in eastern inland Manus. In R.M. Bourke and V. Kesavan (eds), *Proceedings of the second Papua New Guinea food crops conference, part 2*, pp273–278. Papua New Guinea: Department of Primary Industry.

Seacology 2008. <http://www.seacology.org/projects/individualprojects/PNG_manus2008.htm> accessed 1 September 2009.

Southern, W. 1987. *The vegetation history of Lahakai Swamp, Manus Island, Papua New Guinea: A preliminary report.* Unpublished ms.

Specht, J. 2005. Revisiting the Bismarcks: Some alternative views. In A. Pawley, R. Attenborough, J. Golson and R. Hide (eds), *Papuan Pasts: Cultural, linguistic and biological histories of Papuan-speaking peoples*, pp235–288. Canberra: Pacific Linguistics, RSPAS, ANU.

Spriggs, M.J.T. 1997. *The Island Melanesians.* Oxford: Blackwell.

Spriggs, M. 2003. Post-Lapita evolutions in Island Melanesia. In C. Sand (ed.), *Pacific Archaeology: Assessments and prospects. Proceedings of the International conference for the 50th anniversary of the first Lapita excavation. Le Cahiers de l'Archéologie en Nouvelle Calédonie* 15, pp213–220. Nouméa: Départemente Archéologie, Service des Musées et du Patrimoine de Nouvelle-Calédonie.

Stevens, P.F. 1974. A review of *Calophyllum* L. (Guttiferae) in Papuasia. *Australian Journal of Botany* 22:349-411.

Stevens, P.F. 1980. A revision of the Old World species of *Calophyllum* (Guttiferae). *Journal of the Arnold Arboretum* 61:117-699.

Stevens, P.F. 1995. Guttiferae. In B.J. Conn (ed.), *Handbooks of the flora of Papua New Guinea, Volume 3*, pp61-125. Carlton, Victoria: Melbourne University Press.

Stevens, P.F. 2007. Clusiaceae-Guttiferae. In K. Kubitzki, C. Bayer and P. F. Stevens (eds), *The families and genera of vascular plants. Volume 9. Flowering plants: Eudicots: Berberidopsidales, Buxales, Crossosomatales, Fabales, Geraniales, Gunnerales, Myrtales, Proteales, Saxifragales, Vitales, Zygophyllales, Clusiaceae Alliance, Passifloraceae Alliance, Dilleniaceae, Huaceae, Picramniaceae, Sabiaceae*, pp48–66. Springer.

Stokstad, E. 2009. On the origin of ecological structure. *Science* 326:33-35.

Svenning, J.-C., D.A. Kinner, R.F. Stallard, B.M.J. Engelbrecht and S.J. Wright 2004. Ecological determinism in plant community structure across a tropical forest landscape. *Ecology* 85:2526–2538.

Swadling, P. and G.S. Hope 1992. Environmental change in New Guinea since human settlement. In J.R. Dodson (ed.), *The naive lands: Prehistory and environmental change in the South West Pacific*, pp13-42. Melbourne: Longman Cheshire.

Terrell, J.E. 2004. Introduction: 'Austronesia' and the great Austronesian migration. *World Archaeology* 36:586-590.

Terrell, J.E. and R.L. Welsch 1997. Lapita and the temporal geography of Lapita. *Antiquity* 71:548–572.

Torti, S.D., P.D. Coley and T.A. Kursar 2001. Causes and consequences of monodominance in tropical lowland forests. *American Naturalist* 157:141-153.

Wadra, F. 1991. *Around Club Mud: A descriptive catalogue of obsidian artefacts collected from southwest Manus, 1989-1991.* Unpublished thesis, Postgraduate Diploma in Arts. Department of Anthropology and Sociology, University of Papua New Guinea, Waigani.

Walker, D. and Y. Chen 1987. Palynological light on tropical rainforest dynamics. *Quaternary Science Reviews* 6:77-92.

Wheatley, J. 1992. *A guide to the common trees of Vanuatu.* Port Vila, Vanuatu: Department of Forestry.

Whitmore, T.C. 1974. Change with time and the role of cyclones in the tropical rain forest on Kolombangara, Solomon Islands. *Commonwealth Forestry Institute Paper* 46. Oxford: University of Oxford.

Whitmore, T.C. 1984. *Tropical rain forests of the Far East.* 2nd edition. Oxford: Clarendon.

Whitmore, T.C. 1989. Changes over twenty-one years in the Kolombangara rain forests. *Journal of Ecology* 77:469-483.

Willis, K.J., L. Gillson and T.M. Brncic 2004. How 'virgin' is virgin rainforest? *Science* 304:402-403.

Wills, C., K.E. Harms, R. Condit, D. King, J. Thompson, F. He, H.C. Muller-Landau, P. Ashton, E. Losos, L. Comita, S. Hubbell, J. LaFrankie, S. Bunyavejchewin, H.S. Dattaraja, S. Davies, S. Esufali, R. Foster, N. Gunatilleke, S. Gunatilleke, P. Hall, A. Itoh, R. John, S. Kiratiprayoon, S. Loo de Lao, M. Massa, C. Nath, M.N.S. Noor, A.R. Kassim, R. Sukumar, H.S. Suresh, I-F. Sun, S. Tan, T. Yamakura and J. Zimmerman 2006. Nonrandom processes maintain diversity in tropical forests. *Science* 311:527-531.

World Wildlife Fund 2001. http://www.worldwildlife.org/wildworld/profiles/terrestrial/aa/aa0101_full.html

Yen, D.E. 1991. Domestication: The lessons from New Guinea. In A. Pawley (ed) *Man and a half: Essays in Pacific anthropology and ethnobiology in honour of Ralph Bulmer. Memoir* 48, pp558–569. Auckland: The Polynesian Society.

Yen, D.E. 1996. Melanesian arboriculture: Historical perspectives with emphasis on the genus *Canarium.* In M.L. Stevens, R.M. Bourke and B.R. Evans (eds), *South Pacific indigenous nuts.* ACIAR Proceedings 69, pp36–44. Canberra: Australian Centre for International Agricultural Research.

Zimmerman, J.K., J. Thompson and N. Brokaw 2008. Large tropical forest dynamics plots: Testing explanations for the maintenance of species diversity. In W.P. Carson and S.A. Schnitzer (eds), *Tropical forest community ecology*, pp98–117. Chichester, UK: Wiley-Blackwell.

13

Geomorphic and archaeological consequences of human arrival and agricultural expansion on Pacific islands: A reconsideration after 30 years of debate

Matthew Spriggs

School of Archaeology and Anthropology, Australian National University, Canberra, Australia
Matthew.Spriggs@anu.edu.au

Introduction: The distorted history of an idea

The late 1970s was an exciting time to be a PhD student in archaeology at the ANU, with many new ideas flying around and being discussed by staff and students, usually over a beverage or two or three at the ANU Staff Club. Prominent among sometimes-heated discussions were those pertaining to human-environment relations. The dominant paradigm up to that time in the Pacific was a fairly pedestrian environmental determinism, with prehistoric humans having had very little agency (although at the time we did not use, indeed did not know the word in such a context: see Dobres and Robb 1999). Humanly induced, or at least humanly accelerated, landscape change was the latest idea, and the ferment of a paradigm shift was everywhere. Geoff Hope was one of the prominent exponents of the new ideas in those discussions. He and others of like mind were key influences on my thinking as I embarked on PhD fieldwork on the island of Aneityum in southern Vanuatu in mid-1978 (Spriggs 1981).

This time and these intellectual influences are worth reflecting on, as some recent discussions of the history of debate over human-environment relations in the Pacific seriously misrepresent that history. If we are to believe Patrick Nunn, for instance, the 'orthodoxy' among Pacific archaeologists has *always* been (and indeed remains today) that human acts have shaped Pacific environments to the exclusion of natural factors (Nunn 1991, 2001, 2003, 2007:169-71; cf. Kumar et al. 2006:141). To the contrary, during the period in question, the orthodoxy was the opposite to that portrayed by Nunn, and we felt that a battle needed to be fought to establish some human involvement in the story of Pacific environmental change.
Kirch (1997a:3-11) provides a very useful potted history of the issue of humans and their place in island ecosystems (cf. Fosberg 1963), in an edited volume (Kirch and Hunt 1997) that represented the first wide-ranging synthesis exploring pre-industrial human impacts on

Pacific environments. Kirch's story starts with Louis-Antoine de Bougainville arriving in Tahiti in 1769, himself influenced by the 'homme naturel' views of Rousseau. The 'noble savage' trope long held sway until some time after World War II, with only a few early dissenters who saw some reciprocal human influence on the environment of the islands.

The development of the 'New Archaeology' in the United States and parallel developments in the United Kingdom by the late 1960s, with their ecologically based frameworks, led to a change in view among Pacific archaeologists. The general recognition of the role of humans in environmental change in the region only occurred in the mid- to late-1970s as archaeologists and natural scientists began to work together at institutions in places such as Hawaii and Canberra. In 1997, Kirch specifically questioned Nunn's 'curious, and to our mind unfounded, assertion that most writers have attributed environmental change on Pacific islands to human impact' (Kirch 1997a:15, referencing Nunn 1991). Nunn (2003) acknowledges but clearly fails to understand Kirch's point here.

It should be noted that none of the proponents of the argument for an important role for human-induced erosion in Pacific landscape evolution argued that all prehistoric environmental change was human-induced. For example, the interplay of human actions and natural cyclonic and other high-rainfall events was discussed, and major earthquake activity also canvassed, as being significant erosional agents on Aneityum (Spriggs 1981). The differences in landscape susceptibility to erosion between the wetter windward and drier leeward sides of islands were also noted, with the latter seen as far more vulnerable to disturbance (Spriggs 1981: Chapter 5, cf. Spriggs 1985). It was particularly the interplay between human interference with the vegetation and natural catastrophic events that was highlighted.

The 'Spriggs hypothesis'

My 1981 PhD postulated a series of three linked propositions relating to Pacific environmental change. The first, not at all original but crucial to the other two, was that much of the mid to late-Holocene landscape change on Pacific Islands was humanly accelerated, if not humanly induced (Proposition 1). Two further propositions – labelled by the late Roger Green, one of the thesis examiners, the 'Spriggs hypothesis' – developed from this. Proposition 2 was that in certain geological circumstances the effects of such prehistoric landscape changes would make it very difficult for archaeologists to find early cultural sites on volcanic islands, as they would be deeply buried by the products of erosion. This was used to explain putative gaps in Lapita site distribution on many larger Melanesian islands, such as the Solomon Islands chain and in southern Vanuatu. Proposition 3 was that the effects of such early humanly accelerated erosion were beneficial to human settlement on the islands in the medium to long term, by creating large coastal plains where much of the population currently lives. Previously, terms such as 'degradation' had been used to describe such human impacts (as in Hughes et al. 1979). A more tentative corollary was suggested, that at least in some cases the erosion was intentionally induced (Proposition 3a).

A fourth proposition was added post-thesis, that the initial response to (in the short-term sometimes disastrous) humanly accelerated erosion was to move settlement to another area. As Pacific populations grew, this response was no longer possible. Reanalysis of the original Aneityum results suggested a pattern of significant human environmental impacts immediately after initial settlement, followed by a hiatus of several hundred years in occupation or use at particular locations. When use of an area began again, usually for gardening, it was seemingly continuous to the present (Spriggs 1997). Similar findings were being published in the 1980s and early 1990s from other islands such as East Futuna (Frimigacci 1990) and O'ahu, Hawaii (Allen et al. 1987). These suggested a more general model.

Initially, in a land-rich situation with low population densities, conservation measures such as terracing were not considered necessary. Any perceived environmental degradation in the short-term could be alleviated by moving to another valley or even another island. Later, in land-full conditions, this option was not possible. In addition, with larger populations than were present at the earliest stages of settlement, considerable investment in capital works such as soil-retention terraces and storm drains could be undertaken. When catastrophic erosion buried irrigation systems, they were immediately rebuilt (Spriggs 1997:97-98).

A corollary of this model is that conservation measures associated with more intensive agricultural practices developed over time on islands in response to changing subsistence and social demands. People could indeed learn from their earlier mistakes. The investment in erosion-control measures both allowed, and to some extent required, permanent occupation of the affected lands. Indeed, requirements for the maintenance of such capital works may have essentially 'tethered' people to them. The rest of the paper examines how these four linked propositions have stood the test of time since those heady days at the ANU in the late 1970s.

Proposition 1: Human-accelerated landscape change

In the late 1970s, the general view around the ANU among archaeologists and natural scientists was that there were no significant effects in Australia or the Pacific Islands of the Little Ice Age (LIA) and other postulated late-Holocene climatic changes. This again was in reaction to earlier geomorphological thought where minor climatic variations were used to explain any evidence for Holocene deposition of extensive areas of alluvium.

This earlier view had been criticised by Butzer (1974) in relation to the Mediterranean and, closer to home, by Hughes and Sullivan (1981) in relation to alluvial sequences in New South Wales. Noting the lack of correlation between the dating of the postulated climatic changes and sequences of valley fill, they suggested that Aboriginal burning patterns removed leaf litter more frequently than would have happened by natural bushfires and over a longer period of the year. This increased the susceptibility of the soil to rain-splash erosion and led to accelerated erosion. In my thesis, I concluded: 'It is not clear how minor climatic changes could have so altered an established vegetation cover as to lead to the greatly accelerated erosion and valley infilling which have been the object of investigation' (1981:97).

Subsequently, evidence has been marshalled – particularly, but by no means exclusively, by Nunn (1997) – to suggest much greater environmental instability as a result of climatic perturbations affecting our region. We are now in a much better position to evaluate this idea than we would have been in the late 1970s and 1980s. Nunn's main contention seems to be that it was not so much the LIA itself, but the 'AD 1300 event', a period of very rapid climate change marking its start, which led to major landscape instability (Nunn et al. 2007). The supposed effects include a sea-level fall of at least 0.8 m that exposed and killed reefs, and lowered water tables all over the islands, which thus affected agricultural productivity and caused stream incision, which in turn caused slope instability and erosion. This was associated with a sudden cooling and increased storminess.

Nunn claims the AD 1300 event resulted in a catastrophic food crisis for Pacific Island communities, with food resources falling 'by around 80%' (Nunn 2003:224). The human response was suggested to be a move into marginal and inland areas away from the coast, first occupation of many smaller islands, abandonment of other isolated small islands, changes in resource exploitation, reduced interaction between communities except for warfare and cannibalism, the end of long-distance voyaging across the Pacific, the end of pottery making in some areas, the appearance of new belief systems, the construction of megalithic monuments and localised pig extirpation (Nunn 2007: Chapter 6).

If such an event did occur, and if such human responses were indeed linked to it, then it would indeed have been a major period of cultural change in the Pacific. But on every point, from cooling to sea-level change to resource depletion, Nunn's assertions can be challenged.

Allen (2006) answers most of Nunn's assertions decisively, indeed devastatingly. Using recent palaeoclimatic data from the region itself, she shows that the previous Medieval Warm Period (MWP) was a time of colder weather in the equatorial Pacific, while the LIA was warmer - the opposite of the Northern Hemisphere-derived model Nunn is using to underpin his views. Changes there may have been, but there is little to suggest they were as abrupt and of such magnitude. On Nunn's own reading of the preceding MWP period, there were times within it of quite extreme climatic change. This implies that people had long been exposed to sudden changes in climate before any 'event' such as that postulated by Nunn. Suitable responses, therefore, may have already been developed for such eventualities. If a 1300 AD 'event' did occur, it was not a one-off. The sea-level data on which much of the model is based have also been questioned. Gehrels (2001), for instance, points out that the suggested relationship between sea level and climate is compromised by height uncertainties and dating problems, including Nunn's (1998) failure to use calibrated (calendar) radiocarbon dates.

Climatic instability does need to be added into the mix in regards to the degree of humanly accelerated landscape change that has occurred in the region during human occupation. For instance, in the past decade, attention has been drawn to changing cyclone frequencies in particular areas. These variations relate to ENSO cycles and are seen as significant in 'forcing' cultural change (Anderson et al. 2006). Humanly accelerated or induced erosion must always be considered within the context of other significant processes causing slope instability, such as climatic changes leading to altered cyclone frequency, earthquakes leading to uplift or subsidence, and deposition of volcanic ash changing slope susceptibility to catastrophic landslides. But even after these factors are taken into account, it seems absurd to deny that humans, by vegetation clearance for agriculture and other purposes, have had a major effect on the landscape of Pacific islands. To do so is to deny them agency, the ability to act on the land and shape it – for good or ill. Instead, they are seen as passive victims of climatic change, their existence always completely subject to the perturbations of nature they can have no control over and no power to ameliorate. The evidence remains overwhelmingly to the contrary (see below, Proposition 3a).

Propositon 2: Difficulty of finding early sites

The Southern Vanuatu Culture History Project was begun in 1983 specifically to test the idea that humanly accelerated landscape change was affecting early site visibility. The project concentrated on the island of Erromango, where it was predicted that early pottery-bearing sites would be easily located, adjacent to stream mouths on the island and protected from deep alluvial burial by the raised limestone terrace fringe of an otherwise volcanic island (Spriggs and Wickler 1989). The Imponkor limestone formation along the eastern side of the island was targeted, an extensive area of recent raised reef about 2-7 m above sea level (Colley and Ash 1971:48-49). As the formation was backed by older reef terraces, depositional conditions were therefore analogous to those expected to obtain on low coralline islands where site visibility is high and where it had been argued that Lapita sites were preferentially located (Green 1979:32).

Pottery was found at Naen near Ipota along the Imponkor limestone belt, and excavation further south again at Ifo revealed dentate-stamped Lapita pottery (Spriggs and Wickler 1989). This provided strong evidence against the idea that Lapita people preferred low coralline islands for their settlements. This apparent pattern was an artefact of post-depositional changes reducing site visibility on larger volcanic islands. Early sites were thus likely to be deeply buried by the products of erosion. It was a significant issue at the time because Bellwood (1978:262, 264) and Green (1979:47-48) had previously argued for a pre-Lapita occupation of Remote Oceania (specifically Vanuatu and New Caledonia), partly on the basis of the supposedly aceramic nature of the southern Vanuatu archaeological record (see Spriggs 1984 for discussion). Further development of this project was cut short by the general research ban applied by the Vanuatu government in 1985, which lasted for a decade.

Green (1985) was an early commentator on Proposition 2, finding it much more persuasive for areas in Remote Oceania where Lapita is likely to be the founding culture. He noted that the old idea that Lapita sites were found more generally in marginal locations on small islands offshore of islands which already had a population had not entirely stood the test of time, even without the 'alluvial hypothesis'. But this did not stop him from invoking just such a possibility for Near Oceania. He maintained that this was still a viable alternative to the proposition that if there are sites in such marginal locations, then there must be many more on the main islands.

This proposition is still very much contested and is especially pertinent to current debates about Lapita site distribution (or rather the gaps in it) across much of the main Solomons chain. Sheppard and Walter (2006, 2009), supporting Green (1985), would see the lack of Lapita sites as a real absence rather than a sampling problem as would be argued from Proposition 2. They were responding in this instance, however, to the arguments of Felgate (2003, cf. 2007), who has raised the problem of sea-level change obscuring the presence of what were originally stilt-house village sites in the nearshore sub-tidal zone of many islands. Felgate's argument was not so much that early sites were deeply buried, but instead that they were located offshore and so archaeologists had been looking in the wrong places for them, needing to get their feet wet and wade out on to the reefs or even don snorkel and mask to explore the deeper depths.

Such a finding had originally been reported by Stephen Wickler, researching Lapita sites on the island of Buka and adjacent smaller islands (Wickler 1995, 2001a). A recreational wander over the reef at Kessa Plantation had revealed Lapita pottery scattered across and indeed embedded into the reef. His subsequent work on Sohano Island just off Buka revealed similar sites. Lapita stilt-house villages were already known as a site type in the Bismarck archipelago, but 3000 years of coastal progradation had covered up their remains. This meant that excavation of them in 1985, as part of the Lapita Homeland Project (Allen and Gosden 1991), was conducted from what is now dry land (Gosden and Webb 1994; Kirch 1987, 1988a).

Others have now entered the debate. Clark and Bedford (2008:59) use the ecological concept of 'friction zones', landscapes that are 'hostile, fragmented, unfamiliar or difficult to reach' and so are avoided by early colonists. Their study of apparent 'friction zones' within the overall Lapita distribution concluded that an avoidance of the main Solomons chain by early Lapita colonists would be extremely unlikely. Most recently, a linguist, Andy Pawley, has weighed in on the debate (Pawley 2009). His conclusion is that for the northwest Solomons the leap-frog model of Sheppard and Walter could explain the linguistic situation. But it is unlikely on linguistic grounds that the rest of the main Solomons from Guadalcanal to Makira remained empty of Austronesian speakers until 2300-2000 BP. We should expect to find early Lapita sites therefore at points across that region. The problem is thus sampling, and a likely cause of difficulty finding such sites is post-Lapita landscape change on these high volcanic islands.

Systematic research recommenced in Vanuatu in 1994, immediately after the 1985 research ban had been lifted. It has shown that burial by volcanic ash of early sites on many islands in that archipelago is perhaps a more significant factor in obscuring their location than burial by humanly accelerated erosion. Ironically, Aneityum and Erromango are among the few islands where recent volcanism has not had any significant effect in terms of tephra deposition. Early sites on Efate and other central Vanuatu islands, Malakula, Ambae, parts of Santo, and on some of the Banks islands have all been located – sometimes with great difficulty because of depth of burial – beneath such volcanic deposits. The disadvantage of difficulty of site location is compensated for by the often exceptional preservation conditions that result from the sites being emplaced well below the 'plough zone' of traditional agriculture (see Bedford et al. 1998, 2006; Bedford and Spriggs 2000, 2008; Spriggs and Bedford 2001; Bedford 2006).

Human-accelerated erosion has also been blamed for the gaps in the distribution of other early settlement sites in the Western Pacific. Clark and Anderson (2001) invoke such an idea in relation to western Fiji, concluding: 'there is little reason to suspect that within the western group colonisation sites were preferentially distributed on islands adjacent to Viti Levu and Vanua Levu rather than on the major islands themselves' (Clark and Anderson 2001:81).

Kirch and colleagues provided a sophisticated model of landscape change for the Lapita-derived site of Toaga on Ofu in American Samoa, invoking both subsidence and sea-level fall after the mid-Holocene high stand, and slope instability caused by vegetation clearance as operative processes (Kirch and Hunt 1993). The site existed on what was initially a narrow coastal flat below a steep hillside. It was only found after a landfill site was gouged out of the present colluvial slopes of the area, revealing the deeply buried Lapita-derived early pottery site. In places, this is buried by up to 10-15 m of colluvium. Kirch and Hunt (1993:234-236) extend their model to suggest application not only elsewhere within American Samoa, but also in the search for early sites in the southern Cooks, the Australs and the Society Islands. Variants of the model have also been shown to have value for Mangaia in the Cooks (Kirch 1997b), Niuatoputapu in Tonga (Kirch 1988b) and more recently Palau (Wickler 2001b) – but see Dickinson and Athens (2007) for a useful corrective concerning the influence of subsidence for that last island group, perhaps also applicable to other cases.

Subsidence was also established as the reason for the lack of early Lapita sites in Western Samoa, where Dickinson and Green (1998) explained the occurrence of the only true Lapita site yet found in Samoa some 4 m below sea level under a later capping of coral reef growth. In this case, the serendipity of discovery was the need for a deep-water ferry terminal at Mulifanua, requiring a major dredging operation.

It is now clearer that multiple processes can affect site visibility, both natural – as with tephra deposition – and humanly accelerated. Conventional surface survey is unlikely to provide a full picture of Lapita site distribution in either case, and hypotheses based simply on apparent absence of sites, without due consideration of the geomorphological context, are constructed on weak foundations.

Proposition 3: The benefits of human-accelerated erosion

The major benefit was seen in the creation of large coastal plains where most of the population lives today. This is now so well accepted that it is often stated as a given. Leaving aside the cause of the creation of such plains for the moment, it is empirically easy to verify that this is indeed where much of the population lives today and that such areas are often, or were at least until recently, the site of major irrigation or other garden systems providing much of the vegetable produce for growing populations. Kirch (1997b) gives a particularly good illustration of such enhancement in the case of Mangaia in the Cook Islands.

Nunn (1991) has pointed out that in some cases the bedrock substrate for such plains is formed by reef deposits that were raised above sea level by either sea-level fall after the mid-Holocene high stand or by tectonic uplift. But the soil still had to come from somewhere, and natural erosion rates are arguably unlikely to have provided sufficient depth of material across these plains during the past 3000 years to make them productive.

The common presence of dispersed charcoal and sometimes significant charcoal lenses within the alluvial deposits that make up these plains instead suggests another cause: the instability generated by clearing vegetation from slopes upstream using fire, to establish gardens and for other clearance purposes. Natural fire is unlikely as the general explanation, given that fire is rare or absent from pre-human pollen cores in most areas of the tropical Pacific. Even where natural fire is recorded in such cores, there is a quantum increase with the arrival of humans – whether they arrived 3000 years ago or fewer than 1500 years ago. Hope et al. (1999) provide a good summary of examples of these phenomena.

A test of whether natural erosion rates could create these coastal plains is readily to hand. One could compare coastal-plain deposition rates for the past 3000 years on early-occupied islands in island Melanesia and Western Polynesia, with islands settled only 1500 or fewer years ago in Eastern Polynesia. To my knowledge, such a test has not been undertaken.

Propositon 3a: The intentionality issue

The somewhat tentative conclusion was made that, at least in some cases, the erosion leading to the development of such alluvial plains was intentionally induced. This implies strong human agency in the process, and a degree of control of the environment that is quite remarkable for pre-industrial or pre-urban civilisations. Nunn is sceptical, suggesting that 'such understanding of environmental management appears implausible' (Nunn 2003:222). It is, of course, extremely difficult to prove archaeologically.

This sort of control is attested ethnographically, but it could be argued that the evidence may represent influences from the colonial situation (cf. Spriggs 2008). Such influence is possible, for instance, in the practice of 'hydraulicking' to move sediment from one place to another in the construction of new rice terraces in the Ifugao region, as recorded in the later 20th century several hundred years after Spanish colonialism began in the Philippines (Conklin 1980:16-17, discussed in Spriggs 1997). But such an explanation would be more unlikely for the immediate post-contact period in the New Guinea Highlands. There, the directed use of water to erode soil as part of traditional mining operations was recorded at the Ganz-Tsenga stone-axe quarries: 'The quarry was in a clearing on the hillside, and a water race led to the top of it to wash away ground and expose the rocks' (Vial 1940:159). The deliberate and spectacular use of erosion, this time for not fully understood cultural reasons, was also recorded in a situation unlikely to be influenced by colonial practices in the Wantoat area of the Finisterre ranges. Ponds were constructed to hold large amounts of water above steep slopes. At a given sign, the sluices were opened all at the same time and the ensuing flood washed soil, rocks and trees downslope (Vial 1937; Schmitz 1963). The process of deliberately inducing erosion was clearly understood within the Pacific.

There are indeed several possible archaeological examples of the process. Hughes et al. (1979:109) suggested not only that the productive sub-coastal swamplands developed on Lakeba after human arrival as a result of vegetation clearance, but that their creation may have been actively encouraged by the people of the island. Deposition of soil, branches and tree trunks in valley bottoms, assisted by severe cyclones, choked valley outlets, and sediment-charged run-off was ponded behind such barriers to create swamps used for growing taro. Hope et al. (1999:396) have suggested that swamps in the upland area of Evoran in Erromango, Vanuatu, were purposely created by blocking a creek line with earthworks across a valley. A date of 2000 BP has been obtained for the creation of one of these ponds.

From the Keanae Peninsula on Maui comes what seems to be one of the most extreme forms of deliberate soil formation. What became productive irrigated taro lands was originally a barren lava flow. Oral tradition records that a chief, seeking to increase production in his domain, set his people to work 'carrying soil in baskets from the valley down to the lava point. The soil and the banks enclosing the patches were thus, in the course of many years, all transplanted and packed into place' (Handy 1940:110). A quick look at the *Soil Survey of the Territory of Hawaii* maps and accompanying soil descriptions certainly does not rule out this possibility (Cline et al. 1955:347, 607); direct archaeological investigation would settle the matter.

By the time of European contact, the peoples of the Pacific were engaged in sophisticated landscape engineering involving construction of stone-faced terracing, artificial channelling of water through construction of canals, storm drains, stone-lined river channels and earthen aqueducts, large-scale bank and ditch construction for fortifications, and construction of temple platforms and stone sculptures, some of them involving the movement of stones weighing many tonnes (see Kirch 2000 for a summary). Some irrigation canals ran for several kilometres and crossed major watersheds (Spriggs 1990).

Given this sophisticated engineering knowledge, it seems clear that Pacific Islanders could also have known how to shape the landscape in other ways, through deliberately inducing erosion – indeed the New Guinea evidence would suggest they did possess such knowledge. Although perhaps impossible to prove archaeologically, such practices are certainly not beyond their known technological competence.

Proposition 4: A two-stage response to erosion damage

There was an initial response of abandoning the affected site and a later, land-full one of investment in erosion-control measures and rebuilding garden systems in place. The proposition does not seem to have received significant attention since its more explicit formulation in the 1990s. Testing it is extremely difficult with our current lack of understanding of changing settlement patterns in the Pacific and in particular the dearth of detailed agricultural sequences from many islands, particularly in the Western Pacific. Obtaining such understanding should be a priority for the future. There are few new sequences of agricultural change and development additional to those used originally in the formulation of the proposition for Western Polynesia and island Melanesia, where suitably long sequences can be expected – New Caledonia is a notable exception (Sand 1999, 2002; Carlson 2008). In Eastern Polynesia, much more detailed information is available; for instance, most recently for Maui and Hawai'i islands (Ladefoged and Graves 2000; Ladefoged et al. 2003; Kirch et al. 2004; Vitousek et al. 2004), Easter Island (Stevenson et al. 1999, 2006; Mieth and Bork 2005), and New Zealand (Barber 2004, and references therein). Initial settlement dates remain controversial, however, for many Eastern Polynesian archipelagos, making the definitive establishment of cultural sequences difficult (Spriggs In press). Given the probability of pre-European settlement sequences of fewer than 1000 years, the long-term sustainability of these particular agricultural systems cannot easily be assessed.

Palynology and related disciplines are providing further lines of evidence. These are tracking both general vegetation change and, in some cases, individual crop introductions and utilisation (examples include Hope 1996; MacPhail et al. 2001; Haberle 2003; Kennett et al. 2006; Horrocks and Wozniak 2008; Prebble and Dowe 2008; Prebble and Wilmshurst 2009). It would be a pity, however, if archaeologists working in Western Polynesia and island Melanesia were happy to leave the field to these other disciplines. Only archaeology can study the direct field remains of the agricultural systems and attendant erosion-control measures that constitute a large part of the human response to changing environmental conditions over 3000 or more years since first settlement.

Reliable dating of such systems remains a real problem. Another issue is clearly cost: the Hawaiian research constitutes a major interdisciplinary effort, requiring proportionate amounts of funding. Agricultural systems are often spatially extensive and can be stratigraphically complex and differentiated within even small areas. The investigation of swamp systems has its own problems, solved in the case of the Kuk site in the New Guinea Highlands by large-scale modern drainage to create a tea plantation (Golson 1977). Similar opportunities for investigation on this scale have not arisen since elsewhere in the Pacific.

Conclusions

Since first formulated, three of the four main propositions have held up under scrutiny to various degrees. In all cases, there have been supporters and sceptics, and the sceptics have been valuable in helping to give additional processes their proper due. Nunn's (1997) stressing of natural climatic changes has led to a healthy debate about the extent of prehistoric human impacts on Pacific landscapes. The extreme development of this, his AD 1300 'event' scenario, has its fans as well as its detractors. The detractors stress that the people of the Pacific were agents rather than victims of their fate on these scattered and vulnerable islands.

It now seems self-evident that the erosion that created the current coastal plains of the islands has been, in the medium term, beneficial to human existence on them rather than deleterious, whether it was humanly accelerated or largely climate driven – a question that still needs to be resolved through a comparison of early- and late-settled islands across the Pacific.

Proposition 4 on different early and later responses to major erosion episodes has not received as much attention as the others. It requires a greater understanding of settlement patterns and their changes over time than we currently possess, and a different scale of archaeology than is usually attempted in the Pacific Islands.

The most contentious of the propositions remains Proposition 2 about visibility of early sites being obscured by post-settlement landscape changes. It clearly has been demonstrated for some islands – those where often accidental processes have indeed revealed early sites deeply buried under alluvial or colluvial deposition. But its applicability where such sites have not (yet?) been found remains debated – the central Solomons are currently the main focus for such debate. A complicating factor not evident in the early 1980s is the degree of blanketing of landscapes by volcanic tephra deposits across large parts of Vanuatu and several other archipelagos. Subsidence is another factor affecting site visibility in some areas. Given the increasing pace of archaeological discovery of the past few decades, it would be a brave scholar indeed who would suggest that we have already found all there is to be found in this regard. Time will tell.

Acknowledgements

Several teachers and colleagues were instrumental in the development of the ideas presented here during and after preparation of my PhD: Jim Allen, Jack Golson, Philip Hughes, Rhys Jones, Pat Kirch, Betty Meehan, Marjorie Sullivan and Doug Yen, and of course not least Geoff Hope, prominent among them. Since then, colleagues and students at the University of Hawaii and at the ANU have provided stimulating discussions on many of the issues discussed here. During 2009, my RA, Catherine Fitzgerald, has been a tireless 'library hound'. I thank the editors for their patience in waiting for the delayed delivery of this manuscript in May 2009, an anonymous referee whose devastating critique led to recasting in its present form and two later, gentler, referees for their helpful comments.

References

Allen, J., M. Riford, T. Bennett, G. Murakami and M. Kelly 1987. *Five Upland 'Ili: Archaeological and historical investigations in the Kane'ohe Interchange, Interstate Highway H-3, Island of O'ahu*. Report prepared for the Dept of Transportation, State of Hawaii. Honolulu: Dept of Anthropology, B.P. Bishop Museum.

Allen, J. and C. Gosden (eds) 1991. *Report of the Lapita Homeland Project. Occasional Papers in Prehistory* 20. Canberra: Dept of Prehistory, RSPacS, Australian National University.

Allen, M.S. 2006. New ideas about Late Holocene climate variability in the Central Pacific. *Current Anthropology* 47(3):521-535.

Anderson, A., J. Chappell, M. Gagan and J. Grove 2006. Prehistoric maritime migration in the Pacific Islands: an hypothesis of ENSO forcing. *The Holocene* 16:1-6.

Barber, I. 2004. Crops on the Border: the growth of archaeological knowledge of Polynesian cultivation in New Zealand. In L. Furey and S. Holdaway (eds) *Change Through Time: 50 years of New Zealand archaeology*, pp169-192. *New Zealand Archaeological Association Monograph* 26. Auckland.

Bedford, S. 2006. *Pieces of the Vanuatu Puzzle: archaeology of the north, south and centre. Terra Australis* 23. Canberra: Pandanus Books.

Bedford, S. and M. Spriggs 2000. Crossing the Pwanmwou: preliminary report on recent excavations adjacent to and south west of Mangaasi, Efate, Vanuatu. *Archaeology in Oceania* 35:120-126.

Bedford, S. and M. Spriggs 2008. Northern Vanuatu as a Pacific crossroads: the archaeology of discovery, interaction and the emergence of the 'ethnographic present'. *Asian Perspectives* 47(1):95-120.

Bedford, S., M. Spriggs and R. Regenvanu 2006. The Teouma Lapita site and the early human settlement of the Pacific Islands. *Antiquity* 80:812-828.

Bedford, S., M. Spriggs, M. Wilson and R. Regenvanu 1998. The Australian National University-National Museum of Vanuatu Archaeology Project 1994-7: a preliminary report on the establishment of cultural sequences and rock art research. *Asian Perspectives* 37(2):165-193.

Bellwood, P. 1978. *Man's Conquest of the Pacific*. Auckland: W. Collins.

Butzer, K.W. 1974. Accelerated soil erosion: a problem of man-land relationships. In I.R. Manners and M.W. Mikesell (eds) *Perspectives on Environment*, pp47-78. *Association of American Geographers Publication* 13. Washington D.C.: Association of American Geographers.

Carlson, M.T. 2008. Correlation of environmental and cultural chronology in New Caledonia. *Geoarchaeology* 23(5):695-714.

Clark, G. and A. Anderson 2001. The pattern of Lapita settlement in Fiji. *Archaeology in Oceania* 36:77-88.

Clark, G. and S. Bedford 2008. Friction zones in Lapita colonisation. In G. Clark, F. Leach and S. O'Connor (eds) *Islands of Inquiry: Colonisation, Seafaring and the Archaeology of Maritime Landscapes*, pp59-73. *Terra Australis* 29. Canberra: ANU E Press.

Cline, M.G, A.S. Ayres, W. Crosby, P.F. Phillips, R. Elliott, O.G. Magstad, J.C. Ripperton, E.Y. Hosaka, M. Takahashi, G.D. Sherman and C.K. Wentworth 1955. *Soil Survey of the Territory of Hawaii. US Dept of Agriculture Soil Survey Series 1939* 25. Washington, DC: US Government Printing Office.

Colley, H. and R.P. Ash 1971. *The Geology of Erromango*. Port Vila: British Service.

Conklin, H. 1980. *An Ethnographic Atlas of Ifugao*. New Haven: Yale University Press.

Dickinson, W.R. and J.S. Athens 2007. Holocene paleoshoreline and paleoenvironmental history of Palau: implications for human settlement. *Journal of Island and Coastal Archaeology* 2:175-196.

Dickinson, W.R and R.C. Green 1998. Geoarchaeological context of Holocene subsidence at the Ferry Berth Site, Mulifanua, Upolu, Western Samoa. *Geoarchaeology* 13:239-263.

Dobres, M.-A. and J. Robb (eds) 1999. *Agency and Archaeology*. London: Routledge.

Felgate, M.W. 2003. *Reading Lapita in Near-Oceania: Intertidal and shallow-water pottery scatters, Roviana Lagoon, New Georgia, Solomon Islands.* Unpublished PhD thesis. Auckland: University of Auckland.

Felgate, M. 2007. Leap-frogging or limping? Recent evidence from the Lapita littoral fringe, New Georgia, Solomon Islands. In S. Bedford, C. Sand and S.P. Connaughton (eds) *Oceanic Explorations: Lapita and Western Pacific Settlement*, pp123-140. *Terra Australis* 26. Canberra: ANU E Press.

Fosberg, F.R. (ed.) 1963. *Man's Place in the Island Ecosystem: A symposium.* Honolulu: Bishop Museum Press.

Frimigacci, D. 1990. *Aux Temps de la Terre Noire: Ethnoarchéologie des îles Futuna et Alofi. Langues et Cultures du Pacifique* 7. Paris: Peeters.

Gehrels, W.R. 2001. Discussion of Nunn, Patrick D., 1998. Sea-Level Changes over the last 1,000 Years in the Pacific. Journal of Coastal Research 14(1):23-30. *Journal of Coastal Research* 17:244-245.

Golson, J. 1977. No room at the top: agricultural intensification in the New Guinea Highlands. In J. Allen, J. Golson and R. Jones (eds) *Sunda and Sahul: Prehistoric studies in Southeast Asia, Melanesia and Australia*, pp601-638. London: Academic Press.

Gosden, C. and J. Webb 1994. The creation of a Papua New Guinean landscape: archaeological and geomorphological evidence. *Journal of Field Archaeology* 21:29-51.

Green, R.C. 1979. Lapita. In J. Jennings (ed.) *The Prehistory of Polynesia*, pp27-60. Cambridge, Mass.: Harvard University Press.

Green, R.C. 1985. Comment. Spriggs' 'The Lapita Cultural Complex'. *Journal of Pacific History* 20:220-224.

Haberle, S. 2003. Late Quaternary vegetation dynamics and human impact on Alexander Selkirk Island, Chile. *Journal of Biogeography* 20:239-255.

Handy, E.S.C. 1940. *The Hawaiian Planter, Vol. I. His Plants, Methods and Areas of Cultivation. B. P. Bishop Museum Bulletin* 161. Honolulu: Bishop Museum.

Hope, G.S. 1996. Quaternary change and the historical biogeography of Pacific islands. In A. Keast and S.E. Miller (eds) *The Origin and Evolution of Pacific Island Biotas. New Guinea to Eastern Polynesia: patterns and processes*, pp165-190. Amsterdam: SPB Academic Publishing.

Hope, G., D. O'Dea and W. Southern 1999. Holocene vegetation histories in the Western Pacific: alternative records of human impact. In J.-C. Galipaud and I. Lilley (eds) *The Pacific from 5000 to 2000 BP: Colonisation and transformations*, pp387-4-04. Paris: IRD.

Horrocks, M. and J.A. Wozniak 2008. Plant microfossil analysis reveals disturbed forest and a mixed-crop dryland production system at Te Niu, Easter Island. *Journal of Archaeological Science* 35(1):126-142.

Hughes, P.J. and M. Sullivan 1981. Aboriginal burning and Late Holocene geomorphic events in Eastern Australia. *Search* 12:277-278.

Hughes, P.J., G. Hope and M. Latham 1979. Prehistoric man-induced degradation of the Lakeba landscape: evidence from two inland swamps. In H.C. Brookfield (ed.) *Lakeba: Environmental change, population dynamics and resource use*, pp93-110. *UNESCO/ UNFPA Population and Environment Project in the Eastern Islands of Fiji, Island Reports* 5. Canberra: UNESCO.

Kennett, D., A. Anderson, M. Prebble, E. Conte and J. Southon 2006. Prehistoric human impacts on Rapa, French Polynesia. *Antiquity* 80:340-354.

Kirch, P.V. 1987. Lapita and Oceanic cultural origins: excavations in the Mussau Islands, Bismarck Archipelago 1985. *Journal of Field Archaeology* 14:163-180.

Kirch, P.V. 1988a. The Talepakemalai site and Oceanic prehistory. *National Geographic Research* 4:328-342.

Kirch, P.V. 1988b. *Niuatoputapu: The prehistory of a Polynesian chiefdom. Thomas Burke Memorial Washington State Museum Monograph* 5. Seattle: Burke Museum.

Kirch, P.V. 1997a. Introduction: The environmental history of Oceanic Islands. In P.V. Kirch and T.L. Hunt (eds) *Historical Ecology in the Pacific Islands: Prehistoric environmental and landscape change*, pp1-21. New Haven: Yale University Press.

Kirch, P.V. 1997b. Changing landscapes and sociopolitical evolution in Mangaia, Central Polynesia. In P.V. Kirch and T.L. Hunt (eds) *Historical Ecology in the Pacific Islands: Prehistoric environmental and landscape change*, pp147-165. New Haven: Yale University Press.

Kirch, P.V. 2000. *On the Road of the Winds: An archaeological history of the Pacific Islands before European contact.* Berkeley: University of California Press.

Kirch, P.V and T.L. Hunt (eds) 1993. *The To'aga Site: Three millennia of Polynesian occupation in the Manu'a Islands, American Samoa. Contributions of the University of California Archaeological Research Facility* 51. Berkeley: University of California.

Kirch, P.V. and T.L. Hunt (eds) 1997. *Historical Ecology in the Pacific Islands: Prehistoric environmental and landscape change.* New Haven: Yale University Press.

Kirch, P.V., A.S. Hartshorn, O.A. Chadwick, P.M. Vitousek, D. Sherrod, J. Coil, L. Holm and W.D. Sharp 2004. Environment, agriculture, and settlement patterns in a marginal Polynesian landscape. *Proceedings of the National Academy of Sciences* 101(26):9936-9941.

Kumar, R., P.D. Nunn, J.S. Field and A. de Biran 2006. Human responses to climate change around AD1300: A case study of the Sigatoka Valley, Viti Levu, Fiji. *Quaternary International* 151:133-143.

Ladefoged, T.N. and M.W. Graves 2000. Evolutionary theory and the historical development of dry-land agriculture in North Kohala, Hawai'i. *American Antiquity* 65:423-448.

Ladefoged, T.N., M.W. Graves and M.D. McCoy 2003. Archaeological evidence for agricultural development in Kohala, Island of Hawai'i. *Journal of Archaeological Science* 30:923-940.

Macphail, M.K., G.S. Hope and A. Anderson 2001. Plant introductions in the Southwest Pacific: Initial pollen evidence from Norfolk Island. *Records of the Australian Museum, Supplement* 27:123-134.

Mieth, A and H.-R. Bork 2005. History, Origin and Extent of Soil Erosion on Easter Island (Rapa Nui). *Catena* 63:244-260.

Nunn, P.D. 1991. *Keimami sa vakila na liga ni Kalou (Feeling the hand of God): Human and Nonhuman Impacts on Pacific Island Environments. Occasional Papers of the East-West Environment and Policy Institute* 13. Honolulu: East-West Center.

Nunn, P. 1997. *Pacific Island Landscapes.* Suva: Institute of Pacific Studies, University of the South Pacific.

Nunn, P.D. 1998. Sea-Level Changes over the last 1,000 Years in the Pacific. *Journal of Coastal Research* 14(1):23-30.

Nunn, P.D. 2001. Ecological crises or marginal disruptions: the effects of the first humans on Pacific Islands. *New Zealand Geographer* 57(2):11-20.

Nunn, P.D. 2003. Nature-Society interactions in the Pacific Islands. *Geografiska Annaler* 85B:219-229.

Nunn, P.D. 2007. *Climate, Environment and Society in the Pacific during the Last Millennium.* Amsterdam: Elsevier.

Nunn, P.D., R. Hunter-Anderson, M.T. Carson, F. Thomas, S. Ulm and M.J. Rowland 2007. Times of Plenty, Times of Less: Last-Millennium societal disruption in the Pacific Basin. *Human Ecology* 35:385-401.

Pawley, A. 2009. The role of the Solomon Islands in the first settlement of Remote Oceania: bringing linguistic evidence to an archaeological debate. In A. Adelaar and A. Pawley (eds) *Austronesian Historical Linguistics and Culture History: A festschrift for Bob Blust*, pp509-534. *Pacific Linguistics series*. Canberra: ANU E Press.

Prebble, M. and J.L. Dowe 2008. The Late Quaternary decline and extinction of palms on Oceanic Pacific Islands. *Quaternary Science Reviews* 27:2546-2567.

Prebble, M. and J. Wilmshurst 2009. Detecting the initial impact of humans and introduced species on island environments in Remote Oceania using palaeoecology. *Biological Invasions* 11(7):1529-1556.

Sand, C. 1999. From the swamp to the terrace: Intensification of horticultural practices in New Caledonia. In C. Gosden and J. Hather (eds) *The Prehistory of Food: Appetites for Change*, pp252-269. London: Routledge.

Sand, C. 2002. Creations and transformations of prehistoric landscapes in New Caledonia: the southernmost Melanesia islands. In T.N. Ladefoged and M.W. Graves (eds) *Pacific Landscapes: Archaeological approaches*, pp11-34. Los Osos, CA: The Easter Island Foundation, Bearsville Press.

Schmitz, C.A. 1963. *Wantoat: Art and religion of the northeast New Guinea Papuans*. The Hague: Mouton.

Sheppard, P. and R. Walter 2006. A revised model of Solomon Islands culture history. *Journal of the Polynesian Society* 115:47-76.

Sheppard, P. and R. Walter 2009. Inter-tidal late Lapita sites and geotectonics in the Western Solomon Islands. In P. Sheppard, T. Thomas and G. Summerhayes (eds) *Lapita: Ancestors and Descendants*, pp73-100. Auckland: New Zealand Archaeological Association.

Spriggs, M. 1981 [1987]. *Vegetable Kingdoms: Taro irrigation and Pacific prehistory*. PhD thesis, ANU, Canberra [published by University Microfilms, Ann Arbor, Michigan, 1987].

Spriggs, M. 1984. The Lapita cultural complex: origins, distribution, contemporaries and successors. *Journal of Pacific History* 19:202-223.

Spriggs, M. 1985. Prehistoric human-induced landscape enhancement in the Pacific: examples and implications. In I. Farrington (ed.) *Prehistoric Intensive Agriculture in the Tropics*, Vol. II, pp409-34. *British Archaeological Reports, Supplementary Series* 232. Oxford: B.A.R.

Spriggs, M. 1990. Why irrigation matters in Pacific prehistory. In D.E. Yen and J.M.J. Mummery (eds) *Pacific Production Systems: Approaches to economic prehistory*, pp174-189. *Occasional Papers in Prehistory* 18. Canberra: Dept of Prehistory, RSPacS, ANU.

Spriggs, M. 1997. Landscape catastrophe and landscape enhancement: Are either or both true in the Pacific? In P.V. Kirch and T.L. Hunt (eds) *Historical Ecology in the Pacific Islands: Prehistoric environmental and landscape change*, pp80-104. New Haven: Yale University Press.

Spriggs, M. 2008. Ethnographic parallels and the denial of history. *World Archaeology* 40(4):538-552.

Spriggs, M. In press. 'I was so much older then, I'm younger than that now': why the dates keep changing for the spread of Austronesian languages. in J. Bowden, N. Himmelmann and M. Ross (eds) *A Journey through Austronesian and Papuan Linguistic and Cultural Space. Papers in honour of Andrew K. Pawley*. *Pacific Linguistics* series. Canberra.

Spriggs, M. and S. Wickler 1989. Archaeological research on Erromango: recent data on southern Melanesian prehistory. *Bulletin of the Indo-Pacific Prehistory Association* 9:68-91.

Spriggs, M. and S. Bedford 2001. Arapus: a Lapita site at Mangaasi in Central Vanuatu? In G. Clark, A. Anderson and T. Vunidilo (eds) *The Archaeology of Lapita Dispersal in Oceania*, pp93-104. *Terra Australis* 17. Canberra: Pandanus Press, ANU.

Stevenson, C.M., J. Wozniak and S. Haoa 1999. Prehistoric Agricultural Production on Rapa Nui. *Antiquity* 73:801-812.

Stevenson, C.M., T.L. Jackson, A. Mieth, H.-R. Bork and T. Ladefoged 2006. Prehistoric and Early Historic Agriculture at Maunga Orito, Easter Island, Chile. *Antiquity* 80:919-936.

Vial, L.G. 1937. The Dangagamun ceremony of the Wantoat. *Oceania* 7:341-345.

Vial, L.G. 1940. Stone axes of Mount Hagen, New Guinea. *Oceania* 11(2):158-163.

Vitousek, P.M., T.N. Ladefoged, P.V. Kirch, A.S. Hartshorn, M.W. Graves, S.C. Hotchkiss, S. Tuljapurkar and O.A. Chadwick 2004. Soils, agriculture and society in precontact Hawai'i. *Science* 304(5677):1665-1669.

Wickler, S. 1995. *Twenty-Nine Thousand Years on Buka: Long-term cultural change in the Northern Solomon Islands.* Unpublished PhD thesis. Honolulu: University of Hawaii.

Wickler, S. 2001a. *The Prehistory of Buka: A stepping stone island in the Northern Solomons. Terra Australis* 16. Canberra: Dept of Archaeology and Natural History and Centre for Archaeological Research, Australian National University.

Wickler, S. 2001b. The colonisation of western Micronesia and early settlement of Palau. In C.M. Stevenson, G. Lee and F.J. Morrin (eds) *Pacific 2000: Proceedings of the Fifth International Conference on Easter Island and the Pacific*, pp185-196. Los Osos CA: The Easter Island Foundation, Bearsville Press.

14

Pollen evidence for plant introductions in a Polynesian tropical island ecosystem, Kingdom of Tonga

Patricia L. Fall

School of Geographical Sciences, Arizona State University, Tempe, United States
FALL@asu.edu

Introduction

The dynamic nature of tropical Pacific ecosystems results from chance migrations and the evolution of founder species, as well as from physical factors such as changes in sea level, ocean currents, tectonic processes and climate (Hope 2001). In addition, this region's vegetation is constantly adjusting through succession to local perturbations like landslides and tropical cyclones. These ecological and physical processes are compounded by continued immigration of new species, competition, extinctions and extirpation of species. For many island ecosystems the most dramatic impact on species composition results from the arrival of humans and their 'co-voyaging' plants and animals (Hope 2001). The degree to which climatic or other natural variations, versus human impacts, have caused recent environmental fluctuations in island ecosystems is an ongoing debate (cf. Nunn 1994; Burney 1997).

In this paper, I examine palynological evidence for plants introduced to several islands in the three main island groups of Tongatapu, Ha'apai and Vava'u in the Kingdom of Tonga. My purpose is to briefly document the history of the tropical rainforests of Tonga as they can be reconstructed from pollen cores, to understand the role that humans played in the development of the Tongan flora, and to discuss plant introductions to the islands by both Polynesian and later European settlers.

Losses or increases in plant species on remote islands are controlled by many factors, including habitat change, natural factors, and loss of dispersers or pollinators. Habitat change in Pacific tropical ecosystems may be due to human factors, including fire, habitat destruction (deforestation) and the introduction of exotic animals or plants (Southern 1986; Hope and Pask 1998; Stevenson 1998; Hope et al. 1999; Haberle et al. 2001; Stevenson et al. 2001; Haberle 2007). Natural factors such as tropical cyclones (Franklin et al. 2004), sea-level variation (Dickinson et al. 1994; Dickinson 2001, 2003), or the effects of ENSO (Haberle and David 2004) cause disturbances to island ecosystems. The loss of animal pollinators or dispersers, triggered by human or natural causes, can further disrupt reproduction and establishment of plant species (McConkey and Drake 2006, 2007; Steadman 2006; Prebble and Dowe 2008). As Hope (2001) points out, tropical Pacific islands support rain-green forests with relatively high numbers of endemic species. Variations in local species diversity and ecological adaptations between island groups confound the task of understanding present vegetation dynamics, and make deciphering palaeoecological interactions a particularly daunting task. Thus, palaeoecological data often are needed to help unravel the history of an island's ecology. Here, I address one of the historical questions regarding Pacific ecosystems asked by Hope (2001:172): Have human populations caused major changes in species composition and ecological processes?

Kingdom of Tonga

The Kingdom of Tonga, lying southwest of Samoa and east of Fiji, is comprised of more than 170 islands and covers a vast area of the south Pacific Ocean from 15° 30' to 22° 20' S latitude, and between 173° 00' and 177° 15' W longitude (Figure 1). Tonga is formed by a north-south alignment of islands that includes the three main uplifted limestone island groups of Vava'u, Ha'apai and Tongatapu. The Tofua Volcanic Arc to the west of these limestone islands is made up of the younger volcanic islands and sea mounts that run from the northernmost island of Niuafo'ou to 'Ata in the south. The highest point in the archipelago is Koa, an extinct volcano with an elevation of 1046 m. Recent eruptions and earthquakes, including an eruption of a submarine volcano on March 19, 2009, southwest of Tongatapu, demonstrate the ongoing volcanic and tectonic activity of the islands.

The Kingdom of Tonga has a mild tropical maritime climate dominated by the south Pacific trades for most of the year. Annual temperature and precipitation average 21°C and 1780 mm on Tongatapu, and 23.5°C and 2340 mm on Vava'u (Thompson 1986). The climate is slightly wetter in the Austral summer than in the somewhat drier and cooler winter. The islands lie in the track of tropical cyclones, being struck by an average of two tropical cyclones per year (Franklin et al. 2004).

Tonga is part of the southwestern Pacific region of the Indo-Malesian floral realm described by van Steenis (1979) in which floral impoverishment increases with distance from tropical Southeast Asia. Tropical shoreline vegetation is typical of many Pacific islands (Whistler 1992a). More species-rich lowland rainforests are found on the larger islands (Tongatapu, 'Eua and Vava'u), with somewhat more depauperate forests in Ha'apai and on the smaller islands. The islands of 'Eua and 'Uta Vava'u contain more diverse forests due to their larger sizes and greater elevational range (more than 300 m) (Figure 2). A range of botanical inventories document the flora of Tonga (Yuncker 1959; Sykes 1977, 1981; Palmer 1988; Stoddart 1992; Whistler 1992a, 1992b; Drake et al. 1996; Mueller-Dombois and Fosberg 1998; Franklin et al. 1999; Park and Whistler 2001; Wiser et al. 2002; Franklin 2003; Franklin et al. 2006). Park and Whistler (2001) estimate Tonga has about 450 indigenous and 240 introduced plant species. Polynesian introductions (about 40 species) to Tonga tend to be trees or herbaceous cultigens that are dispersed by people or native fauna, whereas European introductions (as many as 200 species) are primarily herbaceous species that are dispersed by wind or epizoochorously by humans or their domestic animals (Fall et al. 2007). Nomenclature (including Tongan and English names where known) follows Smith (1979, 1981, 1985, 1988, 1991) and Whistler (1991) (Table 1).

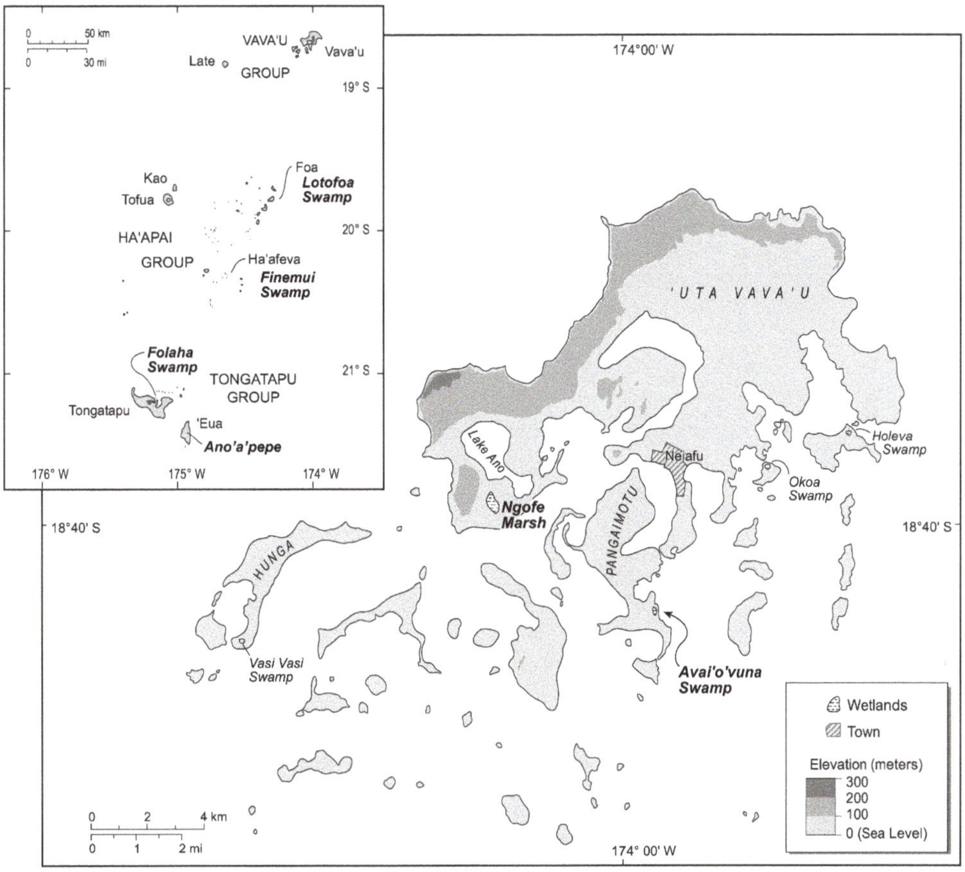

Figure 1. Location of the Kingdom of Tonga. Large map shows the Vava'u island group with the location of the Ngofe Marsh and the Avai'o'vuna Swamp. Inset shows three main island groups (Vava'u, Ha'apai and Tongatapu) where sediment cores were collected. Ha'apai island group: Lotofoa Swamp on Foa, and Finemui Swamp on Ha'afeva. Tongatapu island group: Folaha Swamp in Fanga 'Uta Lagoon on Tongatapu, and Ano'a'pepe on 'Eua (see Table 2 for descriptions of coring localities). Figure by Barbara Trapido-Lurie

Figure 2. Rainforest on 'Uta Vava'u. Photograph by P. Fall

Table 1. Plant taxa mentioned in text including scientific authorities, Tongan and English names (where known). Nomenclature follows Smith (1979; 1981; 1985; 1988; 1991) and Whistler (1991) (Table 1 continues on page 257)

Taxon	(*Tongan*; English names)
Acrostichum aureum L.	(*hakato*; golden leatherfern)
Alphitonia zizyphoides (Spreng.) A. Gray	(*toi*)
Alternanthera sessilis R. Br. ex DC	(*loseli*; joyweed)
Barringtonia asiatica (L.) Kurz	(*futu*; fish poison tree)
Bruguiera gymnorrhiza (L.) Lam.	(*tongo lei*)
Calophyllum inophyllum L.	(*feta'u*; beach mahogany; oil nut tree)
Calophyllum neo-ebudicum Guillaumin	(*tamanu*)
Canarium harveyi var. *harveyi* Leenh.	(*'ai*)
Casuarina equisetifolia J.R. & G. Forst.	(*toa*; ironwood; she-oak)
Cocos nucifera L.	(*nui*; coconut)
Colocasia esculenta (L.) Schott.	(*talo*; taro)
Cordyline fruticosa (L.) A. Chev.	(*si*; ti; good luck plant)
Cyperaceae	(sedge family)
Dysoxylum forsteri (Juss.) DC	(*mo'ota*)
Dysoxylum tongense A.C. Smith	(*mo'ota mea*)
Elaeocarpus tonganus Burk.	(*ma'ama'alava*; blue berry tree)
Eleocharis dulcis (Burm. f.) Trin. ex Hensch.	(*kutu*; water chestnut)
Ellatostachys falcata (Seem.) Radlk.	(*ngatata*)
Erianthus	(plume grass)
Erythrina fusca Lour.	(*ngatae fisi*; Fijian coral tree)
Erythrina variegata L.	(*ngatae*; coral tree)
Excoecaria agallocha L.	(*feta'anu*; blind-your-eye mangrove)
Flacourtiaceae	(Flacourtiaceae family)
Freycentia urvilleana Hombron & Jacqinot	(*kahikahi*)
Gardenia augusta (L.) Merr.	(*siale matalateau*; double gardenia)
Gardenia tannaensis Guillaumin	(*siale lotuma*; gardenia)
Gardenia taitensis DC. Prodr.	(Tahitian gardenia)
Garuga floribunda Decne.	(*manaui*)
Glochidion ramiflorum J.R. & G. Forst.	(*manolo*)
Guettarda speciosa L.	(*puopua*; beach gardenia)
Hedycarya dorstenioides A. Gray	
Hernandia nymphaeifolia (Presl) Kub.	(*fotulona*; lantern tree)
Hibiscus tiliaceus L.	(*fau*; beach hibiscus)
Homalanthus nutans Benth. & Hook. f. ex Drake	(*fonua mamala*; bleeding heart tree)
Homalium	(*moto*)
Inocarpus fagifer (Parkinson) Fosb.	(*ifi*; Tahitian chestnut)
Ipomoea batatas L.	(*kumala*; sweet potato)
Leguminosae	(legume family)
Lepironia articulata (Retz.) Domin	(*kutu kofe*; blue rush)
Ludwigia octovalvis (Jacq.) Raven	(*loaana*; water primrose)
Macaranga haveyana (Muell. Arg.) Muell. Arg.	(*loupata*; giant leaf plant)
Mallotus	(kamala tree)
Malvaceae	(mallow family)
Maniltoa grandiflora (A.Gray) Scheff.	(*pekepeka*; dove tree; handkerchief tree)
Melastomataceae	(Melastomataceae family)
Meliaceae	(Meliaceae family)
Mimosa pudica L.	(*mateloi*; sensitive plant)
Miscanthus floridus (Labill.) Warb.	(*kaho tonga*; reed)
Monimiaceae	(Monimiaceae family)
Moraceae	(fig family)
Morinda citrifolia L.	(*nonu*; Indian mulberry)
Narenga	(genus in the grass family)
Neonauclea fosteri (Seem.) Merr.	(*afa*)
Onagraceae	(primrose family)

Taxon	**(*Tongan*; English names)**
Pandanus tectorius Parkinson	(*fafa*; screw-pine)
Pandanus veitchii Hort.	(*hauai'i*)
Pandanus whitmeeanus Mart.	(*paongo*)
Papillionaceae	(pea family)
Phyllanthus amicorum G.L. Webster	
Pinus caribaea Morelet	(*paini*; Caribbean pine)
Planchonella	(*kalaka*)
Pleiogynium timoriense (DC.) Leehn.	(*tangato* or Burdekin plum)
Poaceae	(grass family)
Podocarpus pallidus N.E. Gray	(*uhiuhi*)
Polygonum dichotomum Bl. Bijar (synonym *Polygonum glabrum*)	(dense-flower knotweed)
Pometia pinnata J.R. & G. Forst.	(*tava*; island lychee)
Psidium guajava L.	(*kuava*; guava)
Rhizophora mangle	(*tongo*; mangrove)
Rhus taitensis Guillemin	(*tavahi*; island sumac)
Saccharum officinarum L.	(*to*; sugarcane)
Saccharum spontaneum L.	(wild sugarcane)
Sapotaceae	(sapote family)
Schizostachyum glaucifolium (Rupr.) Munro	(*kofe*; bamboo; Polynesian 'ohe)
Sclerostachys	(genus in the grass family)
Solanum amicorum Benth.	(*polo tonga*)
Solanum nigrum L.	
Stachytarpeta urticifolia (Salisb.) Sims	(*hiku 'i kuma*; blue rat's-tail)
Stenochlaena palustris (Burm.) Beddome	(*pasivaka*; climbing fern)
Syzygium	(*fekika*)
Thespesia populnea (L.) Solander ex Correa	(*milo*; portia tree)
Trema cannabina Lour.	(*mangele*; poison bush)
Tricale	(grass hybrid of *Triticum* and *Secale*)
Triumfetta rhomboidea Jacq.	(*mo'osipo*; burr bush)
Urticaceae	(nettle family)

Polynesian settlers reached Tonga by 2850 yr BP (Burley 1998). Lapita culture has been documented on Tongatapu (Poulsen 1987; Burley and Dickinson 2001), Ha'apai (Burley 1994; Burley et al. 1999) and Vava'u by about 2700-2800 yr BP (Burley and Connaughton 2007). European contact with Tonga was initiated by Dutch sailors in 1616 and by Abel Tasman in 1643 (Whistler 1991). The next European contact was made by James Cook in 1773, 1774 and 1777; missionary settlement soon followed in the late 1700s. Only 37 of the islands are inhabited today, with the majority of the Tongan people living on the main islands of Tongatapu, Ha'apai, or 'Uta Vava'u, the largest island of Vava'u.

Palynological records from Tonga

Sediment cores have been retrieved from a variety of small wetlands found throughout the Kingdom of Tonga (Table 2). Sediments were collected by the author on several islands in the Vava'u Island Group in 1997 and 1998, including cores from the islands of Pangaimotu (Avai'o'vuna Swamp) and 'Uta Vava'u (Ngofe Marsh) (Fall 2005). In the Ha'apai Islands, Flenley and others (1999) collected cores from two small basins on the islands of Ha'afeva (Finemui Swamp) and Foa (Lotofoa Swamp). In the Tongatapu island group, a sediment core was collected from the island of 'Eua in 1998 from a small lake in the highlands, Ano'a'pepe (Lake of the Butterflies) (Fall unpublished). In addition, Ellison (1989) reported on a series of sediment cores from a mangrove swamp on the main island of Tongatapu.

The evidence for indigenous or pre-Polynesian vegetation, Polynesian plant introductions and later European-introduced plants from these sediment records will be used to outline floral changes in Tonga over about the past 7000 years. Particular attention is directed to Polynesian and European plant introductions, as documented by pollen records in Tonga, and to plant taxa showing substantial declines or increases in abundance associated with Lapita colonisation through later European settlement.

Table 2. Sediment cores collected for pollen analyses in the Kingdom of Tonga

Island Group							
Core locality	**Island**	**Latitude**	**Longitude**	**Elevation (m)**	**Approximate age range**	**Length of core (m)**	**Reference**
Vegetation surrounding coring locality							
Vava'u Island Group							
Ngofe Marsh	'Uta Vava'u	18°39'53" S	174°02'40" W	20	6000-0 yr BP	3.4	Fall unpublished
Cyperaceae (*Eleocharis dulcis and Lepironia articulata*) marsh surrounded by Poaceae, *Stachytarpeta urticifolia, Ludwigia octovalvis, Erythrina variegata*							
Avai'o'vuna Swamp	Pangaimotu	18°41'44" S	173°59'02" W	1.5	4500-0 yr BP	1.5	Fall 2005
Coastal forest of *Pandanus tectorius, Cocos nucifera* and *Hernandia nymphaeifolia, Cocos* plantation, and *Colocasia* swamp							
Ha'apai Island Group							
Lotofoa Swamp	Foa	19°44'48" S	174°18'27" W	3	6700-0 yr BP	3.13	Flenley et al. 1999
Ludwigia octovalvis, Cyperaceae and Poaceae swamp surrounded by *Erythrina variegata*							
Finemui Swamp	Ha'afeva	19°56'45" S	174°42'35" W	6	5800-0 yr BP	3.33	Flenley et al. 1999
Polygonum cf. *glabrum* swamp with *Ludwigia octovalvis*, Poaceae, Cyperaceae, Cocos *nucifera*							
Tongatapu Island Group							
Folaha Swamp	Tongatapu	21°11' S	175°11' W	0.4-0.9	6900-0 yr BP	3.2	Ellison 1989
Rhizophora mangrove at lagoon edge with *Bruguiera, Acrostichum aureum, Excoecaria, Hibiscus tiliaceus* and *Stenochlaena palustris*							
Ano'a'pepe	'Eua	21°24'03" S	174°55'05" W	250	Undated	0.6	Fall unpublished
Small lake (1.5 m deep) in the upland rain forest with Poaceae and Cyperaceae, surrounded by *Calophyllum neo-ebudicum, Dysoxylum tongense* and *Ellatostachys falcata*							

Vava'u Island Group

Vava'u is the northernmost of the three limestone island groups that make up the Kingdom of Tonga. The main island of 'Uta Vava'u forms the northern edge of an uplifted limestone platform, including several embayments and the brackish water Lake Ano. The limestone platform of Vava'u slopes southward, where its subsidence has produced numerous smaller islands (Dickinson and Burley 2007). Several swamps were examined in Vava'u, including the two discussed here – Ngofe Marsh and Avai'o'vuna Swamp – for their potential to provide palaeoecological data on pre-settlement forests and subsequent vegetation change associated with settlement by Lapita and later cultures (Fall 2005). Avai'o'vuna Swamp is a very small basin about 5 m x 10m in size that lies about 1.5 m above mean sea level on the eastern shore of Pangaimotu Island. The swamp is a sedge wetland surrounded by coastal forest trees, *Pandanus tectorius, Hernandia nymphaeifolia* and *Cocos nucifera*, set among more extensive

Cocos plantations on the island. *Colocasia esculenta* and *Ipomoea batatas* are cultivated in wetter soils. Ngofe Marsh is a sedge, grass and reed marsh on the eastern flank of Mount Mo'ungalafa on 'Uta Vava'u, separated from Lake Ano by a low rise. Vegetation on Ngofe Marsh is dominated by *Eleocharis dulcis* and *Lepironia articulata*, with grasses, sedges, *Stachtarpeta urticifolia*, *Polygonum dichotomum*, and the swamp tree *Erythrina variegata* at the edge of the marsh (Figure 3).

Figure 3. Ngofe Marsh on 'Uta Vava'u, Vava'u island group. Photograph by P. Fall

The coastal wetland Avai'o'vuna Swamp produced a 2 m sediment record, documenting sea-level variation, fire history and vegetation change. During a marine high-stand from 4500 to 2600 yr BP, the swamp was dominated by mangroves, including *Rhizophora mangle* and *Excoecaria*, the coastal trees *Barringtonia*, *Pandanus tectorius* and *Cocos nucifera*, and Malvaceae (perhaps *Thespesia populnea*), and was surrounded by lowland rainforest with *Hedycarya*, *Calophyllum* (presumably *C. inophyllum*), *Rhus*, Papillionaceae and *Alphitonia* (Fall 2005). After about 2600 yr BP, sea level dropped and fire, as attested by the appearance of microscopic charcoal in the sediments, became common. The recent vegetation around the swamp is dominated by mangroves (*Excoecaria*), coastal forest (*Pandanus tectorius* and *Cocos nucifera*), successional forest (*Macaranga*), and open vegetation of sedges, grasses and ferns. Past rainforest dynamics have included shifts to *Homalanthus* and *Macaranga* as secondary taxa, with declines in pollen from rainforest trees, including *Hedycarya, Calophyllum, Elaeocarpus,* Monimiaceae, *Neonauclea, Pleiogynium* and Papillionaceae (Fall 2005).

Plant taxa more common at Avai'o'vuna Swamp before Polynesian colonisation included *Hedycarya, Neonauclea, Guettarda* and *Solanum* (possibly *S. amicorum*) (Fall 2005). Following Polynesian colonisation, rainforest trees, particularly *Hedycarya, Calophyllum, Elaeocarpus* and *Rhus,* declined in abundance. Plants that increased and/or were possibly introduced by Polynesian settlers include cultivated grass (Poaceae >40-50 µm), *Casuarina, Erythrina* and *Canarium.*

Ngofe Marsh produced a record indicating that the basin contained a lake between about 7000 and 3000 yr BP. Associated rainforest taxa included *Elaeocarpus, Dysoxylum, Ellatostachys,* Flacourtiaceae, *Garuga, Hedycarya, Maniltoa,* Melastomataceae and Papillionaceae. Microscopic charcoal first appears in sediments from about 2800 yr BP, coincident with the arrival of Lapita people on Vava'u (Burley and Connaughton 2007). Along with the abundant charcoal, the infilling of the lake and the spread of plants on to the marsh is attested by dramatic increases in Poaceae and Cyperaceae pollen and fern spores after 2800 yr BP. The lowland rainforest surrounding Ngofe Marsh became less diverse after this time, with significant losses in most rainforest trees, accompanied by the notable expansion of *Pandanus tectorius* and *Macaranga*. Plant taxa with truncated records at Ngofe Marsh following Polynesian colonisation include *Dysoxylum, Garuga, Homalium, Maniltoa, Pleiogynium, Syzygium, Freycentia* and one type of *Pandanus*. Cultivated grass (Poaceae >40-50 μm; grains often have a double pore, or rarely a triple pore), *Colocasia, Pometia* and *Erythrina* are likely Polynesian introductions.

Ha'apai Island Group

The middle island group of Ha'apai is made up of a multitude of small low limestone islands and atolls to the east and the two more recent, larger and higher volcanic islands of Kao and Tofua at the western edge. Flenley and others (1999) examined two small swamps in Ha'apai on the islands of Ha'afeva and Foa for their potential to provide palaeoecological information applicable to Lapita and later archaeological sites investigated by Burley and colleagues (Burley 1994; Burley et al. 1999, 2001). The two swamps examined, Finemui (on Ha'afeva) and Lotofoa (on Foa), were most likely former lagoons within the atolls (Flenley et al. 1999). Both basins are surrounded by herbaceous and shrub vegetation. Finemui Swamp is dominated by *Polygonum* cf. *glabrum*, with other herbaceous plants, including *Ludwigia octovalvis* and plants in the Poaceae and Cyperaceae families. *Cocos nucifera, Inocarpus fagifer, Morinda citrifolia* and *Erythrina variegata* are marginal trees. Lotofoa Swamp is characterised by *Ludwigia octovalvis*, plants in the Cyperaceae, Poaceae and Leguminosae families, as well as ferns, *Erythrina* and other trees.

Basal organics began to accumulate in Finemui Swamp about 5770 yr BP (Flenley et al. 1999). The indigenous or pre-settlement vegetation at Finemui Swamp is interpreted as a brackish or freshwater wetland, which was replaced by a fern and *Polygonum* swamp, with the coastal trees *Hibiscus* and *Trema* (Flenley et al. 1999). Pollen from *Elaeocarpus, Trema,* Urticaceae/Moraceae and Meliaceae/Sapotaceae is indicative of the surrounding rainforest trees. The record also demonstrates a slight sea-level rise before disturbed sediments that are hypothesised to relate to the Lapita phase. Flenley and others (1999) also associate Lapita occupation with an expansion of *Glochidion* pollen (interpreted as a possible opening in the forest canopy) and increases in *Pandanus, Casuarina* and *Cocos nucifera* pollen, all of which could have been encouraged by planting. Peaks in Cyperaceae and Poaceae pollen and monolete fern spores seen at Finemui Swamp also are associated commonly with forest disturbance on other Polynesian islands (Parkes et al. 1992; Kirch et al. 1995; Kirch 1996; Parkes 1997; Flenley et al. 1999). The modern environment of Finemui Swamp is dominated by *Polygonum*, Cyperaceae and ferns, with *Cocos nucifera* and *Trema* as the most common dryland trees, accompanied by *Pandanus* and *Gardenia*, Elaeocarpaceae and *Macaranga*. Herbaceous plants include Poaceae and weedy taxa. Probable recently introduced species include *Gardenia, Stachytarpheta, Ipomoea* (*Ipomoea* cf. *batatas*) and *Solanum* (*S.* cf. *nigrum*) (Flenley et al. 1999).

Similarly, Flenley and others (1999) interpret the vegetation at Lotofoa Swamp before Polynesian settlement as a fern wetland surrounded by a coastal forest of *Pandanus* and *Hibiscus*, as well as *Podocarpus, Mallotus* and *Planchonella* trees. Lotofoa Swamp became an *Excoecaria* swamp during Polynesian settlement, and provides evidence of *Cocos* pollen. The modern vegetation at the swamp is represented by Cyperaceae and Poaceae pollen and fern spores. Dryland vegetation surrounding Lotofoa Swamp has been deforested, perhaps followed by the planting of *Cocos, Gardenia, Trema* and *Triumfetta*. Probable Polynesian introductions to

Ha'apai include *Ipomoea* cf. *batatas, Cocos* and *Casuarina*, accompanied by an expansion of *Pandanus* (Flenley et al. 1999).

Tongatapu Island Group

Ano'a'pepe, on the island of 'Eua, is a small lake about 1.5 m deep surrounded by plants in the Cyperaceae and Poaceae families. The forest surrounding the lake is dominated by *Calophyllum neo-ebudicum, Dysoxylum tongense* and *Ellatostachys falcata,* as is typical of upland rainforests on 'Eua (Drake et al. 1996). Although the rainforest immediately surrounding the lake is undisturbed, plantations of *Pinus caribaea* grow nearby. Approximately 1 m of organic gyttja was recovered from Ano'a'pepe in 1998; pollen preservation was good in the uppermost 50 cm of the core (Fall unpublished). The presence of microscopic charcoal, the abundance of Poaceae (including grains >40-50 µm; see discussion below), Cyperaceae pollen and fern spores, and the presence of pollen from introduced trees (e.g. *Pinus* and *Mimosa* in the upper few centimetres) suggest a relatively recent age (at least post-Polynesian) for this core.

Ellison (1989) collected a series of sediment cores from Fanga 'Uta Lagoon on Tongatapu to investigate sea-level change. A basal age of about 6900 yr BP demonstrates that the lagoon was a brackish marsh dominated by Cyperaceae and *Acrostichum aureum,* which lay about 1 m above modern mean sea level. This interpretation agrees with Dickinson and Burley's (2007:247) inference of a marine high-stand that peaked about 5000 yr BP, and a subsequent draw-down of sea level beginning about 3000 yr BP. A late Holocene sea-level drop is represented by a mangrove forest designated by pollen from *Hibiscus, Bruguiera, Excoecaria* and *Stenochlaena palustris* spores. The uppermost peat samples contain pollen from the introduced tree *Psidium guajava.*

Introduced plants

A number of tree, crop and ornamental plants was introduced to Tonga and cultivated by its Polynesian and European inhabitants. Analysis of pollen from the small marshes and swamps on Tonga, described above, provides a unique window into the human ecology of these islands. Based on ethnographic information and early botanical collections, Whistler (1991) provides a valuable compendium of the plants of Tonga, their names and their uses. Drawing on Whistler's (1991) observations of cultivated or aboriginal introductions, I discuss useful tree and herbaceous cultigens for which we have a palynological record, as well as inadvertent introductions. Interestingly, many taxa known to be European introductions do not appear in pollen records, perhaps reflecting a combination of palynological invisibility and restricted distributions. Although a wide range of taxa introduced by Polynesians is signalled palynologically, other Polynesian introductions, such as *Colocasia escuelenta* and *Ipomoea batatas,* may be invisible for the same reasons (Haberle 1995; Haberle and Atkins 2005). The subsequent discussions note the species most likely responsible for each pollen type represented in the subfossil record (pollen identification often is limited to genus, type or family), listed by scientific name with authority, followed by the Tongan and English names (in parentheses), as available (after Whistler 1991).

Pollen evidence for Polynesian introductions

Canarium harveyi var. *harveyi* is indigenous to Tonga, where it is found today mostly as a cultivated tree in villages. Its edible almond-like seeds, timber and sap are used by Tongans (Whistler 1991). *Canarium* pollen appeared in small amounts in both Ngofe Marsh and Avai'o'vuna Swamp before Lapita settlement, but is more abundant and consistent in the

Table 3. Plant taxa that first appear (FAD) in sediment cores from Tonga or show an increase in pollen in the uppermost sediments after Lapita (Polynesian) colonisation

Botanical species	Island	Core	Evidence	Reference
Canarium harveyi Leenh.	Pangaimotu, Vava'u	Avai'o'vuna Swamp	Pollen increase c. 2600 BP	Fall 2005
	'Uta Vava'u, Vava'u	Ngofe Marsh	Pollen increase c. 2600 BP	Fall unpublished
	'Eua	Ano'a'pepe	FAD pollen (recent)	Fall unpublished
Casuarina equisetifolia L.	Pangaimotu, Vava'u	Avai'o'vuna Swamp	FAD pollen c. 2600 BP	Fall 2005
	Ha'afeva, Ha'apai	Finemui Swamp	Pollen increase c. 2000 BP	Flenley *et al.* 1999
	'Eua	Ano'a'pepe	FAD pollen (recent)	Fall unpublished
Cocos nucifera L.	Pangaimotu, Vava'u	Avai'o'vuna Swamp	Pollen increase c. 1000 BP	Fall 2005
	'Uta Vava'u, Vava'u	Ngofe Marsh	Pollen increase after 4000 BP	Fall unpublished
	Ha'afeva, Ha'apai	Finemui Swamp	Pollen increase after 1800 BP	Flenley *et al.* 1999
	Foa, Ha'apai	Lotofoa Swamp	FAD pollen c. 2900 BP	Flenley *et al.* 1999
Colocasia esculenta (L.) Schott.	Pangaimotu, Vava'u	Avai'o'vuna Swamp	FAD pollen c. 2600 BP	Fall 2005
	'Uta Vava'u, Vava'u	Ngofe Marsh	FAD pollen c. 2000 BP	Fall unpublished
	'Eua	Ano'a'pepe	FAD pollen (unknown age)	Fall unpublished
Cordyline fruticosa (L.) A. Chev.	Pangaimotu, Vava'u	Avai'o'vuna Swamp	FAD pollen c. 2600 BP	Fall 2005
	'Uta Vava'u, Vava'u	Ngofe Marsh	FAD pollen c. 2200 BP	Fall unpublished
Erythrina variegata L.	Pangaimotu, Vava'u	Avai'o'vuna Swamp	FAD pollen c. 1500 BP	Fall 2005
	'Uta Vava'u, Vava'u	Ngofe Marsh	FAD pollen c. 2400 BP	Fall unpublished
Gardenia tannaensis Guillaumin	Pangaimotu, Vava'u	Avai'o'vuna Swamp	FAD pollen c. 2200 BP	Fall 2005
	Ha'afeva, Ha'apai	Finemui Swamp	Pollen increase c. 2200 BP	Flenley *et al.* 1999
Ludwigia octovalvis (Jacq.) Raven	Foa, Ha'apai	Lotofoa Swamp	FAD pollen c. 2500 BP	Flenley *et al.* 1999
	'Uta Vava'u, Vava'u	Ngofe Marsh	Single Onagraceae grain c. 2000 BP	Fall unpublished
Pandanas tectorius Parkinson	Pangaimotu, Vava'u	Avai'o'vuna Swamp	Pollen increase c. 1500 BP	Fall 2005
	'Uta Vava'u, Vava'u	Ngofe Marsh	Pollen increase c. 2400 BP	Fall unpublished
	Ha'afeva, Ha'apai	Finemui Swamp	Pollen increase c. 2900 BP	Flenley *et al.* 1999
Poaceae (grains >40µm)	Pangaimotu, Vava'u	Avai'o'vuna Swamp	FAD pollen c. 2600 BP	Fall 2005
	'Uta Vava'u, Vava'u	Ngofe Marsh	FAD pollen c. 2800 BP	Fall unpublished
	'Eua	Ano'a'pepe	FAD pollen (unknown age)	Fall unpublished
Poaceae	Foa, Ha'apai	Lotofoa Swamp	Pollen increase c. 3100 BP	Flenley *et al.* 1999
	Ha'afeva, Ha'apai	Finemui Swamp	Pollen increase c. 900 BP	Flenley *et al.* 1999
Polygonum dichotomum Bl. Bijar	Ha'afeva, Ha'apai	Finemui Swamp	Pollen increase c. 1600 BP	Flenley *et al.* 1999
	'Uta Vava'u, Vava'u	Ngofe Marsh	Pollen increase c. 1000 BP	Fall unpublished
Pometia pinnata J.R. & G. Forst.	'Uta Vava'u, Vava'u	Ngofe Marsh	FAD pollen c. 2500 BP	Fall unpublished
Stenochlaena palustris (Burm.) Beddome	Foa, Ha'apai	Lotofoa Swamp	FAD pollen c. 1100 BP	Flenley *et al.* 1999

Note: All ages estimated from sedimentation rates based on radiocarbon ages from each core. Ngofe Marsh and Avai'o'vuna Swamp each have two ^{14}C ages; Finemui and Lotofoa Swamps have one ^{14}C age each.

records following Polynesian settlement. These suggest aboriginal cultivation of this tree, leading to its modern distribution as a cultivated species in villages (Table 3).

Casuarina equisetifolia is suggested to have been either an indigenous tree or an aboriginal introduction to Tonga (Whistler 1991). Haberle (2007) documents the planting of *Casuarina equisetifolia* in New Guinea to aid nitrogen-fixing in the soil, noting that increased *Casuarina* pollen indicates arboriculture. Whistler (1991:125) notes that *Casuarina* often is cultivated in Tonga and that its hard wood is used for posts, tapa mallets, war clubs and other tools. Pollen evidence documents *Casuarina* trees in Tonga before human occupation of the islands, but its abundance and distribution on the islands may have been expanded by Polynesians. Pollen records from Ha'apai (Finemui Swamp) (Flenley et al. 1999) and Vava'u (Ngofe Marsh) (Fall unpublished) suggest that this tree was maintained and probably planted during Lapita occupation. *Casuarina* pollen first appears in Avai'o'vuna Swamp on Pangaimotu Island after Polynesian arrival (Fall 2005). Similarly, *Casuarina* pollen is found in pre-aboriginal sediments on Rapa and Rapanui, then shows a marked increase about 1000 cal. BP (Prebble 2008; Prebble and Wilmshurst 2009).

Cocos nucifera is thought to be an aboriginal introduction to Tonga, where it is cultivated widely in plantations throughout the islands. Coconut provides a liquid that can be substituted for water during periods of drought, meat that can be eaten, and oil that can be made from the nut. Virtually every part of the plant has a use, including its shell, leaves and roots (Whistler 1991:94). While *Cocos nucifera*-type pollen is found in sediments from Tonga prior to colonisation, one record in Ha'apai (Lotofoa Swamp) shows the first *Cocos* pollen coincident with the arrival of Lapita people. Other records from Ha'apai (Finemui Swamp) and Vava'u (Ngofe Marsh and Avai'o'vuna Swamp) document its expansion following Lapita settlement. Similarly, palaeoenvironmental records from Vanuatu (Spriggs 1984), Mo'orea (Parkes 1997) and the Cook Islands (Ellison 1994; Kirch and Ellison 1994) provide evidence for *Cocos nucifera* on Pacific islands prior to human settlement. *Cocos* palms may have been introduced by early Polynesian settlers on the Hawaiian Islands (Athens and Ward 1997) and by European immigrants to French Polynesia (Prebble 2008; Prebble and Wilmshurst 2009).

Colocasia esculenta was brought to Tonga by Polynesian settlers. This starchy cultigen is a staple crop and its many varieties introduced by aboriginal and more recent populations are cultivated widely (Whistler 1991:117). In Tonga, the earliest *Colocasia* pollen is found in Avai'o'vuna Swamp (at about 2600 yr BP), Ngofe Marsh (at about 1800 yr BP), and Ano'a'pepe on the island of 'Eua (in the upper sediments). Although *Colocasia* pollen is not preserved in some sedimentary basins (Haberle 1995), it has been associated with Polynesian settlement and cultivation in other parts of Polynesia, specifically Hawai'i (Athens and Ward 1993, 1997) and French Polynesia (Kennett et al. 2006; Prebble 2008; Prebble and Wilmshurst 2009).

Cordyline fruticosa is an aboriginal introduction to Tonga. The root of *Cordyline fruticosa* was consumed as food in the past; the plant is used today as an ornamental and its leaves are harvested for cooking or medicinal uses (Whistler 1991:111). The first appearance of *Cordyline* pollen at Ngofe Marsh at about 2200 yr BP and at Avai'o'vuna Swamp about 2600 yr BP corroborates the interpretation of *Cordyline fruticosa* as a Polynesian introduction to Vava'u. Prebble (2008) also found *Cordyline* pollen associated with aboriginal colonisation in sedimentary records from subtropical Polynesian islands.

Erythrina variegata, although thought to be an indigenous tree, today is found mostly in cultivation. *Erythrina fusca*, which grows in swamps and marshes on Tonga, may have been introduced to Tonga by Polynesians, perhaps from Fiji, as its Tongan name (*ngatae fisi*, Fijian coral tree) suggests (Whistler 1991:38-39). *Erythrina* trees are commonly found at the edges of wetlands like Ngofe Marsh on Vava'u and at Finemui and Lotofoa swamps on Ha'apai. *Erythrina* pollen appears for the first time in Ngofe Marsh about 2400 yr BP and in Avai'o'vuna Swamp about 1500 yr BP, after Lapita colonisation of Vava'u, supporting

Whistler's suggestion that one of the species of *Erythrina* is a Polynesian introduction, or that *Erythrina* became more prominent in the palynological record.

Gardenia taitensis is an indigenous littoral species in Fiji, which grows on Tonga (Smith 1988). *Gardenia tannaensis* is a tree that is cultivated and naturalised on Niuatoputapu and was probably introduced from Tanna, Vanuatu, where it is indigenous (Whistler 1991:111). *Gardenia augusta* is a European introduction (Whistler 1991:112). *Gardenia* pollen is found in pre-Lapita sediments in both Lotofoa and Finemui swamps (Flenley et al. 1999), strongly supporting Smith's (1988) idea that *Gardenia taitensis* is indigenous to Tonga. *Gardenia* pollen also is present in Lapita and post-Lapita-age sediments at both Finemui and Lotofoa swamps. Increased frequencies of *Gardenia* pollen in Lapita and particularly in post-Lapita sediments at Finemui Swamp (Flenley et al. 1999), and its first appearance in Avai'o'vuna Swamp at about 2200 yr BP (Fall 2005), lend support to the possibility of Polynesian planting or encouragement of *Gardenia* plants, or the increased palynological visibility of *Gardenia* on Tongan Islands.

Ipomoea batatas is an aboriginal introduction to Tonga, commonly found in cultivation today (Whistler 1991). Pollen identified as *Ipomoea* cf. *batatas* is noted in Finemui Swamp at a depth of 0.15-0.24 m (Flenley et al. 1999) and as two grains from Avai'o'vuna Swamp at 0 m and 0.24 m below the surface of the swamp (Fall 2005). This pollen type is very distinctive (Flenley et al. 1999), but normally is not preserved in sediments (Haberle and Atkins 2005). At Avai'o'vuna Swamp the presence of sweet-potato pollen fits the criteria outlined by Haberle and Atkins (2005) for its association with *Casuarina equisetifolia* silviculture, signalled by the first appearance of *Casuarina* pollen in this little swamp.

Ludwigia octovalvis, a pan-tropical species, is a native of tropical America and an introduction to the Pacific islands, where it is common in wetlands and is associated with *Colocasia* cultivation (Whistler 1995:113). *Ludwigia octovalvis* currently dominates the vegetation in some swamps and wetlands in Tonga. Of particular interest here, it grows at the edge of Ngofe Marsh in Vava'u (Fall unpublished) and in both of the swamps investigated by Flenley and colleagues (1999) on Ha'apai, where it dominates the vegetation in Lotofoa Swamp. Although Whistler (1995) suggests that *Ludwigia octovalvis* is a European introduction, the evidence presented below suggests it may have been introduced inadvertently by Polynesians. *Ludwigia* pollen is found associated with *Colocasia* pollen in Polynesian-age deposits in a sediment core from Hawaii (Athens and Ward 1997). *Ludwigia* pollen and seeds document this taxon as an inadvertent Polynesian introduction to Rapa (Prebble 2008). In Tonga, *Ludwigia* pollen first appears in Lapita-age sediments at Lotofoa Swamp (Zone LF 4) (Flenley et al. 1999). At Ngofe Marsh a single Onagraceae (cf. *Ludwigia*) pollen grain is found in association with *Colocasia* pollen in sediments dated to about 2000 yr BP. *Ludwigia octovalvis* is a common weed associated with *Colocasia escuelenta* fields throughout the Pacific (Kirch 1994). The discovery of *Ludwigia* pollen in Lapita-age ponds on Ha'apai and Vava'u suggests that *Ludwigia* was a Polynesian introduction to Tonga.

Pandanus tectorius is an indigenous and cultivated tree in Tonga. Whistler (1991:70-76) describes the many varieties and uses for the indigenous *Pandanus tectorius*, as well as for the Polynesian introduction *Pandanus whitmeeanus* and for *Pandanus veitchii*, which probably was introduced from Hawai'i. *Pandanus tectorius*-type pollen is present in Tonga in sediments deposited before colonisation by Polynesians, as it is on other Polynesian islands (Prebble 2008). However, *Pandaus tectorius*-type pollen is more abundant in Avai'o'vuna Swamp, Ngofe Marsh and Finemui Swamp after Lapita colonisation, supporting the suggestion that people may have planted *Pandanus tectorius* and encouraged its growth throughout the Tongan Islands.

Phyllanthus amicorum, although not mentioned in Whistler (1991), is an endemic tree or small shrub that grows today on 'Eua (Drake et al. 1996) and Vava'u (Steadman et al. 1999). While *Phyllanthus amicorum* or another species of *Phyllanthus* may have grown in

Tonga before human colonisation, *Phyllanthus*-type pollen makes its initial appearance in Avai'o'vuna Swamp in the upper sediments deposited in the past 300 years (Fall 2005). Poaceae pollen increases in all Tongan sediment cores coincident with Lapita settlement and burning. *Saccharum officinarum*, *Schizostachyum glaucifolium* and *Miscanthus floridus* are recognised as Polynesian introductions to Tonga (Whistler 1991). *Miscanthus floridus* was used for arrow shafts, pipes or construction (Whistler 1991:50). The two cane species *Saccharum officinarum* and *Schizostachyum glaucifolium* were cultivated grasses. The sweet stalks of *Saccharum officinarum* provided food and its leaves were used for thatch (Whistler 1991:124-5). *Schizostachyum glaucifolium* occasionally was cultivated or became naturalised, providing materials for poles, vessels, musical instruments and construction (Whistler 1991:57).

While small amounts of Poaceae pollen (1-2%) are found in Ngofe Marsh before Polynesian arrival, increases in Poaceae pollen (up to 270% calculated outside the terrestrial pollen sum), Poaceae pollen >40-50 µm (up to 85% calculated outside the terrestrial pollen sum), and Poaceae pollen with two to three pores (up to 45% calculated outside the terrestrial pollen sum) jump dramatically with the onset of burning of the marsh (denoted by the presence of microscopic charcoal) and the arrival of Polynesian colonists (Figure 4). Similarly, Poaceae pollen, particularly grains >40-50 µm, becomes more abundant in Avai'o'vuna Swamp about 2600 yr BP, and again coincident with the onset of microscopic charcoal deposition (Fall 2005). Ano'a'pepe on 'Eua contains Poaceae pollen >40-50 µm, some Poaceae with multipores, and microscopic charcoal throughout the 1 m core.

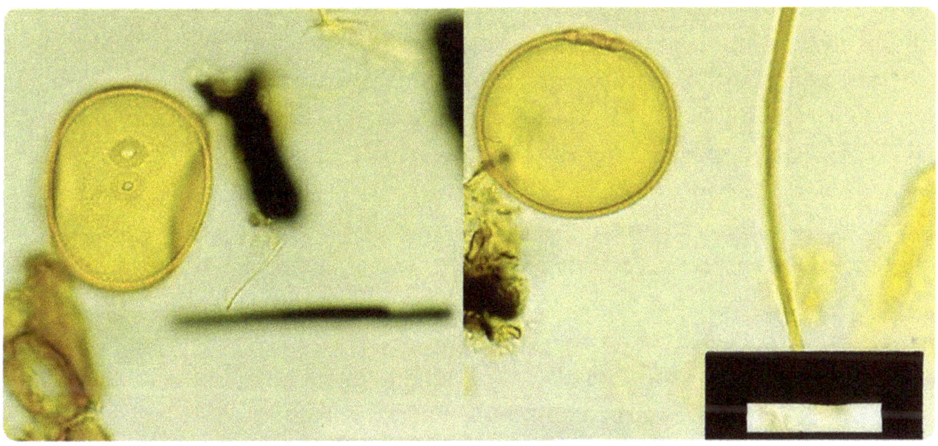

Figure 4. Poaceae pollen from Ngofe Marsh (scale bar = 50 µm [black rectangle]). Photo by P. Fall

The possibility that these Poaceae pollen grains from the Vava'u and 'Eua cores represent a cultivated or hybridised grass is strengthened by their larger size and multiple pores. Erdtman (1969) notes that hybridised cultigens in the Poaceae family may be diporate, multiporate or inapertuate (e.g. *Tricale,* a hybrid of *Triticum* and *Secale,* often has two or more pores). *Saccharum officinarum* hybridises with *Saccharum spontaneum*, as well as other genera, including *Erianthus, Miscanthus, Narenga* and *Sclerostachys* (Aitkens et al. 2007), perhaps leading to the production of these subfossil hybrid pollen grains. Thus, one of the cane or reed grasses brought by Polynesians to Tonga may have been cultivated on Vava'u and 'Eua.

Polygonum dichotomum (synonym *Polygonum glabrum*) has been suggested as a Polynesian introduction to Tonga (Smith 1981). This common wetland plant (Whistler 1992b) currently grows in Tongan swamps, including Finemui Swamp on Ha'afeva (Flenley et al. 1999), and dominates the marsh vegetation at Vasi Vasi Swamp on Hunga Island, Vava'u (Fall unpublished). *Polygonum* pollen first appears at Ngofe Marsh about 4000 yr BP, prior to Polynesian settlement, but is much more common on the marsh after 1000 yr BP. *Polygonum*-type pollen appears throughout the record at Avai'o'vuna Swamp. Although *Polygonum* pollen is found in cores from Lotofoa and Finemui swamps on Ha'apai before Lapita colonisation, *Polygonum* pollen increases in Lapita and subsequent sediments at Finemui Swamp (Flenley et al. 1999).

Pometia pinnata is a large Polynesian introduced tree common in villages and plantations throughout the islands. Fruits of *Pometia pinnata* are similar to litchi in taste and appearance, and are highly valued by Tongans; its wood is used for timber, tapa pounders and other wooden tools (Whistler 1991:121). *Pometia* pollen first appears in Ngofe Marsh at 2500 yr BP, just after the arrival of Polynesians, suggesting that early colonists most likely brought this tree.

Pollen evidence for European introductions

Alternanthera sessilis is a European introduction that is used occasionally as an ornamental around houses (Whistler 1991:70). *Alternanthera* pollen appears in the uppermost sediments from Avai'o'vuna Swamp, Vava'u, attesting to its European introduction (Fall 2005) (Table 4).

Mimosa pudica is an introduced weedy shrub or small tree found in disturbed habitats (Whistler 1991:85). *Mimosa* pollen is quite abundant in a short core collected from the mud-water interface at Ano'a'pepe, the small lake in the upland rainforest, attesting to the recent introduction of *Mimosa pudica* to 'Eua.

Pinus caribaea was introduced to 'Eua, where today it is cultivated in plantations (Whistler 1991:100). Interestingly, although pine species are notoriously prolific pollen producers, and *Pinus caribaea* plantations grow within a few hundred metres of Ano'a'pepe, only three *Pinus* pollen grains have been recovered from a single sample collected at the surface (0 cm) of the lake. *Pinus* pollen is absent from both Avai'o'vuna Swamp and Ngofe Marsh on Vava'u. However, Flenley and others (1999) found small amounts of Pinaceae pollen in the two cores collected on Ha'apai, suggesting long-distance transport to these islands.

Psidium guajava is an introduced fruit tree which has become naturalised in pastures and disturbed areas (Whistler 1991:60). *Psidium guajava* is widespread in Tonga as a cultivated tree and as an adventive species. *Psidium* pollen appears in the upper peats at Folaha Swamp in Fanga 'Uta Lagoon, Tongatapu, dating to the historic period (50-150 yr BP) (Ellison 1989).

Stachytarpeta urticifolia is a weed introduced recently to Tonga (Whistler 1991:46). *Stachytarpeta urticifolia* grows in disturbed soils around swamps and is particularly common on the edge of Ngofe Marsh. *Stachytarpeta* pollen appears in the most recent sediments from Finemui Swamp on Ha'afeva, Ha'apai, in Zone FM 5 (Flenley et al. 1999).

Stenochlaena palustris, an epiphytic fern, grows on *Hibiscus tiliaceous* plants in the *Rhizophora* mangrove at the edge of Folaha Swamp in Fanga 'Uta Lagoon on Tongatapu (Ellison 1989). Whistler (1991:102) notes that it is relatively uncommon in most forests in Tonga. *Stenochlaena* spores were found in Zone LF5, the most recent sediments in the core from Lotofoa Swamp on Foa, Ha'apai (Flenley et al. 1999). Flenley et al. (1999) suggest that *Stenochlaena palustris* was introduced from Southeast Asia, where it is common today, became abundant on Ha'afeva Island, and then was extirpated.

Summary

Dramatic changes follow the settlement and modification of tropical Pacific Island ecosystems by human populations (Hope et al. 1999). The creation of anthropogenic landscapes in Oceania extended from New Guinea eastward with the Lapita culture, and became widespread throughout the Pacific islands (Kirch et al. 1995; Athens et al. 1996; Denham et al. 2003; Kennedy and Clarke 2004; Fairbairn 2005). Vegetation in the Kingdom of Tonga derives from long-distance dispersal (Carlquist 1967, 1974), transport by birds and bats (Rainey et al.

Table 4. Plant taxa that first appear (FAD) in sediment cores from Tonga or show an increase in pollen in the uppermost sediments after European colonisation

Botanical Species	Island	Core	Evidence	Reference
Alternanthera sessilis R.Br. ex DC	Pangaimotu, Vava'u	Avai'o'vuna Swamp	FAD pollen c. 300-0 BP	Fall 2005
Ipomoea batatas L.	Ha'afeva, Ha'apai	Finemui Swamp	FAD pollen historic	Flenley et al. 1999
	Pangaimotu, Vava'u	Avai'o'vuna Swamp	FAD pollen c. 300-0 BP	Fall 2005
Mimosa pudica L.	'Eua	Ano'a'pepe	FAD pollen 0 BP	Fall unpublished
Phyllanthus amicorum G.L. Webster	Pangaimotu, Vava'u	Avai'o'vuna Swamp	FAD pollen c. 300-0 BP	Fall 2005
Pinus caribaea Morelet	'Eua	Ano'a'pepe	FAD pollen 0 BP	Fall unpublished
Psidium guajava L.	Tongatapu	Folaha Swamp	FAD pollen historic (50-150 yr BP)	Ellison 1989
Stachytarpeta urticifolia (Salisb.) Sims	Ha'afeva, Ha'apai	Finemui Swamp	FAD pollen historic	Flenley et al. 1999

Note: All ages estimated from sedimentation rates based on radiocarbon ages from each core. Ngofe Marsh, Avai'o'vuna and Folaha swamps each have two ^{14}C ages; Finemui and Lotofoa swamps have one ^{14}C age each.

1995; Banack 1998) and human introduction and cultivation (Fall et al. 2007). Palynological evidence for Polynesian and later European plant introductions to Tonga comes from Holocene age sediment cores collected from Vava'u (Fall 2005), Ha'apai (Flenley et al. 1999), Tongatapu (Ellison 1989) and 'Eua (Fall unpublished).

Plants introduced commonly by Lapita colonists include both trees and herbaceous cultigens. In addition, indigenous plants may have been planted and cultivated by Polynesian settlers, allowing the expansion of their native habitats. Plants that were cultivated or whose ranges were expanded by Polynesians in Tonga are represented by pollen from the following tree species – *Canarium harveyi, Casuarina equisetifolia, Cocos nucifera, Erythrina variegata, Pandanus tectorius*, and *Pometia pinnata*. Woody shrubs introduced or cultivated by Polynesian settlers include *Cordyline fruticosa* and *Gardenia*. Herbaceous taxa introduced or expanded by early colonists consist of *Colocasia esculenta, Ludwigia octovalvis, Polygonum*, and a cultivated grass species. Pollen deposited in the past few hundred years provides evidence for European cultigens or inadvertent introductions. Pollen from tree species includes the planted *Pinus caribaea* found in plantations on 'Eua and the naturalised weedy species *Mimosa pudica* and *Psidium guajava*, found in very recent-age deposits. Historic-age deposits also contain pollen produced by herbaceous or shrubby vegetation, including *Alternanthera, Ipomoea* cf. *batatas, Phyllanthus* and *Stachytarpeta*, as well as *Stenochlaena* spores.

Pollen data provide a window to understanding many of the changes that have moulded the Tongan landscape. Polynesian colonists significantly modified their environment through the burning and clearing of indigenous rainforests, making way for cultivated root crops, grasses and ferns. These early settlers also brought a number of woody shrubs and useful tree species, and expanded their ranges, thereby encouraging the creation of anthropogenic forests, as well as open landscapes. European settlement of Tonga further added to the number of cultivated plants, promoted forest clearing and introduced new species, particularly weedy or unintentionally introduced plants.

Acknowledgements

This research was funded by a research grant from the National Geographic Society and was conducted with the permission of the government of the Kingdom of Tonga. My research into the history of tropical Pacific rainforests on the islands of Tonga was enhanced greatly during a sabbatical leave in 1998 spent as a Research Fellow in the Department of Archaeology and Natural History, Research School of Pacific and Asian Studies, at the Australian National University, at the invitation of Geoff Hope. The support and camaraderie extended by Geoff Hope, Janelle Stevenson, Simon Haberle, Sophie Bickford, Mike Macphail, Gillian Atkins and Dominique O'Dea at the ANU were invaluable. This paper was modified from a presentation given in Canberra for a workshop organised by Mat Prebble and Geoff Hope, entitled 'A multidisciplinary workshop for defining plant extinctions from island Oceania and Australia', October 11-13, 2006. Steve Falconer and two anonymous reviewers provided valuable comments.

References

Aitkens, K., J. Li, L. Wang, C. Qing, Y. H. Fan and P. Jackson 2007. Characterization of intergeneric hybrids of *Erianthus rockii* and *Saccharum* using molecular markers. *Genetic Resources and Crop Evolution* 54:1395-1405.

Athens, J.S. and J.V. Ward 1993. Environmental change and prehistoric Polynesian settlement in Hawai'i. *Asian Perspectives* 32:205-223.

Athens, J.S. and J.V. Ward 1997. The Maunawili core: prehistoric inland expansion of settlement and agriculture, O'ahu, Hawai'i. *Hawaiian Archaeology* 6:37-51.

Athens, J.S., J.V. Ward and G.M. Murakami 1996. Development of an agroforest on a Micronesian high island: Prehistoric Kosraean agriculture. *Antiquity* 70:834-46.

Banack, S.A. 1998. Diet selection and resource use by flying foxes (Genus *Pteropus*). *Ecology* 79:1949-1967.

Burley, D.V. 1994. Settlement pattern and Tongan prehistory reconsiderations from Ha'apai. *Journal of Polynesian Society* 103:379-411.

Burley, D.V. 1998. Tongan archaeology and the Tongan past, 2850-150 BP. *Journal of World Prehistory* 12:337-92.

Burley, D.V., E. Nelson, and R. Shutler, Jr. 1999. A radiocarbon chronology for the Eastern Lapita frontier in Tonga. *Archaeology in Oceania* 34:59-70.

Burley, D.V. and W.R. Dickinson 2001. Origin and significance of a founding settlement in Polynesia. *Proceedings of the National Academy of Sciences* 98:11829-11831.

Burley, D.V. and S.P. Connaughton 2007. First Lapita settlement and its chronology in Vava'u, Kingdom of Tonga. *Radiocarbon* 49:131-137.

Burney, D.A. 1997. Tropical islands as paleoecological laboratories: Gauging the consequences of human arrival. *Human Ecology* 25:437-457.

Carlquist, S. 1967. The biota of long-distance dispersal. V. Plant dispersal to Pacific Islands. *Bulletin of the Torrey Botanical Club* 94:129-162.

Carlquist, S. 1974. *Island biology*. New York: Columbia University Press.

Denham, T.P., S.G. Haberle, C. Lentfer, R. Fullgar, J. Field, M. Therin, N. Porch and B. Winsborough 2003. Origins of agriculture at Kuk Swamp in the highlands of New Guinea. *Science* 301:189-193.

Dickinson, W.R. 2001. Paleoshoreline record of relative Holocene sea levels on Pacific islands. *Earth-Science Reviews* 55:191-234.

Dickinson, W.R. 2003. Impact of mid-Holocene hydro-isostatic highstand in regional sea level on habitability of islands in Pacific Oceania. *Journal of Coastal Research* 19:489-502.

Dickinson, W.R. and D.V. Burley 2007. Geoarchaeology of Tonga: Geotectonic and geomorphic controls. *Geoarchaeology* 22:229-259.

Dickinson, W.R., D.V. Burley and R. Shutler, Jr. 1994. Impact of hydro-static Holocene sea-level change on the geological context of island archaeological sites. *Geoarchaeology* 9:85-11.

Drake, D.R., W.A. Whistler, T.J. Motley and C.T. Imada 1996. Rain forest vegetation of 'Eua Island, Kingdom of Tonga. *New Zealand Journal of Botany* 34:65-77.

Ellison, J.C. 1989. Pollen analysis of mangrove sediments as a sea-level indicator – assessment from Tongatapu, Tonga. *Palaeogeography, Palaeoclimatology, Palaeoecology* 74:327-341.

Ellison, J.C. 1994. Paleo-lake and swamp stratigraphic records of Holocene vegetation and sea-level changes, Mangaia, Cook Islands. *Pacific Science* 48:1-15.

Erdtman, G. 1969. *Handbook of palynology: Morphology, taxonomy, ecology*. Copenhagen: Munksgaard.

Fall, P.L. 2005. Vegetation change in the coastal-lowland rainforest at Avai'o'vuna Swamp, Vava'u, Kingdom of Tonga. *Quaternary Research* 64:451-459.

Fall, P.L., T.D. Drezner and J. Franklin 2007. Dispersal ecology of the lowland rain forest in the Vava'u island group, Kingdom of Tonga. *New Zealand Journal of Botany* 45:393-417.

Fairbairn, A. 2005. An archaeobotanical perspective on Holocene plant-use practices in lowland northern New Guinea. *World Archaeology* 37:487-502.

Flenley, J.R., C.T. Hannan and M.J. Farelly 1999. Final report on the stratigraphy and palynology of swamps on the islands of Ha'afeva and Foa, Ha'apai, Tonga. *Massey University, Geography Programme Miscellaneous Publication Series* 99/3, Palmerston, New Zealand.

Franklin, J. 2003. Regeneration and growth of pioneer and shade-tolerant rain forest trees in Tonga. *New Zealand Journal of Botany* 41:669-684.

Franklin, J., D.R. Drake, L.A. Bolick, D.S. Smith and T.J. Motley 1999. Rain forest composition and patterns of secondary succession in the Vava'u Island group, Tonga. *Journal of Vegetation Science* 10:51-64.

Franklin, J, D.R. Drake, K.R. McConkey, F. Tonga and L.B. Smith 2004. The effects of Cyclone Waka on the structure of lowland tropical rain forest in Vava'u, Tonga. *Journal of Tropical Ecology* 20:409-420.

Franklin, J., S.K. Wiser, D.R. Drake, L.E. Burrows and W.R. Sykes 2006. Environment, disturbance history and rain forest composition across the islands of Tonga, Western Polynesia. *Journal of Vegetation Science* 17:233-244.

Haberle, S.G. 1995. Identification of cultivated *Pandanus* and *Colocasia* in pollen records and the implications for the study of early agriculture in New Guinea. *Vegetation History and Archaeobotany* 4:195-210.

Haberle, S.G. 2007. Prehistoric human impact on rainforest biodiversity in highland New Guinea. *Philosophical Transactions of the Royal Society B* 362:219-228.

Haberle, S.G. and G. Atkins 2005. Needles in a haystack: Searching for sweet potato (*Ipomoea batatas* (L.) Lam.) in the fossil pollen record. In C. Ballard, P. Brown, R.M. Bourke and T. Harwood (eds), *The sweet potato in Oceania: a reappraisal*, pp25-33, Ethnology Monographs 19. Pittsburgh: University of Pittsburgh Press.

Haberle, S.G. and B. David 2004. Climates of change: Human dimensions of Holocene environmental change in low latitudes of the PEPII transect. *Quaternary International* 118-119:165-179.

Haberle, S.G., G.S. Hope and S. van der Kaars 2001. Biomass burning in Indonesia and Papua New Guinea: Natural and human induced fire events in the fossil record. *Palaeogeography, Palaeoclimatology, Palaeoecology* 171:259-268.

Hope, G.S. 2001. Quaternary change and the historical biogeography of Pacific islands. In A. Keast and S.E. Miller (eds), *The origin and evolution of Pacific Island biotas, New Guinea to Eastern Polynesia: Patterns and processes*, pp165-190. Amsterdam: Academic Publishing.

Hope, G.S., D. O'Dea and W. Southern 1999. Holocene vegetation histories in the Western Pacific- alternative records of human impact. In I. Lilley and J.-C. Galipaud (eds) *The Pacific from 5000 to 2000 BP. Colonisation and Transformations*. Paris: Orstom Publishing.

Hope, G.S. and J. Pask 1998. Tropical vegetational change in the late Pleistocene of New Caledonia. *Palaeogeography, Palaeoclimatology, Palaeoecology*. 142:1-21.

Kennedy, J. and W. Clarke 2004. Cultivated landscapes of the southwest Pacific. *Resource Management in Asia-Pacific Working Paper* No. 50. Canberra: Resource Management in Asia-Pacific Program.

Kennett, D., A. Anderson, M. Prebble, E. Conte and J. Southon 2006. Prehistoric human impacts on Rapa, French Polynesia. *Antiquity* 80:340-354.

Kirch, P.V. 1994. *The wet and the dry: Irrigation and agricultural intensification in Polynesia*. Chicago: University of Chicago Press.

Kirch, P.V. 1996. Late Holocene human-induced modifications to a central Pacific island ecosystem. *Proceedings of the National Academy of Sciences* 93:5296-5300.

Kirch, P.V., J.C. Ellison 1994. Paleoenvironmental evidence for human colonization of remote Oceanic islands. *Antiquity* 68:310-321.

Kirch, P.V., D.W. Steadman, V.L. Butler, J. Hather and M.I. Weisler 1995. Prehistory and human ecology in Eastern Polynesia: Excavations at Tangatatau Rockshelter, Mangaia, Cook Islands. *Archaeology in Oceania* 30:47-65.

McConkey, K.R. and D.R. Drake 2006. Flying foxes cease to function as seed dispersers long before they become rare. *Ecology* 87:271-276.

McConkey, K.R. and D.R. Drake 2007. Indirect evidence that flying foxes track food resources among islands in a Pacific archipelago. *Biotropica* 39(3):436-440.

Mueller-Dombois, D. and F.R. Fosberg 1998. *Vegetation of the tropical Pacific islands*. New York: Springer-Verlag.

Nunn, P. D. 1994. *Oceanic Islands*. Oxford: Blackwell.

Palmer, M.W. 1988. The vegetation and anthropogenic disturbance of Toloa Forest, Tongatapu Island, South Pacific. *Micronesica* 21:279-281.

Park, G. and A. Whistler 2001. The terrestrial ecology and botany of Tofua and Koa Islands in Ha'apai, Kingdom of Tonga. *Report of the botanical survey of the Ha'apai Conservation Area Project of the South Pacific Biodiversity Conservation Programme*, Apia, Samoa.

Parkes, A. 1997. Environmental change and the impact of Polynesian colonization: sedimentary records from central Polynesia. In P.V. Kirch and T.L. Hunt (eds) *Historical ecology in the Pacific islands: Prehistoric environmental and landscape change*, pp166-199. New Haven: Yale University Press.

Parkes, A., J.R. Flenely and J.T. Teller 1992. Environmental history of the Lake Vaihiria drainage basin, Tahiti, French Polynesia. *Journal of Biogeography* 19:431-447.

Poulsen, J. 1987. Early Tongan prehistory: The Lapita Period on Tongatapu and its relationships. *Terra Australis* 12 (2 volumes) 205pp Canberra.

Prebble, M. 2008. No fruit on that beautiful shore: What plants were introduced to the subtropical Polynesian islands prior to European contact? In G. Clark, F. Leach and S. O'Connor (eds) *Islands of Inquiry. Colonisation, seafaring and the archaeology of maritime landscapes*, pp227-251. Canberra: ANU-E Press.

Prebble, M. and J.L. Dowe 2008. The late Quaternary decline and extinction of palms on oceanic Pacific islands. *Quaternary Science Reviews* 27:2546-2567.

Prebble, M. and J.M. Wilmshurst 2009. Detecting initial impacts of humans and introduced species on island environments in Remote Oceania using paleoecology. *Biological Invasions* 11:1529-1556.

Rainey, W.E., E.D. Pierson, T. Elmqvist and P.A. Cox 1995. The role of flying foxes (Pteropodidae) in oceanic island ecosystems of the Pacific. *Symposium of the Zoological Society, London* 67:47-62.

Smith, A.C. 1979, 1981, 1985, 1988, 1991: *Flora Vitiensis Nova: A new flora of Fiji, Vols. 1-5.* Lanai: Pacific Tropical Botanical Gardens.

Southern, W. 1986. The late Quaternary environmental history of Fiji. Unpublished PhD thesis, Australian National University, Canberra.

Spriggs, M. 1984. Early coconut remains from the South Pacific. *Journal of the Polynesian Society* 93:71-76.

Steadman, D.W. 2006. *Extinction and biogeography of tropical Pacific birds.* Chicago, University of Chicago Press.

Steadman, D.W., J. Franklin, D.R. Drake, H.B. Freifeld, L.A. Bolick, D.S. Smith and T.J. Motley 1999. Conservation status of forests and vertebrate communities in the Vava'u island group, Tonga. *Pacific Conservation Biology* 5:191-207.

Stevenson, J., 1998. Human impact from the palaeoenvironmental record in New Caledonia. In J.-C. Galipaud and I. Lilley (eds), *Le Pacifique de 5000 a 2000 BP: Supplements a l'Histoire d'une Colonisation*, pp251-259. Noumea: Orstom.

Stevenson, J., J.R. Dodson and I.P. Prosser 2001. A late Quaternary record of environmental change and human impact from Plum Swamp, New Caledonia. *Palaeogeography, Palaeoclimatology, Palaeoecology* 168:97-123.

Stoddart, D.R. 1992. Biogeography of the tropical Pacific. *Pacific Science* 46:276-293.

Sykes, W.R. 1977. The pteridophytes of 'Eua, southern Tonga. *Bulletin of the Royal Society of New Zealand* 17:119-152.

Sykes, W.R. 1981. The vegetation of Late, Tonga. *Allertonia* 2:323-353.

Thompson, C.S. 1986. The climate and weather of Tonga. New Zealand Meteorological Service Miscellaneous Publication 188. Wellington: New Zealand Meteorological Service.

van Steenis, C.G.G.J. 1979. Plant geography of east Malesia. *Botanical Journal of the Linnean Society* 79:97-178.

Whistler, W.A. 1991. The ethnobotany of Tonga: The plants, their Tongan names, and their uses. *Bernice P. Bishop Museum Series in Botany* 2:1-155.

Whistler, W.A. 1992a. *Flowers of the Pacific Island seashore: a guide to the littoral plants of Hawai'i, Tahiti, Samoa, Tonga, Cook Islands, Fiji, and Micronesia.* Honolulu: University of Hawaii Press.

Whistler, W.A. 1992b. Vegetation of Samoa and Tonga. *Pacific Science* 46:159-178.

Whistler, W.A. 1995. *Wayside plants of the islands: A guide to the lowland flora of the Pacific Islands, including Hawai'i, Samoa, Tonga, Tahiti, Fiji, Guam, Belau.* Honolulu: Isle Botanica.

Wiser, S.K., D.R. Drake, L.E. Burrows and W.R. Sykes 2002. The potential for long-term persistence of forest fragments on Tongatapu, a large island in western Polynesia. *Journal of Biogeography* 29:767-787.

Yuncker, T.G. 1959. Plants of Tonga. *Bishop Museum Bulletin* 220:1-283.

15

Integrating social and environmental change in prehistory: A discussion of the role of landscape as a heuristic in defining prehistoric possibilities in northeast Thailand

William E. Boyd

School of Environmental Science and Management, Southern Cross University, Australia
william.boyd@scu.edu.au

Nigel Chang

School of Anthropology, Archaeology and Sociology, James Cook University, Australia

Introduction

> Just as landscape maps memory and declares identity, so too it offers a key to interpreting society … the land itself, as socially constituted, plays a fundamental role in the ordering of cultural relations … And as a community merges with its habitus, through the actions and activities of its members, the landscape may become a key reference point for expressions of individual as well as group identity. … it is now clear that landscape is neither exclusively natural nor totally cultural; it is a mediation between the two and an integral part of Bourdieu's habitus, the routine social practices within which people experience the world around them. Beyond habitus, however, people actively order, transform, identify with and memorialise the landscape by dwelling within it. The environment manifests itself as landscape only when people create and experience space as a complex of places. (Knapp and Ashmore 1999:16, 20-21)

The archaeological study of past societies is an inherently difficult activity. Relying on extremely small samples of the available evidence, often many millennia after the event, archaeologists have the unenviable task of inferring complex relationships and processes for societies whose social, cultural, political and cognitive characteristics are likely to be very different from those of the archaeologists investigating them. Consequently, archaeologists are developing and adopting an increasingly wide range of analytical and conceptual tools with which to tackle the task of unravelling past social behaviour and history. While palaeoenvironmental study has frequently been called on, it has been largely to provide an assumed passive background, a stage on which people have enacted their social and cultural activity, or a resource providing the essential energy and materials for a successful society, merely at risk of being degraded by the very societies which rely on that resource. However, there is a growing and vibrant literature that promotes the truism that people and environment are intimately linked. Building on this close relationship, geographers, in particular, recognise landscape as a crucial unifying element, a phenomenon that both recognises and reflects the mutual interactivity of people and their environment. In this paper, we discuss the heuristic of the landscape as a source of evidence in assisting in defining social possibilities in the past. By doing this, we recognise both archaeological and geoarchaeological data as equal sources of data reflecting past social behaviour. We use an example from northeast Thailand – the Neolithic to Iron Age occupation of the upper Mun River valley – to demonstrate the potential that such an integrating approach may have for furthering our understanding of past social behaviour. In particular, we focus on a crucial event towards the end of the Iron Age, in which the choice of a re-patterning of settlement across the landscape, while conventionally being understood as an obligatory response to either environmental degradation or an inevitable socio-political progress, can be better explained as a unique and non-obligatory response to a particular set of socio-environmental conditions, emerging from a history of evolving social identity intimately associated with the landscape of the previous 2000 years.

Origins of society in northeast Thailand

The evolution of the human species and subsequent change in social organisation and subsistence practices have long been subjected to explanation based, at least in part, on our relationship to the environment. The clearest example of this is surely the coincidence of the rise of agriculture-based lifestyles and the end of the last Ice Age at a number of locations around the world (e.g. Fagan 2007). In Southeast Asia, for example, Higham and Lu (1998) have suggested that initial rice agriculture in the middle Yangtze River developed as a result of a worldwide return to colder conditions (the Younger Dryas) that disrupted the otherwise relatively steady warming of the planet following about 12,000 BC. However, it is still common to find archaeologists using environment as a non-interactive background that either directs change in human societies or provides context within which change is constrained. What are not always acknowledged are the complex feedback relationships between humans and their environments. This complexity has been brought into sharp focus recently as modern global warming or climate change is increasingly acknowledged as a reality.

To return briefly to Higham and Lu's (1998) discussion of the origins of rice agriculture, we can see a partial attempt to understand this relationship in prehistory. Essentially, their argument is that as the planet warmed, the inhabitants of the middle Yangtze River valley became increasingly reliant on the wild rice that was abundant in the local swampy environment. As temperatures cooled, rice probably grew less abundantly; an important choice had to be made. Should the people move elsewhere (south?) where wild rice presumably continued to grow abundantly, or remain and adapt their subsistence strategy to other food resources? Or, as Higham and Lu conclude, should they adapt in another way, and invest time and resources into encouraging rice to grow? In a sense, they were changing to remain the same. By making this latter choice, the process of domestication begins, and what was a natural swampy environment became a constructed swampy environment that was, eventually, replicated right across Asia. This process is clearly of huge import for understanding past change in Asia and, just as clearly, is the result of a complex interrelationship between people and environment. Implicit in Higham and Lu's discussion is the construction, and subsequently evolution and maintenance, of new landscapes, key elements in reflecting social behaviour and choice and, in turn, in defining social identity (cf. Knapp and Ashmore 1999).

Archaeology and geography: People, place and landscape

As Knapp and Ashmore indicate above, there is good reason to consider people and environment as a single entity, expressed, and potentially interpretable, through the concept of landscape. It is landscape, as a heuristic device, that we discuss here. In particular, we suggest that environmental conditions offer not only a backdrop for past human behaviour, but, more importantly, key evidence to model social change through time. In doing so, we acknowledge that concepts of society, environment and especially landscape are difficult and notably 'unstable' (cf. Tilley 1994), and that the inherent mutual interactivity between people and environment – Basso's 'interanimation' (1996) – makes, unlike conventional use of environmental data as social backdrop or resource, a simple cause-and-effect analysis difficult. Social relations are, however, inherently complex, so their interpretation and analysis must also be expected to be complex.

Here we take as a case study our efforts to integrate environmental and archaeological investigations, research questions and interpretations within an archaeological study in the upper Mun River valley of northeast Thailand. Our study centres on the site of Ban Non Wat and associated projects, most of which have been conducted under the extended Origins of the Civilisation of Angkor project driven by Charles Higham, Rachanie Thosarat and Amphan Kijngam (Higham 2002, 2004; Higham et al. 2007). The considerable palaeoenvironmental research that has already been done in the project will form a central part of the following discussions. An important parallel with Higham and Lu's work (1998) is that while we are not looking at the initial development of agriculture, we are looking at the first introduction of an agricultural economy to an apparently previously unfarmed region (Higham 2002).

The challenge: Integrating palaeosocio-environmental processes

Our specific question is whether we can use an integrated socio-environmental approach to develop a satisfactory explanation of prehistoric conditions. In the context of this case study, the question has been prompted by the need to account for the following broad archaeological observations: there was a long and continuous stable and healthy prehistoric occupation in the Mun River valley; at least during the middle to late Iron Age, there appears to be a major discontinuity in this occupation; there is some evidence of changing social structures over time; new technologies are introduced (bronze and then iron working among others); and the environment has changed, expressed in terms of water, vegetation and sedimentation. The basis for these observations will be discussed below, and they form the basis for a proposed integrated model of change in the region and a strategy for investigating and testing this model. Specifically, we argue that it is important that data from intensive archaeological excavations be extended and integrated with studies focusing on landscape and environment, so as to place Ban Non Wat and other sites more clearly in the wider context of the past social and environmental world(s) of the upper Mun River catchment. This approach may help us to better imagine how the people of Ban Non Wat saw and interacted with this integrated social and environmental world, their landscape (cf. Källén 2004).

Ban Non Wat and the Origins of the Civilisation of Angkor

Context

It is often commented that archaeology in Southeast Asia has but a recent history. However, the pace of research has increased markedly over the past decade or so, particularly reflecting the opening up of China, Burma, Cambodia, Vietnam and Laos to foreign researchers, as well as the growth of cultural heritage and research projects being undertaken by national archaeologists in each country (Higham 2002; Glover and Bellwood 2004; Stark 2006; O'Reilly 2007). The database of excavation and survey has increased to the point that it is increasingly difficult for any one researcher to be familiar with the whole region. One of the many positive outcomes of this situation is that it is now possible (and desirable) to develop detailed regional sequences so that social trajectories in different environments can be compared; one region in which this process is well advanced is the upper Mun River valley in lower northeast Thailand.

An important place in the upper Mun River landscape is the town of Phimai in northeast Thailand, once a provincial northern capital of the Angkorian empire that was based on the northern shores of the Tonle Sap in modern Cambodia, and now a focus for a rich archaeological history (Williams-Hunt 1950; Higham 1977; Welch 1984; Moore 1988a, b; McNeill and Welch 1991; Nitta 1991; O'Reilly 1998, 1999, 2000, 2003). Most recently, a major archaeological project entitled The Origins of the Civilisation of Angkor (hereafter referred to as the Origins project) has examined the prehistoric base from which the Angkorian Empire developed (Higham 2002, 2004; Higham and Thosarat 2005; Higham et al. 2007). The Origins project undertook archaeological excavation and field work at several sites to the northwest of Phimai; the site of Ban Non Wat is the most recent and arguably the most significant. Excavations have also been completed at the sites of Ban Lum Khao (Higham and Thosarat 2005), Noen U-Loke and Non Muang Kao (O'Reilly 1998, 1999; Higham et al. 2007), as well as within the Phimai Prasat (sanctuary) itself (Talbot 2000, 2002).

Archaeological foundations

At Ban Non Wat, eight seasons of excavation (January 2002 to February 2009) have revealed cemetery and occupation evidence for repeated, possibly continuous, use from at least 4000 years ago (Higham 2002, 2004). Initial pre-metal occupation is followed by Bronze and Iron Age settlement. The upper levels include evidence of Angkorian (late first millennium AD) and later use, with a modern living village being the most recent use of the site. The site is thus far unique in having such a long stratigraphic sequence so intensely investigated; more than 550 burials, for example, representing all the prehistoric periods, offer an unparalleled opportunity to understand social change over time.

Details of the archaeological evidence are discussed below. For now, it is important to note that current work at Ban Non Wat represents the latest stage of archaeological discovery that began with the observations of Williams-Hunt (1950), who first alerted the English-reading community to the 'moat-and-mound' sites distributed throughout, principally, the Mun and Chi River valleys of northeast Thailand. The defining characteristic of these sites is the presence of moats or, in Williams-Hunt's words, 'irregular earthworks' (Williams-Hunt 1950; Higham and Kijngam 1982), surrounding the central mound. Initial interpretations of these sites stressed the identity of the earthworks as moats, their association with Iron Age pottery and a likely association with developing chiefdoms in the region. Associations with artefacts of Indian origin or inspiration, the introduction of the plough and therefore possible intensification of agriculture, an assumption that some degree of central organisation was required to construct moats, and the ability to organise the production of iron artefacts, have all been cited as evidence for the development of chiefdoms, along with the appearance of these 'moated' sites (e.g. Higham 1989). Such interpretations bear closer examination and, as we shall see below, can be questioned. For now, they stand as general background to archaeological knowledge before the Origins project field work began. Importantly, in drawing attention to these sites, Williams-Hunt opened a long-standing discussion of the prehistoric landscape of this region, in which many assumptions and preconceptions were brought to bear on interpreting past human behaviour.

The Mun and Chi valley archaeology is characterised by many large sites, now 3 m to 5 m high mounds over several hundred hectares, comprising the remains of substantial and often long-lived settlements, and surrounded by distinctive encircling earthworks and channels commonly known as moats (Boyd 2008). Since the encircling earthworks imply a relationship between hydrological conditions and prehistoric settlement, it is reasonable to assume there was a particular association between environment and society in this region. Environmental examination of these sites and their landscapes (Moore 1988a, b, 1992; Boyd et al. 1999a, b; Boyd and McGrath 2001a, b; McGrath and Boyd 2001; Boyd 2004, 2007, 2008; Boyd and Habberfield-Short 2007; Habberfield-Short and Boyd 2007; McGrath et al. 2008) has progressed to a point where integration with the archaeology is possible (Figure 1).

Calendar C BC/AD	Palaeo-geographic phases	Vegetation phases and thresholds	Hydrological phases and thresholds	Social changes	Regional social influences	Regional environmental influences
9 AD	Phase 6	?modern conditions	?modern conditions	Centralised state landscape	Angkor	Dry, seasonal rainfall, floodplain sedimentation
8 AD					Zhenla	
7 AD						
6 AD						
5 AD		Rapid change	Rapid change	Rapid change		
4 AD	Phase 5C	Phased landscape management	Construction engineering	Change in focus on identity, claim on place, and social intensification	Funan	
3 AD						
2 AD			Adaptive Engineering		Social dislocations in the region	
1 AD						
1 BC						Gradual drying, run-off reduction
2 BC						
3 BC	Phase 5B	Early gradual vegetation change	Settlement beside rivers, gradual hydrological change		Indianisation?	
4 BC				Internal focus on identity and landscape; mixed economy		
5 BC						
6 BC	Phase 5A					
7 BC						
8 BC						
9 BC						
10 BC						
11 BC						
12 BC						
13 BC						
14 BC						
15 BC						
16 BC	Phase 4			Arrival from outside region: establishing a place in the landscape		Warm, humid, lakes and swamps on floodplain
17 BC						
18 BC						
19 BC						
20 BC						

Figure 1. Synthesis of the major social and environmental trends in the upper Mun River valley over the past 4000 years. The boldness of the horizontal lines indicates the importance or significance of the change or difference between phases

We now discuss the geoarchaeological and archaeological materials and data specific to our case study before returning to the wider implications of the case study in later sections of this paper.

Geoarchaeological insights

Overview

There are several strands of evidence that bring us to these conclusions; published in detail elsewhere (summarised by Boyd 2007), they are synthesised here. To open, it is important to understand the geological setting of the archaeological sites (Figure 2; Boyd and Habberfield-Short 2007); the published regional geological history, together with details of field observations, combine in a model of the palaeogeography of the region (Boyd 2007) (Figure 3). That palaeogeographic model identifies several key landscape periods: Palaeogeographic Phases 1 to 3 represent the establishment of the landscape into which people eventually moved; Palaeogeographic Phase 4 represents the early to mid-Holocene period of environmental stability and richness; whereas Palaeogeographic Phase 5 reflects a mid to late-Holocene period of critical environmental

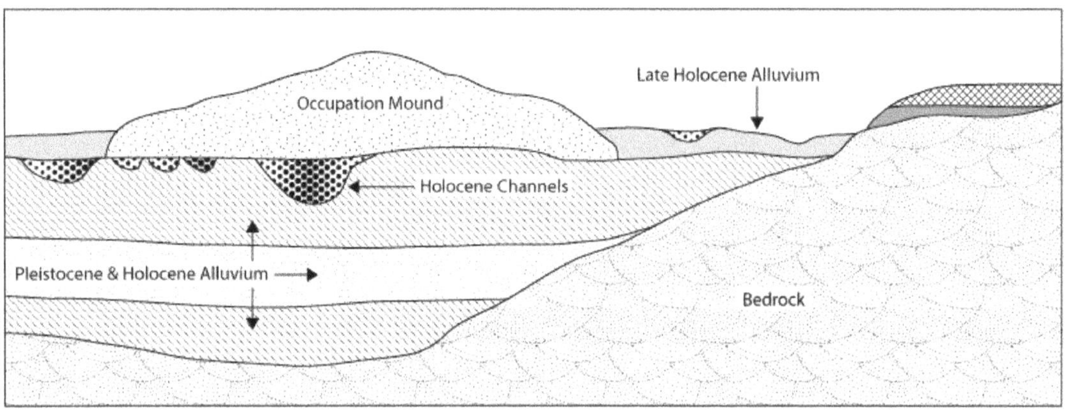

Figure 2. The geological context of the archaeological sites of the upper Mun River valley floodplain

Palaeogeographic phase	Key geographic processes	Social implications
Phase 6: Late Holocene (AD 500-present)	Alluvial conditions include single channels and sheet wash, onset of modern climatic conditions	Establishment of historic and modern modified and degraded landscape
Phase 5: Mid to Late Holocene	**Phase 5C (200 BC-AD 500)** Increasing dryness and seasonality	Increasing landscape management and engineering
	Phase 5B (BC 500-200) Anastomosing channels, reduced runoff, drier floodplain	Mixed economy settlement, tendency towards increased landscape management
	Phase 5A (BC 1500-500) Channel infilling	Settlement by rivers
Phase 4: Early to Mid Holocene (6000-1500 BC)	Floodplain deposition, braided channels to single-string channels, back swamps, lakes	Rich aquatic environment and important resource source for early settlement
Phase 3: Late Pleistocene (12,000-6000 BC)	Hiatus	
Phase 2: Mid to Late Pleistocene (to 12,000 BC)	Regolith stripping, coarse sandy river channels	Establishing the landscape for human settlement
Phase 1: Mesozoic-Tertiary	Bedrock formation and weathering	

Figure 3. A palaeogeographic model for the upper Mun River valley, linking the geological record of the floodplain with its key geographic processes and social implications arising from these

change; and the final Palaeogeographic Phase (6) represents a modified environmental stability, as the changes during the previous phase resulted in a new landscape equilibrium.

Vegetation history

There is pollen evidence for Palaeogeographic Phase 5 (Boyd and McGrath 2001a, b), indicating that by the early Iron Age (c. 200 BC-AD 1), the natural forest of the floodplain had been disturbed and was now partially open; given the archaeological evidence for previous settlement and rice cultivation, this is unsurprising. During the Iron Age, there are cycles of further forest disturbance and the establishment of rice cultivation, woodland management and arboriculture, the latter coming in later in the sequence: during the mid-Iron Age (c. AD 300), forest clearance, rice cultivation and some woodland management is followed by a period of forest and woodland regeneration, and then the resumption of rice cultivation accompanied by arboriculture. The period of forest and woodland regeneration is notable, in that, despite the environmental impact of human activities, local vegetation had the capacity to recover. Finally, however, during the late Iron Age, a decline of agriculture and arboriculture was accompanied not by forest and woodland regeneration but by the establishment of grassland and scrubby vegetation. Linked with other archaeological evidence, for example wild animals and other fauna, we can model the trend of vegetation change during the late Holocene (Figure 1). This model, in particular, charts the gradual and sequential shift from stable Holocene conditions of a floodplain forest, through increasing and varying forms of vegetation disturbance and change, to the eventual conversion to late-Holocene conditions of dry woodland.

Clearly, such a vegetation history could be read largely, and conventionally, in terms of the human impacts on vegetation. However, it perhaps more accurately reflects the continuous and variable interaction between people and the vegetation of their environment. While it is largely the effects of people that best explain the conversion of floodplain pioneer forest to managed forest and, eventually, to agricultural landscapes during the Iron Age, it is also clear that the vegetation had a capacity to regenerate where agricultural pressure declined. This provides evidence that natural ecological processes still prevailed and provided opportunities for both people and their environment to continually evolve in their functioning requirements. The rise of arboriculture later in the sequence also indicates the potential for environmental processes to adapt to changing conditions, and thus while forest was on the decline, it was possible, as in many parts of the world, for this structural element to be replaced by an equivalent vegetation. Moreover, this model of vegetation change suggests there were several long periods of change, reflected in both the vegetation and social histories. In particular, three phases of change are in evidence: a gradual early prehistoric forest change during the Neolithic and Bronze Ages into the early Iron Age (c. 2500 BC to 200 BC); a more rapid period of Iron Age landscape management producing phased and sequential change from c. 200 BC to AD 500; and a rapid immediately post-Iron Age transition at c. AD 500 to AD 600, where the Iron Age agricultural landscapes were replaced by the dry woodlands and grasslands of the late Holocene.

Palaeohydrology

Crucial to these vegetation changes are the shifts in hydrology identified for this floodplain (Figure 1). Initial recognition of the sites was founded on the identification of encircling earthworks interpreted as defensive and water-management features. Subsequent geoarchaeological study (Boyd et al. 1999a, b; McGrath and Boyd 2001; McGrath et al. 2008) developed a model of the earliest settlement being along prehistoric rivers. Importantly, these rivers represented a period where the floodplain was better watered than at present, and where the rivers that did flow across the floodplain were more abundant, of a different form, and differently located to those at present. Importantly, the early rivers provided suitable locations and resources for the human occupation of the floodplain, with settlements becoming established along the banks of the many rivers. Archaeological evidence is clear on this matter, with former river channels being found at the bases of excavations. In due course, these river channels became increasingly managed, with evidence initially for relatively minor modifications, the cutting of channels across oxbow cut-offs to produce bodies of water fully encircling the settlements, and the periodic cleaning out of the channel bottom sediments. This might be described as adaptive engineering, and may represent the need or perceived need to ensure a reliable supply of water to the sites, changing demands on water supply, or a reaction to changes in water supply reliability.

In time, however, the situation had changed to a point where constructive engineering was called upon. By this stage, either demand for a reliable supply of water was increasing beyond the system's capacity or the supply of water was declining. It is also clear that by then individual settlements had grown to such an extent that they were expanding over the former encircling water bodies and infilling them. Whatever the immediate cause, people began to construct new encircling water bodies, probably in imitation of the natural and adapted water bodies that appear to have supplied water for so long. This period of constructive engineering was quite short lived, and did not produce the desired effects; sedimentary evidence demonstrates that during the late Iron Age such channels infilled quickly with sediments. Whereas the earlier channels were filled up with clay best explained as the natural catchment sediment, and thus representing catchment run-off, the later channels were infilled by suspended floodplain sediment, representing in-situ rainfall flooding rather than catchment through-flow. While this engineering response did not manage to reproduce the effects of the previous regime of river flow, the evidence also suggests changing hydrological and possibly climatic conditions, a probable reduction in rainfall, or at least in the reliability of rainfall. The evidence of rainfall flooding may signal the onset of increasing seasonality in climatic conditions (cf. Maxwell 2001; Maxwell and Liu 2002).

Environmental explanations for change

This geoarchaeological evidence would, it has to be acknowledged, provide satisfactory stories of the past, details of the stage on which society acted, accounts of the resources it used, and a history of the changes it wrought on nature. In this way, the various environmental trajectories and especially their conclusion towards the end of the Iron Age can in tandem be explained in a variety of conventional biophysical, climate-change or human-impact ways (Boyd 2008). Typical of such explanation would be discussions of soil and vegetation change, past disruption of regeneration ecology, changes in riverbed loads, critical catchment changes, or specific local biophysical processes such as salt-dome upwelling. Finally, there is always climate change. A key idea in Quaternary geology is that processes such as climate warming or changes in rainfall abundance or patterning provide convincing evidence for changing environmental conditions. While the conclusions of such discussions do provide useful contextual interpretation, we are more interested in the ways in which such stories interact with the people who are the direct subject of the archaeological investigations. Before we turn to that relationship, we describe the archaeological evidence for our case study.

Archaeological insights

Overview

We can now turn to the archaeological material culture and stratigraphic evidence from, in particular, Ban Non Wat and the later Iron Age deposits at the neighbouring site of Noen U-Loke. Reference will be made to other sites as appropriate. It is important to note that analysis of the material from Ban Non Wat is still underway, and that many of the following observations necessarily reflect our (especially Chang's) personal impressions. There is much debate over the analysis and interpretations still to be had.

With these caveats in mind, we return to Figure 1, in which both apparent internal social change and outside (regional) social influences are identified as important in understanding the integrated pattern of social and environmental change over time. Turning first to the internal developments, six characteristics have been identified (Figures 4 and 5): burial practice, burial patterns (cemetery layout), types and materials of grave goods, variety and quantity of grave goods, exchange and contact, and, finally, evidence of warfare.

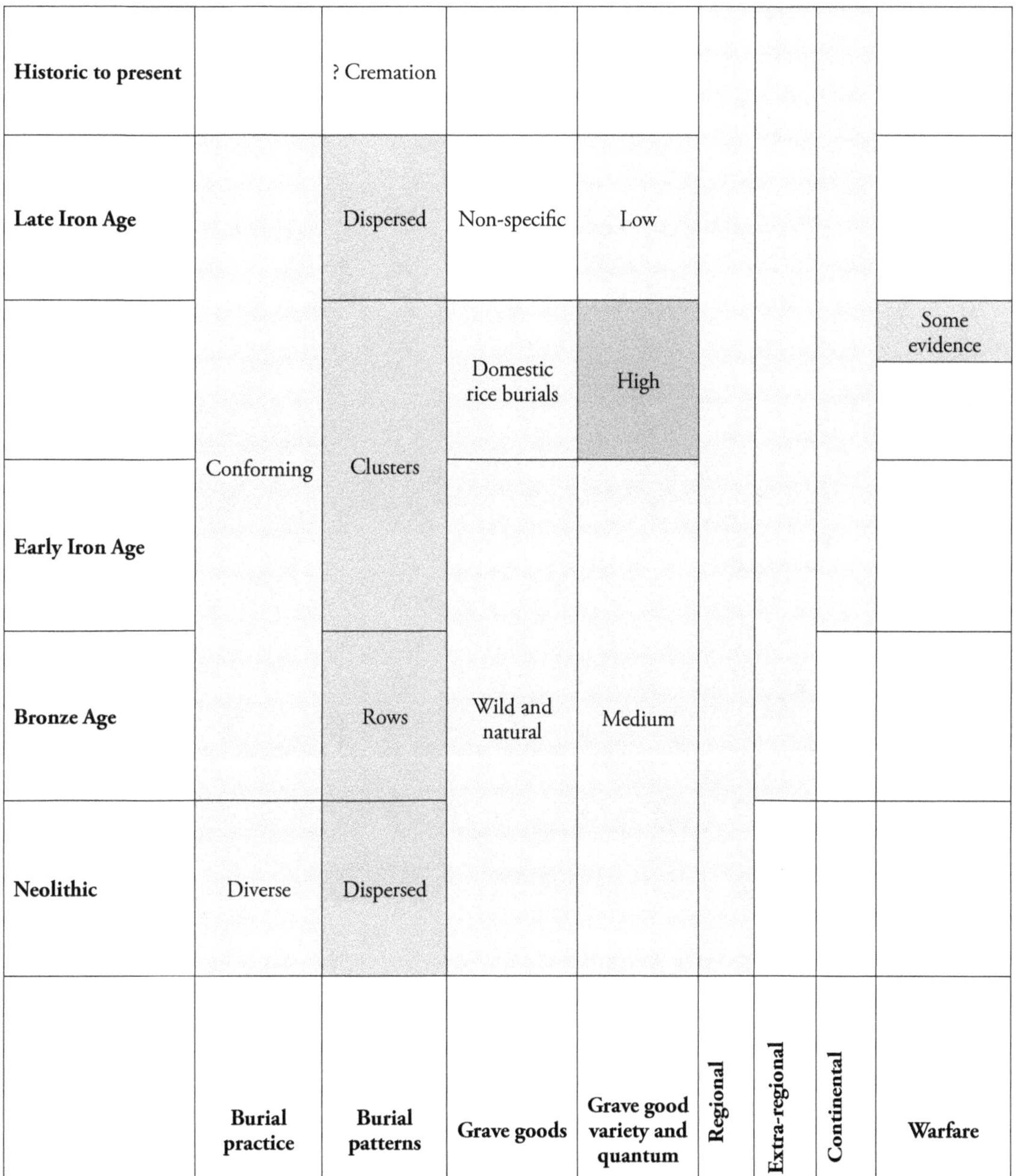

Figure 4. Schematic illustrating the trends in social activity and change throughout the late-Holocene settlement of the Mun Valley; the columns represent broad categories of social activity as witnessed by the archaeological record, which are characterised in broad terms. The schematic indicates that for each broad category of social activity there has been significant change; the heavy lines represent periods of significant change

Figure 5. Photograph of exposed Neolithic and Bronze Age graves at Ban Non Wat, February 2005. Note the different alignments of the overlying rows of Bronze Age burials. The partially disturbed (incomplete) burial running across the lower central part of the picture is Neolithic, while the deeper graves aligned roughly north-south (north is at the top of the picture) are actually dated to the later Bronze Age. Photograph N.J. Chang

Burial practices and patterns

Burial practice and burial pattern at Ban Non Wat reveal a distinction between Neolithic diverse practices and widely dispersed graves, and the more conforming Bronze and early Iron Age graves. During the Neolithic, some individuals were buried crouched in large pots, some in roughly east-west and others in north-south oriented extended graves. Some graves include freshwater gastropods and pig bones, others not. The most obvious feature is the lack of rows or clusters in the distribution of Neolithic graves. One reason for this variety may be that the Neolithic at Ban Non Wat extends over a possible 800 years from roughly 2100 to 1300 BC (Higham pers comm., unpublished radiocarbon dates). If this is the case, then it further emphasises the sparse distribution of graves from this period. Further to this, we should note the relative lack of sediment deposition during this period. In many cases, Bronze Age graves are cut down through Neolithic deposits, leaving the skeletal remains of early Bronze Age individuals deeper below the surface than the earlier Neolithic remains. By the Bronze Age, however, it is clear that graves are being placed in clear and conscious relationship to one another. There is something of a transition evident. The earliest Bronze Age graves are very large and replete with grave goods. They are also relatively widely dispersed across the excavated area. Those that follow are placed in neat rows. Over time, the orientations of the rows vary and the individual graves appear to be placed increasingly close together. As iron working becomes apparent, the graves become clustered, so much so that excavating individuals, and identifying to which grave various grave goods belong, becomes very difficult. Turning to Noen U-Loke for the later Iron Age, we see that clusters remain the major organising principle well into the first millennium AD (Higham 2002; Higham et al. 2007). However, sometime after AD 300, burials become more dispersed. Talbot (2002) has interpreted this as reflecting a breakdown of local social organisational systems.

Grave goods are an important characteristic of the mortuary evidence, as well as being useful indicators of the extent of exchange relationships. There appears to be a general trend at Ban Non Wat and Noen U-Loke towards an increase in quantity and, more clearly, in variety of grave goods over time. A second trend is towards grave goods of increasingly distant origin or, put another way, the participation of the upper Mun River valley people in increasingly extensive exchange networks. The third and most interesting trend seems to be from wild materials and objects to materials and objects of more domestic inspiration. All three of these apparent trends require explanation.

The Origins project began with excavations at the Bronze Age site of Ban Lum Khao (Higham and Thosarat 2005). The remains from this site have been described as relatively poor, and the comparison with the following excavations at Noen U-Loke was clear. Particularly in terms of personal ornaments, the number and variety of grave goods were orders of magnitude greater at the Iron Age site (Chang 2002). However, the discovery of large and well-endowed graves in the early Bronze Age at Ban Non Wat questions this interpretation. The impression that the Bronze Age was significantly poorer or less hierarchical than the Iron Age in the Mun River area may simply be the result of sampling error or bias. At Ban Lum Khao, excavations were undertaken near what was probably the edge of the prehistoric occupation, while at Noen U-Loke and Ban Non Wat, the excavations are closer to the centre of the site.

At Ban Non Wat, there still remains, however, a distinction between the Neolithic graves, which contain relatively fewer and less diverse grave goods, and the more subjectively impressive graves of later periods. In this context, it should also be noted that pottery was an important grave good in all periods. The styles and decorative elements vary, but it is clear that some thought and time was given to providing appropriate ceramic inclusions in mortuary rituals from the time that Ban Non Wat was first occupied. It may be appropriate to consider pottery as a common, background element in mortuary ritual; a base from which variation and diversity develops.

Regional trade and contact

This impression of greater variety over time may derive relatively directly from the wider exchange relationships that were developing. During the Neolithic, there is actually little evidence of extended exchange relationships at Ban Non Wat. The most intriguing exception is the presence of modified cowrie shells with two of the Neolithic burials. The most obvious interpretation is that there was some relationship with parts of China, where such shells were being used as currency about the same time (c. 4000 years ago). During the Bronze Age, large quantities of marine-shell jewellery seem to indicate well-developed networks between the

Khorat Plateau and the ocean. The question remains whether these were overland and via the Chao Phraya river system westwards, or whether they followed the Mun and Mekong route to the east coast. *Tridacna* sp. (giant clam) dominates in the earliest Bronze Age burials, while armbands and bangles of *Trochus* sp. dominate in the later Bronze Age. A continued connection of some sort with China is implied by the presence of T-sectioned bangles (usually made from marble) across Southeast Asia during the Bronze Age, including at Ban Non Wat (Chang 1996, 2002).

During the Iron Age, it is well known that agate, carnelian and glass beads indicate some sort of contact with distant Indian sources and ideas (Glover 1998; Higham 1989, 2002; Bellina 2003; Theunissen 2003). These artefacts are regular finds across northeast Thailand. Other jewellery items at Noen U-Loke, such as gold beads, suggest participation in networks that reach to the mouth of the Mekong, north up the Vietnamese coast, and south into Island Southeast Asia. By the end of the sequence at Noen U-Loke (after AD 300), distinctive finger rings found with one individual suggest direct relationships with the Dvaravati culture sites of the Central Plains of Thailand (Chang 2002). Of course, it is in the first few centuries AD that places such as Oc Eo and the developing state of Funan are thought to begin participating in a 'World Trading System' (Glover 1998). The people of the upper Mun River valley were in some way connected to this system, if at one or two removes, during the latest burial phases at Noen U-Loke (Chang 2002).

Social interaction with natural resources

While quantity and variety of grave goods and the distances from which they are derived are all increasing, there is a parallel movement from 'wild' to 'domestic' grave goods. It may be that the causes of this trend are more closely tied to local environmental and social changes. This change was first identified at Noen U-Loke, where the earliest Iron Age graves include pendants made of pig's tusks and tiger canines (Chang 2002; Talbot 2002) (see also Chang and Voelker 2003 for an argument as to the symbolic significance of personal ornaments as grave goods). These parallel grave goods found at Ban Na Di that include an alligator skin shroud and a dog's tooth necklace (Higham and Kijngam 1982; Higham 2002). Extending back into the Bronze Age at Ban Non Wat, we note again the predominance of *Tridacna* sp. and *Trochus* sp. armbands and bangles. In contrast, as the Iron Age progresses, we see at Noen U-Loke a clear change such that rice becomes a significant grave good itself. This is the most obvious domestic element. Along with the domestic product, tools such as iron spades, knives and possible sickles become common grave goods (Higham 2002; Higham et al. 2007). The important point here is that while it is considered that the first inhabitants at Ban Non Wat (and their contemporaries elsewhere in northeast Thailand) were rice farmers, it is not until the later Iron Age that they chose to stress this part of their life in their mortuary rituals. We suggest, as has Talbot (2002), that this may reflect a social response to a deterioration of the environment. Our argument is as follows (and is reflected in the model we propose in the final section of this paper).

- The initial Neolithic colonists at Ban Non Wat (and presumably elsewhere on the Khorat Plateau) arrived with a mature rice-growing technology as part of their subsistence toolkit.
- Perhaps because during the process of adaptation to the local environment rice agriculture became only a part of their subsistence strategy, or alternatively, perhaps because the local conditions allowed rice to grow so easily, the paraphernalia and products of this activity were not stressed in their ritual practices, certainly not in mortuary ritual. Psychologically, it was not uppermost in their minds. Ritual activities centred on exchange relationships and interactions with the natural world.

- However, during the Iron Age, environmental changes made the growing of rice a more difficult process. This concentrated their minds on what it was they actually did; how they subsisted. Parallel to an increased effort in moat and bank construction to control the increasingly problematic water resource, there was an intensification in ritual activity – for the same reason. Ritual activity was the only opportunity for affecting rainfall and other aspects of the agricultural round, and as these variables began to fail, an obvious response was to increase the ritual response.

In a sense, this parallels Higham and Lu's (1998) arguments about the initial development of rice agriculture in the middle Yangtze River. Environmental change instigates a behavioural response that involves a level of choice. The choice that people make is to change or intensify current practices in order that the overall lifestyle can be maintained. Clearly, the people of Ban Non Wat and Noen U-Loke could have opted to leave the area as agriculture became more difficult or they could have further adapted their subsistence system to the changed natural resources. What they chose to do was to attempt to maintain their current (rice-centred) lifestyle by an intensification of both physical and ritual practices. They changed in order to remain the same.

Warfare and conflict

We turn now to evidence of warfare and conflict in the region, and the relationships with the developing state societies in the lower Mekong River and its delta. At Ban Non Wat and Noen U-Loke the evidence is limited. A single individual at Noen U-Loke in the middle Iron Age was buried face down and with an iron arrowhead embedded in the spine (Higham 2002; Higham et al. 2007). While this is clearly evidence of a violent death, it does not necessarily imply organised or endemic warfare. It is interesting that the early Iron Age burials at Ban Non Wat include large bimetallic spearheads (iron blade and a bronze socket) that are thus far absent from the later Iron Age at Noen U-Loke. In the absence of other martial evidence, it is not unreasonable to hypothesise that these were used for hunting rather than for warfare. A simple contrast can be made with the site of Phum Snay in northwest Cambodia dating to the first centuries AD (that is, broadly contemporary with the middle to later Iron Age in the Mun River system), where iron swords are common grave goods and there are caches of projectile points (O'Reilly and Sytha 2001). Clearly, quite different levels of conflict were apparent there.

It must be remembered that it was also during the first centuries AD that the historically known state of Funan was developing in the Mekong Delta. Phum Snay was considerably closer and potentially in more direct contact with Funan, while any effects from, or response to, such a new form of society in the region for the upper Mun River would have been mitigated by the physical barriers of the Daeng Raek escarpment and the major rapids in the Mekong below its confluence with the Mun. Elsewhere, Chang (2001) has argued that if we accept that the major routes of exchange and contacts were along the rivers, then societies at nodal points, such as the confluence of the Mun and the Mekong, would also have operated as filters through which the traffic of ideas and influence would have been constrained and objects mediated.

Regional influences

Our argument here is that while materials and objects made their way into the Mun River system, they came via various routes and we should not assume a direct connection between, or even any real knowledge of, representatives of Funan (and following states) and the people of Iron Age Ban Non Wat and Noen U-Loke. The obvious question that follows is: what about the development of an Angkorian centre at Phimai, and Dvaravati-inspired centres such as Muang Sema in the upper Mun catchment? We return to this question again below, but essentially our answer is that as environmental changes made agriculture more difficult, an eventual direct impact on the lower Khorat Plateau by the early states would have provided new options and possible responses to these environmental changes. There may have been a very rapid uptake of new religious ideas and organisational strategies if these were seen as having more potential for mitigating environmental change. In a sense, this is an extension of the rice-in-mortuary-ritual argument above. A drive to intensify ritual responses to environmental change could have seen the uptake of new, potentially more powerful, ritual systems such as Buddhism and Hinduism. In the archaeological record, this would be reflected in a sudden truncation of indigenous cultures, adaptations and trajectories. This may be what archaeologists are seeing when they note that there appears to be an abandonment of the moated sites marking the end of the Iron Age at about 500 AD (Welch and McNeill 1991). We have already noted above that this abandonment may be more apparent than real – at Ban Non Wat, for example, there is considerable evidence for continued occupation throughout the ensuing proto-historic and historic periods. Rather than abandonment, we may be seeing a significant and very fast cultural change.

Geoarchaeology and archaeology: Landscapes and people

Conceptual evolution of the prehistory of the study area

Our archaeological and geoarchaeological discussion, in broadly seeking to establish the geography of the landscape before, during and after human occupation (Boyd et al. 1996a, b), has specifically set out to determine reasons for differences between past social and environmental richness and present environmental fragility. This comparative heuristic makes it possible to further understand past intimate human-environment relationships. In particular, we have identified very different past hydrological and climatic conditions as being important influences on social possibilities. This is highlighted in a spectacular, and in some ways conventionally familiar, way by what appears to have been the collapse of, or at least a significant shift in, the environment at the end of the Iron Age. This event may, however, be more apparent than real (as with the idea of site abandonment at the end of the Iron Age), and can, interestingly, be ascribed to a set of ideas in place before our detailed research. For example, at the onset of the research, received thinking assumed that these 'moated' sites represented solely the Iron Age, and marked a distinctive and seemingly evolutionary shift from prior subsistence farming to a more strongly structured and hierarchical society. Because the encircling earthworks had been labelled as 'moats', they took on the identity of defensive structures, thus suggesting conflict (Williams-Hunt 1950; Boyd 2007).

Further presumptions confounded the situation: the initial observation that these moated sites by and large lay distant from present rivers led researchers to assume a complex Iron Age socially and technologically organised canalised landscape, constructed in contrast to any prior landscape. In these conceptualisations, the Iron Age was seen as notably different from previous periods, reflecting contemporary global views of the Iron Age, and presenting a stage in social evolution that both imposes higher-order organisation on the use of resources, and acts as an essential stepping stone towards the emergence of complex societies.

Refocussing the questions

Our views of the prehistory have, however, shifted considerably. We now know that the landscape was occupied since the Neolithic (c. 2500 BC onwards), an occupation that is better characterised largely as a long, continuous, stable and healthy prehistoric occupation. Settlement was by an ancient river system, and represented a continued and dynamic use of rich environment resources, paralleled by dynamic social structures and dynamic technological development. In particular, key signatures of this long-term stability appear to be sophisticated, responsive and evolving water management and changing relationships with other natural resources (plants and animals), reflecting complex interactions with changing environmental conditions. It has become clear that environmental conditions have an important influence in defining patterns, and especially continuity, of settlement and occupation. In particular, the evidence now to hand implies close and complex relationships between social and environmental processes and change.

In this context, it is interesting to contrast the initial intentions of the geoarchaeological research with the outcomes of more than 10 years of such work. Geoarchaeological investigations set out initially to answer questions regarding: the lack of a canalised landscape (where were the canals?); the characteristics of the moats and their use, especially regarding defence (where are the palisades?); the environmental conditions of a unique Iron Age that distinguished it from previous periods (what was special about the Iron Age climate and the seemingly advanced social response to climate?); and the cause of the dramatic end of the Iron Age (what were the climatic and/or social events that caused site and moat abandonment?). As the research progressed, it became increasingly clear that human relationships with the landscape were more consistent than not over time, and that the Iron Age represents a natural, progressive landscape-adaptive response to slowly shifting environmental and social conditions.

Environmental determinism or cultural opportunity?

Therefore, while a regionally significant climatic shift may have provided a critical change at the end of the Iron Age around AD 600, it now seems much more likely that the Iron Age society maintained a much longer social tradition, adapting gradually to long-term environmental (climatic, ecological and hydrological) change, using sophisticated water engineering linked to agricultural change. That is, long periods of environmental constraints were expressed through gradual shifts in social behaviour. Eventually, the point arrived where the established social and engineering solutions that supported social sustainability during periods of environmental uncertainty, declining water supply and supply reliability were no longer able to moderate the effects of environmental change. By the end of the Iron Age, the Dvaravati civilisation of central Thailand was expanding into the region, closely followed by Khmer influences, bringing different and novel social solutions to environmental challenges, and thus providing the opportunities for the creation of new landscapes. But we are running ahead of ourselves; first to the details of our geoarchaeological and archaeological evidence.

Discussion

Overview

What do these seemingly diverse records of environmental and social phenomena mean in terms of the prehistoric human settlement of the upper part of the Mun River valley? Before we answer this, it is important to return to the central conceptual foundation of our approach. Our primary question has been: Can an integrated socio-environmental approach – in other words, a landscape approach – provide good explanation of prehistoric conditions? In part, asking such a question is a response to previous approaches to archaeology in the region, in particular, a tendency towards privileging society over environment that does not sit comfortably with widespread observational evidence of the intimate relationship between people and environment (e.g. Boyd 1988; Knapp and Ashmore 1999).

There are many archaeological models for social change – social, cultural, technological, environmental, migration, etc – but they tend to favour individual processes, and are often socio-politically based. The conventional assumption of an Iron Age in Thailand that is distinctively different from the preceding Bronze Age and is structured more like Iron Age societies elsewhere in the world, with a focus on chiefly structures, warfare and defence, is one such example (for a different view incorporating a heterarchical heuristic in Southeast Asia see, for example, O'Reilly 2003 and White 1995). In many ways, it is relatively easy to consider a single driving force in social or cultural change, especially where evidence – archaeological in this case – is by its very nature limited.

Landscape as culture

However, from a geographical point of view, landscape is fundamentally important. Landscape both provides a foundation for, and represents the synthesis of, all social and biophysical processes, and, by definition, is integrated and complex. It may have many properties – physical, biological, social, cultural – and in character it can be temporary, changing or interlinked. Importantly, geographers understand landscapes as cultural constructions, and use them as a valuable heuristic to examine the complexity of the human condition, the ways in which societies organise their social, technological, economic, cultural, spiritual and ritual behaviours, and thus define themselves. Cultural landscape concepts, for example, emphasise the connectedness rather than the singularity of individual sites (e.g. Head et al. 1994; Ross 1996) and the importance of landscape and environment in understanding past human behaviour (e.g. Butzer 1982; Lasca and Donahue 1990). An individual place, event or person therefore ceases to be isolated, but is relocated within sets of parallel, particular and cognitive landscapes constructed under the influence of many social and cultural parameters (Tuan 1974, 1977; Gould and White 1986). The individual therefore becomes one node within overlapping networks of physical, social, cultural and political linkages, pathways, edges, landmarks and surfaces (Boyd et al. 1996a, 2005). In this way, landscapes remind us of the essential links between people and places, and can only be fully understood in terms of both the physical and social processes enacted on them, and the ideas imposed on them and derived from them. Implicit in such a model are concepts of, for example, mutual influence, process relationships, feedback, change and cultural construction. In short, landscapes not only shape human experience, but are also shaped by that very human experience (Bender 1993).

The role of environment in archaeology

Given this, it remains to comment on the role of environment in studies of ancient social change. It is worth repeating that environment is used in archaeological studies in many ways. Typically, it often provides a passive background (a stage for people) or a deterministic control (people as automatons) for social behaviour. Alternatively, it can be conceptualised as a resource, a source of natural resources (utility), or something to be modified by people (people as controllers). Clearly, we take a different view. While all of these characterisations of environment relative to people can be correct in a non-exclusive way (environment can provide a background for social action, for example, and may in some circumstances provide a deterministic cause-and-effect driver), for a more complete understanding of social change, environment must be viewed as an active element, influence or process in social change. Applying this approach brings archaeological investigations closer to geographic understandings of people as part of the environment and environment as part of society – in other words, to the integrating concept of landscape. Even using landscape, however, fragmentation is possible. Bradley (1997), for example, draws the distinction between landscape archaeology's focus on settlement and subsistence, and social archaeology's interest in ritual and ceremony. Any observation of the human and social diversity of behaviour evident across the world supports claims that the integration of people and environment is a tangible and real entity. Articulated in the concept of landscape, the 'material manifestation of the relationship between humans and the environment' (Crumley 1994), such observation denies any claims that people ignore their environment, are obligated to respond in pre-ordained ways to environmental constraints, operate separate from their environment, or simply quarry it for resources without redefining themselves. In contrast to what Knapp and Ashmore describe as the minimalist situation – 'landscape as the backdrop against which archaeological remains are plotted' – we favour their emphasis on landscape as socio-symbolic: 'landscape is an entity that exists by virtue of its being perceived, experienced, and contextualized by people' (Knapp and Ashmore 1999).

Adopting such an approach, it may be questioned exactly how geoarchaeology, given its strong biophysical or earth-science tradition, can contribute to archaeological interpretations and model building. Here we demonstrate a two-stepped approach. Step one commences with a conventional geoarchaeological question: What was the local geography like before during and after ancient occupation? To answer such a question, we follow standard geoarchaeological procedures (Rapp and Gifford 1986; Waters 1996; Rapp and Hill 1998) in our application of the physical sciences, applying techniques borrowed from geology, geochronology, physical geography and chemistry to help understand the biophysical nature of the landscape and its evolution during the prehistoric period of interest.

Such study is valid, and indeed has provided a detailed set of descriptions of the biophysical conditions in the study area. Importantly, it tends, however, to discount complex human-environment interaction. The second step, therefore, is important for us. Here is where we move on to integrate such biophysical information with the body of emerging archaeological evidence. In doing so, our focus shifts to mutual social and environmental processes, feedback and interactivity, and cultural construction complementing physical and social processes.

Integrating social and environmental dynamics

Integration of social and environmental trajectories (Figures 1, 3 and 4) draws parallels with context and behaviour. Initially, there are clear social implications of the geological and biophysical history, in which environmental opportunities and limitations are important. More importantly, examining the various changes identified in both the geoarchaeological and archaeological studies reinforces an alignment between social and environmental processes. What this comparison does is reinforce a history of evolving relationships between water, vegetation and people. That history traces a model of late-Holocene landscape evolution, influenced by intimate relationships between internal and external social and environmental processes. This evolution spans the development of a landscape of mid-Holocene optimal conditions being settled and adapted, through to a critical change resulting in a new equilibrium landscape. It is in this new landscape, emerging about AD 500, that the Khmer established their important presence in the early historic period.

The whole history can be modelled as a sequence of shifts representing a trajectory from the mid-Holocene period of the Neolithic occupation of the floodplain to the post-Iron Age. As for the histories of vegetation and hydrological change described above, the archaeological trajectory can be modelled in terms of critical changes. Furthermore, there is a relatively long early period (1500 to 300 BC) of gradual river modification, extending throughout the times of Neolithic and Bronze Age settlement, and into the early Iron Age, during which adaptive engineering gradually modified the natural channels around the settlement sites. During the late Iron Age, around AD 300, a second phase of change is evident, with the phased introduction of constructed channels. Importantly, these biophysical and social shifts need not be contemporaneous, and indeed their partial non-synchronicity is testament to the complex interactivity and resilience of both society and the landscape: both can respond and adapt to changing circumstances, and both can operate within broad limits during such change. The end of the Iron Age, a notably rapid period of change around AD 500 to 600, saw significant social and environment changes, the abandonment of sites and constructed channels, and the establishment of a new vegetational regime and the present regime of river flow. Many of these changes could be explained purely in terms of functional process: changing hydrological regimes, for example, coupled with social desire to maintain a comfortable lifestyle based on an adequate and reliable supply of water, encourages sequential technological shifts (adaptive engineering, followed by constructive engineering), which in turn become accommodated with the environment.

However, we can take the relationship between society and environment further than mere functional process. By considering both archaeological and geoarchaeological evidence of natural and agricultural resource use, patterns of change and exchange, forms of landscape patterning and adaptation, technological responses to common problems, etc, we can construct a model in which human and environmental behaviour becomes integrated and expressed through social identity. Social identity links behaviour, process and socio-environmental context, and is based on conceptions and construction of landscape, environment and geography (Boyd et al. 1996a, 2005). In our case study, the model comprises five interlocking process conditions and four characterisations of socio-environmental identity. This evolving social sense of place and self-identity broadly equates with the archaeological chronostratigraphy, and represents the intersection between environmental context, opportunities and constraints, and social behaviour and construction of beliefs, values and culture. Essentially, they provide a unifying statement in which people and their environment are treated equally, in which the processes and characterisations apply equally to both.

Process condition 1: Colonisation (Neolithic)

This phase refers to human entry into a (socially) empty (environmentally) optimal environment. Both society and the environment are adapting to the new conditions, people bringing external adaptations to their needs – farming, technology and social structures – which are suited to, and are absorbed into, the new environmental conditions. The environment, on the other hand, provides a suitable context to allow satisfactory settlement and social adaptation to natural conditions, and neither society nor environment appears to impose significant constraints on the other. Self-identity under these conditions revolved around arrival from outside the region with a prior social identity – the maintenance of classic Southeast Asian Neolithic pottery styles evidences a fully developed social identity, complex social structures and expression of values, and well-established farming techniques (e.g. Higham 2002; Bellwood 2004). The arrivals entered an environment well suited to their lifestyle and, importantly, their self-identity, adapting to it and modifying it to create local socio-environmental balance. Mid-Holocene environmental conditions were optimal and well suited to such settlement, and allowed the new settlers to slowly assume a localised identity associated with the specific conditions of the floodplain – best expressed in the next stage below.

Process condition 2: Stability (Bronze Age and early Iron Age)

This condition resides within the mid-Holocene, allowing long-term mixed social adaptation to optimal environmental conditions. Such a situation allows for gradual change, both of social and natural conditions, without major or significant disruptions. The social and environmental focus is internal, with social change reflecting localised trends and opportunities rather then external ideological shifts, and with environment not needing to respond to any external (e.g. climate) forcing. This condition represents a balanced socio-environmental equilibrium. Self-identity shifts in emphasis, with the establishment of a more localised self-identity within society and geographical place grounded in the regional landscape, evidenced by indications of diversification of communities. Local communities' characteristics reflect the opportunities, resources and restrictions of the local environments, along with the choices made by each community (cf. Mudar and Pigott 2003).

Process condition 3: Forced Adaptation (middle Iron Age)

This condition is characterised by an increasingly strong social use of the landscape, with environmental conditions conducive to adaptive landscape technologies. Change is driven increasingly by both external environmental processes and external social influences, and there is clear evidence for positive socio-environmental feedback on landscape change. Self-identity again shifts in emphasis, towards the establishment of a regional self-identity, based on claims on landscape and place within regional networks. Society became, simultaneously, increasingly outward looking, both locally and regionally (for example, the incorporation of glass and other materials representing broader exchange relationships), and more centrally or locally focused. Evidence of increasing engineering responses to environmental change and increasing partitioning of the landscape point to a stronger sense of social claim on the landscape. Chang (2002) has argued, for example, that the very presence of moats around sites must have affected how the inhabitants saw themselves and their relationships with both neighbouring communities and the wider landscape.

Process conditions 4 and 5: Tipping Point (Crisis) and Resolution-New Equilibrium (late Iron Age)

The Tipping Point condition reflects a period of increasing social and environmental disequilibrium. External social forces provide increasing opportunity and need for social change, while environmental degradation, part internally and part externally driven, induces critical change. Eventually, both society and environment abandon adaptations (both vegetation resilience and engineering effectiveness, for example, decline), and there is further major environmental and social change.

The Resolution-New Equilibrium condition, on the other hand, represents a significant response to prior process conditions and an enforced major change of social behaviour within a non-optimal landscape. New ideas regarding alternative forms of social organisation from outside (Khmer, Dvaravati) become important. Associated with and emerging from these process conditions is an identity subsumed within an emerging centralised state landscape. This represents a significant shift from the trajectory of developing self-identity, and reflects the convergence of need and opportunity for change. The pattern of settlement across the landscape appears to have shifted from dispersed and widespread to centralised and concentrated (McNeill and Welch 1991; Welch and McNeill 1991). This marks a shift from localised identity of smaller communities, marked through attachment to specific areas of land and mediated through regional networks, to identity subsumed into a centralised external state.

How can we best explain these changes between conditions?

This model of integrated social and environmental mutual change and feedback relies more on the fluidity of socio-environmental process than on the primacy of either socio-cultural or environmental process. We are thus able to consider the landscape as a fundamentally socio-environmental construct, in which potentialities emerge, are expressed, and may be influential in changes in social and/or environmental behaviour. This approach is deliberately non-deterministic, in which potentialities may be activated, and in which the most important base condition is that change is only possible where the conception of that action exists. In this way, there may be many possible trajectories – both social and environmental – flowing from any particular set of socio-environmental circumstances, but, importantly, there is no obligation that any of these be necessarily activated.

The power of a new idea: One of how many possible socio-environmental trajectories?

To close, and to illustrate the import of this approach, it is useful to consider one such set of possible trajectories. Our example lies at the end of the archaeological sequence described here, when there was an apparent shift in the patterns of landscape occupation and social conditions, changes identified as highly significant by archaeologists (Moore 1988a, b; Higham 1996b, 2002; Welch and McNeil 1991). These have conventionally been explained in terms of a persuasive and inevitable move towards statehood. On the other hand, the apparent abandonment of sites such as Ban Non Wat at the end of the Iron Age could, equally conventionally, be taken simply as evidence of the deterministic effects of environmental degradation. A third possibility, as noted above, is that the sites were not abandoned at all, and that they were simply used in a different way in a more complex landscape that also included larger central sites.

However, in our model, we start with a situation in which the late Iron Age people were presented with, it seems, the need to continue their adaptive use of natural resources, but in a context of deteriorating environmental conditions – deteriorating, that is, relative to the optimal conditions under which the Iron Age people had inherited and maintained their preferred lifestyle. In the event, the social response was to change settlement pattern and, one presumes, the social conditions and structures that may be influenced by or required by this new landscape organisational arrangement. The vegetation was increasingly less able to cope with the cumulative impacts of agriculture and arboriculture, and the engineering responses to maintaining a reliable and predictable water supply were simply not working. Local rainfall patterns were probably also changing, with good evidence that the water in the rivers ceased to be sourced in the upper catchment, and instead relied on localised rainfall.

Fortunately, a new idea was circulating at this time, a notion of an alternative social structure – centralised settlement characteristic of Indianised ideas entering mainland Southeast Asia at the time. This new idea provided an alternative to the concept of settlement extant for several millennia, a concept of dispersed settlement patterns, and a concept that had suited socio-environmental conditions so far. With a new idea came a new option, thus offering late Iron Age people an alternative social organisation that may have better allowed them to deal with the socio-environmental conditions they now found themselves in. By electing to adopt this option, and thus changing both social organisation and their relationship with the landscape – living in greater numbers and higher concentrations in fewer central towns – society was provided a key possibility that was neither available previously, nor an obligatory pathway into their (then) future.

A valid question still remains. In the absence of the new idea, or, importantly, in a situation where the new idea may not have been taken up, what might have been? Our integrated model of socio-environmental relationship – articulated through and evidenced by the landscape – can provide some suggestions. At an extreme end, a social and population crash and landscape emptying might have been on the cards; such an interpretation is common across the world (cf. Diamond 2005). Alternatively, survival could have relied on one or more of continuing use and adaptation of established technologies and, perhaps, associated social structures within the evolving landscape, or a reordering of the occupation or use of that landscape within the extant social structure. Just because the technologies or social structures that had evolved to date were abandoned at the end of the Iron Age does not mean there is an a priori case that they had become wholly redundant. The experience of human-environment relationships since the Neolithic in this region indicates the resilience and adaptability of both social and environmental conditions within the landscape. Some options may require both the conceptual context for such change, especially if they are stepping outside the received range of possibilities intrinsic in existing technologies and social structures, and the environmental capacity – particular to the history and times of these events – for continued adaptation.

What consideration of such options reinforces, however, is that the trajectory that was adopted represented a unique combination of socio-environmental circumstances, and that its adoption was only one of many possibilities. By considering this unique change event within its landscape, and thus examining it from the perspective of the detailed interrelationships between people and environment, the roles of memory, continuity, discontinuity and transformation (Rowlands 1993; Schama 1995), and of landscape as identity (Knapp and Ashmore 1999), we have discussed a heuristic able to delve deeply into the social behaviour of past people. By doing this, however, we are forced to move from the detailed description of single sites, to consideration of landscape-scale patterning of human behaviour. This next stage demands that field work and research now be aimed at understanding the broader patterning of individual sites as well as their distributions across the landscape. It also demands that we try to understand the mutual relationships and feedbacks between people and environment, that is, the landscapes that were inherited, occupied, constructed and modified throughout prehistoric time.

Acknowledgements

The research we discuss here is part of the Origins of the Civilisation of Angkor Project. Our involvement in the research has been funded and supported by grants from a wide range of sources: Marsden Fund (NZ), Australian Research Council (ARC), Australian Institute for Nuclear Science and Technology (AINSE), Earthwatch Institute (USA), National Geographic (USA), Southern Cross University (Australia), James Cook University (Australia), and Cambridge University (Britain). We acknowledge and thank the directors of the Origins of the Civilisation of Angkor Project, Professor Charles Higham, Dr Rachanie Thosarat and Dr Amphan Kijngam, for their invaluable support. We are also grateful to local villagers for their continuing input to the research, and officials of the National Research Council of Thailand, the Royal Thai Fine Arts Department, and the regional Amphoe Non Sung, for permission to conduct this research. The School of Environmental Science and Management at Southern Cross University has allowed access to its extensive array of specialist equipment and support staff. The opinions expressed in this discussion paper are solely those of the authors, and do not necessarily represent those of the Origins of the Civilisation of Angkor Project.

References

Basso, K.H. 1996. Wisdom sites in places: Notes on a Western Apache landscape. In S. Feld and K.H Basso (eds) *Senses of Place*, pp53-90. Santa Fe: SAR Press.

Bellina, B. 2003. Beads, social change and interaction between India and South-east India. *Antiquity* 77:286-297.

Bellwood, P. 2004. The Origins and Dispersals of Agricultural Communities in Southeast Asia. In I. Glover and P. Bellwood (eds) *Southeast Asia: from Prehistory to History*, pp21-40. Abington: Routledge Curzon.

Bender, B. (ed.) 1993. *Landscape: Politics and Perspectives.* Oxford: Berg.

Boyd, W.E. 1988. Methodological problems in the analysis of fossil non-artefactual wood assemblages from archaeological sites. *Journal of Archaeological Sciences* 15:602-619.

Boyd, W.E. 2004. "As fantastic as the Atlantis": Reading a prehistoric floodscape. In: P. Thom (ed.) *Flood: Writing across the current*, pp65-97. Lismore: Southern Cross University Press.

Boyd, W.E. 2007. The geoarchaeology of the archaeological sites of Non Muang Kao and Noen U-Loke, Mun River valley, northeast Thailand. In C.F.W. Higham, A. Kijngam and S. Talbot (eds) *The Origins of the Civilisation of Angkor Volume 2: The Excavation of Noen U-Loke and Non Muang Kao*, pp29-53. Bangkok: The Fine Arts Department of Thailand.

Boyd, W.E. 2008. Social change in late Holocene mainland SE Asia: A response to gradual climate change or a critical climatic event? *Quaternary International* 184:11-23.

Boyd, W.E., M.M. Cotter, W. O'Connor and D. Sattler 1996a. Cognitive ownership of heritage places: Social construction and the cultural heritage management. *Tempus* 6:123–140.

Boyd, W.E., M.M. Cotter, J. Gardiner and G. Taylor 2005. Rigidity and a Changing Order. Disorder, Degeneracy and Daemonic Repetition: Fluidity of Cultural Values and Cultural Heritage Management. In T. Darvill, C. Mathers, and B. Little (eds) *Heritage of Value, Archaeology of Renown: Reshaping Archaeological Assessment and Significance*, pp89-113. Tampa: University Press of Florida.

Boyd, W.E. and J. Habberfield-Short 2007. Geoarchaeological landscape model of the Iron Age settlements of the upper Mun River floodplain. In C.F.W. Higham, A. Kijngam and S. Talbot (eds) *The Origins of the Civilisation of Angkor Volume 2: The Excavation of Noen U-Loke and Non Muang Kao*, pp1- 27. Bangkok: The Fine Arts Department of Thailand.

Boyd, W.E., C.F.W. Higham and R. Thosarat 1996b. The Holocene palaeogeography of the S.E. margin of the Bangkok Plain, Thailand and its archaeological implications. *Asian Perspectives* 35:139-162.

Boyd, W.E., C.F.W. Higham and R.J. McGrath 1999a. The geoarchaeology of Iron Age "moated" sites of the upper Mae Nam Mun valley, N.E. Thailand. I: Palaeodrainage, site-landscape relationships and the origins of the "moats". *Geoarchaeology: An International Journal* 14:675-716.

Boyd, W.E., R.J. McGrath and C.F.W. Higham 1999b. The geoarchaeology of the prehistoric ditched sites of the upper Mae Nam Mun Valley N.E. Thailand, II: Stratigraphy, and morphological sections of the encircling earthworks. *Indo-Pacific Prehistory Association Bulletin* 18:169-179.

Boyd, W.E. and R.J. McGrath 2001a. The geoarchaeology of the Iron Age ditched sites of the upper Mae Nam Mun valley, N.E. Thailand III: Late Holocene vegetation history. *Palaeogeography, Palaeoclimatology, Palaeoecology* 171:307-328.

Boyd, W.E. and R.J. McGrath, 2001b. Iron Age vegetation dynamics and human impacts on the vegetation of Upper Mun River floodplain, N.E. Thailand. *New Zealand Journal of Geography* 57(2):21-32.

Bradley, R. 1997. *Rock Art and the Prehistory of Atlantic Europe*. London: Routledge.

Butzer, K.W. 1982. *Archaeology as human ecology*. Cambridge: Cambridge University Press.

Chang 2001 MISSING

Chang, N.J. 1996. Personal ornaments in Thai Prehistory: some preliminary observations from Nong Nor, Central Thailand. *Bulletin of the Indo-Pacific Prehistory Association* 14:140-150.

Chang, N.J. 2002. Personal ornaments in Thailand: Nong Nor, Ban Luam Khao and Noen U-Loke. Unpublished PhD Thesis. Dunedin: University of Otago.

Chang, N.J. In press. The Shell, Silver, Gold, Bronze and Bone Personal Ornaments. In C.F.W. Higham, A. Kijngam and S. Talbot (eds) *The Origins of the Civilisation of Angkor Volume 2: The Excavation of Noen U-Loke and Non Muang Kao.* Bangkok: The Fine Arts Department of Thailand.

Chang, N.J. and J. Voelker 2003. Wash and Wear, Disposable and Cheap: Ceramic bangles as everyday personal ornaments not for symbolic use. *Bulletin of the Indo-Pacific Prehistory Association* 23:19-26.

Crumley, C.L. 1994. Cultural ecology: a multidimensional ecological orientation. In C.L. Crumley (ed.) Historical Ecology: *Cultural Knowledge and Changing Landscapes*, pp1-16. Santa Fe: SAR Press.

Diamond, J. 2005. *Collapse: How societies choose to fail or succeed.* New York: Viking Books.

Fagan, B. 2007. *People of the Earth.* Upper Saddle River: Prentice Hall.

Glover, I.C. 1998. The role of India in the late prehistory of Southeast Asia. *Journal of Southeast Asian Archaeology* 18:21-49.

Glover, I. and P. Bellwood (eds) 2004. *Southeast Asia: From Prehistory to History.* Avington: Routledge Curzon.

Gould, P. and R. White 1986. *Mental maps.* London: Allen and Unwin.

Habberfield-Short, J. and W.E. Boyd 2007. The geoarchaeological development of the mounded Iron Age sites of Non Muang Khao and Noen U-Loke, and implications for timing and phasing of the occupation and abandonment of the sites. In C.F.W. Higham, A. Kijngam and S. Talbot (eds), *The Origins of the Civilisation of Angkor Volume 2: The Excavation of Noen U-Loke and Non Muang Khao*, pp55-73. The Fine Arts Department of Thailand, Bangkok.

Harwood, K. and A.G. Brown 1993. Fluvial processes in a forested anastomosing river: flood partitioning and changing flow patterns. *Earth Surface Processes and Landforms* 18:741-748.

Head, L., C. Gosden and J.P. White (eds) 1994. Social landscapes. *Archaeology in Oceania* 29:113-190.

Higham, C.F.W. 1977. The prehistory of the Southern Khorat Plateau, with particular reference to Roi Et Province. *Modern Quaternary Research in Southeast Asia* 3:103-142.

Higham, C.F.W. 1996a. *The Bronze Age of Southeast Asia.* Cambridge: Cambridge University Press.

Higham, C.F.W. 1996b. A review of archaeology in Mainland Southeast Asia. *Journal of Archaeological Research* 4:3-49.

Higham, C.F.W. 2000. *The Origins of the Civilisation of Angkor: the final report to the National Research Council of Thailand.* Dunedin: Department of Anthropology, University Of Otago.

Higham, C. 2002. *Early cultures of Mainland Southeast Asia.* Bangkok: River Books.

Higham, C. 2004. Ban Non Wat: Lessons from the field. In V. Paz (ed.) *Southeast Asian Archaeology: Wilhelm G. Solheim II Festschrift*, pp491-503. Manilla: University of the Philippines Press.

Higham, C.F.W. and A. Kijngam 1982. Irregular earthworks in N.E. Thailand: new insight. *Antiquity* 56:102-110.

Higham, C.F.W. and T.L.-D. Lu 1998. The origins and dispersal of rice cultivation. *Antiquity* 72:867-877.

Higham, C.F.W. and R. Thosarat 1998. *Prehistoric Thailand: From Early Settlement to Sukhothai.* Bangkok: River Books.

Higham, C.F.W. and R. Thosarat 2005. *The Origins of the Civilisation of Angkor: Volume 1: The excavation of Ban Lum Khao.* Bangkok: The Fine Arts Department of Thailand.

Higham, C.F.W., A. Kijngam and S. Talbot 2007. *The Origins of the Civilisation of Angkor Volume 2: The Excavation of Noen U-Loke and Non Muang Kao.* Bangkok: The Fine Arts Department of Thailand.

Källén, A. 2004. *And Through Flows the River: Archaeology and the Pasts of Lao Pako.* Upsalla: Studies in Global Archaeology, 6.

Knapp, A.B. and W. Ashmore 1999. Archaeological Landscapes; constructed, conceptualised, ideational. In W. Ashmore and A.B. Knapp (eds) *Archaeology of Landscape: Contemporary Perspectives*, pp1-30. London: Blackwell.

Lasca, N.P. and J. Donahue (eds) 1990. *Archaeological Geology of North America.* Boulder: The Geological Society of America.

Maxwell, A.L. 2001. Holocene monsoon changes inferred from lakes sediment pollen and carbonate records, northeastern Cambodia. *Quaternary Research* 56:390-400.

Maxwell, A.L. and K.-B. Liu 2002. Late Quaternary pollen and associated records from the monsoonal areas of continental South and SE Asia. In P. Kershaw, B. David, N. Rapper, D. Penny and J. Brown (eds). *Bridging Wallace's Line: The environmental and cultural history and dynamics of the SE-Asian-Australian Region,* pp189-228. Reiskirchen: Catena Verlag.

McGrath, R.J., W.E. Boyd and R.T. Bush 2008. The paleohydrological context of the Iron Age floodplain sites of the Mun River valley, Northeast Thailand. *Geoarchaeology: An International Journal* 23:151-172.

McGrath, R.J. and W.E. Boyd 2001. The chronology of the Iron Age 'moats' of northeast Thailand. *Antiquity* 75:349-360.

McNeill, J.R. and D.J. Welch 1991. Regional and interregional interaction on the Khorat Plateau. *Bulletin of the Indo-Pacific Prehistory Association* 10:327-340.

Moore, E.H. 1988a. *Moated Sites in Early North East Thailand.* Oxford: BAR International Series 400, British Archaeological Reports.

Moore, E.H., 1988b. Notes on two types of moated settlements in North East Thailand. *Journal of the Siam Society* 76:275-287.

Moore, E.H. 1992. Water-enclosed sites: Links between Ban Takhong, northeast Thailand and Cambodia. In J. Rigg (ed.) *The gift of water: Water Management, cosmology and the State of South East Asia,* pp22-25. London: School of Oriental and African Studies.

Mudar, K.M. and V.C. Pigott 2003. Subsistence changes and community based craft production in prehistoric Central Thailand. In A. Karlstrom and A. Kallen (eds) *Fishbones and Glittering Emblems: Southeast Asian Archaeology 2002.* pp149-160. Stockholm: Museum of Far Eastern Antiquities.

Nitta, E. 1991. Archaeological study on the ancient iron-smelting and salt-making industries in the northeast of Thailand: Preliminary report on the excavations of Non Yang and Ban Don Phlong. *Journal of Southeast Asian Archaeology* 11:1-46.

O'Reilly, D.J.W. 1998. The discovery of clay-lined floors at an Iron Age Site in Thailand; Preliminary observations from Non Muang Kao, Nakhon Ratchasima Province. *Journal of the Siam Society* 85:1-14.

O'Reilly, D.J. 1999. A diachronic analysis of social organisation in the Mun River Valley. Unpublished PhD Thesis. Dunedin: University of Otago.

O'Reilly, D.J.W. 2000. From the Bronze Age to the Iron Age in Thailand: Applying the heterarchical approach. *Asian Perspectives* 39:1-19.

O'Reilly, D.J.W. 2003. Further evidence of heterarchy in Bronze Age Thailand. *Current Anthropology* 44:300-306.

O'Reilly, D.J.W. 2007. *Early Civilizations of Southeast Asia.* New York: Alta Mira Press.

O'Reilly, D.J.W. and P. Sytha 2001. Recent excavations in northwest Cambodia. *Antiquity* 75:265-266.

Rapp, G. Jr. and J.A. Gifford (eds) 1986. *Archaeological Geology.* New Haven: Yale University Press.

Rapp, G. Jr. and C.L. Hill 1998. *Geoarchaeology: The Earth Science Approach to Archaeological Interpretation.* New Haven: Yale University Press.

Ross, A. 1996. Landscape as heritage. *Tempus* 5:1-17.

Rowlands, M. 1993. The role of memory in the transmission of culture. *World Archaeology* 25:141-151.

Schama, S. 1995. *Landscape and Memory.* London: Harper Collins.

Stark, M. (ed.) 2006. *Archaeology of Asia.* Malden: Blackwell.

Talbot, S. 2000. Notes from the field: an excavation at a Khmer Temple in Northeast Thailand. *New Zealand Journal of Asian Studies* 11:162-168.

Talbot, S. 2002. From the Iron Age to Angkor in Northeast Thailand. Unpublished Ph.D. Thesis. Dunedin: University of Otago.

Theunissen, R. 2003. Agate and Carnelian Beads and the Dynamics of Social Complexity in Iron Age Mainland Southeast Asia. Unpublished PhD Dissertation. Armidale: University of New England.

Tilley, C. 1994. *A Phenomenology of Landscape: Places, Paths and Monuments.* Oxford: Berg.

Tuan, Y.-F. 1974. *Topophilia: a study of environmental perception, attitudes, and values.* New York: Columbia University Press.

Tuan, Y.-F. 1977. *Space and place: the perspective of experience.* London: Edward Arnold.

Waters, M.R. 1996. *Principles of Geoarchaeology: A North American Perspective.* Tuscon: University of Arizona Press.

Welch, D. 1984. Settlement patterns as an indicator of sociopolitical complexity in the Phimai region. In D.T. Bayard (ed.). *Southeastern Asian Archaeology at the XV Pacific Science Congress,* pp129-151. Dunedin: Otago University Studies in Prehistoric Anthropology 16.

Welch, D.J. and J.R. McNeill 1991. Settlement, agriculture and population changes in the Phimai region, Thailand. *Bulletin of the Indo-Pacific Prehistory Association* 11:210-228.

White, J.C. 1995. Incorporating heterarchy into theory on socio-political development: the case from Southeast Asia. In R. Ehrenreich, C. Crumley and J. Levy (eds) *Heterarchy and the Analysis of Complex Societies: Comments.* pp102-123. Washington DC: American Anthropological Association.

Williams-Hunt, P.D.R. 1950. Irregular earthworks in Eastern Siam: An air survey. *Antiquity* 24:30-37.

16

A 40,000 year wood charcoal record from Carpenter's Gap 1: New insights into palaeovegetation change and indigenous foraging strategies in the Kimberley, Western Australia

Susan Frawley

Archaeology and Natural History, College of Asia and the Pacific, Australian National University, Canberra, Australia

Sue O'Connor

Archaeology and Natural History, College of Asia and the Pacific, Australian National University, Canberra, Australia

sue.oconnor@anu.edu.au

Introduction

This paper presents the results from analysis of wood charcoal from Carpenter's Gap 1, a rock shelter with a human-occupation record spanning more than 40,000 years. The phytolith and macrobotanical remains from this site have been previously studied. They provided proxy records of vegetation change over time but each class of palaeobotanical material has distinct taphonomic biases that affect the likelihood that it will be incorporated in the archaeological deposit, and if it is, how well it preserves. The wood charcoal record provides a new line of evidence and helps build a more holistic profile of palaeovegetation local to the site and human foraging strategies over the 40,000 year span of its occupation.

Background

Carpenter's Gap 1 is a large rock shelter located in the Oscar Napier Range, between 14° S and 19° S latitude and 123° E to 129° E longitude within Windjana Gorge National Park (Figure 1). The shelter is on the north-facing side of the range, approximately 4 km from Windjana Gorge and the Lennard River, c. 25 m above the surrounding plain. Carpenter's

Gap 1 is spacious, with a floor area of more than 50 m^2, and contains a considerable amount of painted rock art and engravings (O'Connor 1995) (Figures 2 and 3).

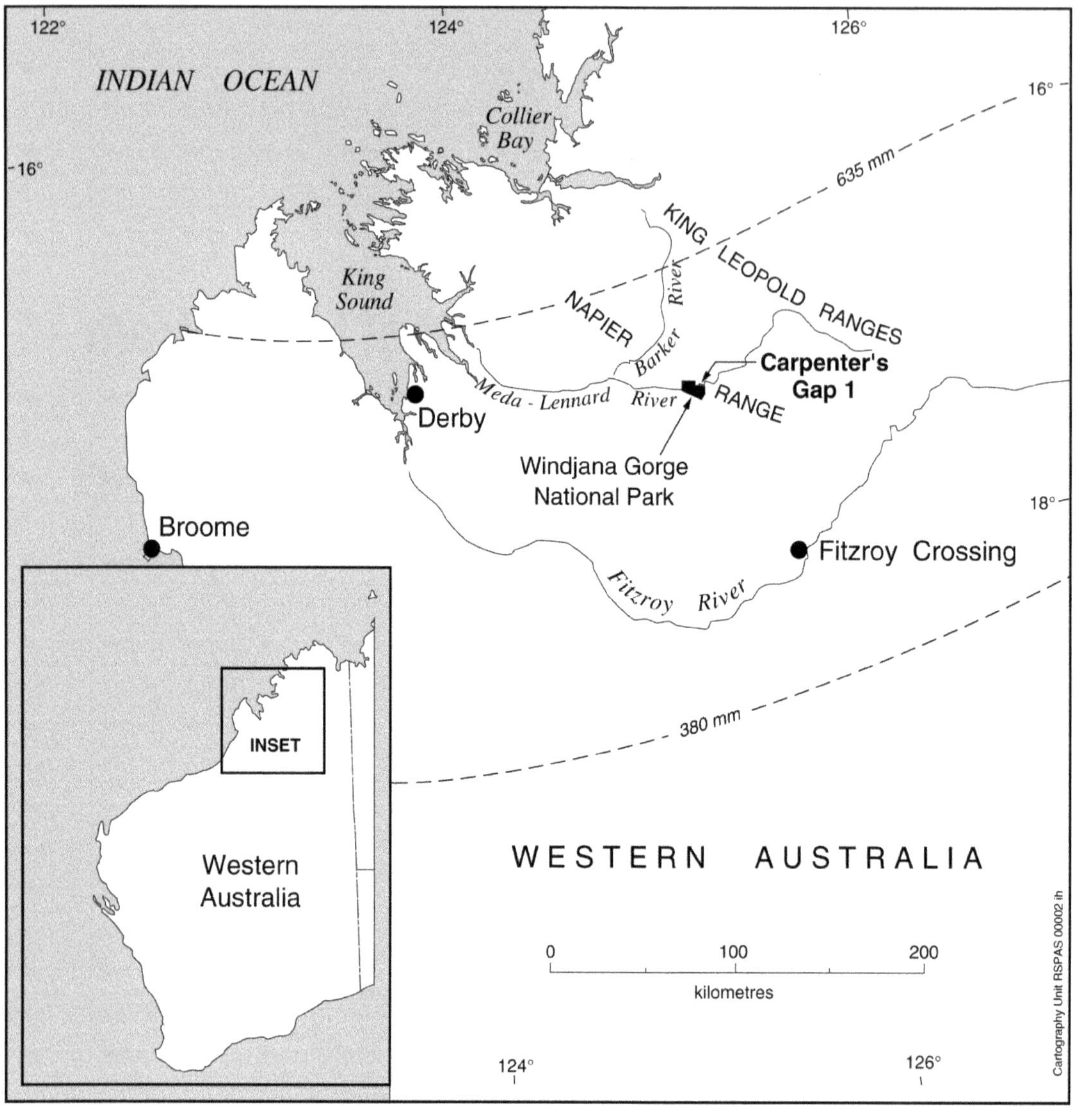

Figure 1. Map showing location of Carpenter's Gap 1

The shelter was excavated by O'Connor over two field seasons between 1993 and 1994 (McConnell and O'Connor 1999; see Figure 3). Its significance lies in its long occupation record and the fact that it preserves a wealth of organic materials, including charcoal and macrobotanic remains, even in the earliest occupation levels. Preliminary reports on the archaeological results have been published elsewhere (O'Connor 1995; McConnell and O'Connor 1999). Radiocarbon dating of the lower levels of the excavated deposit produced a maximum age estimate of 42,800 ± 1850 BP OZD161 (McConnell and O'Connor 1999). Archaeological sequences with this level of preservation and chronological resolution are rare anywhere in Australia, and particularly so in northern Australia, where the monsoon tropical environment usually rapidly destroys organic materials. Well-resolved late Pleistocene-Holocene terrestrial records for palaeovegetation change have also proved elusive, as lakes and other natural deposits preserving continuous pollen records are few in northwest Australia (Hiscock and Wallis 2005).

Carpenter's Gap 1 is unique, as it preserves plant phytoliths, wood charcoal, pollen and macrobotanic remains, as well as organic and non-organic residues of human behaviour and thus provides a window into palaeoenvironmental change and the human behavioural responses

Figure 2. Carpenter's Gap 1, view facing west. Excavation area is behind the large boulder in the shadowed area

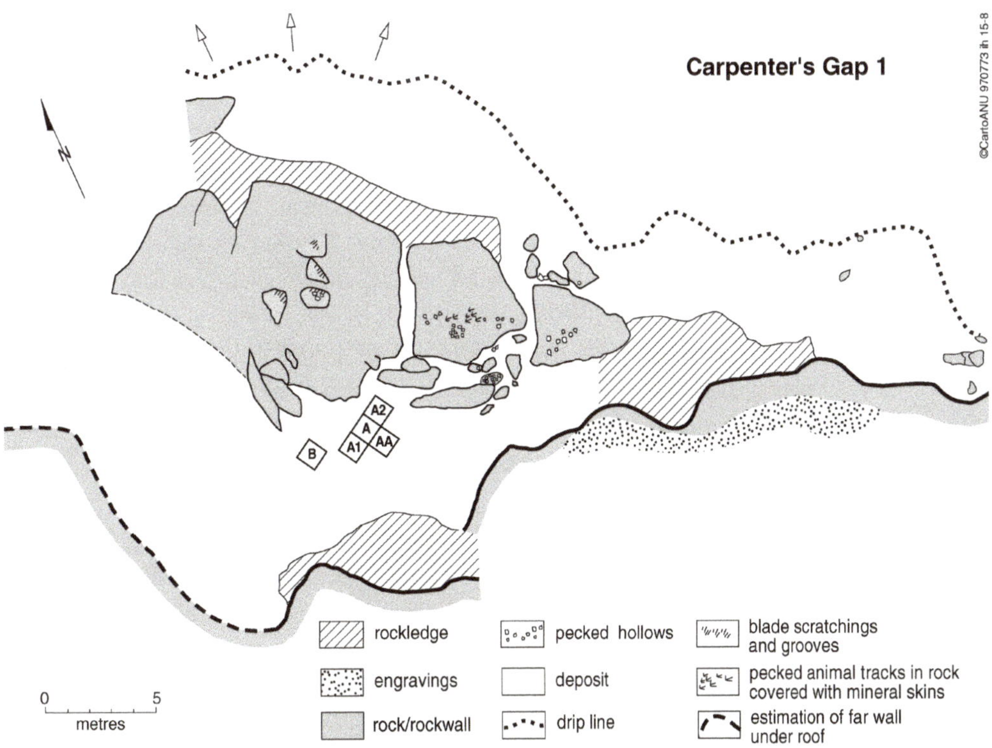

Figure 3. Plan of Carpenter's Gap 1 showing excavation areas

to it, before, during and after the Last Glacial Maximum (McConnell and O'Connor 1997; Wallis 2000, 2001).

Contemporary vegetation in the site surrounds

Four botanical districts are described for the Northern Botanical Province (Kimberley) by Beard (1979:14-17). The Napier Range lies predominantly within the Dampier District, but is in close proximity to the Fitzgerald District. Beard (1979:96) described the regional vegetation on the limestone outcrops as sparse, consisting primarily of the spinifex *Triodia wiseana,* along with some other grasses. Boabs, *Adansonia gregorii,* are scattered on the slopes and plains, along with small trees such as *Sterculia viscidula, Celtis philippensis, Cochlospermum fraseri, Ficus opposita, Terminalia* spp. and *Wrightia* sp. The foot slopes of the limestone ranges have a tree steppe of *Eucalyptus dichromophloia* and *Triodia wiseana* (Beard 1979:96). McConnell (McConnell and O'Connor 1997:20) and Wallis (2000:100-102) carried out field trips to the Napier Range with the specific purpose of collecting and identifying the range of taxa occurring in vicinity of Carpenter's Gap 1, and extended the known taxa for this area. As will be discussed below, Frawley (2010) also carried out a collecting trip to acquire and identify samples to form the basis of the modern wood comparative collection.

Previous palaeobotanical studies at Carpenter's Gap 1

McConnell's (1997) study of the macrobotanic remains at Carpenter's Gap 1 identified significant changes in the species present through time. Almost 2500 plant fragments were analysed, the majority of which are seeds. Other types of material recovered include seed casings, bamboo fibre, tubers, uncharred wood shavings and wood fragments, grass stems, leaves, bark and fibres. McConnell subdivided the assemblage into four time phases which she used to infer a vegetation history for the site surrounds. She downplayed the effects of variable decay, based on a series of taphonomic studies she undertook in the field, arguing instead that the assemblage is a good reflection of the palaeovegetation over time (McConnell and O'Connor 1997:25).

There was little macrobotanic material preserved in the lowest levels between 42,000 BP and 35,000 BP, but *Plectrachne* sp. (grass), Cyperaceae (sedges – aquatic perennial herbs) and occasional seeds of *Terminalia* spp. were recovered. The latter has an edible fruit with a large woody seed that preserves well, and is the most common food seed recovered at Carpenter's Gap 1 (McConnell and O'Connor 1997) and at other Kimberley shelters (e.g. O'Connor 1999:82). *Terminalia* seeds are well represented at Carpenter's Gap 1 in levels dated between 35,000 BP and 30,000 BP and are abundant in the Holocene levels, but are found in very low numbers in all levels dated between 15,000 BP and 22,000 BP. *Plectrachne* and Cyperaceae are found in the levels dated between 30,000 BP and 25,000 BP, along with a variety of fruit and vine species, such as *Terminalia* spp., *Ampelocissus acetosa* and *Vitex glabrata.* However, the fruiting tree species decrease after about 25,000 BP.

Recovered in abundance in levels corresponding with the LGM were Chenopodiaceae (saltbush and bluebush – including saltbush *Atriplex* spp.) and Cyperaceae. The presence of chenopods, grasses and spinifex indicates poor, saline or calcareous soils and dry conditions and would not be unexpected at this time. However, the abundance of the aquatic-associated plants is unexpected. Sedges are found only in moist locations and their closest occurrence to the site today is the edges of the permanently wet pools in the Lennard River, about 4 km to the northwest.

The most recent phase of the site's history, the Holocene, again sees a large input of seeds of broad-leaf deciduous tree and vine fruits of common edible species such as are found around the site today. These include *Ampelocissus acetosa*, *Terminalia* spp., *Vitex glabrata* and *Celtis philippensis*. Fragments of the seed case of boabs, *Adansonia gregorii*, are also well represented in these levels. Significantly, the grasses and sedges prevalent in the Pleistocene layers disappear from the record in the Holocene (McConnell and O'Connor 1997:26).

On the basis of these changes, McConnell inferred major fluctuations in the regional environment. She argued the earliest phase, from 42,000 BP to 35,000 BP, can be characterised 'as a grassland of perennial tall tropical grasses and sedges growing around permanent streams and wetlands, with shifting dunes creating inroads in the wetlands' (McConnell and O'Connor 1997:29). The increase in seeds of deciduous fruit trees between 35,000 BP and 25,000 BP is interpreted by McConnell as indicating that vine thicket/rainforest may have been more extensive, and indicating 'an increase in moisture sufficient to maintain deciduous trees which have higher water requirements than grasses' (McConnell and O'Connor 1997:29). The increase in grasses and shrubs and decline in vine thicket species from 25,000 BP to 11,000 BP is argued to be firm evidence for an arid phase, but with sufficient stands of water to support the Cyperaceae sedges. McConnell and O'Connor (1997) make a strong case that the prevalence of grass seeds in the spits with dates of between 20,000 BP and 18,000 BP reflects their importance as an alternate food or drought food as conditions became more arid. There is a substantial ethnobotanical literature on the use of chenopods and grass seed in arid Australia that provides an empirical basis for this proposition (e.g. Crawford 1982; O'Connell et al. 1983; Goddard and Kalotas 1985; Latz 1995). While there are no ethnographic records for the use of grass seed as food or for seed grinding in this area of the Kimberley, it is possible that grasses were eaten only to expand the food base during more arid times (Gorecki et al. 1997:148). Portable grindstones were not found in the deposit, although grinding patches and pecked and ground hollows were recorded as part of the fabric of the shelter and may have been used for this purpose (O'Connor pers obs.).

McConnell and O'Connor (1997:29) proposed that the macrobotanic record indicates a generally drier environment in the Holocene than during the pre-LGM late Pleistocene. The reappearance of the vine thicket/rainforest taxa in the Holocene is seen as indicating an 'increase in available moisture, but with insufficient surface water to maintain the sedges and perennial grasses' (McConnell and O'Connor 1997:29).

In 2000, Wallis (2000, 2001) made a major study of the phytolith assemblage from Carpenter's Gap 1 for her PhD dissertation. In constructing a modern reference collection, Wallis (2000:151) found that many of the modern tree and shrub plant samples produced no, or limited, phytoliths and hence would not be expected to be represented in the Carpenter's Gap assemblage. However, as grasses are abundant phytolith producers, they are better represented in the assemblage. Like McConnell, she separated the site assemblage into four phases based on the available radiocarbon dates and used the palaeobotanical record to infer a vegetation history for the site. Her findings and interpretations, however, differ quite markedly from McConnell's.

Wallis's Phase 1, dated between c. 40,000 BP and 30,000 BP, was characterised by an assemblage dominated by grassland communities similar to those seen in the study area today, although she identified a wider range of taxa than McConnell. A low representation of phytoliths derived from spinifex and other arid grasses was interpreted as indicating that at this time these arid-tolerant species were spatially restricted to the rocky slopes of the range (Wallis 2001:112). Surprisingly, two distinct types of palm phytoliths were found in considerable abundance between 40,000 BP and 30,000 BP (Wallis 2001:113) and in minor quantities until just before the LGM, after which time they are absent from the record. Palms are dependent on permanent water. They do not occur in the Napier Range today, even in areas that retain freshwater pools throughout the dry season, such as the pools on the Lennard River at Windjana Gorge. Their nearest occurrence to Carpenter's

Gap today is approximately 40 km to the north, on the margins of the King Leopold Range. Palms are very useful plants and recorded as an important food source for local indigenous people. The centre of the palm, known as the 'heart', was eaten and the fronds were used for basketry (Crawford 1982) so it is not surprising that they would be transported some distance to a habitation site. However, 40 km is well outside the limits of a daily foraging range. Therefore, while she recognised a human agency in the transport of palms to the site, Wallis (2001:113) argued that their previous geographic range must have extended southwards, suggesting wetter conditions than at present in the inland southwest Kimberley between 40,000 BP and 30,000 BP.

Vegetation change over time was also evident in the distribution of Ulmaceae phytoliths. As with the palms, two distinct forms differentiated by size were present. These were comparable with the modern leaf and seed phytoliths from *Celtis philippensis* and *Trema tomentosa*. Members of this family are associated with vine thicket as they have reasonably high water requirements. The Ulmaceae phytoliths occur in the 40,000 BP to 30,000 BP deposits, disappear during the LGM, and reappear in the Holocene-aged deposits (Wallis 2001:114). The fruits of these species are an important indigenous food (Crawford 1982) and it seems unlikely they would be neglected if they were locally available. Their complete absence in the LGM levels of the sequence is argued by Wallis (2001:114) to constitute good evidence of their disappearance from the local environment with increasing aridity, and their return in the late Holocene to signal the return to moister conditions.

While the overall phytolith record for the LGM could be interpreted as indicating increased aridity, Wallis identified some undeniable indicators of wet conditions, notably sedges, diatoms and sponge spicules. These appear at precisely the time when they would be least expected in the sequence, as climatic conditions deteriorate. Diatoms are associated with moist habitats such as streams, lakes and swamps, and freshwater sponges also require permanent bodies of standing water for their survival. Unlike McConnell, Wallis did not interpret these remains as a direct reflection of their availability in the immediate site environment, instead seeing human agency and variable preservation as playing a strong part in the formation of the palaeobotanical record. Sedge bulbs are an important late-dry-season food for Aboriginal people in the Kimberley (Crawford 1982) and the stems are also used to cover wooden and bailer-shell water carriers during transport to minimise spillage and evaporation (O'Connor pers obs.). Today, the closest source of standing freshwater in any quantity is the Lennard River about 4 km to the northwest. However, small quantities of freshwater (enough for the daily requirements of a small family group) can be obtained from seepage through the limestone at Carpenter's Gap 1 and at many points along the range (O'Connor pers obs.). Even at the height of the dry season it is not necessary to travel to the Lennard River to obtain a small quantity of freshwater. However, the drier conditions during the LGM may have led to the loss of local seepages and necessitated greater mobility to acquire water (Wallis 2001). People bringing water from the Lennard River may have covered their water containers with sedges, or brought back sedge stems with the bulbs. The diatoms and sponge spicules may also have been inadvertently collected and transported with the aquatic plants.

The wood charcoal sequence at Carpenter's Gap 1

Edgar (2001) undertook a pilot study of a small sample of the wood charcoal remains from Carpenter's Gap 1 for his honours dissertation. He examined 40 archaeological charcoal samples from Square A, of which 25 (62%) were identifiable to at least family level (Edgar 2001:214). He identified four species, four genera and one family. These were Proteaceae including *Grevillea* sp. and *Hakea arborescens*, *Terminalia* spp. including *T. arostrata*,

Eucalyptus spp. including *E. miniata, Grewia breviflora* and *Ficus* sp. He subdivided the sample into the same time periods as Wallis (2000), however his attempts to examine change through time were frustrated by the small sample size and limited range of taxa. The limited modern reference material at his disposal also constrained the number of positive identifications. The prime value of Edgar's study was that it showed the potential of wood charcoal for providing a further line of evidence on the palaeovegetation of the southern Kimberley (Edgar 2001:214).

The current study reports the results of the analysis of a larger quantity of excavated charcoal than analysed by Edgar, as well as the samples identified by Edgar, and thus provides a firmer basis for comparison of the wood charcoal with the macrobotanic and phytolith records.

Creating the modern reference collection

There were two sources for the modern reference materials used in this research. In 2006, Frawley undertook a field trip to the Napier Range to make a collection of modern wood from species growing in the surrounds of Carpenter's Gap 1, with which the archaeological material could be compared for identification. When collecting the samples, a point was made to obtain them from a range of representative environments in the vicinity of the rock shelter. Of note were the different moisture levels and inherent differences in the ecosystems between the open sandy plain below the shelter, the rocky talus slopes outside the shelter entrance, and moister environments such as the seasonal creek lines and within Windjana Gorge. Windjana Gorge is host to plants that rely on a constant water supply, whereas the other more arid talus slopes of the Napier Range were characterised by different plants that were generally more drought tolerant and suited these environments.

Tree species were sampled from Windjana Gorge National Park campground and surrounding areas, on the talus slope immediately in front of the entrance to Carpenter's Gap 1, from the plain below the shelter, from the banks of the river channel within Windjana Gorge, and from the sand dunes and plain to the south of the Napier Range. The reference collection comprises trees from three different ecosystems: the gorge/riverine environment, vine thicket and dry savannah. Samples were collected in accordance with guidelines outlined by Pearsall (1989), and where possible, branches were cut to include seed pods for more accurate identification. In some cases, only a single tree was available for collection.

Specimens were identified using reference books and with the assistance of the CALM Ranger at Windjana Gorge, Rod O'Donnell. The sample branches were preserved by enclosing the specimens in newspaper and placing them between layers of a cardboard press. Thirty-one samples were collected during the 2006 field trip. The branches were transported to Perth where they were identified, or the field identification was confirmed, by botanists at the WA Herbarium.

The identified tree specimens were brought back to the Australian National University (ANU) where the branches were processed to create the modern reference collection (Figures 4 and 5). A small section, approximately 4 cm in length, was cut from each branch, wrapped in aluminium foil and placed in a ceramic crucible. These were then placed in a muffle furnace and heated to 400°C for an hour. The resulting charcoal was then cut for SEM imaging and small sections were prepared for mounting on a stub. The mounted sample was cut along the transverse section across the diameter. This was then split at a 90° angle to produce both a tangential and a radial face, and then mounted on stubs using carbon tape. This reduced the need for multiple samples, as the stage within the SEM could be rotated so that all three faces of the sample could be viewed without removal or remounting. Each stub was coated with a 20 nm coating of gold before viewing. Full information on species collected and all the SEM photographs can be found in Frawley (2010:Appendix 2). A selection of identified species from the comparative collection is presented in Figures 4 and 5.

Comparative Collection

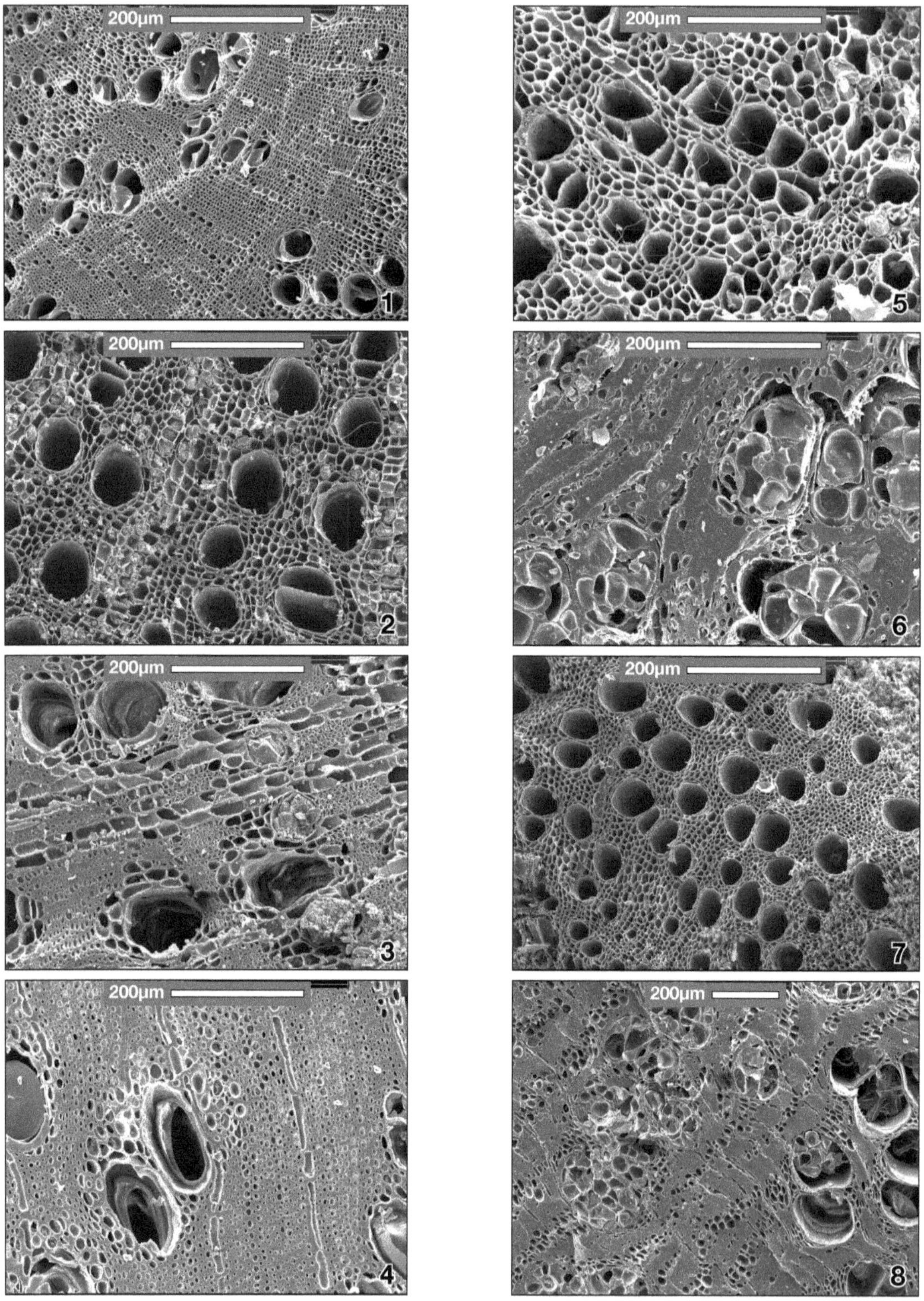

1. *Cochlospermum fraseri*
2. *Andansonia gregorii*
3. *Terminalia arostrata*
4. *Acacia platycarpa*
5. *Ficus platypoda*
6. *Eucalyptus bleeseri*
7. *Eucalyptus camaldulensis*
8. *Eucalyptus ferruginea*

Figure 4. Wood charcoal SEM images for species in the comparative collection

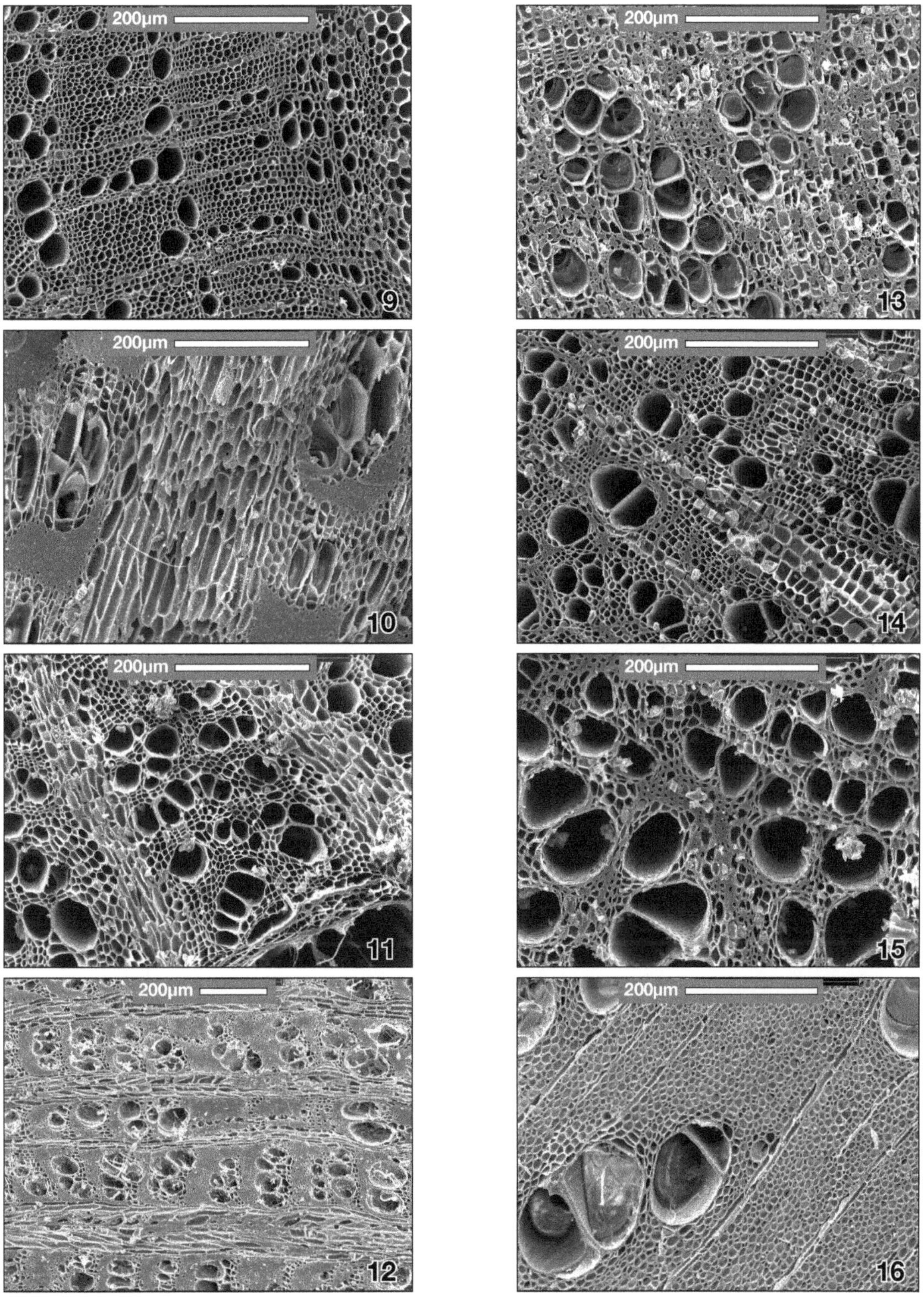

9. Melaluca leucadendra
10. Grevillea heliosperma
11. Grevillea pyramidalis
12. Hakea arborescens
13. Pouteria seracia
14. Brachychiton diversifolius
15. Corymbia dampiere
16. Acacia holosericea

Figure 5. Wood charcoal SEM images for species in the comparative collection

In addition to the samples collected by Frawley, 25 charcoal samples which had been prepared by Edgar (2001) were made available for this study. Combined, this provided a total of 56 samples of species from 22 families.

Sampling the archaeological charcoal

The site yielded a high volume of well-preserved charcoal but only a small sample of it could be processed and analysed for this study.

The charcoal samples selected for analysis were from excavation Square A2 as this excavation area had the greatest number of spits and contained the largest quantities of charcoal (Figure 6). The spits selected for analysis were those that had been dated by O'Connor (O'Connor unpub. data, Table 1). Four additional dates were obtained during this study and are highlighted in Table 1.

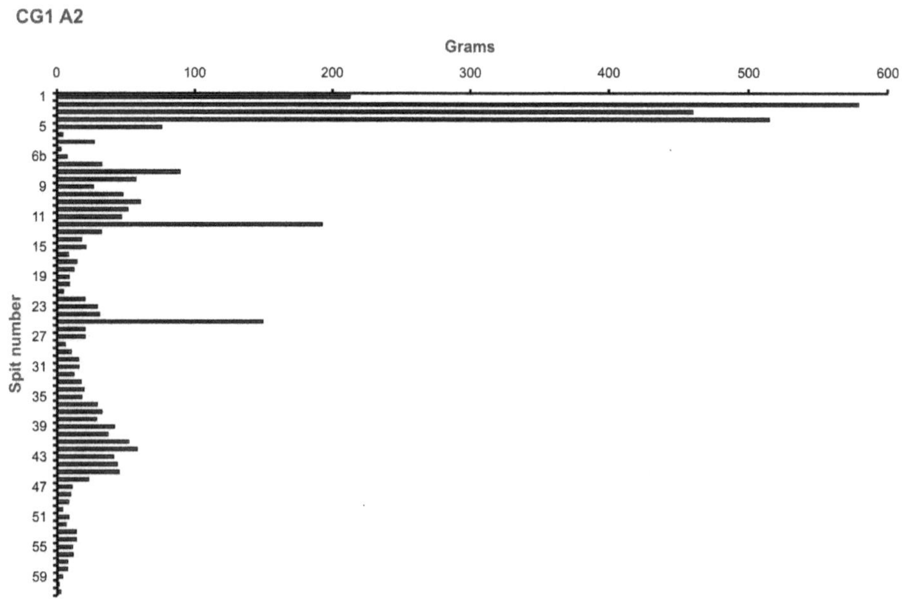

Figure 6. Charcoal weights from Square A2, Carpenter's Gap 1

Spit	C14 Age BP	Delta 13C	Lab. Code
6	3900±70	-26.6±0.1	ANU-11455
5a	**5204±45**	-26.0±0.2	Wk 19723
5a	7470±70	-26.6±0.1	ANU-11454
7	**7468±51**	-25.5±0.2	Wk 19724
7	9600±80	-25.7±0.1	ANU-11458
6b	16,000±710	-24.9±0.1	ANU 11456
6c	17,130±330	-24.0±2.0	ANU-11457
9	17,240±130	-25.2±0.1	ANU-11459
11	18,240±130	-25.0±0.1	ANU-11426
13	18,270±160	-25.0±0.1	ANU-11427
15	18,430±130	-24.0±2.0	ANU-11112
17	**20,292±150**	-25.5±0.2	Wk 19725
25	22,130±130	24.0±0.2 est.	ANU 11113
25	**23,366±150**	-24.4±0.2	Wk 19726
30	23,050±300	-25.5±0.1	ANU-11422
56	33,980±790	-25.9±0.1	ANU-11425
57	40,100±1220	-24.0±0.0	ANU-Abox 340

Table 1. Radiocarbon dates for various spits Square A2, Carpenter's Gap 1. Dates in bold are from Frawley (2010); all others are from O'Connor (unpub. data)

For each spit sampled, 10 charcoal fragments were collected from each of the 6 mm fraction and 3 mm fraction. The 6 mm and 3 mm fractions were sampled by spreading the charcoal on to waxed paper and selecting pieces at random. Both sieve fractions were sampled in order to ensure that species producing less robust, or more easily comminuted, charcoal would stand an equal chance of being represented in the analysis. Two-hundred archaeological charcoal samples were prepared for SEM imaging. Standard wood anatomy identification techniques were employed to determine the species of each sample (Wheeler 2000). It was possible to identify 179 of the 200 samples. Of the 179 that were identified, 136 were identified to the species level. The remainder were identified to the genus or family level.

Chronological division of the archaeological wood charcoal sample

Like Wallis and McConnell, we divided the site into different time phases for analysis:

- 42,000 BP to 30,000 BP (pre-glacial);
- 30,000 BP to 17,000 BP, marking a phase of presumed increasing aridity, with a further subdivision of,
- 20,000 BP to 17,000 BP encompassing the height of the LGM;
- 17,000 BP to 10,000 BP, a time of fluctuation seeing a progressive amelioration; and
- 10,000 BP to the present, encompassing the Holocene.

It should be noted, however, that Square A2 at Carpenter's Gap 1 produced no radiocarbon-dated stratigraphic units within the time period 16,000 BP to 10,000 BP. Spit 6b, dated to 16,000 BP, was grouped for analysis in the 20,000 BP to 17,000 BP phase. Spit 6 is dated to 3900 BP. In the absence of any other dates, we have assumed that Spits 1-5 cover the late Holocene. While there might appear to be inversions in the dates from Spit 6 and Spits 6b and 6c, these spits were not contiguous.

Results of the archaeological wood charcoal analysis

The results of the analysis of the archaeological charcoal samples are presented in Table 2 and discussed below by time period.

Table 2. Counts of identifications by time period (Table 2 continues on page 310)

	Holocene	20,000-17,000	30,000-20,000	42,000-30,000
Acacia holosericea			1	
A. platycarpa				1
Adansonia gregorii	1			
Brachychiton diversifolius	1			
Calthrix exstipulata	1			
Cochlospermum fraseri	2			2
Corymbia dampiere	5	2	1	
Dodonea polyzyga	1			
Eucalyptus bleeseri	1	2	2	2
E. camaldulensis	1	2		
E. conteriflora		1		
E. dendromerenix	2	1	1	1
E. ferruginea	4	4	1	

	Holocene	20,000-17,000	30,000-20,000	42,000-30,000
E. miniata		1		
E. polycarpa	1	1		
E. tectifica			1	
E. tetradonta	3	1	1	
Ficus platypoda	1			
Grevillia heliosperma				1
G. pyramidalis	9	2	1	
Grewia breviflora	1		2	
G. laevigata	1			
G. tiliacfolia	1			
Gyrocarpus americanus	1			
Hakea arborescens	9	19	2	4
Melaluca leucadendra	1			
Lysiphyllum cunninghamii		1		
Petalostigma pubescens	1			
Planchonia careya	3	1		1
Pouteria seracia			1	
Terminalia arostrata	3			
T. canescens	4	4	2	
T. platyphylla	2	1	1	
T. platyptera			2	
Vitex glabrata	2			
Acacia spp.			4	
Terminalia spp.	3	2	6	1
Combretaceae family		1		
Myrtaceae family	3	4	3	1
Proteaceae family	7	3	1	2
Tiliaceae family			1	
Ulmaceae family			1	
Unidentified	5	7	5	4
Identified	76	53	35	16

42,000 BP to 30,000 BP

As noted by McConnell, there is less charcoal and macrobotanic material overall in this earliest phase of occupation than in later periods, and the number of species recovered is restricted. This constrained the sampling to 20 samples. Of the 20 samples from this earliest occupation phase, only seven were able to be identified to the species level. They are: *Acacia platycarpa* (Figure 7, SEM image 2), *Cochlospermum fraseri* (Figure 7, SEM image 3), *Eucalyptus bleeseri* (Figure 7, SEM image 4), *E. dendromerenix* (Figure 7, SEM image 7), *Grevillia heliosperma* (Figure 7, SEM image 5), *Hakea arborescens* (Figure 7, SEM image 1) and *Planchonia careya* (Figure 7, SEM image 6). One sample was identified as belonging to the *Terminalia* genus, though a species could not be assigned. One was identified as belonging to the Myrtaceae family and two as members of the Proteaceae family (Figure 7, SEM image 8). *Acacia platycarpa* and *Grevillia heliosperma* are found exclusively in this period. Both are common today in dry inland areas stretching from the Kimberley through to the Northern Territory (Brock 1988; Wheeler 1992). The other species identified in this period are also indicative of arid to semi-arid environments. The relatively low species diversity in the identified samples for this period and the large percentage of unidentified samples (35%) may potentially indicate that species diversity was restricted in the late Pleistocene. However, as will be discussed below, the small sample size and age of the charcoal at this level suggest that preservation factors must also be taken into consideration.

30,000 BP to 20,000 BP

Of the 40 samples selected for this period, 14 were identified to the species level: *Acacia holosericea* (Figure 8, SEM image 5), *Corymbia dampiere* (Figure 8, SEM image 3), *Eucalyptus bleeseri, E. dendromerenix, E. ferruginea* (Figure 8, SEM image 2), *E. tectifica, E. tetradonta, Grevillea pyramidalis* (Figure 8, SEM image 6), *Grewia breviflora* (Figure 8, SEM image 4), *Hakea arborescens, Pouteria seracia* (Figure 8, SEM image 1), *Terminalia canescens, T. platyphylla* and *T. platyptera*. Four samples were identified as belonging to the *Terminalia* genus, but could not be identified to species level, and six samples were identified as belonging to the *Acacia* genus but again could not be identified beyond this. Three were identified as belonging to the Myrtaceae family, and one each to the Proteaceae, Ulmaceae and Tiliaceae families. The species that are found exclusively in this period are *Acacia holosericea, Eucalyptus tectifica, Pouteria seracia* and *Terminalia platyptera*.

This period is most interesting for its sharp rise in the occurrence of *Terminalia* spp. Of the 40 samples examined in this period, 11 (27.5%) are identified as *Terminalia* (Table 2). *Grewia breviflora, Pouteria seracia* and the sample simply identified as a member of the Tiliaceae family are all trees that grow in association with water courses, vine thickets and gorges (Wheeler 1992). This presents a total of 16 of the 40 samples (40%) for this period identified as trees that are associated with moist vine-thicket habitats or creek edges. It could be argued that this indicates the continued presence of some small vine thicket, even as environmental conditions are presumed to be deteriorating.

LGM 20,000 BP to 17,000 BP

This period was the most intensively sampled, with 60 samples analysed. This was done in order to intensively sample this phase of hypothesised aridity. Despite the larger number of samples analysed, this period does not have the greatest diversity. Fifty-three samples (88%) were able to be identified, with 15 individual species recorded: *Corymbia dampiere, Eucalyptus bleeseri, E. camaldulensis, E. conteriflora* (Figure 9, SEM image 8), *E. dendromerenix, E. ferruginea* (Figure 9, SEM image 1), *E. miniata* (Figure 9, SEM image 2), *E. polycarpa* (Figure 9, SEM image 3), *E. tetradonta* (Figure 9, SEM image 4), *Grevillea pyramidalis* (Figure 9, SEM image 5), *Hakea arborescens, Lysiphyllum cunninghamii* (Figure 9, SEM image 6), *Planchonia careya, Terminalia canescens* and *T. platyphylla*. Two other samples were identified as belonging to the *Terminalia* genus but could not be identified to species. Four were identified as belonging to the Myrtaceae family and three as members of the Proteaceae family. One sample was identified as a member of the Combretaceae family. The species that are found exclusively in this period are *Eucalyptus conteriflora, E. miniata*, and *Lysiphyllum cunninghamii.*

What is instantly apparent from Table 2 is the large amount of *Hakea arborescens* in this period. Nineteen (31.6%) of the samples were identified as this species (see Table 2). Also prolific were *Grevillea pyramidalis* (Figure 9, SEM image 5) and other members of the Proteaceae family (representing a total of 40% of the samples for this period). There is also a sharp rise in *Eucalyptus*, perhaps indicating the wider presence of *Eucalyptus* woodlands. This is supported by the presence of *Planchonia careya,* also found in this environment (Wheeler 1992). There are 19 (31.6%) samples identified as members of the Myrtaceae family. Combining this with those identified only as Myrtaceae undetermined gives a total of 23 samples, which represents 38.3% of the sampled charcoal within this period. If these two families are combined, 78.3%

of the samples are from two families that are predominantly dry woodland species, perhaps indicating a fairly homogenous dry environment.

Surprisingly, two of the *Eucalyptus* samples are identified as red river gum, *Eucalyptus camaldulensis* (see Figure 9, SEM image 7). This tree only grows along the margins of rivers and its presence in the site suggests that, at least for periods of time during the LGM, there was water available in the gorge on the Lennard River. River red gum has other uses besides being a good fuel wood, such as for medicinal infusions and for producing wooden artefacts (Brock 1988; Petheram and Kok 1991).

17,000 BP to the Holocene

There is a chronostratigraphic hiatus in the sedimentary sequence from Square A2 at Carpenter's Gap 1 between 16,000 BP and 9600 BP. The charcoal from Spit 6b was dated to 16,000 BP and was grouped for analysis with the wood charcoal in the 17,000 BP to 20,000 BP phase.

The Holocene

The Holocene sequence in Square A2 at Carpenter's Gap 1 spans from c. 9600 BP to the late Holocene. There is an exponential rise in the quantity of charcoal recovered in this period, as opposed to all earlier periods (Table 2). While superior preservation is no doubt a factor in the overall increase in charcoal, and the commensurate larger number of identified charcoal samples here, the charcoal is also likely due to the marked increase in human use of the site at this time. Stone-artefact numbers and other indicators of human occupation which are less likely to be negatively affected by preservation also show an exponential increase in the Holocene levels (McConnell and O'Connor 1999).

This period has the greatest number of identified specimens, with 76 of the 80 samples identifiable (93.7%), and 26 individual species recorded: *Adansonia gregorii* (Figure 10, SEM image 1), *Brachychiton diversifolius* (Figure 10, SEM image 2), *Calthrix exstipulata* (Figure 10, SEM image 3), *Cochlospermum fraseri, Corymbia dampiere, Dodonea polyzyga* (Figure 10, SEM image 4), *Eucalyptus bleeseri, E. camaldulensis, E. dendromerenix, E. ferruginea, E. polycarpa, E. tetradonta, Ficus platypoda* (Figure 10, SEM image 5), *Grevillea pyramidalis, Grewia breviflora, G. laevigata, G. tiliacfolia* (Figure 10, SEM image 6), *Gyrocarpus americanus, Hakea arborescens, Melaluca leucadendra* (Figure 10, SEM image 8), *Petalostigma pubescens, Planchonia careya, Terminalia arostrata* (Figure 10, SEM image 7), *T. canescens, T. platyphylla* and *Vitex glabrata* (see Figure 10). Of these, *Adansonia gregorii, Brachychiton diversifolius, Calthrix exstipulata, Corymbia bella, Dodonea polyzyga, Ficus platypoda, Grewia tiliacfolia, G. laevigata, Melaluca leucadendra* and *Petalostigma pubescens* are found exclusively in this period. They derive from species found in a wide range of habitats and many are common around the site today. *Gyrocarpus americanus* and *Eucalyptus camaldulensis* both grow in Windjana Gorge. Many, such as *Ficus platypoda, Grewia tiliacfolia, G. laevigata,* the *Terminalia* species and *Vitex glabrata,* are vine thicket/rainforest species. *Melaluca leucadendra* is found around swamps and creek and river courses and can also occur in wet vine thicket. Boabs (*Adansonia gregorii*) and *Brachychiton diversifolius* grow immediately outside the entrance of the shelter and on the plain below.

42,000 BP to 30,000 BP

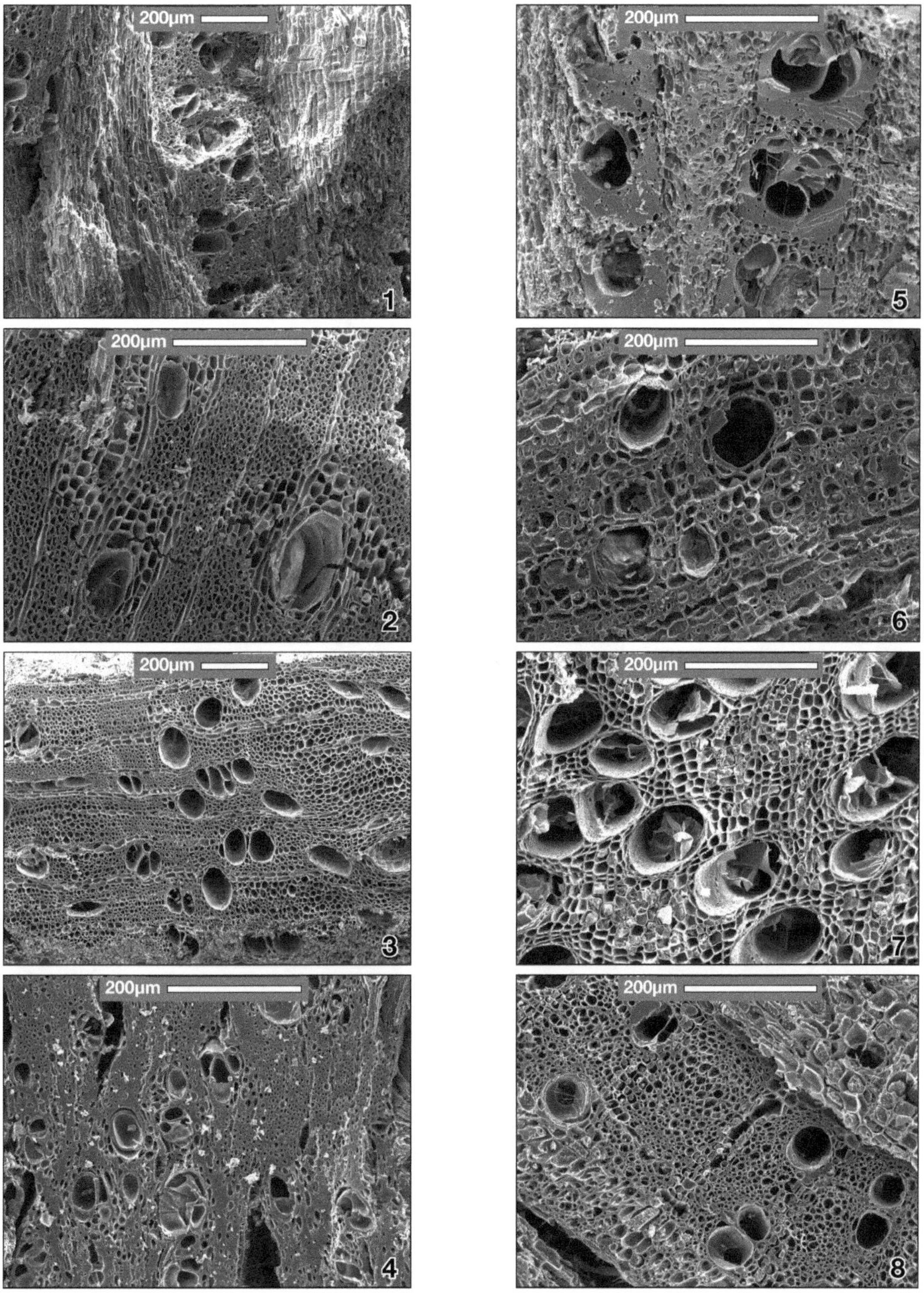

1. Hakea arborescens
2. Acacia platycarpa
3. Cochlospermum fraseri
4. Eucalyptus bleeseri
5. Grevillea heliosperma
6. Planchonia careya
7. Eucalyptus dendromerenix
8. Proteaceae undetermined

Figure 7. SEM images of wood charcoal – species recovered from the 42,000 BP to 30,000 BP levels of Carpenter's Gap 1

30,000 BP to 20,000 BP

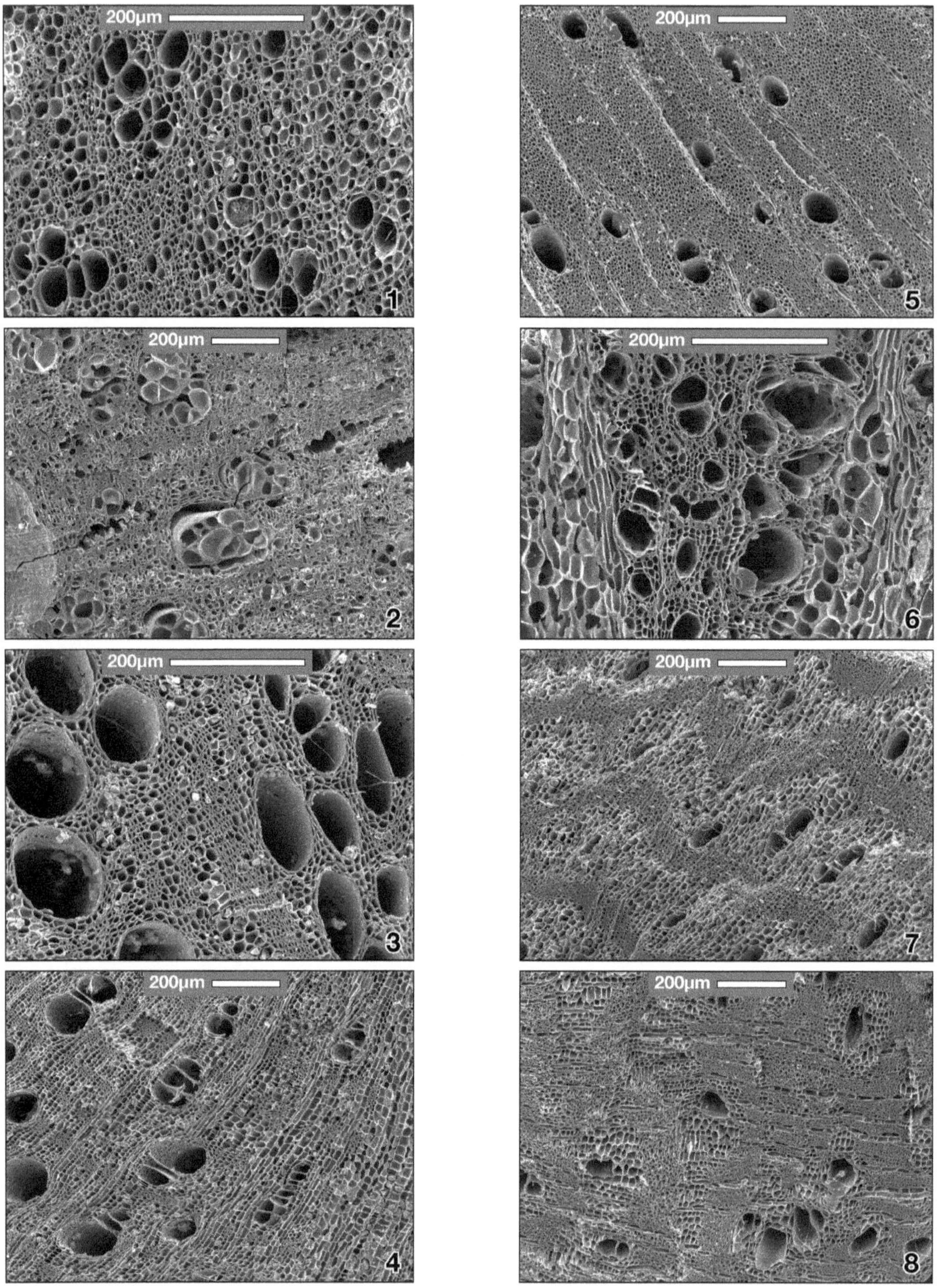

1. *Pouteria seracia*
2. *Eucalyptus ferrunginea*
3. *Corymbia dampiere*
4. *Grewia breviflora*
5. *Acacia holosericea*
6. *Grevillea pyramidalis*
7. *Acacia species*
8. *Ulmaceae unidentified*

Figure 8. SEM images of wood charcoal – species recovered from the 30,000 BP to 20,000 BP levels of Carpenter's Gap 1

20,000 BP to 17,000 BP

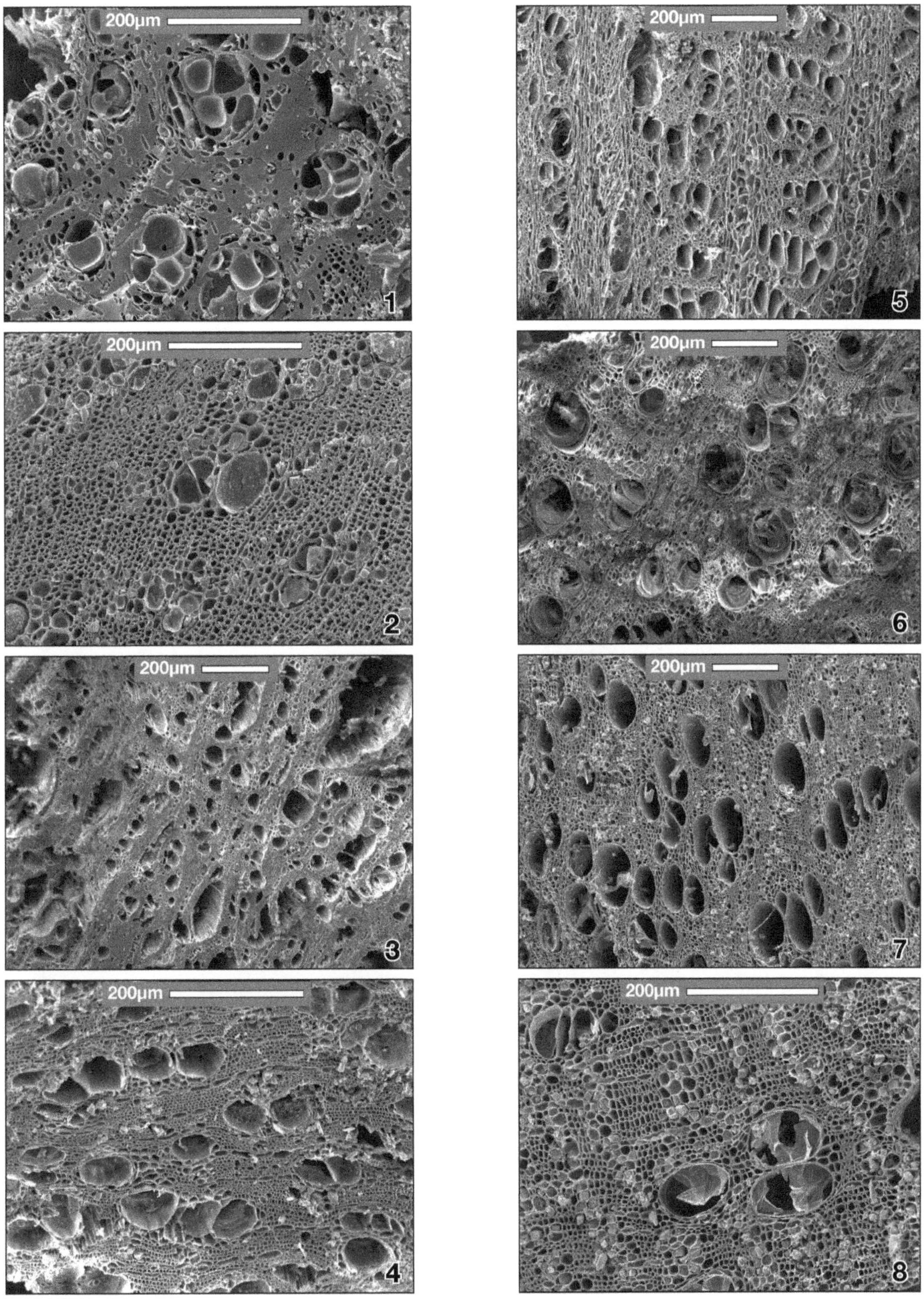

1. Eucalyptus ferruginea
2. Eucalyptus miniata
3. Eucalyptus polycarpa
4. Eucalyptus tetradonta
5. Grevillea pyramidalis
6. Lysiphyllum cunninghamii
7. Eucalyptus camaldulensis
8. Eucalyptus conteriflora

Figure 9. SEM images of wood charcoal – species recovered from the 20,000 BP to 17,000 BP levels of Carpenter's Gap 1

The Holocene

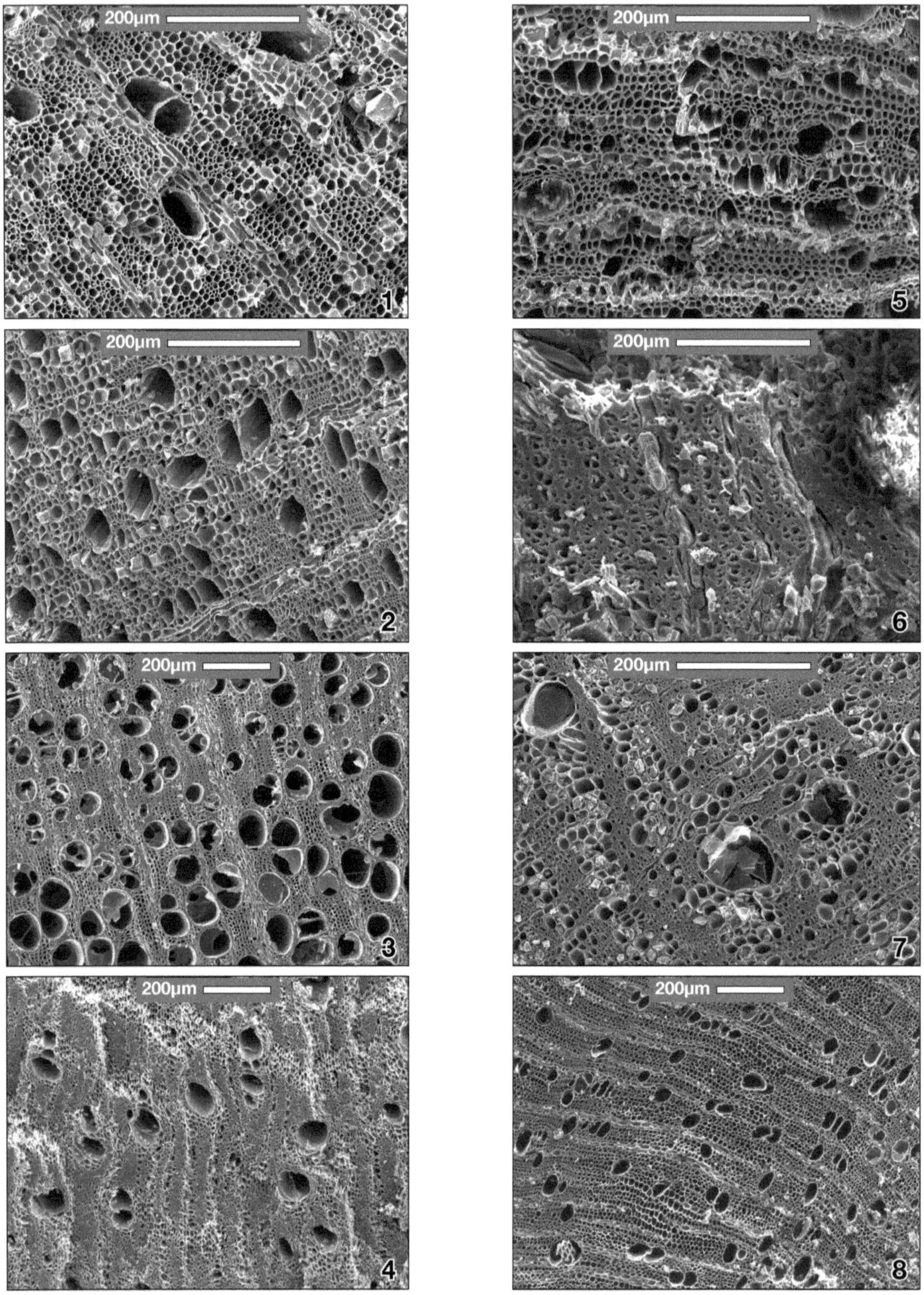

1. *Andansonia gregorii*
2. *Brachychiton diversifolius*
3. *Calthrix exstipulata*
4. *Dodonea polyzyga*
5. *Ficus platypoda*
6. *Grewia tiliacfolia*
7. *Terminalia arostrata*
8. *Melaluca leucadendra*

Figure 10. SEM images of wood charcoal – species recovered from the Holocene levels of Carpenter's Gap 1

Discussion

42,000 BP to 30,000 BP

As in the previous macrobotanic study of McConnell and O'Connor (1997), a restricted range of species was identified in the oldest units of Carpenter's Gap 1 through the wood charcoal analysis. McConnell and O'Connor (1997) noted the small quantity of macrobotanic remains in the earliest levels of Carpenter's Gap 1 and argued that this indicated a lack of preservation with age. This would appear to be corroborated by Wallis's (2000, 2001) study of the phytoliths from the site, which revealed a wider range of plants in the earliest phytolith record than occurs in the macrobotanical record, presumably as a direct outcome of the breakdown of the macrobotanics in the lower spits. While poor preservation would account for the small quantity of charcoal in the lower levels (Figure 6) and mitigate against the survival of some species, the charcoals sampled by Frawley (2010) were well preserved, indicating that the wood charcoal in these lower levels may not have significantly deteriorated over time. The small sample size of macrobotanics in this period is mirrored in the more durable cultural materials, such as stone-artefact numbers, suggesting that occupation at the site in the initial stages was sporadic and of low intensity (McConnell and O'Connor 1999:26). While there will no doubt have been some loss or deterioration of more fragile wood charcoals over time, it may be that less charcoal entered the site.

McConnell and O'Connor (1997:23-26) identified predominantly grassland taxa in this earliest phase, such as *Plectrachne* sp., and *Panicum* sp. Cyperaceae sedges, as well as occasional seeds of fruit-tree species like *Terminalia* sp. and *Ampelocissus acetosa*, were also recovered from the earliest occupation levels. Wallis (2000) also identified a dominance of grasses in this early phase, as well as two distinct forms of Ulmaceae phytolith, possibly from *Celtis philippensis* and *Trema tomentosa*. Members of the Ulmaceae family are associated with vine thicket, thus demonstrating that moister gullies were in reach of the site. As noted above, the Cyperaceae must have been acquired from a waterway or wetland and their presence in the site would not be anticipated in view of the distance to the nearest permanent water on the Lennard River. Surprisingly, Wallis also identified two distinct forms of palm and these were recovered in some quantity. The closest palms today are found in refugia more than 40 km north of Carpenter's Gap in the King Leopold Ranges. From this, Wallis inferred slightly moister conditions than today prevailed between 42,000 BP and 30,000 BP, allowing the palms to extend their range south.

The wood charcoal study makes a valuable contribution to the palaeobotanic record at Carpenter's Gap 1 for this earliest phase of occupation, with the addition of a further seven species, as well as specimens from the *Terminalia* genus and Myrtaceae and Proteaceae families. There is little in the wood charcoal record, however, to support the notion of a moister climatic regime at the time of initial occupation. As noted above, with the exception of the *Terminalia* spp. the tree species identified for this period are indicative of arid to semi-arid environments. Overall, the wood charcoal from this phase supports an interpretation of a semi-arid treed environment, with pockets of economically important fruiting trees such as *Terminalia* spp. restricted to moister vine thickets and seasonal gullies.

30,000 BP to 20,000 BP

Although regionally specific palaeoclimatic records are not available for the inland Kimberley, the LGM generally is characterised as a time of increased aridity, decreased temperatures, greater windiness and reduced effective precipitation (Hesse 1994; Kershaw 1995; Longmore 1997; Hesse and McTainsh 1999; Hiscock and Wallis 2005). It is generally accepted that foraging and social strategies would have required significant modification as major environmental shifts took place and water sources dried up (Hiscock and Wallis 2005; O'Connor and Veth 2006). However, precisely when environmental deterioration set in is more difficult to determine.

The previous palaeobotanical studies at Carpenter's Gap 1 are in some ways difficult to reconcile. McConnell and O'Connor (1997:23) found that seeds of fruiting-tree species such as *Terminalia* spp. are found in some quantity in levels dated between c. 35,000 and 20,000 BP but virtually disappear as the area becomes increasingly arid between c. 20,000 and 16,000 BP. Also recovered in levels dated to the LGM were Chenopodiaceae and Amaranthaceae (saltbush and bluebush, including saltbush *Atriplex* spp.), and perennial grasses such as *Panicum* sp. These species are recorded at the site only for the period between about 30,000 BP and the terminal Pleistocene, during which time they dominate the record (McConnell and O'Connor 1997:29). Cyperaceae sedges also occur in both the macrobotanic and phytolith record from the earliest levels and increase in abundance in layers corresponding with greatest aridity, being found in quantity in levels dated between 30,000 BP and 17,000 BP (McConnell and O'Connor 1997:23; Wallis 2000, 2001). While the presence of wet-environment-associated species at the time of greatest aridity may seem anomalous, a likely explanation can be found in changing human mobility.

One genus that is present in these two earliest periods which does not occur in later periods, *Acacia platycarpa*, is found in the 42,000 BP to 30,000 BP period, and *Acacia holosericea* is found in the 30,000 BP to 20,000 BP phase. Also present in the 30,000 BP to 20,000 BP phase are four samples that are simply identified as belonging to the *Acacia* genus. *Acacia* grows in a variety of habitats. Of interest is that *Acacia* is a colonising species and occurs after fire or other perturbation (Simmons 1999). *Acacia platycarpa* is common in dry inland areas from the Kimberley through to the northern part of the Northern Territory. *Acacia holosericea* is found along creek banks and on well-drained hillsides (Simmons 1987). Another tree found only in this period is *Grevillea pyramidalis*. The common name of this species is caustic bush as there is a substance on the fruits that is acidic and burns the skin when it comes in contact. Although this species does not make good firewood, it is possible that it entered the site as fuel. It is also recorded as used in ceremony and ritual for deliberate scarification of the skin (Petheram and Kok 1991:327). *Grevillea pyramidalis* is found in sparse open woodland and dry regions. Taken together, the samples from this period could be interpreted as indicating a trend towards a drying environment.

The LGM 20,000 BP to 17,000 BP

The wood charcoal record indicates that at the height of the LGM there was a predominance of dry woodland species surrounding Carpenter's Gap 1 (Table 2). The exception to this is the presence of red river gum, which could only have been obtained from the banks of a water course. This is interpreted as indicating that the site's occupants accessed the Lennard River and brought back timber for firewood and/or artefact manufacture. This adds weight to the findings from the phytolith and macrobotanic studies which identified wet-associated species such as sedges, sponge spicules and diatoms during this arid phase, and to the interpretation that there was a reorganisation of mobility and foraging strategies during this phase. Interestingly, tree species which dominate the phytolith record during the LGM, such as *Acacia* spp., do not appear in the wood charcoal record at all. This discrepancy is unlikely to be due to poor preservation or taphonomic factors, as *Acacia* spp. are found in the earlier phases. Wallis (2000) argued that her data indicated that vine thicket/rainforest pockets may have persisted in a reduced state during the height of the LGM, and the persistence of the *Terminalia* species in the wood charcoal record for this period would seem to support this view.

The Holocene

The wood charcoal record from Carpenter's Gap 1 is in good agreement with the phytolith and macrobotanic records for this period. As noted above, 25 tree species were identified for this period, the largest number for any single period. Identified in the wood charcoal record are species from a wide range of habitats and most are common around the site today. All three records indicate the importance of fruiting-tree taxa, including *Vitex glabrata* and *Terminalia* spp. While Wallis (2000) detected *Vitex glabrata* in the phytolith record of the earliest occupation phase and McConnell and O'Connor (1997) reported occasional seeds of *V. glabrata* throughout the sequence, it is only abundant in the palaeobotanic record in the Holocene (McConnell and O'Connor 1997:23). *V. glabrata* is first detected in the wood charcoal record in the Holocene. This may reflect its increased abundance proximal to the site, but could in part be due to preservation, as this species is known to have very porous wood which may not survive as well as the hard seed cases. This economically important species is reported as one of the staple fruits traditionally eaten in the Kimberley during the wet season (Crawford 1982; Scarlett 1985). It was also important because it could be stored. Surplus fruit was often pounded and made into large round cakes which were wrapped in paperbark for later use (Scarlett 1985:7).

Three members of the *Terminalia* genus are identified in the Holocene wood charcoal record. *Terminalia* spp. are recorded as a very important food species in the Kimberley. Of the 12 species recorded as growing in the Kimberley, many have documented ethnographic uses (Kenneally et al. 1996:88). The fruits, which are abundant in the wet season, contain protein and carbohydrate, but most importantly, are enormously high in vitamin C (McConnell and O'Connor 1997:22). Also identified in the Holocene wood charcoal sample is the boab, *Adansonia gregorii*. McConnell and O'Connor (1997:26) reported abundant evidence of pod fragments of *Adansonia gregorii* in the Holocene levels and Wallis (2000) similarly detected high concentrations of boab phytoliths. The white pulp around the seeds was eaten (Crawford 1982), while in historic times the seed pods were collected to be decoratively carved for commercial sale (Akerman 1993:107). *Melaluca leucadendra* was also recorded in the late-Holocene macrobotanic record as fragments of paperbark (McConnell and O'Connor 1997). The soft outer bark of the paperbark was traditionally used for a variety of functions, such as wrapping the bone bundle in secondary burials, and wrapping food and delicate artefacts before placing them in carry bags. It is a very useful species and is still used today by Bunuba people for firewood and the bark sheeting for wrapping (O'Connor pers obs.).

The diversity of species in the wood charcoal record and the fact that many of them currently grow in the vicinity of the shelter suggests a vegetation suite and climate similar to those of today. Many of these species are not found in the macrobotanic record and cannot be identified to species level in the phytolith record. The wood charcoal record also identified species such as *Eucalyptus, Hakea* and *Calthrix*, which may have entered the site as fuel.

Conclusion

Considered in isolation, each of the palaeobotanic studies carried out at Carpenter's Gap 1 has produced a quite different profile of the palaeovegetation, with limited overlap. A variety of taphonomic and preservation factors bias the chances of entry and survival of different plants and plant parts in the archaeological record. For example, Wallis's (2000:151) study of the modern plant specimens demonstrated that Proteaceae and many other locally abundant taxa produced limited or no phytoliths, hence their absence from the Carpenter's Gap 1 phytolith assemblage cannot necessarily be interpreted as evidence of their absence in the landscape. Indeed, Proteaceae are present at the site in the wood charcoal record. The degree to which plants were economically important to the site's occupants will also bias their chances of entering the archaeological record, as well as the degree to which they accurately reflect the vegetation history of the region. Wallis (2000) demonstrated that most of the grass phytoliths were likely to have entered the sedimentary record of the site through the breakdown of macropodid scats or as wind-blown particles, and therefore are arguably a better indicator of changing environmental conditions than residues of plants that were selectively targeted by people. The wood charcoal study is most likely to have sampled tree species that were brought to the site as fuel or for other purposes, however species with porous woods will not preserve well in the wood charcoal record. Poor preservation and intense burning will mitigate against the survival of macrobotanics but may assist with the production of phytoliths. This study has demonstrated that a complete and balanced profile of the vegetation history of the site will only be produced by assessing multiple lines of palaeobotanic evidence in tandem.

In general, the results of the wood charcoal study provide support for the interpretations made by Wallis (2000, 2001) about the vegetation history of the locality around Carpenter's Gap, and challenge some of the earlier assertions made by McConnell and O'Connor (1997). McConnell and O'Connor (1997) interpreted the macrobotanic record as a direct reflection of the surrounding vegetation and paid little regard to the influence of human agency. Thus, they argued that the presence of sedges in the Pleistocene levels indicated moister conditions and more available surface water than the Holocene. Conversely, Wallis (2001) and Frawley (2010) interpreted the presence of the wet-associated species such as the sedges and the red river gum as signalling an increase in human mobility with increased aridity. Together, the data gathered from these three studies provide the best record available of palaeovegetation change in the Kimberley region over the human time span.

References

Akerman, K. 1993. From Boab Nuts to Ilma: Kimberley art and material culture. In J. Ryan with K. Akerman (eds), *Images of Power, Aboriginal Art of the Kimberley*, pp106-117. Melbourne: National Gallery of Victoria.

Beard, J.S. 1979. *The Vegetation of the Kimberley Area: Explanatory Notes to Sheet 1.* Perth: University of Western Australia Press.

Brock, J. 1988. *Top End Native Plants.* Darwin: John Brock.

Crawford, I.M. 1982. Traditional Aboriginal plant resources in the Kalumburu area; Aspects in ethno-economics. *Records of the Western Australian Museum,* Supplement No. 15.

Edgar, J. 2001. Beyond the Black Stump: Macroscopic Charcoal as Palaeobotanical Evidence at Carpenter's Gap 1, Northwest Australia. Unpublished BSocSc(Hons) thesis. Townsville: School of Anthropology, Archaeology and Sociology, James Cook University.

Frawley, S. 2010. Charcoal from Carpenter's Gap 1: Implications for environmental change in the last 42,000 years. Unpublished M. Phil thesis. Department of Archaeology and Natural History, College of Asia and the Pacific, The Australian National University.

Goddard, C. and A. Kalotas (eds) 1985. *Punu, Yankunytjatjara Plant Use.* Alice Springs: Institute for Aboriginal Development Press.

Gorecki, P., M. Grant, S. O'Connor and P. Veth 1997. The morphology, function and antiquity of Australian grinding implements. *Archaeology in Oceania* 32:1-10

Hesse, P.P. 1994. The record of continental dust from Australia in Tasman sea sediments. *Quaternary Science Reviews* 13:257-72.

Hesse, P.P. and G.H. McTainsh 1999. Last Glacial Maximum to early Holocene wind strength in the mid-latitudes of the Southern Hemisphere from Aeolian dust in the Tasman Sea. *Quaternary Research* 52:343-9.

Hiscock, P. and L.A. Wallis 2005. Pleistocene settlement of deserts from an Australian perspective. In P. Veth, M. Smith and P. Hiscock (eds). *Desert Peoples: Archaeological Perspectives*, pp34-57. Carlton: Blackwell Publishing.

Kenneally, K.F., D. Choules Edinger and T. Willing 1996. *Broome and Beyond: Plants and People of the Dampier Peninsula, Kimberley, Western Australia*. Como: Department of Conservation and Land Management.

Kershaw, A.P. 1995. Environmental change in Greater Australia. *Antiquity* 69:656-75.

Latz, P. 1995. *Bushfires and Bushtucker: Aboriginal Plant Use in Central Australia*. Alice Springs: AID Press.

Longmore, M.E. 1997. Quaternary palynological records from perched lake sediments, Fraser Island, Queensland, Australia: Rainforest, forest history and climatic control. *Australian Journal of Botany* 45:507-26.

McConnell, K. 1997. Palaeoethnobotanical remains of Carpenter's Gap Site, the Kimberleys, Western Australia. Unpublished MA thesis. Canberra: Research School of Pacific and Asian Studies, The Australian National University.

McConnell, K. and S. O'Connor 1997. 40,000 year record of food plants in the southern Kimberley ranges, Western Australia. *Australian Archaeology* 45:20-31.

McConnell, K. and S. O'Connor 1999. Carpenter's Gap Shelter 1: A case study for total recovery. In M.J. Mountain and D. Bowdery (eds), *Taphonomy: The Analysis of Processes from Phytoliths to* Megafauna, pp23-34. Research Papers in Archaeology and Natural History 30. Canberra: ANH Publications, Research School of Pacific and Asian Studies, The Australian National University.

O'Connell, J.F., P. Latz and P. Barnett 1983. Traditional and modern plant use among the Alyawara of Central Australia. *Economic Botany* 37(1):80-109.

O'Connor, S. 1995. Carpenter's Gap Rockshelter 1: 40,000 years of Aboriginal occupation in the Napier Ranges, Kimberley, WA. *Australian Archeology* 40:58-59.

O'Connor, S. 1999. *30,000 Years of Aboriginal Occupation in the Kimberley, Northwest Australia*. Terra Australis 14. Canberra: Archaeology and Natural History and Centre for Archaeological Research Publications, The Australian National University.

O'Connor, S. and P. Veth 2006. Revisiting the past: Changing interpretations of Pleistocene settlement subsistence and demography in northern Australia. In I. Lilley (ed.), *Archaeology of Oceania: Australia and the Pacific Islands*, pp31-47. Carlton: Blackwell Publishing.

Pearsall, D.M. 1989. *Paleoethnobotany: A Handbook of Procedures*. San Diego: Academic Press.

Petheram, R.J. and K.B. Kok 1991. *Plants of the Kimberley Region of Western Australia*. Nedlands: University of Western Australia Press.

Scarlett, N.H. 1985. A preliminary account of the Ethnobotany of the Kija people of Bungle Bungle outcamp. East Kimberley Impact Assessment Project, Working Paper no. 6. Canberra: Centre for Resource and Environmental Studies, The Australian National University.

Simmons, M. 1987. *Acacias of Australia*. Melbourne: Nelson Publishers.

Simmons, M. 1999. *Acacia* an introduction. *Eucryphia, Newsletter of the Australian Plants Society (Tasmania)*. Available at http://asgap.org.au/APOL16/dec99-3.html.

Wallis, L.A. 2000. Phytoliths, Late Quaternary Environment and Archaeology in Tropical Semi-arid Northwest Australia. Unpublished PhD thesis. Canberra: Research School of Pacific and Asian Studies, The Australian National University.

Wallis, L.A. 2001. Environmental history of northwest Australia based on phytolith analysis at Carpenter's Gap 1. *Quaternary International* 83-85:103-117.

Wheeler, E. 2000. Wood Anatomy and Properties. http://legacy.ncsu.edu/WPS202/softwood/background.htm. 2000.

Wheeler, J.R. (ed.). 1992. *Flora of the Kimberley Region*. Como: Department of Conservation and Land Management.

17

The burning question: Claims and counter claims on the origin and extent of buttongrass moorland (blanket moor) in southwest Tasmania during the present glacial-interglacial

Mike Macphail

Department of Archaeology and Natural History, Research School of Pacific and Asian Studies, Australian National University, Canberra

mike.macphail@anu.edu.au

Introduction

Claims and counter claims about the origins of the buttongrass moorland (blanket moor) in southwest Tasmania explicitly or implicitly are founded on the 'ecological drift' concept formulated more than 40 years ago by the late Professor W.D. (Bill) Jackson, at the University of Tasmania. Ecological surveys and modelling experiments have provided much valuable information about the dynamic balance between the plant associations forming buttongrass moorland under present-day climates, soil types and fire regimes. However, thus far, fossil pollen provides the only direct evidence about the origins, geographic extent and long-term directions of change in buttongrass moorland in the prehistoric past. For example, fossil pollen demonstrates that the eponymous species of buttongrass moorland, *Gymnoschoenus sphaerocephalus*, was present in the Last Glacial vegetation in southern Tasmania during the late Last Glacial period and provides the first compelling evidence that buttongrass moorland may have been continuously present in the Lake Pedder area of southwest Tasmania throughout the Holocene period. This paper compares various models proposed over the past 40 years to account for the development of buttongrass moorland across southwest Tasmania since the Last Glacial period. How well, and on what time scales, the directions and magnitude of changes predicted by the various models are supported by the palaeoecological evidence is important to understanding the climate-vegetation-fire linkages in present-day buttongrass moorland. How useful these models will be in predicting trends under warmer climates is less clear.

Southwest Tasmania shares with, for example, the Lakes District of Britain, the curious distinction of being a cultural landscape that is widely perceived as natural wilderness. In both instances, burning by humans is seen as responsible for (and one of the key tools to manage) the mosaic of herbaceous and woody communities that dominate these landscapes – grass and sedge-dominated moorland in the Lakes District and buttongrass moorland in southwest Tasmania. The eponymous, often dominant species of the latter vegetation type, the scleromorphic sedge *Gymnoschoenus sphaerocephalus*, occurs in perennially damp sites over a range of elevations in mainland southeast Australia and Tasmania. Jarman et al. (1988) split buttongrass moorland in Tasmania into two groups based on floristics and physiographic characteristics – Eastern Moor, which is confined to poorly drained terrain in northern, central and eastern Tasmania, and Blanket Moor, which is confined to the western half of the state where high rainfall has allowed *Gymnoschoenus sphaerocephalus* to expand beyond the mire niche to cover much of a landscape where rainforest dominance might be expected (see Jackson 1968, 1999). Both types of moorland are dominated by sedges (Cyperaceae) and wire-sedges (Restionaceae), but typically also contain a large proportion of scleromorphic shrubs (Jarman et al. 1988).

Brown (1999) has described buttongrass moorland (blanket moor *sensu* Jarman et al. 1988) as an oligotrophic 'system' which extends over steep-sided hill slopes and ridges, as well as broad, flat valleys, from sea level to about 1050 m across much of western and southwestern Tasmania. These moorlands are developed on ±shallow and highly acidic peat substrates, which overlie typically infertile pre-Carboniferous bedrock or regolith derived from this bedrock. The floristic composition, structure and present-day/short-term dynamics of buttongrass moorland under historical conditions and fire regimes are relatively well researched (references in Brown 1999; Corbett and Balmer 2008; King 2008). In contrast, much less is known about its prehistoric past, including the culturally important question of the origin(s), dynamics, persistence and geographic extent of buttongrass moorland under earlier climatic 'regimes' in the Late Pleistocene (>10 ka BP) and Holocene (10 ka BP to present) in western and southwestern Tasmania.

Reasons for this include a palaeoecological focus on tree- and shrub-dominated communities – in particular, temperate rainforest and highland communities (see Macphail 1979; Markgraf et al. 1986; Colhoun 1996, 2000), as well as an apparent paucity of thick organic deposits within buttongrass moorland that are likely to have escaped periodic truncation by wildfires. Many of the sites pollen-analysed over the past 40 years are either close to or above the present-day upper altitudinal limit of *Gymnoschoenus sphaerocephalus*, or for those at lower elevations, surrounded by forest, e.g. Brown Marsh on the Central Plateau (Macphail 1979), Lake Johnson on Mount Read in northwestern Tasmania (Anker et al. 2001) and the Upper Ringarooma River valley in northeastern Tasmania (Dodson 2001). *Gymnoschoenus* pollen counts usually are excluded from published *summary* pollen diagrams, e.g. for Beattie's Tarn, Lake Dobson and the Tarn Shelf on Mount Field in central southern Tasmania (cf. Macphail 1975a, 1975b, 1979, 1986), or the pollen morphotype is not present or differentiated from other Cyperaceae, e.g. Pulbeena, Mowbray and Broadmeadow Swamps, and Tullabardine Dam in northwestern Tasmania (Colhoun et al. 1982; Colhoun and van de Geer 1986; van de Geer et al. 1986), Newall Creek, King River Valley and Poet's Hill Lake in western Tasmania (van de Geer et al. 1986, 1989, 1991; Colhoun 1992), and Smelter Creek and Governor Bog in southwestern Tasmania (Colhoun et al. 1991, 1992). At sites below about 300 m elevation, *Gymnoschoenus* may have been conflated with *Baumea*, which produces morphologically very similar pollen. Possible examples occur in the Lower Gordon River Valley (Harle et al. 1999) and the Louisa Plains/Melaleuca Inlet area of Bathurst Harbour (Thomas 1995; Macphail et al. 1999).

Only eight sites are known to preserve significant (>2%) relative abundances of *Gymnoschoenus*-type pollen: the three Mount Field sites, Ooze Lake in far southern Tasmania (Macphail 1975a, 1986; Macphail and Colhoun 1985; Colhoun 1996), the 'Tarn Shelf' site and Lake Selina in the Tyndall Ranges area of western Tasmania (Colhoun et al. 1999; M.K. Macphail unpubl. data), the Melaleuca Inlet site (Thomas 1995) and a pond near Lake

Pedder in central southwest Tasmania (Fletcher and Thomas 2007a). The last site is unique at present in that *Gymnoschoenus*-type pollen is more or less frequent throughout the Holocene, assuming continuous deposition at the core site.

This paper discusses fossil pollen evidence for buttongrass moorland in western Tasmania before European settlement, in particular the claim by Fletcher and Thomas (2007a) that the Lake Pedder Pond sequence is evidence that buttongrass moorland has been continuously present in the lowlands of southwest Tasmania since the Last Glacial period. Fossil pollen data from other sites in western Tasmania, including sites from within buttongrass moorland, have been reviewed by Colhoun (1996). The names given to the various ecological models are those used by their authors or informal names coined for the purpose of this paper.

Claims and counter claims

Ecological and pollen analytical studies carried out over the past 40 years have resulted in five models that implicitly or explicitly have (or can be) used to 'explain' the presence of about one million hectares of buttongrass moorland in high rainfall regions capable of supporting temperate rainforest or (subalpine zone) rainforest scrub. All models assume that buttongrass moorland and other sclerophyll-dominated associations in western Tasmania are 'fire-promoting and fire-requiring' vegetation types, resulting from repeated firing of the vegetation by the Tasmanian Aborigines in the prehistoric past. All are based to a lesser or greater degree on the seminal 'ecological-drift' model published by late Professor W.D. (Bill) Jackson (1968, revised 1999) to account for complex relationships between fire-sensitive rainforest and fire-tolerant sclerophyll associations observed in high-rainfall areas of western Tasmania (Figure 1).Given the long history of logging old-growth forests and concomitant debate on the regeneration of commercial eucalypt forests, it is not surprising that earlier research such as that by Jackson (1968), Mount (1979) and Macphail (1980) focused on forest management practices rather than the history of buttongrass moorland *per se.*

Figure 1. Professor Bill Jackson (1975) on the summit of Mount Hazleton, looking across lowland buttongrass moorland below the Norfolk Range, western Tasmania. Photograph by M.K. Macphail

The 'ecologic drift' hypothesis of Jackson (1968)

Jackson (1968) has proposed that buttongrass moorland is a fire-disclimax plant formation that has been able to expand beyond its natural (mire) limits on to steeper terrain, due to a combination of uniformly wet climates and frequent wildfires in the (unspecified) prehistoric past. Critical assumptions are that: (1) ignition sources primarily were Aborigines living on the coast or migrating into the interior along defined trails, not lightning strikes; (2) the arrival of fires in any (inland) locality occurs randomly in time; (3) the vegetation follows a relatively clear successional pattern from sedgeland to *Nothofagus* temperate rainforest via intermediate sclerophyll scrub and wet sclerophyll forest seres; and (4) the *probability* that the vegetation occupying a site will burn is related to its composition. In general, the higher the sclerophyll component, the greater the fire-risk status of the community – i.e. the probability of fire occurring is greater in the earlier, sclerophyll-dominated communities, than in subsequent seral rainforest communities.

The model predicts that as the (chance) interval between successive fires increases, the more likely it is that fire-sensitive vegetation types (chiefly mixed forest and temperate rainforest) will succeed sclerophyll-dominated vegetation types. The reverse situation, characterised by high fire frequencies and an associated loss of soil nutrients, leads to highly inflammable communities, in particular buttongrass moorland. Seral communities are predicted to 'drift' between the two end states (*Nothofagus* temperate rainforest, *Gymnoschoenus* buttongrass moorland) according to the small chance occurrences of very long or very short intervals between fires.

The 'stable fire cycle' model of Mount (1979)

Mount (1979) used a paper discussing short-term processes controlling regeneration in present-day Tasmanian forests (fire, drought, wind-throw, flood and faunal predation) to challenge the 'ecologic drift' model, in particular Jackson's (1968) arrangement of observations made at different locations into the one temporal sequence. Instead, fire was seen as the most important of a spectrum of 'barrier-removing' agencies that prevent forest regeneration. Key assumptions differ from those of Jackson (1968) in that: (1) natural ignition sources are believed to have maintained a constant presence of fire in Tasmania, including in high rainfall areas remote from habitation sites; (2) the probability of a fire occurring depends on the total fuel load, not just the sclerophyll component as proposed by Jackson (1968); and (3) all vegetation types are intrinsically flammable but burning will not occur until sufficient fuel loads have accumulated. The low fuel period (LFP) during which a community cannot catch fire was estimated to vary from about six months in *Gymnoschoenus* communities to about 80 years in *Nothofagus* temperate rainforest. Mount's (1979) observation that burning lichens can cause 'spot' fires in rainforest is supported by a subsequent study of the impact of fire in a drought-impacted rainforest (Hill 1982).

Because frequent fires in communities with short LFPs will only occasionally ignite fires in adjacent vegetation types with long LFPs, boundaries between the major vegetation types are predicted to remain geographically stable over time. Accordingly, Mount (1979) proposed that the mosaic of forest and non-forest communities in western Tasmania was the result of stable fire cycles, rather than the individual associations representing different seral stages towards and away from two ±stable end-member formations (temperate rainforest and buttongrass moorland).

The 'climate reinforcement' model of Macphail (1980)

Macphail (1975a, 1979, 1980, 1986) proposed that postglacial climate change, not fire or humans, is the *primary* determinant of *long-term* trends in the (forest) vegetation, although this control may have been indirect via long-term 'feed-back' relationships between climate and soil fertility, and climate and fire frequency, as proposed by Jackson (1968).

This model is based on fossil pollen sequences preserved in 10 glacial tarns and peat swamps aligned along altitudinal and east-to-west gradients across southern Tasmania and used to reconstruct the history of the regional vegetation and climates in southern Tasmania during the Postglacial period, c. 13 ka BP to present. Despite major variation at the local (individual site) scale, the study indicated that the postglacial forest vegetation in southern Tasmania could be divided into *early* and *late temperate* developmental phases, preceded by a *pre-temperate* phase during which the vegetation at lower elevations was subalpine to alpine in character. Radiocarbon dates show the *early temperate* phase began about 10 ka BP and ended between 6 and 7 ka BP (late-early to middle Holocene). During this interval, *Nothofagus* temperate rainforest associations expanded out of Late Glacial refugia in western Tasmania and on mountains in the centre and south, while *Eucalyptus* sclerophyll forests expanded on the Central Plateau and in eastern Tasmania (cf. Kirkpatrick and Fowler 1998). The *late temperate* phase began about 6 ka BP (middle Holocene). During this interval, many forest and non-forest communities across a range of elevations became more open in structure.

Central to Macphail's (1980) model is the hypothesis that the overall impact of wildfires has varied during the Postglacial period depending on whether the ecological effects of climate change have curtailed (opposed) or reinforced (complemented) the fire-dictated 'drift' away from rainforest. Thus, increasing warm and wet conditions during the early Holocene are seen as having allowed temperate rainforest communities to expand at higher elevations irrespective of (anthropogenic) fire pressure, while increasingly variable (cooler, more drought-prone) conditions during the middle and late Holocene resulted in the regional expansion of sclerophyll vegetation types. This climate-forced reversion (*revertance*) to more open vegetation types during the late Holocene is diachronous in that the trend occurred earlier and is most apparent in the fossil pollen record in central and southeast Tasmania, less so in the perhumid west and far south. For example, fossil pollen data show that temperate rainforest communities have dominated the mid slopes of Frenchmans Cap (Lake Vera glacial trough valley) throughout the Holocene, while infilled glacial tarns at higher elevations on the Denison (Upper Wurawina cirque) and Tyndall Ranges (Tarn Shelf) have supported rainforest scrub over much the same period (Macphail 1986).

At the time of publication, the only lowland site for which fossil pollen data were available were rain-shadowed sites that were (and are) too dry to support *Gymnoschoenus*, viz. Lake Tiberias in the Midlands and thin organic deposits in coastal southeast Tasmania. Nevertheless, if the 'climate reinforcement' concept is extrapolated to the western lowlands, the model predicts that buttongrass moorland will have (1) been least extensive during the early Holocene, (2) begun expanding during the middle Holocene and (3) attained its present-day distribution during the late Holocene as the direct and indirect impacts of Aboriginal fires were reinforced by a higher frequency of short-term 'drought events'.

The 'Iversen glacial-interglacial cycle' model of Colhoun (1996)

Colhoun (1996) accepted that the change from glacial to interglacial climates directed vegetation successions in western Tasmania, but challenged Macphail's (1979, 1980) thesis that climatic change *per se* underlies the late-Holocene *revertance*. Instead, the non-synchronous nature of plant succession recorded at many sites is suggested to be conclusive evidence that biological and physical variables other than climate change were more important. The former (biological factors) include different rates of migration away from Last Glacial refugia; the latter (physical factors) include edaphic phenomena, such as the impact of anthropogenic fires on soil depth, drainage and fertility.

The theoretical basis for these conclusions is the 'glacial-interglacial' model of Iversen (1958). This model, which is widely used in the interpretation of European, British and eastern North American pollen sequences, subdivides glacial-interglacial vegetation successions into four stages, each of which is associated with different stages in the soil maturation-degradation cycle. These are: (1) the Cryocratic Stage, during which frost disturbance prevents herbaceous plant associations (tundra) from completely covering the ground surface; (2) the Protocratic Stage, during which tundra is replaced by grasslands and woodlands growing on unleached calcareous soils; (3) the Mesocratic Stage, during which 'climax' forests occupy sites with deep, slightly acid but relatively fertile soils; and (4) the Telocratic (or Oligocratic) Stage, during which the climax forests revert to lower, more open vegetation types as soils become less fertile, and more poorly drained and acidic due to leaching under colder (and wetter?) climates. In western Tasmania, the Cryocratic and Protocratic Stages encompass the replacement of fell-field and herbfield communities by subalpine-alpine heath and scrub during the latest Pleistocene; the Mesocratic Stage is seen as culminating in the expansion of *Nothofagus* cool temperate rainforest 'throughout' (Colhoun 1996:565) the lowlands sometime before the middle Holocene, while the Telocratic/Oligocratic Stage encompasses the replacement of floristically simple/closed forest types by floristically more complex/open shrub and herbaceous communities, including heath/moorland, during the late Holocene.

The sequence of communities corresponds closely to those observed by Macphail (1979, 1980), but the Holocene reversion towards more lower, open vegetation types is attributed to cultural and environmental factors other than climate, in particular increased burning by Aborigines, soil impoverishment and local sedimentation effects. Colhoun (1996:578) notes that 'the development of [rainforest] vegetation must locally have been retarded by variations in nutrient status of soil types and drainage of the landscape as influenced by topography'. However, he does not argue that the late Holocene forest 'retreat' reflects the inability of tree species to colonise poorly drained/highly acidic blanket peats once these have developed, even though this is seen as a major factor forcing revertance in the Iversen (1958) model (cf. Figures 2 and 3 in Colhoun 1996).

The 'Last Glacial inheritance' hypothesis of Fletcher and Thomas (2007a)

Fletcher and Thomas (2007a) have used fossil pollen preserved in a pond at 312 m elevation near Lake Pedder (hereafter LPP site), supported by published palynological data from other sites within moorland, to challenge two of Colhoun's (1996) key predictions – that rainforest dominated the landscape of western Tasmania during the mid Holocene, and nutrient-leaching, not climate change, was responsible for the decline in rainforest during the late Holocene. Instead, they argue that (1) fire-promoted buttongrass moorland has dominated the landscape since the Last Glacial period, if not earlier, and (2) the mosaic of vegetation types in western Tasmania has remained relatively stable in extent throughout the Holocene whether or not cooler/drier conditions have resulted in increasingly open forest types and an expansion of subalpine communities on adjacent mountains over the past 6000 to 7000 years.

Like Macphail's (1980) hypothesis, the model is underpinned by a detailed analysis of the modern pollen rain and, with some reservations (see Fletcher and Thomas 2007a:670), assumes that the LPP site preserves a mostly continuous record of the local and non-local to regional vegetation during the Holocene. Key observations are: (1) *Gymnoschoenus* has been continuously present since at least 10.3 ka BP; (2) objective comparison of the fossil and modern pollen data (and high concentrations of carbonised particles) confirms that buttongrass moorland has surrounded the site over this period; (3) relatively low pollen percentages for forest taxa show that at no time has rainforest or any other forest type been established around the site or dominated the regional lowland vegetation.

Fletcher and Thomas (2007a:674) concur with previous workers (references in Brown 1999) that the arrival of Aborigines in southwest Tasmania 'provided an ignition source to a perennially wet landscape in which natural ignition sources are rare', and therefore conclude that the absence of forests in inland southwest Tasmania is due to the continuing influence of (Aboriginal) people in the interior of southwest Tasmania. Accordingly, the model challenges archaeological data that imply the spread of rainforest communities during the Holocene was responsible for displacing Aboriginal occupation sites on to the coast (cf. Kiernan et al. 1983; Cosgrove et al. 1990; Thomas 1993; Cosgrove 1999).

Testing the models

All the models discussed above make predictions about the origin and/or geographic extent of rainforest vs. buttongrass moorland that potentially can be tested using tightly AMS-dated palynological evidence (including carbonised particle evidence).

One such test was undertaken by Macphail (1980), who used fossil pollen evidence to compare the 'ecological drift' and 'stable fire cycle' models of Jackson (1968) and Mount (1979), respectively. Macphail (1980) proposed that the critical difference between these two models is that the 'stable fire cycle' model predicts that the boundaries between the major communities in western Tasmania would be 'fixed' in space during the Holocene, whereas the 'ecological drift' model predicts the position of the vegetation boundaries would vary over the same period. The 1980 comparison indicated that the regeneration processes identified by Mount (1979) are important in forest regeneration in high rainfall areas, but each operates on a relatively small scale at any point in time. Conversely, the long-term regional successions recorded by fossil pollen match trends predicted by Jackson (1968). Nonetheless, Macphail (1980) recognised that the very low probability that an area of buttongrass moorland would escape burning for a sufficiently long period for forest species to be established (see Figure 3 in Jackson 1968) gave buttongrass moorland the long-term stability envisaged by Mount (1979) and the fire-disclimax status proposed by Jackson (1968).

In contrast, the 'Last Glacial inheritance' model of Fletcher and Thomas (2007a) explicitly uses fossil pollen and charcoal evidence to infer the regional extent as well as the origins of buttongrass moorland in lowland southwest Tasmania since the Last Glacial. For this reason, a number of questions needs answering before their conclusions, which are based primarily on one site, can be extrapolated to the southwest in general. These include consideration of the strengths and weakness of fossil pollen data in general, as well as specific questions regarding the LPP site. Examples are: (1) Does the LPP site preserve a ±continuous record of the Holocene vegetation? (2) Is the presence of *Gymnoschoenus* pollen essential evidence that buttongrass moorland formed the regional lowland vegetation surrounding the LPP site during the Holocene? (3) What is the evidence for buttongrass moorland in southwest Tasmania during the late Pleistocene? (4) Are there alternative explanations for the low relative abundance of forest tree pollen in the Lake Pedder pond sequence? A related but more general question is whether fossil pollen data can be used to distinguish between climate-forced *revertance* and reversion forced by other cultural and environmental factors, such as the impact of indigenous fire on rates of accumulation of organic matter (peat), and the loss of soil fertility due to the combined effects of burning and leaching in high rainfall areas.

Without answers to these questions, it is premature to use fossil pollen and carbonised particle data to infer buttongrass moorland has dominated the southwest lowlands during the present glacial-interglacial cycle. With answers, it may be possible to use the fossil pollen and charcoal evidence from a network of strategically located sites to predict the longer-term consequences of management strategies centred on planned fires. As elsewhere, the precision with which the palynological data can be used to reconstruct past climate-vegetation-fire linkages is limited by factors intrinsic to the technique, as well as the characteristics of the individual sites.

Fossil pollen evidence – constraints and opportunities

Fortunately, as regards the impact of Aboriginal burning practices, southwest Tasmania is a special case in Australia in having floristically simple major vegetation types, which are dominated often by species that produce morphologically distinctive pollen or spores (cf. Macphail 1979; Colhoun 1996; Tables 13.2, 7.1, 10.4 in Brown 1999; Jarman et al. 1999; and Wells and Hickey 1999).

Nevertheless, interpretation of all fossil sequences is constrained by the following factors: (1) Different plant taxa vary greatly in the amount of pollen or spores (and charcoal) they produce and the distance these are dispersed into the surrounding landscape. (2) Most pollen types can be identified only to family or genus, except under unusually favourable circumstances. (3) Some important canopy tree species can survive as shrubs and produce pollen under unfavourable conditions. For these reasons, all fossil pollen assemblages are a partial record of past floras, with some indication of the relative abundance of the parent plants, and the plant community structure has to be deduced by analogy – e.g. when a fossil flora appears to match the composition of an extant community. Using pollen and spores to reconstruct past environments involves additional assumptions (see Birks and Birks 1980). In many instances, conclusions are based on only a fraction of the fossil data, in some cases the presence of one or two 'indicator' species only. Charcoal preserves well in waterlogged sediments but, unlike pollen, is produced in large amounts only at irregular intervals and also is subject to breakdown from larger (macrocharcoal) to smaller-sized (microcharcoal) particles during release, transport and deposition(Figure 2). Accordingly, the relative abundance of charcoal particles provides evidence of local fire activity, but trends are difficult to interpret in regional terms unless calibrated by modern taphonomic studies complemented by standardised charcoal data from a network of sites (see Patterson et al. 1987; Whitlock and Larsen 2001; Conedara et al. 2009).

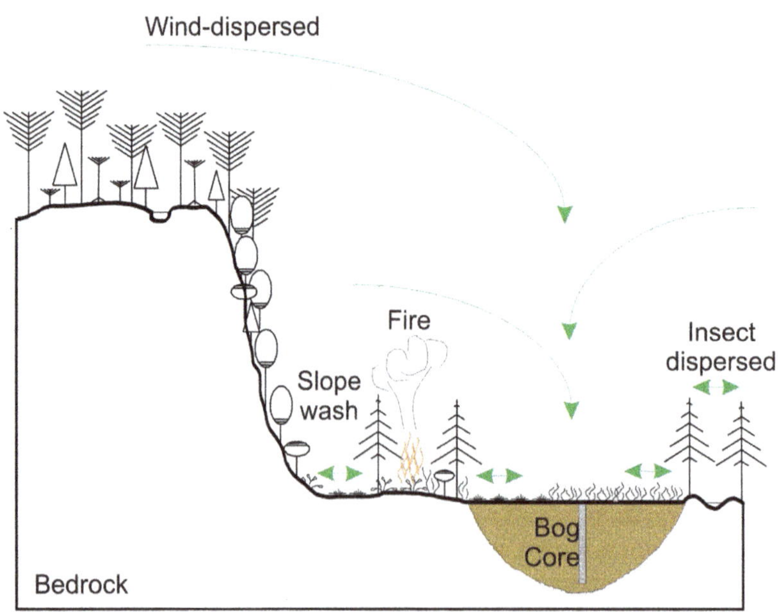

Figure 2. Transport and deposition of pollen and charcoal particles (after Condera et al. 2009)

Does the LPP site pollen sequence represent all or only part of the Holocene?

Macphail and Hope (1985) and more recently Macphail et al. (1999) have demonstrated that many mire sequences include one or more disconformity surfaces that represent all or part of mid-Holocene time. The missing section may be due to wildfires burning into the organic deposits, or simply due to local conditions during this period not supporting the longer-term accumulation of organic matter.

Fletcher and Thomas (2007a:670) note that high concentrations of carbonised particles indicate that fires were a significant feature of the (local) environment at the LPP site throughout the Holocene, and a hiatus resulting from burning of a 'dry pond surface' after about 6.5 ka BP is 'possible'. This can only be tested by additional radiocarbon dates. If the 'spike' in *Lagarostrobos* pollen at 30 cm in the LPP core correlates with similar spikes in the relative abundance of this well-dispersed pollen type elsewhere in southern Tasmania, then this sample accumulated sometime between about 4.5 and 3.5 ka BP (see Figures 13-16 and 19 in Macphail 1979). Underlying sediments at c. 37 cm are dated to about 6.7 ka BP and the combined data are circumstantial evidence that all or part of the middle Holocene is missing in the LPP core. For these reasons, Fletcher and Thomas's (2007a) conclusions regarding the vegetation developments during the late middle Holocene are considered to be premature.

Is the presence of Gymnoschoenus *pollen essential evidence that buttongrass moorland formed the regional lowland vegetation surrounding the LPP site during the Holocene?*

Given the prominence of *Gymnoschoenus* in present-day Blanket Moorland, determining whether this vegetation type formed the regional vegetation in lowland southwest Tasmania would appear to depend on whether *Gymnoschoenus* pollen can be distinguished from other Cyperaceae pollen at the LPP site, and whether these records represent regional as well as local buttongrass moorland.

Many of the shrubs and herbs found in buttongrass moorland cannot be distinguished to species level using fossil pollen, and at the generic level occur in other vegetation types. For these reasons, fossil *Gymnoschoenus* pollen are assumed to be diagnostic of buttongrass moorland. Modern reference material indicates *Gymnoschoenus* pollen can easily be confused with pollen of the unrelated sedge genus *Baumea*, three species of which occur in buttongrass moorland (Table 13.3 in Brown 1999), although none appear to range above elevations of about 500 m (Curtis and Morris 1994). *B. tetragona* is recorded in buttongrass moorland at McPartlan Pass (320-330 m elevation) at the northern end of Lake Pedder (G.J. Jordan pers comm.), although it is uncertain whether this or other *Baumea* species also occur at the LPP site (312 m elevation). The same caveat applies to Macphail et al.'s (1999) study of peat mounds in buttongrass moorland near sea level at Port Davey in southern Tasmania. Conversely, *Gymnoschoenus*-type pollen preserved in mires located above about 500 m elevation is likely to represent *Gymnoschoenus sphaerocephalus*.

Pollen samples from pollen traps, moss polsters and surface muds in western Tasmania show that *Gymnoschoenus sphaerocephalus* is poorly represented by pollen and, not surprisingly, *Gymnoschoenus* pollen was absent in a number of modern samples collected in buttongrass moorland (Table 1 and Figure 3 in Fletcher and Thomas 2007a). Nevertheless, the pollen type is found in trace amounts in surface samples and pollen traps from the lowlands (below 160 m elevation) into the alpine zone (above 1200 m elevation) on Mount Field, as well as in pollen trap and surface samples from the Henty River Plains below the Tyndall Ranges, Yeates Lagoon on the Western Tiers, and Pine Lake and Lake Augusta on the eastern side of the Central Plateau (Macphail 1975a). The maximum value recorded by Macphail (1975a) was 1%, in a mud sample collected in the upper subalpine zone on Mount Field. Since buttongrass moorland occurs in the vicinity of or (as in the alpine zone on Mount Field) below the trap or sample sites, the empirical evidence supports Fletcher and Thomas's (2007a) claim that even low relative abundances of fossil *Gymnoschoenus* pollen are, in fact, reliable evidence for buttongrass moorland in the vicinity of the core site.

Whether 'common' occurrences of fossil *Gymnoschoenus* pollen at the LPP site represent more than locally growing hummocks is less clear. Fletcher and Thomas (2007a) therefore have relied on objective correlation ('finger-printing') techniques to show that fossil pollen assemblages preserved at the LPP site fall within the buttongrass moorland (rather than the alternative rainforest and alpine groups), and cite high charcoal particle values as supporting evidence. In a subsequent paper, they demonstrate that the same objective technique applied to the uppermost samples from Poets Hill Lake, Governor Bog, Smelter Creek and Lake Selina successfully predicted the presence of moorland at these four sites despite the (apparent) absence or non-recognition of *Gymnoschoenus* pollen (Fletcher and Thomas 2007b). The logical corollary is that the presence of *Gymnoschoenus* is helpful but not essential to the recognition of buttongrass moorland in the fossil pollen record, although it should be noted that the objective criterion distinguishing buttongrass moorland from other regional vegetation types is essentially negative viz. the paucity of rainforest taxa compared with values recorded at high elevation sites (see below).

Was Gymnoschoenus-dominated moorland present in southwest Tasmania during the late Glacial?

The oldest known macrofossil evidence of *Gymnoschoenus sphaerocephalus* in southern Tasmania is remains dated at about 3.5 ka BP in the Broad River Valley, Mount Field (Caine 1967). Relative abundances of *Gymnoschoenus*-type pollen (7%) at the base of the LLP core are strong (albeit circumstantial) evidence that buttongrass moorland was present in the region during the Last Glacial (see Figure 6 in Fletcher and Thomas 2007a), and there is little doubt that *Gymnoschoenus*, if not buttongrass moorland, was a persistent element in southern Tasmania, based on trace numbers of its fossil pollen in all sites analysed by M.K. Macphail except for Lake Tiberias (Midlands) and Brown Marsh (eastern Central Plateau) (Macphail 1975a, 1979, 1984, 1986; Macphail and Jackson 1978; Macphail and Colhoun 1985).

Whether the LPP and other palynostratigraphic data support Fletcher and Thomas's (2007a) suggestion that buttongrass moorland extended beyond the mire niche to dominate lowland southwest Tasmania during the late Glacial is debatable for palaeoclimatic reasons. For example, much of southeast Australia and Tasmania was subject to 'cold/arid' conditions during the Last Glacial Maximum (cf. Macphail 1975b; Hope 1994; Colhoun 2000). Nonetheless, at least three upper subalpine sites on Mount Field (Beatties Tarn, Lake Dobson, Tarn Shelf) and two sites in far southern Tasmania (Adamsons Peak, Ooze Lake) preserve frequent (2-5%) numbers of *Gymnoschoenus*-type pollen in radiocarbon-dated sections that are substantially older than 9 ka BP (Macphail 1975a, unpubl. records). In most instances, the sections are dated to about 11.5 ka BP, with the Ooze Lake sequence extending the record of *Gymnoschoenus* back to 14-17 ka BP (cf. Macphail and Colhoun 1985; Colhoun 1996).

These and other high-elevation records are distinctly anomalous compared with the Holocene records, since the pollen types appear to be poorly dispersed (see Fletcher and Thomas 2007b) and the sites were (and still are) at or above the modern upper altitudinal limits of *Gymnoschoenus*. Possible explanations are (1) postglacial warming has caused the extinction of an 'alpine' ecotype, or, preferred, (2) the pollen type has been transported upslope by wind from populations growing in mires at lower elevations. If the latter is correct, then the fact that *Gymnoschoenus* pollen is only sporadically recorded in Holocene assemblages at higher elevations is likely to be due to two related factors. The first is the marked increase in the local pollen influx at about the Pleistocene-Holocene boundary (due to colonisation of the upper subalpine-alpine zone by woody taxa). The second, which is supported by modern pollen trap and surface sample data, is the saturation of the upper subalpine-alpine assemblages by pollen and spores transported upslope from plants that produce/disperse pollen or spores in very large numbers (well-represented taxa) in the montane and lowland forests during the Holocene.

Is the low representation of forest during the Holocene at the LLP site apparent or real?

The relationship between proximity/abundance of a vegetation type and its representation in the fossil pollen record is complex (Figure 2) and depends not only on the pollen production/dispersal characteristics of the dominant taxa, but also on the depositional environment and the way the fossil pollen data are presented.

For example, except for sequences with closely spaced radiocarbon dates (which allow numbers of pollen grains to be estimated in absolute terms: see Birks and Birks 1980), fossil pollen counts are expressed as percentages of a pollen sum comprising all or part of the total pollen counts for that sample (relative pollen counts). These percentage data are subject to a strong inverse correlation effect in that an increase in the relative abundance of one pollen type artefactually decreases the relative value of all other pollen types included in the pollen sum. Accordingly, it can be difficult to determine whether low to moderate values are evidence that the parent plants were present but uncommon at/around a site, or were common but growing at a distance from that site. The problem is aggravated by including pollen of under-represented taxa to make up statistically significant (>250 count) pollen sums. For this reason, wetland taxa are usually excluded from the pollen sum. If pollen counts are sufficiently high (>>250 counts), relative abundance values can be calculated using pollen sums that comprise only well-represented taxa (e.g. Macphail 1975a, 1979; van de Geer et al. 1991; and Colhoun 1996.)

Exclusion of under-represented taxa from the pollen sum minimises the impact that variation in local abundance of the parent plants has on the relative-abundance data for other plants in the pollen source area, especially well-represented taxa growing at some distance from the site (extra-local and regional vegetation). However, in all pond and mire sites, it is difficult, if not impossible, to compensate for effects arising from changes in the pollen deposition regime due to colonisation of the mire surface by plants. This phenomenon, which is analogous to the Neves Effect in large lakes (see Traverse 1988), means that pollen assemblages deposited on the surface of mires (or close to the shoreline of small ponds) usually are dominated by pollen sourced from the local vegetation (high local pollen influx), while assemblages accumulating on unvegetated mire surfaces or in open water in larger lakes will include a higher proportion of pollen and spores derived from well-represented taxa in the extra-local and regional vegetation (non-local pollen influx).

Based on these criteria, pollen assemblages accumulating in the LPP site will have been dominated by locally sourced pollen, except possibly during the earliest Holocene, when *Isoetes* (presumed to be *I. gunnii*), *Myriophyllum* and *Astelia* indicate the site was an alpine-subalpine bog with pools of free-standing water. Subsequently, pollen and spores of these obligate aquatic herbs disappear from the pollen record, and it seems probable that local water tables were at or below the sediment surface, allowing Restionaceae and probably *Gymnoschoenus* to colonise organic soils accumulating within the depression. Whether sclerophyll shrubs (*Bauera*, Epacridaceae, Myrtaceae and? Asteraceae) were able to colonise the mire surface is less clear. If this was the case, then it is plausible that some (much?) of the pollen dispersed from distant stands of *Nothofagus* cool, temperate rainforest and other forest types was filtered out by locally growing shrubs or strongly diluted by the pollen shed by locally growing plants.

If correct, then it follows that the regional abundance of trees such as *Nothofagus cunninghamii* has been underestimated for the LPP site. Because the pollen sum adopted by Fletcher and Thomas (2007a) includes sclerophyll shrubs and herbs, it is difficult to compare the relative-abundance values with those recorded in highland sites analysed by M.K. Macphail and E.A. Colhoun (see Colhoun 1996). Given palaeobotanical evidence that *Nothofagus* populations expanded beyond their present range in the lower Derwent Valley and eastern

Midlands in southeast Tasmania during the early Holocene (Colhoun and Moon 1984; Harle et al. 1993), it would be surprising if forest did not extend on to lowland areas in southwest Tasmania about the same time.

Can fossil pollen data be used to distinguish between climate- and non-climate forcing of late Holocene revertance/reversion in southwest Tasmania?

Palaeobotanical evidence has been widely used to infer climate change in western Tasmania, but attempts to use fossil pollen evidence to separate out the effects of fire pressure and soil deterioration from climate (cf. Colhoun 1996) seem to be futile except at sites where independent evidence allows one or both variables to be excluded. Such ideal sites may occur within rainforest stands (see Macphail 1984; Dodson et al. 1998), but this is improbable in buttongrass moorland. Reasons include: (1) it is highly likely that mean and seasonal rainfall and air temperatures greatly exceeded the minimum requirements for buttongrass (and most forest species) throughout the late Holocene; (2) general agreement exists that the regeneration of *Gymnoschoenus* and other moorland species depends on frequent burning; and (3) modern observations indicate that, at the community level, rainforest and buttongrass are not constrained by regolith geology, although floristic diversity and growth rates (and presumed competitive abilities) of individual taxa will vary with different substrates. For example, *Nothofagus* temperate rainforests are able to thrive on Precambrian quartzites and other highly infertile rock types as well as on more fertile lithologies and organic substrates (see Read 1999). Nutrient cycling in buttongrass moorland has not been studied in detail. However, the efficiency of many moorland plants in recycling phosphorus and potassium from moribund foliage (Bowman et al. 1986) is circumstantial evidence that buttongrass moorland also may be ±independent of edaphic constraints in high rainfall regions. One probable reason is the continuous input of 'cyclic salts' from rainwater, as was proposed for rainforest communities by W.D. Jackson in the early 1970s (Jackson 1977; Brown 1999).

Geographically stable boundaries – a critical test for inferring past fire frequency?

Local-scale evidence continues to support Jackson's (1968) thesis that buttongrass moorland is an intrinsically unstable ecosystem whose boundaries with other communities has and will continue to shift due to changes in fire pressure (whether climate-forced and/or cultural). For example, *Phyllocladus* and other rainforest species were observed to colonise lignotuber 'rafts' around (now submerged) Lake Edgar near Lake Pedder (Macphail and Shepherd 1973) and moribund *Gymnoschoenus* hummocks occur within wet scrub near Mount Anne and elsewhere in southwest Tasmania (G.J. Jordan and M.K. Macphail pers obs.). Moreover, a comparison of the Colhoun et al.'s (1999) Lake Selina sequence (where *Gymnoschoenus* pollen appears relatively late in the poorly dated postglacial section) and the LPP site (where *Gymnoschoenus* pollen is frequent throughout the Holocene) imply the development of buttongrass moorland may have been as diachronous as has been observed for other vegetation types in western Tasmania.

Nevertheless, conclusions reached by Fletcher and Thomas (2007a) regarding buttongrass moorland in southwest Tasmania are remarkably similar to Mount's (1979) prediction that community boundaries are 'fixed in space' due to stable fire cycles in time – the more so if Mount's (1979) pervasive 'natural' ignition sources are replaced by 'cultural' ignition sources. As recognised by Fletcher and Thomas (2007a), their evidence also challenges Kiernan et al.'s (1983) conclusion that the early Holocene expansion of rainforest was responsible for the retreat of Aboriginal communities on to the west coast. Even though there is no archaeological evidence that Aboriginal communities were living away from the coast in southwest Tasmania during the Holocene, such evidence may yet be found. For example, Aboriginal flake tools have been found at Lake Pedder (J. Marsden-Smedley pers comm.) and Macphail et al.

(1975:100) recorded a stone scraper buried under thick sclerophyll scrub growing on Tertiary gravels inland of Lagoon River near Sandy Cape on the West Coast. Hence it might be timely to re-review 19th century journals kept by geologists and prospectors (and contemporary issues of the *Papers and Proceedings of the Royal Society of Tasmania*) for documentary evidence of postglacial occupation sites in inland southwest Tasmania (cf. Jordan and Hill 2002).

The geographic extent of buttongrass moorland and the stability of its boundaries over time potentially can be tested by a combination of fossil pollen, carbonised particle and radiocarbon data from transects of sites located within, and across, the ecotones between forest and buttongrass moorland. Small hollows in outwash fans on the margins of the Arthur, Lake Pedder and Loddon Plains and at Cradle Mountain may be suitable sites (cf. Macphail 1984; Dodson et al. 1998). Whether management concerns justify research across the range of communities making up buttongrass moorland in southwest Tasmania is unknown, but modelling by King (2008) indicates that if boundaries are found to be geographically stable through time at a number of sites located away from Aboriginal trails, this would indicate at least 10% of the buttongrass moorland in southwest Tasmania had been burned on an annual basis. Ecological drift between the major vegetation types would indicate this 'fine mosaic' burning occurred on a smaller geographic scale or over a longer timeframe. Not surprisingly, it should be possible to use the same data to measure the impact of fires on the rate of accumulation of organic soils under buttongrass moorland (see Bowman 2008) and hence the potential value of buttongrass moorland for carbon capture and storage (cf. Hope et al. 2009). Insect remains potentially provide an alternative means of reconstructing past changes within buttongrass moorland (see Driessen 2008; Green 2008; Porch 2008), while the very small amount of organic matter now required to obtain reliable AMS dates means that independent age control can be obtained from the microfossil assemblages, even in sections that have been turbated by crayfish (*Ombrastacoides*, *Spinastacoides*) and other burrowing invertebrates (see Richardson and Doran 2008).

More speculatively, it may be possible to use fossil pollen data to reconstruct longer-term changes in soil pH and fire intensity, which can be correlated to particular climatic phases. For example, Brown et al. (1982) have shown that the distribution of woody and herb species in an 'alkaline pan-acidic peat' mosaic is influenced by soil pH. It is tempting to speculate that changes in the (mostly low to trace) relative pollen abundance of these taxa may document longer-term trends in soil pH that can be related to fire if fossil pollen counts are sufficiently high. Similarly, Brown (1999) has concluded that plant community composition and structure, and the thickness of blanket peat cover are a function of fire intensity. Again (assuming appropriately high pollen counts), it is possible that changes in the relative pollen abundance of 'indicator' taxa are a natural archive of longer-term trends in fire intensity in southwest Tasmania.

Conclusions

Debate is likely to continue regarding the importance of indigenous fire activity relative to other environmental forcing factors at different periods within the Holocene, or in areas that are remote from Aboriginal habitation sites and trails through the southwest. Whether buttongrass moorland's past is a reliable guide to its future is equally debatable since the periods for which we have the most reliable palaeoecologic data were more uniformly wet than the 21st century is predicted to be – a prediction given force by the increase in the number and extent of naturally ignited wildfires in the southwest in recent summers (J. Smedley-Marsden pers comm.). Frequent/extensive planned fires may be helpful in maintaining current vegetation boundaries (and biodiversity). However, 'mosaic' burning may also reduce the value of buttongrass moorland as an important resource for the capture and long-term storage of carbon if Bowman's (2008) prediction that the peat soils will become increasingly flammable proves to be correct. In a similar vein, warmer summer soil temperatures may assist the spread of the plant pathogen *Phytophthora cinnamonii*, which already infests 'tens of thousands of hectares' of buttongrass moorland (Rudman and Balmer 2008).

Accordingly, one task for those responsible for managing the southwest World Heritage Area, which encompasses the greater proportion of buttongrass moorland in Tasmania, will be to identify those natural attributes of buttongrass moorland that can be conserved under warmer/drier conditions. In this process, it is possible that the most important contribution that palaeoecology can make is to redefine humans as the creators as well as the destroyers and protectors of wilderness.

Dedication and Acknowledgements

The paper is dedicated jointly to Bill Jackson (University of Tasmania) and Geoff Hope (Australian National University), whose careers have been devoted unravelling the history of wet vegetation types across the Australasian region. Comments on an earlier draft by Greg Jordan (University of Tasmania) and an anonymous referee were, as always, of great assistance and much appreciated. Dr. Simon Haberle (Australian National University) kindly drafted Figure 2.

References

Anker, S.A., E.A. Colhoun, C.E. Barton, M. Peterson and M. Barbetti 2001. Holocene vegetation and paleoclimatic and palaeomagnetic history from Lake Johnston, Tasmania. *Quaternary Research* 56:264-274.

Birks, H.J.B. and H.H. Birks 1980. *Quaternary Palaeoecology*. Edward Arnold, London.

Bowman, D.M.J.S. 2008. Is global environmental change the end game for prehistoric vegetation legacies? The parallel cases of fire-maintained vegetation mosaics in southwest Tasmania and central Australia. *Australasian Plant Conservation* 16:6-8.

Bowman, D.M.J.S., A.R. Maclean and R.K. Crowden 1986. Vegetation-soil relations in the lowlands of south-west Tasmania. *Austral Ecology* 11:141-153.

Brown, M.J. 1999. Buttongrass moorlands. In J.B. Reid, R.S. Hill, M.J. Brown and M.J. Hovenden (eds) *Vegetation of Tasmania*, pp286-303. Canberra: Australia Biological Resources Study.

Brown, M.J., R.K. Crowden, and S.J. Jarman 1982. Vegetation of an alkaline pan-acidic heath mosaic in the Hardwood River Valley, Tasmania. *Australian Journal of Ecology* 7:3-12.

Caine, N. 1967. Carbon-14 dates from the Broad River Valley, Mt. Field National Park, Tasmania. *Australian Journal of Science* 31:119-120.

Colhoun, E.A. 1992. Late Glacial and Holocene vegetation history at Poet's Hill Lake, western Tasmania. *Australian Geographer* 23:11-23.

Colhoun, E.A. 1996. Application of Inversen's glacial-interglacial cycle to interpretation of the Last Glacial and Holocene vegetation of western Tasmania. *Quaternary Science Reviews* 15:557-580.

Colhoun, E.A. 2000. Vegetation and climate during the Last Interglacial-Glacial cycle in western Tasmania, Australia. *Palaeogeography, Palaeoclimatology, Palaeoecology* 155:195-209.

Colhoun, E.A. and A. Moon 1984. Estuarine sediments at the Bowen Bridge on the Derwent River, southeastern Tasmania. *Search* 15:224-226.

Colhoun, E.A., G. van de Geer, and W.G. Mook 1982. Stratigraphy, pollen analysis and paleoclimatic interpretation of Pulbeena Swamp, northwestern Tasmania. *Quaternary Research* 18:108-126.

Colhoun, E.A. and G. van de Geer, 1986. Holocene to middle last glaciation vegetation history at Tullabardine Dam, western Tasmania. *Proceedings of the Royal Society of London Series B, Biological Sciences* 229:117-207.

Colhoun, E.A., G. van de Geer and S.J. Fitzsimmons 1991. Late Glacial and Holocene vegetation history at Governor Bog King Valley, western Tasmania. *Journal of Quaternary Science* 6:55-66.

Colhoun, E.A., G. van de Geer and S.J. Fitzsimmons 1992. Late Quaternary organic deposits at Smelter Creek and vegetation history of the middle King Valley, western Tasmania. Journal of Biogeography 19: 217-227.

Colhoun, E.A., J.S. Pola, and H. Heijnis 1999. Late Pleistocene vegetation and climate history of Lake Selina, western Tasmania. *Quaternary International* 57/58:5-23.

Conedera, M, W. Tinner, C. Neff, M. Meurer, A.F. Dickens and P. Krebs 2009. Reconstructing past fire regimes: methods, applications and relevance to fire management and conservation. *Quaternary Science Reviews* 28: 555-576.

Corbett, S. and J. Balmer 2008. Buttongrass moorland in Tasmania – what and where? *Australasian Plant Conservation* 16:3-5.

Cosgrove R. 1999. Forty-two degrees south: the archaeology of late Pleistocene Tasmania. *Journal of World Prehistory* 13:357-397.

Cosgrove, R., J. Allen, and B. Marshall, 1990. Palaeo-ecology and Pleistocene human occupation in south-central Tasmania. *Antiquity* 64:39-78.

Curtis, W.M. and D.I. Morris 1994. The Student's Flora of Tasmania, Part 4B. Angiospermae: Alismataceae to Burmanniaceae. Hobart: St David's Park Publishing.

Dodson, J.R. 2001. A vegetation and fire history in a subalpine woodland and rainforest region, Solomons Jewel Lake, Tasmania. *The Holocene* 11: 111-116.

Dodson, J.R., F.J.G. Mitchell, H. Bogeholz and N. Julian 1998. Dynamics of temperate rainforest from fine resolution pollen analysis, Upper Ringarooma River, northeastern Tasmania. *Australian Journal of Ecology* 23:550-561.l

Driessen, M. 2008. Buttongrass moorland fauna. *Australasian Plant Conservation* 16:20-22.

Fletcher, M.-S. and I. Thomas 2007a. Holocene vegetation and climate change from near Lake Pedder, south-west Tasmania, Australia. *Journal of Biogeography* 34:665-677.

Fletcher, M.-S. and I. Thomas 2007b. Modern pollen-vegetation relationships in western Tasmania, Australia. *Review of Palaeobotany and Palynology* 146: 146-168.

Green, D. 2008. The diversity of soil mites in Tasmanian buttongrass moorland in relation to vegetation age. *Australasian Plant Conservation* 16:24-28.

Harle, K.J., A.P. Kershaw, M.K. Macphail and M.G. Neyland 1993. Palaeoecological analysis of an isolated stand of *Nothofagus cunninghamii* (Hook.) Oerst in eastern Tasmania. *Australian Journal of Ecology* 18:161-170.

Harle, K.J., D.A. Hodgson and P. Tyler 1999. Palynological evidence for Holocene palaeoenvironments from the lower Gordon River Valley, in the World Heritage Area of south-west Tasmania. *The Holocene* 9:149-162.

Hill, R.S. 1982. Rainforest fire in western Tasmania. *Australian Journal of Botany* 30 583-589.

Hope, G.S. 1994. Quaternary vegetation. In R.S. Hill (ed) *History of the Australian Vegetation: Cretaceous to Recent*, pp368-389. Cambridge: Cambridge University Press.

Hope, G.S., R. Nanson and I.K. Flett 2009. Technical Report 19. *The peat-forming mires of the Australian Capital Territory*. Canberra: Territory and Municipal Services.

Iversen, J. 1958. The bearing of glacial and interglacial epochs on the formation and extinction of plant taxa. *Uppsala Universiteit Arsskrift* 6:210-215.

Jackson, W.D. 1968. Fire, air, water and earth – an elemental ecology of Tasmania. *Proceedings of the Ecological Society of Australia* 3:9-16.

Jackson, W.D. 1977. Nutrient cycling in Tasmanian oligotrophic environments. In CSIRO (ed) Papers presented at a symposium organized jointly by W.A. Forests Department and CSIRO Division of Land resources Management, pp122-123. CSIRO, Como, W.A.

Jackson, W.D. 1999. The Tasmania legacy of man and fire. *Papers and Proceedings of the Royal Society of Tasmania* 133:1-14.

Jarman, S.J., G. Kantvilas, and M.J. Brown 1988. Buttongrass moorlands in Tasmania. *Tasmanian Forest Research Council Research Report* 2.

Jarman, S.J., G. Kantvilas, and M.J. Brown 1999. Floristic composition of cool temperate rainforest. In J.B. Reid, R.S. Hill, M.J. Brown and M.J. Hovenden (eds) *Vegetation of Tasmania*, pp145-159. Canberra: Australia Biological Resources Study.

Jordan, G.J. and R.S. Hill 2002. Cenozoic plant macrofossil sites of Tasmania. *Papers and Proceedings of the Royal Society of Tasmania* 136:127-139.

Kiernan, K., R. Jones, and D. Ranson 1983. New evidence from Fraser Cave for glacial age man in south-west Tasmania. *Nature* 301:28-32.

King, K.J. 2008. The relative importance of 'fine scale fuel mosaics' in reducing fire risk in southwest Tasmania. *Australasian Plant Conservation* 16:8-10.

Kirkpatrick, J.B. and M. Fowler 1998. Locating likely glacial forest refugia in Tasmania using palynological and ecological information to test alternative climatic models. *Biological Conservation* 85: 171-182.

Macphail, M.K. 1975a. *History of the Vegetation in Southern Tasmania since the Late Pleistocene* (ca. 13,000-0 BP). University of Tasmania: Ph.D. Thesis (2 Vols.).

Macphail, M.K. 1975b. Late Pleistocene environments in Tasmania. *Search* 6:295-300.

Macphail, M.K. 1979. Vegetation and climates in Southern Tasmania since the last glaciation. *Quaternary Research* 11:306-341.

Macphail, M.K. 1980. Natural regeneration processes in Tasmanian forests: a long-term perspective based on pollen analysis. *Search* 11:184-190.

Macphail, M.K. 1984. Small-scale dynamics in an early Holocene wet sclerophyll forest in Tasmania. *New Phytologist* 96:131-147.

Macphail, M.K. 1986. 'Over the top': pollen-based reconstructions of past alpine floras and vegetation in Tasmania. In B.A. Barlow (Editor) *Flora and Fauna of Alpine Australasia: Ages and Origins*, pp173-204. Melbourne: CSIRO.

Macphail, M.K. and R.R. Shepherd 1973. Plant communities at Lake Edgar, South West Tasmania. *Tasmanian Naturalist* 34:1-23.

Macphail, M.K. and W.D. Jackson, 1978. The late Pleistocene and Holocene history of the Midlands of Tasmania, Australia: Pollen evidence from Lake Tiberias. *Papers and Proceedings of the Royal Society of Tasmania* 90:287-300.

Macphail, M.K. and E.A. Colhoun 1985. Late glacial vegetation, climates and fire activity in southwest Tasmania. *Search* 16:43-45.

Macphail, M.K. and G.S. Hope 1985. Late Holocene mire development in montane southeastern Australia: a sensitive climatic indicator. *Search* 15:344-348.

Macphail, M.K., R.R. Shepherd, M.J. Brown and W.D. Jackson 1975. *Norfolk Range Area Environmental Impact Statement.* Devonport: Tasventure Environment Centre.

Macphail, M.K., M. Pemberton and G. Jacobson 1999. Peat mounds of southwest Tasmania: possible origins. *Australian Journal of Earth Sciences* 46:667-677.

Markgraf, V., J.P. Bradbury, and J.R. Busby 1986. Palaeoclimates in Tasmania during the last 13,000 years. *Palaios* 1:368-380.

Mount, A.B. 1979. Natural regeneration processes in Tasmanian forests. *Search* 10:180-186.

Patterson, W.A., K.J. Edwards and D.J. Maguire 1987. Microscopic charcoal as a fossil indicator of fire. *Quaternary Science Reviews* 6:2-23.

Porch, N. 2008. Invertebrates in landscape ecology. In D.B. Thomas (ed) *Handbook of Landscape Archaeology*, pp457-467. Walnut Creek: Left Coast Press.

Read, J. 1999. Rainforest ecology. In J.B. Reid, R.S. Hill, M.J. Brown and M.J. Hovenden (eds) *Vegetation of Tasmania*, pp160-197. Canberra: Australia Biological Resources Study.

Richardson, A.M.M. and N.E. Doran 2008. The role of burrowing crayfish in Tasmanian sedgelands. *Australasian Plant Conservation* 16:22-24.

Rudman, T. and J. Balmer 2008. Death on the moor: the impact of *Phytophthora cinnamonii* on buttongrass moorland. *Australasian Plant Conservation* 16:29-31.

Thomas, I. 1993. Late Pleistocene environments and Aboriginal settlement patterns in Tasmania. *Australian Archaeology* 36:1-11.

Thomas, I. 1995. Where have all the forests gone? New pollen evidence from Melaleuca Inlet in south-western Tasmania. *Monash University Publications in Geography* 295-301.

Traverse, A. 1988. *Paleopalynology.* Boston: Unwin Hyman.

van de Geer, G., E.A. Colhoun, and W.G. Mook 1986. Stratigraphy, pollen analysis and paleoclimatic interpretation of Mowbray and Broadmeadow Swamps, north western Tasmania. *Australian Geographer* 17:121-133.

van de Geer, G., S.J. Fitzsimmons, and E.A. Colhoun 1989. Holocene to middle Last Glaciation vegetation history at Newall Creek, western Tasmania. *New Phytologist* 111:549-558.

van de Geer, G., S.J. Fitzsimmons, and E.A. Colhoun 1991. Holocene vegetation history from King River Railway Bridge, western Tasmania. *Papers and Proceedings of the Royal Society of Tasmania* 125:73-77.

Wells, P. and J. Hickey 1999. Wet sclerophyll, mix and swamp forest. In J.B. Reid, R.S. Hill, M.J. Brown and M.J. Hovenden (eds) *Vegetation of Tasmania*, pp224-243. Canberra: Australia Biological Resources Study.

Whitlock, C. and C.P.S. Larsen 2001. Charcoal as a fire proxy. In J.P. Smol, H.J.B. Birks and W.M. Last (eds) *Tracking Environmental Change Using Lake Sediments, Volume 3 Terrestrial, Algal and Siliceous Indicators*, pp75-97. Dordrecht: Kluwer Academic Publishers.

18

Ecological drift or stable fire cycles in Tasmania: A resolution?

Ian Thomas

Department of Resource Management and Geography, Melbourne School of Land and Environments, University of Melbourne, Victoria, Australia

i.thomas@unimelb.edu.au

Phil Cullen

Natural Resource Planning Pty Ltd, Hobart, Tasmania, Australia

Michael-Shawn Fletcher

Department of Resource Management and Geography, Melbourne School of Land and Environments, University of Melbourne, Victoria, Australia

Introduction

On the Central Plateau of Tasmania, up by Liawenee Moor, the wind whistles with frost-edged harshness. On the shores of Lake Augusta stood the Bernacchi Training Facility of the Australian Antarctic Division in which generations of bearded heroes listened and learned at the foot of gnarly ice veterans. The main room of the facility was barn-like, with exposed beams, coils of rope and a very large and well-used fireplace. The inside of the chimney was damp, soot blackened and slippery. We know this for a fact because in 1984, Phil Cullen and I (Thomas) were wedged inside, trying to squeeze down into the warmth below. Geoff Hope had sent me up to the plateau to re-sample a small lagoon, which he had previously cored but not analysed. It was my Honours year and I was well chuffed. That Geoff had the confidence to simply say, 'Tasmania, Central Plateau, Liawenee, go' gave me additional heart. That he never once hinted at how I should accomplish the task, I took as a compliment as well. Compliments warm the soul but the taste rarely lingers. Phil and I needed something substantial to eat, and up on the Central Plateau in a snowstorm, we needed shelter. Outside the chimney, all was snow and wind. Inside was food and warmth. There was only one solution. We climbed further down the chimney, fell into the fireplace and pillaged a lifetime supply of freeze-dried pineapple before disappearing into the wild and treeless expanses of Liawenee Moor to extract the cores. Geoff encouraged self-reliance.

As it turned out, the data threw up the old castanea of stability versus change. For how long had the treeless expanse of Liawenee been treeless? What were the factors that promoted stability? What were the agents of change? What was the balance? Was there a balance? Not many of these thoughts crossed my mind while trapped in the chimney, but later, back at the ANU under Geoff's guiding hand, the data seemed to indicate that for this little part of Tasmania at least, stability was the answer. Basalt soils and frequent frosts conspired to disadvantage any eucalypt which dared to invade the moorland. For perhaps 8000 years or more, Liawenee Moor was almost certainly a treeless grassland interspersed with sub-alpine heaths. At the time of European settlement, the rapid introduction of sheep in large numbers, along with an annual burning regime that favoured dense communities of fire-loving woody heaths, transformed the moor from a grassland to a grassy shrubland. Changes were apparent but not significant enough to alter the treeless state. The moorland was stable in an overall sense and displayed remarkable resilience, primarily due to edaphic controls exerted by clayey basaltic soils.

Over a number of years, Phil and I wandered our way across some of the great landscapes of Tasmania. We thought about succession and fire. We amused ourselves with plans to shatter what we considered 'eco-myths'. Later, Michael Fletcher and I did exactly the same in western Tasmania. Mostly, we thought about Rhys Jones, Bill Jackson and Tony Mount, but we were always guided by the question; Geoff's question. The question can be applied to every landscape in the world, indeed, perhaps to everything in the world. Geoff simply asked his students to always consider why something was here and not there. Biogeography in a nutshell.

Here was the nub of a great ongoing Tasmanian ecological debate and we were walking through the middle of it. The origin of buttongrass plains. The fact that many grasslands and sedgelands are considered to be pyrogenic disclimaxes provided the botanical clue, which linked vegetation change with cultural traditions. Jones (1968, 1975) recognised the relationship, and in a brilliant example of synthesis, banished forever the notion that Aborigines were children of nature, incapable of affecting the environment. He introduced to Australia the concept that cultural traditions are inevitably interwoven with ecological processes by way of his famous theory of firestick farming. This advance was accomplished with reference to the equally spectacular botanical work of Jackson (1965, 1968), who proposed the first, and so far only, model to explain the nature of the general processes that acted on Tasmanian vegetation during the Holocene.

Alternative theories attracted attention, but failed to win widespread support. Horton (1982) attempted to downplay the effects of Aboriginal fires on the Australian vegetation, but his alternative does not explain the pyrogenic disclimax formations found in Tasmania (Bowman and Brown 1986). Likewise, the static vegetation model of Mount (1979, 1982) does not explain vegetation dynamics on a scale of centuries, much less the longer-term changes determined by pollen analysis (Macphail 1980, 1984, 1988; Colhoun 2000). It is relatively easy to dismiss the arguments of Mount and Horton, especially decadal time scales, but what of Jackson's 'ecological drift', and what happens over centennial and millennial scales? What evidence is out there? Soils, fire and precipitation are the three main factors, but each of these is confounded by local variations and complications. What would Geoff think?

Australia has a history of human occupation which extends for maybe 45,000 to 60,000 years. There has been considerable debate over the exact impact of human activity on the Australian biota over this immense time span. A large body of evidence, gathered from all over the continent, indicates that this impact has been pronounced and largely a result of the regular and sophisticated use of fire by the Aborigines to manipulate their environment (Bowman 1998; Jackson 1999; Lynch et al. 2007). Therefore, an understanding of these impacts and the fire-management practices that produced them should prove very useful in the successful conservation management of many natural ecosystems today.

Southwest Tasmania provides an appropriate and challenging location to test this approach. The area has a long history of human occupation, with much of the vegetation dependent on fire. Other plant communities, especially those in the alpine zone and those dominated by coniferous species, are extremely fire sensitive and great damage has been caused by wildfires in historical times. Furthermore, many of the major land management problems faced in this region are linked to the current fire regime. These include maintenance of habitat diversity, protection of fire sensitive vegetation and soils, control of fuel build up and wildfire, single species management requirements, and the preservation of what can reasonably be argued to be a cultural landscape.

While it has been or will be possible to conduct research to determine the direct impact of Aboriginal fire regimes on contemporary ecosystems in some parts of Australia (Bowman 1998), this will not be possible in Southwest Tasmania as this knowledge has been lost. However, through careful consideration of the historical records and further research in fields such as ecology, geomorphology, palaeoecology and archaeology, it should be possible to devise sustainable management programs that will conserve this rich and unique environment. In the remainder of this chapter, we provide a brief review of the palaeoecology of southwest Tasmania, present some recent palaeoecological findings and suggest further avenues for research that would provide useful information for the pursuit of this goal.

The environment

Southwest Tasmania has a cool, temperate maritime climate with high levels of precipitation, evenly distributed throughout the year, relatively low temperatures, and high humidity levels (Genitilli 1972). Snowfalls and frosts can occur in any month, particularly at higher altitudes.

Many people have described the landforms of the region in detail (e.g. Pemberton 1989). Topography varies from low-lying plains to rugged ranges with extremes of aspect and slope. The geology is relatively complex and encompasses considerable variation in bedrock, with Precambrian quartzites dominating and significant exposures of Cambrian volcanics and mid-Palaeozoic carbonates and clastics. Pleistocene glaciations have shaped the region's erosional landforms, with valley and cirque glacial features evident at higher altitudes. At lower altitudes, deposition forms including outwash tills, and moraines are common (Kiernan 1989). The vegetation of the region is a mosaic of sedgeland-heath, scrub, wet sclerophyll forest and rainforest communities. Alpine vegetation is present on the high peaks and ranges.

Where intact, the soils of the region are dominated by organic horizons. Red/brown fibrous peats overlying mineral soils of varying composition and depth usually occur under forest vegetation. Very dark brown or black fibrous or muck peats overlying sand, gravels and bedrock form large areas of blanket bog (peatland) in the region. These latter soils are found on nutrient-poor Precambrian siliceous substrates which are very slow to weather. They are usually associated with sedgeland-heath and related scrub communities, termed 'moorlands' after the flowering habit of the ubiquitous dominant sedge, buttongrass (*Gymnoschoenus sphaerocephalus*) (Jarman et al. 1988).

There is more than one million ha of buttongrass moorland in Tasmania (Brown 1999), most of which is found in Southwest Tasmania. The moorlands can be found from sea level to the subalpine zone. The vegetation and underlying soils of these ecosystems have been described in detail by Bowman et al. (1986), Pemberton (1989) and Jarman et al. (1988). Marsden-Smedley (1998) estimates there are around 704,200 ha of moorland and associated scrub communities in this region, accounting for 57% of the total area. These ecosystems are of considerable conservation significance and represent some of the largest organic terrains in the Southern Hemisphere (Dixon and Duhig 1996).

Human occupation

Humans have occupied Tasmania for at least 35,000 years (Flood 1995). Most of the oldest cultural sites on the island are found within southwest Tasmania. Radiocarbon dating of cultural material found in caves here indicates that Aborigines occupied inland areas of the region from around 35,000 to around 12,000 years BP. These caves appear to have been vacated as living quarters after this time. There is an unresolved debate as to whether Aborigines generally abandoned inland southwest Tasmania in the face of advancing wet sclerophyll and rainforest vegetation during the late Pleistocene and early Holocene, or whether they occupied the region throughout the Holocene and until European colonisation (Cosgrove et al. 1990, 1994; Thomas 1993, 1995).

Initial pollen investigations suggested that the vegetation of the region was more open in the late Holocene than before (Macphail 1979), possibly as a result of increased anthropogenic firing under a cooler, dryer climate. Further discovery of dateable cultural deposits in the region may provide an answer to this debate. There are certainly a number of historical records which indicate that Aborigines were living in inland southwest Tasmania in the early 1800s (Binks 1980; Thomas 1993, 1994; Marsden-Smedley 1998). These records demonstrate that people were using fire extensively to modify habitat. European exploration and exploitation of the region resulted in high incidences of wildfire in the region until relatively recent times (Marsden-Smedley 1997b).

Ecological drift versus stable boundaries

In a vegetation survey of the Florentine River valley, Gilbert (1959) noted that there was a progression from moorland to rainforest through a set of distinct successional stages. This supported the research by Jackson (1958), who proposed that fire was responsible for moorland regeneration. The dependence of buttongrass moorland on fire stands in stark contrast to the ecology of rainforest vegetation, which is highly sensitive to fire (e.g. Jackson 1968, 1999). Jackson proposed that in the absence of fire, rainforest would displace open moorland. Subsequent work refined this into the 'ecological drift' model, which proposes a predictable set of seral stages in the absence of disturbance.

Overall, Jackson's (1968) model is one in which a succession of communities eventually change to a closed forest community. The model estimates a fire frequency of 10 to 30 years required to maintain an open vegetation community (moorland). As the interval between fires increases, there is an increase in the proportion of woody shrubs. As the fire interval lengthens, taller myrtaceous shrubs, such as *Leptospermum* and *Melaleuca*, become more important and wet scrub develops. The longer the fire-free interval, the more chance there is of trees, such as *Eucalyptus* species, and rainforest undershrubs becoming dominant. A fire-free interval of more than 150 years is sufficient for the development of mixed forest and eventually rainforest. This model holds the incidence of fires as random in space and time.

Mount (1979) agreed with Jackson that fire was important in maintaining vegetation communities. Unlike Jackson, he saw vegetation communities as stable fire climaxes with fixed boundaries. Fire incidence is not random in space and time, according to Mount, but is a reflection of the accumulated fuel loads of each community. As the interval between fires increases, debris accumulates and the risk of fire becomes greater: different decomposition and productivity rates of the plant communities result in differential fuel loads. When adequate fuel is present, a fire will burn, thus resetting the 'fire cycle clock'. Mount placed tentative time limits on his model. Six months was considered sufficient time for fuel accumulation in buttongrass moorlands and 80 years for rainforest (Mount 1979). Neither Jackson nor Mount fully appreciated the effects that extreme climatic events such as the fire weather associated with El Nino oscillations have on vegetation.

Nevertheless, the majority of palynological research points to Jackson's model as the truest reflection of western Tasmanian vegetation succession (Macphail 1980). However, there is some confusion here. At the time of Macphail's exemplary and detailed work and bolstered by Colhoun's wide-ranging additions, nearly all pollen diagrams indicated a post-glacial succession of plant communities starting with invasive *Eucalyptus* and bird-dispersed *Phyllocladus*. This facilitated the expansion of *Nothofagus* rainforest between about 9000 yr BP and 6000 yr BP before a climatic deterioration saw the replacement of some rainforest with *Eucalyptus* forest. Superficially this sounds like succession. It sounds like 'ecological drift'. But the time scales are wrong. According to Jackson, given disturbance free conditions, anywhere in western Tasmania could drift towards and attain rainforest status in as few as 500 years. However, the post-glacial pollen sequences are almost all about 10,000 years long. If they do support ecological drift, then it is a tardy version of what Jackson envisaged and what ecologists can measure over decadal scales.

Pemberton's edaphic model

The major theme of Pemberton's (1990) alternative model was the relationship between edaphic characteristics and vegetation type. He acknowledges that wet sclerophyll forests are the result of rainforest restricted by fire and that temperature and water availability confines subalpine and alpine plant types to their particular niches. His model differs from the two previous models in his emphasis on soil type and its effect on plant communities. He downplays the role played by fire in the regeneration of buttongrass moorland, favouring the notion that they are natural edaphic disclimaxes.

Edaphic disclimaxes develop on nutrient-poor acidic soils that are waterlogged for most of the year. The harsh environmental restraints placed on plants growing in these areas results in low productivity and slow peat accumulation. According to Pemberton, this is reflected in the normally shallow (<30 cm) peat profiles of southwest Tasmania. Burning of the moorland inevitably results in peat destruction, and if this occurs during a period of drought, almost total removal of peat results (Brown and Podger 1982; Pemberton 1990).

Colhoun (1996) places importance on factors other than climate in the regression of regional vegetation back towards that seen in the late glacial or early Holocene. A decrease in climate-sensitive rainforest species (namely *N. cunnghamii*) during this phase is seen as an artefact of nutrient leaching in soils and anthropogenic fire. Nutrient leaching is the vehicle driving an increase in the proportion of sclerophyllous species in the vegetation. Sclerophylly promotes flammability, and the build up of fuel loads increased the effectiveness of Aboriginal burning in the region, expediting the expansion of fire-promoted moorland into areas occupied by rainforest (Colhoun 1996; Jackson 1999).

Thomas (1993) noted an obvious lack of local forest over a late-glacial-Holocene pollen sequence from fire-promoted buttongrass moorland at Melaleuca Inlet in the far southwest of the state. The pollen site is in a coastal moorland close to archaeological sites that display continued coastal occupation through the Holocene (Cosgrove 1999). The record clearly displays the usual regional trends in rainforest pollen present in all western Tasmanian pollen records, but the cumulative values of these pollen types never exceeds 20%. Instead, pollen from Ericaceae, *Leptospermum*, *Melaleuca*, *Gymnoschoenus sphaerocephalus* (buttongrass) and Restionaceae dominate and charcoal concentrations are high throughout the sequence (Thomas 1995).

The southwest near-coastal location of this site exposes it to the westerly winds straight off the ocean and the low rainforest pollen content can be partially explained by the lack of forest upwind from the site. It is clear, however, that this site was occupied by moorland vegetation throughout the Holocene and that no substantial forest was present at any stage in the c. 12,000-year history recorded at the site. Subsequent analysis of a neighbouring pollen record from the same moorland substantiated this interpretation (M.K. Macphail et al. 1999).

Fletcher (2000) analysed a 10,300-year pollen record from moorland vegetation in the interior of southwest Tasmania and came to the same conclusion as Thomas (1993) and Macphail et al. (1999). Fletcher argued that the area around his site (Harlequin Hill near Lake Pedder) had probably been free of forest vegetation since before Oxygen Isotope Stage (OIS) 2.

Whether human activity pre-set the landscape and whether moorland vegetation was of sufficient flammability to ensure a fire regime that perpetuated this vegetation type and to retard the encroachment of forest, despite a climate hostile to natural lightning-strike fires, is less certain. It is clear from long-term (20-year) vegetation studies and space-for-time analyses that moorland succeeds to forest at different rates depending on moisture, geology, vegetation type, slope, aspect and fire history (Brown and Podger 1982; Marsden-Smedley 1997a, b,1998; Brown et al. 2002).

The ecological truism is that rainforest will form in any area of western Tasmania given a sufficient fire-free interval. This may not be true. Furthermore, it appears that moorland vegetation 'requires' a fire interval of fewer than 20 years to maintain ecological diversity (Marsden-Smedley and Kirkpatrick 2000). Long fire-free intervals may in some instances result in hot fires fuelled by thick moorland growth (Pemberton 1988, 1989), resulting in sheet erosion and the loss of fire-sensitive communities (Brown et al. 2002).

The results of Balmer (1990) and Marsden-Smedley et al. (1998) indicate that considerably more research is required into vegetation-soil relationships, peat formation and time scales of successional pathways before we can confidently predict the impact of fire on moorland and forest ecosystems.

Peat accumulation and the formation of blanket bogs

Peat begins to form on a site when the rate of accumulation of organic matter at a site exceeds the rate of decomposition. A cool, moist climate with evenly distributed precipitation throughout the year, low levels of evaporation and nutrient-poor anaerobic soil conditions are conducive to this process. Peat will accumulate in topographic depressions when groundwater tables are high enough to produce soil waterlogging for much of the year. Where precipitation is high enough, soils may remain waterlogged for sufficient time periods for peat accumulation to occur regardless of the groundwater conditions. Peat soils that form under these conditions are called blanket bogs due to the way they blanket the ground surface irrespective of topography. Given sufficiently high and evenly distributed precipitation and low levels of evaporation, blanket bogs can occur on relatively steep slopes.

Blanket bogs are widespread in association with buttongrass moorland in southwest Tasmania. The climatic conditions required for blanket bog development in relation to Tasmania are reviewed by Bridle (1992). Although rainfall is adequate for blanket peat formation (more than 1000 mm per annum), the temperature regime in southwest Tasmania seems to be slightly higher than the maximum (15°C for the warmest month) under which blanket peats form in the Northern Hemisphere. The marginal temperature regime appears to be compensated for by very high and evenly distributed precipitation and high humidity. Although extensive, the blanket bogs of Tasmania are generally shallow compared with those of western Europe and other regions. This fact has led some authors to conclude the climate of western Tasmania is only marginally suitable for blanket bog formation (Jarman et al. 1988).

Balmer (1990) has suggested that rates of peat accumulation and decay reach equilibrium at around 30 cm for much of southwest Tasmania. This view is supported by the apparently low rates of peat accumulation in Tasmania. Tasmanian peat accumulation rates have been deduced by examining sediment cores extracted from a variety of peat deposits. They yield accumulation rates of 1-2 cm/100 years. This is low compared with the 5-6 cm/100 years recorded for the Northern Hemisphere (Bridle 1992). An alternative hypothesis is that frequent burning episodically vaporises and removes surface peat, with freshly burned and exposed peat surfaces recolonised by adjacent, unburned plant communities. This would almost always result in anomalous age/depth relationships if accumulation rates were based on single basal dates.

The impact of fire on peat soils

Large areas of blanket bog (in excess of 100,000 ha) have been degraded by fire and subsequent erosion in southwest Tasmania (Pemberton 1988, 1989; Pemberton and Cullen 1995). Much of this degradation is attributed to wildfires started by humans in the recent times. The direct physical effects of fire on peat soils have been summarised in increasing order of severity (Pemberton and Cullen 1995). They are:

1. Removal of some vegetation with no direct impact on the soil.
2. Removal of vegetation which exposes the soil surface, which may then be vulnerable to wind or sheet erosion.
3. Frequent removal of vegetation and litter. Raw material for organic soil development is lost and soil formation is hindered. Ash may be blown or washed away.
4. Removal of vegetation, litter, soil seed bank, and soil to varying depths. Ash may be blown or washed away. The more soil removed, the longer it is likely to take for the soil to reform because of greater stresses on plant development.
5. Soils burned to bedrock, losing thousands of years worth of soil accumulation.

Even low-intensity fires are likely to influence the rate of peat accumulation and it is possible that they may cause peat losses even when peat is not directly burned. Live plant material and litter that accumulates to form peat will be lost. Removal of vegetation cover is likely to result in raised soil temperatures, causing an increase in organic-matter decomposition. Soil water tables may be lowered as a consequence of increased run-off and increased evaporation resulting from higher soil temperatures and airflows over the soil surface, again causing increased organic-matter decomposition. Elsewhere, studies indicate that peatland ecosystems may switch from being net storages of carbon (organic matter) to net sources of carbon (carbon dioxide) with small changes in soil temperature and soil water table position (Shurpali et al.1995; Johnson et al. 1996; Bubier et al. 1999). On the other hand, such effects may be countered by reductions in evapotranspiration until vegetation regrows following the fire. With time, as vegetation reverts to pre-fire condition, conditions conducive to peat accumulation are likely to re-establish.

Clearly vegetation-soil-fire relationships are not well understood and considerable research is required in this area. If the predictions of CO_2-induced global warming prove correct, then the consequent climate changes may well have a significant impact on peat accumulation. If the climate of southwest Tasmania is only marginal for blanket bog formation at present, then small shifts in the prevailing conditions, such as a rise in mean annual temperatures, particularly if coupled with inappropriate fire regimes, may well cause blanket peats to degrade.

Fire regimes in southwest Tasmania

The association of buttongrass moorland with blanket bog presents an interesting paradox. On one hand, Jackson's (1968) model of 'ecological drift' suggests that moorland vegetation is promoted by, and is resilient to, the impacts of fire, but on the other, the soils that form under this vegetation are readily damaged or destroyed by fire.

Using historical records, Marsden-Smedley (1997b) demonstrates dramatic changes in the fire regime in southwest Tasmania in the past 170 years. He proposes that the pre-European fire regime was one of frequent, low-intensity, small-scale burns used by the Aborigines for habitat management. From the 1850s to the 1930s, the fire regime was one of less frequent, high-intensity, 'landscape-scale' wildfires. From the 1940s to the 1960s, the fire regime was one of medium intensity in spring and autumn. Since the 1970s, there has been a policy of fire exclusion from virtually all of the area. Burning has been largely restricted to small fuel-reduction fires at a few localities, but there have been a few notable exceptions.

Most of the degraded peatland evident on aerial photographs and satellite images falls within the boundaries of fires that have occurred in the region since the 1930s (Cullen unpublished data and fire boundaries mapped by Marsden-Smedley 1997). There also have been dramatic losses of fire-sensitive vegetation in the region in historical times (Cullen and Kirkpatrick 1987; Brown 1988; Gibson and Brown 1991).

The fact that fire-sensitive ecosystems were significantly more widespread before the arrival of Europeans suggests that the burning practices of the Aborigines were sophisticated (Bowman and Brown 1986; Cullen and Kirkpatrick 1987; Thomas 1993; Thomas and Hope 1994; Bowman 1998; Bird et al. 2008). This is not to say that burning by Aborigines did not cause damage to vegetation or organic soils. At Melaleuca, small peat fires during the middle Holocene caused the removal of blanket peat (Macphail et al. 1999), while in the Denison Range, there is clear evidence that plant communities dominated by native pines were severely affected by fires which may have been induced by Aborigines. However, for blanket bogs to form, moorland ecosystems must act as net stores/sinks of carbon (organic matter). This implies that the climate and Aboriginal fire regime over the Holocene has been conducive to peat accumulation, even if rates seem somewhat retarded compared with the Northern Hemisphere. At present, we have little knowledge of Aboriginal occupation and use of fire in southwest Tasmania throughout the Holocene. Future research in this field will depend on detailed research into the ecology, palaeoecology and archaeology of the region.

Buttongrass moorland during the late Pleistocene and Holocene

Pollen sequences obtained from bogs and lakes indicate that the vegetation of lowland southwest Tasmania during the last glacial was open and herbaceous, with alpine and subalpine affinities (see the review in Jackson 1999). By the early to middle Holocene, there is an increasing dominance of pollen associated with closed forest communities. A warmer, wetter climate during the period 9000 to 6000 years ago appears to have favoured the spread of this vegetation. During the late Holocene, decreasing temperatures and precipitation produced an opening of forest vegetation (Macphail 1980), accompanied by increases in charcoal particles in the deposits. This may indicate greater deliberate burning activity by the Aborigines and/or increases in the occurrence of wildfires.

However, pollen evidence from all pollen sites in the southwest indicates that the long-term successional sequence proposed by Jackson does not adequately explain the persistence of buttongrass moorlands for the entire length of the Holocene if people have been largely absent from the southwest during the Holocene.

Colhoun et al. (1991) found fairly consistent levels of charcoal associated with pollen representing a range of vegetation assemblages throughout the Holocene at Govenor Bog in the King River valley and concluded that fire was 'an ever-present ecological factor of the changing alpine, subalpine, temperate vegetation/environmental changes'. It seems likely that at least some of this firing would have resulted from human ignition sources.

Two pollen sequences from Melaleuca in the far southwest provide vegetation records that span the Holocene (Thomas 1995a; Macphail et.al. 1999). These show the persistence of open moorland vegetation and blanket bog from the late Pleistocene until the present in a location that is at low altitude and relatively close to the present coast. Both sequences show that fire was present in the environment throughout this time. Macphail et al. (1999) interpret a hiatus in the sedimentary record as evidence of a peat-burning fire. In addition, neither site provides evidence that the far southwestern coastal plains formed refugia for arboreal taxa. Nor do they indicate any long-term successional changes from moorland to forest. However, the sequences indicate that *Gymnoschoenus* sedgeland coexisted with *Poa* grassland at the end of the last glacial under conditions where fire was present but far less significant than at any subsequent time. In other words, moorland was able to prosper under conditions of both high and low fire pressure (where intensities and frequencies are as yet unknown). If Jackson's model is appropriate and general, the early Holocene should have been a phase of forest expansion similar to the sites investigated by Macphail (1979, 1980). This inconsistency can be resolved by postulating a late last glacial to early Holocene phase in the southwest in which environmental perturbation was not fire but something else, possibly soil instability or increased soil infiltration rates.

A third sequence from the inland southwest taken from a buttongrass moorland near the foot of Mount Anne near the present Lake Pedder also indicates that buttongrass communities have dominated from at least the late last glacial right up to the present day (Fletcher and Thomas 2007a, b). By analysing pollen transport and depositional processes, it was possible to demonstrate that *Gymnoschoenus* pollen, although abundant in the landscape, is rarely presented in pollen diagrams. This has resulted in a severe under-representation of the single major plant association in western Tasmania. When regionally significant pollen types are subtracted from pollen sums, the true nature of local and extra-local plant communities is revealed. These have uniformly been found to be treeless plant communities which have existed for at least the entire Holocene.

This point needs to be reiterated: no moorland site in southwest Tasmania contains pollen values (Thomas 1995a; Fletcher and Thomas 2007b) which indicate the former presence of rainforest on site. It is also clear that areas occupied by closed forest associations today have been forested since the termination of the Last Glacial Maximum. This interpretation is supported by a supplementary meta-ordination of all late glacial and Holocene pollen samples from western Tasmania (Fletcher 2007a). No change in the overall relative proportions of major plant community types compared with that seen today is visible in any pollen record for this period.

Conclusions

It seems that both Tony Mount and Bill Jackson were right. Jackson's 'ecological drift' is rightly seen as the model which best explains changes at decadal and centennial time scales, especially in the ecotones of plant communities in the western Tasmania. Mount's 'stable fire cycle' model cannot explain floristic changes that are demonstrably measurable, either in real time or by space-for-time substitution. However, the data shows quite clearly that major expanses of buttongrass have remained treeless for at least 10,000 years and so stability rather than change is the overriding feature on sub-regional and regional scales. The only way in which Jackson's model can operate over the long time scales involved is to invoke either or both fire and subsequent edaphic modification. However, if fire pressure was so great that a constant state of disclimax was achieved, then we return to a situation where stability is the key and Mount's 'stable fire cycle' model seems rather more attractive than it does for shorter time scales. In other words, scale dependency is paramount.

A great amount of knowledge relevant to current management of southwest Tasmania was lost with the passing of Aboriginal occupation of the region. At present, we have very limited information on which to base future management practices. The extensive losses of fire-sensitive vegetation and organic soils in the region in recent times demonstrate that if we are to manage southwest Tasmania in a sustainable manner, we must proceed with caution. As concluded by Bowman and Brown (1986), experimental investigations and manipulation of fire regimes in association with appropriate palynological studies are likely to provide the best solutions to these problems.

References

Balmer, J. 1990. Two moorland boundaries. *Tasforests* 1:133-141.

Binks, C.J. 1980. *Explorers of Western Tasmania.* Launceston, Tasmania: Mary Fisher Bookshop.

Bliege Bird, R., D.W. Bird, B.F. Codding, C.H. Parker and J.H. Jones 2008. The fire stick farming hypothesis: Australian Aboriginal foraging strategies, biodiversity, and anthropogenic fire mosaics *Proceedings of the National Academy of Science* 105:14796-1480.

Bowman, D.J.M.S. 1998. Tansley Review No. 101: The impact of Aboriginal landscape burning on the Australian biota. *New Phytologist* 140:385-410.

Bowman, D.M.J.S. and M.J. Brown 1986. Bushfires in Tasmania: a botanical approach to anthropological questions. *Archaeology in Oceana* 21:166-171.

Bowman, D.J.M.S. and W.D. Jackson 1981. Vegetation succession in Southwest Tasmania. *Search* 12:358-362.

Bowman, D.J.M.S., A.R. MacLean and J.K. Crowden 1986. Vegetation-soil relations in the lowlands of southwest Tasmania. *Australian Journal of Ecology* 11:141-153.

Bridle, K. 1992. Organic soils on Mt Sprent, south west Tasmania: an analysis of correlations with local climate, microtopography and vegetation. Unpublished Master of Environmental Studies thesis, Department Geography and Environmental Studies, University of Tasmania.

Brown, M.J. 1988. *The distribution and conservation of King Billy Pine.* Forestry Commision, Tasmania.

Brown, M.J. 1999. Buttongrass moorlands. In Reid, J., R.S. Hill, M.J. Brown, and M.J. Hovenden (eds) Vegetation of Tasmania. *Flora of Australia Supplementary Series, Volume 8*, pp286-303. Australian Biological Resources Study, Canberra.

Brown, M.J. and F.D. Podger 1982. Floristics and fire regimes of a vegetation sequence from sedgeland-heath to rainforest at Bathurst Harbour. *Australian Journal of Botany* 30:659-676.

Bubier, J.L., S. Frolking, P.M. Crill and E. Linder 1999. Net ecosystem productivity and its uncertainty in a diverse boreal peatland. *Journal of Geophysical Research* 104(D22):27638-27692.

Colhoun, E.A. 2000. Vegetation and climate during the Last Interglacial-Glacial cycle in western Tasmania, Australia. *PalaeogeogrAphy, Palaeoclimatology, Palaeoecol*ogy 155, 195-209.

Colhoun, E.A., G. van de Geer and S.J. Fitzsimons 1991. Late glacial and Holocene vegetation history at the Governors Bog, King Valley, western Tasmania, Australia. *Journal of Quaternary Science* 6:55-66.

Cosgrove, R., J. Allan and B. Marshall 1990. Palaeo-ecology and Pleistocene human occupation in south central Tasmania. *Antiquity* 64:59-78.

Cosgrove, R., J. Allan and B. Marshall 1994. Late Pleistocene human occupation in Tasmania: a reply to Thomas. *Australian Archaeology* 38:28-35.

Cullen, P.J. and J.B. Kirkpatrick 1987. The ecology, distribution and conservation of *Athrotaxis*: A report to the Australian Heritage Commission. Tasmanian Conservation Trust and Department of Geography and Environmental Studies, University of Tasmania.

Dixon, G. and N. Duhig 1996. Compilation and assessment of some places of geoconservation significance. Report to the Tasmanian RFA Environment and Heritage Committee.

Fletcher, M.S. and I. Thomas 2007a. Modern pollen-vegetation relationships in western Tasmania. *Review of Palaeobotany and Palynology* 146:146-168.

Fletcher, M.S. and I. Thomas 2007b. Holocene vegetation and climate change from near Lake Pedder, south-west Tasmania, Australia. *Journal of Biogeography* 34(4):665-677.

Flood, J. 1995. *Archaeology of the dreamtime. The story of prehistoric Australia and its people.* Australia: Angus and Robertson.

Gentilli, J. 1972. *Australian climatic patterns.* Melbourne, Victoria: Nelson.

Gibson, N. and M.J. Brown 1991. The ecology of *Lagarostrobos franklinii* (Hook.f.) Quinn (Podocarpaceae) in Tasmania. 2. Population structure and spatial pattern. *Australian Journal of Ecology* 16:223-229.

Gilbert, J.M. 1959. Forest Succession in the Florentine Valley, Tasmania. *Papers and Proceedings of the Royal Society of Tasmania.* 93:129-151.

Horton, D. 1982. The burning question: Aborigines, fire and Australian ecosystems. *Mankind* 13(3):237-251.

Jackson, W.D. 1965. Vegetation. In J.L. Davies (ed.) *Atlas of Tasmania*, pp30-35. Hobart: Mercury Press.,

Jackson, W.D. 1968. Fire, air, water and earth-an elemental ecology of Tasmania. *Proceedings of the Ecological Society of Australia* 3:9-16.

Jackson, W.D. 1999. The Tasmanian Environment. In J. Reid, R.S. Hill, M.J. Brown and M.J. Hovenden (eds) Vegetation of Tasmania. *Flora of Australia Supplementary Series* 8, pp11-38. Australian Biological Resources Study, Canberra.

Jackson, W.D. 1999. Palaeohistory of vegetation change: The last 2 million years. In J. Reid, R.S. Hill, M.J. Brown and M.J. Hovenden (eds) Vegetation of Tasmania. *Flora of Australia Supplementary Series* 8, pp286-303. Australian Biological Resources Study, Canberra.

Jarman, S.J., G. Kantvilas and M.J. Brown 1988. Buttongrass moorlands in Tasmania. *Tasmanian Forestry Research Council Research Report No 2*, Hobart, Tasmania.

Johnson, L.C., G.R. Shaver, A.E. Giblin, K.J. Nadelhoffer, E.R. Rasetter, J.A. Laundre and G.L. Murray 1996. Effects of drainage and temperature on carbon balance of tussock tundra microcosms. *Oecologia* 108:737-748.

Jones, R. 1969. Fire stick farming. *Australian Natural Hist*ory 16:224-8.

Jones, R. 1975. The neolithic, paleolithic and the hunting gardeners: man and land in the Antipodes. In R.P. Suggate and M.M. Cresswell (eds) *Quaternary Studies*, pp21-34. Wellington: Royal Society of New Zealand.

Kiernan, K. 1989. Multiple glaciation of the upper Franklin Valley, Western Tasmania Wilderness World Heritage Area. *Australian Geographical Studies* 27(2):208-233.

Lynch A., H., J. Beringer, P. Kershaw, A. Marshall, S. Mooney and N. Tapper 2007. Using the palaeorecord to evaluate climate and fire interactions in Australia. *Annual Review of Earth and Planetary Science* 35:215-39.

Macphail, M.K. 1979. Vegetation and climates in Southern Tasmania since the last glaciation. *Quaternary Research* 11:306-341.

Macphail, M.K. 1980. Natural regeneration processes in Tasmanian forests: a long-term perspective based on pollen analysis. *Search* 11:184-190.

Macphail, M.K., Pemberton, M. and Jacobson, G. 1999. Peat mounds of southwest Tasmania: possible origins. *Australian Journal of Earth Science* 46:667-677.

Marsden-Smedley, J.B. 1997a. Fire and fuel in Tasmanian buttongrass moorlands: regimes, characteristics, behavior and management. Unpublished PhD thesis, Department of Geography and Environmental Studies, University of Tasmania, Hobart, Tasmania.

Marsden-Smedley, J.B. 1997b. *Fire regimes in southwest Tasmania: implications for management.* Unpublished paper, Department of Geography and Environmental Studies, University of Tasmania.

Marsden-Smedley, J.B. 1998. Changes in southwestern Tasmanian fire regimes since the early 1800's. *Papers and Proceedings of the Royal Society of Tasmania* 132:15-29.

Marsden-Smedley, J.B. and J.B. Kirkpatrick 2000. Fire management in Tasmania's Wilderness World Heritage Area: ecosystem restoration using Indigenous-style fire regimes? *Ecological Management and Restoration* 1:195-203.

Mount, A.B. 1979. Natural regeneration processes in Tasmanian forests. *Search* 10:180-6.

Mount, A.B. 1982. Fire-cycles or succession in south-west Tasmania. *Search* 13:174.

Pemberton, M. 1988. Soil erosion between Birches Inlet and Elliot Bay. *Papers and Proceedings of the Royal Society of Tasmania* 122:109-114.

Pemberton, M. 1989. *Land systems survey of Tasmania Region 7- Southwest.* Department of Agriculture, Tasmania.

Pemberton, M. and P. Cullen 1995. Impacts of fire on soils in Tasmania. In *Bushfire '95*: Australian Bushfire Conference 27-30 September 1995, Hobart, Tasmania.

Shurpali, N.J., S.B. Verma and J. Kim 1995. Carbon dioxide exchange in a peatland ecosystem. *Journal of Geophysical Research* 100(D7):14319-14326.

Thomas, I. 1993. Late Pleistocene environments and Aboriginal settlement patterns in Tasmania. *Australian Archaeology* 36:1-11.

Thomas, I. 1995a. Where has all the forest gone: new pollen evidence from Melaleuca Inlet, SW Tasmania. In G. Dixon and G. Aitken (eds) *Proceedings of Institute of Australian Geographers 1993 Conference* pp295-301. Monash University Publications in Geography 45. Melbourne: Monash University.

Thomas, I. 1995b. Models and Prima-donnas in southwest Tasmania. *Australian Archaeology* 41:21-23.

Thomas, I. and G.S. Hope 1994. An example of Holocene vegetation stability from Cameron's Lagoon, a near treeline site on the Central plateau, Tasmania. *Australian Journal of Ecology* 19:150-158.

19

Restoration of mires of the Australian Alps following the 2003 wildfires

Roger Good

1178 Bungendore Road, Bungendore, Australia
rgo03227@bigpond.net.au

Genevieve Wright

NSW National Parks and Wildlife Service, Department of Environment and Climate Change, Queanbeyan, Australia

Jennie Whinam

Biodiversity Conservation Branch, Department of Primary Industries, Parks, Water and Environment, Hobart, Tasmania, Australia

Geoff Hope

Archaeology and Natural History, Australian National University, Canberra, Australia

Introduction

The alpine and subalpine bogs and fens formed some 3000 to 9000 years ago during periods of active growth of *Sphagnum spp.* and other mire plant species. The climate and the wet, acidic and anaerobic soil and environmental conditions during this period provided for the rapid accumulation of partially decomposed organic matter and the formation of deep peatbeds. The bogs and fens at the present time have an underlying peat depth of 50 cm to 150 cm, although the accumulation of organic matter and peat formation has been minimal over the past several thousand years.

Saturated peats of the bogs and fens can store between 100 and 300 litres of water per cubic metre of peat, and make a significant contribution to catchment hydrology in terms of water storage, water quality and stream flow regimes, as well as providing water to other adjacent or associated ecological communities (Ashton and Williams 1989; Good 1992; Grover and Mackenzie 2002).

At the time of European settlement, some 8500 ha of alpine, subalpine and montane mire existed across the Alps in NSW, Victoria and the Australian Capital Territory, but approximately half of this area was lost during the grazing era (early 1800s to the late 1900s) as a result of heavy grazing of the mire vegetation and physical destruction of the peatbeds by cattle and sheep trampling (Costin 1952; Academy of Science 1957).

During the 2003 wildfires, almost all the alpine, subalpine and montane bogs and fens were burnt over, with *Sphagnum* hummocks and other mire plants being severely damaged or destroyed. A further loss of about 15% of the functional bog and fen communities was recorded after the fires. The loss of these mires was the end result of the protracted pre-fire drought which led to drying of almost all surface water pools, desiccation of *Sphagnum* and other mire plant species and dehydration of the peats, providing the opportunity for the fires to burn into the peatbeds.

It was recognised that further bog and fen community losses would occur due to post-fire runoff, leading to peat tunnelling, flowline entrenchment and subsequently peat erosion, unless some remedial works were undertaken in the immediate post-fire months.

The mire restoration program began in March 2003 as a project under the Australian Alps National Parks cooperative management program, with special post-fire funding from the NSW, Victorian and Australian Capital Territory governments. Additional support was provided by research personnel from the Australian National University, the Tasmanian Department of Primary Industries and Water (Biodiversity Conservation Branch) and the Victorian Arthur Rylah Institute.

Restoration works were applied to approximately 130 bog and fen sites burnt by the fires, this number being about one 10th of the number of bogs and fens burnt over by the fires. All the restoration works were underpinned by earlier mire and catchment ecological research (CSIRO) and restoration experience accrued from the post-grazing, alpine-area restoration and revegetation program, carried out over a 15-year period between 1960 and 1974 by the NSW Soil Conservation Service (Good 1976, 2000).

Several mire recovery monitoring and research programs were also initiated immediately post-fire to guide the restoration work and quantify the benefits of the various restoration techniques and programs (Hope et al. 2006). Photo monitoring points were established at all bog and fen restoration sites. Several demonstration sites were also established to provide a visual comparative appreciation of the benefits of mire restoration and as ecological-restoration education and interpretation sites.

Impacts of the 2003 wildfires

At the time of the 2003 fires, many of the bog and fen areas were still recovering from the impacts of stock grazing, some 40 to 50 years after the cessation of grazing in the Alps. Due to the long 2000 to 2003 drought, the peat in many bogs and fens was extremely dry, predisposing the vegetative cover and the underlying peats to burning (Figure 1). Following the wildfires, it was observed that a number of these bogs and fens were so severely impacted that they were beyond restoration, while many others were subject to lesser impact and were identified as being in a condition that would benefit from the implementation of specific restoration works.

The impacts of the fires varied with the intensity and the rate of spread at the time the bogs and fens were burnt. The impacts ranged from minor burning of some *Sphagnum* (moss) hummocks (mainly *Sphagnum cristatum*), to complete destruction of the bog and fen vegetative cover, to partial burning of the underlying peatbeds. The latter resulted in the loss of the functional hydrological role of the peats and as a consequence, the loss of the ecosystem services the bogs and fens provide to catchment water storage, flow regulation and run-off filtering.

Figure 1. A valley bog extensively burnt during the 2003 bushfires

The science underpinning the restoration program

The restoration program has been underpinned by research and field trials commenced in the 1950s by CSIRO personnel (Costin 1952; Costin et al. 1959, 1960, 1964; Wimbush 1970), NSW Soil Conservation Service Alps field research staff (Good 1976, 2000), and several Victorian researchers (Wahren et al. 1996, 1999).

Wimbush (1970) carried out a detailed study of the water relationships and functional role of a bog in the upper Piper's Creek valley, this study being a baseline from which subsequent mire restoration and rehabilitation works were developed. Hahn (2004) revisited the site after the 2003 fires and re-established the hydrological studies before and after restoration works were implemented in the bog.

Mire survey work in the Alps catchments by Hope and Southern (1983), Clark (1980), Clark and Martin (1999), Whinam (1995), Whinam and Chilcott (2002), Whinam et al. (2001, 2003) and Hope (2003) has extended the ecological information base upon which the post-fire restoration works have been carried out. The research work of Hope, Wimbush and Costin also provided the background information for several recent post-graduate mire studies (S. Grover and C. Hahn), which have also contributed to an increased understanding of the ecology and hydrological functioning of bogs and fens and hence their restoration and rehabilitation.

A number of papers and reports have subsequently been published on the mire restoration and research program, (Grover and McKenzie 2002; Good 2004, 2006, 2008; Hahn 2004; Hope et al. 2005, 2006; Grover et al. 2005; Grover 2006; Growcock and Wright 2006; Hope 2006; Hope and Clark 2008). A project outline has also been placed on the Global Restoration Network website (GRN 2009).

Between 1964 and 1982, a number of mire restoration and rehabilitation techniques were trialled and applied to several domestic stock and fire-damaged bogs and fens along the main range of the Snowy Mountains, between Mount Kosciuszko and Mount Twynam, and in the Mount Jagungal area. Several fens were also 'constructed' to determine whether the techniques could be used to re-create 'mire' landscapes in sites where these ecosystems had previously existed but, due to grazing and burning impacts, had been lost from the landscape.

From these early investigations and field trials, it was identified that where the underlying peats of mires exhibited little or no physical damage from stock trampling and grazing, they could readily regenerate a vegetative cover, and their ecosystem function recovered without the implementation of any restoration works.

The recovery of the *Sphagnum spp.* component of the mire vegetation was identified as essential to this, as *Sphagnum* is the dominant contributor to the continuing accumulation of organic matter and peat formation. It was also observed and recognised that *Sphagnum* requires partial shading for its regeneration, this being provided naturally by mire shrubs (*Baeckea, Leptospermum* and *Callistemon spp.*) and taller restionaceous species (*Empodisma* and *Baloskian spp.*).

The many mire studies also identified the need to restore any surface mire pools as quickly as possible after damage or impact. These pools contribute to the rapid recovery of mire vegetation, particularly *Sphagnum, Empodisma* and *Baloskian* species, and are important in the rehydration of any dehydrated peatbeds.

The sharing of research and monitoring data has been maintained through the program of workshops and scientific forums coordinated by the Australian Alps Liaison Committee, which oversees the Alps Parks Cooperative Management Program. A number of workshops on rehabilitation principles and techniques have been conducted over the past six years for some of the above project participants, catchment management authorities, local-government personnel and community groups (Figure 2).

Figure 2. Peat coring during a mire-restoration workshop program

Objectives

The general objectives of the restoration program, drawn from past and current research and restoration programs, have been to:

1. Assist and promote the regrowth of *Sphagnum*, *Empodisma*, *Carex* and other bog and fen vegetation species;
2. Slow the rate of water movements both into and within the bogs and fens to reduce the potential for peat incision channel entrenchment;
3. Restore, where possible, the functional hydrological role of the bog and fen communities in the catchment;
4. Implement sound ecologically based techniques that will ensure sustainability of the restored sites;
5. Implement a research and monitoring program as an integral part of the rehabilitation works such that the success of the program can be assessed;
6. Ensure knowledge transfer between the research personnel, restoration specialists, national-park staff and non-government organisation volunteers involved in the program.

Implementation

There are two main tenets for this ecological restoration work. One is that peats should be prevented from becoming hydrophobic and, as a response, shedding water from the site, as well as ensuring the peats recover their capacity to take up surface and subsurface flows and regain their saturated state. The second is the recovery and restoration of a natural shade cover to provide for widespread recovery of mire plants, particularly *Sphagnum spp.* regeneration.

The most important steps in the restoration and rehabilitation program are, therefore, to slow surface flows to prevent flowline entrenchment; to create pools of surface water where *Sphagnum spp.* can regenerate; and to ensure that the pooled and subsurface water saturates the peats and spreads laterally through the peat profile and adjoining organic soils.

The ultimate objective is to restore the hydrologic regime of the bogs and fens and their functional role in catchment water storage, flow regimes and water quality.

The basic techniques for the restoration of bog and fen ecosystems are relatively simple. They involve the shading and protection of the remnant bog and fen plant populations, particularly *Sphagnum* species from high ultraviolet light levels and desiccation; the construction of straw-bale 'dams' in flowlines to create/restore surface pools; the construction of subsurface organic-matter dams to slow the flow of water from the peats; and the placement of coir and straw-filled jute mesh 'logs' as surface water-spreaders and sediment traps.

Following the placement of straw bales as flow-control structures, sods of *Sphagnum* and *Carex* species have generally been planted into them to hasten the recovery of the species and to assist the integration of the organic materials into the peat complex. In most pools, silt deposition has occurred and *Carex* sods have been planted into the sediment fans. The rehabilitation techniques and materials used in this project and outlined below are only several of a number of techniques and approaches that could be used in bog and fen rehabilitation programs.

1. Spreading and diversion of inflows to the bogs and fens

The initial work at each site has been to identify any entrenched flowlines entering the bogs and fens and to implement water-spreading structures such that flows into the bogs are slowed and spread across the entrant slope. This reduces the potential for flowline incision through the bogs and fens and reduces the rate of flow through any bogs that had already been incised, providing for slow rewetting of the dried or drying peats.Where water-spreading was required, it was achieved through the placement of coir logs and straw bales across the slopes above the bogs.

2. Pool creation and flow-control structures

The fires and a number of post-fire storm flows resulted in destruction and loss of many natural bog and fen pools. To reconstruct the organic-matter dams that naturally provide for the bog and fen pools, sterilised straw bales, some wrapped in jute mesh fabric, were placed in outflow channels to re-create surface ponding and to slow flows from the sites. The ponding provided the opportunity for regeneration of a fringing water-edge *Sphagnum* community and re-saturation of the underlying peatbeds (Figures 3a, 3b). It had previously been noted that *Sphagnum* regenerated quickest where free-standing water existed and shading was provided by other associated pool-fringing bog plants such as *Carex* and *Empodisma* species (Good 1976, 2004, 2006).

Figure 3a. Haybales inserted into a low-flow area to spread water into the surrounding peats and to form a stable pool for the re-establishment of fringing *Sphagnum* hummocks

Figure 3b. Regeneration of *Sphagnum* around the edge of the stable water pool, three years after restoration works

In sites where high flows occurred and substantial vertical incision of the peat had developed, channel depths of 1 m or more were evident. In many places, incision to the underlying gravels and bedrock had occurred, with lateral erosion at the gravel/peat interface up to 2 m either side of the incised flowline. Flows were reduced in these areas through packing straw bales into the incised flowlines, and inserting hessian bags filled with straw into the undermined peatbeds, to provide support for the peat, preventing collapse into the flowlines and subsequent loss from the system.

3. Subsurface straw-bale 'dams'

In several large drained and drying bog areas, narrow trenches were machine cut through the peats down to the underlying gravels or bedrock (1 to 2.5 m) and then filled with one to three levels of straw bales. This was done to provide a subsurface, semi-impervious organic 'dam' to assist the retention of subsurface inflows and the subsequent re-saturation of the peats. After placement, the straw bales were covered with soil and planted with sods of bog vegetation (*Sphagnum, Carex* and *Empodisma spp.*). Eventually, the straw bales will decompose and be incorporated into the peat mass.

This was initially considered to be an extreme measure to impose on several bog areas but was deemed essential if the bogs were to regain their water-holding capacity and have their hydrological role restored. This technique has been very successful, with no identifiable detrimental impact on the structure, function or vegetation of the mires, where these structures were implemented.

4. Vegetation shading

Not all vegetation of the bogs and fens was destroyed by the fire, with some live *Sphagnum* moss hummocks remaining in a number of bogs, post-fire. In other bogs, the *Sphagnum* exhibited small patches of regeneration from within the core of the burnt *Sphagnum* hummocks, with growth being assisted by some shading from overhanging dead shrub material and other dead herbaceous organic matter.

In order to further increase the potential for *Sphagnum* recovery, sterilised straw mulch was initially spread over remnant *Sphagnum* hummocks that exhibited some post-fire recovery. The straw was spread at a rate of approximately two tonnes per hectare, loosely spread to a depth of 3-5 cm. This rate of application provided approximately 70% shading for the underlying vegetation; a level of shading previously identified as optimal for initiating and enhancing *Sphagnum* regeneration (Good 2000; Whinam et al. 2003).

Due to the difficulty of transporting heavy and bulky straw bales to remote sites and the very variable results that accrued, commercial shade cloth was trialled and was found to be a suitable alternative, being easier to apply and longer lasting than the straw shading material (see Whinam et al. in this volume). Some 6000 sq. m of shade cloth have been placed over *Sphagnum* hummocks in bog and fen sites in Kosciuszko and Namadgi National Parks, with noticeable benefits to plant growth and health being evident after only two years (Figure 4).

Figure 4. Shade cloth partly removed to show the benefits of shading to post-fire regeneration of *Empodisma* and *Sphagnum spp.* Photograph by D. Whitfield

Summary

A total of 130 bogs and fens across the Alps have had some restoration works applied to them. Approximately 300 ha of bog and fen ecosystem have been restored to functional and stable mires, with large areas of adjacent organic soils and associated shrub and grassland ecosystems benefiting from the improved soil moisture regime accruing from the mire restoration works. All bogs and fens to which restoration works have been applied/implemented have recovered from a desiccated state to a saturated state, although the full recovery of this condition has taken between 15 months and six years. The full recovery of the functional role of the ecosystem and a complete bog and fen vegetation complex will still take many years to achieve.

The current restoration and rehabilitation work is nearing completion (2010/11) but the associated mire mapping and monitoring programs will continue for at least a further two-to-three years and 15 years respectively.

Important lessons have been learned during the program in terms of the techniques implemented to assist the rewetting of peatlands through peat trenching and damming, peat tunnel blocking, pooling of incised flowlines through the bogs, reconstruction and restoration of permanent bog and fen pools and the re-establishment of *Sphagnum* mossbeds by artificial shading, as well as the development of new rehabilitation/restoration materials (now commercially available).

Importantly, several of the techniques developed, trialled and implemented have been considered suitable and have already been implemented in a number of stable undisturbed bogs and fens, as part of an associated program to enhance mire survival capacity and resilience under predicted climate-change regimes (Good 2008).

New techniques for the mapping and monitoring of mires have also been developed in association with research personnel from the Australian National University, La Trobe

University, Tasmanian DPIWE, ACT Environment and ACTEWAGL research personnel, and NSW Department of Environment and Climate Change (National Parks).

The techniques and approaches to ecosystem restoration have also been taken up by several other management agencies in other bioregions and environments and interest in the program has been expressed by European mire-management personnel.

Acknowledgements

Much of the early bog and fen restoration work in the ACT was planned and coordinated by Amanda Carey. The area now known as Carey's mire will forever remind us of her contribution to bog and fen ecosystem conservation in the Australian Alps. Her work has been continued over the past three years with the same determination and enthusiasm by Dave Whitfield and other Namadgi National Park staff.

This project has been judged one of Australasia's top 25 restoration projects by the *Ecological Management & Restoration Journal*, in partnership with the Global Restoration Network (2009).

See http://www.globalrestorationnetwork.org/countries/australianew-zealand/

References

Academy of Science 1957. A Report on the Condition of the High Mountain Catchments of New South Wales and Victoria. Canberra: Academy of Science. pp62.

Ashton, D.H. and R.J. Williams 1989. Dynamics of the sub-alpine vegetation in the Victorian region pp143-168. In R. Good, (ed) *The Scientific Significance of the Australian Alps.* Proceedings of the First Fenner Conference on the Environment. Canberra: Australian Alps Liaison Committee and the Australian Academy of Science.

Clark, R.L. 1980. *Sphagnum* growth on Ginini Flats, ACT. Report to NSW National Parks and Wildlife Service

Clark, R.L. and A.R.H. Martin 1999. Sphagnum peatlands of Kosciuszko National Park in relation to altitude, time and disturbance. *Australian Journal of Botany* 47:519-536.

Costin, A.B. 1952. Hydrological studies in the Upper Snowy Catchment Area with special reference to the effects of land utilization. *Journal of Soil Conservation Service NSW* 8:5-16.

Costin, A.B., D.J Wimbush, D. Kerr, and L.W. Gay 1959. Studies in catchment hydrology in the Australian Alps 1. Trends in soils and vegetation. Technical Paper 13. Canberra: Division of Plant Industry CSIRO.

Costin. A.B., SD.J. Wimbush, and D. Kerr 1960. Studies in catchment hydrology 2 Surface run-off and soil loss. Technical Paper 14. Canberra: Division of Plant Industry, CSIRO.

Costin, A.B., J.N. Jennings, H.P. Black and B.G. Thom 1964. Snow action on Mount Twynam, Snowy Mountains, Australia. *Journal of Glaciology* 5:219-228.

Dept of the Environment, Water, Heritage and the Arts 2009. *Alpine Sphagnum bogs and associated fens.* In: Community and Species Profile and Threats Database. Canberra: DEWHA.

Good, R.B. 1976. Contrived regeneration of alpine herbfields. Paper presented to the ANZAAS Congress, Hobart, Tasmania, pp8.

Good, R.B. 1992. Kosciusko heritage : the conservation significance of Kosciusko National Park.Hurstville, N.S.W.: National Parks and Wildlife Service of New South Wales, 194p.

Good, R.B. 2000. Rehabilitation and revegetation of the Kosciuszko summit area: an historic review. In J. Mills (ed) Proceedings of the Third Australian Network for Plant Conservation Annual Conference, Albury, pp12-20. Canberra: ANPC.

Good, R.B. 2004. Rehabilitating fire-damaged wetlands in the Snowy Mountains. *Australasian Plant Conservation* 12(4):3-4.

Good, R.B. 2006. Post-fire ecosystems rehabilitation in Namadgi and Kosciuszko National Parks. In K. McCue, S. Lenz, and S. Friedrich (eds) *Caring for Namadgi – Science and people*, pp121-128. Proceedings of ACT NPA Symposium. – Canberra: National Parks of Australia.

Good, R.B. 2008. The impacts of climate change on the Alpine biota: Management Adaptations in the Australian Alps. *International Mountain Forum Bulletin*. January 2008, pp9–11. Mountain Forum Secretariat, Kathmandu, Nepal.

Good, R.B. 2008. Management adaptations to climate change in the alpine area. In K.F. McCue and S. Lenz (eds) *Corridors for survival in a changing world*, pp29-34. Proceedings of ACT National Parks Association Symposium, Canberra.

Global Restoration Network 2009. Top twenty five Australasian restoration projects. www.globalrestorationnetwork.org/countries/aystralianew-zealand/

Grover, S.P.P., and B.M. McKenzie 2002. Catotelms past and present: The characteristics and hydrology of mossland peat soils in the Australian Alps. *Australian Institute of Alpine Studies Newsletter* 14. Dec 2002. AIAS Jindabyne.

Grover, S.P.P., B.M. Mackenzie, J.A. Baldcock and W.A. Papst 2005. Chemical characterisation of bog peat and dried peat of the Australian Alps. *Australian Journal of Soil Research* 43:963-971.

Grover, S.P.P. 2006. Carbon and Water Dynamics of peat soils in the Australian Alps. Unpubl. PhD thesis, Faculty of Science La Trobe Uni. Bundoora. pp186.

Growcock, A. and G. Wright 2006. Mire rehabilitation following the 2003 bushfires in Kosciuszko National Park. – Report of works. Report to NSW Dept Environment and Climate Change, Queanbeyan. pp78.

Hahn, C. 2004. Hydrological study of a subalpine, valley–side bog in the Snowy Mountains, NSW after drought and fire. Report submitted to the Australian National University 2004. pp60.

Hope, G.S. 2003. The mountain mires of southern New South Wales and the Australian Capital Territory: their history and future. In J. Mackay (ed). *Celebrating Mountains*, pp67-79. Proceedings of the International Year of Mountains (2002) Conference, Jindabyne, Australia.

Hope, G.S. 2006. Histories of wetlands in the Australian Capital Territory and the bog recovery program. In J. McCue and S. Lenz (eds) *Caring for Namadgi – Science and people*, pp129-143. Proceedings of ACT NPA Symposium, National Parks of Australia, Canberra .

Hope. G.S and R.L. Clark 2008. A tale of two swamps: subalpine peatlands in the Kelly-Scabby area of Namadgi National Park. In J. McCue and S. Lenz (eds) *Corridors for Survival in a Changing World*, pp61-75. Proceedings of ACT NPA Symposium, National Parks of Australia, Canberra.

Hope, G.S., M. Macphail, and B. Keaney 2006. Mires of the Australian Capital Territory region. Notes on comparative sites for Palaeoenvironmental reconstruction students GE3029. Dept of Archaeology and Natural History, Australian National University, Canberra.

Hope, G.S. and W. Southern 1983. Organic deposits of the Southern Tablelands Region, New South Wales. Unpublished report to the NSW National Parks and Wildlife Service, Sydney.

Hope, G.S., J. Whinam and R.B. Good 2005. Methods and preliminary results of post-fire experimental trials of restoration techniques in the peatlands of Namadgi (ACT) and Kosciuszko National Parks (NSW). *Ecological Management and Restoration* 6(3):214-217.

Wahren, C.-H.A., R.J. Williams and W.A. Papst 1996. The ecology of wetlands and snow patches on the Bogong High Plains. Unpublished report to the Australian Heritage Commission and Victorian Dept. of Natural Resources and Environment. Melbourne.

Wahren, C-H.A., R.J. Williams and W.A. Papst 1999. Alpine and subalpine wetland vegetation on the Bogong High Plains southeastern Australia. *Australian Journal of Botany* 47:165-185.

Whinam, J. 1995. Effects of fire on Tasmanian Sphagnum peatlands. *Bushfire '95*. Australian Bushfire Conference, September 1995. Presented papers 133pp.

Whinam J., G.S. Hope, P. Adam, B.R. Clarkson, P.A. Alspach and R.P. Buxton 2003. Sphagnum peatlands of Australasia: the resource, its utilisation and management. *Wetlands Ecology and Management* 11:37-49.

Whinam, J., L. Barmuta and A. Chilcott 2001. Floristic description and environmental relationships of Tasmanian *Sphagnum* communities and their conservation management. *Australian Journal of Botany* 49(6):673-685.

Whinam, J., G.S. Hope, B. Clarkson, R. Buxton, P. Alspach and P. Adam 2003. *Sphagnum* in peatlands of Australasia: Their distribution utilization and management. *Wetlands Ecology and Management* 11:37-49.

Whinam, J. and N. Chilcott 2002. Floristic description and environmental relationships of Sphagnum communities in NSW and ACT and their conservation management. *Cunninghamia* 7(3):463-500.

Wimbush. D.J. 1970. Hydrological studies on Sphagnum bogs in the Snowy Mountains, New South Wales. MSc. Thesis, Department of Botany, University of Sydney. pp57.

20

Post-fire experimental trials of vegetation restoration techniques in the peatlands of Namadgi (ACT) and Kosciuszko National Parks (NSW), Australia

Jennie Whinam

Biodiversity Conservation Branch, Department of Primary Industries, Parks, Water and Environment, Hobart Tasmania, Australia

Jennie.Whinam@dpipwe.tas.gov.au

Geoffrey Hope

Archaeology and Natural History, Australian National University, Canberra, Australia

Roger Good

Research Associate, Australian National Botanic Gardens

Genevieve Wright

NSW Parks and Wildlife Service, Department of Climate Change, Queanbeyan, NSW, Australia

Introduction

In January 2003, wildfires, ignited by lightning strikes, burnt approximately 2.1 million ha in the ACT, NSW and Victoria. These wildfires burnt a high proportion of subalpine peatlands, (both *Sphagnum*-shrub bog and sedge fens) across the Snowy Mountains, in New South Wales and in the Brindabella Ranges, in the ACT. These communities are widely distributed and cover about 5500 ha in the alpine and subalpine zones of Kosciuszko National Park, but are of limited area and occurrence in Namadgi National Park, where they total 320 ha. The bogs and fens occur above approximately 1000 m altitude at the heads of streams and along stream floodplains where annual precipitation is greater than 850mm and mean annual temperatures below 12°C (Hope 2003).

As *Sphagnum* and its underlying peats are known to be fire-sensitive (Good 2000; Whinam and Chilcott 2002; Walsh and McDougall 2004), there was considerable concern about the long-term impacts of the fires on the bog and fen ecosystems. This concern, coupled with observation of active peat tunnelling and incision, as well as erosion of natural peat 'dams' in streams, prompted a program to monitor the recovery of the vegetation communities and to trial several techniques to initiate and enhance bog and fen vegetation recovery at selected bogs in Kosciuszko National Park and Namadgi National Park (Figure 1).

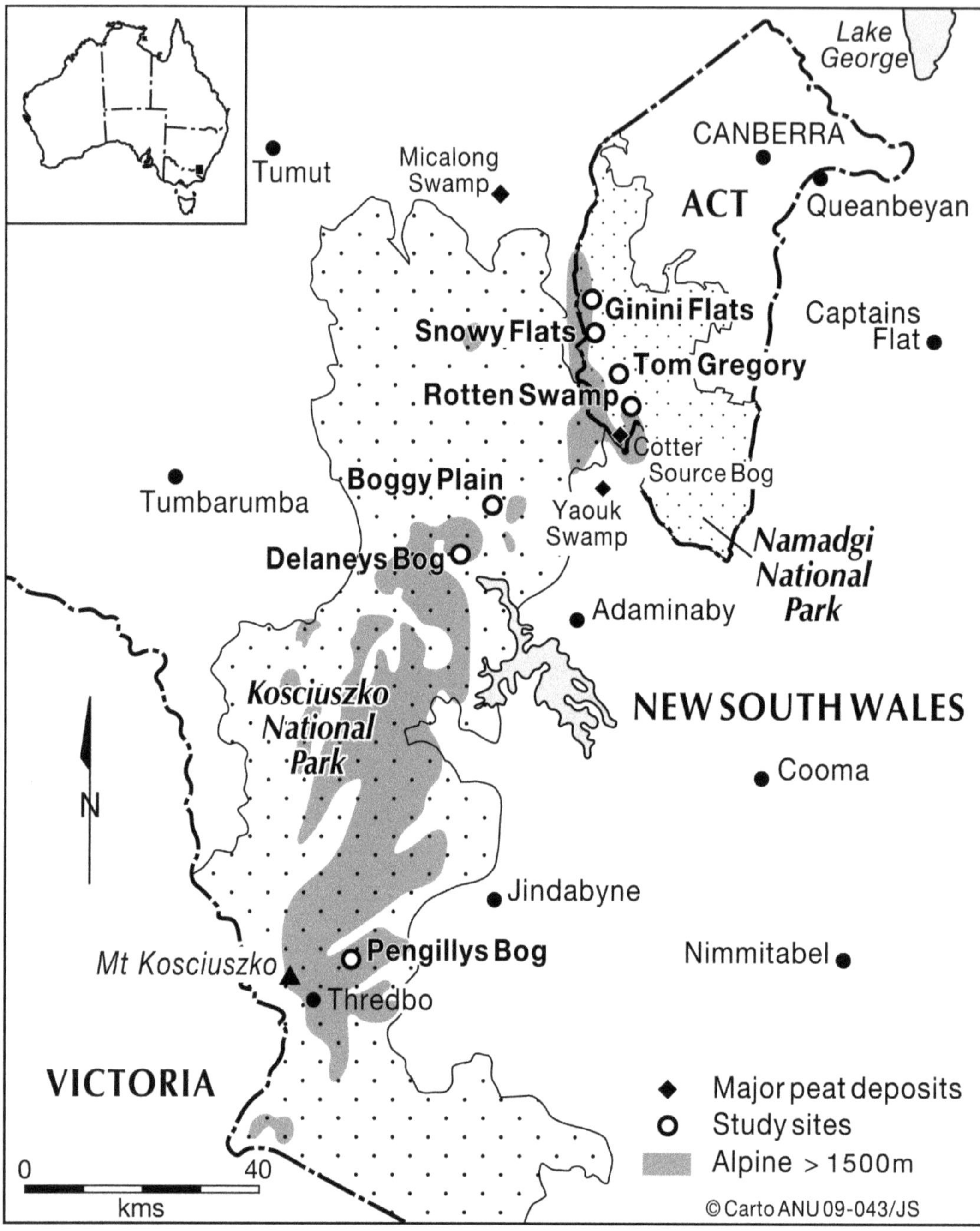

Figure 1. Location of sites mentioned in the text

The peatlands that burnt are an important functional component of the water supply to Canberra and to the Snowy Mountains Hydroelectric Scheme as they retain sediment and solutes and moderate flow discharges. The higher altitude sites also act as snow accumulation sites in winter, with the snow thawing in spring and saturating the underlying peatbeds. Most of these peatlands were recovering gradually from the effects of grazing that ended more than 50 years ago (Costin et al 1959; Wimbush and Costin 1979; Good 1992, 2000), but the January 2003 fires reversed these gains. The few long-term studies of bog recovery trends in Australia have shown that recovery after fires and grazing takes decades (Carr and Turner 1959; Wahren et al. 1996; Wahren and Walsh 2000). *Sphagnum*-shrub bog-with-pool complexes are very slow to recover from disturbance (Ashton and Hargreaves 1993; Good 2000; Whinam and Chilcott 2002; Whinam et al. 2003). Although *Sphagnum* can grow rapidly in favourable circumstances, most undisturbed bogs are currently only accumulating organic matter slowly, or may be oxidising with net loss of bulk and carbon (Clark 1980; Clarke and Martin 1999).

Post fire, while the vegetation in the montane peatlands had been severely damaged, most underlying peat surfaces were scorched by the fires but remained intact. Residual moisture in the peat prevented burning of the peat except along channel edges and other dry sites. In these sites, peat burnt to depths of 5-20 cm, leaving sterile ash surfaces with a pH 6, leading to increased weed invasion, frost heave and peat and flowline erosion.

The *Sphagnum* mires of the Australian Alps are at the climatic limit for these peatlands (Whinam et al. 2003). Evapotranspiration in the hottest month is the limiting factor for the distribution of Australian *Sphagnum* peatlands (Whinam et al. 2003). Up to 70% natural shading by shrubs has been shown to favour *Sphagnum* moss growth (Whinam 1990; Whinam and Buxton 1997). A combination of reseeding with moss and the application of small amounts of low phosphorous fertiliser have been shown to improve recovery rates after *Sphagnum* moss harvesting (Rochefort et al. 2003; Cobbaert et al. 2004) and as a strategy in post-grazing alpine rehabilitation works (Good 1976, 1992, 2000).

The results after four years of trials of vegetation recovery techniques and post-fire monitoring of the peatland vegetation in selected sites in Namadgi and Kosciuszko National Parks are presented here (Figure 1, Table 1). Experimental treatments include (a) different types of shading, (b) addition of fertiliser, and (c) transplants of *Sphagnum* moss. The aim of these trials is to assess additional peatland revegetation techniques which can be incorporated into the extensive restoration and rehabilitation program underway in Namadgi and Kosciuszko National Parks. The restoration program is attempting to restore and rehabilitate the hydrological functioning of these disturbed peatlands (Hope et al. 2005; Good et al. In press). The results of these trials may help guide future peatland restoration and management across the Alps.

Table 1. Sites for post-fire regeneration plots

Site Name	Locality	State	Altitude m	Lat Long
Boggy Plain	Northern Kosciuszko NP	NSW	1370	35°52.7' S, 148°36.0' E
Delaneys Bog	Northern Kosciuszko NP	NSW	1480	35°54.5' S, 148°33.7' E
Ginini Bog	Mt Ginini, Namadgi NP	ACT	1590	35°31.1' S, 148°46.3' E
Pengillys Bog	Central Kosciuszko NP	NSW	1680	36° 22.8' S, 148° 24.8' E
Rotten Swamp	Mt Kelly, Namadgi NP	ACT	1445	35°42.3' S, 148°53.2' E
Snowy Flat	Mt Gingera, Namadgi NP	ACT	1618	35°33.9' S, 148°51.85' E
Tom Gregory Bog	Upper Cotter, Namadgi NP	ACT	1024	35°38.9' S, 148°49.8' E

Methods

Bogs were visited between February and May 2003 to set up plots and assess the 2003 fire damage. As these sites are also localities for the threatened frog species (corroboree frog *Pseudophryne corroboree* and alpine tree frog *Litorria verreauxii alpina*), both of which are sensitive to Chytrid (*Batrachochytrium dendrobatiidis*) fungal attack (Lintermans and Osborne 2002), all boots and tools were sterilised in bleach before entering each peatland. Care was also taken to avoid the use of zinc-coated products, given the known long-term problems of zinc toxicity in alpine humus soils (Johnston and Good 1996). Fire damage was measured in terms of the degree of preservation of the biomass and peat profile. The categories of damage were identified as (i) unburnt, (ii) burnt but moss cushion intact, (iii) burnt cushion margins, (iv) charred but root mat intact and (v) burnt and incised. Sketch maps of burn classes were made and photo survey points marked with wooden stakes. These are being used to prepare post-fire maps from rectified air photography flown a few weeks after the fires (Nanson and Hope unpubl. data).

Permanent plots

Permanent plots were established at seven sites (Ginini Bog, Snowy Flat, Tom Gregory Bog and Rotten Swamp in the ACT, and Pengillys Bog, Delaneys Bog and Boggy Plain in Kosciuszko) in April-November 2003 (Figure 1).

The plots sample a range of altitudes (1050-1740 m) in granidiorite catchments, except for Ginini and Tom Gregory bogs, which have shaley metasediment basements. All are located in national parks and catchment areas in which commercial grazing has been excluded for more than 50 years. Some feral stock has been present more recently at some localities (horses at Rotten Swamp and Tom Gregory until 1986 and at Boggy Plain currently). Before the fire, all plots were a mosaic of actively growing *Sphagnum*-epacrid shrub bog in which scattered areas of *Carex* and *Empodisma* fens and *Poa* tussock-dominated herbland (the *Baeckea gunniana-Callistemon pityoides-Sphagnum cristatum* wet heathland and subalpine valley grassland communities of MacDougall and Walsh 2007). They were actively grazed by kangaroos, rabbits and rodents such as *Mastacomys fuscus.* The plots are located in open valley sites on gentle slopes on peats of medium depth (95% are on 40-125 cm of *Sphagnum* or humic peat with pH 4-5), subject to both surface-water run-on and runoff. Active streamlines and pools were avoided but waterlogging due to drainage lines varied between quadrat sites. Snow-patch areas were avoided but snow lie varied from a few months at Pengillys to shortlived cover at the other plots.

All plots were damaged severely by the 2003 fire (Table 2) but individual quadrats had variable damage, ranging from being unburnt to quadrats in which the surface peats had been ashed. Almost all quadrats were in damage classes 3 (36%), 4 (34%) and 5 (26%), hence had experienced complete death of vegetation above the surface. It is impossible to standardise quadrats from different localities, but similar settings and substrates are intended to allow comparisons for treatments, damage and other factors between plots.

Table 2. Distribution of damage classes by site

	Quadrats by damage class by site						
Damage class	**Boggy Plain**	**Delaneys**	**Ginini Bog**	**Pengillys Bog**	**Rotten Swamp**	**Snowy Flat**	**Tom Gregory**
0	0	0	1	0	0	0	0
1	2	1	0	1	0	0	0
2	1	0	0	0	0	2	0
3	4	4	3	21	7	10	4
4	2	4	2	29	7	4	1
5	1	1	4	24	1	4	0
Total number of quadrats	10	10	10	75	15	20	5

At each bog, between five and 75 x 0.25 m^2 quadrats were established and located by fixed stainless-steel pins driven into the basal gravels. This quadrat size has been used in *Sphagnum* peatlands to monitor natality and mortality rates (Whinam 1995). The quadrats were located in a range of sites with different degrees of fire damage, and sited on hummocks or in hollows. Each plot was assessed for fire damage, surface pH, peat depth and species regeneration. Individual plants were located by 5 cm squares within the quadrat to allow estimates of survival. Quadrats were remeasured twice a year for five years between 2004 and 2009 for plant cover, floristic diversity, peat accumulation/depth and pH values.

Fertilisation and transplants

At Snowy Flat and Pengillys Bog, additional quadrats were set up adjacent to the monitoring plots. Several had fertiliser applications (80 g) of low-phosphorous, slow-release pellets (Osmocote) to determine whether fertiliser would increase moss growth rates (Whinam and Buxton 1997; Rochefort et al. 2003). In March 2003 at Snowy Flat, transplants of fist-sized plugs of living *Sphagnum* cushion were placed in some quadrats to test whether they could act as nucleii for future growth, and some received both fertiliser and a transplant. In October-November 2003, larger transplants, approximately 20 x 20 x 30 cm, cut with a spade from living *Sphagnum* and the underlying peat, were placed in holes in shaded hollows at Tom Gregory and Pengillys bogs. These were hand-fertilised with 10 g of Osmocote at the time of transplant.

Shading

At Pengillys Bog, three shading treatments were established in burnt and partially burnt peat areas during October 2003, using 70% shade-cloth material and loosely spread straw (Figure 2, overleaf). Five plots were established, each one including a 20 m long and 1.6 m high 70% shade-cloth fence aligned east-west. Four quadrats were placed on each side of the vertical shade fence and solar radiation levels on northern and southern sides were recorded over the 2004 summer. The effect of shading was compared with a second treatment, in which 15 m lengths of shade cloth were pegged down horizontally, to lie loosely over the bog surface. A third shade treatment was established using sterilised grass straw spread at a rate equivalent to two tonnes per hectare (70% cover equivalent).

At Rotten Swamp and Ginini Bog, several lengths of shade cloth were laid out across areas of hummock which were exhibiting a small amount of natural regeneration, to enable subjective comparison with adjacent unshaded areas. At Rotten Swamp, five shaded quadrats can be compared with five unshaded quadrats placed 2-4 m away from the shade cloth. The shade treatments will demonstrate whether shading has a beneficial effect on recovery and allow us to differentiate between shade alone (vertical shade cloth) and shade with humidity/evapotranspiration control (mulch and horizontal shade cloth).

Data analysis

The number of plots and number of quadrats per plot at each site varied, largely due to logistical constraints (Table 3). In general, quadrats were remeasured each autumn and spring from 2003 until spring 2005, and then in autumn 2006 and 2007 (Table 4 and Table 5). At Delaneys Bog, which is immediately adjacent to the Snowy Mountains Highway in the north of Kosciuszko National Park, three of the quadrat markers had disappeared by autumn 2007.

Virtually all quadrats had no, or little, vegetative cover post fire, while some showed little regrowth over the entire period of observations. Even in quadrats where regrowth has occurred, there were taxa that showed little or no recovery in any quadrats. Formal statistical analysis was restricted to the analyses of data from those taxa, families, lifeforms and regeneration types for

Figure 2. Set up of shade, fertiliser and transplant plots at Pengillys Bog October 2003

Table 3. Number of plots and number of quadrats per plot at each site

	Site						
Plot	**Boggy Plain**	**Delaneys**	**Ginini Bog**	**Pengillys Bog**	**Rotten Swamp**	**Snowy Flat**	**Tom Gregory**
1	1	1	1	14	5	4	1
2	1	1	1	14	5	4	1
3	1	1	1	14	5	4	1
4	1	1	1	14		4	1
5	1	1	1	14		4	1
6	1	1	1	5			
7	1	1	1				
8	1	1	1				
9	1	1	1				
10	1	1	1				
Total number of quadrats	10	10	10	75	15	20	5

Table 4. Number of quadrats at which observations were made in each survey time

	Site							
Year season	**Boggy Plain**	**Delaneys**	**Ginini Bog**	**Pengillys Bog**	**Rotten Swamp**	**Snowy Flat**	**Tom Gregory**	**Total**
2003 Autumn	10	10	10	5	5	20	5	65
Spring			10	70	5	20	5	110
2004 Autumn			10	55	15	20	5	105
Spring			10	75	15	20	5	125
2005 Autumn	10	10	10	55	15	20	5	125
Spring			10	70	15	20	5	120
2006 Autumn			10	70	15	20	5	120
2007 Autumn	10	7	10	75	15	20	5	142
Total	**30**	**27**	**80**	**475**	**100**	**160**	**40**	**912**

The experimental design of treatments based on combinations of shading, fertiliser and moss transplants is shown in Table 5.

Table 5. Treatments based on combinations of physical support and regrowth aids employed at different sites, with the number of replications of factor combinations employed at individual sites

		Site							
Regrowth aid	**Cover**	**Boggy Plain**	**Delaneys**	**Ginini Bog**	**Pengillys Bog**	**Rotten Swamp**	**Snowy Flat**	**Tom Gregory**	**Total**
Fertiliser	**Horizontal shade cloth**				**5**				**5**
	Mulch				**5**				**5**
	None						**5**		**5**
	Vertical shade cloth north side				10				10
	Vertical shade cloth south side				**10**				**510**
Fertiliser and moss transplant	**Horizontal shade cloth**				**5**				**5**
	Mulch				5				5
	None						**5**		**5**
	Vertical shade cloth north side				5				5
	Vertical shade cloth south side				**5**				**5**
Moss transplant	None						5		5
Nil	**Horizontal shade cloth**				**5**	**5**			**10**
	Mulch				**5**				**5**
	None	**10**	**10**	**10**	**5**	**10**	**5**	**5**	**55**
	Vertical shade cloth north side				**5**				**5**
	Vertical shade cloth south side				5				5
Grand total		**10**	**10**	**10**	**75**	**15**	**20**	**5**	**145**

which there is at least one time of observation when more than 50% of observations is greater than zero (G. McPherson, pers comm.). Even with this restriction, it was necessary to group taxa into families and lifeforms to obtain sufficient non-zero data for treatment comparisons to be possible.

There was substantial variance heterogeneity and evidence of inconsistency among like-treated plots. Statistical analysis was based on the use of a linear, additive and apply repeated-measures analysis of variance (ANOVA) across the time periods for which observations are available, employing the Huynh-Feldt correction for serial correlation. To allow for possible violations in model assumptions, a conservative approach is taken in the interpretation of findings. Only where p-values are less than 0.01 are significant differences assumed, with p-values less than 0.001 assumed to identify strong evidence of differences. It is not judged appropriate to apply any pairwise comparison because the p-values are highly sensitive to model violation.

Results

For illustrative purposes, cover values of plant species are shown for different treatments over time (Figures 5, 6, 8). These cover values have been calculated by dividing the sum of cover percentages of a species in a set of quadrats by the number of quadrats in the set multiplied by 100 in order to standardise the cover estimates derived from variable numbers of quadrats.

Climate of the study period

During the study period (between 2003 and 2007), daily rainfall at the Bureau of Meteorology station at the Perisher Valley Ski Centre, 1712 m, provided a reasonable estimate of precipitation at Pengillys Bog, which is 2.78 km to the north of the centre, at 1680 m, and is the highest, wettest and most southerly site. Annual precipitation values for all years of the study were below the local average (1754 mm per annum), with 2008 recording the lowest annual rainfall (1061 mm) since the local station was established in 1976 (Figure 3). Since the fire, only 2005 recorded precipitation above 1500 mm. Further, the period of study had the most consecutive years with below-average rainfall since records commenced, indicating a very dry period for the local area. The rainfall was predominantly received in spring and winter, with the spring months receiving the most. The temperature for the same period did not show any significant trends, although the mean maximum temperature for the summer months for 2003 and 2006 was three degrees above average (Bureau of Meteorology, 2009), which could have hindered the rate of *Sphagnum* recovery through bleaching. However, there is evidence that the temperature in the alpine areas in Australia has been increasing over the past 35 years (Hennessy et al. 2003).

Daily rainfall figures from automated gauges are available from the ACT mountains, and annual precipitation from Tom Gregory Bog, the lowest and driest study site, is also shown in Figure 3. Rainfall has been below the 1960-1990 mean (981 mm) since 1997 at this site, but the 2002-2003 pre-fire year was not as markedly dry as at Perisher. The other ACT sites track these results, with reasonable spring rain but dry, late summers occurring in most years.

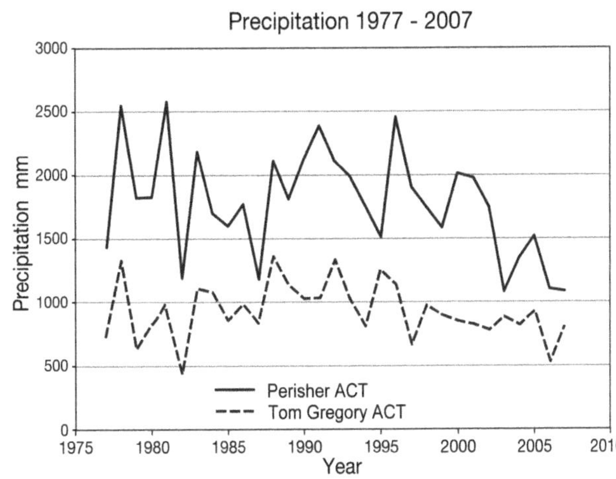

Figure 3. Annual precipitation (mm) 1977-2007 at the wettest site (Perisher Valley near Pengillys) and driest site (Tom Gregory, ACT). Long-term means are 1960-1990

Vegetation response

The data are characterised by the sparseness of presence of most taxa – of the 125 taxa recorded, only five taxa have more than 50% non-zero readings for at least one of the monitoring periods. Only by grouping taxa (families, lifeforms) is there sufficient data for meaningful analysis, and even after this grouping, there tends to be many zero readings. Figure 4 shows that mean floristic diversity (flowering plants, ferns and mosses) increased from three species/quadrat after the fire to eight species/quadrat after 32 months, and has remained stable from 32 to 50 months.

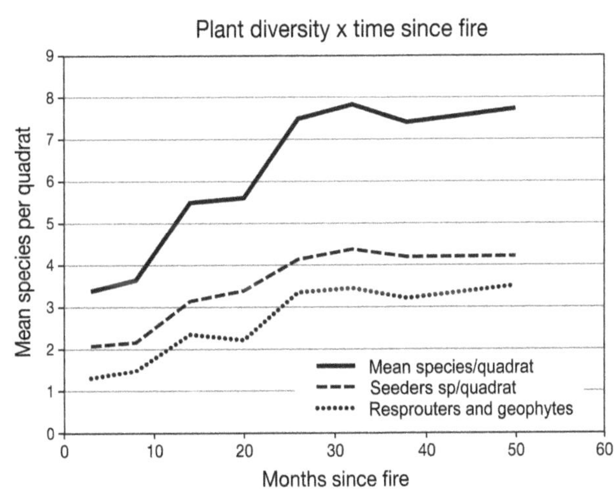

Figure 4. Floristic diversity change through time in mean species per quadrat for seeders and resprouters

Over four years, almost all quadrats recorded increases in cover due to both seed germination of grasses, sedges and epacrids and resprouting by *Empodisma minus* and myrtaceous shrubs. The ratio of resprouters to seeders has declined slightly over this time. Across all sites and treatments, the greatest number of post-fire observations of live taxa on our plots was of graminoids (32% of observations) in autumn 2003, and they retained that dominance throughout the monitoring until autumn 2007 (96% of observations).

There is insufficient data to quantitatively analyse the benefits of the restoration techniques on the recovery of *Sphagnum* moss (predominantly *S. cristatum)*. However, *Sphagnum* recovery does provide some broad indication of trends (Figure 5), with total *Sphagnum* cover in non-transplant plots under all other treatments increasing from 100 m²/ha to 850 m²/ha after 32 months, and stabilising after this to 50 months.

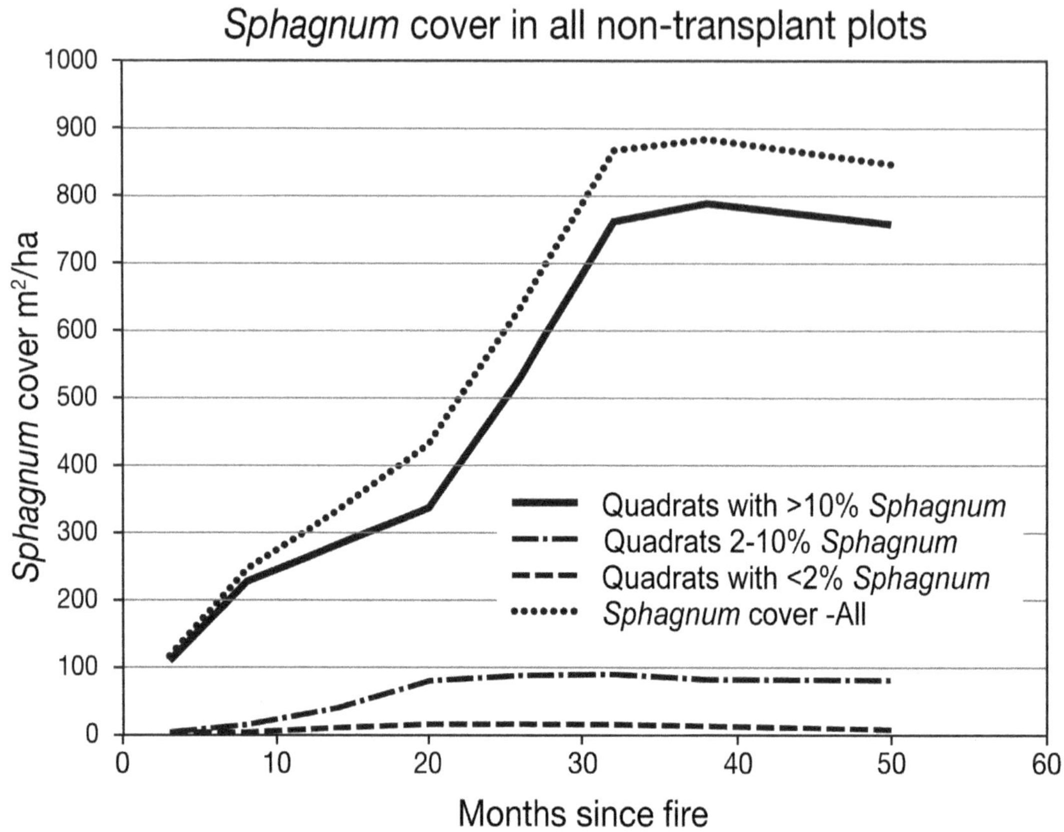

Figure 5. Total non-transplanted *Sphagnum* moss cover through time, shown as m²/ha. The contribution of plots with >10% and plots with <2% is shown

Figure 5 shows that most *Sphagnum* regeneration occurs by the expansion of the larger patches (>10% cover) of *Sphagnum* after 20 months. Many quadrats did not regenerate over 50 months, and low-cover patches (<2%) did not always persist. At 20 months, some sheltered and moist sites showed signs of regeneration and recruitment of moss, but exposed burnt hummocks remained dead. The proportion of non-transplant quadrats containing live *Sphagnum* rose from 10% after the fire to 62% at 32 months and has since declined to 50% at 50 months. This may reflect continuing below-average rainfall at all sites.

There is no strong evidence that between-site variability in respect of changes in percentage cover is greater than within-site variability for any taxa or broader grouping that could be analysed ($p>0.01$). Statistical analysis does indicate that the pattern and extent of recovery is dependent on the treatment or treatments employed for some taxa and broader (family and/or lifeform) groupings. Based on total percentage vegetation cover for all native species (not graphed here), there is evidence of treatment difference ($p<0.001$), with the earliest and most extensive recovery associated with horizontal shade cloth combined with the addition of fertiliser. This treatment favoured species that can resprout after fire when compared with species that are seed obligates. While different treatments have favoured different taxa, in general revegetation treatments that include the addition of fertiliser provide the fastest rate of recovery of vegetation cover. A range of taxa contributed to increases in vegetation cover, including *Poa costiniana,* other grasses, graminoids, bryophytes and shrubs. The combination of fertiliser and moss transplants had the greatest impact on the cover of *Poa costiniana,* however horizontal shade cloth alone provided for the greatest increase in the vegetation cover of *Empodisma minus* and *Baloskion australe*, while assisting the early establishment and regeneration of remnant *Sphagnum* plants. Straw mulch favoured only the Cyperaceous graminoids.

Figure 6 shows the response of *Sphagnum* to the shading treatments, with controls for comparison. Horizontal shade appears to maintain *Sphagnum* cover, in comparison with the small increases in cover values seen in the controls after 30 months.

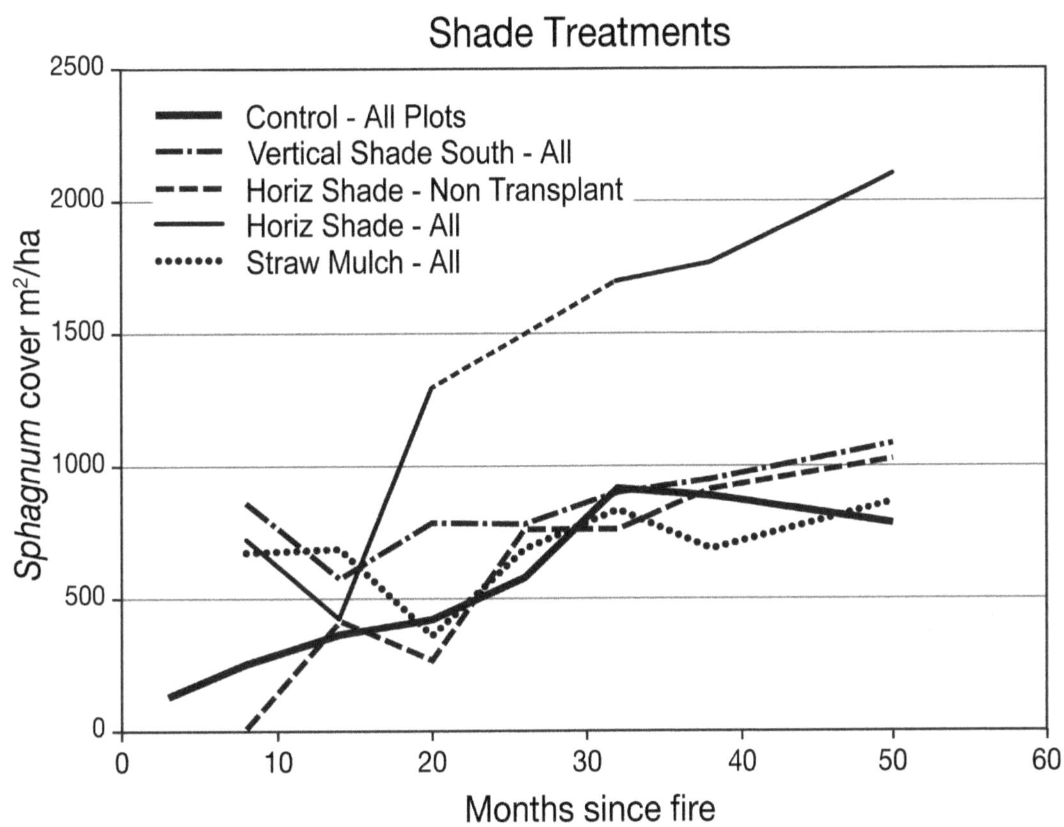

Figure 6. Comparison of treatments in terms of *Sphagnum* cover m^2/ha: 1. Control, 2. Horizontal shade without transplants, 3. Mulch straw (no transplants), 4. Fertiliser, transplants and all shade treatments

Where plots were treated with horizontal shade cloth, there was good recruitment and the greatest expansion of transplants after 20 months. Both horizontal and vertical shade treatments showed declines in some transplant cover in the first six months, but then a continuous increase in cover to 50 months. When shade cloth was removed, the exposed vegetation was thicker, taller and greener than adjacent unshaded plots, but further measurements will reveal whether these effects persist over time (Figure 7, overleaf).

The assessment of the effects of all horizontal shade-cloth-only treatments on species cover combines Rotten Swamp and Pengillys Bog data for comparison with controls at these sites. When compared with unshaded plots, there is significant difference in recovery of all native plant species (percentage vegetation cover) treated with horizontal shade cloth over time ($p=0.001$), although forbs are the only lifeform that show significant increases in cover over time ($p=0.005$). Although the overall results from Rotten and Pengillys are similar, the data from Rotten Swamp are obtained from limited plot data and so results rely heavily on an assumption that the observations there are typical of the entire site. There was considerable inter-plot variability at Pengillys Bog, showing that not all shaded quadrats shared a positive response and that some unshaded quadrats recovered strongly on their own. Straw mulch, used only at Pengillys, had less effect than the controls, and had largely disappeared by 32 months and was undetectable at 50 months. Strong winds removed the mulch from some quadrats.

Figure 7. Shade cloth at Rotten Swamp removed for survey March 2007

The effects of fertiliser, transplants and shade on *Sphagnum* cover (m^2/ha) are shown in Figure 8 (opposite). Low overall survival rates of the fist-sized transplants at Snowy Flat prevent a comparison of transplant-only and fertilised transplants. Only the larger transplants were successful, but all transplants at Tom Gregory and Pengillys received fertiliser.

For all fertilised plots, the continuing rise in moss cover after 32 months is largely due to moss expansion in quadrats with shade, but unshaded plots showed stable or declining cover. Hence, fertiliser on its own may not outperform the controls. However, fertiliser seems to support transplant growth, with most transplants also responding to shade. Analysis of plots that had no treatment applied showed there were no differences in cover of any species over time by site. However, when analysing changes in vegetation cover by only the time variable, *Empodisma minus, Baloskion australe* and *Luzula* sp. all showed significant increases in changes of percentage cover over time ($p<0.01$).

Discussion

In our study, tussock grasses (*Poa* spp.) and rhizomatous restiads (*Empodisma minus* and *Baloskion australe*) have rapidly regenerated from stools and underground rhizomes and they are replacing areas of dead *Sphagnum* moss. Shrubs of *Baeckea gunniana* and *Callistemon sieberi* were resprouting within 20 months, but epacrids, particularly *Richea continentis,* only started to appear in our plots after 20 months, suggesting that they will only regenerate from seed, as also noted by Keith (2002). In several plots, moss species, notably *Polytrichum* sp., are growing on burnt peat and dead *Sphagnum* cushions, a phenomenon noted in other burnt peatlands (Huber and Markgraf 2003).

Sphagnum regeneration has been shown to be complex and delayed. Long-term recovery to pre-fire *Sphagnum* cover will require significant colonisation and spread of *Sphagnum* beyond

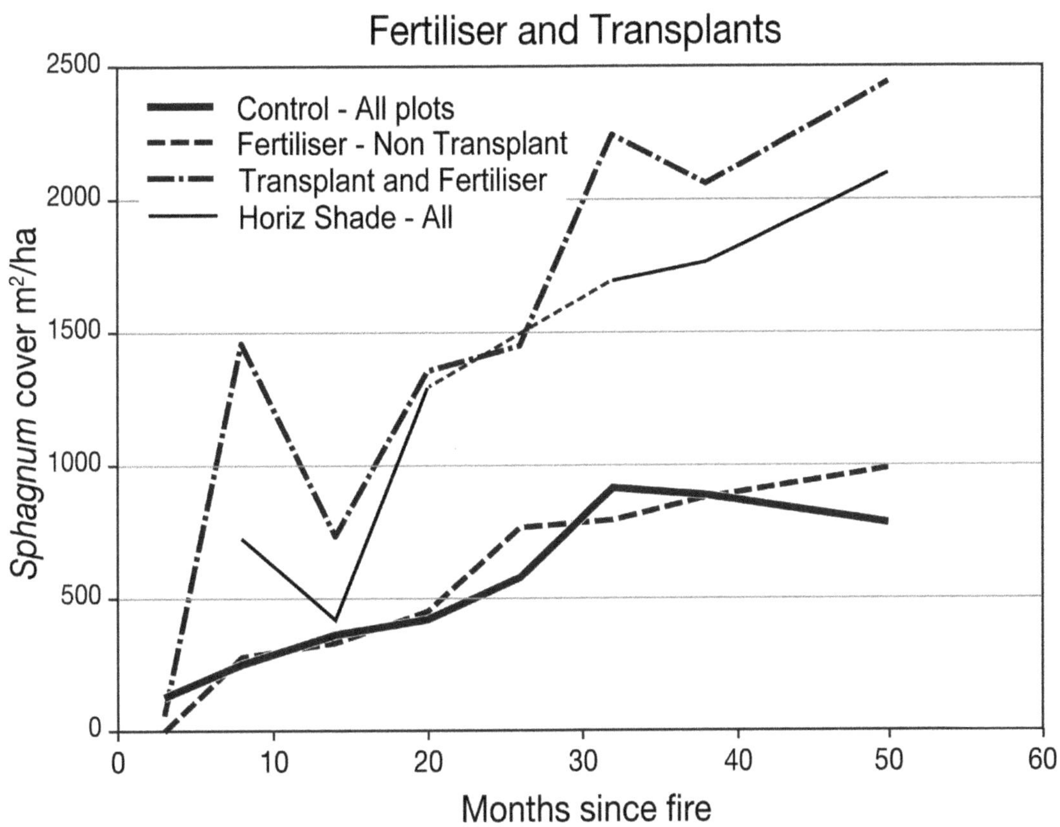

Figure 8. Effects of fertiliser and transplants on *Sphagnum* cover by comparison with controls. The quadrats include both shaded and non-shaded values. The initial steep rise is an artifact, as the earliest transplants were small and the larger transplants were not placed until month eight

the regeneration achieved in the first 32 months. Drought may in part explain the levelling out of the overall extent of *Sphagnum* cover after 32 months, but hummocks with shaded flanks and in moist shaded hollows have often continued to expand. These favourable micro-habitats seem to have enhanced survival and subsequent recovery of large patches of *Sphagnum* moss, in some cases not until more than 25 months after the fire. It is also clear that *Sphagnum* has not been able to regenerate on some plots. Such plots are typically severely burnt, or located on the exposed tops of dead hummocks, or located on drying sites where water flow has gone underground. However, after 18 months, many dead hummocks became crusted and hydrophobic and some had completely eroded down to hollow level by 50 months. These results indicate that very long recovery times are needed to restore *Sphagnum* bog structure. Hence, implementing a specific and appropriate fire-protection program for bogs and fens, particularly for high to very high intensity wildfires, may be necessary if long-term loss of bogs is to be avoided.

Peat where *Sphagnum* moss has been completely killed by the fire reached pH values of 5-7 shortly after the fire, but most moist sites returned to normal long-term values (pH 4-4.5) in the four years after the fire. The success rate of the large moss transplants at Tom Gregory is already less than 40% and it is recognised that site choice for transplants is critical, with shaded wet hollows offering the best chance for establishment and growth. The combination of transplants and fertiliser appears to be the most successful treatment at this time (2010). Ninety percent of the transplants at Pengillys Bog were successful in 2007, but site setting in wet hollows appeared to be more critical than any shade treatment. Figure 9 shows *Empodisma* expanding above a healthy transplant in a badly burnt plot with little other regeneration. In some transplants *Sphagnum* was outcompeted by tussock grasses or sedges that responded to fertiliser. These results show that transplants have some potential for re-establishing *Sphagnum* in areas from which it has been largely excluded, but are subject to a high rate of failure and probably not worthwhile where early natural recovery is observed.

Figure 9. Successful *Sphagnum* transplant and fertiliser treatment in severely damaged area at Pengillys

Differences were observed between horizontal and vertical shading. The horizontal shade cloth and litter not only provided shade protection, but also reduced evaporation and wind movement over the burnt vegetation. The vertical shade cloth benefited species number, but only provided a benefit for a small area adjacent to the shade structure. Straw mulch was reasonably persistent after one year and effective in shielding moss hummocks, but after 12 months had packed down into mats, and after two to three years had decomposed. Neither shade treatment seems to be effective in encouraging moss development when compared with controls (Figure 6). However, shading had beneficial effects for most other lifeforms, with vegetation under shade cloth observed to have increased vigour and growth when compared with unshaded plots. Straw commonly has been used in restoration projects as a protective mulch and was initially used in these trials. However, the difficulty of transport and effective spreading of the straw at the required rate make it difficult to use in large field restoration programs. The use of shade cloth was found to be a suitable alternative and is now recommended, as it provided a uniform 70% shade cover and could be readily rolled out over large areas, or laid out by slinging under a helicopter.

Recent field rehabilitation and restoration work (post 2003 fires) in the NSW and Victorian alps (Good 2006; Good et al. In press) and the field experiments reported here indicate the importance of shading for *Sphagnum* growth and recovery. Shrubs (*Epacris* spp. *Baeckea gunniana* and *Richea continentis*) and taller herbs (*Empodisma minus* and *Baloskion australe)* provide shading in stable natural mires. *Sphagnum* regenerates very slowly, if at all, after severe burning of *Sphagnum* hummocks, but artificial shading provides the critical level of protection from solar radiation (high UV levels) and wind desiccation for any surviving plants, and a considerable enhancement of the rate of regeneration of *Sphagnum*. From the results of this study, *Empodisma minus* has also benefited from artificial shading, but is not dependent on it for survival and regeneration.

While there were a significant number of weed species recorded post fire, none of the treatments appears to have differentially favoured the weeds, although some plots have become dominated by weed species, such as *Acetosella vulgaris, Cirsium vulgare, Holcus lanatus, Hypochaeris radicata* and *Taraxacum officinal.* These are largely the most damaged plots where surface peats were burnt. The enhancement of *Sphagnum* regeneration and its water-retention capacity is likely to minimise colonisation of the fire-damaged bogs and fens by particular weed species. Dead *Sphagnum* and the drying of the underlying peat potentially provide a suitable environment for active colonisation by weeds and native herbaceous species from the surrounding grasslands. Our data, and observations post fire, indicate that bog and fen ecosystems can be readily transformed to grassland if the ecosystem dries out, indicating the importance of ensuring appropriate moisture/water regimes are maintained or recovered.

The trials sought to establish the potential and value of early intervention to accelerate the natural succession to mature *Sphagnum*-shrub bogs. It is still too early to fully quantify the value of artificial shading, transplants and fertiliser applications to the regeneration of bog communities. Although considerably altered from the pre-disturbance state, regeneration of damaged bog and fen vegetation is possible, and can be enhanced by rehabilitation techniques, such as the addition of fertiliser and the provision of shading. Transplants may have a limited role in some cases. Our experimental and monitoring programs aim to support the choice of techniques that will enhance regeneration and guide future management of these peatlands.

This analysis after only four years indicates that bog and fen vegetation, particularly the significant and functionally important *Sphagnum,* can be enhanced in its rate of recovery, thus shortening the long regeneration time of bog and fen ecosystems that would occur under 'natural' regeneration. The restoration of the water-retention capacity and the functional hydrological role of the underlying peats is an extremely long process (up to 30 years). The implementation of vegetation restoration techniques is only the starting point to full recovery of these systems, which will require sensitive ecologically based management for many years. Remeasurement of plots in the future will allow the longer-term efficacy of the trials to be assessed more fully.

Bogs and fens have a fundamental role in catchment hydrology in the Alps and should be maintained in a condition that ensures catchment stability, water yield, water quality and stream flow regimes. This is likely to become an even greater management issue as the importance of bogs and fens for catchment management becomes clearer, with climate-change predictions – including increases in bushfire intensity and occurrence – for the Alps being realised in the next decade or two (Zylstra 2008).

Acknowledgements

This paper is dedicated to the memory of our friend and colleague Amanda Carey, who was an energetic, professional and cheerful advocate for the conservation and restoration of montane peatlands. We thank Brett MacNamara and Dave Whitfield, Namadgi National Park, ACT; Murray Evans, ACT Parks Conservation and Lands; Ken Green, NSW NPWS; and Colin de Paget, Alan Wade and Bren Weatherstone for their active encouragement and help. Trish McDonald, ACT Parks Conservation and Lands, assisted with most of the fieldwork and data analyses and provided cheerful project support. The analyses would not have been possible without the understanding, determination and resolve of our biometrician, Glen McPherson. The research monitoring was supported by Parks Conservation and Lands (ACT TAMS), the Australian National University, the Biodiversity Conservation Branch (Tasmanian DPIW), Namadgi National Park and the NSW National Parks and Wildlife Service. Shading, replanting and water spreading measures have been put in place by Namadgi and Kosciuszko National Parks staff as part of the post-2003 fire catchment rehabilitation program, with the welcome assistance of numerous park-care volunteers. This paper has been improved by addressing the comments of an anonymous referee.

References

Ashton, D.H. and G.R. Hargreaves 1993. Dynamics of subalpine vegetation at Echo Flat, Lake Mountain, Victoria. *Australian Journal of Ecology* 12:35-60.

Carr, S.G.M. and J.S. Turner 1959. The ecology of the Bogong High Plains, II. Fencing experiments in grassland. *Australian Journal of Botany* 7:34-63.

Clark, R.L. 1980. *Sphagnum* growth on Ginini Flats. Canberra: ACT Report to NSW National Parks and Wildlife Service.

Clarke, P.J. and A.R.H. Martin 1999. *Sphagnum* peatlands of Kosciuszko National Park in relation to altitude, time and disturbance. *Australian Journal of Botany* 47:519-536.

Cobbaert, D., L. Rochefort and J.S. Price 2004. Experimental restoration of a fen plant community after peat mining. *Applied Vegetation Science* 7:209-220.

Costin, A.B., D.J., Wimbush, D. Kerr and L.W. Gay 1959. Catchment hydrology in the Australian Alps 1. Div. Plant Industry CSIRO Technical Paper N013, pp36. Canberra: CSIRO.

Good, R.B. 1976. Contrived regeneration of alpine herbfields. Paper presented to the ANZAAS Congress, Hobart, Tasmania. pp8.

Good, R.B. 1992. *Kosciusko Heritage – The conservation significance of Kosciusko National Park.* Sydney: National Parks and Wildlife Service and Surrey Beatty & Sons.

Good, R.B. 2000. Rehabilitation and revegetation of the Kosciuszko summit area: an historic review. In J. Mill (ed.) *Proceedings of the Third Australian Network for Plant Conservation Annual Conference*, pp12-20. Albury: Australian Network for Plant Conservation.

Good, R.B. 2004. Rehabilitating fire-damaged wetlands in the Snowy Mountains. *Australasian Plant Conservation* 12:3-4.

Good, R.B. 2006. Post-fire ecosystem rehabilitation in Namadgi and Kosciuszko National Parks. In K. McCue, S. Lenz and S. Friedrich (eds) *Caring for Namadgi, Science and People*, () pp121-128. Canberra: National Parks Assoc (ACT).

Good, R.B., G. Wright, J. Whinam and G. Hope 2010. Restoration of mires of the Australian Alps following the 2003 wildfires. In S.G. Haberle, J. Stevenson and M. Prebble (eds), *Altered Ecologies: Fire, Climate and Human Influence on Terrestrial Landscapes*, pp353-362 Terra Australis 32. The Australian National University, Canberra: ANU E-Press.

Hennessy, K., P. Whetton, I. Smith, J. Bathols, M. Hutchinson and J. Sharples 2003. The impact of climate change on snow conditions in mainland Australia. Melbourne: CSIRO Atmospheric Research.

Hope, G.S. 2003. The mountain mires of southern New South Wales and the Australian Capital Territory: their history and future. In J. Mackay and Assoc. (eds) *Celebrating Mountains. Proceedings of an International Year of the Mountains Conference,* pp67-79. Jindabyne: Australian Alps Liason Committee.

Hope, G.S., J. Whinam and R. Good 2005. Methods and preliminary results of post-fire experimental trials of restoration techniques in the peatlands of Namadgi (ACT) and Kosciuszko National Parks (NSW). *Ecological Management and Restoration* 6:215-218.

Huber, U.M. and V. Markgraf 2003. European impact on fire regimes and vegetation dynamics at the steppe-forest ecotone of southern Patagonia. *The Holocene* 13:567-579.

Johnston, S.W. and R.B. Good 1996. The impacts of exogenous Zinc on the soils and plant communities of Carruthers Peak, Kosciuszko National Park NSW. *Proceedings of ASSSI and NZSSS National Soils Conference, Melbourne*. 3:117-118.

Keith, D. 2002. Population dynamics of an endangered heathland shrub, *Epacris stuartii* (Epacridaceae). Recruitment, establishment and survival. *Austral Ecology* 28:67-76.

Lintermans, M. and W. Osborne 2002. *Wet and Wild; a Field Guide to the Freshwater Animals of the Southern Tablelands and High Country of the ACT and NSW.* Canberra: Environment ACT.

McDougall, K. and N. Walsh 2007. Treeless vegetation of the Australian Alps. *Cunninghamia* 10(1):1-57.

Rochefort, L., F. Quinty, S. Campeau, K. Johnson and T. Malterer 2003. North American approach to the restoration of *Sphagnum* dominated peatlands. *Wetlands Ecology and Management* 11:3-20.

Wahren, C-H.A. and N.G. Walsh 2000. Impact of fire in treeless sub-alpine vegetation at Mt Buffalo National Park, 1982-1999. Report to the Australian Alps Liaison Committee.

Wahren, C-H.A., R.J. Williams and W.A. Papst 1996. The ecology of wetlands and snow patches on the Bogong High Plains. Melbourne: Unpublished report to Australian Heritage Commission and Dept of Natural Resources and Environment.

Walsh, N.G. and K.L. McDougall 2004. Progress in the recovery of the flora of treeless subalpine vegetation in Kosciuszko National Park after the 2003 fires. *Cunninghamia* 8:439–452.

Whinam, J. 1990. A study of the ecology of Tasmanian *Sphagnum* peatlands. Ph.D. thesis, University of Tasmania, Australia.

Whinam, J. 1995. Effects of fire on Tasmanian *Sphagnum* peatlands. *Bushfire '95*. Australian Bushfire Conference 27-30 September 1995, presented papers 133pp.

Whinam, J. and R. Buxton 1997. *Sphagnum* peatlands of Australasia: an assessment of harvesting sustainability. *Biological Conservation* 82:21-29.

Whinam, J. and N. Chilcott 2002. Floristic description and environmental relationships of *Sphagnum* communities in NSW and the ACT and their conservation management. *Cunninghamia* 7:463-500.

Whinam, J., G.S. Hope, P. Adam, B.R. Clarkson, P.A. Alspach and R.P. Buxton 2003. *Sphagnum* peatlands of Australasia: the resource, its utilisation and management. *Wetlands Ecology and Management* 11:37-49.

Wimbush, D.J. and A.B. Costin 1979. Trends in vegetation at Kosciuszko. I. Grazing trials in the subalpine zone, 1957-1971. *Australian Journal of Botany* 27:741-787.

Zylstra, P. 2008. Climate change, forest flammability and strategic management options in Kosciuszko National Park. In K. McCue and S. Lenz (eds) *Corridors for Survival in a Changing World*, pp9-20. Canberra: National Parks Assoc. ACT.

21

The archaic and puzzling record of Lake Xere Wapo, New Caledonia

Janelle Stevenson

Archaeology and Natural History, ANU College of Asia and the Pacific, Australian National University, Canberra

Janelle.Stevenson@anu.edu.au

Richard Gillespie

Archaeology and Natural History, ANU College of Asia and the Pacific, Australian National University, Canberra; School of Earth and Environmental Sciences, University of Wollongong, Wollongong

Geoff Hope

Archaeology and Natural History, ANU College of Asia and the Pacific, Australian National University, Canberra

Geraldine Jacobsen

Institute for Environmental Research, Australian Nuclear Science and Technology Organisation, Menai

Stewart Fallon

Research School of Earth Sciences, ANU College of Physical Sciences, Australian National University, Canberra

Vladimir Levchenko

Institute for Environmental Research, Australian Nuclear Science and Technology Organisation, Menai

Introduction

Research into the palaeoenvironmental history of New Caledonia was begun independently by Hope and Stevenson in the early 1990s. While the original work of Hope and colleagues was centred around questions of the long-term vegetation dynamics of maquis and rainforest within the ultramafic terrain of New Caledonia (Hope and Pask 1998; Read et al. 2000), Stevenson and colleagues were exploring questions of human impact and the detection of initial human settlement (Stevenson and Dodson 1995; Stevenson 1998; Stevenson et al. 2001; Stevenson 2004). Hope and Stevenson later came together to work on the longest record so far recovered from the tropical southwest Pacific, Lake Xere Wapo in southwest New Caledonia.

Having published the initial findings from this site (Stevenson and Hope 2005), a major problem remained, that of a robust chronology. The 12 m core XW-B reached radiocarbon background shortly after 300 cm, had several significant age inversions and had what appeared to be a very shallow Holocene sequence of fewer than 20 cm. Stevenson and Hope revisited Lake Xere Wapo in 2005, collecting new material for further dating aimed at untangling the chronology. Results from this new dating program are reported here, with a summary of previous radiocarbon determinations.

Environmental setting

New Caledonia (20-23° S and 164-167° E; Figure 1) has a tropical to subtropical climate influenced by the prevailing southeast trade winds. The average annual rainfall for the region is around 3000 mm yr^{-1}, with the warmest and wettest months being from December to April and the driest from August to November. Annual rainfall is also highly variable and years of rainfall shortage are linked to the El Niño – Southern Oscillation (ENSO) phenomenon (Morliere and Rebert 1986).

The archipelago is probably best known for its rich and distinctive flora, with the high rate of endemism (80% of an estimated 3000 species) thought to occur as a result of its Gondwanan origins and the unusual ultramafic terrain that covers approximately one third of the main island (Morat et al. 1984). Within this terrain at the southeast end of the island is the Plaine des Lacs, an old plateau around 180-250 m in altitude and crossed by a series of low ridges separated by gently sloping areas with numerous lakes. Within this terrain lies Lake Xere Wapo (22° 17.5′ S, 166° 58.5′ E) at an altitude of 220 m (Figures 1 and 2). The lake is roughly triangular in plan, is approximately 0.85 km^2 in area and has a catchment consisting of the area within ~100 m of the lake. The lake is usually shallow, only 1-2 m in depth, with shelves of laterite on the northern shoreline flooded on occasion. There are no feeder streams and it is thought that the basin formed by solution, possibly influenced by some fault control, with the lake rising after heavy rain and draining underground to the Wajana River.

The gentle slopes that surround Lake Xere Wapo are an ultramafic complex of hartzbergite and serpentinite mostly covered in a ferritic soil mantle, although in places characterised by an iron pan crust known as 'sols cuirasse'. These soils are high in iron, magnesium, manganese and nickel, and plant growth is challenged by a lack of phosphorous, potassium and nitrogen. Vegetation on ultramafic substrates therefore tends to have a distinct species composition, rich in local endemics and lacking species from adjacent substrates. In many parts of the world, the ultramafic floras are species-poor (Brooks 1987), but in New Caledonia the species diversity of the ultramafic terrain is high and the vegetation ranges from stunted maquis (heath-shrubland) to structurally complex rainforest, with many of the rainforest species not confined to the ultramafic substrate.

The vegetation surrounding Lake Xere Wapo is characterised by a bushy maquis dominated by either *Gymnostoma deplancheanum* or *Dacrydium araucarioides*, with various Myrtaceae forming a denser, closed scrub down to the water's edge (examples: *Austromyrtus altemifolius*, *Babingtonia lerattii*, *Melaleuca gnidioides*, *Syzygium ngoyensis*, *Tristaniopsis glauca*, *T. guillaimii*, *Uromyrtus myrtoides*, *U. emarginata*). Two sedges (*Costularia xyridioides* and *Schoenus brevifolius*) grow in the shallow water margins. The low nutrient status for all lakes and streams within this landscape is evident from the sparseness of aquatic flora and fauna.

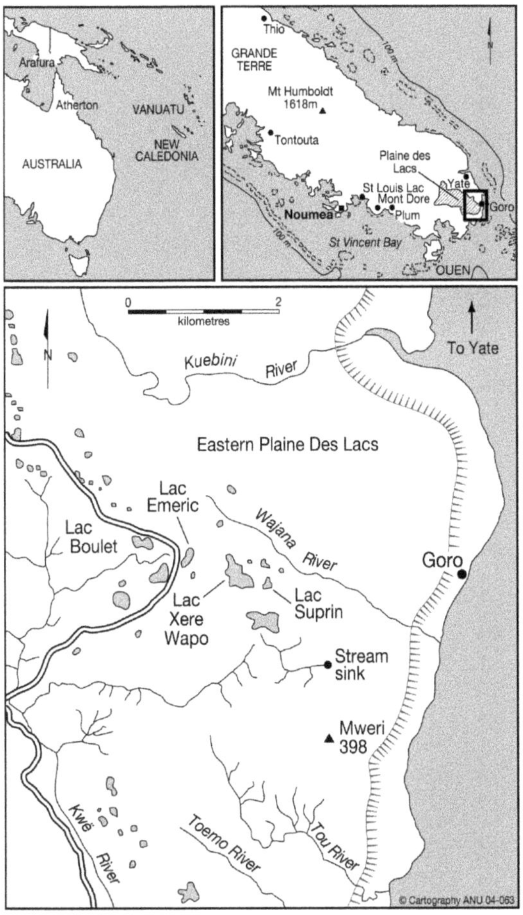

Figure 1. Location map of Lake Xere Wapo within the Plaine des Lacs region of New Caledonia

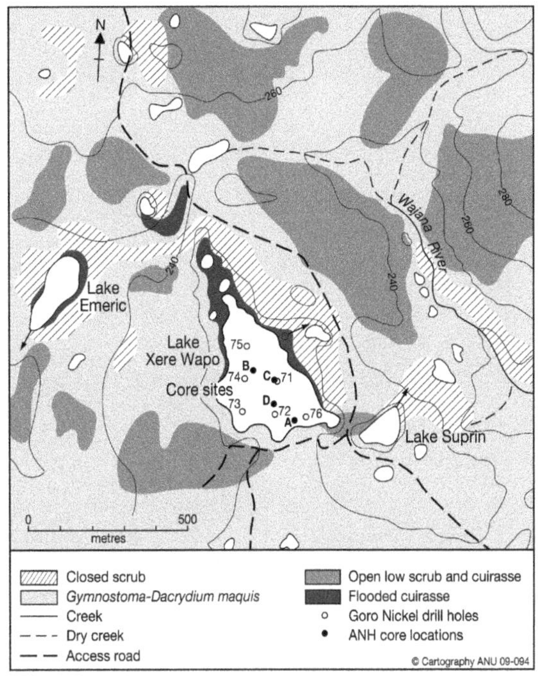

Figure 2. Site map of Lake Xere Wapo showing coring locations

Of note is that, while the diversity of the dicotyledonous flora of the Plaines des Lacs region is high, grasses are absent.

Methods and materials

Lake sediment coring

Sediment core locations for Lake Xere Wapo are shown in Figure 2. A 50 cm core, XW-A, was taken in June 1992, about 25 m from the southern edge of the lake, and a 1280 cm core, XW-B, was taken in December 1994 from the centre of the lake in a water depth of ~60 cm. In both cases, sediment was collected with a D-section corer, except for the interval 1000 cm to 1280 cm in XW-B, where a Livingston piston corer was used. Core XW-B was chosen for the pollen and charcoal analyses, published in Stevenson and Hope (2005), given the greater depth of sediment recovered. Some difficulty was encountered in collecting material from the upper 3 m, however, as several quite fluid horizons were encountered.

At location XW-C in 2005, a 60 cm mud-water interface core was collected and extruded in the field into 0.5 cm slices. Lake sediments were also sampled from 30-230 cm at this location with a D-Section corer as the upper sediments were too loose for collection with a Livingstone corer. An attempt was made to collect sediments from 200 cm onwards (giving a 30 cm overlap with the D-Section cores) with a Livingston corer. However, two major woody layers at 290 cm and 340 cm prevented further collection using either coring system. Operations were therefore moved to site XW-D (see Figure 2), where sediments from 150-850 cm were recovered using a combination of D-Section and Livingstone corers.

In 2001, the company Vale Inco Nouvelle Calédonie drilled Lake Xere Wapo as part of a mining survey for the Goro Nickel plant (Figure 2). This drilling survey reinforced how difficult the sediments are to collect, as all five drill locations had extremely poor recovery down to 7-9 m. Below this, however, the drillers were able to recover more compact lake muds to a maximum depth of 25 m. Drill hole 01RG-76 from the Goro Nickel drilling program had the best and deepest recovery, at 25.2 m, with this and the other cores held in cool storage at the Grand Lac mining camp. Because an aim of the 2005 field season was to extend the record of XW-B, 01RG-76 was sampled at 10 cm intervals from 6.3 m to 20.35 m.

At the time of the 2005 field season, the layout of the mining survey was not known. That information became available at a later date, revealing that core location XW-C is in close proximity to drill hole 01RG-71, with the drilling notes revealing that three attempts were made to collect material from the site and only on the third attempt was there reasonable recovery.

Radiocarbon dating

So far, 31 samples from five Xere Wapo cores have been analysed by four radiocarbon laboratories, one using radiometric techniques (ANU) and three using AMS (OZ, SANU and Wk codes), along with a variety of physical and chemical pre-treatments. Radiocarbon and stable-isotope measurements were made using standard procedures at the respective laboratories (e.g. Fink et al. 2004; Hogg et al. 2006)

For the bulk sediment sample from 115-123 cm in core XW-A, the <500 µm fraction was solvent extracted, then given standard hot acid-base-acid (ABA) chemistry. Samples from core XW-B comprised three wood and three 250-600 µm bulk sediment fractions, all treated with hot HCl only. For later sediment samples from core XW-B, the 10-250 µm pollen size fractions were treated with cold HCl, cold HF, warm KOH and cold HCl. Similar pre-treatment was applied to samples from XW-C and XW-D, but on the 10-125 µm fraction.

Four paired sediment and wood samples from cores XW-C and 01RG-76 were processed in 2009. Sediment samples were treated with cold NaOCl, HCl, warm NaOH and HCl (OxABA), then sieved to produce a pollen-rich size fraction of 38-50 µm; the dominant pollen type was *Dacrydium*, which varied from 30% to 70% of the total fraction. Wood samples were treated with two cycles of cold NaOCl, hot HCl, hot pH3 $NaClO_2$, hot NaOH, HCl (OxAOxBA). Because these waterlogged wood samples were cut into matchstick-size slices with a scalpel, complete decontamination may not have been achieved; reference standard New Zealand kauri wood samples, ground to <1 mm and treated with two cycles of cold NaOCl, hot pH3 $NaClO_2$ and HCl (OxAOxA), produced clean white cellulose which yielded consistent results (OZL-485 and 486). The hypochlorite and chlorite oxidation (bleaching) steps are based on chemistry discussed in Gillespie et al. (2008).

Results

Stratigraphy

The Goro Nickel drilling program provides an overview of the depth of organic sediment within the lake (Figure 3). That drilling program and the hand coring carried out by Stevenson and Hope reveal that the surface sediments of Lake Xere Wapo are firm and the lake bed appears to have no discernable topography. Underlying this, however, are many fluid horizons, possibly resulting from voids within and around tree debris (Figure 4). Wood fragments have been

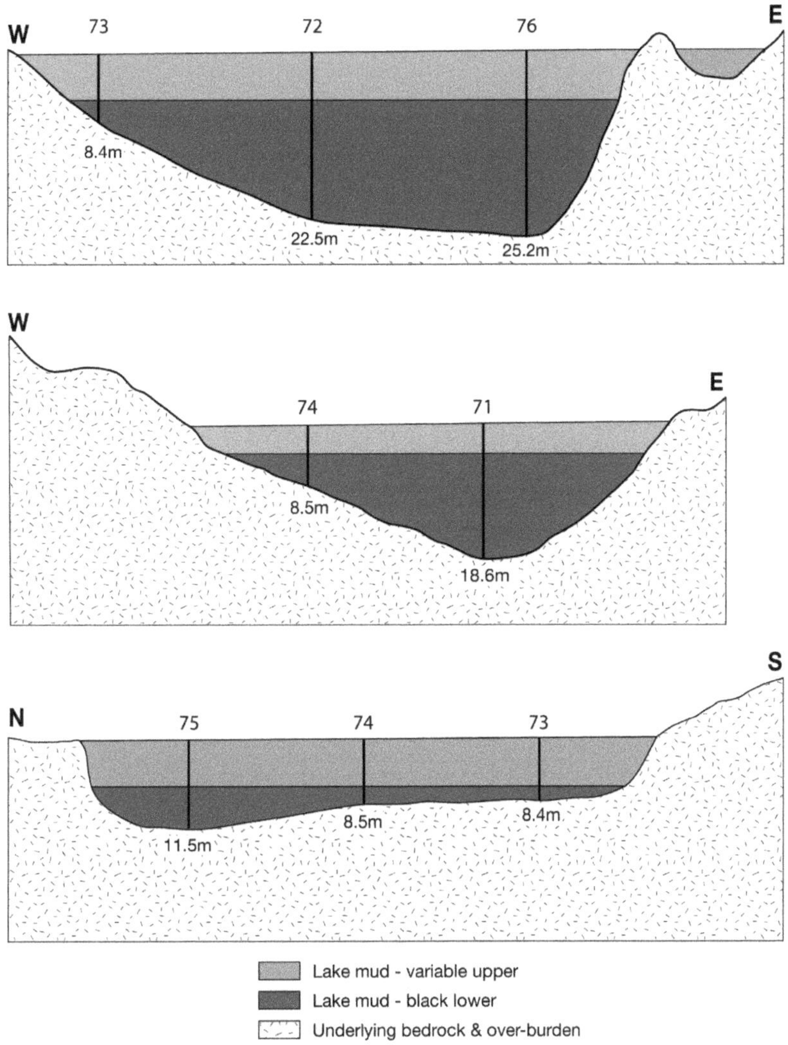

Figure 3. Goro Nickel survey sections

noted at various depths in the Lake Xere Wapo sediments. Wood from a similar context in Lake Suprin (see Figure 1), but with better preservation, has been identified as the freshwater mangrove conifer *Retrophyllum minor* (Hope and Pask 1998). The wood from Lake Xere Wapo is from several taxa, including *R. minor*, and is indicative of considerable fluctuations in water level.

Sediment descriptions of XW-A to XW-D as well as 01RG-76 are set out in Table 1 and a stratigraphic diagram illustrating the major changes in sediment type is shown in Figure 5. Across the cores there are several correlating units. Prominent among these are the black organic mud layers at 150 -175 cm in XW-B, 101-172 cm in XW-C, 150 -160 cm in XW-D, 35-100 cm in XW-A and from 60-70 cm in 01RG-76. In all cases this unit is underlain by dark yellowish brown clay (10YR 4/3-4/6) and, depending on location within the lake, this unit can be extremely fluid. In the original cores collected by Hope, this layer was described (without a Munsell chart description) as an orange clay. The next stratigraphic unit common to all the cores is another dark yellowish brown clay unit, which like the unit above, is also very fluid. However, the major sedimentary unit common to all of the deeper cores is the black organic mud that commences at 300 cm in XW-B, 356 cm in XW-D, and 275 cm in 01RG-76. From 660-2030 cm in the 01RG-76 core, the sediment is predominantly this very dark grey mud, with punctuations of lighter coloured sediment from 1530-1580 cm (dark grey), 1660-1720 cm (dark olive brown), and 1945-2030 (dark grey brown-dark olive brown). These black/grey muds and clays comprise the bulk of the sediment in the lake and

Table 1. Sediment core descriptions (see Figure 2 for locations) (Table 1 continues on page 389)

Depth (cm)	**XW-A**
0-15	Sloppy-brown organic mud with wood fragments
15-35	Orange-brown mud
35-100	Black organic mud
100-138	Orange-brown mud
138-290	Brown organic mud with thin white clay band at 160 cm
290-302	Brown-orange mud
302-455	Brown-organic mud with wood at 310-330, 430-440
455->500	Grey silty clay and coarser mineral particles
	XW-B
0-25	Yellow-brown organic clay
25-40	Dark-brown clayey organic mud – fine yellow band 32 cm
40-45	Brown organic clay
45-125	Black organic mud – fine yellow clay band at 90 cm, wood from 90-125 cm
125-140	Brown organic mud
140-150	Wood – in matrix of black mud
150-165	**Black organic mud**
162-175	Wood – in matrix of black mud
175-180	Orange clay
180-200	Brown organic mud
200-240	Dark-brown organic mud, wood and roots at 222 and 238
240-265	Light-brown/orange clay with wood fragments
265-300	Dark-brown organic mud – lighter colour + wood 275-280
300-610	Black org. mud – coarse, grading into fine from 460-480 (grey clay 410-415; wood @ 470-480, 600)
610-650	Grey mud
650-1100	Black organic mud
0-14	Dark yellowish-brown mud
Depth (cm)	**XW-C**
14-18	Transition to very dark-brown mud
18-26	Very dark-brown mud
26-28	Transition to very dark grey-brown mud
28-30	Very dark grey-brown mud
30-50	Very dark-brown organic mud – wood 31-36 cm, rootlets at 42 cm, wood 50-53 cm
50-53	Dark yellowish-brown clay – loose and watery with lots of wood
53-60	Very dark-brown mud – voids filled with dark yellowish-brown mud
60-101	Dark yellowish-brown mud with rootlets
101-172	Black organic mud – lots of wood and other organic debris – yellow bands at 149 and 156
172-185	Wood in a matrix of dark yellowish-brown clay – very fluid
185-187	Dark yellowish-brown clay – no wood
187 -230	Grades into very dark grey-brown organic clay – wood at 190, 193-196, 199-205, 214-223
	XW-D
150-159	Black organic mud
159-173	Dark yellowish-brown clay – gravels at 163-167
173-280	Dark-brown organic clay – very fluid
280-335	Dark yellowish-brown clay
335-356	Dark greyish-brown mud

356-850	Black organic mud – wood at 358-362 and 635-650
	01RG-76
0-58	Dark yellowish-brown clay
58-63	Black organic mud
63-265	Very dark-brown mud – wood and roots at 180-185
265-275	Dark yellowish-brown mud – wood throughout – very fluid
275-670	Black – wood throughout from 295
670-1660	V. dark grey mud – fine roots and wood throughout – olive-brown mud with roots and small wood fragments from 710-740
1660-1720	Dark olive-brown slightly organic clay
1720-1835	Changes gradually to very dark grey clay – wood and roots from 1825-1835 – less consolidated
1835-1950	Black organic mud – unconsolidated at time of collection
1950-1990	Very dark grey-brown – wood from 1980-1990
1990-2035	Dark olive-brown, changing to very dark grey clay

Note: 01RG-76 description from Goro Nickel core logs. Colours descriptions are based on the Munsell colour chart.

are seen in the other Goro Nickel drill holes (Figure 3). Many of the cores have extremely poor recovery for the upper 2-4 m due to the highly fluid nature of the sediments, resulting in the uppermost black unit missing from many of these locations.

Wood layers are noted throughout the cores, but the greatest concentration is in cores closer to the current shoreline. XW-D contains the least woody debris, suggesting that this may be the location where greater water depth has been most consistent over time.

Chronology

All ^{14}C determinations for Lake Xere Wapo are listed in Table 2, and are shown against depth in the schematic core stratigraphy for XW-A to D and 01RG-76 (Figure 5).

Results for cores XW-A and XW-B were discussed by Stevenson and Hope (2005). The sample from 96-107 cm, originally reported as a bulk sediment sample, was in fact the >600 μm

Figure 4. Composite showing the fluid layers within the upper sediments of Lake Xere Wapo. A) XW-D 300-358 cm. B) Zero recovery or voids in drill core 01RG74

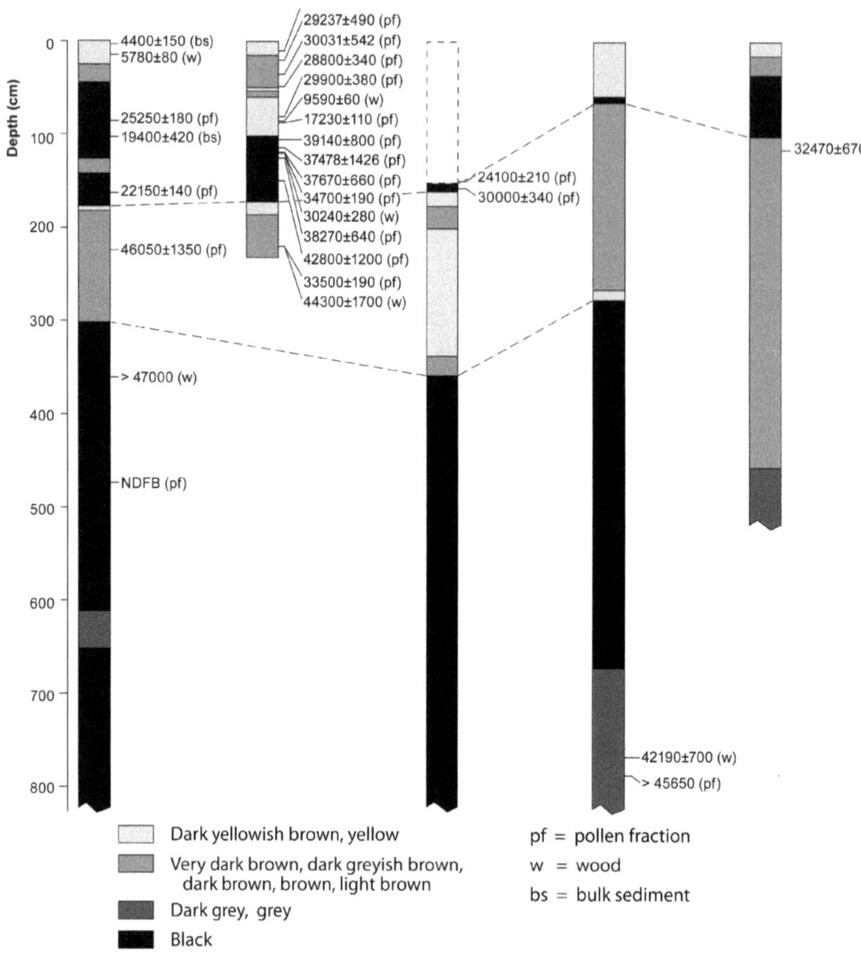

Figure 5. Schematic of core stratigraphy for XW-A to XW-D and 01RG-76, with age determinations

fraction, composed largely of small twigs; it is therefore now viewed as a wood sample. Extreme age inversions were also noted at 250-270 cm and 290-300 cm, and laboratory records indicate these two samples were unusually small for a conventional radiocarbon date; they were dropped from any discussion concerning the chronology. Further results from pollen-size fractions produced another age inversion, with background reached at 360 cm in the core XW-B.

Because the fluid yellow-brown layers have always been considered to stem from disturbance within the catchment, the focus for fresh material collected in 2005 was on the black, highly organic layers, as they were likely to be indicative of greater stability. Four samples were dated from field-extruded 5 mm slices of the XW-C mud-water interface section in an attempt to construct a better Holocene chronology. The sample from 10 cm returned an age of 8792 ± 53 (Wk-18065), but four samples covering 20 cm to 81 cm depth are surprisingly old and statistically indistinguishable at 1σ, with an error-weighted mean age of 29,540 ± 220 BP. Four pollen size fractions from the black organic mud layer spanning 101 cm to 126 cm in XW-C are also statistically indistinguishable at 1σ, with an error-weighted mean age of 37,450 ± 380 BP, and the sample from 150-151 cm at the base of this black organic mud unit in XW-C returned an age of 42,800 ± 1200 (OZJ-298). Two samples from the bottom 9 cm of the corresponding unit in core XW-D returned ages in stratigraphic order at 24,100 ± 210 (OZJ-293) and 30,000 ± 340 (OZJ-294).

The latest round of dating on the Lake Xere Wapo sediments, using paired wood cellulose and pollen-rich preparations, adds further disarray to an already confusing set of radiocarbon ages. For the three sample pairs from core XW-C, each pair has significantly different ages and both the 85-87 cm and 119-121 cm pairs are out of sequence with the two pooled means noted above. The much deeper pair from drill core 01RG-76 also returned significantly different ages, with the wood cellulose at 42,190 ± 700 BP (OZL-484) and the pollen-rich fraction (SANU-8320) not distinguishable from background.

The graph in Figure 6 plots ^{14}C age against $\delta^{13}C$, showing the expected difference between wood cellulose at $\delta^{13}C$ = -20 to -22‰ and the pollen-rich organics, most of which have $\delta^{13}C$ = -27 to -30‰. The four samples at $\delta^{13}C$ = -23 to -25‰ appear anomalous, less negative than other pollen-rich samples, but the oldest three of these were given hypochlorite oxidation plus ABA chemistry, which is expected to significantly reduce humic acids and other lignin-derived organics, thus moving $\delta^{13}C$ of residual non-pollen plant debris towards the cellulose values.

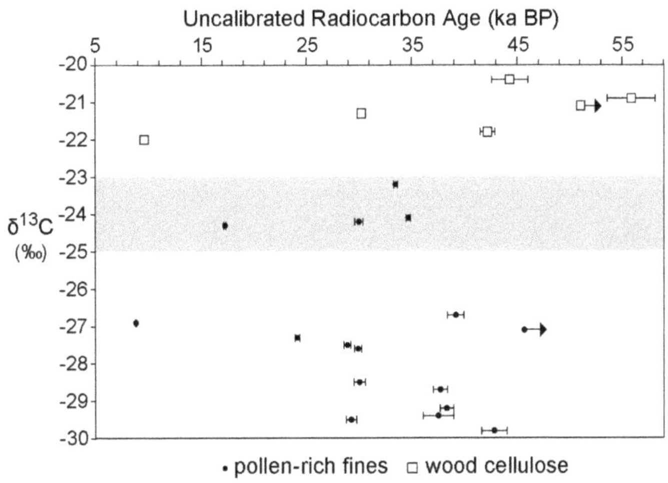

Figure 6. Radiocarbon age against $\delta^{13}C$ (‰)

Discussion

The most consistent unit dated between cores is the yellow-brown lake surface sediment that has a varying depth of 15-58 cm across the basin. In XW-B, this unit is 25 cm thick and returned two mid-Holocene dates. In XW-C, this surface layer is 14 cm thick and a sample from 10 cm returned an early Holocene age. These Holocene dates suggest that this most recent sedimentary unit may have started accumulating around 10,000-12,000 BP, and be indicative of (1) how slowly these sediments accumulate, and (2) the potential age of all underlying sediments. The next most consistent unit within the lake sediment is the upper black organic mud unit in cores XW-B, XW-C and XW-D. However, samples dated from the base of this unit in the three different cores (163 cm, 150 cm, 158 cm) returned ages of 22,150 BP, 42,800 BP and 29,910 BP respectively, suggesting no horizontal correlation in age. Another curiosity in the dating of Xere Wapo is that by 20 cm, the XW-C ages have already reached 29,540 BP, although there is no discernable stratigraphic hiatus in the sediments to explain this sudden shift in age.

The paired dating of wood and pollen-rich samples from these layers was carried out to resolve this issue. Vandergoes and Prior (2003) and a follow up study Newnham et al. (2007) highlighted the difficulties of dating the organic fraction in old sediments from several lakes and swamps in a high rainfall area of southern New Zealand. They concluded that it was the accompanying fine organic fraction in some pollen concentrates that caused young carbon contamination, or in some cases, the incomplete removal of humic acids that may have entered the older sediments through root penetration. The paired dating exercise in this study produced mixed results, with three out of the four pollen-rich dates being older than

Table 2. Carbon isotope analyses. Radiocarbon determinations by Australian National University (ANU radiometric, SANU accelerator), University of Waikato (Wk accelerator (in collaboration with UC Irvine)), Australian Nuclear Science and Technology Organisation (OZ accelerator). Pre-treatment chemistry: A = acid only; ABA = acid, base, acid; OxABA = hypochlorite, acid, base, acid; OxAOxBA = hypochlorite, acid, chlorite, base, acid; OxAOxA = hypochlorite, acid, chlorite, acid; all with distilled water washes between reagents. Reference standards are ancient NZ kauri wood samples courtesy of Alan Hogg, Waikato. NDFB = not different from background

Core	Depth (cm)	Fraction dated	Chemistry	$\delta^{13}C$ (‰)	^{14}C Age (yr BP)	Lab number
XW-A	115-123	bulk sediment <500 µm	solvent, ABA	-24.0 (est.)	32,470 ± 670	ANU-8420
XW-B	3-4	bulk sediment 250-600 µm	A	-24.0 (est.)	4400 ± 150	ANU-9797
	10-20	whole wood	A	-24.0 (est.)	5780 ± 80	ANU-9793
	83-85	pollen-rich 10-250 µm	ABA	N/A	22,150 ± 140	OZF-756
	96-107	whole wood >600 µm	A	-24.0 (est.)	19,400 ± 420	ANU-9794
	163-164	pollen-rich 10-250 µm	ABA	N/A	25,250 ± 180	OZF-755
	223-225	pollen-rich 10-250 µm	ABA	N/A	46,050 ± 1350	OZF-757
	250-270	bulk sediment 250-600 µm	A	-24.0 (est.)	8500 ± 420	ANU-9795
	290-300	bulk sediment 250-600 µm	A	-24.0 (est.)	14,660 ± 660	ANU-9796
	360	whole wood	ABA	N/A	> 47,000	OZE-449
	475	pollen-rich 10-250 µm	ABA	N/A	NDFB	OZE-448
XW-C	10-10.5	pollen-rich 10-125 µm	ABA	-26.9	8792 ± 53	Wk-18065
	20-20.5	pollen-rich 10-125 µm	ABA	-29.5	29,237 ± 490	Wk-18066
	35-35.5	pollen-rich 10-125 µm	ABA	-28.5	30,031 ± 542	Wk-18067
	48-49	pollen-rich 10-125 µm	ABA	-27.5	28,880 ± 340	OZJ-295
	80-81	pollen-rich 10-125 µm	ABA	-24.2	29,990 ± 380	OZJ-296
	85-86	wood cellulose	OxAOxBA	-22.0	9590 ± 60	OZL-481
	86-87	pollen-rich 38-50 µm	OxABA	-24.3	17,250 ± 110	SANU-8316
	105-106	pollen-rich 10-125 µm	ABA	-26.7	39,140 ± 800	OZJ-297
	110-111	pollen-rich 10-125 µm	ABA	-29.4	37,478 ± 1426	Wk-17760
	115-116	pollen-rich 10-125 µm	ABA	-28.7	37,670 ± 660	OZJ-291
	119-120	pollen-rich 38-50 µm	OxABA	-24.1	34,700 ± 190	SANU-8317
	120-121	wood cellulose	OxAOxBA	-21.3	30,240 ± 280	OZL-482
	125-126	pollen-rich 10-125 µm	ABA	-29.2	38,270 ± 640	OZJ-292
	150-151	pollen-rich 10-125 µm	ABA	-29.8	42,800 ± 1200	OZJ-298
	219-220	pollen-rich 38-50 µm	OxABA	-23.2	33,500 ± 190	SANU-8318
	220-221	wood cellulose	OxAOxBA	-20.4	44,300 ± 1700	OZL-483
XW-D	150-151	pollen-rich 10-125 µm	ABA	-27.3	24,100 ± 210	OZJ-293
	158-159	pollen-rich 10-125 µm	ABA	-27.6	29,910 ± 340	OZJ-294
01RG-76	770	wood cellulose	OxAOxBA	-21.8	42,190 ± 700	OZL-484
	790	pollen-rich 38-50 µm	OxABA	-27.1	NDFB	SANU-8320
Ref. std.	NZK-1	wood cellulose	OxAOxA	-21.1	>51,100	OZL-485(1)
		wood cellulose	OxAOxA	-20.9	55,900 ± 2300	OZL-485(2)
	NZK-2	wood cellulose	OxAOxA	-22.3	NDFB	OZL-486(1)
		wood cellulose	OxAOxA	-22.2	NDFB	OZL-486(2)

wood cellulose dates and several more age inversions, suggesting that contamination by young carbon through the fine organic fraction is not the main factor operating here.

The sediment fractions in this latest trial were treated with hypochlorite oxidation in an attempt to destroy organic contamination not removed by standard ABA chemistry (e.g. Gillespie 1990), and then concentrated to a size fraction of 38-50 µm. The paired wood samples were converted to cellulose by a combination of alkaline hypochlorite and acidic chlorite oxidations, which produced ages consistently younger than the existing dataset. The stable isotope results presented in Table 2 are another way of looking this data. One interpretation is that they verify that the latest dating exercise has removed organic contamination (most likely humic acids and other lignin degradation products) not removed by previous methods and by doing this, has increased the cellulose content derived from non-pollen plant debris. However, the results might also indicate a significant amount of non-pollen fine plant debris in the samples. The source of this may be coarse non-pollen plant debris sinking through the quite fluid muds, contributing ages that are apparently too young. However, without doing more detailed work, we can't draw any firm conclusions on the reliability of the various dated fractions.

Stevenson and Hope (2005) suggested that the upper 250 cm of Lake Xere Wapo were most likely disturbed, possibly by in-washing of catchment sediment and erosion of shallow sediments, combined with phases of aquatic woodland interrupting deposition over the past 20,000 years. It is worth noting that a number of neighbouring lakes, such as Lake Suprin and Lake Emeric, have similar sedimentation histories for the uppermost sediments, with units of dark-yellow to reddish-brown sediments alternating with darker organic units (Hope and Pask 1998). Dating of these sediments also produced age inversions and finite dates in sediments assumed to date from beyond the radiocarbon age limit.

Given the similarity in some aspects of their pollen records, Stevenson and Hope (2005) compared the Lake Xere Wapo record with that from Lynch's Crater in northern Queensland (Kershaw 1986). Radiocarbon dating of the organic-rich Lynch's Crater sediments also produced a number of age inversions, although none as severe as those encountered in the Lake Xere Wapo dating exercises, and simple ABA pre-treatment yielded more consistent results than ABOX-SC treated samples (Acid-Base-Oxidation-Step Combustion) (Turney et al. 2001a, b). The decline of *Araucaria* within the past 50 ka and increase in fire in the Australian record has been attributed variously to human impact and climate change, and remains largely unresolved. New Caledonia was not inhabited by people until c. 3000 years ago and so does not have the complicating factor of human agency, hence the importance of the record. However, a robust chronology remains elusive.

The dating of sediment cores collected by the ANU team and Goro Nickel drilling program has found that the underlying sedimentation history of the lake is more complex than first imagined, and the discussion here suggests that sediment disturbance may well be a significant factor. Fluctuations in water level and changes in sedimentary type over the past 50,000 years or more have at various times led to forest growth on the lake-bed surface. Sediments recovered in core XW-D suggest that this may be a location that had less forest incursion than other coring sites. It is this core that will now receive most attention with regard to pollen analysis and further dating. Due to the lack of free quartz in this environment, we have not been able to pursue alternative dating techniques such as optically stimulated luminescence. One avenue that may be explored is the detection of peaks in volcanic glass shards. Several peaks in volcanic glass were detected in Marine Isotope Stages 2, 3 and 4 sediments in deep-sea core Fr7/01:GC4 from the New Caledonia Basin off southern New Caledonia (Gretton 2002). If similar peaks occur in the Lake Xere Wapo sediments, then it may be possible to link and correlate the sediments with the marine stages.

Conclusion

Although some of the problems encountered in dating the Lake Xere Wapo sediments may stem from the variety of sample type and pre-treatment protocols used, later dating exercises have attempted to keep the material dated and chemistry performed more consistent. However, given the frequency of age inversions in the record – at least one in each set of radiocarbon results obtained – it is unlikely that sample contamination can explain the disparate age-depth relationships at the five core locations. Disturbance of the sediments must therefore be considered. The most recent dating programs have not resolved the problems originally reported; indeed, they add new levels of complexity and confusion to interpretations of the tempo and mode of lake-sediment formation in the Plaines des Lacs, New Caledonia.

Acknowledgements

We thank the late John Head for ANU results, the Australian Institute for Nuclear Science and Engineering for grants AINGRA01/074 (Hope), AINGRA05/157 (Stevenson), AINGRA 090/035 (Gillespie), and the Centre for Archaeological Research, ANU (Stevenson). Thanks also to Christian Tessarolo and Stephane McCoy of Goro Nickel for providing Stevenson and Hope with all their field needs during 2005, including field assistants Tom, Gab, Max, Maggie and Rachel Lestideau.

References

Brooks, R.R. 1987. New Caledonia. In R.R. Brooks (ed.) *Serpentine and its vegetation: a multidisciplinary approach*. pp330-353. London.

Fink, D., M. Hotchkis, Q. Hua, G. Jacobsen, A.M. Smith, U. Zoppi, D. Child, C. Mifsud, H. van der Gaast, A. Williams and M. Williams 2004. The ANTARES AMS facility at ANSTO. *Nuclear Instruments and Methods* B 223-224:109-115.

Gillespie, R. 1990. On the use of oxidation for AMS sample pretreatment: *Nuclear Instruments & Methods* B52:345-347.

Gillespie, R., L.K. Fifield, V. Levchenko and R. Wells 2008. New ^{14}C ages on cellulose from *Diprotodon* gut contents: explorations in oxidation chemistry and combustion. *Radiocarbon* 50:75-81.

Gretton, E. J. 2002. *Palaeoceanographic changes offshore New Caledonia for the past 140,000 years*. Unpublished Honours Thesis. Canberra: The Australian National University.

Hogg, A.G., L.K. Fifield, C.S.M. Turney, J.G. Palmer, R. Galbraith and M.G.K. Baillie 2006. Dating ancient wood by high sensitivity liquid scintillation counting and accelerator mass spectrometry – Pushing the boundaries. *Quaternary Geochronology* 1:241-248.

Hope, G.S. and J. Pask 1998. Tropical vegetational change in the late Pleistocene of New Caledonia. *Palaeogeography, Palaeoclimatology, Palaeoecology* 142:1-21.

Kershaw, A.P. 1986. Climatic change and Aboriginal burning in north-east Australia during the last two glacial/interglacial cycles. *Nature* 322:47-49.

Morat, Ph., J-M. Veillon and H.S. Mackee 1984. Floristic relationships of New Caledonian rainforest phanerogams. In F.J. Radovsky, P.H. Raven and S.H. Sohmer (eds) *Biogeography of the Tropical Pacific: Proceedings of a Symposium*, pp71-128. Honolulu: Bernice P. Bishop Museum.,

Morliere, A. and J.P. Rebert 1986. Rainfall shortage and El Niño-Southern Oscillation in New Caledonia, Southwestern Pacific. *Monthly Weather Review* 114:1131-1137.

Newnham, R.M., M.J. Vandergoes, M.H. Garnett, D.J. Lowe, C. Prior and P.C. Almond 2007. Test of AMS 14C dating of pollen concentrates using tephrochronology. *Journal of Quaternary Science* 22:37-51.

Read, J., T. Jaffré, E. Godrie, G.S. Hope and J-M. Veillon 2000. Structural and floristic characteristics of some monodominant and adjacent mixed rainforests in New Caledonia. *Journal of Biogeography* 27:233-250.

Stevenson, J. 2004. Late Holocene environmental change at Lac Saint Louis, New Caledonia. *The Holocene* 14(6):888-898.

Stevenson, J. 1998. *Late Quaternary Environmental Change and the Impact of Melanesian Colonisation in New Caledonia*. Unpublished PhD Thesis. Kensington: University of New South Wales.

Stevenson, J. and J.R. Dodson 1995. Palaeoenvironmental evidence for human settlement on New Caledonia. *Archaeology in Oceania*. 30:36-41.

Stevenson, J., J.R. Dodson and I.P. Prosser 2001. A late Quaternary record of environmental change and human impact from New Caledonia. Palaeogeography, *Palaeoclimatology, Palaeoecology* 168:97-123.

Stevenson, J and G.S. Hope 2005. A comparison of late Quaternary forest changes in New Caledonia and northeastern Australia. *Quaternary Research* 64(3):372-383.

Turney, C.S.M., M.I. Bird, L.K. Fifield, A.P. Kershaw, R.G. Cresswell, G.M. Santos, M.L. di Tada, P.A. Hausladen and Z. Youping 2001a. Development of a robust C-14 chronology for Lynch's Crater (North Queensland, Australia) using different pretreatment strategies. *Radiocarbon* 43:45-54.

Turney, C.S.M., A.P. Kershaw, P. Moss, M.I. Bird, L.K. Fifield, R.G. Cresswell, G.M. Santos, M.L. di Tada, P. A. Hausladen and Y. Zhou 2001b. Redating the onset of burning at Lynch's Crater (North Queensland): implications for human settlement in Australia. *Journal of Quaternary Science* 16:767-771.

Vandergoes, M.J. and C.A. Prior 2003. AMS dating of pollen concentrates: a methodological study of Late Quaternary sediments from south Westland, New Zealand. *Radiocarbon* 45:479-491.

22

Comparative AMS ^{14}C dating of plant macrofossils, beetles and pollen preparations from two late Pleistocene sites in southeastern Australia

Nick Porch

Archaeology and Natural History, Australian National University, Canberra, Australia
Nicholas.Porch@anu.edu.au

A. Peter Kershaw

School of Geography and Environmental Science, Monash University, Melbourne, Australia

Introduction

In the Northern Hemisphere, late Quaternary chronologies are commonly constructed using AMS ^{14}C dated plant macrofossils because they are generally argued to provide the most reliable chronology (MacDonald et al. 1991; Törnqvist et al. 1992; Snyder et al. 1994; Birks and Birks 2000; Hatté and Jull 2007). Although terrestrial macrofossils are often relatively abundant and well preserved in a variety of site types, they are still potentially subject to a range of complications that include reworking, movement of dissolved organic carbon, contamination by modern carbon due to inappropriate storage and analysis, and potentially, measurement effects relating to small sample sizes (Wohlfarth et al. 1998; Turney et al. 2000; Nilsson et al. 2001; Oswald et al. 2005; Hatté and Jull 2007).

Insects, likewise, are common and often abundant in a range of sediment types (Elias 1994; Porch 2008). Examination of their utility as dating materials has focused on beetle sclerites (Elias and Toolin 1989; Elias et al. 1991; Cong et al. 1996; Hodgins et al. 2001; Walker et al. 2001; Hormes et al. 2004; Tripp et al. 2004) and chironomid head capsules (Jones et al. 1993; Wolfe et al. 2001; Fallu et al. 2004; Hormes et al. 2007). Results suggest that insect determinations may, in some cases, be younger than determinations on other fractions, including plant macrofossils (Hodgins et al. 2001; Walker et al. 2001), bulk sediments – gyttja (Fallu et al. 2004) and humic acid fractions (Wolfe et al. 2001). The results of Child and Werner (1999) comparing bulk sediment, chironomid/cladoceran and plant macrofossil ages indicated concordance of bulk sediment and invertebrate ages with, conversely, younger macrofossil ages. In contrast, Tripp et al. (2004) report similar ages from both uncharred and charred plant macrofossils and beetle sclerites. In essence, there is a general assumption that insect sclerites may not be reliable for AMS radiocarbon dating, although this is poorly founded. Site-specific differences in insect dates and dates from other fractions suggest the observed variation may relate to the taphonomy of the various components of the sedimentary column rather than specific problems with any particular fraction.

Although the perceived advantages of dating pollen preparations (ubiquity and chemical stability), rather than bulk sediment organics, have been made explicit (Brown et al. 1989, 1992; Regnéll and Everitt 1996; Prior and Chester 2001; Vandergoes and Prior 2003), little consideration has been given to assessing the general suitability of pollen preparations in a variety of contexts. In Australia and New Zealand, there have been limited efforts to examine late Quaternary plant or insect macrofossils for dating potential. However, there has been a trend towards dating lake, swamp and playa sequences using pollen preparations or bulk sediment dates on relatively inorganic materials (e.g. Gillespie et al. 1991; Magee 1995; Chester and Prior 2004; Vandergoes et al. 2005; Cupper 2006; Kershaw et al. 2007; Newnham et al. 2007). In the latter case, the carbon being dated in organic poor sediments is likely to be effectively similar to a pollen preparation concentrate.

Almost all the studies comparing sediment or pollen and plant or insect macrofossil ages have been based on late glacial or Holocene material from extended lake and/or peat-bog sequences. In contrast, many important palaeoecological sites in the Australian region are near-surface exposures, characterised by slow rates of sediment accumulation, or relatively inorganic. Depositionally old sediments may be preserved in such contexts. Here we report on the results of a dating exercise in which subfossil seeds, beetle sclerites and 'pollen preparations' from two such sites, Spring Creek in western Victoria and Pulbeena Swamp in northwestern Tasmania, were dated. Results suggest that pollen preparations from these two sites may yield aberrantly young ages and that both beetle and plant macrofossils give more comparable and potentially more reliable results. Although sample sizes utilised here are much larger than usually available for most Quaternary sequences, they reflect the requirements of sampling for insect assemblages rather than material for radiocarbon dating. Plant and insect macrofossils are frequently preserved and often abundant in many late-Quaternary records from mesic Australia (Porch pers obs.) and may, in some cases, provide a more reliable approach for establishing robust radiocarbon-based chronologies.

Site descriptions and previous research

Spring Creek

Spring Creek, western Victoria, is a mega-fauna-bearing site, initially argued to provide evidence for late survival of a number of extinct taxa (Flannery and Gott 1984). Fossil bone has been recovered and still erodes from the edge of a thin exposure of heterogeneous gravelly silt. Flannery and Gott (1984:391) reported a single conventional radiocarbon determination of 19,800 ± 390 (Teledyne 1-11,018) on 'plant remains (leaf, seed and stem fragments)' from sediments associated with the extinct form of the macropod *Macropus giganteus*, and seven other extinct vertebrate taxa. Subsequently, four AMS radiocarbon determinations (White and Flannery 1995) were obtained on bone to test the concern that the original radiocarbon determination was not indicative of the age of megafauna – two determinations on two macropod bones: 27,268 ± 330 BP (NZA 3202) and 27,863 ± 333 BP (NZA 3201), and two further determinations on halves of a longitudinally sectioned humerus of an extinct mega-fauna taxon, *Palorchestes*, the marsupial tapir: 36,500 ± 2400 BP (NZA 3871) and 25,440 ± 150 BP (CAMS 9559/BETA 67231). Results were interpreted by White and Flannery

(1995:16) to indicate contamination of the bone samples, although the pair made no further attempt to date the depositional context of the fauna.

Pulbeena Swamp

Pulbeena Swamp, northwestern Tasmania, has been the subject of a number of palaeoecological research projects, including analysis of vertebrate fossils (Banks et al. 1976), pollen and chronology (Colhoun et al. 1982), ostracods (De Deckker 1982) and non-marine molluscs. More recently, it has become a focus of subfossil insect studies (Porch 2007). The swamp is characterised by inter-bedded peaty and marly sediments and generally very slow rates of sediment accumulation (<<10 cm/1000 years). Peats have been argued to be deposited during periods of low groundwater flux resulting from regionally reduced effective precipitation, and conversely, marls have been argued to represent periods of high groundwater flow (Colhoun et al. 1982). Several laterally continuous peat beds within the range of radiocarbon dating occur in the upper 1-1.5 m of the Pulbeena sequence.

Sample descriptions

Spring Creek SC97/5

A sample of approximately 9 kg of laminated, organic sandy silt derived from a short sequence of bedded silts immediately downstream from the concentration of bone. It represents an equivalent of the disturbed-bone-containing layers upstream, but is difficult, because of cover, to physically trace laterally: both layers at least rest upon Miocene marl and are overlain by the distinctive ferruginous gravel referred to above. The sample is 40-50 mm in thickness. Four samples taken from above SC97/5 are in total 160-180 mm in thickness, and another 150 mm of coarse relatively inorganic gravels were removed before beginning sampling. A further 250 mm of bedded organic silts underlie the sample.

Spring Creek SC97/6

A 5 cm thick, approximately 11 kg sample of heterogeneous gravelly, organic, clayey, sandy silts excavated upstream from the sampling locality of SC97/1-5. Here, there is no evidence of bedding; rather, a suggestion of mixing of several sediment types at some time in the past, although these sediments are capped with a distinctive ferruginous gravel. On top of the Tertiary marls lies a layer (50-200 mm) of silty and sandy gravel containing large rock clasts, mega-fauna bones, occasional wood and material from the marl below. Overlying this are blocky structured, gravelly, clayey, silty sands that contain occasional mega-fauna bone fragments. Interpretation of these deposits suggests they may represent several depositional and reworking episodes, almost certainly within a relatively short period, given the excellent preservation of most of the mega-fauna material before being covered by gravels and, later, Holocene alluvium.

Depositional environment of the Spring Creek samples

Based on the nature of the sediments and interpretation of the insect and plant macrofossil fauna, both Spring Creek samples were deposited at the head of a pool in a flowing creek, probably very similar to the arrangement that occurs at the site today. Fossil material in the deposit is almost certainly derived from the creek catchment above the waterfall. Significantly, the pollen and macrofossil floras and the insect assemblages, to be described elsewhere, are almost identical in both samples.

Pulbeena Swamp P3/D

Sample P3/D is a 13 kg sample of moderately humified, shelly peat from 104-107 cm below the present-day surface. The top of this peat layer elsewhere in the site has been dated previously. The resulting determination was 27,300 ± 200 (Beta-70065). Three further conventional determinations were obtained on the sample sequence referred to here. These dates, which are essentially identical because of the wide errors, range from 29,820 ± 1350 (GX-25690) at the top of the peat bed, to 30,400 ± 1450 (GX-25691) for the middle and 29,380 ± 1280 (GX-25692) for the base on the bed. There is no obvious evidence of physical disturbance of the sequence.

Depositional environment of the Pulbeena sample

The abundance of both aquatic and terrestrial shells in the Pulbeena sample, its organic matrix, the presence of sedge seeds and the remains of micro-sclerophyllous Myrtaceae (*Leptospermum* and/or *Melaleuca*) all imply deposition in a shallow swamp with flowing water. The snail assemblage is dominated by the small aquatic snail *Fluvidona* (Hydrobiidae), commonly found in springs and creeks. Terrestrial snails, especially the presence of *Victaphanta milligani,* indicate there was a vegetation canopy over the swamp surface.

Sample preparation

Plant and insect macrofossils were recovered by disaggregating approximately 10 kg of sediment in sodium pyrophosphate and washing the residues through a series of sieve screens; the Pulbeena Swamp sample was treated with hydrochloric acid before pyrophosphate to remove carbonate. Material for dating was picked from the 1 mm sieve, dried and stored in glass vials. Pollen samples measuring 5 cm^3 were processed using standard pollen preparation methods, although without acetolysis, which has the potential to introduce modern carbon. Samples were dated at ANSTO using conventional acid-base-acid pre-treatment, graphitisation and analysis procedures.

Results

All samples yielded sufficient carbon for AMS radiocarbon dating and the results are presented in Table 1. Because the majority of samples are from a context that is at the limit of the radiocarbon method and beyond the range of standard calibration curves, the results are presented in radiocarbon years BP. Two samples from Spring Creek gave results that were not different from background (OZD380 on beetle sclerites and OZD387 on *Pimelia* seeds). The other beetle-sclerites determination from Spring Creek yielded an age >44,100 BP. All other samples gave finite results.

Table 1. Results of comparative AMS ^{14}C dating of seeds, insects and pollen from Spring Creek (Series 1 and 2) and Pulbeena Swamp (Series 3)

Lab. code	Material	Sample series and identity	pmc±1SD	^{14}C yr BP±1SD	δ(^{13}C) per mil
Spring Creek					
OZD379	*Pimelia sp.* seeds	**1**-SC97/6A	0.37 ± 0.16	44,900 ± 3,600	-26.86
OZD380	beetle sclerites	**1**-SC97/6B	0.10 ± 0.11	NDFB[1]	-23.6
OZD381	pollen preparation	**1**-SC97/6C	0.72 ± 0.12	39,600 ± 1,400	-27.4
OZD382	Asteraceae seeds	**1**-SC97/6D	0.45 ± 0.11	43,400 ± 2,100	-27.9
OZD386	*Pimelia sp.* seeds	**2**-SC97/5A	-0.09 ± 0.07	NDFB[2]	-24.2
OZD387	beetle sclerites	**2**-SC97/5B	0.18 ± 0.12	>44,100	-26.0
OZD388	pollen preparation	**2**-SC97/5C	0.95 ± 0.12	37,400 ± 1,100	-26.1
Pulbeena Swamp					
OZD383	Cyperaceae seeds	**3**-P3D/A	3.33 ± 0.11	27,350 ± 260	-26.5
OZD384	beetle sclerites	**3**-P3D/B	3.03 ± 0.16	28,100 ± 410	-26.0
OZD385	pollen preparation	**3**-P3D/C	5.05 ± 0.18	24,000 ± 300	-26.1

[1]Not Distinguishable from Background (NDFB) – background for this sample is 43,400 years

[2]Not Distinguishable from Background (NDFB) – background for this sample is 46,300 years

Discussion

Reliability of macrofossil versus pollen preparations

The trend in the results of the ^{14}C AMS determinations is the general concordance of the plant and beetle macrofossil determinations and the fact that pollen preparations gave finite or younger ages; in the case of sample Series 1 from Spring Creek (SC97/6A-D in Table 1), the large errors of the two plant-macrofossil samples mean, however, they overlap the pollen age at two standard deviations. Given that the two Spring Creek samples represent laterally equivalent layers and that the flora and fauna from the samples is very similar, the two series can be considered as dating the same unit. If taken together, there are two finite dates on seeds that are close to background (OZD379, OZD382), two dates not different from background (OZD380, OZD386), a single date >44,100 (OZD387) and two pollen dates markedly younger than background. It is therefore likely the two finite plant macrofossil ages from Spring Creek are minimum ages rather than reflecting the real age of the Spring Creek sequence. For the Pulbeena Swamp sequence, the difference in the macrofossil ages relative to the single pollen age is much clearer.

In Australia, the use of pollen preparations for dating late-Quaternary sequences often stems from the apparent lack of terrestrial macrofossils in sites, particularly during the last glacial. In some sites, this absence may, however, be more apparent than real, with few Australian researchers actively searching for macrofossils (see, however, Kershaw et al. 2007). The potential problems of dating pollen samples need to be assessed, and the data presented here points to one of them, contamination by modern or younger fine organic material. Another, contamination by derived older charcoal or uncharred organics, is more likely to be significant in lacustrine sites dominated by minerogenic sedimentation, especially during periods of instability (Gillespie et al. 1991).

In the Australian context, few pollen preparations derived from Quaternary sediments are composed exclusively or even principally of pollen. Much of the material remaining after physical and chemical pollen separation techniques consists of resistant plant fragments, including fine charcoal, even after extensive treatment. The Spring Creek samples are estimated to contain <20% pollen and Pulbeena Swamp <10%, the majority of the remaining material being fine, uncarbonised plant matter, and to a lesser extent, opaque organics, possibly charcoal. For the determinations reported here, the pollen samples are biased towards fine insoluble organics other than pollen. If contamination problems by older organic material from the catchment, or younger penetrative root contamination are to be avoided, methods for the production of relatively pure pollen samples should be routinely employed (Long et al. 1992; Regnéll 1992; Richardson and Hall 1994; Regnéll and Everitt 1996; Prior 1998). Alternatively, more attention needs to be paid to sampling and processing for terrestrial macrofossils.

It is possible the finite ages for the Spring Creek pollen samples reflect contamination by modern and/or Holocene root penetration, a factor noted by Newham et al. (2007) to be a potential problem in bog contexts. A small amount of fine modern organic matter added to the Spring Creek samples would be enough to give a finite age. However, in the case of Pulbeena Swamp, a significantly greater amount of contamination would be required. The samples at both sites are derived from near-surface contexts, and in the case of Pulbeena Swamp, swamp forest occurred on the site before clearing in the late 19th century. Presumably, such contamination issues become less problematic in more rapidly accumulated sites and sites dominated by organic rather than minerogenic sedimentation, where dates on bulk organics or pollen are less likely to be influenced by the inclusion of younger carbon.

Utility of insect sclerites for AMS dating

The most interesting result from this study was the apparent consistency of the AMS ages of beetle sclerites in relation to the plant-macrofossil ages. The Spring Creek results were background (OZD380) and infinite (OZD387), in line with the postulated age of the Spring Creek sequence, noted above. The Pulbeena Swamp determination (OZD384) was completely consistent with the plant-macrofossil age from the same sample and the conventional peat age from the same layer (GX-25690 29,820 ± 1350 BP: Porch unpublished) – all dates overlap at two standard deviations.

Elsewhere, comparative AMS radiocarbon dating of insect sclerites has shown they can yield younger dates (than other materials), which have remained unexplained (Hodgins et al. 2001; Walker et al. 2001), or younger ages that have been assumed to be accurate relative to the older, contaminated, bulk sediment or humic fractions (Wolfe et al. 2001; Fallu et al. 2004), or in the case of the palaeosol dating of Hormes et al. (2004), reflect the extended period of palaeosol formation and accumulation of its constituent fractions over an extended period. Other studies have demonstrated that beetle sclerites yield ages that are comparable to ages from plant macrofossils, in line with the results reported here (Tripp et al. 2004). Similarly, there are conflicting results when comparing simple pre-treatments like ABA with more complex pre-treatments involving the isolation and purification of chitin or its derivative, chitosan. Finally, Tripp et al. (2004), however, reported no difference in ages derived from fractions that had only received a simple acid-wash pre-treatment and those that had been chemically purified to almost pure chitin.

There is clearly a need for more research into the utility of insect sclerites for AMS radiocarbon dating. This research should examine alternative pre-treatment methods, fossil insect taphonomy, whether taxonomic differences in ecology are related to differences in age, and whether these issues can explain differences between sites in the comparability of insect, plant and other determinations. These issues are important because direct dating of insect sclerites is an ideal way to demonstrate whether selected taxa are indigenous or introduced under contentious circumstances and to demonstrate the age of insect sclerites in sites with complex depositional histories.

Acknowledgements

Ursula Pietrzak for pollen preparation and other laboratory assistance. The Whiteheads of 'Spring Creek' and John Head of the now defunct Pulbeena Limeworks Company for access to the sites on their properties. This study was supported by Australian Nuclear Science and Technology Organisation ANISE/ARC Grant No. 97/200R. Andrew Jenkinson (ANSTO) checked and recalculated the sample ages. Reviews from Henk Heinjis and Janelle Stevenson improved the paper.

References

Banks, M.R., E.A. Colhoun and G. van de Geer 1976. Late Quaternary *Palorchestes azael* (Mammalia: Diprotodontidae) from northwestern Tasmania. *Alcheringa* 1:159-166.

Birks, H.H. and H.J.B. Birks 2000. Future uses of pollen analysis must include plant macrofossils. *Journal of Biogeography* 27:31-35.

Brown, T.A., D.E. Nelson, R.W. Mathewes, J.S. Vogel and J.R. Southon 1989. Radiocarbon dating of pollen by accelerator mass spectrometry. *Quaternary Research* 32:204-212.

Brown, T.A., G.W. Farwell, P.M. Grootes and F.H. Schmidt 1992. Radiocarbon AMS dating of pollen extracted from peat samples. *Radiocarbon* 34:550-556.

Chester, P. I. and C. Prior 2004. An AMS ^{14}C pollen-dated sediment and pollen sequence from the late Holocene, southern coastal Hawke's Bay, New Zealand. *Radiocarbon* 46:721-731.

Child, J.K. and A. Werner 1999. Evidence for a hardwater radiocarbon dating effect, Wonder Lake, Denali national park and preserve, Alaska, U.S.A. *Géographie physique et Quaternaire* 53:407-411.

Colhoun, E.A., G. van de Geer and W.G. Mook. 1982. Stratigraphy, pollen analysis and paleoclimatic interpretation of Pulbeena Swamp, northwestern Tasmania. *Quaternary Research* 18:108-126.

Cong, S., A.C. Ashworth, D.P. Schwert and S.M. Totten 1996. Fossil evidence for a short warm interval near 40,000 yr B.P. at Titusville, Pennsylvania. *Quaternary Research* 45:216-225.

Cupper, M.L. 2006. Luminescence and radiocarbon chronologies of playa sedimentation in the Murray Basin, southeastern Australia. *Quaternary Science Reviews* 25:2594–2607.

De Deckker, P. 1982. Non-marine ostracods from two Quaternary profiles at Pulbeena and Mowbray Swamps, Tasmania. *Alcheringa* 6:249-274.

Elias, S.A. 1994. *Quaternary Beetles and Their Environments.* Washington: Smithsonian Institution Press.

Elias, S.A., P.E. Carrara, L.J. Toolin and A.J.T. Jull 1991. New radiocarbon and macrofossil analyses from Lake Emma, Colorado: Revision of the age of deglaciation. *Quaternary Research* 36:307-321.

Elias, S.A. and L.J. Toolin 1989. Accelerator dating of a mixed assemblage of late Pleistocene insect fossils from the Lamb Spring site, Colorado. *Quaternary Research* 33:122-126.

Fallu, M.-A., R. Pienitz, I.R. Walker and J.T. Andrews 2004. AMS ^{14}C dating of tundra lake sediments using chironomid head capsules. *Journal of Paleolimnology* 31:11-22.

Flannery, T.F., and B. Gott 1984. The Spring Creek locality, southwestern Victoria, a late surviving megafaunal assemblage. *Australian Zoologist* 21:385-422.

Gillespie, R., J.W. Magee, J.G. Luly, E. Dlugokencky, R.J. Sparks and G. Wallace 1991. AMS radiocarbon dating in the study of arid environments: examples from Lake Eyre, South Australia. *Palaeogeography, Palaeoclimatology, Palaeoecology* 84:333-338.

Hatté, C. and A.J.T. Jull 2007. Radiocarbon Dating: Plant Macrofossils. In S.A. Elias, (ed.) *Encylopedia of Quaternary Science*, pp2958-2965. Amsterdam: Elsevier Scientific Publishing.

Hodgins, G.W.L., J.L. Thorpe, G.R. Coope and R.E.M. Hedges 2001. Protocol development for purification and characterisation of sub-fossil insect chitin for stable isotopic analysis and radiocarbon dating. *Radiocarbon* 43:199-208.

Hormes, A., W. Karlen and G. Possnert 2004. Radiocarbon dating of palaeosol components in moraines in Lapland, northern Sweden. *Quaternary Science Reviews* 23:2031-2043.

Hormes, A., M. Blaauw, S.O. Dahl, A. Nesje and G. Possnert 2007. A radiocarbon wiggle-match dated age-model for the proglacial lake Hervavatnet, western Norway. *Geophysical Research Abstracts* 9, European Geosciences Union.

Jones V.J., R.W. Battarbee and R.E.M. Hedges 1993. The use of chironomid remains for AMS 14C dating of lake sediments. *The Holocene* 3:161–163.

Kilian M.R., J. van der Plicht, B. van Geel and T. Goslar 2002. Problematic 14C-AMS dates of pollen concentrates from Lake Gosciaz (Poland). *Quaternary International* 88:21–26.

Long, A., O.K. Davis and J. Delanois 1992. Separation and 14C dating of pure pollen from lake sediments: nanofossil AMS dating. *Radiocarbon* 34:557-560.

Kershaw, A.P., G.M. McKenzie, N. Porch, R.G. Roberts, J. Brown, H. Heinjis, M.L. Orr, G. Jacobsen and P.R. Newall 2007. A high resolution record of vegetation and climate through the last glacial cycle from Caledonia Fen, southeastern highlands of Australia. *Journal of Quaternary Science* 22:481-500.

MacDonald, G.M., R.P. Beukens and W.E. Kieser 1991. Radiocarbon dating of limnic sediments: a comparative analysis and discussion. *Ecology* 72:1150-1155.

Magee, J.W., J.M. Bowler, G.H. Miller and D.L.G. Williams 1995. Stratigraphy, sedimentology, chronology and palaeohydrology of Quaternary lacustrine deposits at Madigan Gulf, Lake Eyre, South Australia. *Palaeogeography, Palaeoclimatology, Palaeoecology* 113:3-42.

Mensing S.A. and J.R. Southon 1999. A simple method to separate pollen for AMS radiocarbon dating and its application to lacustrine and marine sediments. *Radiocarbon* 41:1-8.

Newnham, R.M., M.J. Vandergoes, M.H. Garnett, D.J. Lowe, C. Prior and P.C. Almond 2007. Test of AMS 14C dating of pollen concentrates using tephrochronology. *Journal of Quaternary Science* 22:37-51.

Nilsson, M., M. Klarqvist, E. Bohlin and G. Possnert 2001. Variation in radiocarbon age of macrofossils and different fractions of minute peat samples dates by AMS. *The Holocene* 11:579-586.

Oswald, W.W., P.M. Anderson, T.A. Brown, L.B. Brubaker, F.S. Hu, A.V. Lozhkin, W. Tinner and P. Kaltenrieder 2005. Effects of sample mass and macrofossil type on radiocarbon dating of arctic and boreal lake sediments. *The Holocene* 15:758-767.

Porch, N. 2007. Beetle Records: Late Pleistocene of Australia. In S.A. Elias (ed.) *Encyclopedia of Quaternary Science*, Volume 1, pp179-190. Amsterdam: Elsevier Scientific Publishing.

Porch, N. 2008. Invertebrates in landscape archaeology. In B. David, and J. Thomas, (eds) *Handbook of Landscape Archaeology*, pp457-467. World Archaeological Congress, Research Handbook Series. California: Left Coast Press.

Prior, C. 1998. A new method in the preparation of pollen for AMS radiocarbon dating. Society of American Archaeology, 63rd Annual Meeting (PLACE), *Abstracts*, 244.

Prior, C. and P. I. Chester 2001. Precision radiocarbon dating of a late Holocene vegetation history. In M. Jones and P. Sheppard (eds), *Australian Connections and New Directions: Proceedings of the 7th Australasian Archaeometry Conference,* pp285-294. Auckland: University of Auckland.

Regnéll, J. 1992. Preparing concentrates for AMS dating – a methodological study from a hard-water lake in southern Sweden. *Boreas* 21:373-377.

Regnéll, J. and E. Everitt 1996. Preparative centrifugation – a new method for preparing pollen concentrates suitable for radiocarbon dating by AMS. *Vegetation History & Archaeobotany* 5:201-205.

Richardson, F. and V.A. Hall 1994. Pollen concentrate preparation from highly organic Holocene peat and lake deposits for AMS dating. *Radiocarbon* 36:407-412.

Snyder, J.A., G.H. Miller, A. Werner, A.J.T. Jull and T.W. Stafford Jr. 1994. AMS-radiocarbon dating of organic-poor lake sediment, an example from Linnévatnet, Spitsbergen, Svalbard. *The Holocene* 4:413-421.

Törnqvist, T.E., A.F.M. de Jong and K. van der Borg 1992. Accurate dating of organic deposits by AMS 14C measurement of macrofossils. *Radiocarbon* 34:566-577.

Tripp, J.A., T.F.G. Higham and R.E.M. Hedges 2004. A pretreatment procedure for the AMS radiocarbon dating of sub-fossil insect remains. *Radiocarbon* 46:147-154.

Turney, C.S.M., G.R. Coope, D.D. Harkness, J.J. Lowe and M.J.C. Walker 2000. Implications for the dating of Wisconsinan (Weichselian) late-glacial events of systematic radiocarbon age differences between terrestrial plant macrofossils from a site in SW Ireland. *Quaternary Research* 53:114-121.

Vandergoes, M.J., R.W. Newnham, F. Preusser, C.H. Hendy, T.V. Lowell, S.J. Fitzsimons, A.G. Hogg, H.U. Kasper and C. Schlucter 2005. Regional insolation forcing of late Quaternary climate change in the Southern Hemisphere. *Nature* 436:242-245.

Vandergoes, M.J. and C.A. Prior 2003. AMS dating of pollen concentrates. A methodological study of Late Quaternary sediments from south Westland, New Zealand. *Radiocarbon* 45:479-491.

Walker, M.J.C., C. Bryant, G.R. Coope, D.D. Harkness, J.J. Lowe and E.M. Scott 2001. Towards a radiocarbon chronology of the Late-Glacial: sample selection strategies. *Radiocarbon* 43:1007-1019.

White, J.P. and T. Flannery 1995. Late Pleistocene fauna at Spring Creek, Victoria: A re-evaluation. *Australian Archaeology* 40:13-17.

Wolfe, A.P., G.H. Miller and S.L. Forman 2001. A cautionary notice regarding radiocarbon dating of humic acid extracted from arctic sediments. *31st Annual Arctic Workshop*, p106. Amherst, MA, USA: University of Massachusetts.

Wohlfarth, B., G. Skog, G. Possnert and B. Holmquist 1998. Pitfalls in the AMS radiocarbon-dating of terrestrial macrofossils. *Journal of Quaternary Science* 13:137-145.

23

Can Myrtaceae pollen of the Holocene from Bega Swamp (New South Wales, Australia) be compared with extant taxa?

Andrew H. Thornhill

Botany and Zoology, Research School of Biology, Australian National University, Canberra, Australia
Andrew.Thornhill@anu.edu.au

Introduction

Bega Swamp (Figure 1) is a seldom disturbed restiad-shrub bog (Hope et al. 2000) located 50 km inland at the eastern side of the Southern Tablelands in Wadbilliga National Park, New South Wales (36° 31' S, 149° 30' E) at an altitude of around 1080 m and with a mean annual rainfall of 800 mm to 1200 mm (Polach and Singh 1980). Rainfall is a limiting factor to plant growth in the region (Donders et al. 2007). It is thought that the swamp originated as a valley fill between Yankee Creek and Bemboka River (Polach and Singh 1980) and occupies an elongated north-south valley on a heavily forested granitic plateau (Hope et al. 2000). The site was selected by Dr Gurdhip Singh to represent montane southeastern Australia as it was a Holocene site with continuous sedimentation that would allow near annual resolution analysis as well as being surrounded by relatively undisturbed natural vegetation, providing the possibility of relating the modern pollen deposition to climatic conditions (Hope 1995).

The surrounding catchment area of Bega Swamp is predominantly a wet, tall, open sclerophyll forest, with dominant *Eucalyptus* tree species being *E. fraxinoides*, *E. fastigata*, *E. radiata* and *E. dalrympleana*, along with a rich shrub and herb understorey containing genera such as *Epacris*, *Leptospermum*, Asteraceae, *Acaena*, *Ranunculus*, *Hydrocotyle*, *Plantago*, *Wahlenbergia*, *Gonocarpus*, *Blechnum* and *Pteridium* (Green et al. 1988; Hope et al. 2000). The swamp area has a zone of *E. pauciflora* (snowgum) and *E. rubida* woodlands with a grassy shrubland suggesting that this is a frost hollow (Hope et al. 2000). *Poa* grasslands occur in the western area of the swamp, while the majority of it is thickly overgrown with *Empodisma* and *Restio*, with scattered aquatic sedges and ericaceous and myrtaceous shrubs (Hope et al. 2000). It was noted by Hope et al. (2000) that *Leptospermum lanigerum* would probably occupy parts of the swamp but had been partly removed by fire.

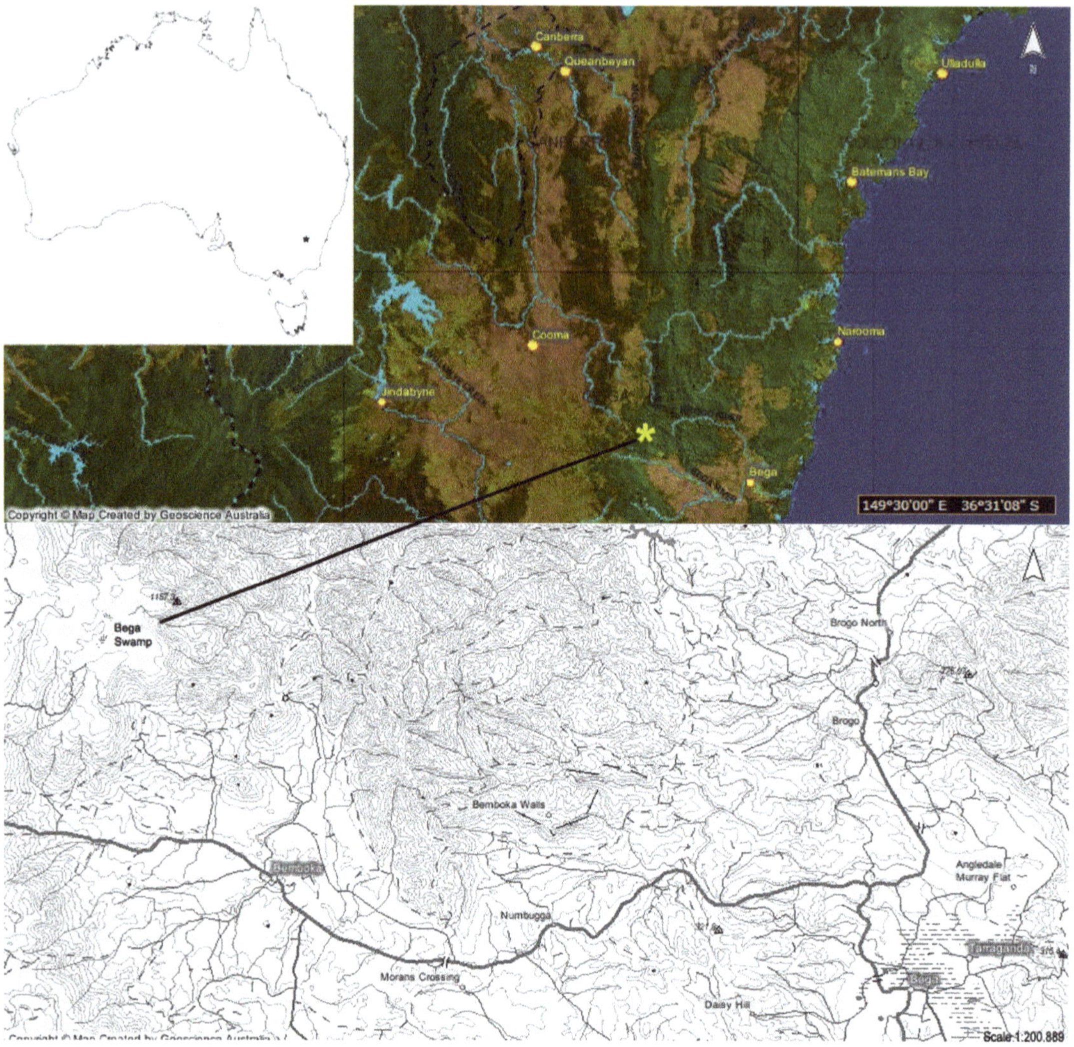

Figure 1. Locality map of Bega Swamp

Detailed pollen counts from the past 12,000 years have previously been conducted (Green et al. 1988; Hope 1995; Hope et al. 2000, 2004; Donders et al. 2007), from which 35 Myrtaceae-type pollen grains have been identified (Hope et al. 2000), with these various pollen types appearing at different times (Figure 2). Interestingly, the number of recorded Myrtaceae species housed at the Australian National Herbarium that come from the Wadbilliga National Park region is 37 (see Table 1 for species list). Various assumptions have been made about the identities of the pollen types, first by Gurdhip Singh (unpublished data) and then by Geoff Hope (Figure 2). One notable assumption is that *E. pauciflora* was a dominant eucalypt in the past, associated with a cooler and wetter climate at Bega Swamp. This is illustrated in Figure 2, which distinguishes five groups of *Eucalyptus* appearing and disappearing through the past 12,000 years. It was assumed that *Eucalyptus* 2 represented a high altitude group such as *E. pauciflora*, while *Eucalyptus* 3 and 4 were wet forest groups such as *E. fastigata* and *E. fraxinoides* (Hope 1995). The present study examined extant Myrtaceae taxa from the Bega Swamp area with light microscopy and compared images of Myrtaceae pollen grains from various sections of Singh's 1980 core using visual character comparison and a Lucid key of extant Myrtaceae pollen morphology (Thornhill unpublished) with the aim of correlating extant and fossil pollen.

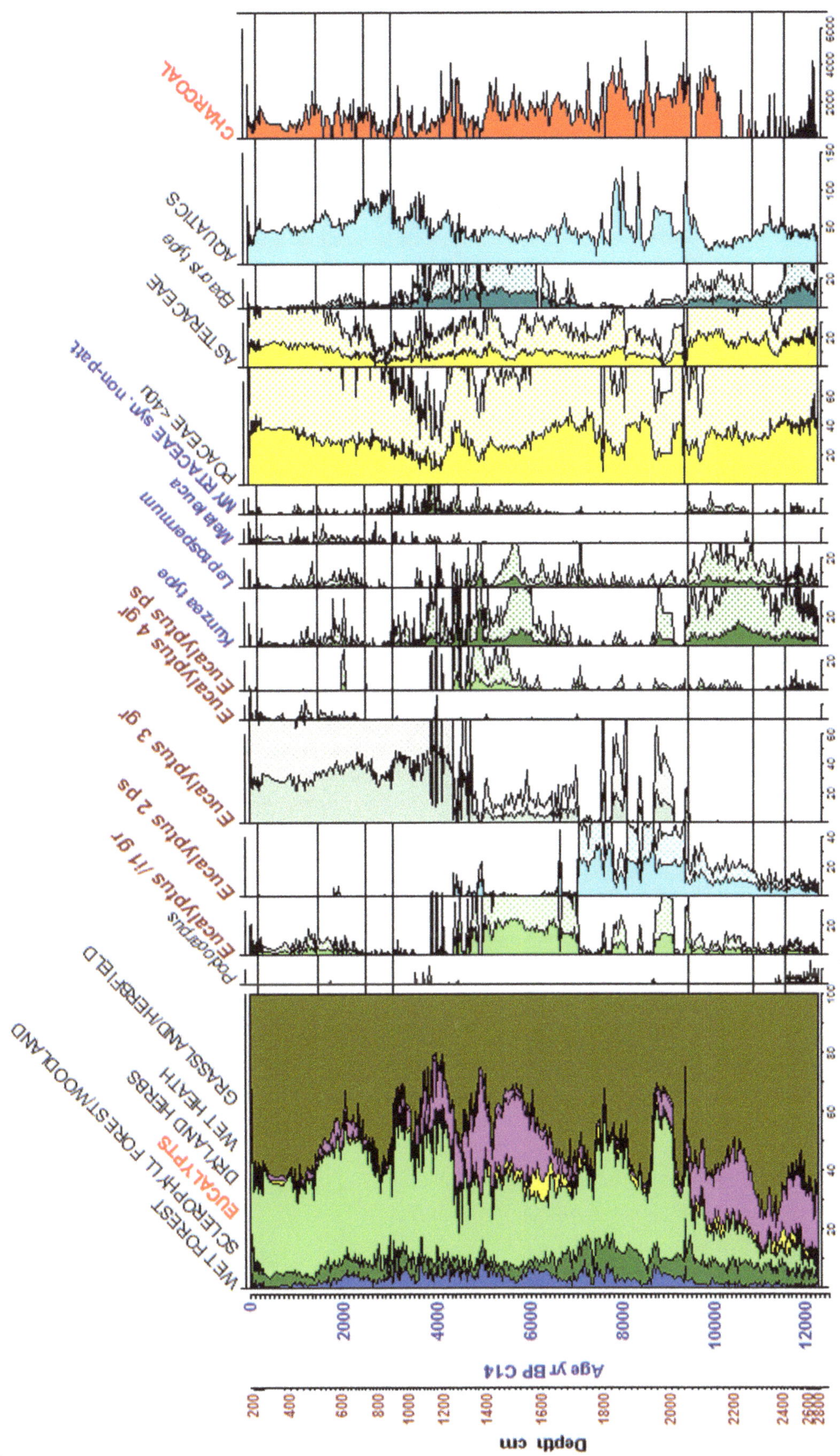

Figure 2a. Pollen diagrams from a 1980 core of Bega Swamp showing various plant families and genera found in the core

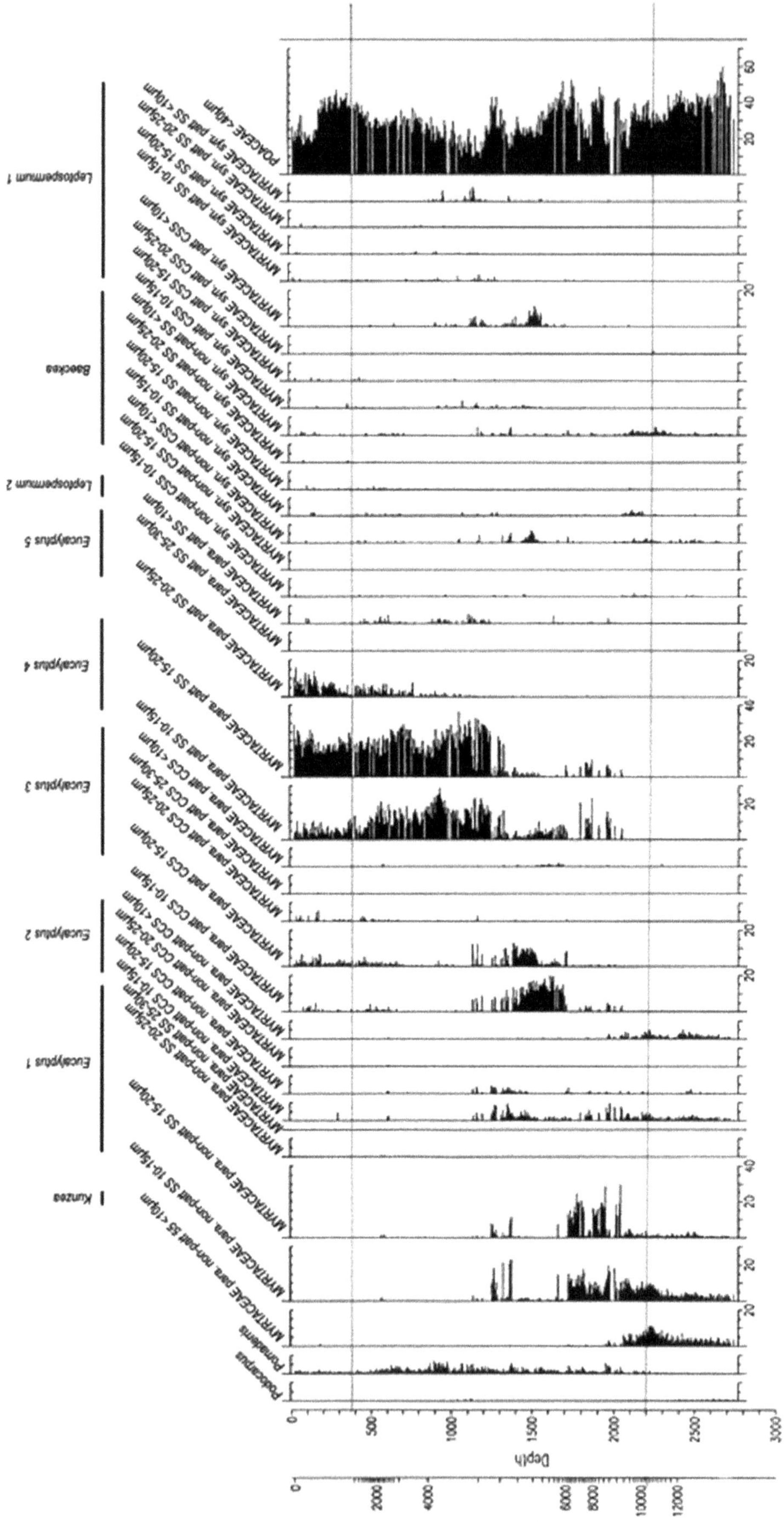

Figure 2b. Pollen diagrams from a 1980 core of Bega Swamp showing the presence of 35 Myrtaceae pollen grains and how they were grouped by Hope et al. in 2000

Table 1. Species list of Myrtaceae taxa from the area surrounding Bega Swamp (Table 1 continues on page 410)

Myrtaceae taxa	Taxa from Wadbilliga National Park recorded in Australian herbaria	Other taxa listed to occur in the area	Taxa that Gurdhip Singh suggested resembled fossil pollen
Acmena smithii (Poir.) Merr. & L.M.Perry	x		
Angophora floribunda (Sm.) Sweet			x
Babingtonia pluriflora (F.Muell.) A.R.Bean ex Govaerts & al.		x	
Backhousia myrtifolia Hook. & Harv.	x		
Baeckea gunniana Schauer ex Walp., Repert			x
Baeckea linifolia Rudge		x	
Baeckea utilis F.Muell. ex Miq.	x		
Baeckea virgata (J.R.Forst. & G.Forst.) Andrews (now Sannantha)		x	
Callistemon citrinus (Curtis) Skeels		x	
Callistemon salignus (Sm.) Colv. ex Sweet		x	
Callistemon sieberi DC.	x		
Callistemon subulatus Cheel		x	
Calytrix tetragona Labill.		x	
Corymbia gummifera (Gaertn.) K.D.Hill & L.A.S.Johnson		x	
Corymbia maculata (Hook.) K.D.Hill & L.A.S.Johnson		x	
Eucalyptus badjensis Beuzev. & M.B.Welch	x		
Eucalyptus baeuerlenii F.Muell.	x		
Eucalyptus dalrympleana Maiden	x		
Eucalyptus elata Dehnh.	x		
Eucalyptus fastigata H.Deane & Maiden			x
Eucalyptus fraxinoides H.Deane & Maiden	x		x
Eucalyptus kybeanensis Maiden & Cambage	x		
Eucalyptus moorei Maiden & Cambage subsp. *moorei*	x		
Eucalyptus nitens (H.Deane & Maiden) Maiden	x		
Eucalyptus olsenii L.A.S.Johnson & Blaxell	x		
Eucalyptus paliformis L.A.S.Johnson & Blaxell	x		
Eucalyptus parvula L.A.S.Johnson & K.D.Hill	x		
Eucalyptus pauciflora Sieber ex Spreng.	x		x
Eucalyptus radiata Sieber ex DC. subsp. *radiata*	x		
Eucalyptus smithii R.T.Baker	x		
Eucalyptus stellulata Sieber ex DC.	x		x

Myrtaceae taxa	Taxa from Wadbilliga National Park recorded in Australian herbaria	Other taxa listed to occur in the area	Taxa that Gurdhip Singh suggested resembled fossil pollen
Eucalyptus stenostoma L.A.S.Johnson & Blaxell	x		
Eucalyptus viminalis Labill.			x
Eucalyptus wilcoxii Boland & Kleinig	x		
Euryomyrtus denticulata (Maiden & Betche) Trudgen	x		
Kunzea ambigua (Sm.) Druce		x	
Kunzea ericoides (A.Rich.) Joy Thomps.		x	
Kunzea parvifolia Schauer		x	
Kunzea sp. 'Wadbilliga' (Rodd 6168)	x		
Leptospermum brevipes F.Muell.	x		
Leptospermum grandifolium Sm.	x		
Leptospermum morrisonii Joy Thomps.	x		
Leptospermum myrtifolium Sieber ex DC.	x		x
Leptospermum obovatum Sweet	x		
Leptospermum polygalifolium Salisb. subsp. *polygalifolium*	x		
Leptospermum scoparium J.R.Forst. & G.Forst.			x
Melaleuca armillaris (Sol. ex Gaertn.) Sm. subsp. *armillaris*	x		
Melaleuca citrina (Curtis) Dum. Cours.	x		
Melaleuca hypericifolia Sm.	x		
Melaleuca pityoides (F.Muell.) Craven	x		
Melaleuca subulata (Cheel) Craven	x		
Tristania neriifolia (Sieber ex Sims) R.Br.			x
Tristaniopsis laurina (Sm.) Peter G.Wilson & J.T.Waterh.	x		x

Pollen of the plant family Myrtaceae is characterised by a syncolpate or parasyncolpate appearance, although some Myrtaceae tribes are brevicolpate and asyncolpate (Thornhill unpublished data), with many studies conducted on the pollen of the family or specific groups within (Pike 1956; McIntyre 1963; Churchill 1968; Barth and Barbosa 1973; Graham 1980; Gadek and Martin 1981; Gadek and Martin 1982; Patel et al. 1984; Van Wyk and Dedekind 1985; Martin and Gadek 1988; Chalson and Martin 1995; Pickett and Newsome 1997; Parnell 2003; Eliseu and Dinis 2008). It has been said that at the generic and species levels, separation of taxa can be difficult or impossible (Pickett and Newsome 1997), but a number of useful and distinctive characters have been created to tackle the task (Dodson 1974; Chalson and Martin 1995). The high significance that is placed on Myrtaceae, especially *Eucalyptus*, when identifying pollen to interpret past vegetation and climatic conditions means that the ability to differentiate Myrtaceae into distinct pollen groups is of great value. Most important, as Pickett and Newsome (1997) noted, is whether *Eucalyptus* pollen might be identifiable to species, or groups of species.

Methods

As many extant Myrtaceae genera and species as were readily available were imaged with light microscopy. Images were taken from existing slides sourced from the pollen slide collection in the Department of Archaeology and Natural History (ANH), Australian National University, and from slides made by Kathleen Pike in the 1940s and '50s (many of which are line illustrated in Pike 1956) housed at the School of Botany, University of Melbourne, and the School of Land and Environment, University of Melbourne. Taxa used for extant reference pollen were selected by searching a species list of Wadbilliga National Park housed in the Australian National Herbarium, Canberra, and from a Bega Swamp species list compiled by Gurdhip Singh (unpublished). New slides were created for the Bega Swamp fossil pollen from archived residues, as the slides used in previous studies had been lost. The original core was taken in 1980 by Gurdhip Singh who froze the core on site and then sliced it with a fine bandsaw into a series of 500 discs of 20 cm diameter and approximately 2.5 mm thickness, resulting in a sample every 4 mm (Hope 1995). The archived acetolysed samples are stored in silicone oil at ANH. Peaks in concentration of fossil Myrtaceae pollen grains were recognised from the pollen diagram (Figure 2), and from these peaks archived samples were selected for the new slides. The chronology used in this study comes from several previous published studies (Hope et al. 2000; Wheeler et al. 2001), as well as unpublished dates provided by Dr Nick Porch from ANH. This chronology is summarised in Table 2 and it is worth noting that the chronology is now the highest resolution sequence in Australia. Specimens for new slides were mounted in silicone oil and all images were captured using a Zeiss (Oberkochen) Microscope, objective lens x63 and the image capturing software Axiovision.

Table 2. Depth and age of core sample numbers

Sample number	Depth (mm)	Age
54, 55, 56	302–316	~ 400 years
81, 82, 83	400-419	850 years
115, 116, 117, 118	583-606	2300 years
194, 195	980-1002	3400 years
202	1006-1009	3500 years
238, 239, 240	1197-1213	4300 years
269, 270, 271	1369-1386	4700 years
274	1399-1404	4800 years
290, 291	1494-1505	5000 years
344, 345, 346	1796-1814	8500 years
374	1971-1975	8800 years
440, 441, 442	2199-2213	10,900 years
498, 499, 500	2497-2511	12,500 years

Both extant and fossil pollen grains were scored using the characters described by Chalson and Martin (1995) and descriptions followed the terminology of Punt et al. (2007). Pollen measurements were also made in Adobe Photoshop. A Lucid key was created compiling character scores of extant Myrtaceae pollen. The key, which contains 329 Myrtaceae taxa that have been imaged with light microscopy (and another 220 with scanning electron microscopy), was then used to search for similarities with the fossil pollen by entering the characters of each individual fossil pollen grain (more than 400 grains in total).

Results

Extant Myrtaceae

In total, 25 Myrtaceae taxa representing 10 genera were used as a reference for extant pollen, and in most instances, pollen grains could be sorted down to a species level using a total of 15 morphological characters and image measurements (Figure 3).

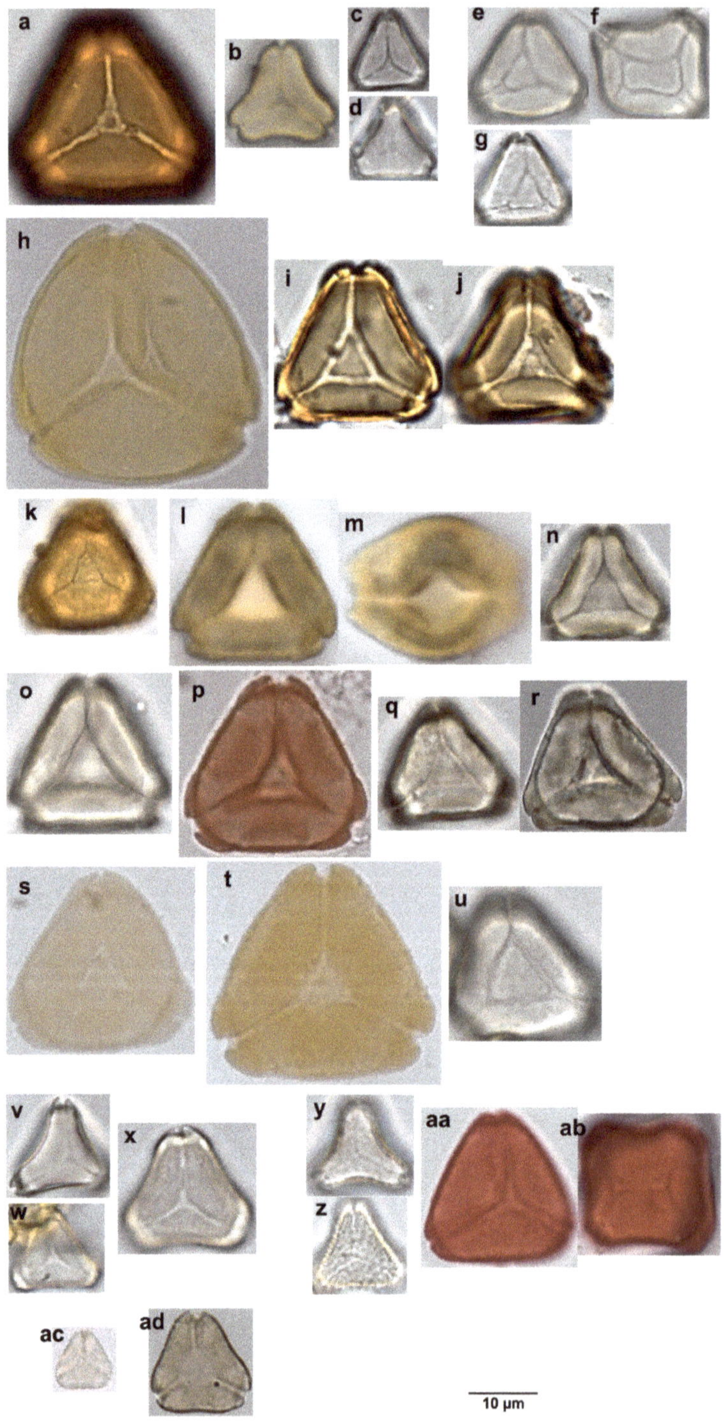

Figure 3. Extant Myrtaceae pollen of Bega Swamp (a) *Angophora floribunda*; (b) *Backhousia myrtifolia*; (c) *Baeckea gunniana*; (d) *Baeckea utilis*; (e) and (f) *Callistemon citrinus*; (g) *Callistemon sieberi*; (h) *Corymbia gummifera*; (i) and (j) *Corymbia maculata*; (k) *Eucalyptus dalrympleana*; (l) and (m) *E. fastigata*; (n) *E. fraxinoides*; (o) *E. pauciflora*; (p) *E. radiata*; (q) and (r) *E. rubida*; (s) *E. smithii*; (t) *E. stellulata*; (u) *E. viminalis*; (v) *Kunzea ambigua*; (w) *K. ericoides*; (x) *K. parvifolia*; (y) *Leptospermum grandifolium*; (z) *L. lanigerum*; (aa) and (ab) *L. scoparium*; (ac) *Tristania neriifolia*; (ad) *Tristaniopsis laurina*

Scale bar - 10 µm

Eucalyptus and allies

Nine *Eucalyptus* species (Figure 3 k-u) were sampled, representing five eucalypt groups: peppermints (*E. radiata*), longitudinale (*E. stellulata*), symphomyrtus (*E. smithii, E. viminalis, E. rubida* and *E. dalrympleana*), snowgum (*E. pauciflora*) and ashes (*E. fastigata* and *E. fraxinoides*), showing a variation in pollen length among species (16-25.7 µm). *Eucalyptus pauciflora* and *E. fraxinoides* exhibited pollen grains which were parasyncolpate with straight sides and had psilate/scabrate exine patterning, with *E. pauciflora* appearing slightly larger on average (20.4 µm compared with 16 µm). *Eucalyptus fastigata* was also straight sided, but had psilate exine patterning. *Eucalyptus rubida* showed presence of an apocolpial island, as did *E. viminalis. Eucalyptus dalrympleana* had scabrate exine patterning, while *E. stellulata* was large in size with convex pollen sides and psilate/scabrate patterning. The pollen grains of genera *Angophora* and *Corymbia* (Figure 3 a, h-j), which are in the same Myrtaceae tribe as *Eucalyptus*, could be distinguished by their larger size (pollen length of 18-34 µm) and arcuate-shaped colpi. The majority of *Angophora* and *Corymbia* also had apocolpial islands present.

Other Myrtaceae

Nine other Myrtaceae genera were used as reference pollen. Grains of *Callistemon* (now *Melaleuca*) were parasyncolpate with straight edges and a large closely fitting island present and had a pollen length range of 12-15.6 µm (Figure 3 e-g). The majority of other Myrtaceae genera exhibited small pollen grains (pollen length <15 µm). *Kunzea* and *Leptospermum*, which are both in the Leptospermeae tribe, had syncolpate grains with straight or concave edges and arcuate colpi, the difference being that *Leptospermum* pollen had granulate/scabrate exine patterning while *Kunzea* was psilate (Figure 3 v-ab). *Baeckea* pollen (Figure 3 c and d) was also syncolpate with arcuate colpi making it difficult to separate from *Kunzea* pollen, the most notable difference being that *Kunzea* were less syncolpate than *Baeckea*. *Backhousia myrtifolia* (Figure 3 b) and *Tristaniopsis laurina* pollen (Figure 3 ad) looked similar, both being parasyncolpate with concave sides, while pollen from *Tristania neriifolia* (Figure 3 ac) was extremely small, on average being less than 8 µm in length.

Fossil Myrtaceae

A total of 428 Myrtaceae pollen grains was recognised from the 32 core sections sampled. Comparison of the fossil pollen with extant specimens showed trends in the genera and species that were present at Bega Swamp at various times (Table 3). *Eucalyptus* pollen grains were found in every sample examined but the total number of eucalypt grains per sample, as well as the number of eucalypt 'species' identified per sample, varied through time. The results given by entering pollen characters into the Lucid key and by simply looking at the grains were compared and found to be compatible. Sixty-nine (16%) Lucid key suggestions did not match visual assumptions. Three-hundred-and-fifty-nine (84%) Lucid suggestions matched to at least genera visual judgements; of these, 119 (28%) Lucid suggestions gave matching species.

Core samples 54-56, depth 302-316 mm, age ~400 BP

52 grains, 8 unmatched between Lucid and visual, 34 generically matched, 10 fully matched.

The youngest core samples investigated, approximately 400 years old, presented a variety of pollen types, the most dominant being similar to *E. rubida* (Figure 4 h) and *E. pauciflora* (Figure 4 a, g and j). Seven other *Eucalyptus* pollen types were identified in smaller numbers. One diporate grain (Figure 4 d) was found and attributed to *E. fastigata*, although it is possible that other *Eucalyptus* species have diporate pollen grains. Pollen thought to belong to *Callistemon* (Figure 4 m) was present, as were *Kunzea* (Figure 4 i), *Tristania* (Figure 4 n), *Baeckea* (Figure 4 i) and *Leptospermum*.

Table 3. Grain numbers of various core sections attributed to Myrtaceous pollen types using visual judgement

	Core sample numbers												
Genus	**54-56**	**81-83**	**115-118**	**194-195**	**202**	**238-240**	**269-271**	**274**	**290-291**	**344-346**	**374**	**440-442**	**498-500**
Eucalyptus fraxinoides	3	2	11	4	0	11	14	5	5	5	0	5	3
Eucalyptus pauciflora	8	0	8	2	0	2	7	1	5	7	1	10	7
Eucalyptus fastigata	4	4	9	0	1	3	6	1	6	5	1	1	7
Eucalyptus rubida	14	3	12	3	0	5	6	2	6	1	1	4	3
Eucalyptus dalrympleana	2	2	3	0	0	1	3	1	1	0	0	1	0
Eucalyptus viminalis	6	1	2	0	0	1	1	1	0	2	0	2	1
Eucalyptus smithii	1	2	5	1	0	2	2	0	0	0	0	0	1
Eucalyptus radiata	1	0	0	0	0	0	0	0	0	0	0	0	
Eucalyptus stellulata	1	1	2	1	0	1	2	0	0	1	0	0	1
Angophora/Corymbia	0	0	2	0	0	0	0	0	0	0	0	0	0
Callistemon	7	1	3	1	1	2	6	0	0	0	1	1	1
Kunzea	4	1	6	0	1	5	9	2	15	0	3	1	3
Leptospermum	1	0	1	0	0	6	8	1	1	0	2	1	1
Baeckea	1	0	5	1	1	5	6	1	8	0	1	1	3
Tristania neriifolia	1	1	2	0	0	2	1	0	1	0	0	0	1
Tristaniopsis laurina	0	0	1	0	0	1	0	0	0	0	0	0	0

Core samples 81-83, depth 400-419 mm, age 850 BP

22 grains, 2 unmatched, 16 generically matched, 4 fully matched.

Seven *Eucalyptus* pollen grains were identified but no species appeared dominant and grains from *E. pauciflora* were not seen. One pollen grain from an *Angophora* or *Corymbia* (Figure 4 t) was observed, as well as grains from *Callistemon* (Figure 4 v), *Kunzea* (Figure 4 w) and *Tristania* (Figure 4 s, overleaf).

Core samples 115-118, depth 583-606 mm, age 2300 BP

74 grains, 9 unmatched, 43 generically matched, 22 fully matched.

Eucalyptus rubida (Figure 5 o, overleaf), *E. fraxinoides* (Figure 5 c, d i, s and t), *E. fastigata* (Figure 5 h) and *E. pauciflora* (Figure 5 k and n) all had many grains attributed to them during this time period. Another four *Eucalyptus* species had pollen grains attributed to them and grains from *Angophora* or *Corymbia* (Figure 5 q) were identified. *Kunzea* and/or *Baeckea* pollen grains (Figure 5 f, p and u) showed an increased number, while grains from *Tristania, Leptospermum* and *Tristaniopsis* were found.

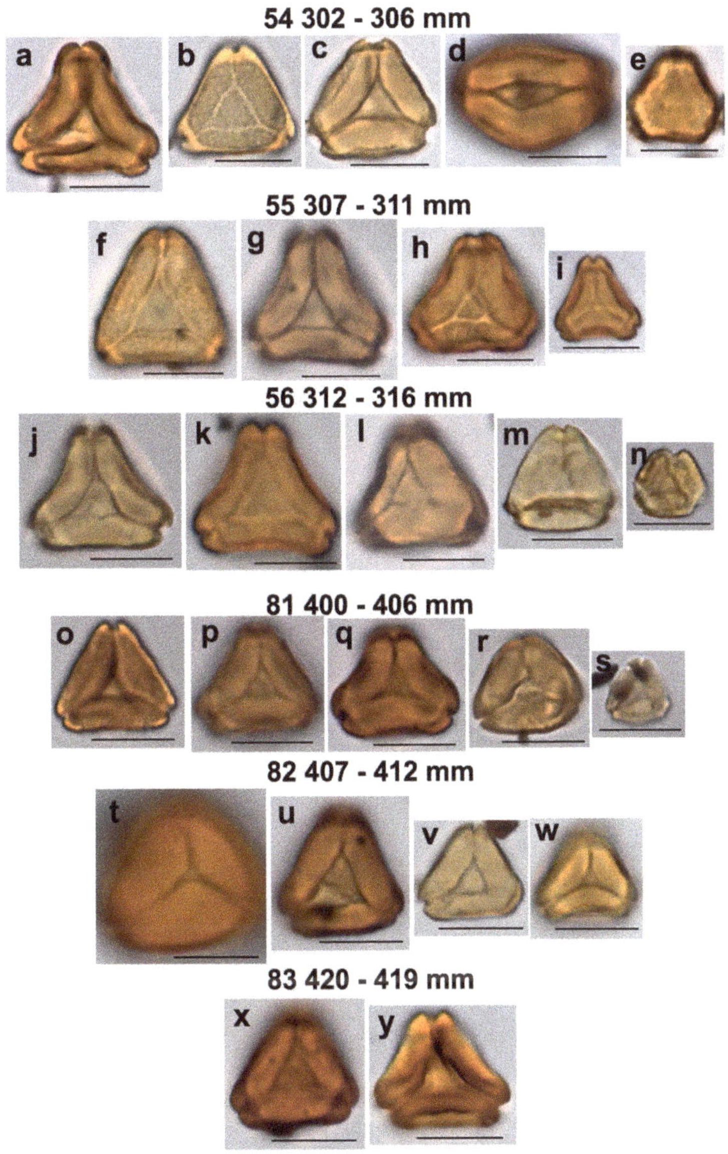

Figure 4. Examples of fossil pollen types from Bega Swamp core sections 54-56 and 81-83 with related core depth. 54 (a) *E. pauciflora*; (b); (c) *E. fastigata*; (d) Diporate *E. fastigata*; (e) *Kunzea*; 55 (f) *E. viminalis*; (g) *E. pauciflora*; (h) *E. rubida*; (i) *Baeckea*; 56 (j) *E. pauciflora*; (k) *E. fastigata*; (l) *E. rubida*; (m) *Callistemon*; (n) *Tristania*; 81 (o) *E. fraxinoides*; (p) *E. dalrympleana*; (q) *E. pauciflora*; (r) *E. rubida*; (s) *Tristania*; 82 (t) *Angophora/Corymbia*; (u) *E. viminalis*; (v) *Callistemon*; (w) *Kunzea*; 83 (x) *E. smithii*; (y) *E. fastigata*

Scale bar – 10 μm

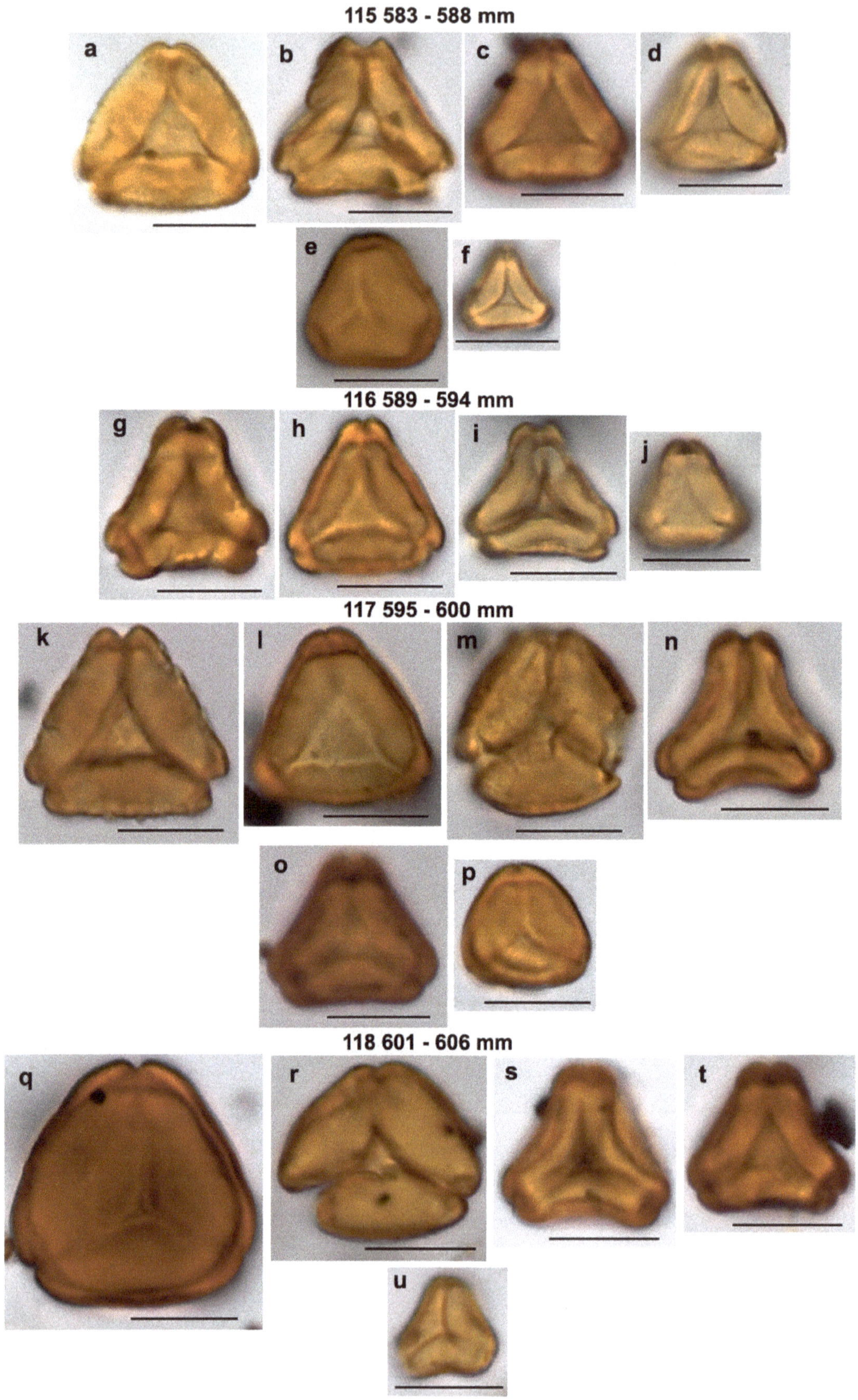

Figure 5. Examples of fossil pollen from Bega Swamp core sections 115-118 with related core depth. 115 (a) *E. stellulata*; (b) *E. pauciflora*; (c) and (d) *E. fraxinoides*; (e) *Kunzea*; (f) *Baeckea*; 116 (g) *E. pauciflora*; (h) *E. fastigata*; (i) *E. fraxinoides*; (j) *Callistemon*; 117 (k) *E. pauciflora*; (l) *E. viminalis*; (m) *Angophora/Corymbia*; (n) *E. pauciflora*; (o) *E. rubida*; (p) *Kunzea/Baeckea*; 118 (q) *Angophora/Corymbia*; (r) *Eucalyptus* sp.; (s) and (t) *E. fraxinoides*; (u) *Kunzea/Baeckea*

Scale bar – 10 μm

Core samples 194, 195, depth 980-1002 mm, age 3400 BP

14 grains, 1 unmatched, 7 generically matched, 6 fully matched.

Five *Eucalyptus* species were identified in samples from 3400 years ago (Figure 6), with one pollen grain each being attributed to *Callistemon* (Figure 6 e) and *Baeckea.*

Core sample 202, depth 1006-1009 mm, age 3500 BP

3 grains, 1 unmatched, 1 generically matched, 1 fully matched.

Three grains in total were found, with one thought to be similar to *E. fastigata* (Figure 6 g), one *Callistemon* and one *Kunzea* or *Baeckea.*

Core samples 238-240, depth 1197-1213 mm, age 4300 BP

47 grains, 6 unmatched, 19 generically matched, 22 fully matched.

A marked increase in smaller grains was seen from 4300 years ago. *Leptospermum* (Figure 6 q and w), *Kunzea* (Figure 6 r and x) and *Baeckea* (Figure 6 l and p) grain types all appeared more prevalent, suggesting that vegetation composition was different during this time. *Eucalyptus fraxinoides* (Figure 6 i, m and t) pollen type was abundant. Seven other *Eucalyptus* grains were identified, as well as *Callistemon*, *Tristania* and *Tristaniopsis.*

Core samples 269-271, depth 1369-1386 mm, age 4700 BP

69 grains, 15 unmatched, 33 generically matched, 21 fully matched.

Eucalyptus fraxinoides (Figure 7 c and k) and *E. pauciflora* (Figure 7 p) type pollen were the most common seen in samples from 4700 years. *Eucalyptus fastigata* (Figure 7 a) and *E. rubida* (Figure 7 r) also had a number of pollen grains identified, while four other *Eucalyptus* species were found. The number of *Callistemon* (Figure 7 g and m) pollen grains increased in this time period and many smaller pollen grains from *Kunzea* (Figure 7 o and u), *Leptospermum* (Figure 7 f, n, s and t) and *Baeckea* (Figure 7 v) were seen.

Core sample 274, depth 1399-1404 mm, age 4800 BP

15 grains, 2 unmatched, 9 generically matched, 4 fully matched.

Eucalyptus fraxinoides type grains (Figure 7 y, overleaf) were most prevalent and five other *Eucalyptus* species were found in small numbers. Smaller grains from *Kunzea* (Figure 7 aa), *Baeckea* (Figure 7 z) and *Leptospermum* (Figure ab) were found in smaller numbers.

Core samples 290, 291, depth 1494-1505 mm, age 5000 BP

42 grains, 9 unmatched, 18 generically matched, 15 fully matched.

Four *Eucalyptus* species, *E. fraxinoides* (Figure 8 h), *E. pauciflora* (Figure 8 a), *E. fastigata* (Figure 8 c) and *E. rubida* had similar numbers of pollen types identified. One tetraporate grain (Figure 8 j) was found and attributed to *Eucalyptus.* The most common pollen grains were small syncolpate grains of *Kunzea* (Figure 8 m) and *Baeckea* (Figure 8 n), while one grain each of *Tristania* (Figure 8 g) and *Leptospermum* was found.

Core samples 344-346, depth 1796-1814 mm, age 8500 BP

19 grains, 2 unmatched, 15 generically matched, 2 fully matched.

Grains from *Eucalyptus* species were the only pollen identified in this period, *Eucalyptus fraxinoides* (Figure 8 v), *E. pauciflora* (Figure 8 s) and *E. fastigata* (Figure 8 o) being the most abundant, with three other species also being recognised.

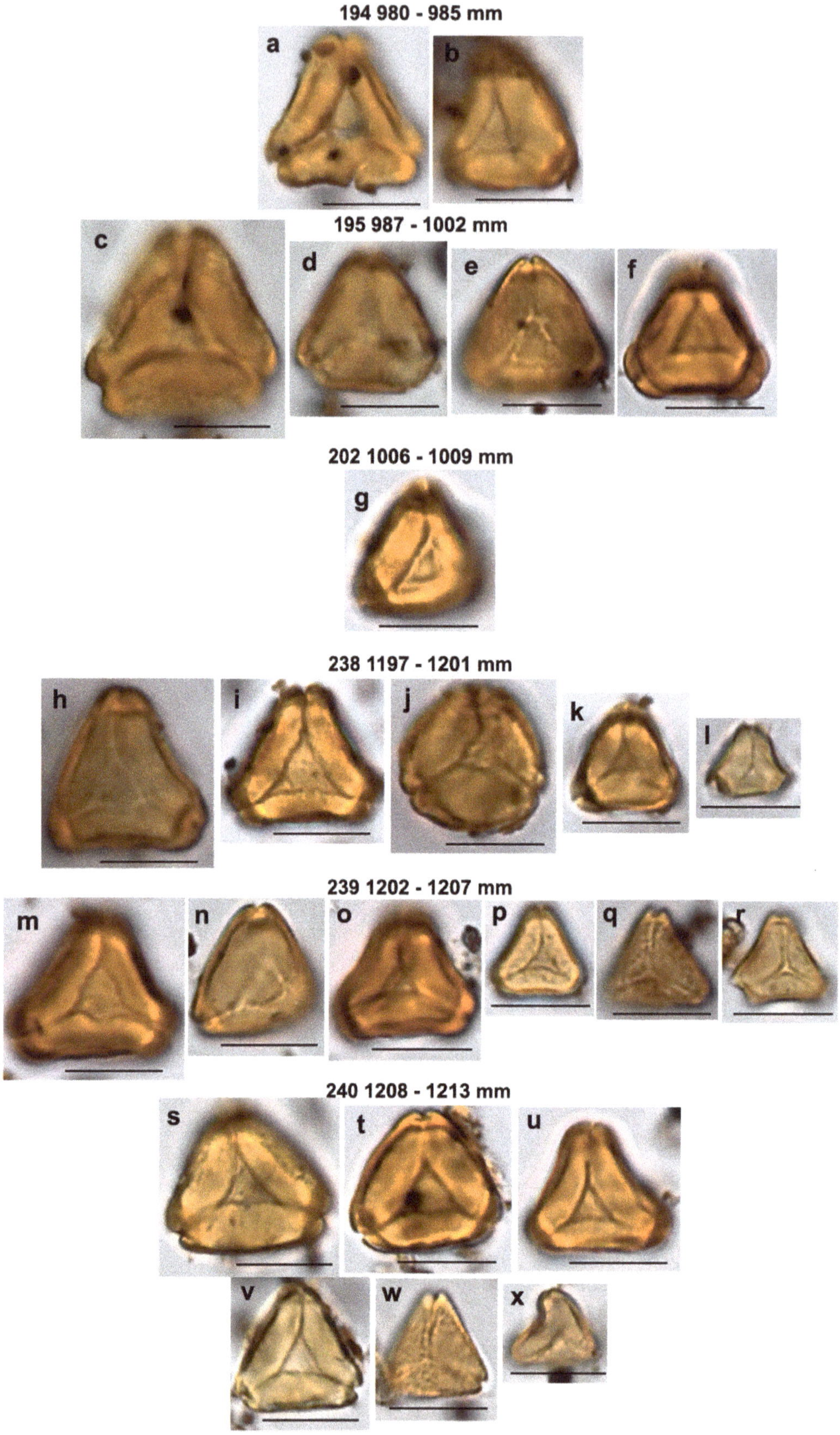

Figure 6. Examples of fossil pollen types from Bega Swamp core sections 194, 195, 202 and 238-240 with related core depth. 194 (a) *E. pauciflora*; (b) *E. fraxinoides*; 195 (c) *E. stellulata*; (d) *E. rubida*; (e) *Callistemon*; (f) *E. smithii*; 202 (g) *E. fastigata*; 238 (h) *E. viminalis*; (i) *E. fraxinoides*; (j) *E. dalrympleana*; (k) *E. radiata/rubida*; (l) *Baeckea*; 239 (m) *E. fraxinoides*; (n) *E. viminalis*; (o) *E. rubida*; (p) *Baeckea*; (q) *Leptospermum*; (r) *Kunzea*; 240 (s) *E. stellulata*; (t) *E. fastigata*; (u) and (v) *E. fraxinoides*; (w) *Leptospermum*; (x) *Baeckea/Kunzea*

Scale bar – 10 µm

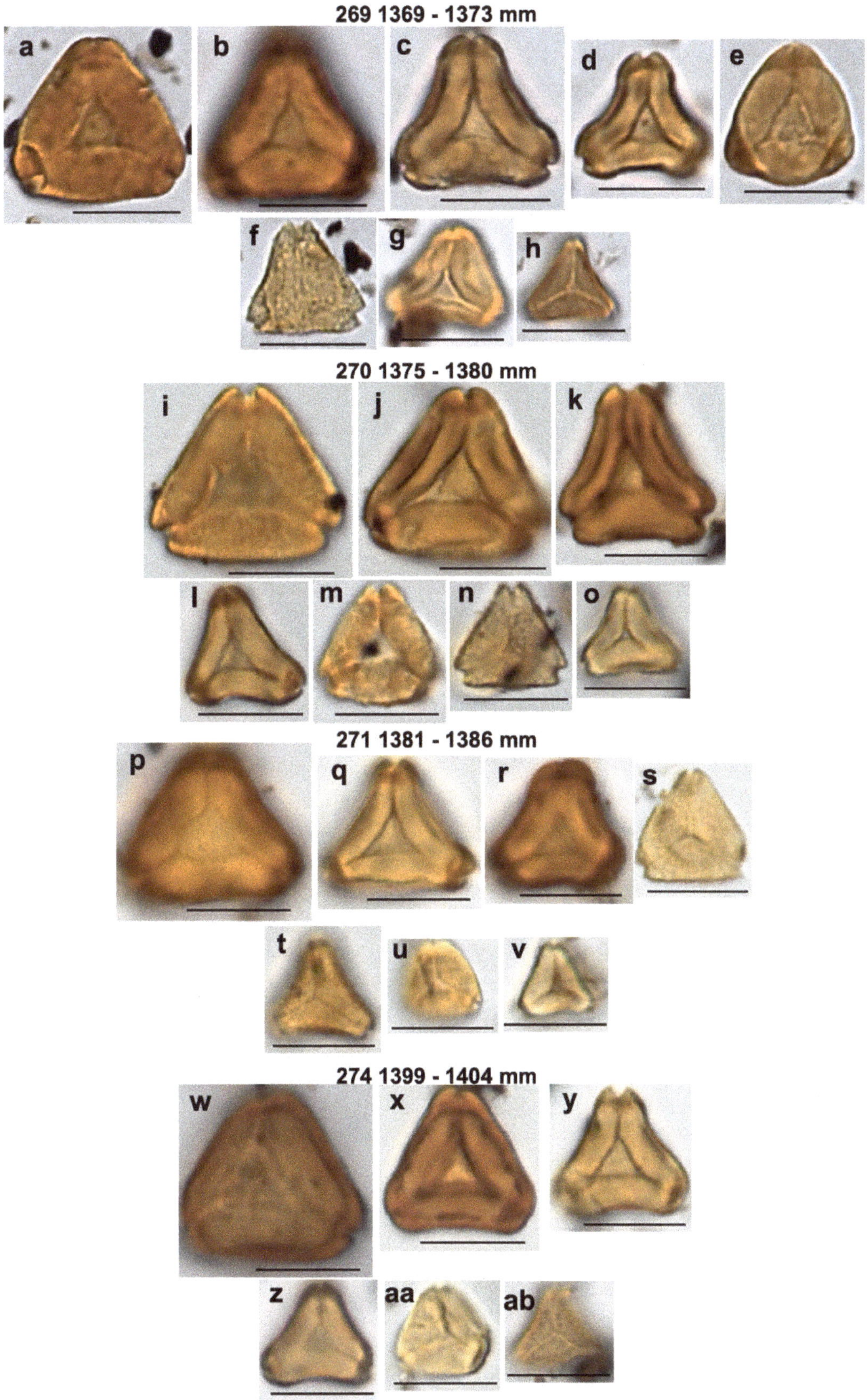

Figure 7. Examples of fossil pollen types from Bega Swamp core sections 269-271 and 274 with related core depth. 269 (a) *E. fastigata*; (b) *E. dalrympleana*; (c) *E. fraxinoides/pauciflora*; (d) *E. fastigata*; (e) *E. smithii*; (f) *Leptospermum*; (g) *Callistemon*; (h) *Kunzea/Baeckea*; 270 (i) *E. stellulata*; (j) *E. viminalis*; (k) *E. fraxinoides*; (l) *Kunzea*; (m) *Callistemon*; (n) *Leptospermum*; (o) *Kunzea*; 271 (p) & (q) *E. pauciflora*; (r) *E. smithii*; (s) *Leptospermum*; (t) *Leptospermum*; (u) *Kunzea*; (v) *Baeckea*; 274 (w) *E. viminalis*; (x) *E. fastigata*; (y) *E. fraxinoides*; (z) *Baeckea*; (aa) *Kunzea*; (ab) *Leptospermum*

Scale bar – 10 µm

Core sample 374, depth 1971-1975 mm, age 8800 BP

12 grains, 4 unmatched, 5 generically matched, 3 fully matched.

Around half the grains identified in this section were small syncolpate grains of *Kunzea* (Figure 8 z), *Baeckea* and *Leptospermum* (Figure 8 aa, overleaf). Three *Eucalyptus* species were identified, as was one *Callistemon* (Figure 8 y) grain.

Core samples 440-442, depth 2199-2213 mm, age 10,900 BP

33 grains, 4 unmatched, 24 generically matched, 5 fully matched.

Six *Eucalyptus* species were found, *E. pauciflora* (Figure 9 i, overleaf) and *E. fraxinoides* (Figure 9 j) types being the most common. A tetraporate pollen grain with an apocolpial island (Figure 9 b) was found and appeared similar to extant tetraporate grains of *Callistemon* (Figure 3 f). Smaller syncolpate grains were less abundant.

Core samples 498-500, depth 2497-2511 mm, age 12,500 BP

26 grains, 6 unmatched, 16 generically matched, 4 fully matched.

Eucalyptus viminalis (Figure 9 q) and *E. fastigata* (Figure 9 n) grains were identified as the most common grains from the oldest samples analysed. *Eucalyptus* type pollen was more diverse in this section, with five other pollen types being identified. The number for *Kunzea* (Figure 9 u) and *Baeckea* (Figure 9 w) pollen was slightly higher, and one grain each of *Callistemon* (Figure 9 z) and *Leptospermum* was found.

Validating fossil data

Images of fossil pollen were shown to Geoff Hope to establish which pollen types represented Myrtaceae groups from the Hope et al. (2000) pollen diagram (Figure 2). *Eucalyptus* type 1 grains were best illustrated by Figure 5 k. *Eucalyptus* type 2 pollen were similar to Figure 4 a and j. *Eucalyptus* type 3 pollen was similar to Figure 5 n. *Eucalyptus* type 4 was most like Figure 4 b and *Eucalyptus* type 5 like Figure 4 f. The smaller pollen grains were attributed to *Kunzea* (Figure 5 e), *Baeckea* (Figure 5 f) and *Leptospermum* (Figure 6 q). When comparing the Hope et al. (2000) *Eucalyptus* types to extant grains it was seen that *Eucalyptus* 1 looked similar to *E. pauciflora* (Figure 3 o). *Eucalyptus* 2 grains, which were thought to belong to *E. pauciflora,* looked similar to *E. fraxinoides* (Figure 3 n), *E. pauciflora* (Figure 3 o) and *E. fastigata* (Figure 3 m). *Eucalyptus* type 3 grains appeared most like *E. dalrympleana* (Figure 3 k*), E. radiata* (Figure 3 p) and *E. smithii* (Figure 3 s). *Eucalyptus* type 4 was very similar to *E. viminalis* (Figure 3 u), as was *Eucalyptus* type 5.

Discussion

The vegetation history of Bega Swamp interpreted from pollen diagrams by Hope et al. (2000) observes a transition from shrubby daisy-grass steppe to a low eucalypt (presumed snowgum) woodland by about 11,800 BP and then a phase of herbfield development. Eucalypt forest appeared abruptly around 9400 BP and increased in diversity. Changes in eucalypt dominance occurred around 7000 and 4000 BP, which may represent shifts to the *E. fastigata*/*E. dalrympleana* forest of the present day. The centre of a possible 'wet phase' happened between 6000 BP and 3300 BP. The present study searched selected core samples for Myrtaceae pollen which could correlate with this vegetation history interpretation. It was found that younger core samples contained higher concentrations and a wider variety of *Eucalyptus* type pollen than older sections. The dominant pollen type fluctuated through the core and this was also stated by Hope et al. (2004), who suggested that this may represent quite minor transitions across environmental limits or even autogenic processes that are not driven by external changes.

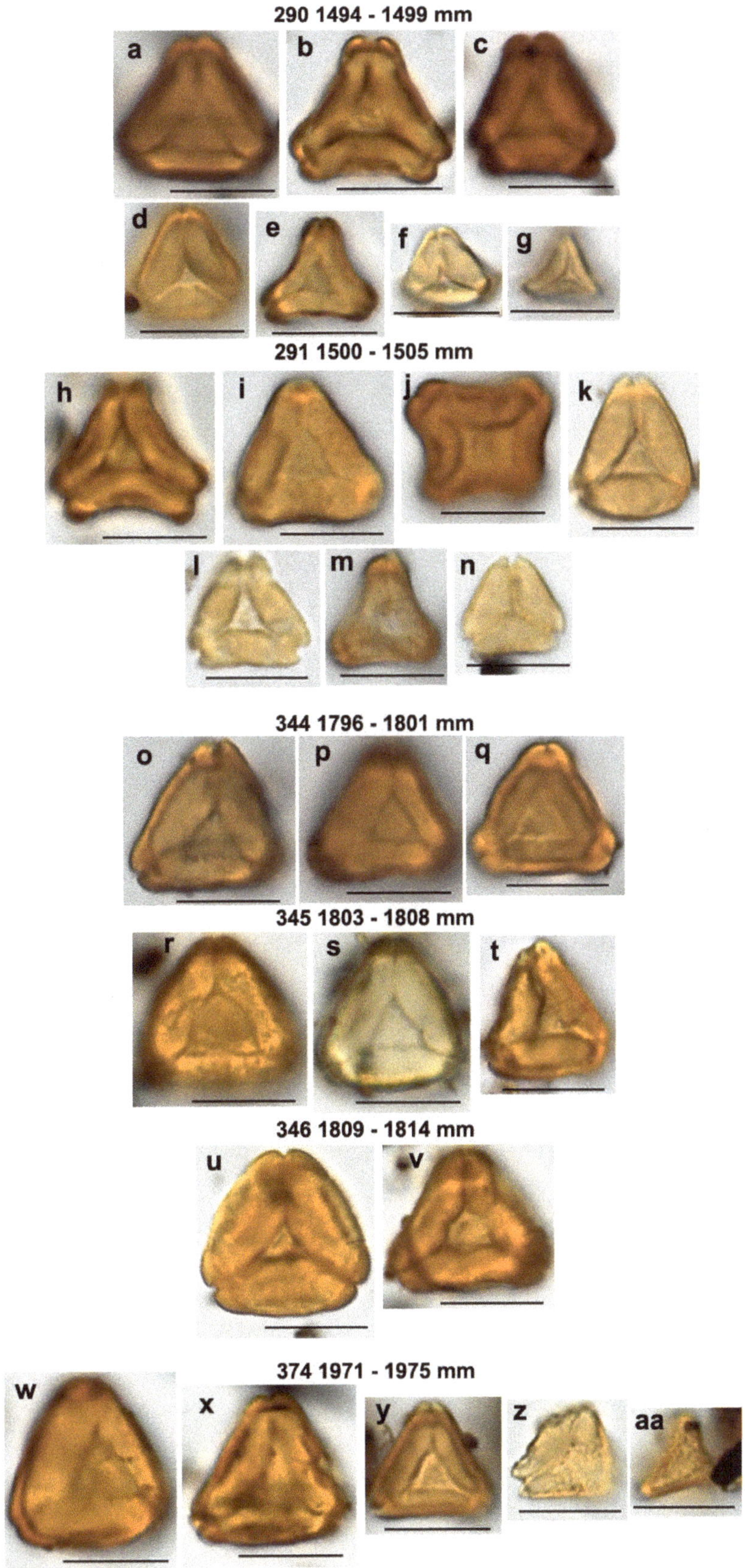

Figure 8. Examples of fossil pollen types from Bega Swamp core sections 290, 291, 344-346 and 374 with related core depth. 290 (a) *E. pauciflora*; (b) *E. fraxinoides*; (c) *E. fastigata*; (d) and (e) *Kunzea*; (f) *Baeckea*; (g) *Baeckea/Kunzea*; 291 (h) *E. pauciflora*; (i) *E. dalrympleana*; (j) Tetraporate *Eucalyptus*; (k) *E. smithii*; (l) *E. pauciflora*; (m) *Kunzea*; (n) *Baeckea*; 344 (o) *E. fastigata*; (p) *E. viminalis*; (q) *E. rubida*; 345 (r) *E. fastigata*; (s) *E. pauciflora*; (t) *E. fastigata*; 346 (u) *E. stellulata*; (v) *E. fraxinoides*; 374 (w) *E. fastigata*; (x) *E. pauciflora*; (y) *Callistemon*; (z) *Kunzea/Leptospermum*; (aa) *Leptospermum*

Scale bar – 10 µm

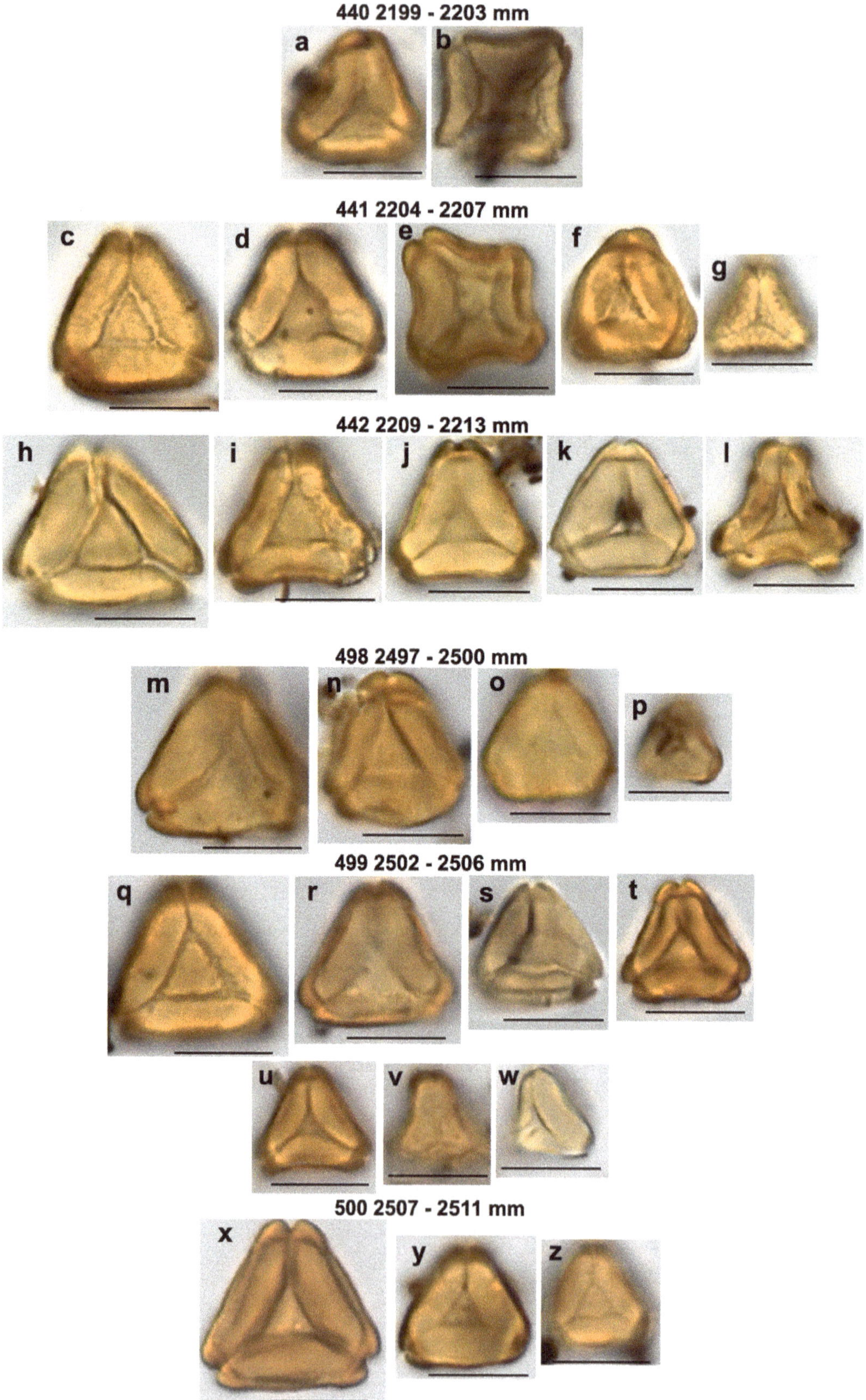

Figure 9. Examples of fossil pollen types from Bega Swamp core sections 440-442 and 498-500 with related core depth. 440 (a) *E. rubida*; (b) Tetraporate *Callistemon*; 441 (c) *E. viminalis*; (d) *E. fastigata*; (e) Tetraporate *Eucalyptus*; (f) *E. rubida*; (g) *Leptospermum*; 442 (h) and (i) *E. pauciflora*; (j) *E. fraxinoides*; (k) *E. pauciflora*; (l) *E. fraxinoides*; 498 (m) *E. stellulata*; (n) *E. fastigata*; (o) *E. fraxinoides*; (p) *Baeckea/Kunzea*; 499 (q) *E. viminalis*; (r) *E. fastigata*; (s) *Eucalyptus* sp.; (t) *E. fraxinoides*; (u) *Baeckea/Kunzea*; (v) *Kunzea*; (w) *Baeckea*; 500 (x) *E. fastigata*; (y) *E. pauciflora*; (z) *Callistemon*

Scale bar – 10 μm

Eucalyptus pauciflora (snowgum) pollen was found in all but two of the sections sampled and in higher concentrations in older and younger samples, with lower values in between. However, *E. fraxinoides (*white ash) type pollen, which looks very similar to *E. pauciflora* but is smaller, was found in high concentrations when the larger *E. pauciflora* type was not as abundant (4300 to 4700 BP). If these fossil grains are indeed *E. fraxinoides*, it would suggest a wet, tall eucalypt forest present at Bega Swamp during this time. *E. dalrympleana*, *E. rubida*, *E. smithii* and *E. viminalis* pollen types all began showing higher concentrations in mid-core sections, starting around 5000 BP, matching suggestions that dominant present-day vegetation emerged somewhere between 6000 BP and 3300 BP. *Eucalyptus fastigata*, however, was found in high concentrations in the oldest core samples, differing from previous pollen studies, suggesting that components of a tall eucalypt forest have existed in the Bega Swamp area for longer than has been assumed.

Comparing the suggested eucalypt groups of Hope et al. (2000) with extant pollen presented problems with the interpretation. Nine extant *Eucalyptus* species were used as reference material in the present study, which contrasted with Hope et al. (2000) who had five distinct eucalypt pollen groups. While the idea of having five groups would concur with the morphological and ecological groups that occur in the Bega Swamp area (i.e. snowgum, ashes, peppermints etc) and create a neat vegetation transition history as shown in the pollen diagram (Figure 2), the extant pollen morphology did not give full support to the morphological eucalypt pollen groups created by Hope et al. (2000). Eucalypt pollen types 2, 3 and 5 have similarities with extant *Eucalyptus* pollen across multiple morphological groups – for instance, eucalypt group 2 pollen could belong to the snowgums or ashes, and eucalypt type 3 could belong to ashes or peppermints. To create a vegetation history based on the occurrence of various eucalypt pollen type groups would thus seem implausible.

There have been previous attempts to find distinctive *Eucalyptus*/Myrtaceae pollen groups (Dodson 1974; Chalson and Martin 1995; Pickett and Newsome 1997). Dodson (1974) identified four types of *Eucalyptus* pollen from Lake Leake in South Australia and related these pollen types to eucalypt groups comprising 11 extant *Eucalyptus* species, although how this was done is not exactly clear as the results are published as an appendix. Chalson and Martin (1995) created the characters used in this study and worked only on extant Myrtaceae pollen of the Blue Mountains, finding that all but two taxa could be sorted to a species level. This study showed that there is the ability to separate Myrtaceae pollen to a species level if a wide range of genera and a low number of species from each genus is selected. Pickett and Newsome (1997) used nine *Eucalyptus* species and fossils from a Holocene swamp at Walpole, Western Australia, finding that extant species could be differentiated to pollen types with some confidence so long as suites of morphological characters, rather than single characters, were used, and that much of the fossil pollen could be allocated to a pollen type if the necessary features were preserved in the pollen grain. However, they also stated that 'a basic pre-requisite is that the area under study has a relatively small number of *Eucalyptus* species and that these can be separated to a satisfactory extent, on the basis of pollen morphology' (Pickett and Newsome 1997:203). This study also showed that when a small number of known taxa is used as a pollen reference, comparisons with fossil pollen can be more confidently made as there is likely to have been minimal floral composition change during the Holocene and known pollen synapomorphies can be discounted. However, it was not *Eucalyptus* pollen, but other representatives of the Myrtaceae family that best illustrated this concept.

Sections sampled between 4300 and 5000 BP showed a dramatic increase in the number of smaller syncolpate pollen grains found, representing the wet phase that has been suggested by previous studies (Hope et al. 2000; Donders et al. 2007). While much focus at Bega Swamp has been on what type of *Eucalyptus* was present throughout the Holocene, little has been mentioned about the smaller Myrtaceae pollen grains present in the core, which may be more indicative and less difficult to interpret than the various *Eucalyptus* species. The smallest of these grains most likely belongs to *Tristania neriifolia* (water gum), which occurs along rocky creek beds and banks on sandstone. The presence of this grain in core sections suggests that there was running water at or into Bega Swamp at various stages. Pollen grains sized 10-15 μm that were psilate and syncolpate most likely came from the genera *Baeckea* and *Kunzea*. Differentiating between the grains of these two genera is not easy, so most fossil grains with these characters were assigned to a combined genera type. However, they are both myrtaceous shrubs that occur in bogs and smaller heathland communities and the presence of these pollen grains suggests a more open vegetation type community. The quantity of *Eucalyptus* pollen types was lower in sections where *Baeckea*/*Kunzea* and *Leptospermum* pollen grains show a marked increase. While this could be interpreted as a reduction in the area of the surrounding tree forest community during these phases, it could be attributed alternatively to pollen filtration caused by shrub cover. Hope (1968) showed from traps in Wilson's Promontory that shrub cover intercepts pollen from nearby communities, which is then presumably destroyed by sunlight. This was also demonstrated by Tauber (1967) in shrubland surrounding a pond in Denmark. This suggests that while the Bega Swamp bog area may have been dominated by an open heathland community during this time, the surrounding vegetation would most likely have remained a eucalypt forest type.

Pollen grains sized 10-15 μm that were granulate or scabrate and syncolpate closely matched extant grains of *Leptospermum* species. The presence of these grains also suggests a more open heathland type community. It can also be noted from the Bega Swamp pollen diagram (Figure 2) that there is a decrease in charcoal found around the same time of the *Leptospermum* and *Baeckea*/*Kunzea* increase. Bickford and Gell (2005) suggested that the presence of *Leptospermum* at around 6500 BP in a core from the Fleurieu Peninsula, South Australia, combined with the presence of Cyperaceae, *Allocasuarina* (*paludosa* or *robusta*) and *Banksia marginata* represented a wet-heath community. This concurs with the previous studies on Bega Swamp that suggest a mid-Holocene wet phase, characterised by expansion of wet heath and ferns, starting around 7500 BP (Donders et al. 2007). Bickford and Gell (2005) also suggested that the dramatic increase in *Leptospermum* pollen could be accounted for in part by the reduction of *Acacia* present in the wetlands and surrounding vegetation following clearance and the cessation of regular burning. Other research has shown that *Leptospermum* benefits from a lack of fire (Bennet 1994; Johnson 2001), while it has also been suggested that fire benefits *Leptospermum*, which is tolerant of high-intensity fires (Morrison and Renwick 2000; Bond et al. 2004). This result suggests that *Leptospermum* and *Kunzea*/*Baeckea* pollen are better vegetation and climate indicators than separating *Eucalyptus* pollen into different sub-genera or species types.

Using size as a sorting characteristic did not prove efficient when using the Lucid key. Entering morphological characteristics of fossil pollen into the Lucid key often resulted in close matches with extant Myrtaceae pollen, but when measurements were combined with morphology no matches were commonly encountered. This may be explained by the use of different mounting media. The Bega Swamp samples were mounted in silicone oil; extant samples in a variety of substrates such as syn-matrix and glycerine, and differing sizes in relation to mounting media and acetolysis treatment has been noted before (Deuse 1960; Drugg 1962). To rely on size as a characteristic when searching for possible fossil-extant matches thus involves greater ranges depending on how many mounting substrates have been used, and creates less certainty.

Conclusion

Eucalyptus pollen morphology of the extant taxa did not consistently match morphological or ecological groups. Pollen grains of the symphomyrtus group did not look the same (*E.*

dalrympleana, *E. rubida*, *E. smithii* and *E. viminalis*), but similarities could be found within the group, such as the presence of a large apocolpial island (*E. rubida* and *E. viminalis*). Another important pollen morphology similarity in *Eucalyptus* was that seen between *E. pauciflora* (snowgum) and *E. fraxinoides* (white ash).

This raises doubts about using *Eucalyptus* pollen types when interpreting past vegetation and climatic conditions. On the other hand, Myrtaceae genera such as *Leptospermum*, *Kunzea*, *Baeckea*, *Melaleuca* (*Callistemon*), *Angophora*/*Corymbia*, *Tristaniopsis* and *Tristania* are more easily identifiable, especially when known pollen synapomorphies do not occur in the area. While it would be foolish to announce with complete certainty that fossil pollen grains match extant pollen grains, sentiments that have been expressed before by Joosten and de Klerk (2002), it is not unreasonable to suggest that pollen grains from the Holocene that display similar characteristics as extant pollen are most likely the same genus, or more safely, tribe within the Myrtaceae family. The confidence in comparing fossil and extant pollen increases when an extant taxa sample size is small and the extant taxa have distinct pollen characteristic suites such as observed in *Leptospermum*, which has small, granulate/scabrate patterned grains, *Kunzea*, which is syncolpate and less patterned than *Leptospermum*, or *Callistemon*, which always has a polar island, and *Tristania*, which has very small pollen grains (less than 8 µm).

Acknowledgements

The author would like to thank Gillian Brown and Andrew Drinnan from the School of Botany, University Melbourne, and Ian Thomas from the School of Land and Environment for loaning slides from the Kathleen Pike collection that were used as reference for extant Myrtaceae pollen. Finally, Geoff Hope provided his entire Bega Swamp reference material, including past articles, correspondence, maps and images. Two anonymous reviewers provided useful comments on the manuscript.

References

Barth, O.M. and A.F. Barbosa 1973. Catalogo sistematico do polens das plantas arborea do Brasil meridional XV Myrtaceae. *Memorias do Instituto Oswaldo Cruz* 70:467-497.

Bennet, L.T. 1994. The expansion of *Leptospermum laevigatum* on the Yanakie Isthmus, Wilson's Promontory, under changes in the burning and grazing regimes. *Australian Journal of Botany* 42(5):555-564.

Bickford, S. and P. Gell 2005. Holocene vegetation change, Aboriginal wetland use and the impact of European settlement on the Fleurieu Peninsula, South Australia. *Holocene* 15(2):200-215.

Bond, W.J., K.J.M. Dickinson and A.F. Mark 2004. What limits the spread of fire-dependent vegetation? Evidence from geographic variation of serotiny in a New Zealand shrub. *Global Ecology and Biogeography* 13(2):115-127.

Chalson, J.M. and H.A. Martin 1995. The Pollen Morphology of Some Co-occurring Species of the Family Myrtaceae from the Sydney Region. *Proceedings of the Linnean Society of New South Wales* 115:163-191.

Churchill, D.M. 1968. Distribution and Prehistory of *Eucalyptus diversicolor* F Muell E Marginata Donn Ex Sm and E Calophylla R Br in Relation to Rainfall. *Australian Journal of Botany* 16(1):125-151.

Deuse, K.F.P. 1960. Size variations in pollen grains with different treatments. *Pollen et Spores* 2:293-298.

Dodson, J.R. 1974. Vegetation history and water fluctuations at Lake Leake, south-eastern South Australia. I. 10 000 BP to present. *Australian Journal of Botany* 22(4):719-741.

Donders, T., S. Haberle, G. Hope, F. Wagner and H. Visscher. 2007. Pollen evidence for the transition of the eastern Australian climate system from the post-glacial to the present-day ENSO mode. *Quaternary Science Reviews* 26(11-12):1621-1637.

Drugg, W.S. 1962. Pollen Morphology of Lennoaceae. *American Journal of Botany* 49(10):1027-1032.

Eliseu, S.A. and A.M. Dinis. 2008. Ultrastructure and cytochemistry of *Eucalyptus globulus* (Myrtaceae) pollen grain. *Grana* 47:39-51.

Gadek, P.A. and H.A. Martin 1981. Pollen Morphology in the Subtribe Metrosiderinae of the Leptospermoideae (Myrtaceae) and Its Taxonomic Significance. *Australian Journal of Botany* 29(2):159-184.

Gadek, P.A. and H.A. Martin 1982. Exine Ultrastructure of Myrtaceous Pollen. *Australian Journal of Botany* 30(1):75-86.

Graham, A. 1980. Morfologia del polen de Eugenia-Myrcia (Myrtaceae) y Combretum-Terminalia (Combretaceae) en relacion a su alcance estratigrafico en el Terciario del Caribe. *Biotica* 5(1):5-14.

Green, D., G. Singh, H. Polach, D. Moss, J. Banks and E. A. Geissler 1988. A fine-resolution palaeoecology and palaeoclimatology from south-eastern Australia. *Journal of Ecology* 76(3):790-806.

Hope, G. 1968. Pollen studies at Wilson's Promontory. Unpublished M.Sc.. Melbourne: University of Melbourne.

Hope, G. 1995. Robustness and sensitivity of pollen data –A high resolution record from Bega Swamp, New South Wales. *Climate Change Retrospect and Prospect; Quantifying the Past to Understand the Future.* University of Melbourne November 6-8 1995.

Hope, G., A. Kershaw, S. van der Kaars, S. Xiangjun, P. Liew, L. Heusser, H. Takahara, M. McGlone, N. Miyoshi and P. Moss 2004. History of vegetation and habitat change in the Austral-Asian region. *Quaternary International* 118-19:103-126.

Hope, G., G. Singh, E. Geissler, L. Glover and D. O'Dea 2000. A Detailed Pleistocene-Holocene Vegetation Record from Bega Swamp, southern New South Wales. Quaternary Studies Meeting. Regional analysis of Australian Quaternary Studies: strengths, gaps and future directions. Canberra: Department of Geology, The Australian National University.

Johnson, P.N. 2001. Vegetation recovery after fire on a southern New Zealand peatland. *New Zealand Journal of Botany* 39(2):251-267.

Joosten, H. and P. de Klerk 2002. What's in a name? Some thoughts on pollen classification, identification, and nomenclature in Quaternary palynology. *Review of Palaeobotany and Palynology* 122(1-2):29-45.

Martin, H.A. and P.A. Gadek 1988. Identification of *Eucalyptus spathulata* pollen and its presence in the fossil record. *Memoirs of the Association of Australasian Palaeontologists* 5:311-327.

McIntyre, D.J. 1963. Pollen Morphology of New Zealand Species of Myrtaceae. *Transactions of the Royal Society of New Zealand* 2(7):83-107.

Morrison, D.A. and J.A. Renwick 2000. Effects of variation in fire intensity on regeneration of co-occurring species of small trees in the Sydney region. *Australian Journal of Botany* 48(1):71-79.

Parnell, J. 2003. Pollen of *Syzygium* (Myrtaceae) from SE Asia, especially Thailand. *Blumea* 48(2):303-317.

Patel, V.C., J.J. Skvarla and P.H. Raven 1984. Pollen characters in relation to the delimitation of Myrtales. *Annals of the Missouri Botanical Garden* 71(3):858-969.

Pickett, E.J. and J.C. Newsome 1997. *Eucalyptus* (Myrtaceae) pollen and its potential role in investigations of Holocene environments in southwestern Australia. *Review of Palaeobotany and Palynology* 98(3-4):187-205.

Pike, K.M. 1956. Pollen morphology of Myrtaceae from the south-west Pacific area. *Australian Journal of Botany* 4:13-53.

Polach, H. and G. Singh 1980. Contemporary 14C levels and their significance to sedimentary history of Bega Swamp, New South Wales. *Radiocarbon* 22(2):398-409.

Punt, W., P.P. Hoen, S. Blackmore, S. Nilsson and A. Le Thomas 2007. Glossary of pollen and spore terminology. *Review of Palaeobotany and Palynology* 143(1-2):1-81.

Tauber, H. 1967. Investigations of the mode of pollen transfer in forested areas. *Review of Palaeobotany and Palynology* 3:277-286.

Van Wyk, A.E. and I. Dedekind 1985. The genus *Eugenia* (Myrtaceae) in Southern-Africa: morphology and taxonomic value of pollen. *South African Journal of Botany* 51(5):371-378.

Wheeler, D., A.R. Chivas and G. Hope 2001. Isotope Stratigraphy of Bega Swamp, Southern NSW. *Australasian Quaternary Association meeting*, Southcombe Lodge, Port Fairy.

24

The evolution of a coastal peatland at Byron Bay, Australia: Multi-proxy evidence from the microfossil record

Kathryn H. Taffs

School of Environmental Science and Management, Southern Cross University, Lismore, NSW, Australia
kathryn.taffs@scu.edu.au

Brendan Logan

School of Environmental Science and Management, Southern Cross University, Lismore, NSW, Australia

Jeff F. Parr

School of Environmental Science and Management, Southern Cross University, Lismore, NSW, Australia

Geraldine E. Jacobsen

Australian Nuclear Science and Technology Organisation, Lucas Heights, Sydney, NSW, Australia

Introduction

Peatlands are highly valuable ecosystems for their ecological functions as well as their economic and societal values (Charman 2002). Yet they are also highly vulnerable to degradation by a range of anthropogenic activities and climate change (Charman 2002; Gorham and Rochefort 2003; O'Connell 2003; Rochefort et al. 2003; Vasander et al. 2003). In Australia, peatlands are an unusual and infrequent component of the landscape (Whinam et al. 2003), mostly distributed in the alpine areas of the southeast of the continent (Clarke and Martin 1999). However, areas of peat also occur in the coastal lowlands, often in dune swales, both on the east coast of Australia and on the southwest coast of Western Australia (Whinam et al. 2003). Coastal peatlands are unique environments and can be very informative about environmental and sea-level change in the coastal zone. Peat deposits in coastal areas of the Northern Hemisphere have provided informative records of water-level changes, which in turn can be used to complement existing palaeoclimatic and sea-level reconstructions (e.g. Eronen et al. 1987; Denys 1989; Campeau et al. 2000; Finkelstein and Davis 2005). At present, there is a lack of research focusing on Australian coastal peatlands. A deeper understanding of the response of coastal areas to climatic and anthropogenic influences is required to maximise conservation efforts of these unique environments.

Coastal peat formation is induced by waterlogging of silted-up areas when there is sufficient supply of freshwater (Kumano et al. 1990). In eastern Australia, many coastal peatlands occur naturally as thin strips behind the barrier dune systems. Australian coastal peatlands are dominated by Restionaceae, Cyperaceae and Epacridaceae vegetation (Whinan et al. 2003). They were generally formed in the past 4000 to 6000 years, since the last high-sea-level stand (Lewis et al. 2008). However, they have been heavily impacted by anthropogenic activities in the densely populated, resource-rich, coastal zone (Taffs et al. 2006). As a result, many coastal peatlands have been extensively damaged and/or removed by agriculture and urbanisation land uses.

Peatland conservation is Australia is very ad hoc and mostly aims to conserve representative portions of different peatland types (Whinan et al. 2003). Yet coastal peatlands are under-represented compared with other peatland types because of their location within the densely populated coastal zone. Developing a full understanding of the pre-anthropogenic impacts and natural variability of coastal peatlands is required to improve their management and conservation. Palaeoecological techniques provide ideal tools to identify natural benchmark conditions and variability such as vegetational change and response to anthropogenic impacts (Gaiser and Ruhland 2010). This provides essential information for managers of peatland systems to maximise their conservation.

Plant macrofossils such as phytoliths (e.g. Lewis 1981; Piperno 1988; Carter 1998; Kealhofer and Penny 1998; Parr 2003; Taffs et al. 2006) and pollen (e.g. von Post 1916; Godwin 1940; Wein et al. 1987; Gignac et al. 1991; Jasinski et al. 1998) are the traditionally used microfossils in peatland palaeoecological studies. They provide important information on vegetation composition and change over time. Phytoliths and pollen complement each other in a number of ways. Firstly, due to its airborne nature, pollen provides an indicator of regional vegetation changes, while phytoliths are for the most part deposited in situ and are useful for examining changes in vegetation at a specific site. Secondly, not all plants are prolific producers of phytoliths and/or pollen, and thus using both of these microfossils for palaeoecological reconstruction provides the best opportunity to identify a broad range of vegetation types (Kealhofer and Penny 1998; Taffs et al. 2006).

In coastal peatlands, the changing interface between the marine and terrestrial landscape strongly influences peatland water levels and water quality (Freund et al. 2004). Diatoms are regarded as very sensitive indicators of changes in water quality (Stoermer and Smol 1999) and are useful tools to describe palaeoenvironmental conditions in humiferous as well as in clastic lithofacies, such as those that accumulate in coastal lagoons. However, comprehensive studies of diatom flora from peatland systems remain scarce (Pienitz 2001; Gaiser and Ruhland 2010).

While peats are useful archives of environmental information, they are difficult environments in which to study siliceous microfossils such as diatoms and phytoliths (Bennett et al. 1991). Biogenic silica cycling causes breakdown of siliceous materials in anoxic, organic-rich, neutral-pH environments that are typical of coastal peatlands (Bennett et al. 1991). The fragility of diatom frustules makes them particularly susceptible to degradation and breakdown. Yet diatom dissolution data can aid interpretation of sedimentary profiles as records of limnological, ecological and environmental change (Ryves et al. 2009).

There are some coastal peatland sites that have sufficiently well-preserved siliceous microfossils to identify environmental change. These include the Beaufort Sea (Campeau et al. 2000), the western Belgian coast (Denys 1989), the western coastline of the United States and Canada (Eronen et al. 1987), the Lower Saxony, Germany (Freund et al. 2004), and the coastal plains of Japan (Kumano et al. 1990). In some of these studies, diatom dissolution occurred to the extent that full counts of each sample could not always be obtained (Eronen et al. 1987; Denys 1989; Campeau et al. 2000). Yet the extent of dissolution or fragmentation was used to aid environmental interpretation.

Given the strengths and weaknesses of any given environmental indicator on its own, the advantages of using multiple proxy indicators have been underscored in numerous peatland palaeoecological papers (e.g. Korhola 1990; Jaskinski et al. 1998; Rühland et al. 2000, 2006; Langdon et al. 2003; Blundell and Barber 2005; Myers-Smith et al. 2008). The aim of this paper is to examine the preservation of siliceous microfossils (diatoms and phytoliths) in a coastal peatland in eastern Australia and determine the environmental history that these fossils are able to reveal in combination with pollen, lithological and chronological information. A more detailed understanding of the evolution of this coastal plain and the response of the ecosystem to anthropogenic impacts is a priority to inform management strategies for the region's coastal peatlands.

Study area

Byron Bay (Figure 1) is located on the northern coast of New South Wales, Australia. It is a quartz-dominated sand province that has had its geomorphology shaped by pre-Holocene estuarine sediment deposits, the onshore movement of sands which accompanied the Holocene sea-level rise, and the reworking of coastal deposits by wind, wave and current actions after sea levels stabilised around 6000 yr BP (Chapman et al. 1982). The adjacent Belongil estuary has developed as a result of tidal forces in the relatively short time available during the Holocene for the estuary to infill under a steady sea level (Harris et al. 2002).

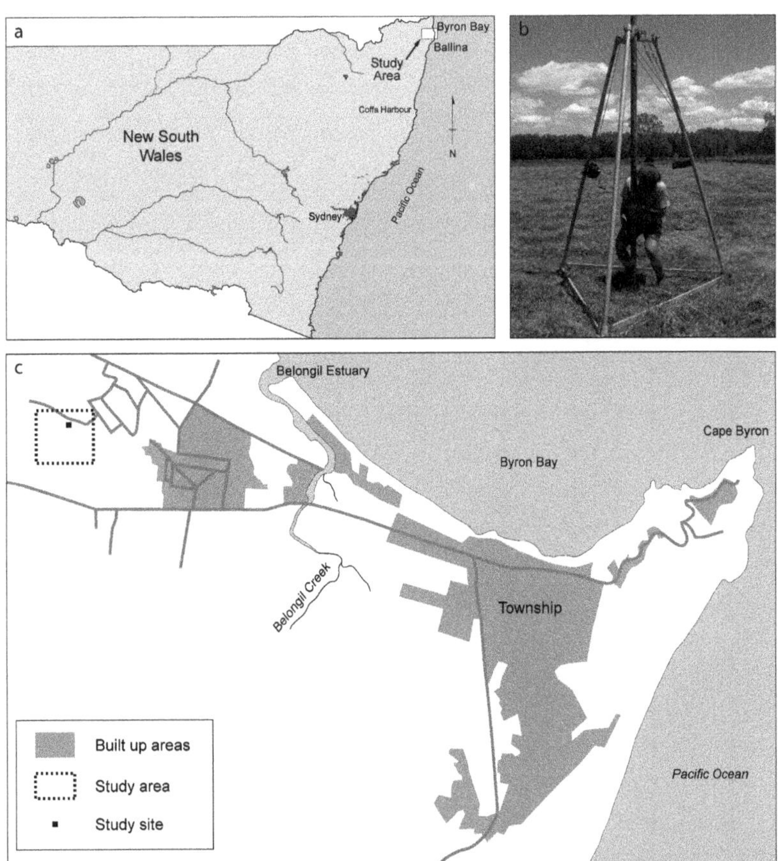

Figure 1. Location of Byron Bay: (a) location of Byron Bay within New South Wales, Australia; (b) photo of the site where the sediment core was extracted; (c) location of the sediment core extraction site

The site used for this study is 2.6 m asl (Byron Shire Council 2003). It is a 24 ha wetland located in the Belongil Swamp, which is a back barrier swamp on an estuarine/alluvial plain, bordered by bedrock hills to the west, and by a series of up to 13 Pleistocene beach ridges with an elevation of 5-6 m to the east (Geomarine 1997). The swamp is connected to an intermittently open estuary, Belongil Creek, via artificial drains. The Belongil Creek catchment drains an area of approximately 2840 ha (Geomarine 1997). The wetland is dominated by peat, and is bordered with *Melaleuca quinquinervia* and *Blechnum indicum* dominated vegetation. The surrounding area also contains significant proportions of dry sclerophyll, sedgeland and littoral rainforest associations.

Methods

In 2003, three sediment cores were extracted along a transect that incorporated the major depositional environments of a coastal peatland complex near Byron Bay (28° 38' 00" S, 135° 34' 25" E). We present here the results of analysis of a 5.2 m long sediment core retrieved in 1.5 m sections using a Vibracorer. Results of the two smaller sediment cores were reported in Taffs et al. (2006). The sediment-core sections were split vertically, with one half frozen for lithological analyses (sediment grain size, loss on ignition and pyrite content) and the other half sub-sampled for microfossil analysis (pollen, phytoliths and diatoms). In addition, pollen was extracted from sediment samples for AMS ^{14}C analysis using the analytical facilities of the Australian Nuclear Science and Technology Organisation (ANSTO).

Organic content was determined according to the loss-on-ignition method of Bengsston and Enell (1986) to characterise the sediment. Samples were oven-dried then ignited in a muffle furnace at 550°C for two hours. After cooling within a dessicator, samples were weighed and the lost material expressed as a percentage of dry weight.

The reduced inorganic sulfur fraction (which includes pyrite and other iron disulfides, elemental sulfur, and acid volatile sulfides) was measured in duplicate on 20 samples using the chromium reducible sulfur technique of Sullivan et al. (2000) in order to provide a measure of pyrite content of the sediments. After peroxide oxidation, all residues were washed three times with distilled water, centrifuged, and oven-dried at 60°C before chromium reducible sulfur analysis. Pyrite can indicate the prevalence of a reducing environment occurring under anoxic conditions during deposition (Bajpai et al. 2001). Thus, the occurrence of pyrite in sedimentary sequences indicates marine origins and can have significant palaeoenvironmental value.

Diatom samples were processed according to the method of Parr et al. (2004), and were inspected under an Olympus CX40 compound light microscope fitted with an Olympus DP10 digital camera at 1000X magnification. A minimum of 300 diatom frustules were identified and counted from each sample to determine the diatom assemblage where possible. The diatoms were identified using the photographs of Foged (1978) and Gell et al. (1999). Only species that had a relative abundance greater than 2% were included in the data set. For rapid assessment of diatom preservation, the method of Dong et al. (2008) was adopted. This method estimates diatom preservation in three categories: (1) slides with good diatom preservation (the assemblages are almost pristine with the whole community preserved well without signs of dissolution and fragmentation); (2) medium fragmentation (assemblages partially dissolved but with most species well preserved); and (3) poor preservation (most species dissolved, only robust species remaining).

Phytoliths were extracted from sediment using the method of Parr (2002). Phytoliths extracted from each sample were weighed, mounted on to microscope slides and scanned at 400x magnification on an Olympus BH2 microscope. A total of 150 diagnostic phytolith types were counted for each slide. A collection of plant species was made in and around the study site and phytoliths extracted using the method of Parr et al. (2001) for use in a comparative database. Identification of phytoliths was based on a probability method (Parr 2004), in which fossil phytolith morphotypes were compared with those represented in a regionally specific phytolith database; the more commonly a plant taxon is represented by different morphotypes, the more likely it is considered that the taxon in the database is represented.

Pollen was extracted from 1 cu. cm of sediment using the method described in Cotter and Boyd (1998), excluding the hydrofluoric acid treatment that was considered unnecessary due to the low silica content of the sediment. Pollen extracted from each sample was mounted on to microscope slides and scanned at 600x and 1000x magnification on an Olympus BH2 microscope. A total pollen sum of 500 grains was counted per sample.

Ten samples were prepared for AMS ^{14}C dating using the pollen fraction. These were processed and measured at the Australian Nuclear Science and Technology Organisation (ANSTO, sample numbers OZH042 to OZH051, Fink et al. 2004). Radiocarbon dates were calibrated using CALIIB 5.0 (Stuiver and Reimer 1993). Dates less than 11 ka were calibrated using the Southern Hemisphere radiocarbon calibration data set ShCal04 (McCormac et al. 2004). Dates 11-26 ka were calibrated using the international radiocarbon calibration data set IntCal04 (Reimer et al. 2004), while dates older than 26 ka are uncalibrated (Reimer et al. 2004).

All results were graphed using the package C2 (Juggins 2004) and displayed using a stratigraphic diagram. Biostratigraphic zones were established, based on the distinct lithological sequence observed in the sediment core.

Results

Chronology

Table 1 shows the AMS ^{14}C results with radiocarbon and calibrated ages. Radiocarbon ages ranged from 4300 ± 50 at 50-55 cm depth to 41,800 ± 1200 at 510-515 cm depth. Figure 2 shows the age-depth model for the sediment core and the core lithology. The basal material is fine, muddy sand, extending from 520 cm to 300 cm and corresponds to an uncalibrated radiocarbon age of ~ 42,000 yr BP to ~ 33,000 yr BP. A sand layer was deposited ~ 33,000 yr BP, which grades back into sandy mud from 260 cm to 130 cm, or ~ 33,000 yr BP to 9000-9396 cal. yr BP. There is a small horizon representing a distinct colour change at 170-175 cm, that is 10,309-10,763 cal. yr BP. The Sandy Mud and Muddy Sand sediments are typical of Pleistocene estuarine deposits on the north-coast coastal plain (McGrath and Boyd 1998; Taffs et al. 2008). The upper peat deposit is 130 cm deep and extends from approximately 5327-5709 cal. yr BP to the present.

Table 1. Radiocarbon ages from the Byron Bay sediment core

Laboratory No.[a]	Core depth (cm)	Radiocarbon age (yr BP)	Calibrated age (cal yr BP)[b]	Dated material
OZH042	50-55	4300±50	4581-4959	pollen
OZH043	120-125	4880±60	5327-5709	pollen
OZH044	140-145	8240±60	9000-9396	pollen
OZH045	160-165	9430±60	10304-10763	pollen
OZH046	170-175	15,110±110	18062-18713	pollen
OZH047	190-195	15,480±100	18647-18936	pollen
OZH048	250-255	33,530±320		pollen
OZH049	310-315	33,050±300		pollen
OZH050	410-415	42,900±1600		pollen
OZH051	510-515	41,800±1200		pollen

a. Australian Nuclear Science and Technology Organisation (ANSTO) laboratory code.
b. Calibrated using program CALIB 5.0, reported using the 2 sigma range.

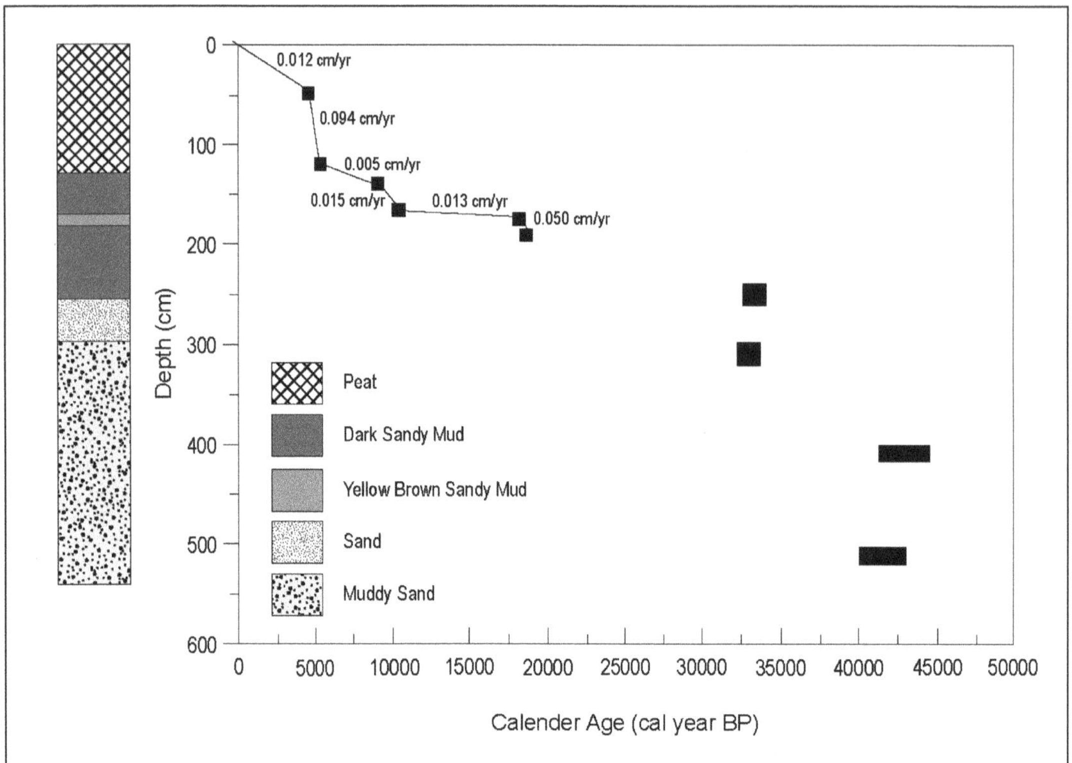

Figure 2. Age-depth curve of the Byron Bay sediment core. Note samples >26 ka have not been calibrated and hence sedimentation rate for this portion of the core has not been calculated

Sedimentation rates are very low throughout the core. Accumulation rates have not been calculated for the basal estuarine muddy-sand substrates as the dates are not able to be calibrated accurately and hence compared with the calibrated dates of the upper sediments. The upper Sandy Mud layer has a very low average accumulation rate of 0.018 cm/yr (range 0.0013–0.05 cm/yr). The accumulation of the peat layer reduces considerably towards the surface, with a sedimentation rate of 0.094 cm/yr for the lower section of the peat and 0.012 cm/yr for the upper section of peat. The general stratigraphic consistency within the core and the gradual boundaries between sediment layers suggest continuous sediment accumulation with only minor reworking of the sediments between boundaries.

Diatoms

Diatom preservation deteriorated with increasing depth in the core (Figure 3). Diatoms were abundant and well preserved in the surface peat. However, there were no diatoms present in zone 3, in the sand layer, and diatom preservation was very poor in zone 4, with only frustles of the most robust species, *Nitzschia fonticola*, present. Only six diatom species occurred in the sediment core with a relative abundance of >2%, representing a very low diversity. The dominant genera were *Nitzschia, Eunotia* and *Pinnularia. Nitzschia fonticola* was dominant throughout the core. This species is known to be quite cosmopolitan and in some coastal sites widespread and abundant (Foged 1978; Vyverman et al. 1995). *Nitzschia paleaeformis* was present only in the surface peats, suggesting a change in the depositional environment occurred around 5000 cal. yr BP, marking the commencement of peat formation. *Eunotia soleroii, Eunotia flexuosa* and *Pinnularia gibba* were found throughout zone 1 and 2, suggesting a continuously acidic habitat (Sterrenburg et al. 2007). Pyrite, also displayed in Figure 3, was present throughout the core but was detected at very high levels within zone 4, indicating marine conditions were prevalent during deposition of these sediments. The organic content was high throughout the surface peat and in zone 2, negligible in zone 3 and a minor component in zone 4.

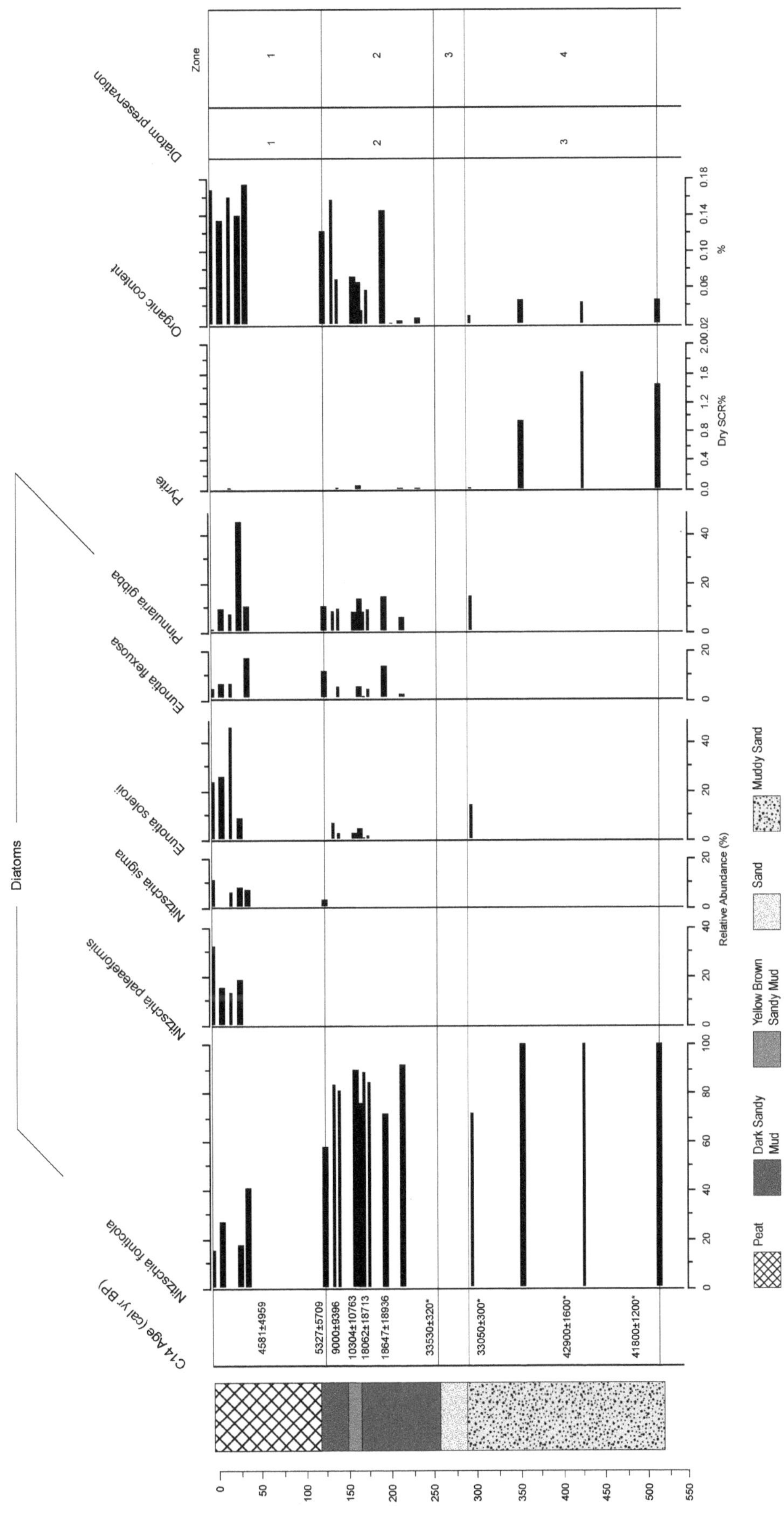

Figure 3. Diatom record of the coastal peat sediment core from Byron Bay. Radiocarbon ages >26 ka have not been calibrated

Phytoliths and pollen

The phytolith record (Figure 4) indicates the vegetation surrounding the site has been a grassland fringed with *Melaleuca* forest for the past ~33,500 yr BP. The understorey is dominated by Poaceae, Restionaceae and Cyperaceae species. The tree component was dominated by Myrtaceae *Melaleuca* species type, possibly indicating a single-species tree component of the vegetation community. The tree component is significant in the lower sections of the core until 4581-4959 cal. yr BP, after which it reduces considerably until relatively recently. In the more recent sediments, there are low counts of Myrtaceae phytolith types. The understorey of shrub and grass species dominates the vegetation, forming 80-90% of the vegetation mass throughout the period of peat deposition. Restionaceae and Cyperaceae are peat-forming vegetation types and their abundance increases within the peat litological zones.

The pollen record provides further evidence of a grassland community with fringing *Melaleuca* forests within the catchment area. The tree component is dominated by Myrtaceae *Melaleuca* type and Casuarinaceae. Cyantheaceae is a dominant understorey shrub, most likely on the margins of the peatland. The pollen data suggest that herbaceous taxa have been the dominant vegetation component of this peat environment for the past ~33,000 yr BP. The upperstorey component is minor but shows distinct fluctuations over time.

Discussion

The coastal peat layer at Byron Bay is 130 cm deep and has accumulated over the past 5000 cal. yr BP, demonstrating a very slow accumulation rate typical of peatlands (Charman 2002). The presence of peat of this age in such close proximity to the ocean indicates that sea levels stabilised in this area approximately 5000 cal. yr BP and the peatland has since been infilling with terrestrially derived sediment. Underlying the peat are sediments of estuarine origin with fine mud particles and very low organic component, indicating higher sea levels in the period preceding 5000 cal. yr BP. The presence of pyrite in these sedimentary layers confirms a strong marine influence on the sediments deposited during this time (Bajpai et al. 2001). This correlates with evidence from Larcombe et al. (1995) and Lewis et al. (2008) on eastern-Australian sea-level fluctuations. The distinct colour change of the estuarine muds at 10,300 cal. yr BP suggests a fluctuation of sea level at this time and a change in sediment origin. Stronger marine influence as indicated by the deposition of sand around 33,000 yr BP indicates a slight increase of sea level or change in coastal geomorphology and hence the proximity of Belongil estuary. However, before this time, up to 42,000 yr BP, the site experienced stable estuarine conditions.

Unfortunately, the diatom record has not contributed significant evidence for the reconstruction of environmental conditions of the coastal plain due to degradation of the silica frustules. Even in the surface peats, only six species were present, indicating very low species diversity. This is typical of peats with their harsh acidic environment (Sterrenburg et al. 2007). However, even six species is a very low diversity for a peatland (Campeau et al. 2000). The species present do indicate acidic freshwater environments (Foged 1978; Vyverman et al. 1995). The introduction of *Nitzschia paleaeformis* when peat accumulation commences is indicative of a change of trophic status that would be expected with the commencement of the accumulation of organic matter.

The diatom record does show a gradual deterioration in the preservation of frustules with increasing depth. This does indicate that diatoms were present at the time of sediment deposition but silica cycling has caused gradual deterioration of frustules present. This is typical of both peat and marine environments. With the two processes occurring at the one location due to both the acid environment of the peat and the saline marine origins, this deterioration

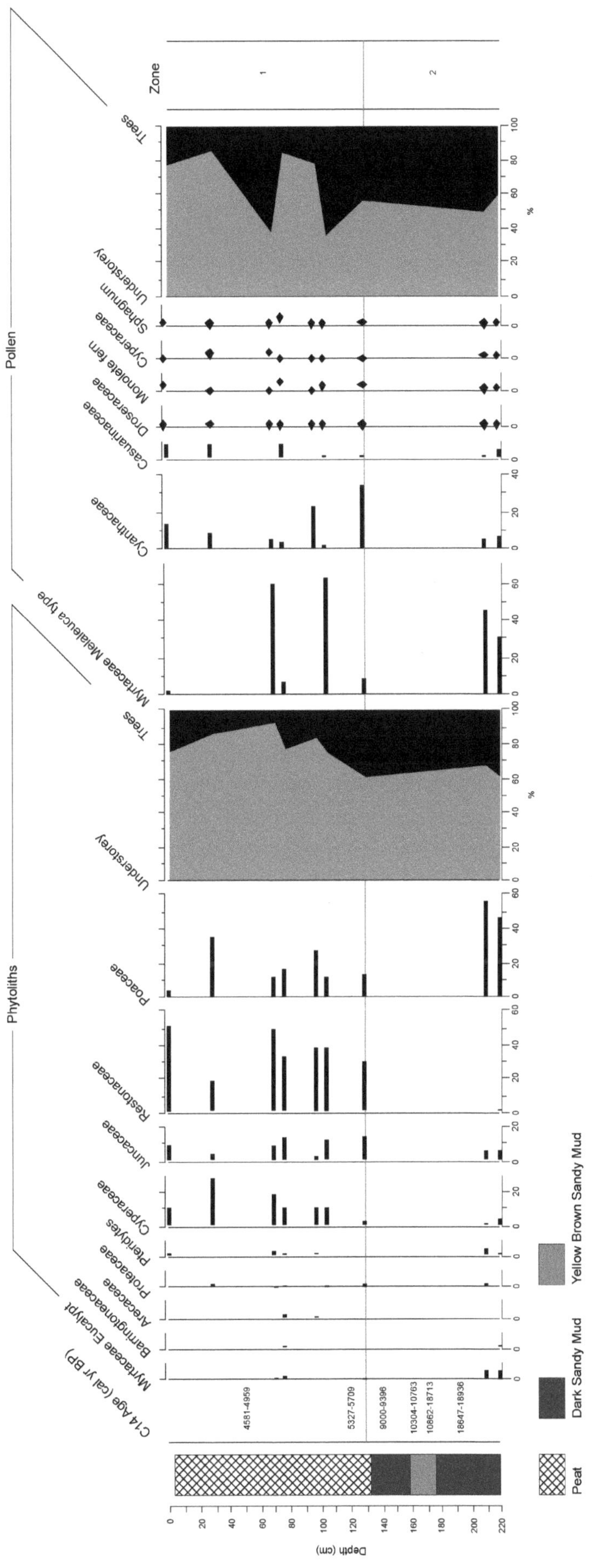

Figure 4. Phytolith and pollen record of the coastal peat sediment core from Byron Bay

is not surprising (Lewin 1960; Bennett et al. 1991). Clearly, more robust species were better preserved and had a higher relative abundance in the sediment core.

In contrast, the phytoliths were well preserved in the peat layer. Phytoliths are very robust silica structures and less prone to deterioration than diatoms (Parr and Sullivan 2005). Comparison of the phytolith and pollen record indicates the strength of using the two indicators together, with the phytolith record indicating the local site vegetation characteristics and the pollen record reflecting vegetation composition changes within the catchment area. Both indicators show varying levels of dominance of the understorey vegetation over the period of peat accumulation. Throughout the past 5000 cal. yr BP, the peatland has been a grassland surrounded by *Melaleuca* forest. The dominant peat-forming vegetation is Restionaceae and Cyperaceae species, which is typical of Australian coastal peatlands (Whinan et al. 2003).

Conclusion

This palaeoecological study found no detectable evidence of sea-level change for the past 5000 cal. yr BP, which indicates that sea levels have possibly been relatively stable on the coastal plain adjacent to Byron Bay during this period, allowing terrestrial processes to dominate and the accumulation of peat to occur. Before this time, the sea level was slightly higher than at present, and the site was shaped by estuarine processes, as evidenced by the constant deposition of estuarine muds. There was a period of very strong marine influence around 33,000 yr BP, which indicates either higher sea levels or a change in the proximity of the Belongil estuary to the core location. This period of estuarine dominance continues until at least 42,000 yr BP.

During the period of peat accumulation, the vegetation was dominated by grassland, with Restionaceae and Cyperaceae being the dominant peat-forming vegetation. The grassland was surrounded by *Melaleuca* forest that contributed significantly to the regional pollen record. Throughout the period of peat deposition, the quality of the interstitial water was freshwater and acidic. These conditions are not conducive to the preservation of diatom frustules and hence the diatom record showed deteriorating preservation throughout the core. This is especially so in the estuarine-influenced section of the core and is typical of marine deposits (Lewin 1960; Dixet et al. 2001).

This study indicated that a multi-proxy methodology in a peatland environment has been able to provide insightful information on the development of a coastal peatland. In combination with an excellent dating record, diatom, phytolith, pollen and lithological changes have contributed to a better understanding of the development of the peatland, and baseline information has been ascertained from which more effective conservation and management decisions can be made.

Acknowledgements

Funds for the project were provided by a Southern Cross University Internal Research grant. The AMS ^{14}C dating was made possible by an Australian Institute of Nuclear Science and Engineering (AINSE) grant to Taffs (04/150). Many thanks to Steve Abbott for assistance with field work and Keith Bolton for development of site understanding.

References

Bajpai, U., M. Kumar, M. Shukla, A. Prakash and G.P. Srivastava 2001. Nature and composition of pyrite framboids and organic substrate from degraded leaf cuticles of Late Tertiary sediments, Mahuadanr Valley, Palamu, Bihar. *Current Science* 81:102-106.

Bengsston, L. and M. Enell 1986. Chemical Analysis. In B.E. Berglund (ed) *Handbook of Holocene Paleoecology and Paleohydrology*, pp423-454. New Jersey: Blackburn Press.

Bennett, P.C., D.I. Siegel, B.M. Hill, and P.H. Glaser 1991. Fate of silicate minerals in a peat bog. *Geology* 19:328-331.

Blundell, A. and K. Barber 2005. A 2800-year palaeoclimatic record from Tore Hill Moss, Strathspey Scotland: the need for a multi-proxy approach to peat-based climate reconstructions. *Quaternary Science Reviews* 24:1261-1277.

Byron Shire Council 2003. *Byron Shire Council Mean Sea Level Data.* Byron Bay, NSW.

Campeau, S., A. Héquette and R. Pienitz 2000. Late Holocene diatom biostratigraphy and sealevel changes in the southeastern Beaufort Sea. *Canadian Journal of Earth Sci*ences 37:63-80.

Carter, J.A. 1998. Palaeoenvironmental reconstruction from the last interglacial using phytolith analysis, southeastern North Island, New Zealand. In D.M. Hart, G.S. Humphreys and R.J. Field (eds) p46. *Australasian Phytolith Workshop*. Sydney: Macquarie University.

Chapman, D.M., M. Geary, P.S. Roy and B.G. Thom 1982. *Coastal evolution and coastal erosion in New South Wales.* Darlinghurst, Sydney: David Ell Press.

Charman, D. 2002. *Peatlands and Environmental Change.* Chichester: John Wiley and Sons.

Clarke, P.J. and A.R.H. Martin 1999. Sphagnum Peatlands of Kosciuzko in relation to altitude, time and disturbance. *Australian Journal of Botany* 47:519-536.

Cotter, M.M. and W.E. Boyd 1998. *Fossil pollen extraction for palaeoenvironmental reconstruction. A method manual.* Lismore, Australia: School of Resource Science and Management, Southern Cross University.

Denys, L. 1989. The transition from Calais deposits to surface peat in the western Belgian coastal plain. Results of a paleo-environmental diatom study. *Professional Paper Belgische Geologische Dienst* 241:20-43.

Dixet, S., P. Van Cappellen and J. van Bennekm 2001. Processes controlling solubility of biogenic silica and pore water build up of silicic acid in marine sediments. *Marine Chemistry* 73:333-352.

Dong, X., H. Bennion, R.W. Battarbee, X. Yang, H. Yang and E. Liu 2008. Tracking eutrophication in Taihu Lake using the diatom record: Potential and problems. *Journal of Paleolimnology* 40:413-429.

Eronen, M., T. Kankainen and M. Tsukada 1987. Late Holocene sea-level record in a core from the Puget Lowland, Washington. *Quaternary Research* 27:147-159.

Fink, D., M. Hotchkis, Q. Hua, G. Jacobsen, A.M. Smith, U. Zoppi, D. Child, C. Mifsud, H. van der Gaast, A. Williams and M. Williams 2004. The ANTARES AMS Facility at ANSTO. *Nuclear Instruments and Methods B* 223-224:109-115.

Foged, N. 1978. Diatoms In Eastern Australia. *Bibliotheca Phycologica* 47. Stuttgart: J. Cramer.

Finkelstein, S.A. and A.M. Davis 2005. Paleoenvironmental records of water level and climatic changes from the middle to late Holocene at a Lake Erie coastal wetland, Ontario, Canada. *Quaternary Research* 65:33-43.

Freund, H., G. Gerdes, H. Steif, O. Dellwig and F. Watermann 2004. The indicative meaning of diatoms, pollen and botanical macro fossils for the reconstruction of palaeoenvironments and sea-level fluctuations along the coast of Lower Saxony; Germany. *Quaternary International* 112:71-87.

Gaiser, E. and K. Ruhland 2010. Diatoms as indicators of environmental change in wetlands and Peatlands. In J.P. Smol and F. Stoermer (eds) *The Diatoms: Applications for the Environmental and Earth Sciences*. Second edition. Cambridge, UK: Cambridge University Press.

Gell P.A., J.A. Sonneman, M.A. Reid, M.A. Illman and A.J. Sincock 1999. *An illustrated key to the common diatom genera from southern Australia*. Cooperative Research Centre for Freshwater Ecology – Identification Guide No. 26.

Geomarine 1997. *Belongil Creek: Estuarine Processes Study*. Final Report, December 1996. Report prepared for the Belongil Creek Estuarine Management Committee.

Gignac, L.D., D.H. Vitt, S.C. Zoltai and S.E. Bayley 1991. Bryophyte response surfaces along climatic, chemical, and physical gradients in peatlands of western Canada. *Nova Hedwigia* 53:27-71.

Godwin, H. 1940. Pollen analysis and forest history of England and Wales. *New Phytologist* 39:370-400.

Gorham, E. and L. Rochefort 2003. Peatland Restoration: A brief assessment with special reference to *Sphagnum* bogs. *Wetlands Ecology and Management* 11:109-119.

Harris, P.T., A.D. Heap, S.M. Bryce, R. Porter-Smith, D.A. Ryan and D.T. Heggie 2002. Classification of Australian clastic coastal depositional environments based upon a quantitative analysis of wave, tidal and river power. *Journal of Sedimentary Research* 72:858-870.

Jasinski, J.P.P., B.G. Barner, A.A. Andreev, R. Aravena, S.E. Gilbert, B.A. Zeeb, J.P. Smol and A.A. Velichko 1998. Holocene environmental history of a peatland in the Lena River valley, Siberia. *Canadian Journal of Earth Sciences* 35:637-648.

Juggins, S. 2004. *C2 Data Analysis*. Version 1.4 Beta (build 3.2). University of Newcastle, UK.

Kealhofer, L. and D. Penny 1998. A combined pollen and phytolith record for fourteen thousand years of vegetation change in northeastern Thailand. *Review of Palaeobotany and Palynology* 103:83-93.

Korhola, A. 1990. Paleolimnology and hydroseral development of the Kotasuo bog, southern Finland, with special reference to the Cladocera. *Annales Academie Scientiarum Fennicae A III 155*, 40pp.

Kumano, S., M. Ihira, Y. Maeda, M. Yamauchi, E. Matsumoto and L. Matsuda 1990. Holocene sedimentary history of some coastal plains in Hokkaiso, Japan. IV. Diatom assemblages in the sediments from Kushiro Moor (2). *Ecological Restoration* 5:221-235.

Langdon, P.G., K.E. Barber and P.D.M. Hughes 2003. A 7500-year peat-based palaeoclimatic reconstruction and evidence for an 1100-year cyclicity in bog surface wetness from Temple Hill Moss, Pentland Hills, southeast Scotland. *Quaternary Science Reviews* 22:259-274.

Larcombe, P., R.M. Carter, J. Dye, M.K. Gagan and D.P. Johnson 1995. New evidence for episodic postglacial sea-level rise, central great-barrier-reef, Australia. *Marine Geology* 127:1-44.

Lewin, J. 1960. The dissolution of silica from diatom walls. *Geochima et Geochimica Acta* 21:182-198.

Lewis, R.O. 1981. Use of Opal Phytoliths in Palaeoenvironmental Reconstruction. *Journal of Ethnobiology* 1:175-181.

Lewis, S. E., R.A. Wust, J.M. Webster and G.A. Shields 2008. Mid-late Holocene sea-level variability in eastern Australia. *Terra Nova* 20:74-81.

McCormac, F.G., A.G. Hogg, P.G. Blackwell, C.E. Buck, T.F.G. Higham and P.J. Reimer 2004. SHCal04 Southern Hemisphere Calibration 0 - 11.0 cal kyr BP. *Radiocarbon* 46:1087-1092.

McGrath, R. and W.E. Boyd 1998. Holocene vegetation history of Bundjalung National Park and Bungawalbin Creek, north eastern New South Wales. *Australian Geographer* 29:205-221.

Myers-Smith, I.H., J.W. Harden, M. Wilmking, C.C. Fuller, A.D. McGuire and F.S. Chapin III 2008. Wetland succession in a permafrost collapse: interactions between fire and thermokarst. *Biogeosciences* 5:1273-1286.

O'Connell, M.J. 2003. Detecting, measuring and reversing changes to wetlands. *Wetlands Ecology and Management* 11: 397- 401.

Parr, J.F. 2002. A comparison of heavy liquid flotation and microwave digestion techniques for the extraction of fossil phytoliths from sediments. *Review of Palaeobotany and Palynology* 120:315-336.

Parr, J.F. 2003. *A study of Palaeo-Landscapes in the Numundo region of West New Britain, Papua New Guinea, as indicated by Fossil Phytolith Analysis.* Unpublished PhD Thesis. Lismore: Southern Cross University.

Parr, J.F. 2004. Morphometric and visual fossil phytolith identification using a regionally specific digital database. *Phytolitharian* 16(2):2-10.

Parr, J.F. and L.A. Sullivan 2005. Soil carbon sequestration in Phytoliths. *Soil Biology and Biochemistry* 37(1):117-124.

Parr, J.F., V. Dolic, G. Lancaster and W.E. Boyd 2001. A microwave digestion method for the extraction of phytoliths from herbarium specimens. *Review of Palaeobotany and Palynology* 116:203 - 212.

Parr, J.F., K.H. Taffs and C.M. Lane 2004. A microwave digestion technique for the extraction of fossil diatoms from coastal lake and swamp sediments. *Journal of Paleolimnology* 31(3):383-390.

Pienitz, R. 2001. Analyse des microrestes végétaux : diatomées. In S. Payette and L. Rochefort (eds)*Écologie des tourbières du Québec-Labrador*, pp311-326. Québec: Les Presses de l'Université Laval.

Piperno, D.R. 1988. Phytolith Analysis: An Archaeological and Geological Perspective. London: Academic Press.

Reimer, P.J., M.G.L. Baillie, E. Bard, A. Bayliss, J.W. Beck, C.J.H. Bertrand, P.G. Blackwell, C.E. Buck, G.S. Burr, K.B. Cutler, P.E. Damon, R.L. Edwards, R.G. Fairbanks, M. Friedrich, T.P . Guilderson, A.G. Hogg, K.A. Hughen, B. Kromer, G. McCormac, S. Manning, C.B. Ramsey, R.W. Reimer, S. Remmele, J. R. Southon, M. Stuiver, S. Talamo, F.W. Taylor, J. van der Plicht and C.E. Weyhenmeyer 2004. Intcal04 terrestrial radiocarbon age calibration, 0-26 cal kyr BP. *Radiocarbon* 46:1029-1058.

Rochefort, L., F. Quinty, S. Campeau, K. Johnson and T. Malterer 2003. North American approach to the restoration of Sphagnum dominated Peatlands. *Wetlands Ecology and Management* 11(1-2):3-20.

Rühland, K., J.P. Smol, J.P.P. Jasinski and B. Warner 2000. Response of diatoms and other siliceous indicators to the developmental history of a peatland in the Tiksi Forest, Siberia, Russia. *Arctic, Antarctic and Alpine Research* 32:167-178.

Rühland, K., N.R. Phadtare, R.K. Pant, S.J. Sangode and J.P. Smol 2006. Accelerated melting of Himalayan snow and ice triggers pronounced changes in a valley peatland from northern India. *Geophysical Research Letters 33 (15):*L15709.

Ryves, D., R. Battarbee and S. Fritz 2009. The dilemma of disappearing diatoms: Incorporating diatom dissolution data into palaeoenvironmental modelling and reconstruction. *Quaternary Science Reviews* 28**:**120-136.

Sterrenburg, F., R. Gordon, M.A. Tiffany and S.S. Nagy 2007. Diatoms: living in a constructal environment. In J. Seckbach (ed) *Algae and Cyanobacteria in extreme environments*, pp143-172. Dordrecht, Netherlands: Springer Press.

Stoermer, F. and J.P. Smol 1999. *The Diatoms: Applications for the Environmental and Earth Sciences*. Cambridge, UK: Cambridge University Press.

Stuiver, M. and P.J. Reimer 1993. Extended C14 database and revised CALIB 3.0 C14 age calibration program. *Radiocarbon* 35:215-230.

Sullivan, L.A., R.T. Bush and D.M. McConchie 2000. A modified chromium-reducible sulfur method for reduced inorganic sulfur: Optimum reaction time for acid sulfate soil. *Australian Journal of Soil Research* 38:729-734.

Taffs, K.H., J.F. Parr and K.G.E. Bolton 2006. Using Palaeobotany to resolve ecological disasters in East Australian peatlands. *Ecological Management and Restoration* 7(2):132-135.

Taffs, K.H., L.J. Farago, H. Heijnis and G. Jacobsen 2008. A diatom-based holocene record of human impact from a coastal environment: Tuckean Swamp, eastern Australia. *Journal of Paleolimnology* 39:71-82.

Vasander, H., A.S. Tuittila and E. Lode 2003. Status and restoration of Peatlands in northern Europe. *Wetlands Ecology and Management* 11(1-2):51-63.

von Post, L. 1916. Om Skogstradspollen i sydsvendka torfmosselagerfoljder (foredragsreferat). *Geologiska Föreningen i Stockholm Förhandlinga* 38:384-394.

Vyverman, W., R. Vyverman, D. Hodgson and P. Tyler 1995. Diatoms from Tasmanian mountain lakes: a reference data-set (TASDIAT) for environmental reconstruction and autoecological study. *Bibliotheca Diatolologica, Band 33*. Berlin: J. Cramer.

Wein, R.W., M.P. Burzynski, B.A. Sreenivasa and K. Tolonen 1987. Bog profile evidence of fire and vegetation dynamics since 3000 years BP in the Acadian Forest. *Canadian Journal of Botany* 65:1180-1186.

Whinam, J., G.S. Hope, B.R. Clarkson, B.P. Buxton, P.A. Alspach and P. Ada 2003. Sphagnum in peatlands of Australasia: their distribution, utilisation and management. *Wetlands Ecology and Management* 11(1-2):37-49.

25

Development of mountain peatlands in stable equilibrium with open-channel hydraulics: A new concept in peatland formation and maintenance

Rachel Nanson

Geography and Environmental Studies, Level 9 Napier Building, University of Adelaide, Adelaide, SA
rachel.nanson@adelaide.edu.au

Introduction

Peatlands are often perceived to be relatively fragile ecosystems that are sensitive to fluctuating water tables and disrupted by channel formation. The Barrington Tops region of New South Wales exhibits peatlands and channels that have coevolved to form stable equilibrium systems not previously recognised in upland systems. The channels have adjusted their cross-sections, bedforms and planform to optimise flow efficiency and limit vertical peatland development. This research summarises these conditions for several peatland channels.

High channel bank strength is afforded by dense vegetation that facilitates optimal hydraulic efficiencies with unusually low channel width/depth ratios of ~2. Such channels frequently flow at or close to bankfull but rarely flow overbank, an unlikely coincidence unless the channels and the peatlands coevolved. Shear stress calculations suggest that bankfull flows are more than capable of transporting the scant sediment load but planform and bedform adjustments consume surplus energy and minimise erosion, resulting in highly stable, flow-efficient channels.

Polblue Creek, Barrington River and Edwards Creek have each developed sinuous planforms, with some examples of very tight bends. The reduction in energy gradient resulting from sinuous planform development and the substantial turbulence that these tight bends generate combine to partly counter the exceptional hydraulic efficiency of the narrow and deep channel cross-sections.

Bedform magnitude (steepness) is strongly linked to flow resistance and there are large variations in the scale of bedforms in the channels studied here. In the smaller but more sinuous Polblue Creek, power is moderate and bed features display only moderate steepness values. In contrast, the larger but less sinuous Barrington River and Edwards Creek channels have higher stream powers and their armoured bedforms have developed with much greater steepness. The considerable turbulence associated with sinuous planforms and large well-armoured bedforms are sufficient to generate the energy losses required to maintain channel and peatland stability.

Once peatland channels have attained their most efficient cross-sections for the evacuation of water from the peatland, any further (vertical) growth of the peatlands is restricted by enhanced flow velocities and limited flow depths; the volume of peat in such peatlands is determined by the hydraulic efficiency of their channels. The development and maintenance of the geometry of these peatland channels is bio-geomorphic and bio-hydraulic in nature and yet accords to the same optimising principles that govern the formation of self-adjusting channels with floodplains formed of clastic alluvium.

Channelled peatlands

Streams that flow through wetlands are usually sinuous and are bound by vertical organic and/or peat banks. Their beds are comprised of sand and gravel, peat or basement material (bedrock) and they tend to exhibit slot-like cross-sections. Some research has focused on the cross-sectional and planform geometry of such organic streams (Jurmu and Andrle 1997; Jurmu 2002), however little research has examined their geometry-flow interactions and the relationship between the channels and their associated peatlands.

The expected response of a peatland to channel incision is that as the water table is lowered, existing vegetation is greatly modified and fluctuations within the peatland water table increase (Price et al. 2003). Peatlands can become dewatered and compacted following such incision (Johnston and Brierley 2006), with consequent impacts on peatland hydrology, floristics (e.g. Southern 1982; Hope 2003) and sedimentology. The recent history of *Sphagnum* peatlands in southern Australia shows that increased drainage can convert *Sphagnum* bog to grasslands or sedgelands (Good 1992). However, while peatland vegetation might be replaced by more terrestrial species, over a longer timeframe the cycle of peatland development can recommence (Young 1986b). While peatland ecology is likely to be severely altered by the development of channelled conditions (Price et al. 2003), the development of channelled peatlands does not necessarily imply non-equilibrium systems. Some peatlands appear to have co-evolved with the channels that occupy them and the palynological record is hence a record of both peatland geomorphology and climate.

In order to maintain a highly organic peatland, bankfull flows need to be relatively frequent or the vegetation at the surface of the peatland will die; peatland vegetation is highly dependent on the level of the water table. At Barrington Tops, New South Wales, bankfull flows were observed many times each year, during the 2002-2004 field seasons, and channels showed remarkably uniform maximum capacities along their length. Furthermore, significant overbank flooding during these peak rainfall events was rare and spatially isolated. This would be a remarkable coincidence if the channels and peatlands had not co-evolved. Once the channels have attained an optimal cross-section through the vertical accretion and growth of a system of inset floodplains (Nanson 2009), further growth of the peatland vertically, which is highly water dependent, is limited by bankfull stage height (Nanson et al. 2010).

This research summarises recent investigations into the evolution, morphology and flow hydraulics of several peatland streams that flow through two upland peatlands on the Barrington Tops plateau (Figure 1a).

a.

b.

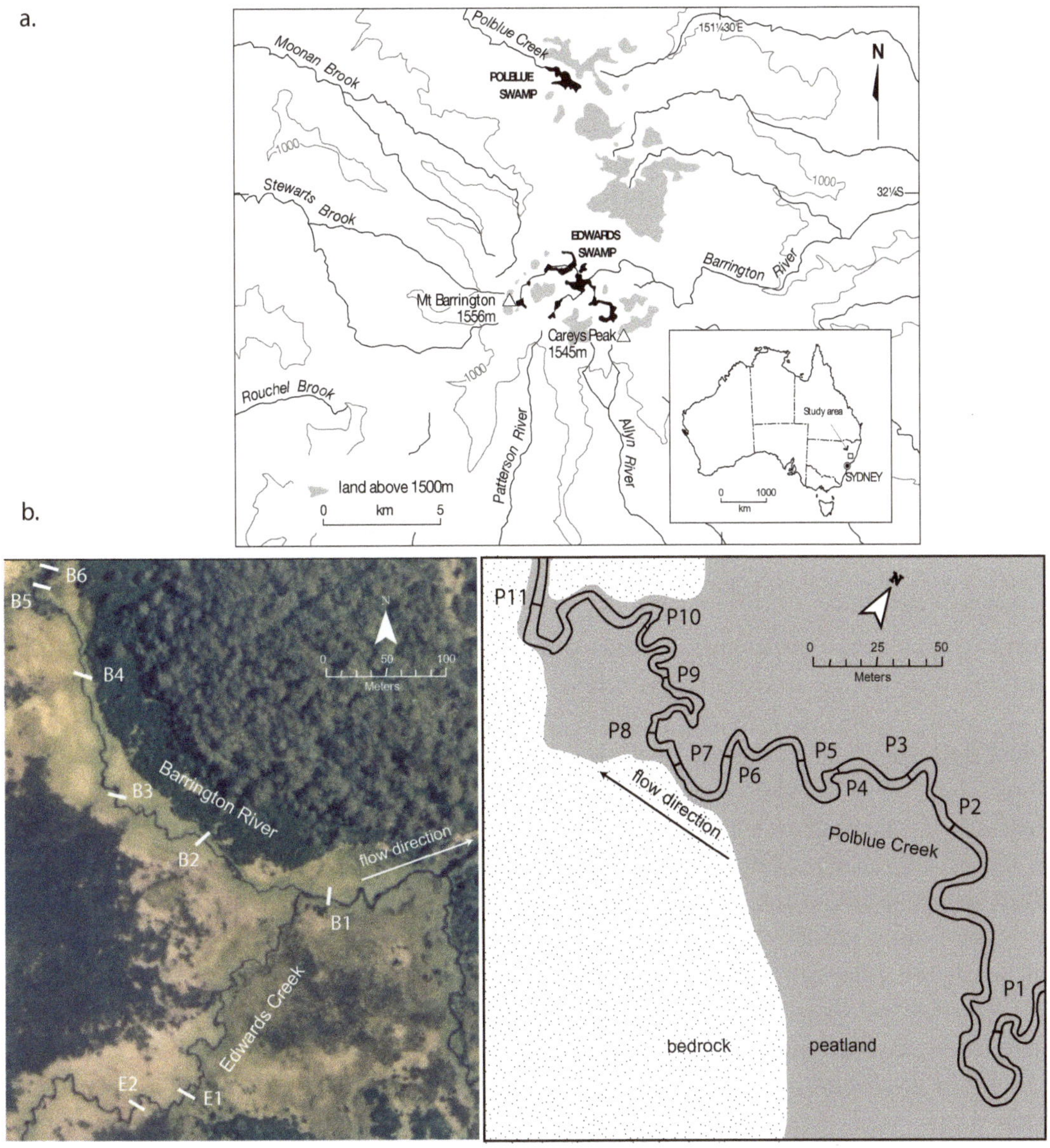

Figure 1. Location of (a) the Barrington Tops (modified from Dodson 1987) and (b) the peatland channels. Hydraulic geometry stations are indicated for Polblue Creek (P), Barrington River (B) and Edwards Creek (E)

Barrington Tops

Regional setting

The channels and peatlands of the Barrington Tops are situated on a 1000 km^2 upland plateau ~1500 m above sea level. The plateau is situated approximately 200 km north of Sydney and incorporates both the Barrington Tops National Park (New South Wales National Parks and Wildlife Service) and the Barrington Tops Reserve (New South Wales State Forests). Mean annual precipitation, in the form of rain, fog and snow, reaches 2000 mm in areas over 700 m asl. (Zoete 2000). The western slopes rise ~1000 m in fewer than 8 km. As a result of its high altitude and steep terrain, the region is subject to severe weather fluctuations. At 1300 m asl, average July temperatures range from -2.3°C to 8.8°C and January temperatures from 9.1°C to 22.8°C (Tweedie 1963). Areas above 1100 m asl experience up to five snowfalls each year, generally up to ~0.2 m deep (Zoete 2000).

Montane peatlands and swamps have formed in the broad, flat hollows (Dodson 1987) of this sub-alpine environment (Figure 1b) and are listed as Threatened Ecological Communities by the Department of Environment and Conservation (NSW). Many of the Barrington peatlands have developed meandering channels and this research specifically examines three such channels in the Polblue (1480-1500 m asl) and Edwards (1400-1420 m asl) swamps. These are Polblue Creek, Barrington River and Edwards Creek (Figure 1b). Each of these channels drain sub-catchments approximating 4 km^2, and Edwards Swamp is collectively supplied by 8 km^2 of catchment as it is situated at the confluence of both Edwards Creek and the Barrington River. Polblue and Edwards swamps cover areas of approximately 0.5 km^2. However, elongated arms of Edwards Swamp extend further up both tributaries of that swamp. The vegetation of the peatlands are either *Sphagnum* hummock-hollow communities or Cyperaceae-Restionaceae sedgelands, surrounded by *Poa sieberana* and *Epacris paludosa* in better-drained locales (Dodson et al. 1986; Dodson and Myers 1986). The beds of the channels are sparsely vegetated with several ribbon-weed species, but these were not collected for taxonomic identification.

Channel evolution

Peatland channels in peatlands at Barrington Tops typically have narrow and deep cross-sections, near-vertical resistant banks, relatively flat beds and almost perfect symmetry (Figure 2). Despite remarkably frequent bankfull flow, they transport almost no bedload (Nanson 2006) and therefore exhibit low width/depth ratios close to the optimum for transporting water alone (Chow 1959; Huang et al. 2002; Huang et al. 2004). That these channels frequently accommodate bankfull flow and rarely flood overbank prompted Nanson (2006) to suggest that they may be self-forming systems adjusted to contemporary flow conditions. Many peatland environments have long been subject to direct (e.g. draining) and indirect (e.g. fires) land-use pressures (Mitsch and Gosselink 1986). Furthermore, the draining of many peatlands and valley fills post-European settlement in Australia is well documented (e.g. Eyles 1977; Fryirs and Brierley 1998) and the possibility has been considered that the peatland channels of the Barrington Tops may be anthropogenically caused. However, Nanson (2009) dated *inset floodplain* (peat) *units* and presented palynological data which demonstrated that, while the peatlands themselves (*upper floodplains*) commenced accumulation before at least 5.4 ka (Polblue Swamp, Dodson et al. 1986) and 3.6 cal. BP (Edwards Swamp, Nanson 2009), over the past ~1000 years the channels have reduced in size from larger channels or gullies that were formed during a different flow regime. These channels are flanked by inset floodplains. While in places the channels are somewhat depth-constrained by a resistant basement to the peatlands, they are otherwise essentially free to adjust their cross-sectional and planform geometry within a matrix of peat. For this reason, these channels and inset floodplains are considered to adopt the self-adjusting equilibrium conditions widely described for alluvial channels and their floodplains.

Channel hydrology

Further support for the evolutionary link between the Barrington peatlands and the streams that flow through them is indicated by channel-flow hydrograph analyses.

Andeera Water Level Recorders were deployed in each peatland during two three-week field trips, the first during November/December 2003 (early summer) and the second during May/June 2004 (early winter). The most significant water-level rises were, not unexpectedly, directly linked to rainfall events, the steep rising limbs of the hydrographs confirming this observation. When the flow-stage time-series data for summer flows are detrended (using an *additive decomposition*), a dominant (statistical) seasonal-cycle of diurnal fluctuations is exhibited by both peatland channels (Figure 3). Channel flow stages fluctuate through 0.06 m and 0.05 m

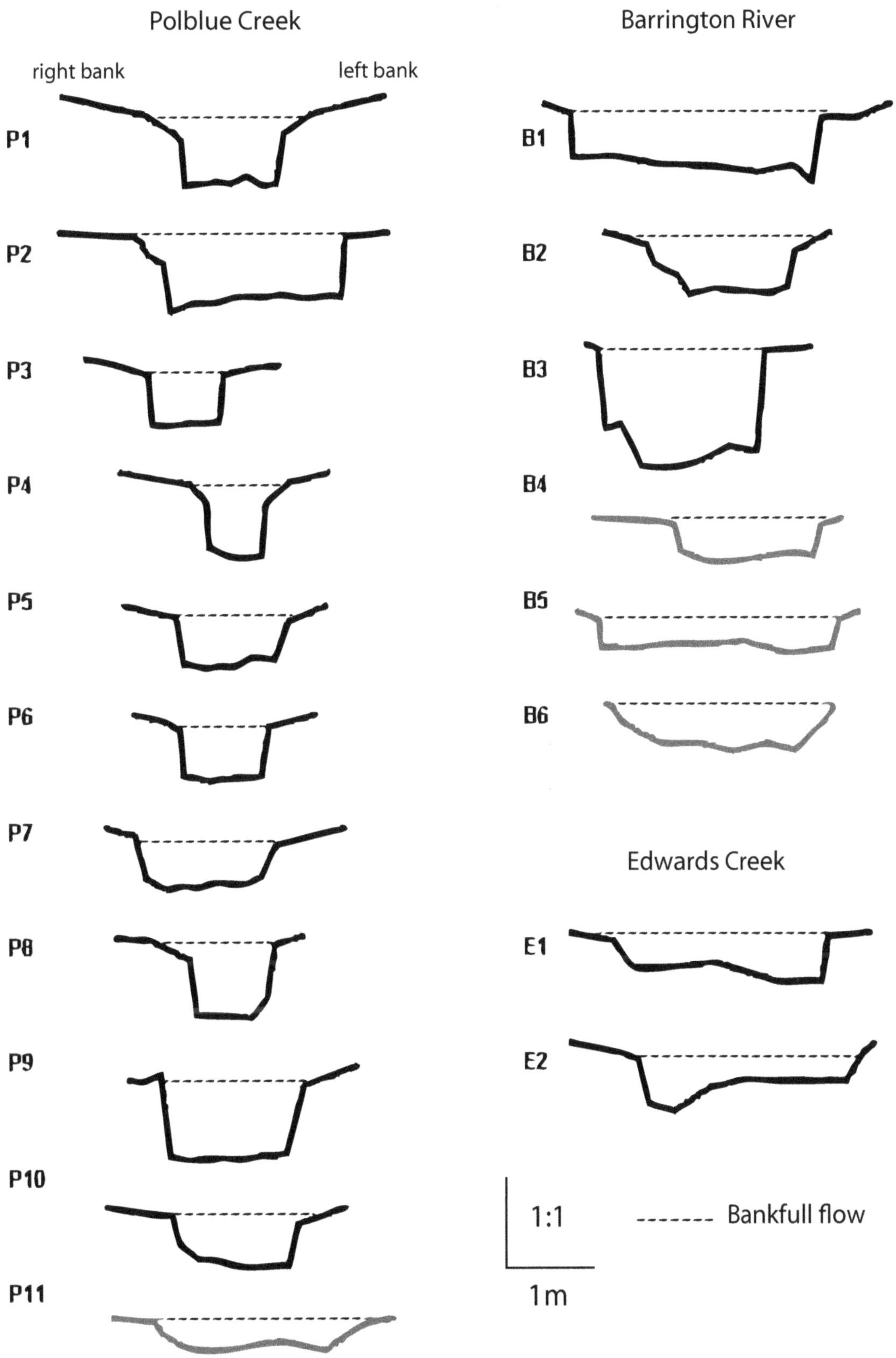

Figure 2. Hydraulic geometry station cross-sections. Flow is towards the reader, out of the page. Grey cross-sections indicate those near entry and exit locations to each peatland

seasonal cycles in Polblue Creek and the Barrington River, respectively. In both channels, flow reached its maximum stage at approximately 6am, immediately before sunrise, from which time it dropped until ~2pm, levelled off and then rose again until ~10pm. From then until midnight, the pattern plateaued, then rose again until 6am.

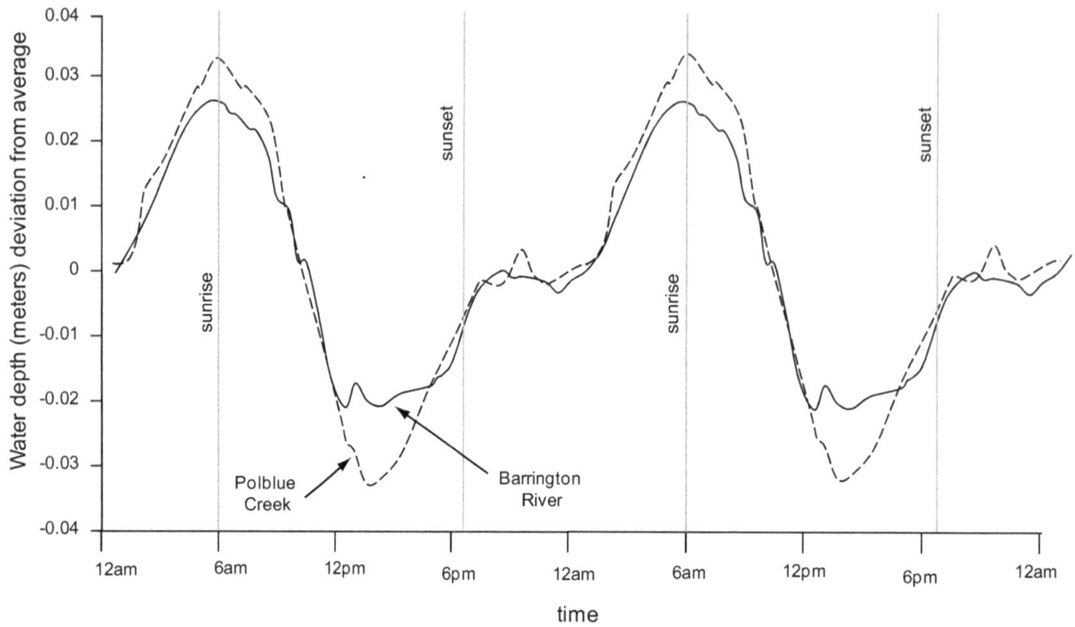

Figure 3. Average seasonal (daily detrended) flow stage variations in Polblue Creek and the Barrington River for selected days recorded in summer (2003)

Daylight and/or temperature has a clear impact on summer flow fluctuations in the Barrington peatland channels, consistent with the (statistically) seasonal variations described by Todd (1964). Day-time losses in channel flow may well be related to the drawing of water from the channels into the surrounding peatlands by plants during daylight hours, when evapotranspiration is at a maximum (Todd 1964; Mitsch and Gosselink 1986). As the plants' need for water reduces during night-time hours, particularly between the hours of midnight and 4am (Todd 1964) (or 6am in the Barrington peatlands), the pressure gradient is reversed towards the channel, thereby increasing flow-stage.

Furthermore, the Barrington Tops channelled peatlands illustrate healthy vegetation, and systematic exchange has been indicated between the channels and peatland-floodplain water table. If the channel water levels were not balanced with the peatland water table, then it is expected that the peatland vegetation would die. That the hydrological balance between the peatlands and the channels is so satisfactory to both peatland vegetation and bankfull flow frequency implies the stability of both features. Furthermore, this balance indicates the importance of channel morphology in determining peatland depth. Channel morphology determines flow-stage height, which, in turn, determines the height of the peatland water table. The peatland water-table height controls peatland growth, in a manner that closely imitates the link between more conventional alluvial channels and their adjacent floodplain growth by overbank sedimentation.

Barrington peatland channel geometry and hydraulics

It is widely accepted that self-forming alluvial streams are adjusted towards what are broadly termed 'equilibrium conditions'. G.K. Gilbert recognised this in his seminal work (1914) but Mackin (1948) enunciated the concept clearly in his classic paper, focusing primarily on the river's ability to adjust channel slope, and described such adjusted rivers as 'graded'. His definition was subsequently expanded by Leopold and Bull (1979) to summarily include adjustments in velocity, roughness, channel pattern and cross-sectional morphology and slope.

The evolution of a channel towards a stable endpoint, therefore, involves the adjustment of all three morphological elements – cross-section (examined through hydraulic geometry), bedform and planform – and can be assessed through an examination of water surface slope (Figure 4).

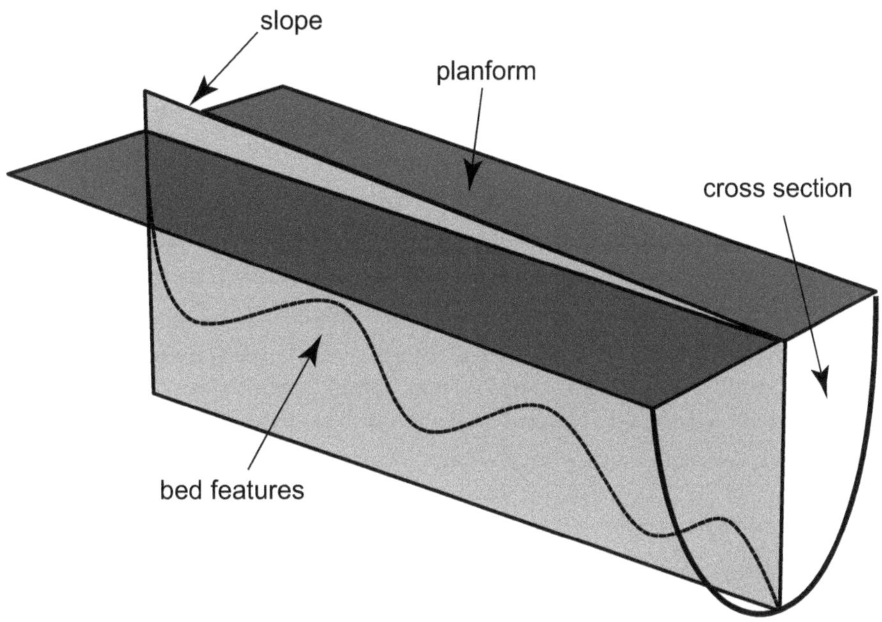

Figure 4. Morphological elements and planes of channel adjustment, modified from Knighton (1998)

Hydraulic geometry

Hydraulic geometry is essentially used to represent the rate of change in channel width, depth and flow velocity with increasing flow discharge, i.e. the proportions of these three variables that accommodate increased discharge. This is expressed as:

$$w = a\,Q^{b}$$

$$d = c\,Q^{f}$$

$$v = k\,Q^{m}$$

where

$$Q^{1} = aQ^{b}\,cQ^{f}\,kQ^{m}$$

(w = width, d = depth, v = velocity and Q = discharge (m^3/s)).

Hydraulic geometry can be considered at a station as a rising flow fills a channel cross-section, or in a downstream direction as the bankfull channel dimensions expand to accommodate increasing catchment runoff (termed bankfull hydraulic geometry). Such measures have traditionally been used to understand or characterise the regional setting (Park 1977; Rhodes 1977), and the sediment transport dynamics or bank strength of a fluvial system, or to investigate more complex theoretical considerations of equilibrium fluvial forms (e.g. Yang 1971; Pickup 1976; Nanson and Huang 2008).

Nanson et al. (2010) measured at-a-station and bankfull hydraulic geometry at 19 temporary gaging stations along three of the Barrington peatland channels. They reported average at-a-station hydraulic geometry exponents for the study streams for width (b), depth (f) and velocity (m) of 0.05, 0.35 and 0.60, respectively. These results indicate that with changing at-a-station discharge, rapid changes in velocity are accompanied by moderate changes in depth and almost no change in width.

Despite some constraint on channel depth (f = 0.35), caused by a limited depth to basement at a number of locations where periglacial cobble lobes extend into the peatland mid-basin, most of the gaged stations have near vertical banks, at-a-station and bankfull width exponents approaching zero and width/depth ratios of ~2. The geometry of these channels is certainly partly due to remarkably high bank strength, and hence the botanical influence on channel form is unusually high compared with alluvial channels. While in most alluvial systems increases in discharge are accommodated primarily by increases in width and to a lesser extent depth, in the Barrington peatland channels, these variables are more constrained and discharge adjustments are accommodated largely by high rates of increase in velocity (m = 0.60).

The Barrington peatland channels are essentially free of sediment transport. While Huang et al. (2004) suggested that, despite such narrow geometries of ~2 being theoretically optimal for sediment-free rectangular channels in equilibrium, insufficient bank strength in natural streams greatly limits their occurrence. Nanson et al. (2010) have shown that peatland channels can indeed provide such optimal conditions. At some locations along the Barrington peatland streams, large bedforms of sand and pebbles do occur, but even at high flows there is almost no bed-sediment movement. The role that these bedforms play in dampening such high rates of velocity increase is discussed below.

Bed features

Nanson (2006) analysed survey data along the channel thalweg to investigate the role bed features play in regulating channel energy. Bed feature steepness (wavelength/height, St) is proportionate to flow resistance (Robert 2003); high St indicates high flow turbulence and hence high flow resistance. Three bed-feature types were identified on the basis of St values.

Entry and exit reaches to each of the peatland systems exhibit wider and shallower cross-sections as the channels flow over shallow basalt and granite outcrops and demonstrate the lowest hydraulic efficiency of the sample channels. Nevertheless, bed shear stresses through these reaches were shown to exceed the critical values required for the transport of sand and silt, if any were present. St values were not calculated for these reaches as bed feature spacing was unsystematic.

In deeper peat reaches of the peatlands, such as along much of Polblue Creek, a second type of low amplitude pool-riffle bedform (St <0.03), mostly derived of massive and cohesive clay, has developed. These provide little resistance to flow (Nanson 2006), and are present throughout reaches with highly efficient channel cross-sections. Through such exceptionally narrow channel reaches (width/depth ~2), the majority of the flow resistance is provided by the banks, and these were shown to severely reduce the sediment-transporting ability of the channels (Nanson 2006). Nevertheless, sufficiently high rates of increase in velocity through these reaches (Nanson et al. 2010) indicate high sediment-transport capacities and sufficient bed shear stress to transport bedload, should there be any; however, very little was observed. These channels are therefore supply, rather than transport, limited. The strength of the channel banks is, therefore, clearly sufficient to resist flow forces without the necessity of developing energy-reducing bedforms in these reaches.

The third category of bedform was identified along the larger Edwards Creek and mid-Barrington River reaches, where sediment supply has, at some stage or over a long enough period of time, been sufficient to construct very large bedforms that are intermediate between dunes and riffles (Nanson 2006). These features are associated with the tightest bends and are scaled to channel planform and width, consistent with pool-riffle sequences. However, they are also very steep features (St >0.06), more consistent with mobile bedforms such as dunes. These reaches are otherwise very hydraulically efficient (width/depth ~2). Nanson (2006) calculated that the shear stresses available for sediment transport (and channel erosion) were severely reduced in these higher-discharge channels by the presence of these bed features. High bank strengths, aquatic vegetation and armouring (of fine sediment by coarser surficial sediment) of their beds has ensured the stability of these bed features despite the large turbulent flow structures they produce (Nanson 2006). Their effect on hydraulic geometry is subdued rates of velocity increase with increasing at-a-station discharge (Nanson et al. 2010).

The apparent lack of sediment supply and the armouring and partial aquatic vegetative cover of the steepest bed features indicate that these bedforms could be vulnerable to disturbance. These bed features presumably formed rapidly following channel incision into the peatlands. Loss of bed armouring or disturbance of the vegetation could result in greatly increased mobility of the bed sediment, as subsurface materials are more easily entrained by the grain shear stresses available. Because these catchments do not have a ready supply of sediment, disturbance to these steep bed features would very likely result in a decrease in flow resistance and a concomitant increase in bed, and possibly bank, erosion.

Planform

Using the concept of minimum stream power, Bettess and White (1983) described how channel planform adjusts from straight to meandering (to braided) in order to create a balance between the regime channel slope, defined as that slope required for the equilibrium transport of the supplied sediment, and the valley slope. Channels in which regime slope coincidentally equals valley slope are straight and very rare, whereas valley slopes in excess of the required regime slope are common and result in either meandering or braiding planforms. Channels meander for two reasons, both of which directly impact on channel energy: one, to reduce channel slope and hence the energy gradient (Langbein and Leopold 1966; Bettess and White 1983); and two, to provide bend flow resistance through the development of secondary circulation and turbulent flow structures and thereby consume energy (Leopold et al. 1960; Langbein and Leopold 1966) and facilitate the most probable distribution of channel energy in accordance with the theory of minimum variance (Langbein and Leopold 1966). The more sinuous the channel, the greater the reduction in slope (potential energy) and the greater the loss of (kinetic) energy due to internal distortion of the flow-through bends.

Channel sinuosity (P) is a measure of the length of channel divided by the straight line length of valley; the Barrington peatland channels vary from P = 1.05 (very straight) to P = 1.98 (very sinuous) (Nanson 2006). Bend radius of curvature to width ratios also provide an indication of *sharpness* of channel bend curvature relative to channel size and Nanson (2010) examined bends varying from 2.4 to as low as 0.6, demonstrating unusual three-dimensional bend flow patterns resulting from interactions between various combinations of particularly low width-depth ratio cross-sections, steep bedforms and over-tightened bends. Most importantly, very low radius of curvature to width ratios indicate exceptionally high energy losses. The narrow deep peatland channels at Barrington Tops offer maximally efficient cross-sections and such extreme sinuosity is necessary to consume the resulting surplus energy and maintain channel stability.

A relationship between a proposed flow-resistance proxy, the product of sinuosity (P) and bedform steepness (St), and a proposed flow-efficiency term, the depth-width ratio, is presented in Figure 5. The uppermost point in this figure represents two overlapping stations from Edwards Creek and, for reasons discussed below, these plot as outliers. If these are removed from the dataset, the Barrington River and Polblue Creek stations illustrate a linear

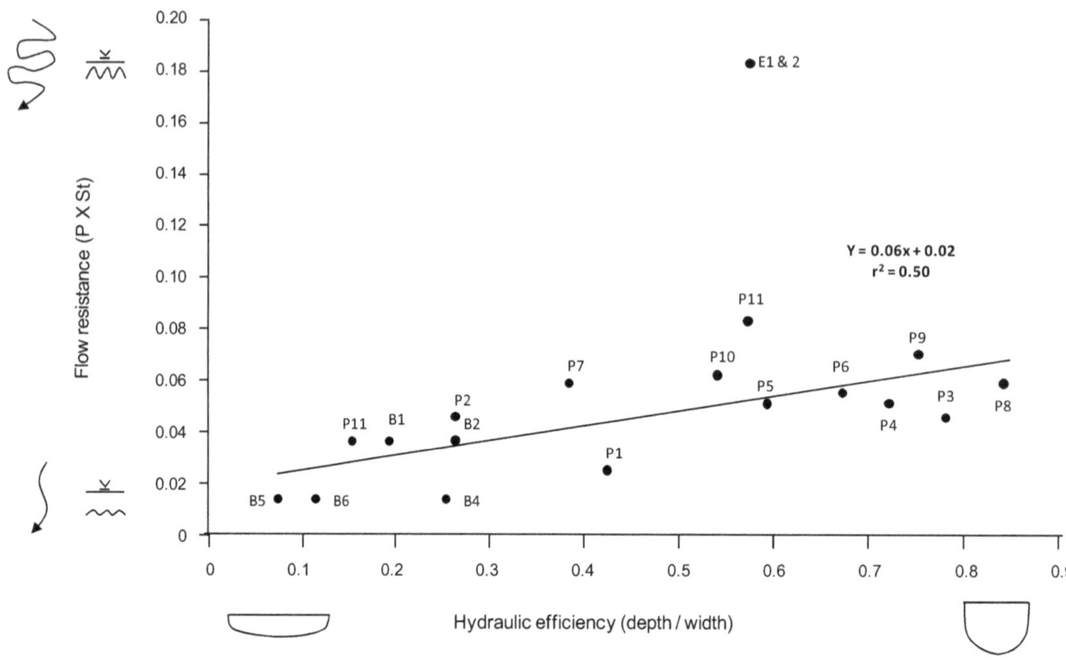

Figure 5. Stations with low hydraulic efficiency have similarly low flow resistance (bottom left of plot), whereas channel sections with higher hydraulic efficiency (upper right) illustrate that flow resistance elements have developed to partially counter such efficiency

relationship, with an r^2 of 0.50. Cross-sectional efficiencies of the channel stations are positively related to their associated measures of flow resistance, at a rate of 0.06 times efficiency.

Stations that plot in the lower-left region of the graph are located at entry and exit points of the peatlands – they are wide and shallow with low cross-sectional efficiency. Their planforms are therefore straight, as there is little surplus energy to consume, and they have only subtle bedforms that generate little roughness. Stations towards the right-hand side of the plot experience greater cross-sectional hydraulic efficiency and the channel response to consume excess energy and maintain stability is to increase sinuosity or bedform development, or both. Outlier Stations (E1 and E2) have very steep bed features (St = 0.11) and are located along highly sinuous reaches, resulting in exceptionally high flow resistance.

Assessment of the graded profile

Entropy is a thermodynamic quantity representing the amount of energy in a system that is no longer available for doing work (Leopold and Langbein 1962; Chorley and Kennedy 1971) and considers the distribution of available energy within a system. The principle was first applied to fluvial environments by Leopold and Langbein (1962), in an effort to produce definitive equations for the solution of hydraulic geometry through the application of two principles: least work and maximum probability. They stated that (pp11 and pp7, respectively): 'the *equilibrium profile* of the graded river is the profile of maximum probability and the one in which entropy is equally distributed' and 'the most probable distribution of energy exists when the rate of gain of entropy in each interval of length along the river is equal'. Over an unconstrained length of channel, they concluded that the most likely form is exponential in long profile. However, over a more moderate length of stable channel that is neither aggrading (which might cause backwater effects) nor degrading, it follows that the energy of an equilibrium river (which is represented by the energy grade line and approximated by the bankfull water surface slope) should be evenly distributed and represented by a straight line.

Mackin (1948) stated that the length of channel constituting a graded (equilibrium) system is not fixed in length. Herein, the graded condition is considered within the channelled reaches of each peatland that lie between constrictive lobes of shallow basement material, as it is these reaches, and their interaction with the peatland, that have the ability to self-adjust as equilibrium systems.

Figure 6 illustrates longitudinal profiles of the bankfull water surface for each of the three peatland channels. The entire study reach for more than 460 m of Edwards Creek shows a water-surface slope with an almost linear relationship and an r^2 of 0.99. The 180 m reach of the Barrington River, downstream of its confluence with Edwards Creek, also demonstrates a very even water-surface profile with an r^2 value of 0.98, and the middle reach of 370 m of Polblue Creek has a similarly constant gradient with an r^2 value of 0.99.

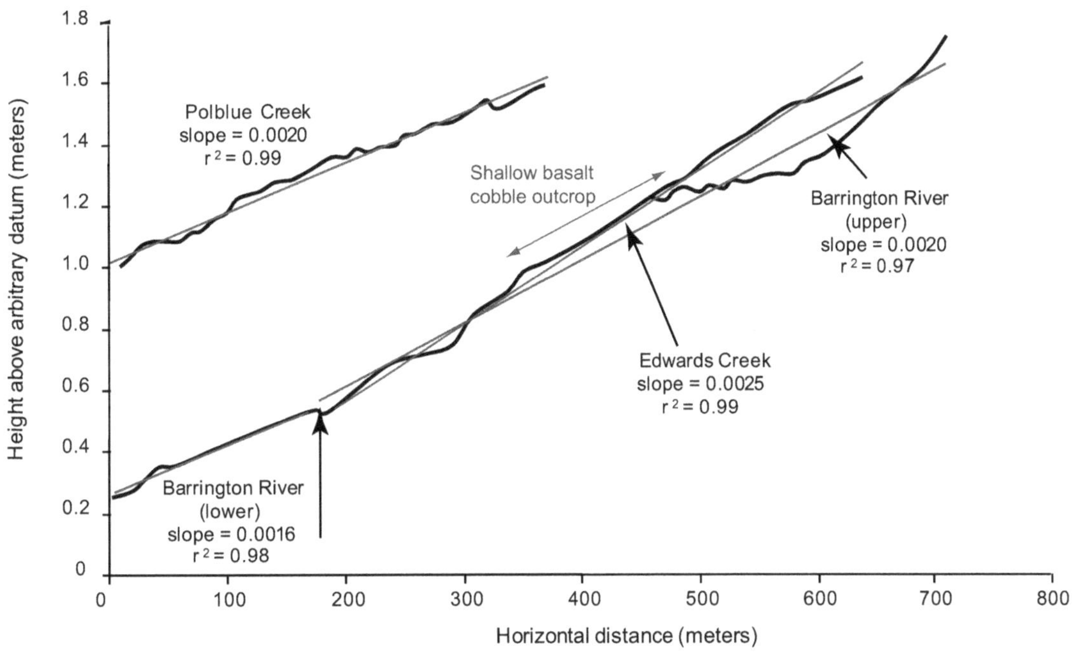

Figure 6. Bankfull (longitudinal) water-surface profiles. The regression lines along the Barrington River Upper and Edwards Creek terminate to the right of the vertical arrow

Nanson (2006) demonstrated that deviations from linear longitudinal water-surface profiles along the Barrington peatland channel water-surface profiles are the direct result of limiting peat depths at entry and exit reaches to the peatlands, or where the channels flow over shallow lobes of basalt or granite cobbles and boulders. It would appear that, when not constrained by a shallow basement, peatland channels are able to adjust to the most probable distribution of energy whereby variance is held at a minimum and the rate of gain of entropy in each interval of channel length is equal.

Conclusions

In the Barrington peatlands, narrow, deep and steep-banked channels provide optimal conditions for the hydraulically efficient flow of sediment-free water, consistent with the principal of minimum energy. The development of meanders in these peatland channels appears to have been sufficient, in the absence of bedforms in most reaches, for the attainment of channel stability. Leopold et al. (1960 pp131 and pp132, respectively) stated that 'a river which is deficient of bed material of large enough size to stabilise a straight channel may become stabilised in an irregular channel merely by creating random bank projections', and 'the meander form may represent one of the channel patterns to which the above generalisation applies, only the random bank projections are replaced by somewhat symmetrical channel curves'.

In the higher stream power channels (Edwards Creek and Barrington River), where very steep bedforms have developed, sediment must have either been supplied more readily at some time in the past, perhaps immediately following channel incision, or accumulated very slowly from minimal supply and transport rates. Where sediment has been sufficient, and where additional resistance has been required for channel stability in higher stream power systems, bedforms ranging from negligible (in most instances) to extremely large indicate a tendency for the channels to form flow-resisting elements in the vertical plane. Furthermore, channel sinuosities have developed from nearly straight to intensely sinuous, thereby illustrating the freedom for these channels to adjust in the lateral plane to achieve equilibrium. However, the presence of exceptionally tightly curving and stable meander bends at many locations indicates that the banks of the Barrington streams must be sufficiently resistant to erosion to maintain stability. Channels with more mobile banks would presumably erode and release sediment, form bedforms and consume energy more evenly between the bed roughness and the distortion of flow in bends.

Graded, almost linear, water-surface long profiles that extend over hundreds of metres in channels with almost no bedforms (Polblue), and higher stream power channels with exceptionally large bedforms (Edwards and Barrington channels) suggest that planform is closely integrated with other flow-resisting elements in order to maintain a condition of minimum variance and a uniform increase in entropy. Clearly, self-forming equilibrium channels can be generated and maintained in these densely vegetated, sediment-supply-limited and highly erosion-resistant peatland environments.

This research has demonstrated that the cross-sectional form of these peatland channels results in highly efficient flow conduits with velocities that increase sharply with increasing stage. The stability of such velocity-responsive peatland channels is dependent on two major factors: (1) high bank strength, provided by vegetation; (2) bedform strength, provided by sediment armouring and aquatic vegetation. The preservation of these two controlling parameters is integral to the continuing stability of both the channels and peatlands. If the very steep bedforms were to erode through the loss of armouring, limited sediment supply from upstream of the peatlands would inhibit their redevelopment. This research and that elsewhere (Nanson 2006; Nanson 2009; Nanson 2010; Nanson et al. 2010) suggests that loss of bedforms would result in a marked reduction in flow resistance and the destabilisation of the channel geometry. Likewise, the loss or significant alteration of riparian vegetation would have major implications for bank and bed strength and would also encourage changes in channel form and peatland hydrology. A feedback between stable, equilibrium channels and the surrounding peatlands is evident; there cannot be one without the stability of the other and the disruption of either would likely be to the detriment of both.

Acknowledgements

I would like to acknowledge funding from the following sources: the University of New South Wales at the Australian Defence Force Academy (UNSW@ADFA) University College Postgraduate Scholarship, the UNSW@ADFA Faculty Research Grants scheme, the School of Physical, Environmental and Mathematical Sciences (UNSW@ADFA) and the Australian Institute of Nuclear Science and Engineers (AINSE). I thank Geoff Hope, Gerald Nanson, Tim Cohen and Roger McLean for rewarding discussions and suggestions, Ben Keaney for pollen counting, and James Boyle, Paul Grevinitz and Tim Robinson for their assistance in the field. I also thank an anonymous reviewer for improving the manuscript. I acknowledge the National Parks and Wildlife Service (NSW) for issuing a scientific permit.

References

Bettess, R and W.R. White 1983. Meandering and braiding of alluvial channels. *Proceedings of the Institution of Civil Engineers, Part 1: Design and Construction* 75:525-538.

Chorley, R.J and B. Kennedy 1971. *Physical geography, a systems approach*. London: Prentice Hall.

Chow, V.T. 1959. *Open-channel hydraulics*. Singapore: McGraw-Hill Book Company.

Cohen, T.J. and G.C. Nanson 2007. Mind the gap: valley deposits and independant proxy climate data identifying the Holocene hypsithermal period of enhanced flow regime in southeastern Australia. *Holocene* 17:1-8.

Dodson, J.R. 1987. Mire development and environmental change, Barrington Tops, New South Wales, Australia. Quaternary Research 27:73-81.

Dodson, J.R., P.W. Greenwood and R.L. Jones 1986. Holocene forest and wetland vegetation dynamics at Barrington Tops, New South Wales. *Journal of Biogeography* 13(6):561-585.

Dodson, J.R. and C.A. Myers 1986. Vegetation and modern pollen rain from the Barrington Tops and Upper Hunter River regions of New South Wales. *Australian Journal of Botany* 34(3):293-304.

Eyles, R.J. 1977. Changes in drainage networks since 1820, Southern Tablelands, N.S.W. *Australian Geographer* 13(6):377-386.

Fryirs, K. and G. Brierley 1998. The character and age structure of valley fills in Upper Wolumla Creek catchment, South Coast, New South Wales, Australia. *Earth Surface Processes and Landforms* 23(3):271-287.

Gilbert, G.K. 1914. The transportation of debris by running water. *United States Geological Survey Professional Paper* 86:1-263.

Good, R.B. 1992. *Kosciusko Heritage - The conservation significance of Kosciusko National Park*. National Parks and Wildlife Service and Surrey Beatty and Sons, Chipping Norton, NSW.

Hope, G.S. 2003. The mountain mires of southern New South Wales and the Australian Capital Territory: their history and future. In: Janet Mackay and Associates (Ed.), *Celebrating mountains: Proceedings of an International Year of Mountains conference*, Jindabyne, Australia. Canberra: Australian Alps Liason Committee.

Huang, H.Q, H.H. Chang and G.C. Nanson 2004. Minimum energy as the general form of critical flow and maximum flow efficiency and for explaining variations in river channel pattern. *Water Resources Research* 40(4):1-13.

Huang, H.Q., G.C. Nanson and S.D. Fagan 2002. Hydraulic geometry of straight alluvial channels and the principal of least action. *Journal of Hydraulic Research* 40(2):152-160.

Johnston, P. and G.J. Brierley 2006. Late Quaternary river evolution of floodplain pockets along Mulloon Creek, New South Wales, Australia. *Holocene* 16:661-674.

Jurmu, M.C. 2002. A morphological comparison of narrow, low-gradient streams traversing wetland environments to alluvial streams. *Environmental Management* 30(6):831-856.

Jurmu, M.C. and R. Andrle 1997. Morphology of a wetland stream. *Environmental Management* 21(6): 921-941.

Knighton, A.D. 1998. *Fluvial forms and processes: A new perspective*. London: Arnold.

Langbein, W.B. and L.B. Leopold 1966. River meanders - theory of minimum variance. *United States Geological Survey Professional Paper* 422-H:H1-H15.

Leopold, L.B., R.A. Bagnold, M.G. Wolman and L.M. Brush (Jnr) 1960. Flow resistance in sinuous or irregular channels. *United States Geological Survey Professional Paper* 282D:108-134.

Leopold, L.B. and W.B. Bull 1979. Base level, aggradation and grade. *Proceedings of the American Philosophical Society* 123(3):168-202.

Leopold, L.B. and W.B. Langbein 1962. The concept of entropy in landscape evolution. *United States Geological Survey Professional Paper* 500A:A1-A20.

Mackin, J.H. 1948. Concept of the graded river. *Geological Society of America Bulletin* 59(5):463-511.

Mitsch, W.J. and J.G. Gosselink 1986. *Wetlands*. New York: Van Nostrand Reinhold Company Inc.

Nanson, G.C. and H.Q. Huang 2008. Least action principle, equilibrium states, iterative adjustment and the stability of alluvial channels. *Earth Surface Processes and Landforms* 33(6):923-942.

Nanson, R.A. 2006. Stream channel adjustment in upland swamps, Barrington Tops, New South Wales, Australia. Unpublished Ph.D. Thesis, University of New South Wales at the Australian Defence Force Academy, Canberra.

Nanson, R.A. 2009. The evolution of peat-swamp channels and organic floodplains, Barrington Tops, New South Wales, Australia. *Geographical Research* 47(4):434-448.

Nanson, R.A. 2010. Flow fields in tightly curving meander bends of low width-depth ratio. *Earth Surface Processes and Landforms* 35:119-135.

Nanson, R.A., G.C. Nanson and H.Q. Huang 2010. Bio-geomorphology and hydraulic geometry of narrow and deep channels in peatland floodplains; evidence for optimisation. *Geomorphology* 117(1-2):143-154.

Park, C.C. 1977. World-wide variation in hydraulic geometry exponents of stream channels: an analysis and some observations. *Journal of Hydrology* 33(1-2):133-146.

Pickup, G. 1976. Alternative measures of river channel shape and their significance. *Journal of Hydrology (NZ)* 15(1):9-16.

Price, J.S., A.L. Heathwaite and A.J. Baird 2003. Hydrological processes in abandoned and restored peatlands: an overview of management approaches. *Wetlands Ecology and Management* 11:65-83.

Randell, F., D. Ellsmore, A. Messner, I. Brown, G. Bubb and Low and Hooke 2003. *Contextual history for the Barrington Tops National Park and heritage action statements for five huts within the park*, NSW National Parks and Wildlife Service.

Rhodes, D.D. 1977. The b-f-m diagram: graphical representation and interpretation of at-a-station hydraulic geometry. *American Journal of Science* 277(1):73-96.

Robert, A. 2003. *River Processes: An introduction to fluvial dynamics*. London: Arnold.

Southern, W. 1982. Late Quaternary vegetation and environments of Jackson's Bog and the Monaro Tablelands, New South Wales. Masters Thesis, Monash University, Melbourne.

Todd, D.K. 1964. Groundwater. In V.T. Chow (ed.) *Handbook of applied hydrology: A compendium of water-resources technology*, pp13/1-55. New York: McGraw-Hill.

Tweedie, A.D. 1963. Climate of the Hunter Valley. *CSIRO Australian Land Research Series* 8:62-70.

Yang, C.T. 1971. Potential energy and stream morphology. *Water Resources Research* 7(2):311-322.

Young, A.R.M. 1986a. The geomorphic development of dells (Upland Swamps) on the Woronora Plateau, N.S.W., Australia. *Zeitschrift fur Geomorphologie* 30(3):317-327.

Young, A.R.M. 1986b. Quaternary sedimentation on the Woronora Plateau and its implications for climatic change. *Australian Geographer* 17:1-5.

Zoete, T. 2000. Vegetation survey of the Barrington Tops and Mount Royal National Parks for use in fire management. *Cunninghamia* 6(3):511-578.

26

Glacier crippling and the rise of the snowline in western New Guinea (Papua Province, Indonesia) from 1972 to 2000

Michael L. Prentice

Indiana Geological Survey and Department of Geology, Indiana University Bloomington, Indiana, United States

mlprenti@indiana.edu

S. Glidden

Institute for Earth, Oceans and Space, University of New Hampshire, Durham, United States

Introduction

Whereas surface temperatures in the tropics (20°N-20°S) have increased ~0.13°C/decade between 1979 and 2005 (Trenberth et al. 2007), the smaller warming of the lower tropical troposphere over this interval, ~0.06°C/decade, is within the error of the measurements (Karl et al. 2006). This situation is problematic because it calls into question climate model results that show vertical amplification of tropical surface warming (Karl et al. 2006). More specifically, climate models, with natural and anthropogenic forcing, show a decadal-scale warming trend that increases with elevation in the troposphere. On the other hand, several types of observation in the tropics show less warming aloft than at surface (Karl et al. 2006), though the uncertainties are considerable (Fu and Johanson 2005). This uncertainty hinders our understanding of important climate feedback processes, such as water vapour and lapse-rate feedbacks, that contribute significantly to the uncertainty in global climate model predictions of the greenhouse effect. Observations of tropical tropospheric temperature before 1979 are even more uncertain (Gaffen et al. 2000; Lazante et al. 2003, 2006).

Tropical mountain glaciers can supplement the instrumental record of tropical tropospheric climate change because they have receded drastically since the late 1800s and this recession can be inverted for the climate forcing (Kaser and Osmaston 2002; Hastenrath 2005; Oerlemans 2005; Lemke et al. 2007). Tropical glaciers exist in three widespread regions: the Andes, the East African highlands and western New Guinea. The fact that glaciers nearly everywhere have receded argues that the cause has a global footprint. This favours atmospheric warming as the primary driver (Thompson 2000) because changes in atmospheric moisture are more localised. Evidence supporting this notion comes from Thompson et al. (2002), who, based on ice-core data, suggested that glaciers atop Mount Kilimanjaro survived the severe droughts of the Holocene and have not receded in the past 11,000 years. This implies that the recent Kilimanjaro ice loss was due not to drying, but primarily to atmospheric warming. Taylor et al. (2006) suggested that a warming trend has been the primary driver of glacier recession in the Rwenzori Mountains, matching the extent of recession to local meteorological records.

On the other hand, local to regional modelling studies of East African glacier recession indicate that changes in atmospheric moisture content have been the primary driver. For instance, Kruss (1983) and Hastenrath (1984) attributed the glacier recession on Mount Kenya from the end of the 19th century until the early 1960s primarily to a significant decrease in precipitation starting around 1880. The decrease caused a decrease in both cloudiness and glacier albedo and this, in turn, increased absorption of solar radiation at the glacier surface, which drove up ablation. Hastenrath and Kruss (1992) proposed that continued glacier recession from 1963 until 1987 was mainly caused by an increase in atmospheric humidity that forced more efficient use of available energy to increase melting. More recently, Mölg et al. (2003), Kaser et al. (2004), Cullen et al. (2006), and Mölg et al. (2009) proposed that the 20th century recession of Kilimanjaro's glaciers was driven entirely by decreasing precipitation and atmospheric drying. As part of the proof, Mölg and Hardy (2004), Cullen et al. (2007) and Mölg et al. (2008, 2009) showed, using energy and mass-balance modelling, that the retreat of Kilimanjaro glaciers since 2000 reflects atmospheric aridity and the associated increased receipt of solar radiation.

In the tropical Andes, Francou et al. (2004) showed that the cumulative loss of ~25 m in water equivalent from Antizana Glacier in Ecuador over the eight-year period from 1995 to 2002 was due to the intensity of El Niño events in this interval. During El Niño events, glacier ablation increased because of atmospheric warming and decreased precipitation. The warming resulted in more of the precipitation being rain. The precipitation decrease accompanied reduced cloud cover and greater solar-radiation receipt at the glacier surface. During La Niña events, the opposite conditions occurred and mass balance increased and remained positive for a few months. This evidence implies that tropical Andean glaciers respond to changes in both air temperature and atmospheric moisture related to the El Niño-Southern Oscillation (ENSO).

Before the availability of mass-balance data, changes in the altitude of the equilibrium line (ELA) of tropical Andean glaciers were used to evaluate climate forcing of glacier recession. Kaser and Osmaston (2002) proposed that the ELA of glaciers in the Cordillera Blanca, Peru, rose 95 ± 5 m from the 1920s to 1970. Based on modelling and independent climate records, they suggested that the majority of this rise was due to a decrease in atmospheric moisture, with warming playing a small role.

In New Guinea, analysis of the climate forcing of the 20th century glacier recession has been hampered by the lack of field studies on the glaciers, owing to their remoteness and also because of conflicting evidence from studies based on remote sensing. The principal New Guinea ice bodies of the late 20th century are the Carstensz Glaciers in Indonesia (Figure 1). Since the last and only field study of glacier mass balance was conducted in 1971-1973 (Allison 1974, 1976), several estimates of the ELA of the glaciers have been published (Figure 2). Fluctuations in the ELA of a glacier are closely correlated to its mass balance and can be estimated without mass balance data by several means. Van Ufford and Sedgwick (1998) estimated that the ELA of the Carstensz Glaciers increased 60 ± 10 m between 1972 and 1995. Klein and Kincaid (2006) estimated that the ELA of the Carstensz Glaciers rose at least 100 m between 1972 and 2002 (Figure 2). Taken together, these estimates represent a 'low road' ELA scenario. Based on other data, Prentice and Hope (2007) suggested that the ELAs were higher in altitude and rose considerably more from 1972 to 1997 (Figure 2). This constitutes a 'high road' ELA scenario. The purpose of this paper is to present mass-balance data that was the basis for the high-road ELA hypothesis of Prentice and Hope (2007) in the context of glacier recession to 2006. Further resolution of the New Guinea ELA narrative will aid resolution of the regional and, perhaps, global tropical climatic forcing of the retreat of tropical glaciers.

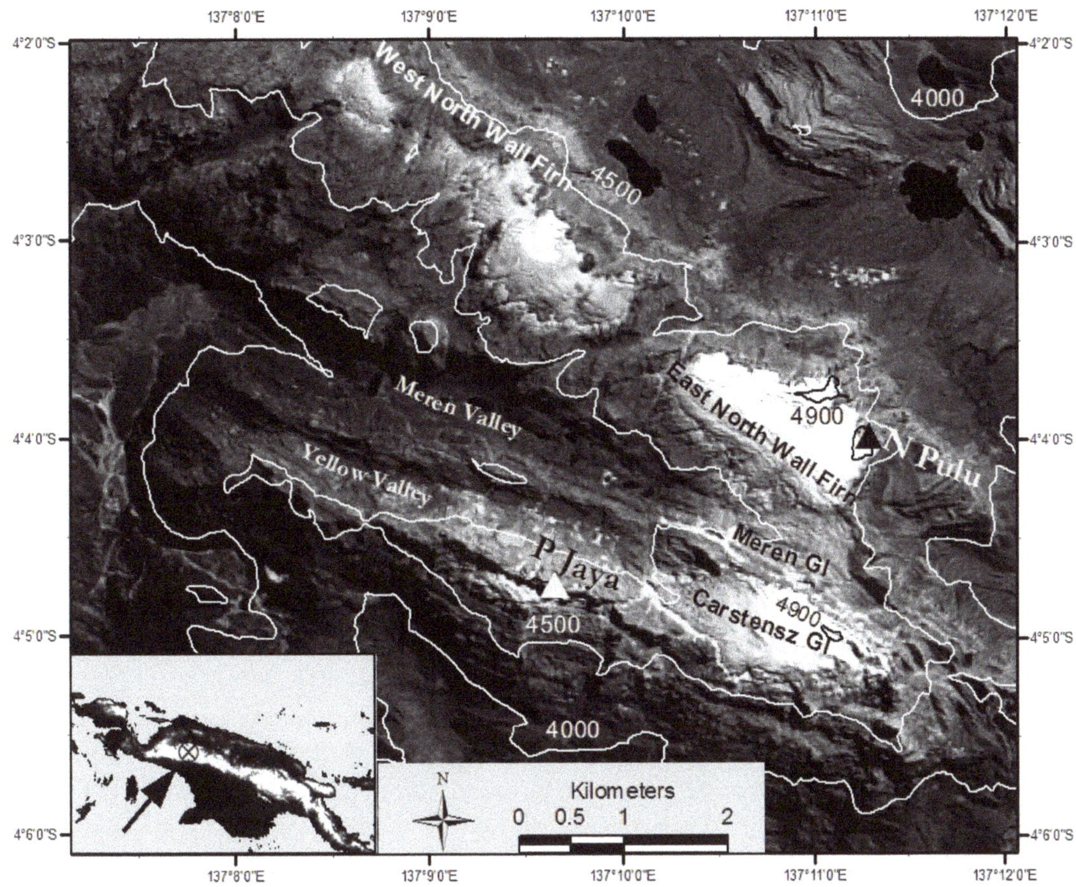

Figure 1. Satellite image of the Carstensz Glaciers, Papua Province, Indonesia, in 1987. Contours are in metres above sea level. The image was taken in July and shows the transient snowline separating snow from firn on the glaciers. The inset shows New Guinea location

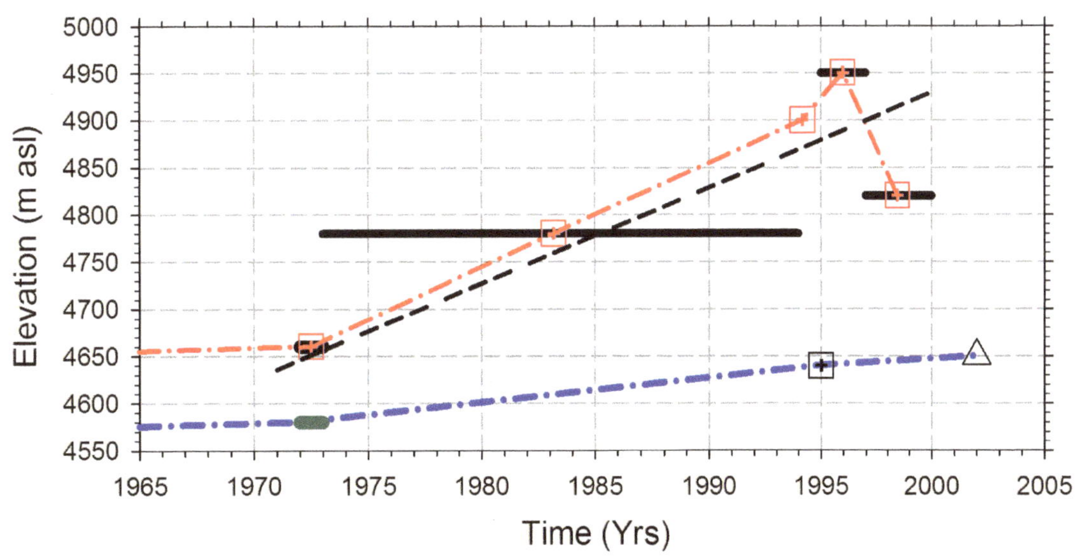

Figure 2. ELA estimates for the Meren Glacier System between 1965 and 2002. Black horizontal lines depict ELAs based on mass balance; their length depicts the mass-balance time interval. The 1972/73 ELA is from Allison (1976), increased by 77 m as explained in the text; others are from this paper. The green line is the unmodified 1972/73 ELA from Allison. The red squares and dash-dot line show the high-road ELA hypothesis of Prentice and Hope (2007). The blue dash-dot line represents a low-road ELA hypothesis that uses the Van Ufford and Sedgwick (1998) ELA (Ð) and the Klein and Kincaid (2006) minimum ELA for 2002 (Δ). The dashed black line shows the high-road ELA trend based on high-road ELAs with that centered on 1972/72, decreased by 30 m to offset assumed interannual bias in the Allison (1976) value

New Guinea glacier recession

At the time of their first survey (Dozy 1938), the Carstensz Glaciers consisted of two valley glaciers, the Meren and Carstensz in the Meren and Yellow valleys, respectively, as well as a plateau glacier, the Northwall Firn (Figure 1). The Northwall Plateau trends northwest-southeast and, along its northeastern edge, reaches peak elevations of nearly 5000 m asl. The plateau dips relatively gently to the southwest before dropping off into the Meren Valley. The landmark study of the Carstensz Glaciers was accomplished over the course of two Australian Universities Carstensz Glacier Expeditions (CGE I and II) between December 1971 and February 1973. The results of CGE I and II as well as previous study of the glaciers are detailed in Hope et al. (1976). By the time of the CGE I, the Northwall Firn had separated into western and eastern ice masses. The eastern Northwall Firn (ENWF) provided ice flow to, and so was part of, the Meren Glacier in the Meren Valley.

Glacier positions from 1936 until 1972 were described by Peterson et al. (1973) and Allison (1974). Allison and Peterson (1989) and Peterson and Peterson (1995) further chronicled the recession of the glaciers using 1987 SPOT imagery. Van Ufford and Sedgwick (1998) contributed a map with 1993 glacier boundaries from oblique aerial photography and a few 1994 boundaries from field surveys. Klein and Kincaid (2006) updated the recession of the Carstensz Glaciers to 2002 by contributing ice areas measured on IKONOS satellite images from 2000 and 2002. Prentice and Hope (2007) presented maps of the recession of the Carstensz Glaciers to 2000. Their work was based on orthophotographs of the glaciers in 1995, 1997 and 2000, as well as topographic data supplied by PT Freeport Indonesia (PTFI, Table 1). For their results, Prentice and Hope (2007) transformed all pre-1995 maps to the 1995 basemap.

In this paper, we focus on the Meren Glacier System (MGS), in which we include the Meren Glacier and the ENWF (Figure 3). Until the late 1980s, there were two lobes of the Meren Glacier. The lobe that flowed westward down the Meren Valley is commonly referred to as the Meren Glacier but, in this paper, is referred to as the West Meren Lobe to distinguish it from the eastward-flowing lobe. Hope at al. (1976) produced a topographic map of the MGS with 10 m contours on the West Meren Lobe and 50 m contours on the ENWF (Hope et al. 1976). This map is crucial to estimating changes in the ELA since 1972. By the early1990s, the Meren Glacier had effectively separated from the ENWF, which hastened complete melting of the former between 1997 and 2000 (Figure 3). Before summarising the mass-balance data that supports the high-road ELA hypothesis from 1972 to 2000, we update the record of the recession of the MGS to 2006.

Methods

MGS lengths, areas and front elevations were computed based on a series of digital maps of the glacier boundaries that were derived from the sources listed in Table 1. For 1995, 1997, 2000 and 2006, boundaries of perennial, semi-permanent ice were drawn on the digital orthophotographs provided by PT Freeport Indonesia (PTFI). For 1987, boundaries were drawn using two nearly simultaneous SPOT images that were projected by PTFI. For 1972, the boundaries were derived from the topographic map of the glaciers and immediate surroundings produced by CGE I and II (Hope et al. 1976) after that map was digitised and transformed to the PTFI 1995 digital topographic basemap.

We calculated the mass balance for the MGS for three periods, 1973/74-1994/95 (hereafter, 1973-1995), 1995/96-1996/97 (i.e. 1995-1997), and 1997/98-1999/00 (i.e. 1997-2000) from DEMs of the MGS and surrounding bedrock, following the method of Østrem and Stanley (1969). Input to the DEMs is listed in Table 2. Surface topography of the MGS for 1995, 1997 and 2000 was provided by PTFI. The topography for 1995 was produced using photogrammetric methods from 1995 aerial photographs with a scale of 1:30,000.

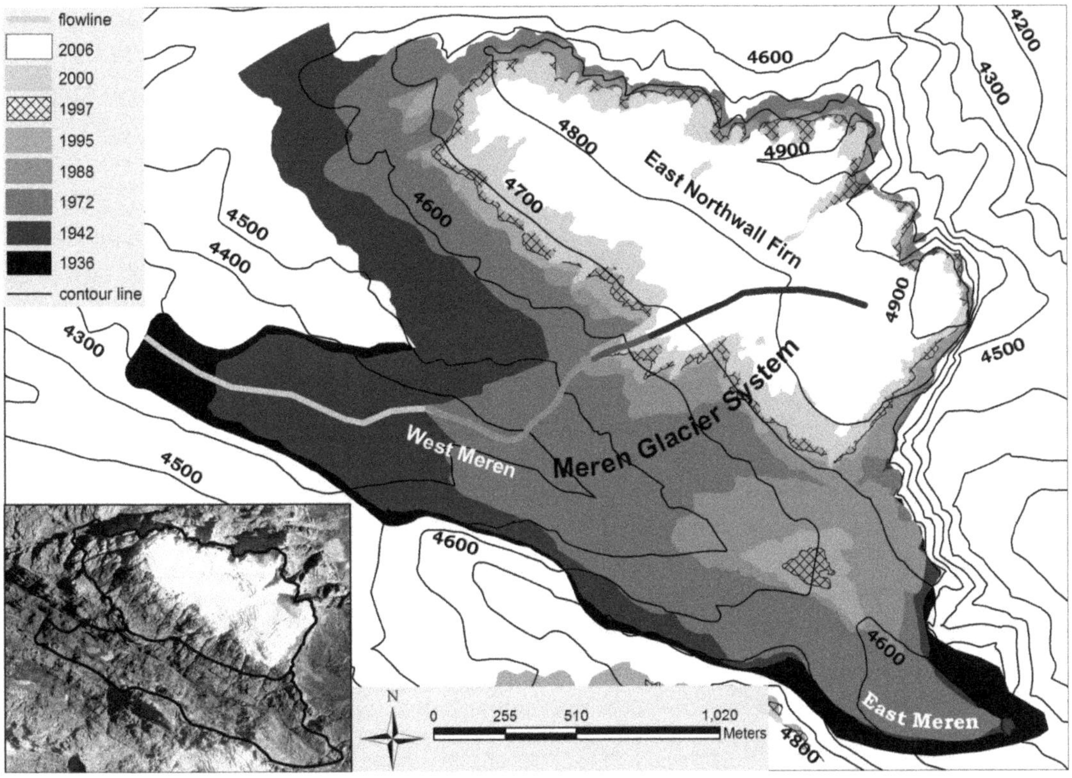

Figure 3. Map of the Meren Glacier System between 1936 and 2006. The inset map shows the 1936 extent on the 2000 orthophotograph. The Meren central flowline is in multiple gray shades. Topography in metres above sea level is for 1995. The East Meren Lobe extent is based on trimlines. Projection is UTM Zone 53S

Table 1. Sources of glacier boundary information

Date	Data source
1936	Peterson and Peterson (1995)
1942	Peterson and Peterson (1995)
1962	Peterson and Peterson (1995)
Dec,1971-Feb,1973	Hope et al. 1976
10-21 July 1987	SPOT Pan and XS
4 Feb 1995	PTFI* orthophoto (1:30,000)
8 May 1997	PTFI* orthophoto (1:25,000)
5 Feb 2000	PTFI* orthophoto (1:25,000)
6 Apr 2006	PTFI* orthophoto

* PTFI: PT Freeport Indonesia

Table 2. Input to the glacier and bedrock DEMS

Date	Data source
1972-1973	1972/73 glacier surface elevation increased by 77 m, 1972/73 glacier boundary
4 Feb 1995	1995 glacier surface elevations, 1995 glacier boundary, 1995 surface topography
8 May 1997	1997 glacier surface elevations, 1995 surface topography decreased by 5-10 m between the 1995 and 1997 glacier boundaries
5 Feb 2000	2000 glacier surface elevations, 2000 glacier boundary, 1995 surface topography decreased by 5-10 m between the 1995 and 2000 glacier boundaries

In order to calculate mass balance for 1995-1997 and 1997-2000, we used an estimate of bedrock topography between the 1995 and 2000 ice limits. This is required in order to estimate ice thickness change outside the MGS boundary at the end of those mass-balance periods. Bedrock topographic contours were inferred in those areas by subtracting 5-10 m from the 1995 MGS surface topography.

We restricted our calculation of mass balance for 1973-1995 to the West Meren Lobe because the surface topography of the ENWF is considered too low in resolution (50 m contours) and accuracy for meaningful results. After transforming the topographic map of Hope et al. (1976) to the 1995 PTFI map, we found that bedrock surfaces were lower than the same surfaces in the 1995 PTFI map by an average of 77 m. Therefore, all 1973 ice elevations were increased by 77 m.

We computed grids of ice thickness change or net specific balance, b_n for the three time intervals from the DEMS. We contoured b_n at both 5 m and 10 m intervals and then converted the areas between the contours into non-overlapping polygons that terminate at the glacier edge. Each polygon was assigned an average b_n. These polygons were merged with topographic contour lines to produce mass-balance polygons within each contour interval.

We calculated total net mass balance, B_n, in m^3 for each contour interval by multiplying the area of each mass-balance polygon by its average b_n and summing across each 10 m contour interval. Dividing B_n by the area of the elevation zone gives average b_n for the elevation zone. We added B_n for each contour interval to obtain total B_n for the entire glacier.

To convert average b_n in m to conventional units of mm of water equivalent (mm we), we assumed that firn with an average density of 500 kg/m^3 was lost. We converted B_n and b_n for 1995-2000 into annual budgets by dividing by the number of months in the balance period and then multiplying by 12. For the 1973-1995 interval, we divided by 22 years to derive an average B_n and b_n in mm/year during that time.

Results

Glacier length and area changes

Relative to its length in 1936 established by the Colijn Expedition (Dozy 1938), the length of the MGS system has decreased by 72 per cent from 1936 to 2006 (Figure 4, Table 3). The area of the MGS decreased by 77 per cent between 1942 and 2006 (Figure 4, Table 3). We measured length along a flowline that we infer to be central to and therefore representative of the MGS from 1936 to 2006 (Figure 3). The flowline is anchored at a spot in the northeast sector of the MGS, approximately on top of a persistent ice dome. The flowline bisects the ENWF ice lobe that occupied a shallow channel that connects the Northwall Plateau to the Meren Valley. In the Meren Valley, the flowline turns west and bisects the front of the former West Meren Lobe as it receded from 1936 to 1972. Along this flowline, the MGS front receded 2114 m up the Meren Valley and the Northwall Plateau.

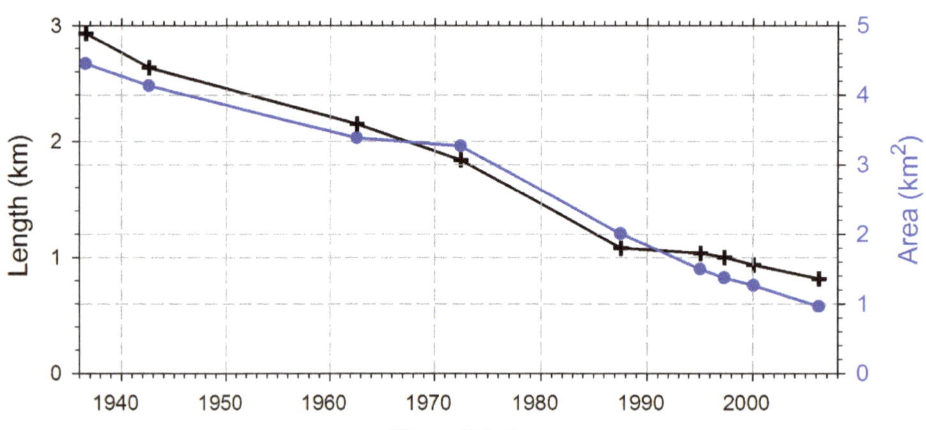

Figure 4. Changes in the length and area (blue dots) of the Meren Glacier System between 1936 and 2006. Length was measured on a map along the central flowline and is not adjusted for slope

Table 3. Length, area, front elevation and ELA of the Meren Glacier system for 1936-2006

Year	MGS length (m)	MGS area (km²)	MGS front elevation (m asl)
1936	2931		4305
1942	2633	4.13	4335
1962	2150	3.38	4
1972	1838	3.26	4355
1987	1081	2.03	4532
1995	1036	1.49	4540
1997		1.37	
2000	934	1.26	
2006	817	0.96	4658

Both the rate of ice-front recession and areal shrinkage appear to have been relatively high between 1972 and 1987 (Figure 4). Whereas the length record may reflect the switch of the flowline from the Meren Valley to the steeper-sloped Northwall Plateau, the increased areal loss rate must partially reflect the rapid melting of West Meren Lobe after it separated from the ENWF. The MGS retreat rate was slightly less between 1995 and 2006 than between 1972 and 1987.

Glacier front elevation

Between 1936 and 2006, the altitude of the retreating MGS front increased by ~350 m along the central MGS flowline (Figure 5). The big step between 1972 and 1987 in part reflects the steep topography between the 1972 front position, which is in the Meren Valley, and the 1987 position, which is on the Northwall Plateau. Figure 6 also shows the elevation of the front of the West Meren Lobe in the Meren Valley in 1987 and 1995. Regardless of which front is followed, the rate of front altitude increase was higher after 1972 than between 1936 and 1972, as was the case with glacier area.

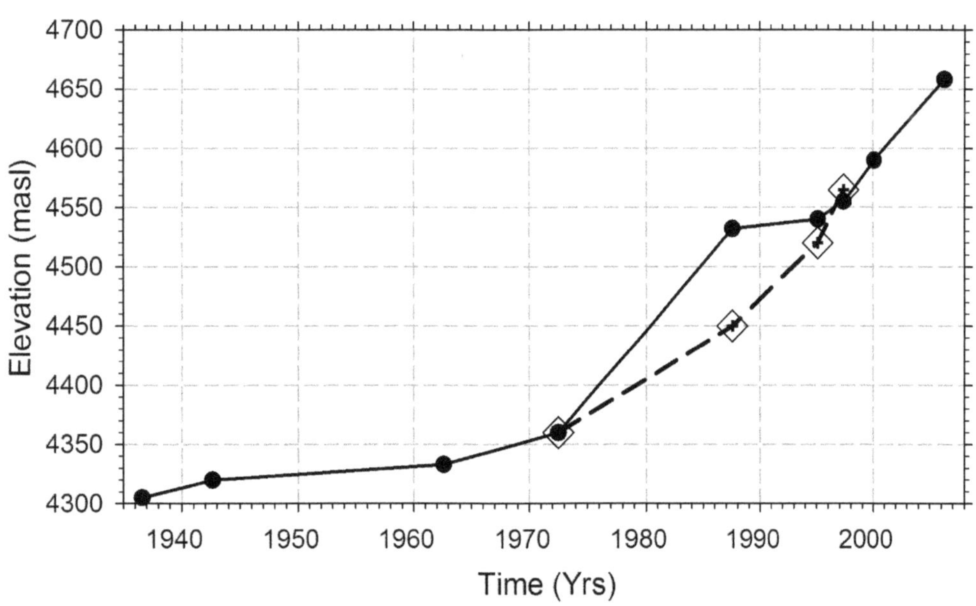

Figure 5. Front elevation of the Meren Glacier System along the central flowline between 1936 and 2006. The dashed line shows the front elevation of the West Meren Lobe in the Meren Valley

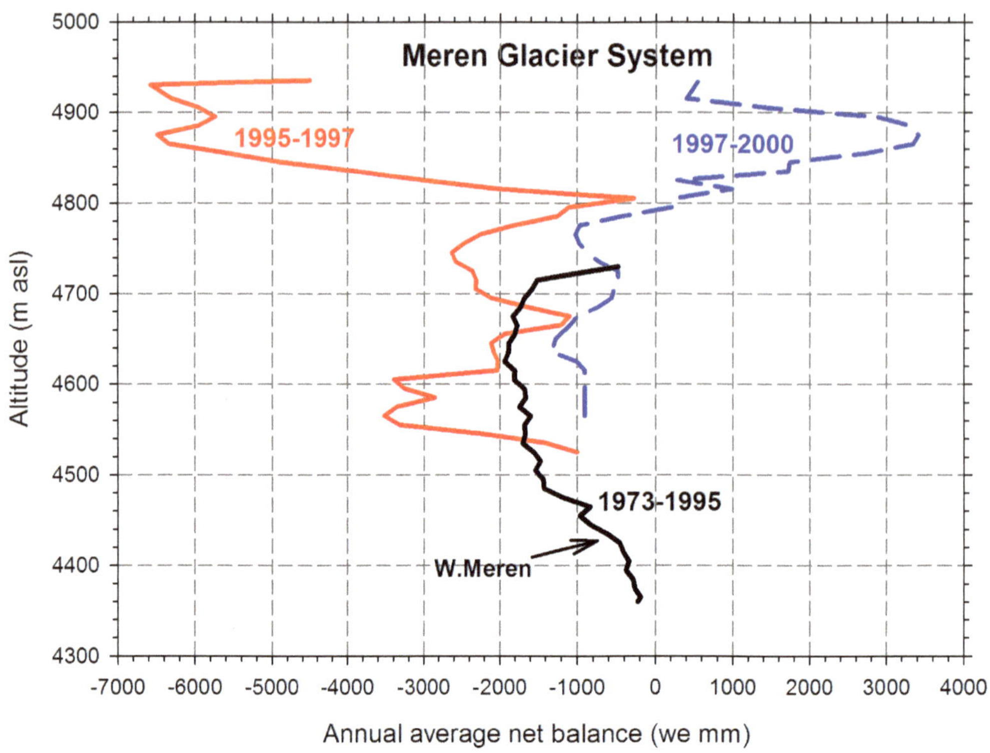

Figure 6. Variation of the annual net specific mass budget of the Meren Glacier System with altitude. The data are annual averages for 1973-1995, 1995-1997 and 1997-2000

Our ice-front elevations are considerably higher than those reported by Klein and Kincaid (2006). They reported that the lower margin of the MGS rose 165 m from 1972 until 2002. We estimate that the rise over this interval was 230 m.

Glacier mass balance and hypsometry

The annual specific net balance, b_n, for the West Meren Lobe over the 22-year interval 1973-1995 was -1517 mm we (Table 4). Figure 6 shows that the entire glacier experienced net melting and that the peak melting rate was -1900 mm we/yr at ~4620 m asl. As much as 130-140 m of ice thickness were lost from the West Meren Lobe over the entire interval. The annual total budget, B_n, for the West Meren Lobe over this interval averaged -1278 x 10^3 m^3 we (Table 4). Variation of annual B_n with altitude shows that the volume losses were highest at ~4700 m asl (Figure 7). Though B_n and b_n were not calculated for the ENWF over this interval, it is clear from comparison of the hypsometries for the MGS in 1972/73 and in 1994/95 (Figure 8) that a considerable area of ENWF was lost up to an altitude of 4780 m asl, above which there does not appear to have been much loss.

Table 4. Annual average total balance and specific net balance for the Meren Glacier system: 1971-2000

	East Northwall Firn		West Meren Lobe		Meren Glacier System		
Balance Yr	B_n (10^3 m^3 we /yr *)	b_n (mm we/yr)	B_n (10^3 m^3 we/ yr *)	b_n (mm we/yr)	B_n (10^3 m^3 we/yr *)	b_n (mm we/yr)	Average ELA (m asl)
1972/73			-990 #	-510 #	-57 #	-60 #	4657†
1973–1995			-1278	-1517			4780
1995–1997	-4213	-2999	-218	-3298	-4,430	-3012	~4950
1997–2000	452	336	~ -8^	~ -545^	452	336	4820

* Assumes that the snow and ice has an average density of 500 kg/m^3
\# Mass budgets measured between 1971 and 1973 by CGE I and II (Allison, 1976)
† From Allison (1976) increased by 77 m as explained in text
^ Assumes average thickness of 3 m in May 1997 and that ice loss spanned 33 months

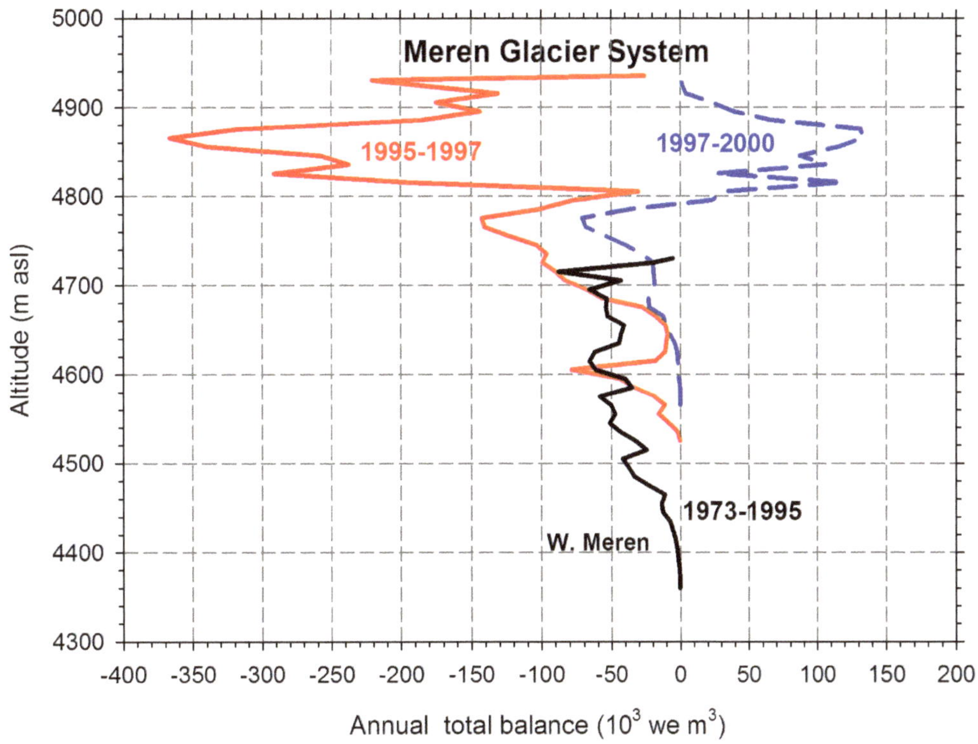

Figure 7. Variation of the annual total mass budget of the Meren Glacier System with altitude. The data are annual averages for 1973-1995, 1995-1997 and 1997-2000

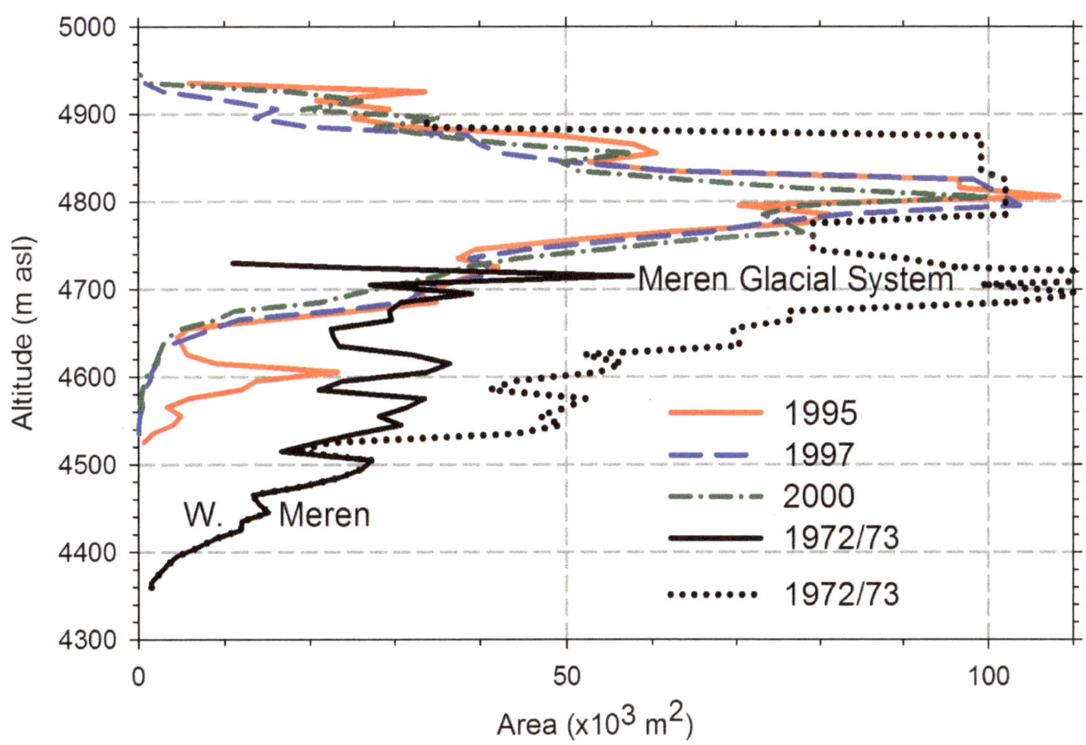

Figure 8. Hypsometry of the Meren Glacier System between 1972 and 2000. For 1972/73, the hypsometries of the West Meren Lobe and the entire MGS are shown with solid line and dotted lines, respectively

The annual b_n for the MGS over the 2.25 year interval 1995-1997 averaged -3012 mm we (Table 4), which is quite negative. The entire MGS had a negative mass budget, with melting rates peaking at more than 6 m we/yr at ~4900 m asl (Figures 6, 7). The net budget was dominated by losses from the ENWF, though there is a secondary B_n peak at 4600 m asl that represents the deflating West Meren Lobe (Figure 7).

In a significant turn-around, the annual b_n for the MGS over the 2.75-year interval 1997-2000 averaged +336 mm we (Table 4). Annual average b_n was negative in all elevation zones below 4800 m asl but positive above this elevation (Figure 6). It was between May 1997 and February 2000 that the West Meren Lobe completely disintegrated but its contributions to the budget are negligible. The greatest mass loss occurred between altitudes of 4860 m and 4940 m asl.

Glacier ELA

The aforementioned mass-balance data provide the basis for inferring the average ELA for the three time intervals of interest and were used to generate the Prentice and Hope (2007) hypothesis (Figure 2). For 1995-1997, we infer that the average ELA was above 4940 m asl, the highest point on the MGS, and arbitrarily choose 4950 m asl as a convenient value above 4940 m (Table 4). For 1997-2000, we suggest that the ELA averaged 4820 m asl because b_n is significantly positive above this altitude (Figure 6).

Allison (1974) determined that the ELA of the MGS for 1972/73 was at 4580 m asl. Given the 77 m offset mentioned above between the 1972 Hope et al. (1976) and the 1995 PTFI basemaps, we estimate that the 1972/73 ELA is actually at 4657 m asl.

We suggest that the average ELA for the 22-year interval 1973-1995 was at 4780 m asl. This is based on the hypsometry (Figure 8) and mass-balance information that indicate net melting below and net accumulation above this altitude. Given this information, the 1972/73 ELA at 4657 m asl, and also that the 1995-97 ELA was above 4940 m asl, we suggest that, on average, the annual ELA rose over the 22-year interval. The simplest hypothesis is a linear rise in the annual ELA that started at 4657 m asl, the 1972/73 ELA, and passed through 4780 m asl, the ELA for the 22-year interval, in 1983, the midpoint of the interval. This hypothesis projects the annual ELA for 1994/1995 to be at 4900 m asl. We stress that the annual ELA for 1994/95 could have been lower or higher than 4900 m asl.

Prentice and Hope (2007) estimated that, in 1942, the ELA was at 4639 ± 10 m asl, based on Allison and Kruss (1977), who used a 1-D flowline glacier model to constrain the rate of the rise in the ELA before 1972.

Discussion

Based on area and ice-front elevation, it is clear that the Meren Glacier System was crippled by climate change between 1972 and 1995. In order to sort out the climatic causes of the drastic ice loss, it is essential to sort out the few clues available for the mass balance and ELA narratives. We focus here on the ELA record because it is related to mass balance but, additionally, captures the elevational component of glacier change. Besides, there are various published estimates for the ELA based on independent data. We first examine how the mass-balance-based ELAs proposed by Prentice and Hope (2007) and supplemented here compare with other published estimates.

Our hypothesis places the average ELA for the 2.25-year interval 1995-1997 at or above 4950 m asl, which is about 300 m above the 1994/95 ELA proposed by Van Ufford and Sedgwick (1998) (Figure 2). Van Ufford and Sedgwick (1998) observed a snowline on the Northwall Firn in May of 1995 at 4640 ± 10 m asl and inferred that it was a good approximation for the annual ELA. This observation corresponds temporally to the first few months of our

2.25-year interval. Whereas 300 m vertical variation in the transient (daily) ELA is plausible, 300 m variation in the annual ELA over at most 2.25 years is unlikely. It would require an extreme climate change and none are apparent in local to regional climate records. The time interval in question is between the major El Niño events of 1992/93 and 1997/98. Even if the Van Ufford and Sedgwick estimate was shifted upwards by 77 m, the offset between the Hope et al. (1976) and the PTFI 1995 maps, we suggest that the resulting value, 4720 m asl, is still incompatible with our 2.25-year mean ELA.

Another way to evaluate the Van Ufford and Sedgwick estimate is to assume that our 1995-1997 high ELA is incorrect and ask whether our average ELA for 1973-1995 is compatible with the Van Ufford and Sedgwick estimate. It does not appear so for the following reason. The length, area and ice-front elevation retreat records all show continuous retreat (Figures 4, 5). The balance-based ELA for 1973-1995, 4780 m asl, is higher than the adjusted estimate of Van Ufford and Sedgwick, which means that, if both are considered, then the ELA decreased in elevation by at least 40 m at some point between 1973 and early 1995. This would be a minimum because the ELA for a considerable portion of the 22-year interval must have been above 4780 m asl in order to generate an interval mean of 4780 m asl. The lowering required by this scenario does not agree with continuous ice retreat through the interval. In sum, the Van Ufford and Sedgwick ELA is not compatible with either of our mass-balance-based ELAs.

So how should we regard the Van Ufford and Sedgwick (1998) ELA? Our best interpretation is that it should be increased by 77 m and considered as a transient, multi-monthly snowline, but not an annual ELA. If this is correct, it implies that there is significant interannual variability in the ELA of the MGS and, thereby, also the Carstensz Glaciers. This seems reasonable given the significant variability in the monthly mass balance of the single intensely studied tropical glacier on the eastern side of the Pacific, Antizana in Ecuador (Francou et al. 2004). In the Ecuadoran case, the monthly variability translated into 290 m of variation (4 sigma) in the annual ELA between 1994 and 2005 (Haeberli et al. 2007). If interannual ELA variability is indeed significant in western New Guinea, then it is complex to analyse decadal ELA trends using ELAs that are themselves based on widely varying time intervals.

Based on analysis of IKONOS satellite images from 2000 and 2002, as well as a DEM based on data from the Shuttle Radar Topography Mission (SRTM), Klein and Kincaid (2006) suggested that the ELA for the Carstensz Glaciers as a group rose about 100 m from 1972 to ~2000. This would put the absolute value of the ELA, according to Klein and Kincaid (2006), at 4680 m asl using SRTM topography. This number is too low to be compatible with our ELA estimates for the same reasons described above regarding the Van Ufford and Sedgwick ELA estimate.

If the Klein and Kincaid (2006) ELA estimate is increased by 77 m, the resulting value, 4740 m asl, fits our ELA narrative better. However, the adjusted Klein and Kincaid value is still lower than our ELA estimate for the 2.75-year interval between 1997 and 2000, implying further lowering of the ELA after 1997-2000. This is somewhat at odds with continued retreat of the MGS as referred to above. Klein and Kincaid did suggest that their ELA estimate represents a minimum value and this seems to be the best interpretation.

Our own hypothesis calls for more than 100 m of ELA lowering between 1995-1997 and 1997-2000 (Figure 2). Because the 1997-2000 ELA is based on a 2.75-year mass balance, the ELA of 4820 m asl cannot be considered as an interannual extreme. As with the other estimates discussed above, this lowering is also at odds with the records of continuous ice retreat in this interval. However, in this case, we side with the ELA lowering primarily because it is based on a mass-balance calculation. It appears as if the increase in mass occurred between the altitudes of 4840 m and 4900 m asl in the eastern half of the MGS (Figures 3, 7). This is well above the ice front along our designated flowline, which was then at ~4590 m asl. Much of the ice between those elevations could have been stagnant so that they were effectively decoupled.

We close with an assessment of the trend in the ELA for the MGS between 1970 and 2000. Given the potential for significant interannual variability in the ELA, ELAs used to assess decadal trends should be based on mass balances that cover at least two years, though an average ENSO cycle is preferable. In this regard, the 1972/73 ELA is probably high relative to a multi-year ELA centered on 1972/73. This is because 1972/73 coincided with a large El Niño event which put western New Guinea in an extreme drought. For the sake of discussion, we suggest that the multi-year ELA centered on 1972/73 was 30 m below 4657 m, at, say ~4625 m asl. With this value as the starting point, we suggest that the MGS ELA rose at ~100 m per decade between 1970 and 2000, 294 m total (Figure 2).

Conclusions

1) From 1936 to 2006, the area of the Meren Glacial System (MGS) decreased by 78 per cent, while the length of a prominent flowline decreased by 72 per cent. Of this ice loss, 66 per cent of the area was lost since 1972/73. The elevation of the MGS snout rose 353 m between 1936 and 2006, with 84 per cent of this rise coming after 1972.

2) Based on the photogrammetry method, we estimate the specific mass balance of the MGS for 1995-1997 (2.25 years) and 1997-2000 (2.75 years) at -3012 mm we/yr and +336 mm we/yr, respectively. For the 22-year period 1973-1995, we estimate the specific mass balance of the West Meren Lobe of the MGS at -1517 mm we/yr.

3) The ELA of the MGS rose ~100 m per decade between 1970 and 2000. For 1995-1997, the average ELA was above the highest part of the MGS and we estimate it at 4950 m asl. For 1997-2000, the average ELA was 4820 m asl. For 1973 to 1995, the ELA averaged 4780 m asl.

Acknowledgements

We thank PT Freeport Indonesia for topographic data as well as imagery of the glaciers. Dr D. Susanto provided the 2006 image. The work was partially funded by PTFI and the National Science Foundation, Paleoclimate Program. We thank two anonymous reviewers for thoughtful comments that improved the manuscript. Professor Geoffrey Hope has been the principal catalyst for our investigation of New Guinea glaciation and climate history and for that we are grateful. Without Geoff, our involvement would not have happened. By virtue of his interdisciplinary approach, Geoff recognised the importance of understanding New Guinea glaciation for an appreciation of climate and cultural change. Geoff's spirit, generosity and friendship have made our New Guinea-related work highly rewarding.

References

Allison, I. 1974. Morphology and dynamics of the tropical glaciers of Irian Jaya: *Zeitschrift fur Gletscherkunde und Glazialgeologie* 10:129-152.

Allison, I. 1976. Glacier regimes and dynamics. In G.S. Hope, J.A. Peterson, U. Radok and I. Allison (eds) *The Equatorial Glaciers of New Guinea*, pp39-59. Rotterdam: A.A. Balkema.

Allison, I., and Peterson, J.A. 1976. Ice areas on Mt. Jaya: Their extent and recent history, In G.S. Hope, J.A. Peterson, U. Radok and I. Allison (eds) *The Equatorial Glaciers of New Guinea*, pp27-38. Rotterdam: A.A. Balkema.

Allison, I. and J.A. Peterson 1989. Glaciers of Irian Jaya, Indonesia In R. Williams and J., Ferrigno (eds) United States Geological Survey Professional Paper 1386-H, pp1-20. Washington, DC, U.S.: Government Printing Office.

Cullen, N.J., T. Mölg, G. Kaser, K. Hussein, K. Steffen and D.R. Hardy 2006. Kilimanjaro Glaciers: Recent areal extent from satellite data and new interpretation of observed 20th century retreat rates. *Geophysical Research Letters* 33, L16502, doi:10.1029/2006GL027084.

Cullen, N.J., T. Mölg, G. Kaser, K. Steffen and D.R. Hardy 2007. Energy-balance model validation on the top of Kilimanjaro, Tanzania, using eddy covariance data: *Annals of Glaciology* 46:227-233.

Dozy, J.J. 1938. Eine gletscherwelt in Niederlandisch Neuguinea: *Zeitschrift für Gletscherkunde* 26:45-51.

Francou, B., M. Vuille, V. Favier and B. Cáceres 2004. New evidence for an ENSO impact on low-latitude glaciers: Antizana 15, Andes of Ecuador, 0°28'S. *Journal of Geophysical Research* 109, D18106, doi:10.1029/2003JD004484.

Fu, Q. and C.M. Johanson 2005. Satellite-derived vertical dependence of tropical tropospheric temperature trends. Geophysical Research Letters 32, L10703, doi:10.1029/2004GL022266.

Gaffen, D.J., B.D. Santer, J.S. Boyle, J.R. Christy, N.E. Graham and R.J. Ross 2000. Multidecadal changes in the vertical structure of the tropical troposphere. *Science* 287:1242-1245.

Georges, C. 2004. The 20th century glacier fluctuations in the Cordillera Blanca (Peru). *Arctic, Antarctic, and Alpine Research* 36:100-107.

Haeberli, W., M. Hoelzle and M. Zemp 2007. Glacier Mass Balance Bulletin. No. 9 (2004-2005), p100. Zurich: Staffel Druck AG.

Hastenrath, S. 1984. *The Glaciers of Equatorial East Africa.* 353p. Dordrecht: Reidel.

Hastenrath, S. 2005. The glaciers of Mt. Kenya 1899-2004. Erdkunde 59:120-125.

Hastenrath, S. and A. Ames 1995. Recession of Yanamarey Glacier in Cordillera Blanca, Peru, during the 20th century. *Journal of Glaciology* 41:191-196.

Hastenrath, S. and P.D. Kruss 1992. The dramatic retreat of Mount Kenya's glaciers between 1963 and 1987: greenhouse forcing. *Annals of Glaciology* 16:127-133.

Hastenrath, S. and D. Polzin 2004. Volume decrease of Lewis Glacier, Mount Kenya, 1978-2004. *Zeitschrift für Gletscherkunde und Glazialgeologie* 39:133-139.

Hope, G.S., J.A. Peterson, U. Radok and I. Allison 1976. *The Equatorial Glaciers of New Guinea.* 244p. Rotterdam: A. A. Balkema.

Karl, T.R., S.J. Hassol, C.D. Miller and W.L. Murray 2006. Temperature Trends in the Lower Atmosphere: Steps for Understanding and Reconciling Differences. A Report by the Climate Change Science Program and the Subcommittee on Global Change Research: Washington, DC.

Kaser, G., D.R. Hardy, T. Mölg, R.S. Bradley and T.M. Hyera 2004. Modern glacier retreat on Kilimanjaro as evidence of climate change: observations and facts. *International Journal of Climatology* 24:329-339.

Kaser, G. and H.A. Osmaston 2002. *Tropical Glaciers.* 207p. Cambridge: Cambridge University Press.

Klein, A.G. and J.L. Kincaid 2006. Retreat of glaciers on the Puncak Jaya, Irian Jaya, determined from 2000 and 2002 IKONOS satellite images. *Journal of Glaciology* 52:65-79.

Kruss, P.D. 1983. Climate change in East Africa: a numerical simulation from the 100 years of terminus record at Lewis Glacier, Mount Kenya. *Zeitschrift für Gletscherkunde und Glazialgeologie* 19:43-60.

Lanzante, J.R., S.A. Klein and D.J. Seidel 2003. Temporal homogenization of monthly radiosonde temperature data. Part II: trends, sensitivities, and MSU comparison. *Journal of Climate* 16:241-262.

Lanzante, J.R., T.C. Peterson, F.J. Wentz and K.Y. Vinnikov 2006. What do observations indicate about the change of temperatures in the atmosphere and at the surface since the advent of measuring temperatures vertically?. In T.R. Karl, S.J. Hassol, C.D. Miller and W.L. Murray (eds) *Temperature Trends in the Lower Atmosphere: Steps for Understanding and Reconciling Differences.* pp47-70. A Report by the Climate Change Science Program and the Subcommittee on Global Change Research,Washington, DC.

Lemke, P., J. Ren, R.B. Alley, I. Allison, J. Carrasco, G. Flato, Y. Fujii, G. Kaser, P. Mote, R.H. Thomas and T. Zhang 2007. Observations: Changes in snow, ice, and frozen ground. In S. Solomon, D. Quin, M. Manning, Z. Chen, M. Marquis, K.B. Averyt, M. Tignor and H.L. Miller (eds) *Climate Change 2007: The Physical Science Basis. Contribution of Working Group 1 to the Fourth Assessment Report of the Intergovernmental Panel on Climate Change.* pp337-377. Cambridge, United Kingdom: Cambridge University Press.

Mölg, T., N.J. Cullen, D.R. Hardy, G. Kaser and E.J. Klok 2008. Mass balance of a slope glacier on Kilimanjaro and its sensitivity to climate. *International Journal of Climatology* 28:881-892.

Mölg, T., N.J. Cullen and G. Kaser 2009. Quantifying climate change in the tropical midtroposphere over East Africa from glacier shrinkage on Kilimanjaro. *Journal of Climate* 22:4162-4181.

Mölg, T., C. Georges and G. Kaser 2003. Solar-radiation-maintained glacier recession on Kilimanjaro drawn from combined ice-radiation geometry modeling. *Journal of Geophysical Research* 108, doi:10.1029/2003JD003546.

Mölg, T. and D.R. Hardy 2004. Ablation and associated energy balance of a horizontal glacier surface on Kilimanjaro. *Journal of Geophysical Reasearch* 109, doi:10.1029/2003JD004338.

Oerlemans, J. 2005. Extracting a climate signal from 169 glacier records. *Science* 308:675-677.

Østrem, G. and A. Stanley, 1969. Glacier Mass Balance Measurements, A Manual for Field and Office Work. p111. Oslo: Canadian Department of Energy, Mines, and Resources and Norwegian Water Resources and Electricity Board.

Peterson, J.A., G.S. Hope and R. Mitton 1973. Recession of snow and ice fields of Irian Jaya, Republic of Indonesia. *Zeitschrift für Gletscherkunde und Glazialgeologie* 9:73-87.

Peterson, J.A. and L.F. Peterson 1995. Ice retreat from the Neoglacial Maximum in the Puncak Jaya kusuma region, Republic of Indonesia. *Zeitschrift für Gletscherkunde und Glazialgeologie* 30:2.

Prentice, M.L. and G.S. Hope 2007. Climate of Papua and its recent changes. In A.J. Marshall and B.M. Beehler (eds) *The Ecology of Papua*, pp177-195. Singapore: Periplus Editions.

Taylor, R.G., L. Mileham, C. Tindimugaya, A. Majugu, A. Muwanga and B. Nakileza 2006. Recent glacial recession in the Rwenzori Mountains of East Africa due to rising air temperature.*Geophysical Research Letters* 33:L10402, doi:10.1029/2006GL025962.

Thompson, L.G. 2000. Ice core evidence for climate change in the Tropics: implications for our future. *Quaternary Science Reviews* 19:19-35.

Thompson, L.G., E. Mosley-Thompson, M.E. Davis, K.A. Henderson, H.H. Brecher, V.N. Zagorodnov, D.R. Hardy and J. Beer 2002. Kilimanjaro ice core records: Evidence of Holocene climate change in tropical Africa. *Science* 298:589-593.

Trenberth, K.E. 1990. Recent observed interdecadal climate changes in the northern hemisphere *Bulletin of the American Meteorological Society* 71:988-993.

Trenberth, K.E., P.D. Jones, P. Ambenje, R. Bojariu, D. Easterling, A. Klein Tank, D. Parker, F. Rahimzadeh, J.A. Renwick, M. Rusticucci, B. Soden and P. Zhai 2007. Observations: Surface and atmospheric climate change In S. Solomon, D. Quin, M. Manning, Z. Chen, M. Marquis, K.B. Averyt, M. Tignor and H.L. Miller (eds) Climate Change 2007: The Physical Science Basis. Contribution of Working Group 1 to the Fourth Assessment Report of the Intergovernmental Panel on Climate Change, pp235-335. Cambridge, United Kingdom and New York City, NY, USA: Cambridge University Press.

Van Ufford, A.Q. and P. Sedgwick 1998. Recession of the equatorial Puncak Jaya glaciers (~1985 to 1995), Irian Jaya (western New Guinea), Indonesia. *Zeitschrift für Gletscherkunde und Glazialgeologie* 34:131-14.

27

Altitudinal limits of 230 economic crop species in Papua New Guinea

R. Michael Bourke

State Society and Governance in Melanesia, College of Asia and the Pacific, Australian National University, ACT, Australia
mike.bourke@anu.edu.au

Introduction

Temperature extremes set limits on the growth of all crop species. In mountainous regions, such as Papua New Guinea (PNG), there is a regular and linear decrease in temperature with increasing altitude. This regularity is known as the lapse rate (McAlpine et al. 1983:92).[1] This relationship is sufficiently precise to enable altitude data to be substituted for temperature data. Furthermore, in regions located at low latitudes, the temperature differences from north to south at a given altitude are small, and similarly, seasonal variation in temperature is very small.[2] This means that estimates of average yearly temperature can be made from altitude with a high degree of precision.

In this paper, I report on the altitudinal limits of 230 crop species based on numerous field observations. It is relatively easy to observe altitudinal limits in PNG because of the rugged topography and the large altitudinal contrasts over short distances. These data were mostly recorded over a three-year period (1980-1982) and the observations were made during extensive driving trips and walking traverses. The main economic product of most of the species (182) is food. The main products of the other species are export commodities (13 species), shade and timber (11), stimulants (6), decoration (6), body covering (6), cover crops, fish poison and weeds (6).

1 This is treated in more detail in Appendix 1 where lapse rates for mean maximum, mean minimum and mean annual temperatures for non-coastal locations in PNG are given.

2 Most of PNG lies in the latitude range 1-9 °S where seasonal temperature differences are very small (typically 1-2°C). A small proportion of the land mass is at 9-12 °S and the wettest part of the year coincides with the Southern Hemisphere winter. There seasonal temperature differences are a little greater (typically 2-3°C) but, even there, the most important environmental influence on temperature is altitude.

The altitudinal limits of crops are of interest for a number of reasons:

1. Agricultural planning. Such data provide basic information for planning and research for agricultural researchers, planners and extension officers.
2. Agricultural technology transfer. Those involved in transferring agricultural information require basic information about where certain crops will grow. For example, an extension officer promoting peanuts, winged bean or pigeon pea as a rotational crop with sweet potato for a particular location must be certain that these species will grow at that location, as sweet potato is grown almost 1000 m higher than the upper altitudinal limit for these three species.
3. Prehistoric agriculture inferences. Information on altitudinal limits is needed to make deductions about prehistoric agriculture, particularly the limits to cultivation before the adoption of sweet potato in the PNG highlands about 300 years ago (Bayliss-Smith 1988; Brookfield 1991; Haberle 1993). These data can also be used to make inferences about species which may have been used for food during the Pleistocene.
4. Assessing climate change. Because of the close relationship between temperature and altitude in PNG, data on crop limits are a surrogate for temperature recordings. Historical data on where crops grow provide a baseline to gauge the impact of temperature changes associated with global climate change.

Altitude, as a surrogate for temperature, also influences the rate of crop development, as well as setting limits to growth. Only a limited amount of information exists on the influence of altitude on crop development and yield in PNG. The available information for five crops is summarised in Appendix 2.

Previous observations

Many villagers in PNG have extensive knowledge about crop altitudinal limits in their territory. This comes from their planting crops in a number of locations in their territory and noting where crops fail to bear, or bear very poorly because of altitude. When a new species is introduced, villagers typically plant it in many environments, including some that are unsuitable because it is too cold, or occasionally, because it is too hot. This experience teaches them the altitudinal range of a species in their territory. This can be illustrated by the introduction of cardamom as a cash crop in the intermediate altitude zone (600-1200 m) and the lower highland valleys (1200-1400 m) in the mid to late 1970s. Some people living in the main highland valleys attempted to grow cardamom over a range of altitudes, many of which were much higher than the upper altitudinal limit recommended by the Department of Agriculture. In the process, villagers discovered that cardamom would bear successfully as high as 1700 m, and occasionally as high as 1890 m, but not above these altitudes.

Knowledge on crop limits is more detailed where altitudinal contrast is greatest; for example, in much of Enga and Simbu provinces in the highlands. This knowledge has not been recorded systematically, although I used it when enquiring about the productivity of species at particular locations. In a number of locations in the highlands, villagers distinguish between 'warmer' and 'colder' zones and this altitudinal break generally coincides with the upper or lower altitudinal limits of a number of important species. This is discussed later.
Researchers working in villages, particularly anthropologists and geographers, commonly report the altitudinal limits of some species for specific communities or localities. For example,

some observations on crop altitudinal limits are given for the upper Lai Valley, Enga Province (Meggitt 1958:314-320); Baiyer-Lai valleys, Western Highlands Province (Bulmer 1960:63); Asaro Valley, Eastern Highlands Province (Howlett 1962:81-82, 102); and the Sinasina area of Simbu Province (Hide et al. 1979). In particular, a lower altitudinal limit of 1800-1850 m for *karuka* nut pandanus (*Pandanus jiulianettii* complex) is reported by numerous field workers in the highlands. Papua New Guinea-wide limits for several food crops have been presented by Treide (1967:118-119) and reproduced in a modified form by Brookfield with Hart (1971:171). Bayliss-Smith (1985), French (1986) and Haberle (1993:301) have also given altitudinal limits for various food crops for all of PNG, based in part on my unpublished recordings.

A major data source for crop altitudinal limits in PNG is the collection records and publications of the Division of Botany, Department of Forests in Lae (now Forest Research Institute of the PNG Forest Authority). The collections have been made by Division of Botany staff and by others seeking botanical identification. For example, the altitudinal distribution of various species of grasses, weeds, legumes and nut-bearing plants are given by Henty (1969), Henty and Pritchard (1975), Verdcourt (1979) and Henty (1982) respectively. Some botanists provide data on other species. Hynes (1974), for example, gives the altitudinal range of 13 species of *Nothofagus* in New Guinea, while Smith (1977a, 1977b) presents data on the altitudinal distribution of both food and non-food species in the Mount Wilhelm area. Gardner (2003) gives limits for certain *Piper* species.

However, the published information should be used cautiously, for the following reasons:

1. Before the 1:100,000 map sheets becoming available in the late 1970s and early 1980s, most quoted figures for locations in PNG underestimated the true altitudes. This is because these were derived from altimeter readings when the instrument was zeroed at sea level, generally using aircraft instruments. Because of air-pressure surface flows into the mountains, these instrument readings give systematic errors. This is discussed more fully in Appendix 3.
2. Botanical specimens identified by Division of Botany staff were collected for purposes other than defining limits. Rarely is there an indication that the specimen was collected at or near the species' altitudinal limit. As well, the collection records do not indicate whether the specimen was found within or outside the usual altitudinal range for that species.

Provided these limitations are recognised, the Division of Botany records are a valuable data source. I have incorporated some of their recordings in data presented here, particularly when the Division of Botany records extend the extreme ranges that I recorded.

I have published some of my observations in other papers. These include observations for Enga Province only (Bourke and Lea 1982); citrus species (Bourke and Tarepe 1982); limits of some introduced vegetables (Bull and Bourke 1983); coffee and associated shade species (Bourke 1984); for Gumine District only (Bourke 1988); 22 selected species (Allen and Bourke 2009, Table 1.13.3); fruits and edible nuts (Bourke 2009a); and starchy food species (Bourke In press). Some of the figures presented here differ slightly from those I have published previously. This reflects additional observations made since earlier papers, or reinterpretation of some data.

Methods

Observations on crop altitudinal limits were made during numerous traverses by foot and vehicle throughout PNG (Figure 1). For any given geographical unit, such as a valley, I recorded the altitude at which each species was growing near the previously observed limit of that species. The altitude was recorded with an accurate pocket altimeter (Thommen 2000) and appropriate adjustments were made so the readings were within plus or minus 20 m of the true altitude (see Appendix 3 for details of necessary corrections). The highest recording for a species in any geographical unit was considered the upper altitudinal limit, provided that the species ceased to be grown below the highest point of the traverse. The same technique was used for lower limits. An example of a traverse by car and foot in the Marigl Valley, Gumine District, Simbu Province, over the range 1450-2450 m illustrates the technique. The highest recordings for triploid banana (2310 m), Arabica coffee (2170 m) and guava (2020 m) were taken as upper altitudinal limits for these species in this valley. However, the lower recordings (1450-1500 m) were not taken as the lower limits because observations elsewhere in PNG indicated these species grow below the lowest point of this traverse.

A distinction was made between the 'usual' and 'extreme' altitudinal limits for any species[3]. Most limits were considered as usual, but limits were classed as extreme under the following circumstances:

1. Isolated plantings that were well above or below the usual limit for that species in that location.
2. Plantings that were clearly experimental and were not bearing a product of generally accepted quality.
3. Plantings made under extraordinary conditions, such as in a hothouse.

For example, recordings for pineapple plants at Margarima (2200 m) and Kandep (2380 m) producing tiny fruit and grown well above other pineapple plants in nearby locations (1800-1900 m) were considered as extreme limits. Similarly, a recording for pineapples grown in a glasshouse at Tambul (2320 m) was taken as an extreme value.

The distinction between usual and extreme limits is to some degree subjective, but it is an important one. It is not uncommon for small plantings of a species to be made well outside the usual altitudinal range of that species. These extreme limits are of some interest, but are less important than the usual limits. A distinction between usual and extreme altitudinal limits has also been made by authors working in other countries, although the terminology differs among authors. In Ethiopia, Alkämper (1973) classed altitudinal zones as main cultivation, sporadic cultivation and extreme cultivation. Brush (1976) distinguished between the 'effective' crop limits and 'absolute' limits in Peru, while in Ecuador Stadel (1986) used the terms 'major belt' and 'minor belt' for altitudinal ranges.

Plants of some species grow above (or below) where they produce their main economic product. This is particularly the case for some fruit- and nut-bearing species such as mango, elderberry and coconut. All limits presented here are for plants that are yielding their main economic product. When it was not obvious whether the economic product was being produced, I asked local villagers about the behaviour of individual plants. Responses indicating

3 An alternative term to 'usual limit' is 'optimal limit'. However the altitude at which people usually cease growing a species is often somewhat lower (or higher) than the optimal altitude. For example, the usual altitudinal range of pawpaw (*Carica papaya*) was recorded from sea level (0 m) to 1700 m (n=30; Standard Deviation 100 m for the upper limit). However the altitude range where the best fruit are produced is from sea level to about 1400 m in PNG. The difference between the optimal limit (which was not recorded systematically) and the usual limit (data presented here) varies between species, but is typically about 200 m (equivalent to 1°C temperature difference).

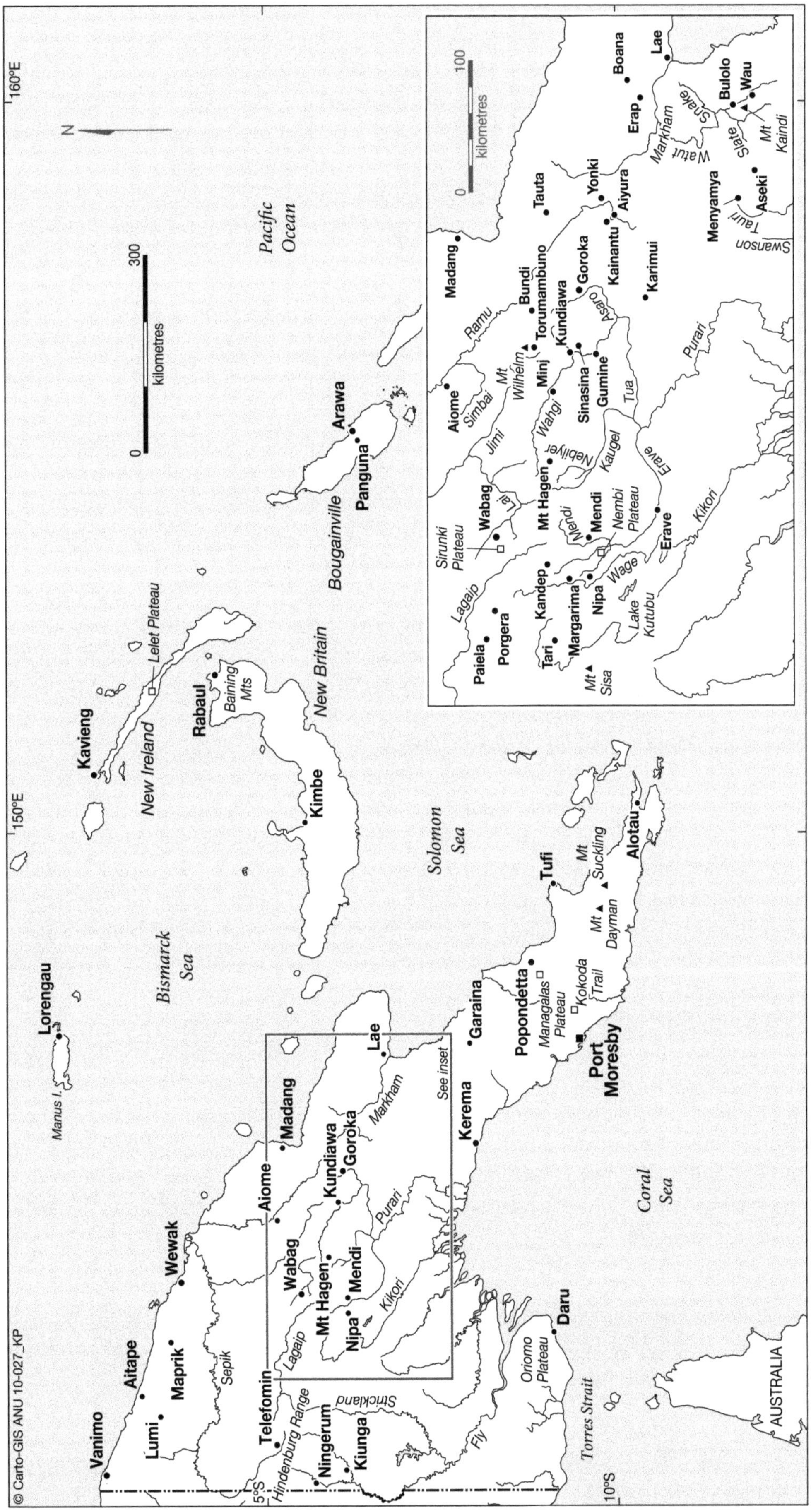

Figure 1. Locations referred to in the text

that a particular plant or group of plants did not bear were taken as indicating that they were outside their usual range. A response indicating that plants bore sparsely or yielded poor quality products was also useful as this suggested that plants were growing near their limit.[4] Similarly, comments by villagers (or my observations) that plants had failed to bear their economic product was a useful indicator that they had been planted above (or below) their altitudinal limit.[5]

Most observations were made over a three-year period (1980-1982). Recordings commenced in mid 1978, but when more accurate topographic maps became available in 1979, I realised that I was systematically underestimating altitude using an altimeter (as others had previously done). I adjusted the 1978 and 1979 observations with more accurate recordings of altitude. Some additional recordings were made between 1983 and 1995, particularly in 1991 when new observations were made at various locations on the Huon Peninsula of Morobe and Madang provinces, including the Nankina Valley, Teptep area, Pindiu area and Kabwum area; and in the Telefomin area of Sandaun Province.

I made most of the observations, but some were made by others (see Acknowledgements). I also examined the collection records for many species held by the Division of Botany in Lae. This was useful in guiding field observations. Where one of the identifications made by Division of Botany staff was from a location with an altitude outside the altitude range that I recorded, this is indicated in the footnotes of the tables.

Observations were made in 18 of the 19 provinces in PNG; the exception was Manus Province, which is a lowland province where people occupy a small altitude range and altitudinal limits are of minor importance. More observations were made in the central highlands than elsewhere in PNG because of the high density of people and agriculture and the good road network. Observations covered all districts of the Highlands region and most districts in the Islands and Momase regions. Traverses were made along virtually all vehicle roads in PNG that extend more than 100 m above sea level. All major mountainous areas were visited. There were numerous lowland areas where recordings were not made; for example, in Western Province, I only made observations in the Kiunga to Ningerum area. The major foot traverses were as follows: Menyamya (Morobe Province) to Swanson River (Gulf); Baining Mountains of New Britain; Nakanai Mountains of New Britain; east coast to west coast of New Ireland, via the Lelet Plateau; Managalas Plateau (Oro) to Kokoda Trail (Central); Chimbu Valley (Simbu) to Ramu Valley via Bundi (Madang); a series of walking traverses on the Huon Peninsula of Morobe and Madang provinces; Golgobip to Olsobip area (Western Province); and Bimin to Oksapmin (Sandaun Province).

The geographical unit used as a sampling framework was usually a valley or plateau, but sometimes it was an administrative one such as a census division. The areas of the units differ and are smaller where population density is greater. For each geographical unit, there are four possible values for any species: the usual upper and lower altitudes, and the extreme upper and lower altitudes. The mean usual limits were derived by averaging the values from the different geographical units. This technique provides an imperfect estimate of mean altitudinal limits. Sources of error are:

1. Erroneous altimeter readings. Such errors were largely eliminated in this study by setting the instrument at a point of known altitude and using a number of corrections where necessary (see Appendix 3 for details).

4 Where plants did not bear, this was recorded as outside the usual and the extreme limits for that location. Recordings of extreme limits are for plants which bore the economic product.

5 Black pepper provides an example of this process (See Footnote 12, Table 6). Pepper is not commonly grown in PNG except at low altitudes, so records of where pepper failed to bear suggest that the usual upper limit is likely to be below 700 m.

2. An inconsistent sampling framework and varying numbers of observations per species. For any particular geographical unit, observations are available only for some species and the mix of locations used to derive mean figures is different for each species. Observations are dictated by where crops are grown and road or walking access to the altitude where they cease to be grown. To overcome this limitation in part, data were gathered on 12 important economic crops at 10 locations between 1600 m and 2500 m; and for five other crops at 15 locations between 700 m and 1400 m.
3. Slope aspects. Cultivation tends to be higher on the east- and north-facing slopes in PNG (Smith 1978). Differences in slope aspect contribute to the error component of these data, but are not considered as great as climatic differences between different locations.
4. Global climate change. Most observations were made over a relatively short period in the early 1980s, but some were made in the late 1980s and early 1990s when temperatures were somewhat higher than in the early 1980s (Bourke et al. 2002).
5. Poorly defined limits. There is a large range in the upper limit for a number of species, particularly green vegetables, as many different species can be grown and villagers are not tempted to grow a species as high as possible. This results in a high error term for some species, particularly green leafy vegetables.

Results and discussion

Data on the usual and extreme altitudinal ranges of 230 crop species and one plant pathogen are presented in Tables 1 to 6. These data are grouped depending on the main economic product, as follows: starchy foods, traditional vegetables, introduced vegetables, fruits, nuts and non-food species. Traditional vegetables are those that were grown, at least in some parts of PNG, before permanent settlement by Europeans (1870 onwards). Introduced vegetables are those introduced into PNG since 1870. For each species the following information is given: botanical family, scientific and common name; mean usual altitudinal range (mean lower and mean upper limits); extreme altitudinal range; number of observations and standard deviation for usual limits. Mean figures for the usual limits are derived from different geographical units and are rounded to the nearest 50 m. Figures for extremes are single observations and are rounded to the nearest 10 m, as are standard deviations. The extreme figures are the highest and lowest obtained for either usual or extreme growing conditions. Common names generally follow French (1986).

Where the usual or extreme lower limit is near sea level (<100 m), this is given as 0 m rather than a mean of actual recordings. Virtually all species that grow below 100 m grow near the ocean, the exception being *marita* pandanus (*Pandanus conoideus*), which is not usually planted near the ocean (Table 4).

Means have been derived for a usual limit where I have a minimum of three observations in different parts of PNG. Means based on three or four observations only must be regarded as approximate. The exception is where the mean upper limit is about 2700 m. This is because the upper limit of arable agriculture in PNG may be taken as 2700 m and food gardens are uncommon above this altitude. Because gardens above 2700 m are planted in a limited number of locations, means for crops that grow up to the limit of agriculture are generally based on only three to five recordings. These crops are potato (Table 1), *Ficus dammaropsis*, *Rorippa schlechteri*, *Rungia klossii*, *Setaria palmifolia* and *Solanum americanum* (Table 2), leek, shallot, spring onion, turnip, carrot, pea and broad bean (Table 3), and pyrethrum and *Eleocharis dulcis* (Table 6). This upper limit of arable agriculture is imposed by the upper limit of sweet potato, the staple food in the highlands. If agriculture was practised at higher altitudes, many of these 15 species would have a higher usual upper limit.

For many vegetables, both traditional and introduced, it is sometimes difficult to define altitudinal limits with much precision, the reason being that many substitute species are available and people tend not to plant a species close to its altitudinal limit. This is reflected in the high standard deviations (>150 m) for the lower limit for many vegetables; for example, *Dicliptera papuana*, *Ficus dammaropsis*, *Oenanthe javanica* (Table 2) and silverbeet (Table 3).

Table 1. The altitudinal range of indigenous and introduced starchy food species in PNG[1]

Scientific name	Family name	Common name	Mean usual altitudinal range (m)	Extreme altitudinal range (m)	Number of observations/ standard deviation	
					Usual min.	Usual max.
Alocasia macrorrhizos[2]	Araceae	Giant taro	0–?	0–400	–	–
Amorphophallus paeoniifolius	Araceae	Elephant-foot yam	0–700	0–1230	–	4/50
**Canna edulis*	Cannaceae	Queensland arrowroot	0–?	0–1620	–	–
Colocasia esculenta[3]	Araceae	Taro	0–2400	0–2760	–	17/150
Dioscorea alata	Dioscoreaceae	Greater yam	0–1900	0–2100	–	15/80
Dioscorea bulbifera	=	Aerial yam	0–1900	0–2110	–	12/110
Dioscorea esculenta[4]	=	Lesser yam	0–1550	0–1670	–	4/120
Dioscorea nummularia	=	Nummularia yam	0–1900	0–2050	–	7/90
Dioscorea pentaphylla	=	Five-leaflet yam	0–1500	0–1620	–	3/40
**Ipomoea batatas*[5]	Convolvulaceae	Sweet potato	0–2700	0–2850	–	10/150
**Manihot esculenta*	Euphorbiaceae	Cassava	0–1800	0–2210	–	32/120
Metroxylon sagu	Arecaceae	Sago	0–1150	0–1250	–	10/60
Musa cvs	Musaceae	Fe'i banana	0–1750	0–2060	–	8/160
Musa cvs	=	Diploid banana	0–1800	0–2030	–	19/70
Musa cvs	=	Triploid banana	0–2150	0–2580	–	30/130
**Oryza sativa*	Poaceae	Rice	0–?	0–2360	–	–
Pueraria lobata	Fabaceae	Pueraria (*kudzu*)	0–2300	0–2740	–	5/180
Saccharum officinarum	Poaceae	Sugar cane	0–2600	0–2760	–	8/160
**Solanum tuberosum*[6]	Solanaceae	Potato	700–2750	0–2850	7/90	4/60
**Xanthosoma sagittifolium*	Araceae	Chinese taro	0–2000	0–2460	–	25/100
**Zea mays*	Poaceae	Maize (corn)	0–2450	0–2680	–	13/130

1. Indigenous species denotes those present in PNG before permanent settlement by Europeans (c. 1870). Introduced species were bought to PNG after 1870 (or after c. 1700 in the case of sweet potato and cassava). Introduced species are denoted by *. The period when food crops were introduced into PNG is given by Bourke (2009b).
2. Native inedible *Alocasia nicolsonii* grows as high as 2640 m, but *Alocasia macrorrhizos* is planted as a food crop only at low altitudes.
3. Stands of *Colocasia* taro are not planted as high as individual plants. The highest observed plot was at 2370 m (Sau Valley, Enga Province) and the mean of six observations was 2250 m (SD 100 m).
4. The lesser yam is not common above 900 m.
5. The usual upper limit of sweet potato (2700 m) is derived from the 10 highest observations, not all observations (Table 7).
6. Potato grown by villagers for subsistence only is usually planted above about 1900 m. Below 1900 m, most crops are intended for both sale and subsistence. Smith (1977a:189) records that he grew potato on Mount Wilhelm at 3580 m.

Table 2. The altitudinal range of indigenous vegetable species in PNG (Table 2 continues on page 482)

Scientific name	Family name	Common name	Mean usual altitudinal range (m)	Extreme altitudinal range (m)	Number of observations/ standard deviation	
					Usual min.	Usual max.
Abelmoschus manihot	Malvaceae	*Aibika*	0–1900	0–2110	–	20/110
Amaranthus tricolor	Amaranthaceae	Amaranth	0–1950	0–2050	–	11/50
Caryota rumphiana	Arecaceae	Fishtail palm	0–1250	0–1600	–	10/70
Commelina diffusa[1]	Commelinaceae	-	0–?	0–2390	–	–
Cucumis sativus	Cucurbitaceae	Cucumber	0–1950	0–2210	–	19/80
Cyanotis moluccana	Commelinaceae	Cyanotis	0–?	0–2410	–	–
Cymbopogon citratus	Poaceae	Lemon grass	0–?	0–2140	–	–
Desmodium repandum[2]	Fabaceae	–	?–2250	1100–2350	–	6/80
Dicliptera papuana[3]	Acathaceae	Dicliptera	1000–2000	720–2660	6/220	5/170
Erythrina variegata	Fabaceae	Coral tree	0–1550	0–2210	–	9/170
Ficus copiosa	Moraceae	*Kumu musong*	0–2200	0–2450	–	6/140
Ficus dammaropsis	=	Highland *kapiak*	800–2750	0–2820	5/300	3/60
Ficus pungens	=	–	0–1850	0–1900	–	5/60
Ficus wassa[4]	=	–	0–?	0–2520	–	–
Gnetum gnemon[5]	Gnetaceae	*Tulip*	0–1100	0–1330	–	10/150
Graptophyllum pictum	Acanthaceae	Caricature plant	0–?	0–1730	–	–
Ipomoea aquatica	Convolvulaceae	*Kangkong*	0–600	0–760	–	5/100
Lablab purpureus	Fabaceae	Hyacinth bean	0–2000	0–2430	–	15/170
Lagenaria siceraria[6]	Cucurbitaceae	Bottle gourd	0–?	0–2670	–	–
Oenanthe javanica[7]	Apiaceae	Java water dropwort	1050–2700	0–3400	9/220	3/40
Pipturus argenteus	Urticaceae	–	0–1800	0–1950	–	5/90
Polyscias sp.	Araliaceae	*Valangur*	0–1200	0–1230	–	4/30
Psophocarpus tetragonolobus[8]	Fabaceae	Winged bean	0–1900	0–2070	–	28/90
Ricinus communis	Euphorbiaceae	Castor	0–2350	0–2760	–	5/100
Rorippa schlechteri[9]	Brassicaceae	Rorippa	750–2700	180–2850	4/120	4/120
Rungia klossii[10]	Acanthaceae	Rungia	950–2700	0–2760	10/170	4/40
Saccharum edule	Poaceae	Lowland *pitpit*	0–1800	0–2270	–	23/160
Setaria palmifolia[11]	=	Highland *pitpit*	0–2700	0–2760	–	4/40
Solanum americanum	Solanaceae	Nightshade (*karakap*)	0–2550	0–2800	–	6/190
Trichosanthes pulleana	Cucurbitaceae	–	0–2000	0–2200	–	3/80
Zingiber officinale	Zingiberaceae	Ginger	0–1950	0–2200	–	15/80

Table 2. *notes*

1. The extreme upper limit for *Commelina diffusa* is a Division of Botany (now PNG Forest Research Institute) identification from the Kandep area in Enga Province.
2. *Desmodium repandum* has been identified from Mount Wilhelm at 3350 m by staff from the Division of Botany.
3. *Dicliptera papuana* is given as *Hemigraphis* sp. by some authors.
4. The extreme upper limit for *Ficus wassa* is a Division of Botany identification from the Kaugel Valley in Western Highlands Province. There is insufficient data to establish the usual upper altitudinal limit for *F. wassa*, but it is probably about 2200 m.
5. *Gnetum gnemon* produces both edible nuts and leaves over these altitudinal ranges.
6. The extreme upper limit of 2670 m for bottle gourd is from the upper Wage Valley in Enga Province (P. Wohlt, pers comm.). I have not seen bottle gourd at over 2200 m.
7. *Oenanthe* is planted in food gardens up to the altitudinal limit of gardening (c. 2700 m) in Enga and Simbu provinces. Self-sown plants occur as high as 3000 m in Enga Province and 3400 m in Simbu Province. The mean usual lower limit of 1050 m is for locations where *Oenanthe* was traditionally grown. It is now commonly planted in coastal locations by highland migrants.
8. The usual range for all plantings of winged bean is from sea level to 1900 m. This species is planted mainly for tuber production over the range 1200-1900 m.
9. *Rorippa schlechteri* is sometimes given as *Nasturtium schlechteri*.
10. Rungia is planted up to about 2700 m; for example, on the Sirunki Plateau, the Kaugel Valley, the Wage Valley (Bowers 1968:89; Wohlt 1978:132; Allen 1982:112) and in the Chimbu Valley. It is not common above about 2300 m.
11. While highland *pitpit* is grown between sea level and 2700 m, it is more commonly planted above about 500 m.

Table 3. The altitudinal range of introduced vegetable species in PNG (Table 3 continues on pages 483-484)

Scientific name	Family name	Common name	Mean usual altitudinal range (m)	Extreme altitudinal range (m)	Number of observations/ standard deviation	
					Usual min.	Usual max.
Abelmoschus esculentus	Malvaceae	Okra	0–?	0–1600	–	–
Allium ampeloprasum cv. group Leek	Alliaceae	Leek	750–2700	0–2760	5/150	3/70
Allium cepa cv. group Aggregatum	=	Shallot	0–2600	0–2740	–	3/150
Allium cepa cv. group Aggregatum	=	Spring onion	0–2700	0–2850	–	5/60
Allium cepa cv. group Common Onion	=	Bulb onion	?–?	0–2630	–	–
Allium sativum cv. group Common Garlic	=	Garlic	?–?	1520–2760	–	–
Amaranthus caudatus[1]	Amaranthaceae	Amaranth	1600–2400	0–2520	5/90	4/150
Amaranthus cruentus[2]	=	Amaranth	1350–2300	1000–2760	6/230	7/110
Amaranthus dubius	=	Amaranth	0–1800	0–2610	–	4/110
Amaranthus blitum	=	Amaranth	0–2050	0–2320	–	9/140
Apium graveolens var. *dulce*	Apiaceae	Celery	950–2300	870–2800	4/40	8/80
Asparagus officinalis	Asparagaceae	Asparagus	1150–2400	0–2630	4/170	3/220
Basella alba	Basellaceae	Ceylon spinach	0–?	0–1670	–	–
Beta vulgaris cv. group Garden Beet	Amaranthaceae	Beetroot	?–2350	0–2720	–	5/190
Beta vulgaris cv. group Spinach Beet	=	Silverbeet	750–2350	0–2760	4/250	7/140
Brassica oleracea cv. group Broccoli	Brassicaceae	Broccoli	1100–?	0–2760	6/140	–
Brassica oleracea cv. group Brussels Sprouts	=	Brussels sprout	?–?	1620–2630	–	–

Scientific name	Family name	Common name	Mean usual altitudinal range (m)	Extreme altitudinal range (m)	Number of observations/ standard deviation	
					Usual min.	Usual max.
Brassica oleracea cv. group Cauliflower	=	Cauliflower	1100–?	0–2760	4/110	–
Brassica oleracea cv. group Chinese kale	=	Chinese kale	?–?	500–2450	–	–
Brassica oleracea cv. group Kohlrabi	=	Kohlrabi	?–?	0–2630	–	–
Brassica oleracea cv. group White Headed Cabbage[3]	=	Head cabbage	700–2700	0–2850	11/80	5/80
Brassica rapa cv. group Chinese cabbage	=	Chinese cabbage (wong bok)	0–2300	0–2720	–	4/140
Brassica rapa cv.group Pak Choi	=	Chinese cabbage (pak choi)	0–2550	0–2800	–	7/160
Brassica rapa cv. group Vegetable Turnip	=	Turnip	?–2700	720–2790	–	3/80
Cajanus cajan	Fabaceae	Pigeon pea	0–1950	0–2200	–	6/200
Canavalia ensiformis	=	Jack bean	0–?	0–1630	–	–
Canavalia gladiata	=	Sword bean	0–?	0–1780	–	–
Capsicum annuum cv. group Grossum	Solanaceae	Capsicum	0–2300	0–2410	–	4/100
Capsicum annuum cv. group Longum	=	Long cayenne pepper	0–?	0–2380	–	–
Cucurbita moschata[4]	Cucurbitaceae	Pumpkin	0–2350	0–2760	–	14/130
Cucurbita pepo	=	Zucchini	1050–2050	0–2760	3/110	5/180
Daucus carota[5]	Apiaceae	Carrot	0–2650	0–2760	–	4/70
Foeniculum vulgare	=	Fennel	?–?	1410–2240	–	–
Glycine max	Fabaceae	Soya bean	0–2150	0–2600	–	9/110
Lactuca sativa[6]	Asteraceae	Lettuce	700–2350	0–2800	4/40	7/210
Lycopersicon esculentum	Solanaceae	Tomato	0–2250	0–2630	–	13/110
Mentha spicata[7]	Lamiaceae	Mint	0–?	0–3580	–	–
Pastinaca sativa	Apiaceae	Parsnip	?–?	1160–2790	–	–
Petroselinum crispum	Apiaceae	Parsley	0–?	0–2760	–	–
Phaseolus lunatus	Fabaceae	Lima bean	1050–2000	0–2770	8/110	18/170
Phaseolus vulgaris[8]	=	Common bean	0–2350	0–2760	–	12/120
Pisum sativum[9]	=	Pea	900–2700	0–2800	6/310	5/70
Raphanus sativus cv. group Small Radish[10]	Brassicaceae	Radish	0–?	0–2630	–	–
Rheum × cultorum	Polygonaceae	Rhubarb	1000–?	750–2760	4/220	–
Rorippa islandica[11]	Brassicaceae	–	2100–2750	1990–2850	4/100	4/20

Scientific name	**Family name**	**Common name**	**Mean usual altitudinal range (m)**	**Extreme altitudinal range (m)**	**Number of observations/ standard deviation**	
					Usual min.	Usual max.
Rorippa nasturtium-aquaticum	=	Watercress	0–2900	0–3580	–	3/70
Sechium edule	Cucurbitaceae	Choko	0–2300	0–2680	–	16/120
Solanum melongena cv. group Common Eggplant	Solanaceae	Eggplant	0–1800	0–2260	–	7/100
Spinacia oleracea	Amaranthaceae	Spinach	1000–?	870–2630	4/130	–
Symphytum asperrimum	Boraginaceae	Russian comfrey	0–2350	0–2630	–	5/170
Talinum triangulare	Portulacaceae	–	0–?	0–1600	–	–
Vicia faba	Fabaceae	Broad bean	2050–2650	1620–2760	4/90	4/80
Vigna radiata	=	Mung bean	0–?	0–1620	–	–
Vigna umbellata	=	Rice bean	0–?	0–2070	–	–
Vigna unguiculata cv. group Sesquipedalis	=	Snake bean	0–1600	0–1890	–	10/200
Vigna unguiculata cv. group Unguiculata	=	Cowpea	0–1750	0–1840	–	6/70

1. *Amaranthus caudatus* was recorded by Peekel (1984:164) as growing near sea level on New Ireland Province. It is rarely seen in the lowlands, but I have seen it near Angoram in East Sepik Province at 60 m.
2. *Amaranthus cruentus* has been recorded by Peekel (1984:164) from near sea level on New Ireland Province and identified by Division of Botany staff from specimens from Malalau (3 m) and Maprik (200 m). The lowest altitude I have seen plants growing was at 1000 m at Bomai in Simbu Province.
3. Cabbage grown by villagers for subsistence only is usually planted above about 1700 m. Smith (1977a:189) records that he grew cabbage on Mount Wilhelm at 3580 m.
4. Records for pumpkin are for plants producing both fruit and leaves. I have seen another species of pumpkin (*C. pepo*?) growing at 2680 m on the Sirunki Plateau in Enga Province.
5. Carrot generally produces better-quality roots at 400 m and above.
6. Lettuce can be grown at sea level in PNG, but the quality is often poor. Acceptable quality lettuce is grown at 700 m and above.
7. Mint is grown at about 2600 m in Simbu and Enga provinces, and this is probably the usual upper limit.
8. Common bean grows poorly at sea level and is usually grown above 400 m.
9. Peas grown by villagers for subsistence only are usually planted above about 1700 m.
10. Radish was grown by Smith (1977a:189) at 3580 m on Mount Wilhelm.
11. *Rorippa islandica* has been identified by Division of Botany staff from specimens at lower altitudes in the Baliem Valley, Papua (West New Guinea) (1600 m) and in south Simbu Province (1680 m).

Table 4. The altitudinal range of indigenous and introduced fruit species in PNG[1] (Table 4 continues on page 486)

Scientific name	Family name	Common name	Mean usual altitudinal range (m)	Extreme altitudinal range (m)	Number of observations/ standard deviation	
					Usual min.	Usual max.
**Ananas comosus*[2]	Bromeliaceae	Pineapple	0–1800	0–2380	–	25/110
**Annona cherimolia*	Annonaceae	Cherimoya	?–?	750–2200	–	–
**Annona muricata*	=	Soursop	0–1000	0–1460	–	7/100
**Annona reticulata*	=	Bullock's heart	0–?	0–1210	–	–
**Annona squamosa*	=	Custard apple (sweetsop)	0–?	0–1210	–	–
**Artocarpus heterophyllus*	Moraceae	Jackfruit	0–?	0–1230	–	–
**Averrhoa bilimbi*	Oxalidaceae	Tree cucumber	0–?	0–750	–	–
**Averrhoa carambola*	=	Five corner (carambola)	0–1300	0–1430	–	4/120
Burckella obovata	Sapotaceae	*Bukabuk*	0–?	0–390	–	–
**Carica pubescens*	Anacardiaceae	Mountain pawpaw	?–?	1750–2760	–	–
**Carica papaya*	=	Pawpaw	0–1700	0–1950	–	30/100
**Citrullus lanatus*[3]	Cucurbitaceae	Watermelon	0–1700	0–1980	–	6/180
**Citrus aurantifolia*	Rutaceae	Lime	0–1800	0–2260	–	5/80
**Citrus limon*[4]	=	Lemon	0–2150	0–2240	–	6/90
**Citrus maxima*	=	Pomelo	0–1300	0–1640	–	4/70
**Citrus paradisi*	=	Grapefruit	0–1800	0–1980	–	6/100
**Citrus paradisi* × *Citrus reticulata?*	=	*Ugli*	0–1800	0–1830	–	3/50
**Citrus reticulata*	=	Mandarin	0–1800	0–2260	–	9/50
**Citrus sinensis*	=	Orange	0–1800	0–2280	–	12/80
**Cucumis melo*	Cucurbitaceae	Rockmelon	0–?	0–2180	–	–
**Cyphomandra betacea*	Solanaceae	Tree tomato (tamarillo)	1050–2300	0–2600	8/160	7/170
**Durio zibethinus*[5]	Malvaceae	Durian	0–?	0–?	–	–
**Eriobotrya japonica*	Rosaceae	Loquat	850–1800	0–2410	3/170	6/70
**Eugenia uniflora*	Myrtaceae	Brazil cherry	0–1750	0–1880	–	3/140
**Fortunella japonica*[6]	Rutaceae	Cumquat	0–?	0–1160	–	–
**Fragaria* sp.	Rosaceae	Strawberry	800–2450	660–2800	7/100	5/240
**Fragaria vesca*	=	Alpine strawberry	?–?	1740–3580	–	–
**Garcinia mangostana*[7]	Clusiaceae	Mangosteen	0–?	0–?	–	–
**Hibiscus sabdariffa*	Malvaceae	Rosella	0–1700	0–2220	–	8/160
**Malus domestica*	=	Apple	?–?	600–2670	–	–
**Mangifera indica*[8]	Anacardiaceae	Mango	0–1600	0–1820	–	13/110
Mangifera minor	=	Traditional mango	0–1750	0–1900	–	6/100
**Monstera deliciosa*	Araceae	Ceriman	?–2200	0–2330	–	3/180
**Morus nigra*	Moraceae	Mulberry	800–2200	0–2760	6/100	8/150
**Nephelium lappaceum*	Sapindaceae	Rambutan	0–?	0–750	–	–
Pandanus conoideus[9]	Pandanaceae	*Marita*	0–1700	0–1980	–	37/90
**Passiflora edulis* f. *edulis*	Passifloraceae	Purple passionfruit	800–2300	700–2520	7/90	13/90

Scientific name	Family name	Common name	Mean usual altitudinal range (m)	Extreme altitudinal range (m)	Number of observations/ standard deviation	
					Usual min.	Usual max.
**Passiflora edulis* f. *flavicarpa*	=	Lowland yellow passionfruit	0–850	0–960	–	5/80
**Passiflora ligularis*	=	*Suga prut* (highland yellow passionfruit)	1350–2350	1300–2460	3/80	4/140
**Passiflora mollissima*[10]	=	Banana passionfruit	1850–2800	1640–2920	7/110	5/100
**Passiflora quadrangularis*	=	Granadilla	0–1000	0–1520	–	10/300
**Persea americana*	Lauraceae	Avocado	0–2050	0–2430	–	16/160
**Physalis peruviana*	Solanaceae	Cape gooseberry	950–2800	750–2870	5/190	6/60
Pometia pinnata[11]	Sapindaceae	*Ton* (*taun*)	0–800	0–1120	–	7/160
**Prunus* sp.	Rosaceae	Plum	?–?	1590–2600	–	–
**Psidium cattleianum*	Myrtaceae	Cherry guava	0–1850	0–1900	–	3/90
**Psidium guajava*	=	Guava	0–1850	0–2020	–	19/110
**Punica granatum*	Rosaceae	Pomegranate	0–?	0–1620	–	–
**Rubus lasiocarpus*	=	Black raspberry	950–2250	760–2830	5/150	11/180
Rubus moluccanus	=	Red raspberry	0–2150	0–2250	–	3/120
Rubus rosifolius[12]	=	Red raspberry	950–2800	700–2900	8/180	5/60
**Sambucus nigra*[13]	Adoxaceae	Elderberry	450–1900	0–2150	3/110	11/140
Spondias cytherea	Anacardiaceae	Golden apple	0–950	0–1070	–	4/110
**Syzygium aqueum*	Myrtaceae	Watery rose apple	0–1600	0–1640	–	3/50
Syzygium malaccense	=	Malay apple	0–850	0–1580	–	5/80

1. Indigenous species denotes those present in PNG before permanent settlement by Europeans (c. 1870). Introduced species were bought to PNG after 1870. Introduced species are denoted by *. The period when food crops were introduced into PNG is given by Bourke (2009b).
2. Smooth-leaf and rough-leaf pineapple have the same usual upper altitudinal limit (1800 m). The rough leaf is less common above about 1500 m and produces better-quality fruit between about 400 m and 1200 m.
3. Watermelon is not common above about 1200 m, but is grown up to a mean usual upper limit of 1700 m.
4. Lemon grows better above about 400 m. For a fuller discussion on the optimum altitudinal range for citrus species in PNG, see Bourke and Tarepe (1982).
5. Durian has failed to establish at Bulolo (750 m) (Simpson and Arentz 1982), suggesting that its upper limit may be below 750 m.
6. Cumquat would almost certainly bear at higher altitudes in PNG as it is cold tolerant, but it has not been recorded growing above 1160 m.
7. Mangosteen failed to bear fruit at 550 m on the Managalas Plateau, suggesting that its upper limit may be below this.
8. Mango fruit quality is poor above about 1200 m and bearing is irregular above 1600 m.
9. *Marita* pandanus is not usually planted near the ocean, but it is grown inland at altitudes below 100 m; for example, near Kiunga, Popondetta, Gogol Valley and Aitape. It is more commonly planted above about 500 m.
10. Banana passionfruit plants grow as high as 3580 m, but the highest that I recorded fruit was at 2920 m (Chimbu Valley) and 2850 m (Sirunki Plateau). Self-sown plants are not common below about 2000 m, although planted vines bear as low as 1640 m (Aiyura).
11. *Pometia pinnata* bears edible fruit up to a mean upper limit of 800 m, but the tree grows at higher altitudes. It has been recorded at about 1700 m in the Nipa area by Sillitoe (1983:115).
12. Peekel (1984:202) implies that *Rubus rosifolius* has been recorded as low as 300 m in New Ireland and New Britain. The Division of Botany has one identification from 200 m (Tufi), but all other specimens in their collections come from over 1000 m. My observations are of a mean minimum of 950 m (eight recordings).
13. Elderberry grows up to about 2650 m, but does not usually bear fruit above a mean of 1900 m.

Table 5. The altitudinal range of indigenous and introduced edible nut species in PNG[1]

Scientific name	Family name	Common name	Mean usual altitudinal range (m)	Extreme altitudinal range (m)	Number of observations/ standard deviation	
					Usual min.	Usual max.
Aleurites moluccana	Euphorbiaceae	Candle nut	0–1800	0–2160	–	9/140
**Arachis hypogaea*	Fabaceae	Peanut	0–1850	0–1940	–	21/70
Artocarpus altilis	Moraceae	Breadfruit	0–1250	0–1450	–	23/130
**Anacardium occidentale*	Anacardiaceae	Cashew	0–?	0–1400	–	–
Barringtonia procera	Barringtoniaceae	*Pao*	0–500	0–620	–	4/90
Canarium indicum	Burseraceae	*Galip*	0–700	0–930	–	5/160
*Carya illinoensis		Pecan	?–?	1390–1640	–	–
Castanopsis acuminatissima[2]	Fagaceae	Castanopsis	700–2350	570–2440	6/80	8/110
Cocos nucifera[3]	Arecaceae	Coconut	0–950	0–1310	–	20/190
Finschia chloroxantha[4]		Finschia	0–1850	0–2000	–	4/110
Inocarpus fagifer	Fabaceae	Polynesian chestnut (*aila*)	0–400	0–870	–	4/90
**Macadamia integrifolia/* **M. tetraphylla*[5]	Proteaceae	Macadamia	0–1750	0–1810	–	3/60
Pandanus antaresensis	Pandanaceae	Wild *karuka*	1000–2350	850–2460	4/110	9/90
Pandanus brosimos	=	Wild *karuka*	2400–3100	1800–3300	20/150	6/60
Pandanus jiulianettii	=	Planted *karuka*	1800–2600	1450–2800	50/110	18/100
Pangium edule		*Sis* (*solomon*)	0–1050	0–1380	–	11/120
Terminalia catappa	Combretaceae	Sea almond (*talis*)	0–300	0–460	–	4/100
Terminalia impediens	=	*Okari*	0–1000	0–1100	–	3/110
Terminalia kaernbachii	=	*Okari*	0–1100	0–1260	–	11/90

1. Data for tulip nuts (*Gnetum gnemon*) are given in Table 2. Indigenous species denotes those present in PNG before permanent settlement by Europeans (c. 1870). Introduced species were bought to PNG after 1870. Introduced species are denoted by *. The period when food crops were introduced into PNG is given by Bourke (2009b).
2. Self-sown castanopsis is more common above about 1100 m, although the usual mean lower limit is 700 m.
3. Coconut palms grow as high as 1760 m, but the highest palms that bore nuts in the period 1980-1982 were at Yonki, EHP (1310 m) and the Baiyer Valley, WHP (1220 m). By 1999, coconuts were bearing as high as 1370 m (Benabena Valley, EHP), 1420 m (Korofeigu, EHP) and 1450 m (Wahgi Valley, WHP). By mid-2009, coconut palms were bearing as high as 1560 m (near Goroka, EHP).
4. The highest recording for finschia (2000 m) is a Division of Botany record from Aseki in Morobe Province.
5. In their natural range in Australia, *Macadamia tetraphylla* occurs in a slightly cooler climate than *M. integrifolia. M. tetraphylla* has also proven more cold tolerant than *M. integrifolia* in California (Cull and Trochoulias 1982). The limited numbers of observations from PNG do not indicate a separate range for the two species; both species bear from sea level up to 1700-1800 m.

Table 6. The altitudinal range of some indigenous and introduced non-food species in PNG[1] (Table 6 continues on page 489)

Scientific name	Family name	Common name	Mean usual altitudinal range (m)	Extreme altitudinal range (m)	Number of observations/ standard deviation		Main uses[2]
					Usual min.	Usual max.	
Albizia chinensis	Fabaceae	Albizia	0–1900	0–2060	–	15/110	S
Araucaria cunninghamii[3]	Araucariaceae	Hoop pine	0–?	0–2140	–	–	T
Araucaria hunsteinii	=	Klinki pine	0–?	0–2140	–	–	T
Areca catechu	Arecaceae	Betel nut	0–1100	0–1390	–	17/140	N
Areca macrocalyx	=	Highland betel nut	1100–1950	770–2120	8/160	14/110	N
**Bixa orellana*	Bixaceae	Bixa	0–1650	0–1910	–	12/140	B
Broussonetia papyrifera	Moraceae	Paper mulberry	?–?	0–2440	–	–	B
**Camellia sinensis*[4]	Theaceae	Tea	?–?	1210–2670	–	–	E
Campnosperma brevipetiolata[5]	Anacardiaceae	*Tigaso* oil tree	0–850	0–?	–	–	B
**Cannabis sativa*	Cannabaceae	Marijuana	0–?	0–2570	–	–	N
**Capsicum frutescens*	Solanaceae	Birdseye chilli	0–1800	0–2420	–	10/110	E
Casuarina oligodon	Casuarinaceae	Casuarina	700–2600	120–2820	8/100	10/150	TS
**Ceiba pentandra*	Malvaceae	Kapok	0–1250	0–1550	–	6/200	E
**Chrysanthemum cinerariaefolium*	Asteraceae	Pyrethrum	2400–2800	1630–2850	8/20	5/50	E
**Codiaeum variegatum*	Euborbiaceae	Croton	0–1700	0–1920	–	16/100	D
Coix lachryma-jobi	Poaceae	Job's tears	0–?	0–2460	–	–	B
**Coffea arabica*[6]	Rubiaceae	Arabica coffee	700–2050	80–2380	15/150	30/90	E
**Coffea canephora* var. *robusta*	=	Robusta coffee	0–550	0–1700	–	5/50	E
Cordyline fruticosa[7]	Laxmanniaceae	Tanget	0–3100	0–3250	–	4/180	BD
**Crotalaria lunata*	Fabaceae	Crotalaria	1450–2400	1300–2740	4/100	10/140	S
**Crotalaria micans*	=	Crotalaria	0–1850	0–2090	4/100	10/90	S
Derris sp.	=	Derris	0–?	0–1110	–	–	F
**Elaeis guineensis*	Arecaceae	Oil palm	0–?	0–920	–	–	E
Eleocharis dulcis	Cyperaceae	Reeds	0–2750	0–2820	–	3/80	B
**Elettaria cardamomum*	Zingiberaceae	Cardamom	550–1700	0–1890	4/110	11/140	E
**Flemingia macrophylla*	Fabaceae	Flemingia	0–1850	0–2140	–	9/130	D
Gymnostoma papuanum	Casuarinaceae	Casuarina	0–2000	0–2260	–	13/130	T
**Helianthus annuus*	Asteraceae	Sunflower	0–2200	0–2630	–	5/120	E
**Hevea brasiliensis*	Euphorbiaceae	Rubber (para)	0–?	0–700	–	–	E
**Hibiscus rosa-sinensis*	Malvaceae	Hibiscus	0–?	0–2360	–	–	D

Scientific name	Family name	Common name	Mean usual altitudinal range (m)	Extreme altitudinal range (m)	Number of observations/ standard deviation		Main uses[2]
					Usual min.	Usual max.	
Lantana camara	Verbenaceae	Lantana	0–2100	0–2200	–	5/100	DW
**Leucaena leucocephala*	Fabaceae	Leucaena	0–1800	0–1890	–	15/70	ST
**Lupinus* cv. *Russell Hybrid*[8]	=	Blue lupin	1850–?	1530–2920	8/80	–	D
**Mimosa invisa*	=	Mimosa	0–1650	0–1770	–	5/90	W
Mucuna novo-guineensis	=	–	0–1500	0–1890	–	4/270	D
**Mucuna pruriens*	=	Velvet bean	0–?	0–1620	–	–	C
Musa ingens	Musaceae	Wild banana	1300–1850	1150–1930	4/130	6/70	D
**Nicotiana tabacum*[9]	Solanaceae	Tobacco	0–2400	0–2790	–	14/200	N
**Phytophthora colocasiae*[10]	Phythiaceae	Taro blight	0–1300	0–1620	–	11/180	P
**Piper aduncum*	Piperaceae	–	0–1850	0–1920	–	4/90	W
Piper betle	=	Lowland betel pepper	0–1000	0–1140	–	7/90	N
Piper gibbilimbum[11]	=	Highland betel pepper	1150–2300	780–2580	4/260	10/110	N
**Piper nigrum*[12]	=	Black pepper	0–?	0–1080	–	–	E
Ptychococcus sp.	Arecaceae	Black palm	?–1900	?–1990	–	4/90	T
**Spathodea campanulata*	Bignoniaceae	West African tulip	0–?	0–1600	–	–	T
**Stylosanthes humilis*	Fabaceae	Townsville stylo	0–?	0–1620	–	–	C
**Tephrosia vogelii*	=	Tephrosia	1050–1950	0–2540	3/100	20/130	S
**Theobroma cacao*	Malvaceae	Cocoa	0–800	0–1390	–	7/140	E
**Vanilla planifolia*	Orchidaceae	Vanilla	0–1000	0–1080	–	4/100	E

1. Indigenous species denotes those present in PNG before permanent settlement by Europeans (c. 1870). Introduced species were bought to PNG after 1870 (or after c. 1700 in the case of tobacco and bixa). Introduced species are denoted by *. The period when stimulants were introduced into PNG is given by Bourke (2009b).
2. The main use of these species in PNG is as follows:
   B. Body covering (*Cordyline fruticosa* and *Eleocharis dulcis*), a dye (*Bixa*), a source of beads (Job's tears), tapa cloth (*Broussonetia papyrifera*) or body oil (*tigaso* tree)
   C. Cover or pasture crop
   D. Decorative plant
   E. Export crop (or potential one)
   F. Fish poison
   N. Stimulant or used with a stimulant (betel pepper)
   P. Plant pathogen
   S. Shade for export crop
   T. Timber source
   W. Weed
3. Natural stands of hoop pine occur as high as 2800 m on Mount Suckling and Mount Dayman according to Whitmore (1984:212).
4. Plantation tea is grown over a range of 1210-1960 m in Western Highlands Province. It grows at about 750 m in the Garaina area in Morobe Province.
5. The upper limit for *tigaso* oil tree is based on the distribution in Southern Highlands Province (C. Ballard, pers comm. 1994).
6. Plantation Arabica coffee occurs over a range of 930 m (Wau area) to 1820 m (Asaro Valley and Wahgi Valley). For a fuller discussion on the distribution of coffee see Bourke (1984).
7. *Cordyline fruticosa* has been recorded on Mount Wilhelm at 3580 m by Smith (1977a:192).
8. The mean usual lower limit of 1850 m for blue lupin is for self-sown plants.
9. Tobacco is commonly grown under the eves of houses towards the top of its altitudinal range. However, the range for tobacco grown in the open and under shelter is the same and observations on plants grown in all situations have been combined.
10. All specimens of the fungus that causes taro blight were identified by Department of Primary Industry plant pathologists.
11. The Division of Botany has two records for highland betel pepper from about 200 m (at Aitape and Maprik), indicating that the extreme lower limit is less than the 780 m that I recorded. Gardner (2003) records that *P. gibbilimbum* grows to about 3000 m.
12. Black pepper does not bear at Bulolo (750 m), Garaina (760 m) or Boana (920 m), although it has borne at Pindiu (930 m) and at Erave (1080 m). Thus, the mean usual upper limit is likely to be less than 700 m.

Usual altitudinal ranges

Considerable variation exists in the usual altitudinal ranges of different species. The ranges may be classed as small (<1000 m), medium (1000-2000 m), and large (2000-3000 m). This is illustrated with data on 22 crop species (Figure 2).

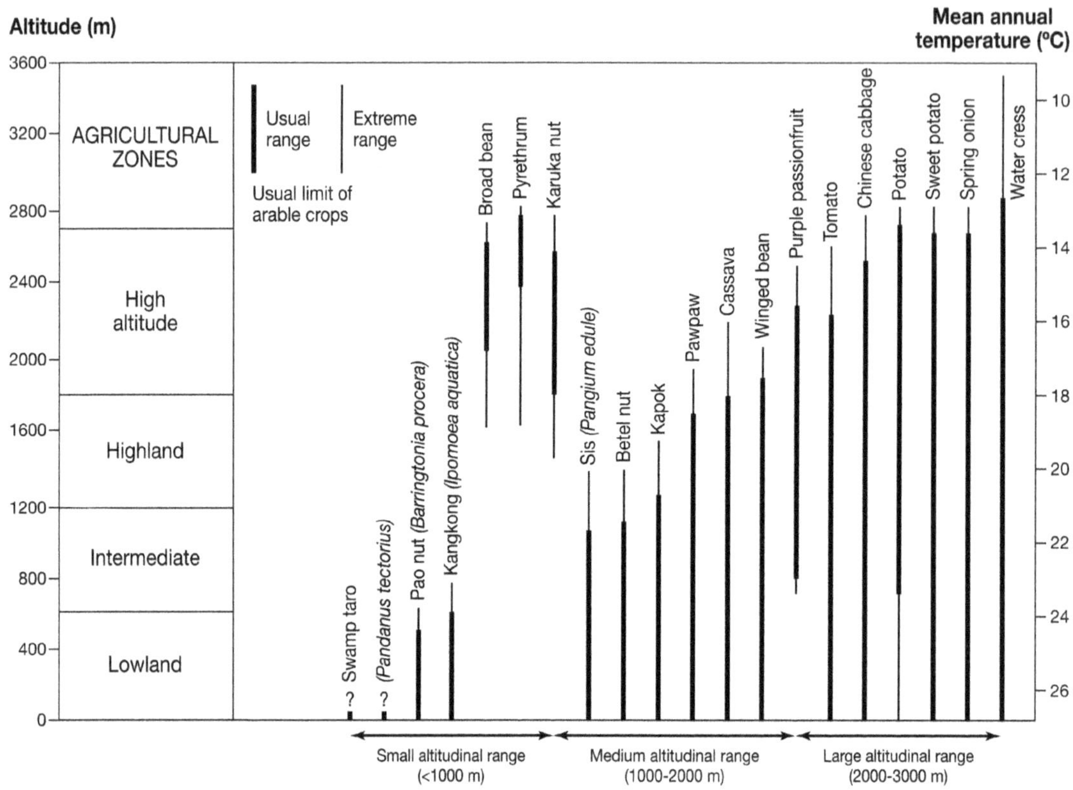

Figure 2. The usual and extreme altitudinal ranges of 22 crop species in PNG

Small range (<1000 m). A number of crops are grown from sea level to altitudes not above 100 m. These include swamp taro (*Cyrtosperma chamissonis*), lotus (*Nelumbo nucifera*), water lily (*Nymphaea pubescens*) and *Pandanus tectorius*.[6] It is likely that the upper limit of these species is determined by factors other than temperature.

A number of other species have restricted altitudinal ranges (400-600 m). These either grow near the lower limit of agriculture in PNG, that is, near sea level, or near the upper limit (2700 m). Species that grow at low altitudes only include kangkong (*Ipomea aquatica*) (Table 2), durian (*Durio zibethinus*), mangosteen (*Garcinia mangostana*) (Table 4) and *pao* nut (*Barringtonia procera*) (Table 5). Presumably these species would grow in warmer conditions than those in PNG if other environmental conditions were favourable. At the other extreme are cold-tolerant species that grow only near the upper altitudinal agricultural limit in PNG. These include *Rorippa islandica*, broad bean (*Vicia faba*) (Table 3) and pyrethrum (*Chrysanthemum cinerariaefolium*) (Table 6). One species with a restricted altitudinal range does not grow near sea level, nor at the usual upper limit of agriculture. This is the cultivated karuka nut (*Pandanus jiulianettii*), which grows over a usual range of 800 m (Figure 2).

Medium range (1000-2000 m). Species that have a usual range of 1000-2000 m mostly fall into two groups. The first are tree species that bear over an altitudinal range of 1000-1250 m; for example, sago (*Metroxylon sagu*) (Table 1), breadfruit (*Artocarpus altilis*), *sis* nut (*Pangium edule*), *okari* nut (*Terminalia kaernbachii*) (Table 5) and kapok (*Ceiba pentandra*) (Table 6).

6 The limits for these four species are not presented in the tables as they range in altitude from sea level to less than 100 m altitude.

The second group contains many species and these have a usual range from sea level to 1700-1900 m. They include three yam species (*Dioscorea alata*, *D. bulbifera* and *D. nummularia*), cassava (*Manihot esculenta*) (Table 1), cucumber (*Cucumis sativus*), winged bean (*Psophocarpus tetragonolobus*) (Table 2), pawpaw (*Carica papaya*) and five citrus species (Table 4).

Large range (2000-3000 m). Again, there are two main groups of species in this category, although the grouping is not as marked as for the medium-range species. The first group contains species that grow from sea level up to 2200-2400 m, such as pumpkin (*Cucurbita moschata*), Chinese cabbage (*Brassica rapa* cv. group Chinese cabbage) and tomato (*Lycopersicon esculentum*) (Table 3). The upper limit of this group coincides with the altitude above which frosts are common (c. 2200 m) and their upper limit is probably determined by annual frosts. Species in the other group have a usual range of about 2700 m because they grow from sea level to the usual limit of agriculture. These include sweet potato (*Ipomoea batatas*) (Table 1), highland *pitpit* (*Setaria palmifolia*) (Table 2), spring onion (*Allium cepa* cv. group Aggregatum), carrot (*Daucus carota*) (Table 3) and reed (*Eleocharis dulcis*) (Table 6).

Several species have an even larger range. Watercress (*Rorippa nasturtium-aquaticum*) grows from sea level to 2900 m (Table 3, Figure 3). It was not possible to define the usual upper limit of mint (*Mentha spicata*), but it has an extreme range of about 3600 m (Table 3).

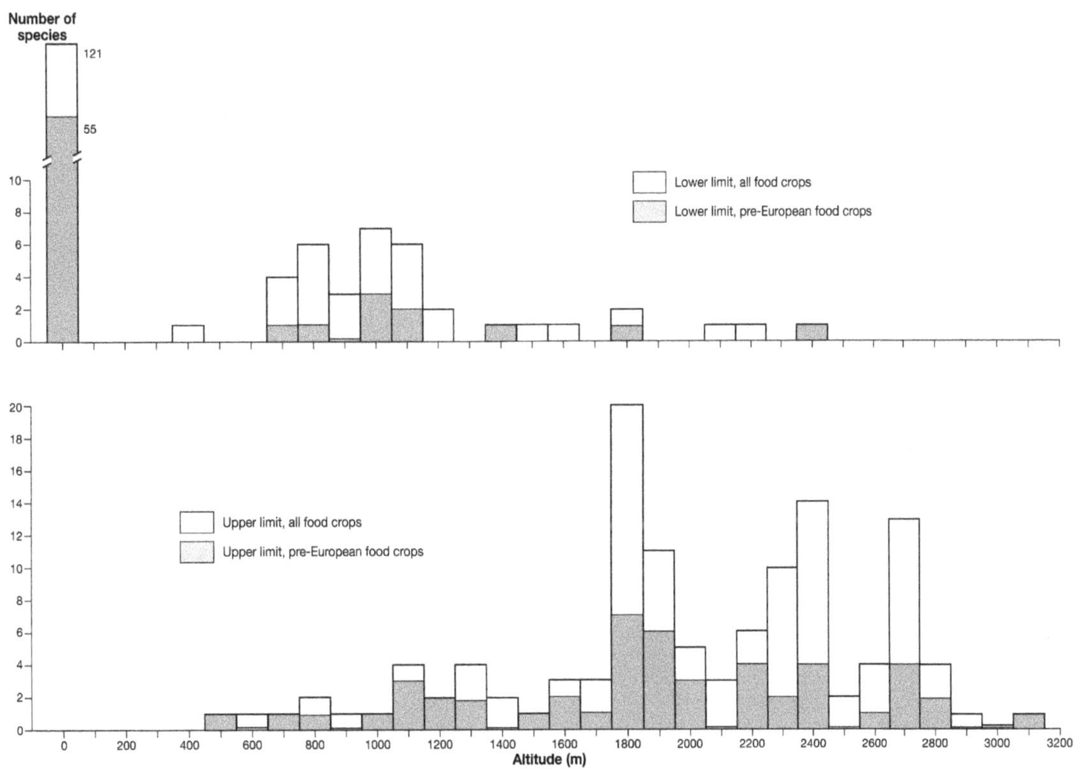

Figure 3. Distribution of the upper and lower altitudinal limits of 160 food crops in PNG

Distribution of crop limits

The distribution of the upper and lower usual altitudinal limits of 160 food crops by 100 m altitudinal classes is plotted in Figure 3. This plot illustrates the grouping of limits within a number of relatively narrow bands.

The most common single altitudinal limit is sea level, as 76 per cent (121 species) of food crops for which observations are available have their usual lower limit at sea level. A number of species have their usual lower limit between 650 m and 1150 m (28 species). Upper altitudinal limits are concentrated in three bands: 1750-1900 m, 2250-2350 m and at about 2700 m (Figure 3).

The concentration of crop altitudinal limits provides a basis for the altitudinal zones frequently recognised in PNG. These are the lowlands (0-600 m), the intermediate zone (600-1200 m), the highlands (1200-1800 m) and the high altitude zone (1800-2700 m) (Figure 2). The lower altitudinal limits of many cold-tolerant species and the upper limits of a number of important lowland tree species occur in the intermediate zone. The large number of species with an upper altitudinal limit of 1750-1900 m defines the upper limit of the highland zone (Figure 3).

The distinction is made in Figure-3 between food crops that pre-date European settlement in PNG in the 1870s and those that post-date European settlement. The same general pattern applies to both pre-European and post-European species (Figure 3). However, before 1870, only a limited number of food crops had a usual lower limit other than sea level. The increase in the number of species with lower limits other than sea level reflects the adoption of cold-tolerant species for use in the highlands. These originated in both Asia and Europe (for example, cabbage, silverbeet and mulberry), and in America (for example, potato and purple passionfruit).

Usual upper limit of food gardens

Near the upper limit of agriculture in PNG, sweet-potato gardens interplanted with other food crops constitute the major agricultural activity. Sweet-potato gardens are not common above about 2700 m and this may be taken as the usual upper limit of arable agriculture. Nevertheless, in parts of the highlands, some food gardens are planted at a higher altitude. The highest garden in PNG was observed between the Sirunki Plateau and the Lagaip Valley in Enga at 2850 m (Table 7). Persistent cloud cover and atmospheric saturations rather than low temperatures probably limit cultivation in parts of the highlands, particularly on the outer faces of the main mountain ranges (Brookfield 1964). Thus, the upper limit of food gardens in parts of the region occurs below 2700 m (Table 7).

Table 7. Highest subsistence sweet-potato (*Ipomoea batatas*) gardens observed in PNG[1, 2]

Location	Altitude (m)
Sirunki Plateau, Enga	2850
Upper Wage Valley, Enga	2840
Upper Lagaip Valley, Enga	2830
Tambul Basin, Western Highlands	2800
Chimbu Valley, Simbu	2770
Lai-Ambum Valleys divide, Enga	2750
Nebilyer Valley, Western Highlands	2570
Teptep area, Morobe	2560
Minyamb Valley, Enga	2500
Mendi Valley, Southern Highlands	2490
Gumine area, Simbu	2460
Margarima area, Southern Highlands	2460
Sinasina area, Simbu	2370
Asaro Valley, Eastern Highlands	2300
Porgera area, Enga[3]	2300
Bundi area, Madang	2200
Paiela area, Enga	2200

1. Houses and cleared land (and presumably food gardens) are shown at somewhat higher altitudes on the 1:100 000 map sheets. For example, they occur as high as 2640 m in the Margarima area in Southern Highlands Province. The highest dwelling and cleared land is shown at about 2900 m on the divide between the Lai and Ambum valleys in Enga Province.
2. The highest sweet-potato gardens recorded in Papua (West New Guinea) are at Kwiyawagi (4° 04' S, 138° 10' E) at c. 2800 m in the West Baliem Valley. There are higher gardens in the East Baliem Valley at Iniuni at c. 2900 m, where cabbage and *Solanum* potato, but not sweet potato, are grown (G. Hope, pers comm. 2009).
3. The highest food gardens observed in the Porgera area in Enga Province in 1980-1981 were at 2300 m. By 1993, sweet-potato gardens were being planted as high as 2650 m (G. Banks, pers comm. 1993).

Based on observations in the early 1960s, Brookfield (1964:30-31) reported on the limits to cultivation in several parts of the highlands. After correcting for systematic underestimation of altitudes, the figures reported by Brookfield are similar to the upper limits that I observed. For example, Brookfield noted abandoned gardens at 9000 feet (2880 m corrected figure) in the Chimbu Valley and upper Wage Valley (Wage-Kandep uplands) of Enga Province. This is similar to my observations in the early 1980s from these locations (Table 7).

In many parts of the highlands, two main types of food gardens may be distinguished. These are sweet-potato gardens and mixed gardens. The latter contain numerous species of food crops planted in a mixed arrangement. Sweet potato is unusual in these gardens. Cropping intensity is low and they are located on more fertile sites. I made 12 observations on the usual upper limit of mixed gardens. The mean was 2050 m (n = 12, SD = 130 m). The highest garden seen was in the Minyamb Valley at 2480 m. Some mixed gardens are also planted as high as 2500 m in the Chimbu Valley (Wohlt and Goie 1986:35-39). The mean upper limit of this garden type coincides with the upper limit of most of the supplementary food crops that are planted in mixed gardens (Figure 3).

The upper altitudinal limit of *Colocasia* taro is of particular interest because it was probably the main staple food for high-altitude dwellers before the adoption of sweet potato about 300 years ago. The mean usual upper limit for all taro planting was 2400 m (Table 1). The mean upper limit for monospecific taro planting is lower, at 2250 m. Thus, the mean upper limit of arable agriculture before the introduction of sweet potato into the highlands was probably about 2200 m, assuming climatic conditions were similar to those prevailing today. Triploid banana cultivars had a mean usual limit almost as high, at 2150 m, in the early 1980s (Table 1). As banana is likely to have been an important supplementary food to taro, this reinforces the suggestion that the upper limit for agriculture before the adoption of sweet potato was about 2200 m.

Variation between locations

These data on crop altitudinal limits confirm earlier studies on temperature and vegetation boundaries (McAlpine et al. 1983:94) that indicate that the mass mountain heating effect operates in PNG. This is the phenomenon, also known as Massenerhebung, whereby large mountains and the central parts of large ranges are warmer at a given elevation than small mountains and outlying spurs (Van Steenis 1961; Whitmore 1984:253).

Data on the usual altitudinal limits for 12 crop species at 10 locations in the central highlands between 1500 m and 2500 m are presented in Table 8. The usual upper limits for five crop species at 15 locations between 700 m and 1400 m are given in Table 9. The same data for selected species and locations are given in Figures 4 and 5. Within the central highlands, the usual limits for these species do not vary greatly among locations. Upper limits are lower for most species at the two locations in Southern Highlands Province. The same phenomenon occurs throughout that province and probably reflects high cloud cover and infertile soils.[7] The upper limits for most species are particularly high in the upper Chimbu Valley (Table 8) and many of the extreme upper altitudinal figures in Tables 1-6 were observed in this valley.

7 This suggests that there is an interaction between certain environmental factors, such as temperature and soil fertility, with a good value in one factor able to compensate to some degree for a poor value in another.

Table 8. The usual altitudinal limit of 12 crops at 10 locations between 1600 m and 2500 m in PNG[1]

Scientific name	Common name	PNG mean[2]	NAN	KTU	ASA	SIN	GUM	U/S	HGN	LAI	MEN	NEM
Arachis hypogaea	Peanut	1840	1750	1940	1880	1760	1770	1920	1800	1870	1810	1810
Carica papaya	Pawpaw	1700	1690	1730	1730	1790	1770	1790	1740	1780	1790	1750
Coffea arabica	Arabica coffee	2070	2280	2080	2050	2060	2170	2390	2080	2150	2030	1990
Manihot esculenta	Cassava	1820	1800	1830	2030	1950	2010	2000	1990	1900	1880	1700
Musa cvs	Banana	2160	2370	2120	2010	2260	2310	2460	2280	2220	2260	2280
Pandanus conoideus	*Marita* pandanus	1720	1800	1670	1780	1870	1830	1940	1820	1780	1640	1670
Pandanus jiulianettii	*Karuka* pandanus	1780	2020	1690	1850	1730	1870	1960	1820	1860	1700	1820
Persea americana	Avocado	2060	2280	2090	1900	1960	2190	2430	2010	2260	2030	1920
Psidium guajava	Guava	1840	1800	1780	1730	1940	2020	1930	1820	1900	1790	1700
Psophocarpus tetragonolobus	Winged bean	1910	2050	2040	2020	1870	2020	2050	2000	1910	1810	1810
Saccharum edule	Lowland *pitpit*	1810	1750	1840	1670	1920	2120	1920	1910	1870	1790	1710
Xanthosoma sagittifolium	Chinese taro	2020	2040	2150	2050	1970	2060	2130	2130	2000	2030	1980
Mean		1890	1970	1910	1890	1920	2010	2080	1950	1960	1880	1850

1. Locations are: Nankina Valley and Teptep area, Huon Peninsula, Madang and Morobe provinces; Kainantu area, Eastern Highlands Province; Asaro Valley, Eastern Highlands Province; Sinasina area, Simbu Province; Gumine area, Simbu Province; upper Chimbu Valley, Simbu Province; Mount Hagen and Bukapena areas, Western Highlands Province; Lai Valley, Enga Province; Mendi Valley, Southern Highlands Province; Nembi Plateau and Nipa area, Southern Highlands Province.
2. The PNG mean is the mean usual limit, with means rounded to 10 m, not 50 m as in Tables 1 to 6. Figures for *Pandanus jiulianettii* are the usual lower limit. Figures for the other species are the usual upper limit.

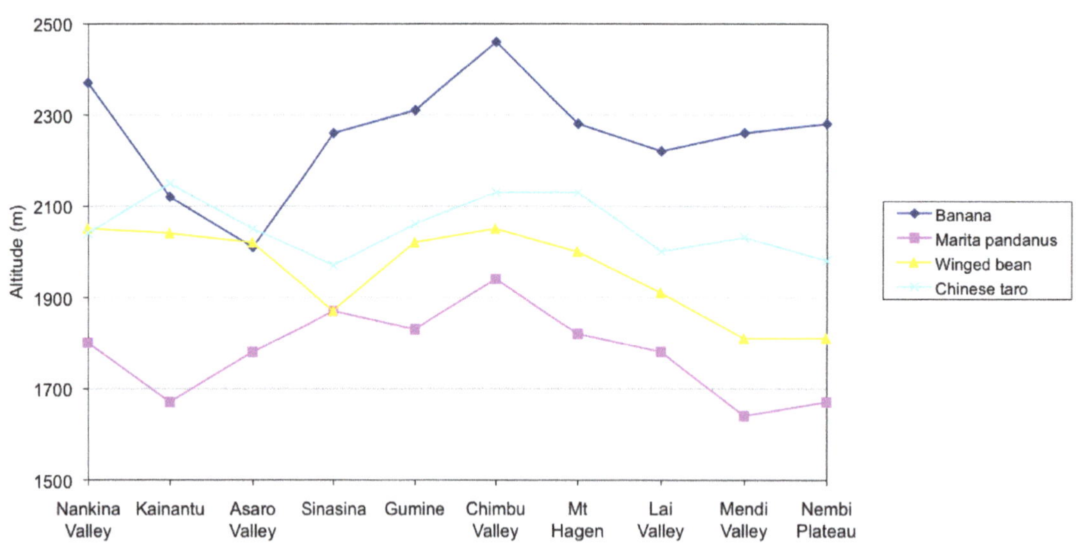

Figure 4. The usual upper altitudinal limits of four food crops at 10 locations in PNG

Table 9. The usual upper altitudinal limit of five crops at 15 locations between 700 m and 1400 m in PNG

Scientific name	Areca catechu	Artocarpus altilis	Cocos nucifera	Pangium edule	Terminalia kaernbachii
Common name	Betel nut	Breadfruit	Coconut	*Sis*	*Okari*
PNG mean[1]	1080	1240	960	1060	1080
Bougainville Island	–	–	730	–	–
Baining Mts, ENB	840	–	750	–	–
Lelet Plateau, NIP[2]	960	–	<960	–	–
Managalas Plateau, ORO	940	–	920	890	980
Kokoda Trail, CEN	–	970	830	970	–
Pindiu area, MOR	1190	1300	1190	–	1020
Kabwum area, MOR	1180	1180	970	–	–
Snake Valley, MOR	1000	1020	1020	–	950
Wau area, MOR	1160	1210	1160	–	1000
Slate Creek Valley, MOR	1170	1280	1130	1230	1260
Tauri Valley, GUL	930	1330	<1200	1180	1130
Tauta, MAD	1110	1110	1010	1040	<1060
Bundi area, MAD	1390	1390	1250	1380	–
Erave area, SHP	1080	1250	1080	–	1110
Elip Valley, SAN	1270	1310	<1270	–	–

1. The PNG mean is the mean usual upper limit given in Tables 1, 5 and 6, rounded to 10 m, not 50 m.
2. Data from the Lelet Plateau differ from those from other locations because observations there are only from the narrow altitudinal range where hamlets and food gardens are located (950-970 m). Thus, betel nut bears at 960 m on Lelet Plateau, but it would possibly bear if planted at a higher altitude. Coconut bears only infrequently at 960 m there, thus the usual altitudinal limit is given as less than 960 m. For the other locations, observations were made over a range of altitudes greater than the local limits for these crops. However, settlement and food gardens are less dense here than in the central highlands of PNG and this is why the limits for any locality for these crops often tend to be grouped in a more restricted range in this dataset compared with locations in Table 8.

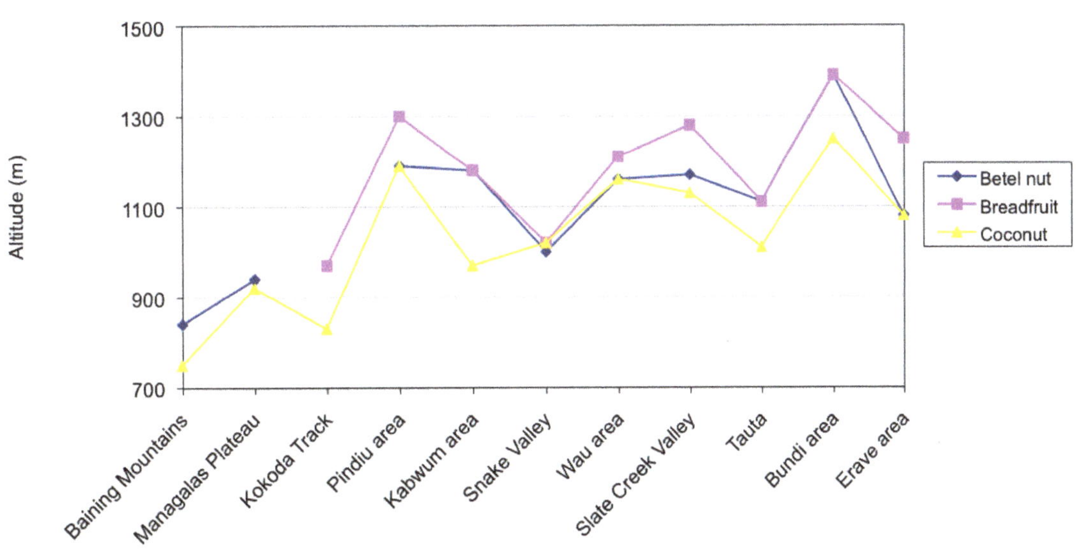

Figure 5. The usual upper altitudinal limits of three crops at 11 locations in PNG

There is a greater difference in usual upper altitudinal limits among locations on smaller island ranges and the main ranges of New Guinea (Table 9, Figure 5). On the small islands of Bougainville, New Britain and New Ireland, the limits are lower than on the large island of New Guinea. Limits are also lower on the Managalas Plateau and Kokoda Trail. This probably reflects the fact that these two places are further south (about 9° S) and thus experience somewhat lower temperatures in the middle of the year.

The comparison between altitudinal limits in the Snake Valley, Wau area and Slate Creek Valley illustrates microclimatic differences. These locations are close to one another (Figure 1). The Slate Creek Valley and Wau area are broad valleys, with predominantly grasslands on the valley floor. In contrast, Snake Valley is narrower and the dominant vegetation is forest. The upper altitudinal limits for crops are generally lower in Snake Valley than in the other two (Table 9, Figure 5). Presumably, the broader grassland valleys have a slightly higher maximum temperature than the narrower forested valley and this is reflected in the crop limits.

Influence of cultivars on species limits

Data presented here are generally at the species level. It is implicit that these observations are for the cultivars presently grown in PNG. However, other cultivars may have somewhat different limits, as it is well established that cultivars do not have the same tolerance of temperature extremes. See, for example, Laing et al. (1983:243) on common bean (*Phaseolus vulgaris*) and Wu et al. (1974) on sweet potato. Data presented here give some indication of cultivar-temperature interactions in PNG. For example, triploid bananas grow to a mean of 350 m higher than diploid cultivars (Table 1).

Recognition of altitudinal divisions

For a number of locations in the highlands and highlands fringe, it has been documented that villagers distinguish with separate terms 'warmer' and 'cooler' zones, based in part on crop altitudinal limits. Some examples of these breaks are: in the Karimui area of Simbu Province, where people recognise a break at about 1100 m (Hide et al. 1984:17, 211); in the Bundi area on the other side of the main ranges, where a break is known at 1450 m (author's field work, 1984); in the Marigl Valley of Simbu Province, where people speak of a break at about 1850 m (author's field work, 1988); in the Lai Valley in Enga Province, where villagers recognise an ecological break also at 1850 m (Meggitt 1958:314); and in the Sinasina area of Simbu Province, where the villagers note an ecological break at about 2100 m (Hide et al. 1979:5).

It is likely that this perception of a warmer and cooler zone dichotomy based on crop performance and altitudinal limits is widespread within the PNG highlands and that the actual altitude of the ecological break varies between locations. Such breaks are probably more commonly recognised where a number of important species have usual upper or lower limits in a narrow range, particularly at about 1800 m (Figure 3).

Crop altitudinal limits in PNG and some other equatorial countries

It is difficult to compare altitudinal limits between an equatorial location, such as PNG, and a non-equatorial location because of the confounding influence of latitude on temperature and crop growth. However, cautious comparisons are possible with other equatorial locations; for example, maize and sweet potato (Table 10) and a number of crop species that are common to both PNG and Indonesia (Table 11).[8]

8 Sources for data from Indonesia do not indicate where the data comes from, but it is almost certainly not from Papua (Indonesian New Guinea).

Table 10. The usual and extreme upper altitudinal limits of maize and sweet potato in a number of countries

Country	Maize		Sweet potato		Source
	Usual	Extreme	Usual	Extreme	
Kenya[1]	1850	2450	1850	–	Morgan (1968:274)
Rwanda	2000	–	–	–	Ford (1990:45)
Ethiopia	2100	2400	–	–	Alkämper (1973:30)
East Africa[2]	–	–	2100	2400	Jana (1982:68)
Papua New Guinea	2450	2700	2700	2850	This paper
Peru	3100	>3100	–	–	Brush (1976:158)
Ecuador	3200	–	–	–	Stadel (1986:57)
Bolivia[1,3]	2450	3650	2750	>2750	Yen (1974:151)
Tibet Plateau, China	3200	–	–	–	Yu and Sun (1981)

1. Figures given by Morgan (1968) and Yen (1974) were in thousands of feet.
2. Jana's discussion covers Kenya, Tanzania and Uganda.
3. The extreme altitudinal limit for maize given by Yen (1974) is for Bolivia and Peru. His extreme limit for sweet potato is for Bolivia and Ecuador.

Table 11. The usual altitudinal range of eight crops in Indonesia and PNG

Scientific name	Common name	Usual altitudinal range (m)		Indonesian source
		Indonesia	PNG	
Dioscorea alata	Greater yam	0–800	0–1900	IBPGR (1981a:53) Ochse and Brink (1977:231)
Artocarpus altilis	Breadfruit	0–1000	0–1250	IBPGR (1980:21)
Psidium guajava	Guava	0–1200	0–1850	IBPGR (1980:109)
Leucaena leucocephala	Leucaena	0–1500	0–1800	Ochse and Brink (1977:393)
Psophocarpus tetragonolobus	Winged bean	0–1500	0–1900	IBPGR (1981b:89)
Cyphomandra betacea	Tree tomato	250–1700	1050–2300	IPBGR (1980:53)
Zingiber officinale	Ginger	0–1700	0–1950	IBPGR (1981a:101)
Vicia faba	Broad bean	>1300	2050–2650	IBPGR (1981b:111)

For maize, sweet potato and some other crops for which data are available (such as cassava), altitudinal limits are generally higher in PNG than in East African locations. The PNG limits are generally lower than those in the Andes of South America (Table 10). Limits are generally much higher in PNG than in Indonesia (Table 11). These differences are again a likely reflection of the mass mountain heating effect. The mountains of the Andes are more massive than the New Guinea mountains, which in turn are larger than the ranges of the Indonesian islands. This is consistent with observations from within PNG, where altitudinal limits are higher on the island of New Guinea than on the smaller islands of Bougainville, New Britain and New Ireland (Table 9).

Impact of climate change on crop limits

Most observations reported here were made in the early 1980s. Thus, they provide baseline data on crop growth that can be used in the future to assess the impact of increasing temperature associated with global climate change.

An analysis of changes in temperatures up to 1999 was made for nine coastal stations by Bourke et al. (2002). They found that maximum, mean and minimum temperatures increased in the PNG lowlands by a mean of 0.6°C between 1970 and 1999; that is 0.2°C per decade. Datasets that covered a longer period, from 1904 in the case of Port Moresby, indicated that increases in temperature had been greater from the mid 1970s onwards, which is consistent with the global pattern. The only highland station for which good quality long-term data were available was Aiyura (1640 m) in Eastern Highlands Province. There, the maximum temperature increased by 0.75°C over the period 1977-2001, a rate of 0.3°C per decade.

In the period 1980-1982, coconut palms grew at altitudes as high as 1760 m, but the highest palms that bore nuts were at 1310 m. By 1999, coconut palms were bearing as high as 1450 m.[9] The increase in the altitude at which coconuts bore nuts was 140 m over a 20-year period, which suggests that temperatures increased by about 0.7°C over this period. This is consistent with the recorded increase in temperature at Aiyura between 1977 and 2001. By mid-2009, coconut palms were bearing small nuts near Goroka, at 1560 m. This was 110 m higher than the highest recorded bearing palms 10 years earlier. Presuming this difference was caused by increasing temperatures, and this is the most likely explanation, this implies a temperature increase of 0.57°C in 10 years – an extremely high rate of increase. This suggestion needs to be checked against long-term temperature data, which are available from Aiyura.

The recorded increase in the altitude at which coconut palms bear is also consistent with observations by many villagers that coconut, betel nut, breadfruit and other tree crops were bearing at higher altitudes in the 1990s and 2000s than in the 1970s and 1980s. At Nokopo village on the Huon Peninsula in Morobe Province, people claim that the climate warmed during the 1980s. They cite the facts that they could plant *marita* pandanus (*Pandanus conoideus* complex) in this period for the first time; that certain species of birds occupied higher-altitude locations; and that there had been an increase in altitude at which a certain palm could be grown (Kocher Schmid 1991:27).

Bananas also provide evidence of the impact of increasing temperature in the highlands. Up to the end of 1978, bananas were not grown at Tambul (2300 m) in Western Highlands Province. From the mid-1980s onwards, people started to plant bananas, which bore fruit in that area (M. Gunter pers comm. 2001).

These limited observations indicate increasing temperature in the New Guinea highlands. As temperatures continue to increase in the coming decades, the observations reported here, particularly the detailed observations at named locations, can be used to assess the rate of increase (Tables 8 and 9).

Acknowledgements

Most of the observations reported here were made while I was employed by the PNG Department of Primary Industry (1978-1983). Some were made while I was supported by the Department of Human Geography in the Research School of Pacific Studies at the Australian National University. This Department provided support for compilation of the work.

Many villagers shared their often extensive knowledge of where plants grow. A number of agriculturalists and other observers also gave information on the distribution of crop plants, particularly Bruce French, Robin Hide and Tevo Tarepe. Others who assisted were Will Akus,

9 See Footnote 3, Table 5 for more detail.

Chris Ballard, Vern Berley, Bruce Carrad, Euclid D'Souza, Geoff Hope, Ger Reesink, Jack Simpson, Clement Tumana and Steve Woodhouse. David Evesson of the National Weather Bureau in Port Moresby advised on air pressure in PNG. Staff of the Division of Botany in Lae gave access to altitude data in their collection records and were unfailingly helpful, particularly Jim Croft, Ted Henty and Karl Kerenga, in identifying plants. Diseased specimens were identified by plant pathologists of the Department of Primary Industry. Robin Hide and Geoff Humphreys commented on an earlier draft, and two referees offered helpful suggestions. Tracy Harwood and Sandra Davenport edited the paper. It is a pleasure to acknowledge the assistance of all those named above.

Appendix 1. Lapse rates in Papua New Guinea

Temperature in PNG declines with increasing altitude (McAlpine et al. 1983:92). The rate of decline is termed the lapse rate. A number of authors have calculated lapse rates for PNG. I have recalculated them here because more accurate information is now available on the altitude of weather stations. (Earlier figures underestimated true altitudes by up to 100 m.) As well, more long-term data has became available for some stations, particularly in the altitude range of 2000-3000 m.

Previously published information on lapse rates for PNG is firstly summarised. Fitzpatrick (1965) and McAlpine (1970) quoted a general lapse rate of 5.5°C per 1000 m. McAlpine et al. (1983:91-95) gave a lapse rate of 6.7°C for mean maximum temperature and 5.4°C for mean minimum temperature for stations above 500 m altitude. The altitudes used by these authors were generally too low. Humphreys (1981a), using corrected altitude figures, calculated 36 lapse rates for non-coastal locations of mainland PNG. In a later publication, he gave lapse rates for a narrow strip that covers Simbu Province (Humphreys 1984:26-30). Researchers at the Centre for Resource and Environmental Studies at the Australian National University devised a computer program that generates lapse rate surfaces for all of the island of New Guinea. This provides estimates of maximum and minimum temperatures as a function of latitude, longitude and altitude (H. Nix, pers comm. 1988).

Humphreys (1981b) calculated a theoretical lapse rate for soil at 50 cm depth in PNG. Dronia (1983) recorded soil temperatures at various depths and altitudes in three regions of PNG. He gave the lapse rate at a depth of 50 cm as 4.6°C per 1000 m between sea level and 2600 m and commented that this value is probably one of the lowest soil lapse rates on earth.

Temperature data covering a minimum of three years are available from 31 non-coastal locations in PNG. All except Panguna are located in the eastern half of the island of New Guinea and are a minimum of 20 km from the ocean. The 30 stations on New Guinea lie between latitudes of 3°30' S and 8°40' S and between longitudes of 141°40' E and 148°20' E. Mean temperatures for these non-coastal stations are presented in Table A1.

The relationship between altitude and mean maximum/minimum temperature is shown in Figure A1. The moderating influence of the ocean is apparent for coastal locations. The mean maximum temperatures for coastal locations are less than those for non-coastal low-altitude locations. The mean minimum temperatures are greater than those for non-coastal low-altitude locations. For non-coastal locations, there is a regular linear decline in temperature with increasing altitude. This decline in temperature may be expressed as a series of equations, as follows:

$$Y_{max.} = 32.67 - 0.0052\,x \qquad r = 0.958^{***}$$

$$Y_{ann.} = 27.32 - 0.0052\,x \qquad r = 0.983^{***}$$

$$Y_{min.} = 22.08 - 0.0052\,x \qquad r = 0.979^{***}$$

where: $Y_{max.}$, $Y_{ann.}$ and $Y_{min.}$ are mean maximum, mean annual and mean minimum air temperatures (°C) respectively (31 locations)
x is the altitude in metres
r is the regression coefficient.

All three regressions are highly significant ($P<0.001$) and explain between 92 and 97 per cent of the variation. These regressions are similar to those derived by Humphreys (1981a). The lapse rates are 5.2°C per 1000 m. This is somewhat lower than many other locations on the globe, as the mean global lapse rate is about 6°C per 1000 m (Giddings 1980:17).

For the two non-coastal stations located in either coastal ranges (Lumi in Sandaun Province) or on an island range (Panguna on Bougainville Island), the mean maximum temperature is lower than for locations on the fringe of the central highlands of New Guinea (Figure A1). McAlpine et al. (1983:94) suggested that data from these two locations indicate that maximum (but not minimum) temperatures are depressed in isolated coastal and island ranges. However, observations on crop altitudinal limits from locations on the northern side of the mountainous Huon Peninsula, including the Nankina Valley and Teptep area, suggest that temperatures are high for the altitude in those locations (Table 8). Thus the observations from the Huon Peninsula contradict the hypothesis of McAlpine et al. (1983). Observations made on crop altitudinal limits in island ranges on New Britain, New Ireland and Bougainville islands indicate that temperatures are lower at any given altitude than on the island of New Guinea, and this is consistent with the temperature recordings from Panguna (Table 9).

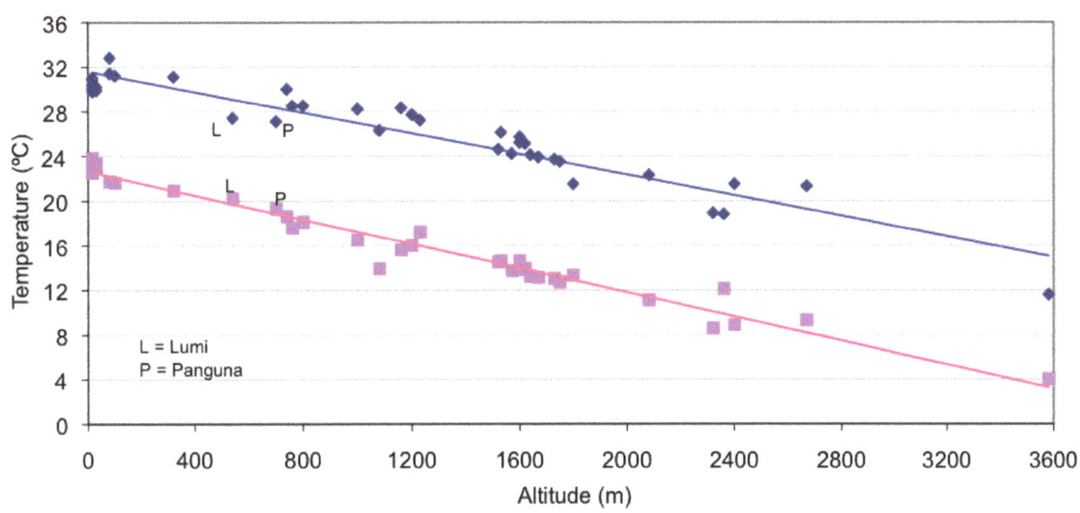

Figure A1. Mean maximum and minimum temperatures for seven coastal and 31 non-coastal locations in PNG, by altitude

Appendix 2. The influence of altitude on crop development and yield in Papua New Guinea

As well as setting limits to crop growth, altitude (that is, temperature) influences the rate of crop development and yield. There is very little information from PNG about the relationship between altitude and crop development or yield. Detailed studies, such as that done by Cooper (1979) which examined the influence of altitude on maize growth in Kenya, and that done by Laing et al. (1983) on the effect of altitude on *Phaseolus* beans in Colombia, have yet to be done in PNG.

It is widely recognised within PNG that crops require more time to attain maturity in the cooler highlands than in lowland locations. It is not generally recognised that the final crop yield per unit area is commonly higher in the highlands than in the lowlands. For example, an experimental yield for sweet potato of 20 tonnes per hectare (t/ha) would be considered high in the lowlands. In contrast, experimental yields at highland locations of 20-30 t/ha are not uncommon. This occurs because there is a longer period available for the plant to accumulate carbohydrate during its growth cycle in the highlands.

In this appendix, the available information on the influence of altitude on crop development and yield is brought together and summarised. Information is provided for the following crop species:

Arabica coffee	*Coffea arabica*
Marita pandanus	*Pandanus conoideus*
Potato	*Solanum tuberosum*
Pyrethrum	*Chrysanthemum cinerariaefolium*
Sweet potato	*Ipomoea batatas*

Arabica coffee

Despite the importance of Arabica coffee to the PNG economy and the wide altitudinal range over which it is grown (700-2050 m), very little is known about the influence of altitude on its growth. It has been observed by some that the harvesting season starts later at higher altitudes. For example, Joughin and Thistleton (1987:23), following a rapid rural appraisal, state that this is the case in the Jimi Valley over an altitudinal range of about 1400-2000 m. In that location, coffee harvesting is reported to start in April-May in the lower part of the valley, in May-June in the middle valley, and in July-August at higher locations. Similarly, in the Asaro Valley west of Goroka, the main coffee flush is reported as July-August at 1600 m, but in August-September at 1800-2000 m (B. Carrad, pers comm. 1990).

Marita pandanus

Marita pandanus is grown from just above sea level to a mean upper limit of 1700 m in PNG (although it is generally not grown adjacent to the ocean) (Table 4). Information on fruiting seasons of *marita* pandanus at 25 different locations in PNG was assembled by Bourke et al. (2004:155, 198). These data are qualitative not quantitative, except for those from the four highest locations. As well, the timing and duration of the fruiting season varies somewhat between years. Despite these limitations, the available information indicates a strong negative linear relationship between the length of the fruiting season and altitude. Near sea level, the fruit are available year round. The cropping season starts later in the year and is shorter at higher altitudes. Towards the top of the crop altitudinal range, the fruiting season is limited to four months.

The relationship between the length of the fruiting season and altitude can be expressed as follows:

$$Y = 10.01 - 0.004x \qquad r = 0.6542^{**}$$

where: Y is the fruiting season in months
x is the altitude in metres
r is the regression coefficient.

The relationship is highly significant and explains 43 per cent of the variation. When two outlying data points are excluded, the relationship becomes much stronger and explains 89 per cent of the variation. See Figure 92 in Bourke et al. (2004) and its reproduction in a modified form in Figure 1.13.4 by Allen and Bourke (2009).

Potato

In Simbu Province, Goodbody (1982) compared the performance of six cultivars of *Solanum* potato at two altitudes. At 1500 m, the crops needed 81-93 days to mature, whereas at a higher-altitude location (2400 m), they required a longer period (101 days). For all six cultivars, mean tuber weight was greater at 1500 m than at 2400 m. However, for five of the six cultivars, more tubers developed per plant at the higher-altitude location. This resulted in a higher yield per hectare at the higher location.

Pyrethrum

By 1980, pyrethrum cultivation was restricted to locations above 2400 m, but when the crop was first introduced to PNG villagers in the late 1960s, it was grown as low as 1800 m. Data on yields from village plantings in the 1960s from five locations in Enga were presented by Quinlan (1968). The mean yield of flowers harvested per eight months has been plotted against corrected altitudes (Figure A2). This shows a striking positive relationship between flower yield and altitude. These data show that mean yields at 2700 m are about three times greater than those at 1900 m. This suggests that the contraction of pyrethrum plantings by villagers from the 1970s onwards to locations above 2400 m occurred because yields are unacceptably low below this altitude.

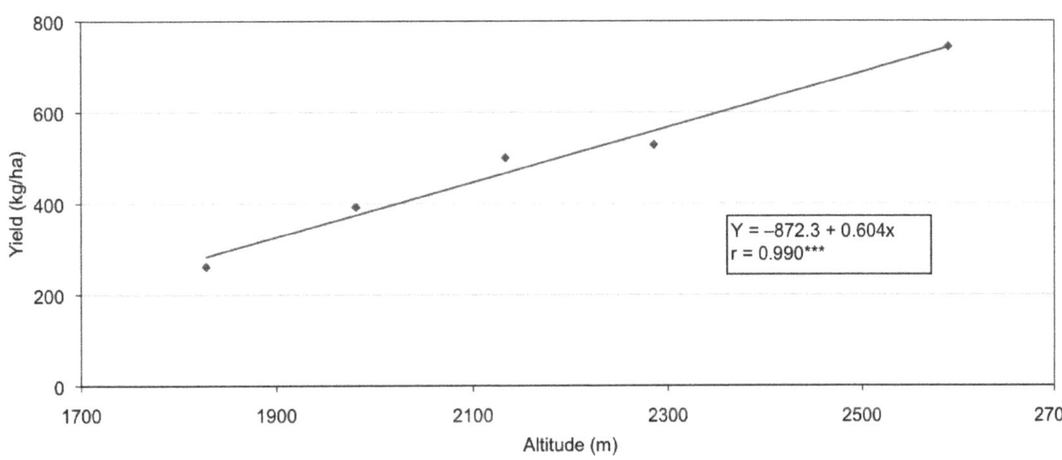

Figure A2. Yield of pyrethrum flowers over eight months in village plantings, Enga Province vs altitude (after Quinlan 1968)

The linear regression between pyrethrum flower yield and altitude is:

$$Y = -872.3 + 0.604x \qquad r = 0.990^{***}$$

where: Y is the yield of dried flowers in kg/ha
x is the altitude in metres
r is the regression coefficient.

The statistical relationship is very highly significant ($P < 0.001$) and explains 98 per cent of the variation.

Sweet potato

In Simbu Province, Goodbody (1984) recorded the time to first harvest and tuber yield from 72 village plantings of sweet potato over an altitudinal range of 1400 m to 2600 m. He found that the tuber yield was not related to altitude. However, the period to first harvest increased with increasing altitude. The relationship was as follows:

$$Y = -149 + 0.19x \qquad r = 0.86^{**}$$

where: Y is the number of days to first harvest
x is the altitude in metres
r is the regression coefficient.

The relationship suggests that sweet potato would mature at sea level (x = 0) at 149 days before it was planted. Clearly this is not true and the linear regression found by Goodbody does not hold over the entire altitudinal range of sweet potato in PNG (sea level to 2850 m).

Appendix 3. Recording the altitude of locations in Papua New Guinea

The purpose of this note is to explain how to record the altitude of locations within PNG with reasonable accuracy. The most convenient method for recording altitude is to use an altimeter. However, altimeters measure air pressure, not altitude. Hence, it is necessary to adjust the instrument before it is used. It is also helpful to make certain corrections to ensure that the reading taken from the instrument is as accurate as possible.

The altitudes given on older maps and on signposts for many locations in PNG are frequently incorrect. This is because the readings are based on uncorrected readings taken from aircraft altimeters.

Adjusting the instrument

Altimeters can be adjusted so that the instrument reading is the same as the altitude of a known point. This is known as 'zeroing'. The altitudes of locations in PNG can be obtained from trigometric points or spot heights marked on the 1:100 000 topographic maps. For example, the trigometric point at Kundiawa airstrip is at exactly 1525 m above sea level. So persons working out of Kundiawa should adjust their altimeters to 1525 m at the airstrip or the government offices in Kundiawa before leaving town. The altitudes of some urban areas in the highlands are given in Table A2.

Correcting the reading

A good quality altimeter gives a reasonably accurate measure of the altitude of a location, provided that it has been correctly adjusted at a point of known altitude. However, there are still other sources of error that may influence the reading. It is possible to make corrections for these so that the reading is accurate to within plus or minus 20 m of the actual altitude of a location.

There are three sources of error in altimeter readings:

- Pressure surface flows into mountains;
- The effect of daily variation in air temperature on air pressure;
- Rapid changes in air pressure, often associated with storms.

Pressure surface flows

Air is constantly being forced into the highlands. The effect of this is to increase air pressure slightly. Hence, altimeter recordings are somewhat lower than they would be if this was not occurring. So even if an altimeter is adjusted accurately at sea level, the instrument will give a slightly low reading when it is taken into the highlands. This is the reason that many altitudes quoted for highland locations are too low.

I have found that it is possible to correct for pressure surface flows using a simple formula. The formula is accurate at least between sea level and 3000 m. The correction factor follows.

For locations higher than where an instrument was adjusted, add 10 m to the reading for every 200 m you have risen above where the instrument was zeroed. For example, on a flight between Lae and Goroka, an altimeter is adjusted to zero at Lae airstrip. When the aircraft reaches the hangar in Goroka, the altimeter is reading 1520 m. The correction factor is:

$$\frac{(1520 - 0) \times 10}{200} = 80 \text{ m}$$

So add 80 m to the reading of 1520 m to obtain the corrected reading of 1600 m. (The correction factor is rounded to the nearest 10 m.)

For places lower than where the instrument was adjusted, subtract 10 m from the reading for every 200 m you have descended since the instrument was zeroed. For example, if you zero your instrument at 1740 m on Mendi airstrip and then fly by helicopter into a village where the instrument reads 1320 m, the correction factor is:

$$\frac{(1740 - 1320) \times 10}{200} = 20 \text{ m}$$

So the corrected reading is 1320 – 20 = 1300 m.

When travelling between locations of known altitude over short periods, I have found that use of this correction factor gives a corrected altimeter reading that is generally accurate to within plus or minus 10 m of the true altitude.

Daily variation in air temperature

Changes in air temperature also influence air pressure and hence altimeter readings. Because temperature varies during the day, there is generally a daily cycle in instrument recordings. If an altimeter is kept in one location and read at hourly intervals, the pattern will be similar to that shown in Figure A3. (These recordings were made in Kundiawa at 1550 m. The instrument was not moved during the day.) During the early morning, there is little change in the instrument readings. Between late morning and early afternoon the reading rises fairly rapidly. Typically, the greatest change occurs between about 11 am and 2 pm. The reading is fairly steady between mid and late afternoon. It then declines steadily until late at night. The instrument reading increases and decreases by 20 m to 30 m during a typical day in the highlands.

It is possible to adjust for this source of error by adjusting the instrument at a location of known altitude. This is best done in the early afternoon when the effects of changes in air temperature have stabilised. In practice, it is often not possible to adjust the instrument during field work because the exact altitude of locations will not be known. However, it is still possible to make a crude adjustment.

To do this, record the time of day that any altitude recordings are made. Also record the instrument reading when you return to your base at the end of the day. The reading will typically be 20-30 m greater than when you adjusted the instrument at the base in the morning. Assume that this increase occurred evenly between 11 am and 1-2 pm and adjust the field recordings accordingly. For example, you set your instrument to 2080 m at Wabag at 8 am. When you return to Wabag at 4 pm it is reading 2110 m; that is, the reading is 30 m higher. So deduct

10 m from any recordings made between 11 am and noon, deduct 20 m from any recordings made between noon and 1 pm and deduct 30 m from any recordings made after 1 pm.

Rapid changes in air pressure

In general, air pressure is quite stable in PNG. However, large changes in air pressure sometimes occur, often associated with low-pressure fronts and storms. These rapid changes in air pressure will markedly affect an altimeter reading. Just before a storm there is sometimes a rapid drop in air pressure and this results in an increase in the instrument reading. There is no simple correction for this unless one maintains a base recording station. In practice, however, sudden changes in air pressure and hence in altimeter readings are not common in PNG. But these changes in air pressure highlight the need for instruments to be adjusted correctly at a point where the altitude is known before they can be used to record the altitude of other places.

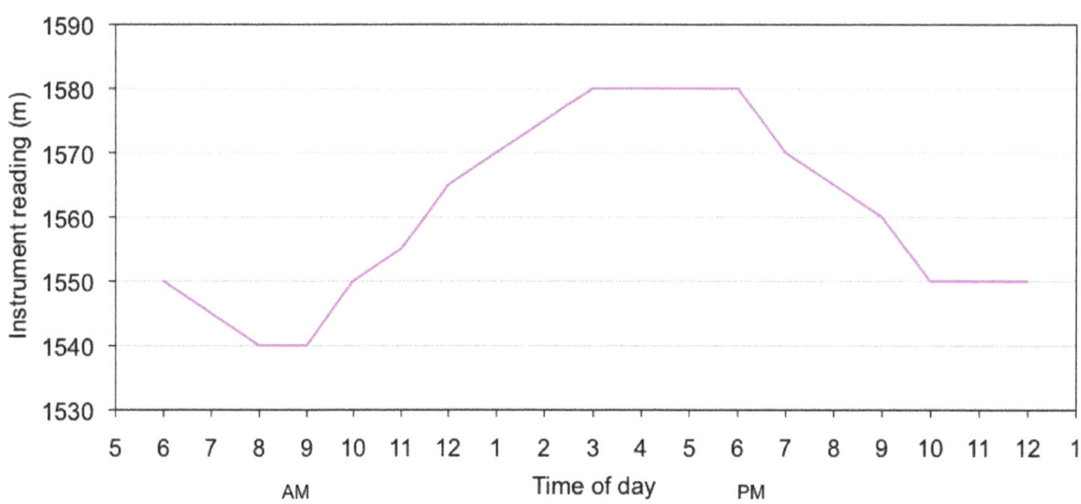

Figure A3. A typical pattern for hourly altimeter readings for a highland location during the course of the day (where the instrument is not moved)

Conclusion

An altimeter can be used in PNG to record the altitude of locations usually within an accuracy of plus or minus 20 m of the actual altitude. It is necessary for the instrument to be adjusted before use at a location where the altitude is known accurately.

The recordings obtained are highly accurate only when taken soon after adjusting and only at similar altitudes to the point where the instrument was adjusted. When the instrument is used several hours after adjustment and more than 200 m higher or lower than the point of adjustment, it is necessary to make certain corrections to allow for this. When these corrections have been made, recordings are generally within plus or minus 20 m of a location's actual altitude.

Table A1. Mean annual temperatures for non-coastal stations in PNG

Station[1]	Altitude (m asl)	Annual temperature (°C)			Source[2]
		Mean max.	Mean annual	Mean min.	
Aiome	80	31.4	26.5	21.7	1
Erap	80	32.8	27.2	21.7	1
Popondetta	100	31.2	26.4	21.6	1
Kaiapit	320	31.1	26.0	20.9	1
Lumi	540	27.4	23.8	20.2	1
Panguna	700	27.1	23.2	19.3	2
Bulolo Forestry College	740	30.0	24.3	18.6	1
Garaina	760	28.5	23.0	17.6	1
Lake Kutubu	800	28.5	23.3	18.1	1
Wau Forestry	1000	28.2	22.3	16.5	1
Erave	1080	26.3	20.1	13.9	1
Baiyer River	1160	28.3	21.9	15.6	1
Menyamya	1200	27.7	21.8	16.0	1
Wau Ecology Institute	1230	27.2	21.1	17.2	3
Telefomin	1520	24.6	19.6	14.5	1
Kundiawa	1530	26.1	20.4	14.6	1
Kainantu	1570	24.2	19.0	13.7	1
Goroka	1600	25.7	20.1	14.6	1
Minj	1600	25.2	19.5	13.8	1
Kagamuga	1620	25.1	19.5	13.9	1
Aiyura	1640	24.1	18.6	13.2	1
Tari	1670	23.9	18.5	13.1	1
Mount Hagen	1730	23.7	18.3	13.0	1
Mendi	1750	23.5	18.1	12.7	1
Mount Kaindi	1800	21.5	16.4	13.3	3
Wabag	2080	22.3	16.7	11.1	1
Tambul	2320	18.9	13.8	8.6	4
Mount Kaindi	2360	18.8	14.7	12.1	3
Toromambuno	2400	21.5	15.2	8.9	5
Yumbisa (Wage V, Enga)	2670	21.3	15.4	9.3	6
Mount Wilhelm	3580	11.6	7.8	4.0	7

1 The altitudes of stations were obtained from the 1:100 000 map sheets or adjusted altimeter recordings made by the author.

2 Sources used for temperature data are as follows:

1. McAlpine, Keig and Short (1975)
2. N.A. Marshman, Bougainville Copper Ltd (pers comm. 1982)
3. Gressitt and Nadkarni (1978:8)
4. A. Menge-Nang (pers comm. 1982)
5. Humphreys (1984:104)
6. Wohlt and Humphreys (1981)
7. Hnatiuk, Smith and McVean (1976)

Table A2. Altitudes of some urban areas and airstrips in the PNG highlands

Location[1]	Altitude (m)
Aiyura airstrip	1640
Goroka airstrip	1600
Kagamuga airstrip (Mount Hagen)	1620
Kainantu airstrip	1570
Kandep airstrip	2360
Kundiawa airstrip	1525
Mendi airstrip	1740
Tambul (DPI station)	2320
Tari airstrip	1670
Wabag airstrip	2080
Wau (Katherine Lehman School)	1000

[1] The altitudes given for airstrips are for the higher end for all locations. Some of these airstrips are no longer in use.

References

Alkämper, J. 1973. *Die Getreideproduktion Athiopiens: Stand, Leistung und Möglichkeiten für eine bessere Versorgung der Bevölkerung.* Abteilung Pflanzenbau und Pflanzenzüchtung in den Tropen und Subtropen. Gie⊠en: Tropeninstitut der Justus Liebig-Universität.

Allen, B.J. 1982. Subsistence agriculture: three case studies. In B. Carrad, D.A.M. Lea and K.K. Talyaga (eds) *Enga: Foundations for Development: Volume 3,* pp93-127. Armidale: University of New England.

Allen, B.J. and R.M. Bourke 2009. Part 1. People, Land and Environment. In R.M. Bourke and T. Harwood (eds) *Food and Agriculture in Papua New Guinea,* pp27-127. Canberra: The Australian National University.

Bayliss-Smith, T. 1985. Pre-Ipomoean agriculture in the New Guinea highlands above 2000 metres: some experimental data on taro cultivation. In I.S. Farrington (ed.) *Prehistoric Intensive Agriculture in the Tropics,* pp285-320. Oxford: British Archaeological Reports International Series 232.

Bayliss-Smith, T. 1988. Prehistoric agriculture in the New Guinea highlands: problems in defining the altitudinal limits to growth. In J.L. Bintliff, D.A. Davidson and E.G. Grant (eds) *Conceptual Issues in Environmental Archaeology,* pp153-160. Edinburgh: Edinburgh University Press.

Bourke, R.M. 1984. The altitudinal range of coffee and some associated shade crops in Papua New Guinea. *Coffee Industry Board Research Newsletter* 3(1):7-12. Goroka: Coffee Industry Board.

Bourke, R.M. 1988. Gumine Land Use Workshop: Consultant's Report. Kundiawa. South Simbu Rural Development Project.

Bourke, R.M. 2009a. An overview of edible fruit and nuts in Papua New Guinea. *Proceedings of Papua New Guinea Fruit and Nut Workshop.* NARI Conference Proceedings No 9. Lae: National Agricultural Research Institute.

Bourke, R.M. 2009b. History of agriculture in Papua New Guinea. In R.M. Bourke and T. Harwood (eds) *Food and Agriculture in Papua New Guinea,* pp10-26. Canberra: ANU E Press, The Australian National University. http://epress.anu.edu.au/food_agriculture_citation.html

Bourke, R.M. In press. Environmental influences on food crop distribution and production in Papua New Guinea. In J. Golson, T. Denham, P. Swadling and J. Muke (eds) *Ten Thousand Years of Gardening in the New Guinea Highlands.*

Bourke, R.M. and D.A.M. Lea 1982. Subsistence horticulture. In B. Carrad, D.A.M. Lea and K.K. Talyaga (eds) *Enga: Foundations for Development: Volume 3*, pp76-92. Armidale: University of New England.

Bourke, R.M. and T.N. Tarepe 1982. Locations for commercial citrus production in Papua New Guinea. *Harvest* 8(4):147-155. Port Moresby: Department of Primary Industry.

Bourke, R.M., G.S. Humphreys and M. Hart 2002. Warming in Papua New Guinea: some implications for food productivity. Unpublished paper. Canberra: Department of Human Geography, Research School of Pacific and Asian Studies, The Australian National University.

Bourke, R.M., C. Camarotto, E.J. D'Souza, K. Nema, T.N. Tarepe and S. Woodhouse 2004. *Production Patterns of 180 Economic Crops in Papua New Guinea.* Canberra: Coombs Academic Publishing, The Australian National University, 213pp.

Bowers, N. 1968. The ascending grasslands: an anthropological study of ecological succession in a high mountain valley of New Guinea. Unpublished PhD thesis. New York: Columbia University.

Brookfield, H.C. 1964. The ecology of Highland settlement: some suggestions. *American Anthropologist* 66(4):20-38.

Brookfield, H. 1991. Research in the mountains of the island of New Guinea. *Mountain Research and Development* 11(3):203-211.

Brookfield, H.C. with D. Hart 1971. *Melanesia: A Geographical Interpretation of an Island World.* London: Methuen.

Brush, S.B. 1976. Man's use of an Andean ecosystem. *Human Ecology* 4(2):147-166.

Bull, P.B. and R.M. Bourke 1983. Growing introduced vegetables in the lowlands. [Articles 2 and 3]. *Harvest* 9(2):68-79. Port Moresby: Department of Primary Industry.

Bulmer, R.N.H. 1960. Leadership and social structure among the Kyaka people of the Western Highlands District of New Guinea. Unpublished PhD thesis. Canberra: Australian National University.

Cooper, P.J.M. 1979. The association between altitude, environmental variables, maize growth and yields in Kenya. *Journal of Agricultural Science* 93:635–649.

Cull, B.W. and T. Trochoulias 1982. Macadamias – environmental range for commercial production. In D. Noel (ed.) *Tree Crops: The Third Component. Proceedings of the First Australasian Conference on Tree and Nut Crops, Perth, Western Australia, 1982*, pp54-61. Perth: Cornucopia Press.

Dronia, H. 1983. Bodentemperaturmessungen in tropischen gebirgen: Virunga-Vulkane (Ruanda), Sierra Nevada (Kolumbien), Papua-Neuguinea. [Ground (soil) temperature measurements in tropical mountains]. *Erdkunde* 37:292-295.

Fitzpatrick, E.A. 1965. Climate of the Wabag-Tari area. In *General Report on Lands of the Wabag-Tari area, Territory of Papua and New Guinea, 1960-61*, pp56-69. Melbourne: Land Research Series No. 15. CSIRO.

Ford, R.E. 1990. The dynamics of human-environment interactions in the tropical montane agrosystems of Rwanda: implications for economic development and environmental stability. *Mountain Research and Development* 10(1):43–63.

French, B.R. 1986. *Food Plants of Papua New Guinea. A Compendium.* Tasmania: Published privately, Sheffield.

Gardner, R.O. 2003. Piper (Piperaceae) in New Guinea: the non-climbing species. *Blumea* 48(1):47-68.

Giddings, L.E. 1980. The physical environment of the highlands. In M. S. Mani and L. E. Giddings (eds) *Ecology of Highlands*. Monographicae Biologicae 40, pp11-34. The Hague, Boston and London: Dr W Junk.

Goodbody, S. 1982. The performance of Irish potato in the highlands. *Harvest* 8(3):123-125. Port Moresby: Department of Primary Industry.

Goodbody, S. 1984. Research Report of the Simbu Land Use Project Volume 3. Agronomy. Unpublished report. Kundiawa: Simbu Land Use Project.

Gressitt, J.L. and N. Nadkarni 1978. *Guide to Mt Kaindi: Background to Montane New Guinea Ecology*. Handbook No. 5. Wau: Wau Ecology Institute.

Haberle, S.G. 1993. Late Quaternary environmental history of the Tari Basin, Papua New Guinea. Unpublished PhD thesis. Canberra: The Australian National University.

Henty, E.E. 1969. *A Manual of the Grasses of New Guinea*. Botany Bulletin No. 1. Lae: Division of Botany, Department of Forests.

Henty, E.E. 1982. Some nut-bearing plants in Papua New Guinea. In R.M. Bourke and V. Kesavan (eds) *Proceedings of the Second Papua New Guinea Food Crops Conference. Part 3,* pp78-85. Port Moresby: Department of Primary Industry.

Henty, E.E. and G.H. Pritchard 1975. *Weeds of New Guinea and their Control*. Botany Bulletin No. 7. Lae: Division of Botany, Department of Forests.

Hide, R.L. with S. Goodbody and G. Gertru 1984. Agriculture. In R.L. Hide (ed.) *Research Report of the Simbu Land Use Project Volume 6. South Simbu: Studies in Demography, Nutrition, and Subsistence*, pp207-289. Port Moresby: Institute of Applied Social and Economic Research.

Hide, R., M. Kimin, A. Kora, G. Kua and K. Kua 1979. A Checklist of Some Plants in the Territory of the Sinasina Nimai (Simbu Province, Papua New Guinea), with Notes on their Uses. Working Paper No. 54. Auckland: Department of Anthropology, University of Auckland.

Hnatiuk, R.J., J.M.B. Smith and D.N. McVean 1976. *The Climate of Mt Wilhelm*. Mt Wilhelm Studies 2. Canberra: Department of Biogeography and Geomorphology Publication BG/4. Australian National University.

Howlett, D.R. 1962. A decade of change in the Goroka Valley, New Guinea: land use and development in the 1950s. Unpublished PhD thesis. Canberra: Australian National University.

Humphreys, G.S. 1981a. Lapse rates for Papua New Guinea. Unpublished paper. Kundiawa: Simbu Land Use Project.

Humphreys, G.S. 1981b. Soil temperature data for Papua New Guinea. Unpublished paper. Kundiawa: Simbu Land Use Project.

Humphreys, G.S. 1984. *The Environment and Soils of Chimbu Province, Papua New Guinea with particular reference to soil erosion*. Research Bulletin No. 35. Port Moresby: Department of Primary Industry.

Hynes, R.A. 1974. Altitudinal zonation in New Guinea Nothofagus forests. In J.R. Flenley (ed.). *Altitudinal Zonation in Malesia*, pp75-120. Department of Geography Miscellaneous Series No. 16. Hull: University of Hull.

IBPGR 1980. Fruits [in Indonesia]. Rome: International Board for Plant Genetic Resources.

IBPGR 1981a. Roots and Tuber Crops [in Indonesia]. Rome: International Board for Plant Genetic Resources.

IBPGR 1981b. Vegetables [in Indonesia]. Rome: International Board for Plant Genetic Resources.

Jana, R.K. 1982. Status of sweet potato cultivation in East Africa. In R.L. Villareal and T.D. Griggs (eds). *Sweet Potato. Proceedings of the First International Symposium*, pp63-72. Tainan: Asian Vegetable Research and Development Centre.

Joughin, J. and B. Thistleton 1987. *A Rapid RuralAappraisal in the Jimi Valley, Western Highlands Province, Papua New Guinea.* DAL Discussion Paper 87/3. Port Moresby: Department of Agriculture and Livestock.

Kocher Schmid, C. 1991. *Of People and Plants: A Botanical Ethnography of Nokopo Village, Madang and Morobe Provinces, Papua New Guinea.* Basler Beiträge zur Ethnologie, Band 33. Basel: Ethnologisches Seminar der Universität und Museum für Völkerkunde.

Laing, D.R., P.J. Kretchmer, S. Zuluaga and P.G. Jones 1983. Field beans. In *Potential Productivity of Field Crops Under Different Environments*, pp227-248. Los Baños: International Rice Research Institute.

McAlpine, J.R. 1970. Climate of the Goroka-Mount Hagen area. In *Lands of the Goroka-Mount Hagen Area, Territory of Papua and New Guinea*, pp66-79. Land Research Series No. 27. Australia: CSIRO.

McAlpine, J.R. and G. Keig with R. Falls 1983. *Climate of Papua New Guinea.* Canberra: Australian National University Press.

McAlpine, J.R., G. Keig and K. Short 1975. *Climatic Tables for Papua New Guinea.* Division of Land Use Research Technical Paper No. 37. Canberra: CSIRO.

Meggitt, M.J. 1958. The Enga of the New Guinea highlands: some preliminary observations. *Oceania* 28(4):253-330.

Morgan, W.T.W. 1968. The role of temperate crops in the Kenya highlands. *Acta Geographica* 20:273-278.

Ochse, J.J. and R.C. Bakhuizen van der Brink 1977. *Vegetables of the Dutch East Indies.* Canberra: Australian National University Press.

Peekel, P.G. 1984. *Flora of the Bismarck Archipelago for Naturalists.* E.E. Henty (translator). Lae: Office of Forests, Division of Botany.

Quinlan, T.J. 1968. Cultivation of pyrethrum in Papua and New Guinea. *Rural Digest* 9(5): 7-13. Port Moresby: Department of Agriculture, Stock and Fisheries.

Sillitoe, P. 1983. Natural resources exploited by the Wola in the manufacture of artifacts. *Science in New Guinea* 10(2):111-132.

Simpson, J.A. and F. Arentz 1982. Notes on tree fruits cultivated in the Bulolo and Wau Valleys with particular reference to diseases and pests. In R.M. Bourke and V. Kesavan (eds) *Proceedings of the Second Papua New Guinea Food Crops Conference. Part 3*, pp458-466. Port Moresby: Department of Primary Industry.

Smith, J.M.B. 1977a. Man's impact upon some New Guinea mountain ecosystems. In T.P. Bayliss-Smith and R.G. Feachem (eds) *Subsistence and Survival. Rural Ecology in the Pacific*, pp185-214. London: Academic Press.

Smith, J.M.B. 1977b. Vegetation and microclimate of east- and west-facing slopes in the grasslands of Mt. Wilhelm, Papua New Guinea. *Journal of Ecology* 65:39-53.

Smith, J.M.B. 1978. Relationship of slope aspect to deforestation in highland Papua New Guinea. *Queensland Geographical Journal* (3rd Series) 4:69-75.

Stadel, C. 1986. Del valle al monte: altitudinal patterns of agricultural activities in the Patate-Pelileo area of Ecuador. *Mountain Research and Development* 6(1):53-64.

Treide, B. 1967. Wildpflanzen in der Ernährung der Grundbevölkerung Melanesiens. Veröffentlichungen des Museums für Völkerkunde zu Leipzig Heft 16. Berlin: Akademie-Verlag.

Van Steenis, C.G.G.J. 1961. An attempt towards an explanation of the effect of mountain mass elevation. Proceedings of the Koninklijke Nederlandse Akademie van Wetenschobben. Series C. *Biological and Medical Sciences* 64:435-442.

Verdcourt, B. 1979. *A Manual of New Guinea Legumes.* Botany Bulletin No. 11. Lae: Division of Botany, Department of Forests.

Whitmore, T.C. 1984. *Tropical Rain Forests of the Far East.* Second edition. Oxford: Oxford University Press.

Wohlt, P.B. 1978. Ecology, agriculture and social organization: the dynamics of group composition in the highlands of Papua New Guinea. Unpubished PhD thesis. Minneapolis: University of Minnesota.

Wohlt, P.B. and A. Goie 1986. *Research Report of the Simbu Land Use Project Volume 5. North Simbu Land Use*. Port Moresby: Institute of Applied Social and Economic Research.

Wohlt, P.B. and G.S. Humphreys 1981. A summary of some climate data for Yumbisa (2670 m a.s.l.), Enga Province. Unpublished paper. Kundiawa: Simbu Land Use Project.

Wu, H.B-F., T-T. Yu and T-D. Liou 1974. Physiological and biochemical comparisons of sweet potato varieties sensitive (Tai-long 57) and insensitive (Red-tuber-tail) to chilling temperature. In R.L. Bieleski, A.R. Ferguson and M.M. Creswell (eds) *Mechanisms of Regulation of Plant Growth*, pp483-486. New Zealand: Royal Society of New Zealand Bulletin 12.

Yen, D.E. 1974. *The Sweet Potato and Oceania. An Essay in Ethnobotany*. Bernice P. Bishop Museum Bulletin 236. Honolulu: Bishop Museum Press.

Yu Xiao-gan and Sun Shang-zhi 1981. The upper limit of agriculture in Xizang and its factor analysis. In *Geological and Ecological Studies of Qinghai-Xizang [Tibet] Plateau. Volume 2. Proceedings of the Symposium on Quinghai-Xizang [Tibet] Plateau*, pp2033-2044. Beijing: Science Press.

www.ingramcontent.com/pod-product-compliance
Lightning Source LLC
LaVergne TN
LVHW071202210926
848482LV00015B/710
9781921666803